MODERN HEATING AND VENTILATING SYSTEMS DESIGN

MODERN HEATING AND VENTILATING SYSTEMS DESIGN

George E. Clifford

Regents/Prentice Hall
Englewood Cliffs, New Jersey 07632

Library of Congress Cataloging-in-Publication Data

Clifford, George E.
    Modern heating and ventilating systems design / George E.
Clifford.
      p. cm.
    Includes index.
    ISBN 0-13-602830-6
    1. Heating. 2. Ventilation. 3. Air conditioning. I. Title.
TH7223.C55 1992
697—dc20   91-30038
                                 CIP

Acquisitions editor: Robert Koehler
Editorial/production supervision and
  interior design: Tally Morgan, WordCrafters Editorial Services, Inc.
Prepress buyer: Ilene Levy
Manufacturing buyer: Ed O'Dougherty

Limited portions of this book are reprinted with permission from
Modern Heating, Ventilating and Air Conditioning, © 1990 by
Prentice-Hall, Inc. and previously published as *Heating, Ventilat-
ing, and Air Conditioning,* © 1984 by Prentice-Hall, Inc.

  © 1993 by Regents/Prentice Hall, Inc.
A Division of Simon & Schuster
Englewood Cliffs, New Jersey 07632

Printed in the United States of America
10 9 8 7 6 5 4 3 2 1

0-13-602830-6

Prentice-Hall International (UK) Limited, *London*
Prentice-Hall of Australia Pty. Limited, *Sydney*
Prentice-Hall Canada Inc., *Toronto*
Prentice-Hall Hispanoamericana, S.A., *Mexico*
Prentice-Hall of India Private Limited, *New Delhi*
Prentice-Hall of Japan, Inc., *Tokyo*
Simon & Schuster Asia Pte. Ltd., *Singapore*
Editora Prentice-Hall do Brasil, Ltda., *Rio de Janeiro*

To my wife, Sally

Contents

Preface

Modern Heating and Ventilating Systems Design was written specifically for use in Engineering Technology courses where emphasis is usually place on *practical* system design. However, the text material and method of presentation make it adaptable for use in junior- and senior-level Mechanical Engineering courses, as well as in adult evening classes, on-the-job training, self-instruction, and as a reference.

It has been the author's experience that a year of classroom study is necessary to cover even the basic principles of the design of heating, ventilating, refrigeration, and comfort cooling systems. Many schools are limited to a shorter period of time—a semester or one or two terms—for study in these areas. Therefore, it was the author's aim in preparing this book to bring together in logical order and in a small volume the data necessary to design heating and ventilating systems for buildings. In doing this, he has been guided by his experience as a teacher and as a consulting engineer and by consulting not only the valuable data prepared by the American Society of Heating, Refrigerating, and Air Conditioning Engineers in the *ASHRAE Handbook* series, but also that from the many design manuals and catalogs available from industry. When these data have been used, credit has been given to the author or the company. These data, which, in general, are scattered among various books, manuals, pamphlets, and catalogs and must be used by design engineers in their work,

have been brought together and placed in a logical position in a single book.

This book assumes that previous study has been accomplished by the student in the field of mathematics through introduction to calculus; however, the application of calculus is not required. Formal courses in thermodynamics, fluid mechanics, and heat transfer are desirable, but sufficient review of these subject areas is provided in the text.

A thorough understanding of psychrometrics, the psychrometric chart, and applied psychrometrics is one of the important considerations in the design of heating and ventilating systems. Psychrometrics is discussed in Chapters 3 and 4. A well-designed heating and ventilating system requires the best possible automatic control system, within cost limits, to obtain the best system operating characteristics and provide for economical operation. Chapters 15 and 16 provide information for an understanding of automatic control systems.

The author is grateful for the permission to use certain data and written information from the publications of the American Society of Heating, Refrigerating, and Air Conditioning Engineers, The Trane Company, Honeywell, Johnson Controls, ITT Fluid Transfer Division, Taco, Armstrong Machine Works, Lima Register Company, and many others.

George E. Clifford

MODERN HEATING AND VENTILATING SYSTEMS DESIGN

1

Fundamental Concepts

1-1 INTRODUCTION

Air conditioning as an inclusive term means *the control of temperature, moisture content, circulation, and purity of the air within a space* to produce desired effects on the occupants of that space or on the products and materials manufactured or stored there. *Complete air conditioning* involves the control of all four factors and implies the establishment of year-round conditions for the comfort and health of people and for the stability of products and manufacturing processes. In addition to this term are terms such as *summer air conditioning* or *summer cooling* and *winter air conditioning* or *winter heating*.

Summer cooling implies the maintenance of a space temperature *below* that of the surroundings and may include *dehumidification* of the air. Both cooling and dehumidification of the air normally require some type of refrigeration system. Summer cooling should also include means for controlling air motion and distribution and air purity within the conditioned space.

Winter heating implies the maintenance of a space temperature *above* that of the surroundings and normally should include *humidification* of the air. It should also include a means of controlling air motion and distribution and air purity within the conditioned space.

The distribution of the air must be uniform with gentle motion in all parts of the occupied zone so that a sensation of comfort is produced for the greatest number of individuals.

Ventilation is usually required in order to produce and to maintain either healthful, comfortable, or specified air conditions within a structure and involves *the introduction into and removal from the spaces of the structure a definite amount of air per unit of time.* This change of air may be accomplished by either natural or mechanical means. The air supplied may or may not be processed to change its properties. The most successful ventilation systems embody a part or all of the features of air conditioning.

Industrial air conditioning is often required during the storage of material and products. It is often a necessity during some of the manufacturing operations in the following industries: brewing, chemicals and drugs, engines, electrical products, foods, furs, glass, precision instruments, machine shops, printing, rubber goods, steel, textiles, and tobacco in the form of cigarettes, cigars, and other materials. Proper air temperatures, humidity, and cleanliness are of extreme importance in a number of the foregoing operations; otherwise, the materials cannot be handled or the finished products are inferior or worthless. Industrial air conditioning frequently, but not always, produces comfort conditions for the personnel employed.

We are concerned in this book with heating and ventilating systems design. The term *air conditioning* may be used as an inclusive term in discussions of year-round control of environmental space conditions, regardless of the time or season of the year. More and more applications of complete year-round air conditioning to industrial and commercial buildings are being employed.

Since heating, ventilating, and air conditioning (HVAC) systems are component parts of a building and are designed to overcome building heat losses and gains, the *entire building system* should be analyzed, including the basic structure itself. It is desirable, then, for the designer of air conditioning systems to carefully study the complete building system in order to design the best possible mechanical systems.

Energy is generally used in buildings to perform functions of heating, lighting, mechanical drives, cooling, and special applications. The energy is available to the building in limited forms, such as electricity, fossil fuels, and solar energy, and these energy forms must be converted within the building to serve the end use of the various functions. A loss of energy is associated with any conversion process. In energy conservation efforts, there are two avenues of approach: (1) reducing the amount of use and/or (2) reducing conversion losses. The latter is an unfortunate situation inherent in conversion processes. For example, the furnace that heats the building produces unusable and toxic flue gas, which must be vented to the outside of the building, and thus part of the energy is lost. (This "lost" energy is not destroyed; it simply ends up as low-grade heat energy.)

The efficient use of energy in buildings can only be achieved through implementation of optimized energy designs, well-developed policies on energy use, and dedication by management backed up by a properly trained and motivated operating staff. Optimum energy conservation is obtained when the least amount of energy is used to achieve a desired result. If this is not fully realizable, the next best method is to move the excess energy from where it is not wanted to where it can be used or stored for future use, which generally results in a minimum expenditure of new energy. If possible, a system should be designed so that it cannot heat and cool the same locations simultaneously.

ASHRAE Standard 90.1-1989, *Energy-Efficient Design of New Buildings Except Low-Rise Residential Buildings,* and the 100 series, *Energy Conservation in Existing Buildings,* provide minimum guidelines for energy conservation design and operation. They incorporate the following types of energy standards:

1. *Prescriptive standards* specify the materials and methods for design and construction of buildings.
2. *System performance standards* set requirements for each component, system, or subsystem within a building.
3. *Building energy standards* consider the performance of the building as a whole.

The design goal is set for the annual energy requirements of the entire building on a unit basis—for example, kW-hr/m²·yr (Btu/ft²·yr). Any combination of materials, systems, and operating procedures can be applied as long as design energy usage does not exceed the building's annual energy budget.

This kind of approach allows greater flexibility while promoting the goals of energy efficiency. It also allows and encourages the use of innovative techniques and the development of new methods of saving energy. Means for its implementation are still being developed. They are different for new and for existing buildings; in both cases, an accurate database is required as well as an accurate, verifiable means of measuring consumption.

From an energy management viewpoint, there are essentially three classes of buildings: (1) residential/domiciliary, which use about 34% of all energy for buildings; (2) commercial/institutional, which use about 25%; and (3) manufacturing/industrial, which use about 38%. Overall, buildings use about 30% of the energy consumed in the United States, excluding energy for industrial processes.

The Arab oil embargo of 1973 triggered the push toward energy management in buildings and the continuing fervor to make buildings in this country use less energy; office buildings, hospitals, hotels, schools, and industrial plants have been thoroughly scrutinized. Although born in crisis, an era of energy-efficient building design was long overdue. As energy prices continue to rise, lower losses and higher efficiencies have become a raging concern of building owners and managers. In addition to employing recognized methods of improving the building envelope and equipment performance, more sophisticated schemes of slashing energy consumption have been conceived. Included in such schemes are the following: cogeneration, energy management systems (EMS), direct digital control (DDC), daylighting, heat pumps, variable air volume (VAV), thermal storage, and heat recovery in commercial and institutional buildings and industrial plants.

1-2 DIMENSION AND UNITS

A *physical quantity* is something that can be measured. A *dimension* is a property or quality of the quantity or entity that characterizes or describes that quantity. Thus, distance has the characteristic property, or dimension, of length (L). Area is a function of length and breadth, or dimensionally, $A = L^2$. Also, volume is a function of length × breadth × height, or dimensionally, $V = L^3$.

The dimension of length may be expressed in various *units,* such as inches, feet, yards, miles, centimeters, meters, and kilometers. Any physical quantity is exactly specified by a dimension and by a multiple or fraction of a defined unit, such as a length of 14 ft.

A dimensional system that will completely describe an event can be constructed from a relatively small number of fundamental dimensions. The usual fundamental dimensions are time (T), length (L), force (F), and mass (M).

For many years, engineering calculations in the United States have been made using the unit *second* (s) for time, the unit foot (ft) for length, the unit *pound force* (lbf) for force, and the unit *pound mass* (lbm) for mass. This system of units is called the *English Engineering System* (EES).

Another system of units, the Engineering Dynamical System, makes use of second for time, foot for length, pound mass for mass, and poundal for force. The English Gravitational System makes use of second for time, foot for length, slug for mass, and pound force for force. The cgs system uses the centimeter (cm) for length, second for time, gram for mass, and dyne for force.

As world technology grew, it became apparent that there was a need for international standardization of units. The official name of the international standardization of units is *Le Système International d'Unités,* abbreviated SI. Do not call it the "SI system" because the *S* in SI means "system."

SI is an absolute system. The basic unit of length is the *meter* (m), the unit of mass is the *kilogram* (kg), and the unit of time is the *second* (s). The unit of force is derived and is called the *newton* (N) to distinguish it from kilogram, which, as indicated, is the unit of mass.

The SI base units, with their symbols, are shown in Table 1-1. These are dimensionally independent. Lowercase letters are used for the symbols unless they are derived from a proper name, in which case a capital is used for the first letter of the symbol. Note that the unit of mass uses the prefix kilo; this is the only base unit having a prefix. Table 1-1 shows that the SI unit of temperature is the kelvin. The Celsius temperature scale

TABLE 1-1

SI base units

Quantity	Name	Symbol
Length	meter	m
Mass	kilogram	kg
Time	second	s
Electric current	ampere	A
Thermodynamic temperature	kelvin	K
Amount of matter	mole	mol
Luminous intensity	candela	cd

(once called "centigrade") is not part of SI, but a difference of one degree on the Celsius scale equals one on the Kelvin scale.

A second class of SI units comprises the derived units, many of which have special names. Table 1-2 is a list of derived units of primary concern in our work. Note that each unit is spelled with an initial lowercase letter, except when it occurs at the beginning of a sentence. The first letter is then capitalized. If the unit is named for an individual, the first letter of the symbol is capitalized; otherwise, the symbol is lowercase.

1-3 FORCE, MASS, AND WEIGHT

In Section 1-2, different systems of units were discussed briefly. In this text, we will be concerned primarily with the English Engineering System of units and the SI units. A brief discussion of what is meant by force, mass, and weight will help us to understand the relationships and conversions from one unit system to the other.

A *force* may be defined as a push or pull tending to change the motion or direction of a body on which it acts. A *mass* is a quantity of matter. A given mass has the same value any place in the universe. A body has the tendency to resist a change in motion, and we say that it has *inertia.* Force and mass are related by Newton's second law of motion. *Weight* is a force exerted by a mass due to the pull of gravity, which varies with geographical location.

Newton's second law of motion states that if an unbalanced force acts on a body, the body will accelerate in the direction of the force, and the acceleration is directly proportional to the unbalanced force and inversely proportional to the body mass. Stated mathematically,

$$a \propto \frac{F}{m}$$

TABLE 1-2

SI derived units

Quantity	Name	Symbol	Formula	Expressed in Base Units
Acceleration	meter per second squared	m/s^2	m/s^2	m/s^2
Area	square meter	m^2	m^2	m^2
Density	kilogram per cubic meter	—	kg/m^3	$kg \cdot m^{-3}$
Energy or work	joule	J	$N \cdot m$	$m^2 \cdot kg \cdot s^{-2}$
Force	newton	N	$m \cdot kg \cdot s^{-2}$	$m \cdot kg \cdot s^{-2}$
Moment	newton-meter	$N \cdot m$	$N \cdot m$	$m^2 \cdot kg \cdot^{-2}$
Plane angle	radian	rad	rad	rad
Power	watt	W	J/s	$m^2 \cdot kg \cdot s^{-3}$
Pressure	pascal	Pa	N/m^2	$m^{-1} \cdot kg \cdot s^{-2}$
Rotational frequency	revolution per second	r/s	s^{-1}	s^{-1}
Velocity (speed)	meter per second	m/s	m/s	$m \cdot s^{-1}$
Volume	cubic meter	—	m^3	m^3

or

$$a = g_c \left(\frac{F}{m}\right) \quad \text{and} \quad F = \frac{m}{g_c}a \qquad (1\text{-}1)$$

where

F = unbalanced force
m = body mass
a = acceleration
g_c = proportionality constant

The numerical value of g_c depends on the units used for the terms in Eq. (1-1). A newton (N) is a unit of force that will accelerate 1 kg of mass at the rate of 1 m/s² (meter per second squared). Using Eq. (1-1) and solving for the units of g_c, we have

$$g_c = \frac{ma}{F} = \frac{(1 \text{ kg})(m/s^2)}{N} = 1 \frac{\text{kg-m}}{\text{N-s}^2} \qquad (1\text{-}2)$$

Equation (1-2) gives the value and units of g_c for SI.

The English Engineering System (EES) uses the unit of pound mass (lbm), which is equal to 0.453 592 37 kg. The pound force (lbf) is defined as the weight of one pound mass when subjected to standard acceleration of gravity of 32.174 048 56 ft/s². Substitution into Eq. (1-1) gives us

$$g_c = \frac{ma}{F} = \frac{(1 \text{ lbm})(32.174\ 048\ 56 \text{ ft/s}^2)}{\text{lbf}}$$

or

$$g_c = 32.2 \text{ lbm-ft/lbf-s}^2 \text{ (approximate)} \qquad (1\text{-}3)$$

Equation (1-3) gives the value and units of g_c for EES.

If the only force acting on a body is the force of gravity (its weight, w), the resulting acceleration will be the local acceleration of gravity, g. From Eq. (1-1), we would have

$$w = \frac{1}{g_c}mg \quad \text{or} \quad \frac{w}{g} = \frac{m}{g_c} \qquad (1\text{-}4)$$

Combining Eqs. (1-1) and (1-4) produces the following important relationship:

$$F = \frac{w}{g}a \qquad (1\text{-}5)$$

Equation (1-4) is important because the ratio of g/g_c appears in many equations for the purpose of obtaining proper engineering units. If the average acceleration of gravity, g, on the surface of the earth is assumed to be 32.2 ft/s², the ratio becomes

$$\frac{g}{g_c} = \frac{32.2 \text{ ft/s}^2}{32.2 \text{ lbm-ft/lbf-s}^2}$$

$$= 1 \text{ lbf/lbm (approximate)}$$

and, if standard gravity is 9.806 m/s² or about 9.80 m/s² using SI units,

$$\frac{g}{g_c} = \frac{9.80 \text{ m/s}^2}{1 \text{ kg-m/N-s}^2} = 9.80 \text{ N/kg}$$

1-4 MATTER

The molecule is the smallest division of matter that has all the chemical and physical properties of a large quantity of matter. Molecules, in turn, are composed of smaller particles known as electrons, protons, and neutrons. The study of atoms and subatomic particles is beyond the scope of this book, and the discussion will be limited for the most part to the study of molecules and their behavior.

When chemical reactions occur, the atoms are regrouped to form molecules of different substances without basic changes in the atoms themselves. In the combustion process, the rearrangement of atoms into different molecules is accompanied by the release of energy, part of which may be converted into work in a suitable engine. This energy could also have been used to heat a building. When energy is released by chemical reactions, there is a very slight decrease in mass, but the change is too small to be measured. Accordingly, chemical reactions may be said to take place in accordance with the *law of conservation of matter,* which states that matter is indestructible, that the mass of material is the same before and after the chemical reactions occur, and that the mass of each chemical element remains unchanged during the chemical reactions. Thus, if fuel oil and air are supplied to the combustion chamber of a furnace, the mass of flue gas leaving the furnace in a given time period equals the mass of fuel and air supplied; also, the mass of oxygen, hydrogen, carbon, nitrogen, and other elements entering the combustion chamber is equal to the mass of those same elements leaving, although an entirely new set of chemical compounds may result from the combustion process.

Matter ordinarily exists as a solid, liquid, or gas. These are called *phases* of matter. In the solid phase or condition, the molecules are held in fixed positions by powerful forces. However, the atoms may vibrate about mean positions within the molecules. Solids, therefore, have definite shape and volume, and relatively great external forces are required to deform them. In the liquid phase, the molecules have a motion of translation with frequent collisions from which they rebound as perfectly elastic bodies. A liquid will occupy a definite volume at a given temperature but conforms to the shape of the confining vessel. In the gaseous phase, the molecules are far apart in comparison to their size, move in straight lines between collisions, and will fill a confining vessel regardless of the amount of gas present.

Many substances, such as water, may exist in the solid, liquid, and gaseous phase, depending upon the pressure and temperature. Thus, ice may melt to form water, and water may evaporate to form steam (vapor).

When a substance is in the gaseous phase, but at a temperature and pressure not far removed from the temperature and pressure at which it can be liquefied, it is usually referred to as a *vapor.* A *perfect gas* is at a temperature far removed from the pressure and temperature at which it can be liquefied.

1-5 PROPERTIES AND STATE OF A SUBSTANCE

Work is obtained from an engine by causing a working substance (fluid) such as steam or a gas to undergo heating, cooling, expansion, and compression processes, or changes, in a suitable mechanism. Heat is added to a fluid, such as water in a water boiler, and the heated water is forced to circulate through a system of piping to terminal heat transfer devices where some of its heat is given up to heat the building. A refrigerant absorbs heat as it boils at low temperature. The resulting vapor is compressed to a high pressure and temperature and is then condensed back to liquid by the rejection of heat.

It is important that the *state* or condition of a fluid be known while it is being subjected to changing conditions.

A property of a substance is a quantitative characteristic of that substance. For most purposes, the properties of a fluid may be classified as (1) *state properties,* which define the physical conditions of the fluid; (2) *thermodynamic properties,* which define the thermal energy conditions of the fluid; or (3) *transport properties,* which measure the diffusion within the fluid resulting from molecular activity. These are as follows:

State Properties

- Density, ρ
- Specific weight, γ
- Specific volume, v
- Pressure, p
- Temperature, T

Thermodynamic Properties

- Specific internal energy, u
- Displacement energy, pv
- Enthalpy, h
- Entropy, s
- Constant-volume specific heat, c_v
- Constant-pressure specific heat, c_p

Transport Properties

- Viscosity, μ
- Thermal conductivity, k

The state properties are easily determined by direct observation or simple measurement. The thermodynamic properties are more complex in their conception. Internal energy, enthalpy, and entropy are used almost entirely as differences, and the numerical values are relative to arbitrary datum levels chosen for their simplifications of problems. State properties and thermodynamic properties are discussed briefly in the following paragraphs. The transport properties, viscosity and thermal conductivity, are reviewed where encountered in other sections of the book.

In general, if a proper combination of two of these properties is known, the state or molecular condition of the fluid is defined, and the values of the other properties are fixed. Also, if a given mass of fluid initially occupying some known volume at a known pressure and temperature undergoes some change or process after which the final volume, pressure, and temperature can be determined, then the *change* in volume, pressure, and temperature may be computed by subtracting the final values of these properties from the initial and final states.

Referring to Figure 1-1, let the coordinates of point 1 be pressure p_1 and volume V_1 representing to some scale the pressure and volume of a given mass of gas to the left of the gas-tight frictionless piston in the cylinder. If the piston moves to the right to a final position, point 2, the final pressure and volume would be p_2 and V_2. During the expansion process of the gas within the cylinder, the relationship between the pressure and volume at each successive position of the piston might be represented by some curve such as *a* or *b,* which is called the *path* of the process. The increase in volume

($V_2 - V_1$) during the process depends only on the initial and final positions of the piston and is independent of how the pressure varied with the volume—that is, the path. Also, the change in pressure for the process is dependent only on the difference between the initial and final pressures and is independent of the path. Such properties of matter as pressure and volume are known as *point functions* since the change in their values during any process depends upon the initial and final values of the properties and are independent of the process path.

1-6   DENSITY

Density is a fluid property and is defined as the mass per unit of volume. It is calculated by

$$\rho \text{ (rho)} = \frac{m}{V} \tag{1-6}$$

where m is the total mass and V is the total volume occupied by the mass.

> *Definition:*   Mass per unit of volume
> *Symbol:* ρ (rho)
> *Units:*   EES: lbm/ft^3; SI: kg/m^3
> *Conversion:*   To convert lbm/ft^3 to kg/m^3, multiply by 16.02.

1-7   SPECIFIC WEIGHT

Specific weight is defined as weight (force) per unit of volume. It is calculated by

$$\gamma \text{ (gamma)} = \frac{w}{V} \tag{1-7}$$

where w is the total weight and V is the total volume.

> *Definition:*   Weight (force) per unit of volume
> *Symbol:* γ (gamma)
> *Units:*   EES: lbf/ft^3; SI: N/m^3
> *Conversion:*   To convert lbf/ft^3 to N/m^3, multiply by 157.

Relationship to Density.   By definition, specific weight is defined by Eq. (1-7). By Eq. (1-4),

$$\frac{w}{g} = \frac{m}{g_c} \quad \text{or} \quad w = m\left(\frac{g}{g_c}\right)$$

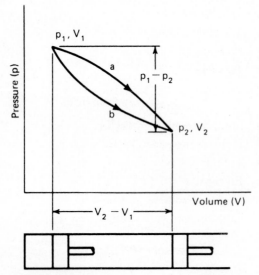

FIGURE 1-1   Change in pressure and volume during expansion of a gas.

If the latter expression is substituted into Eq. (1-7), we have

$$\gamma = \frac{m(g/g_c)}{V}$$

But, $\rho = m/V$; therefore,

$$\gamma = \rho \left(\frac{g}{g_c}\right) \qquad (1\text{-}7a)$$

It should be remembered that the ratio of g/g_c has units of lbf/lbm; therefore, if the local acceleration of gravity, g, is equal to 32.2 ft/s² and g_c is constant at 32.174 (approximately), then the numerical value of specific weight (γ) and density (ρ) are equal. It must be remembered, however, that the *units are different.*

1-8 SPECIFIC VOLUME

Specific volume is defined as the volume per unit of mass. It may be calculated by

$$v = \frac{V}{m} \qquad (1\text{-}8)$$

In many references, specific volume is defined as the reciprocal of density, or $v = 1/\rho$.

Definition: Volume per unit of mass
Symbol: v
Units: EES: ft³/lbm; SI: m³/kg
Conversion: To convert ft³/lbm to m³/kg, multiply by 1/16.02.

Relationship to Specific Weight. If $v = 1/\rho$, we may substitute into Eq. (1-7a) and obtain

$$\gamma = \frac{1}{v}\left(\frac{g}{g_c}\right) \quad \text{or} \quad v = \frac{1}{\gamma}\left(\frac{g}{g_c}\right) \qquad (1\text{-}8a)$$

To repeat, it must be remembered that the ratio of g/g_c has units of lbf/lbm and, if the ratio is unity, $v = 1/\gamma$, numerically.

1-9 PRESSURE

Definition: Force per unit of area
Symbol: p
Units: EES: lbf/ft² or lbf/in.² (psf or psi); SI: N/m² or Pa

Conversion: To convert lbf/in.² to N/m², multiply by 6895.

Pressure is the force exerted per unit of area. It may be defined as a measure of the intensity of a force at a given point on the contact surface. Whenever a force is evenly distributed over a given area, the pressure at all points on the contact surface is the same and may be calculated by dividing the total force exerted by the total area over which the force is applied, or

$$p = \frac{F}{A} \qquad (1\text{-}9)$$

where

p = pressure expressed in units of force per unit of area
F = total force in units of force
A = total area in units of area

The most frequently used units of pressure are lbf/in.² (psi). However, since the units of force and area may be any appropriate unit, the unit of pressure may be lbf/ft² (psf), tons/ft², and, in SI units, N/m² or Pa. Use is also made of a pressure unit called the *bar*, defined as 100,000 N/m². Since the standard atmospheric pressure is 101,325 Pa (101.325 kPa), the bar can be seen to be 0.98066 times as large as the standard atmosphere, at 14.5038 psi, and, of course, 100 kPa.

We will find in our work in the design of HVAC systems that it is often convenient to express pressure in terms of the height of a column of liquid such as feet of water, inches of water, millimeters of mercury, and so forth. It is well known from the principles of hydrostatics that the unit of pressure exerted by a column of liquid is a function of its height and its specific weight and is independent of the cross section of the column, or $p = \gamma h$, where γ is the specific weight and h is the height of the column of liquid. Water itself can be used as a measuring liquid, but mercury has a density of 0.4912 lbm/in.³ (1.3596×10^4 kg/m³) measured at 0°C. At 20°C (68°F), the corresponding values are 0.4894 lbm/in.³ (1.3546 kg/m³).

To change fluid pressure units, the following obvious relationship exists:

$$p_1 = p_2$$

or

$$h_1\gamma_1 = h_2\gamma_2$$

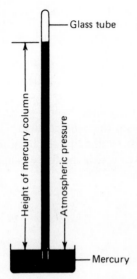

FIGURE 1-2   Barometer.

or

$$h_1\rho_1\left(\frac{g}{g_c}\right) = h_2\rho_2\left(\frac{g}{g_c}\right) \qquad (1\text{-}10)$$

where the subscripts refer to two different fluids.

Atmospheric pressure is usually measured by a barometer and may be expressed in either inches of mercury, millimeters of mercury, or pounds per square inch. Atmospheric pressure is exerted by the air that extends outward from the earth's surface. The pressure of the atmosphere does not remain constant but will vary from hour to hour depending on the air temperature, water-vapor content, and elevation above sea level. Fig-

ure 1-2 illustrates the basic principle of a mercury barometer.

Gauge pressure is the difference between the pressure of a fluid in a container, such as a tank or pipe, and that of the atmosphere. Gauge pressure may be above (positive) local atmospheric pressure or below (negative) atmospheric pressure. Negative gauge pressure is usually called *vacuum*.

Absolute pressure is the algebraic sum of gauge pressure and atmospheric pressure. Figure 1-3 illustrates the relationship between gauge pressure and absolute pressure. A useful conversion factor for converting inches of mercury to pounds per square inch is 0.491 psi per inch of mercury.

1-10   TEMPERATURE

Definition: A measure of the level of heat intensity

Symbols and units: EES: t_F (degree Fahrenheit, °F), T_R (degree Rankine, °R); SI: t_C (degree Celsius, °C), T_K (kelvin, K)

Conversions: To convert °F to °R, use

$$T_R = t_F + 460$$

To convert °C to K, use

$$T_K = t_C + 273$$

To convert °R to K, use

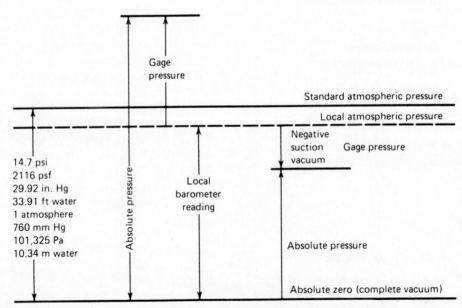

FIGURE 1-3   Units and scales for absolute and gauge pressures.

$$T_K = \frac{T_R}{1.8}$$

To convert °F to °C, use

$$t_C = \frac{t_F - 32}{1.8}$$

Temperature Scales

Fahrenheit temperature scale: This scale has been used in the United States for many years for ordinary temperature measurements. On this scale, water freezes at 32°F, sometimes called the *ice point,* and boils at 212°F, the *steam point,* when the pressure above the liquid is at standard barometric pressure of 14.696 psia.

Celsius temperature scale: This scale (formally centigrade) has an ice point of 0°C and a steam point of 100°C.

Kelvin temperature scale: This scale is the absolute Celsius scale. The kelvin (K with no degree sign) is defined as the SI unit of temperature and is 1/273.16 of the fraction of the thermodynamic temperature of the triple point of water. The International Practical Temperature Scale of 1968 assigns a value of 0.01°C to the triple point of water.

Rankine temperature scale: This scale is the absolute Fahrenheit scale.

The relationship of the four temperature scales is shown in Figure 1-4. The Celsius scale has 100° between the ice and steam points; the Fahrenheit scale has 180° between the same two points.

1-11 SPECIFIC GRAVITY

Definition: Fluid density/reference fluid density
Symbol: SG
Dimensions: Dimensionless ratio
Units: None
Reference fluids: Water for solids and liquids; air for gases

Since the density of water changes with temperature and, at very high pressures, for a precise definition of specific gravity, the temperatures and pressures of the fluid and reference fluid should be stated. In practice, two temperatures are stated—for example, 60/60°F where the upper temperature refers to the fluid and the lower to water. If no temperatures are stated, it must be assumed that reference is made to water at its maximum density. The maximum density of water at atmospheric pressure is at 3.98°C and has a value of 999.97 kg/m³ or 62.4 lbm/ft³.

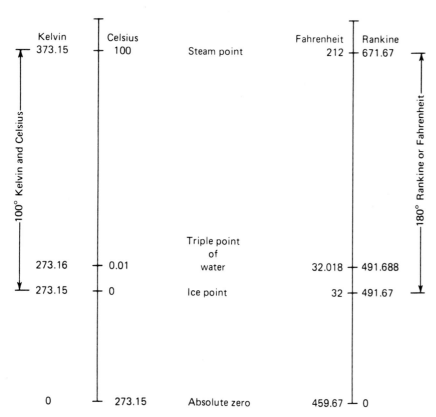

FIGURE 1-4 Temperature scales.

For gases, it is common practice to use the ratio of the molecular weight of the gas to that of air (28.96), thus eliminating the necessity of stating the pressure and temperature for ideal gases.

Hydrometer scales have been established by industry that have arbitrary graduations. In the petroleum and chemical industries, the Baumé (°B) and the American Petroleum Institute (°API) scales are used.

Conversions: Baumé scale (heavier than water):

$$SG_{60/60} = \frac{145}{145 - °B} \qquad (1\text{-}11)$$

Baumé scale (lighter than water):

$$SG_{60/60} = \frac{140}{140 + °B} \qquad (1\text{-}12)$$

American Petroleum Institute scale:

$$SG_{60/60} = \frac{141.5}{131.5 + °API} \qquad (1\text{-}13)$$

Since the specific gravity of a liquid is greatly influenced by temperature, the following equation may be used to determine the specific gravity of a liquid at some temperature, *t,* other than 60°F:

$$(SG)_t = SG_{60/60} - 0.00035(t - 60) \qquad (1\text{-}14)$$

1-12 ENERGY

Definition: Capacity to produce an effect; capacity to do work
Symbol: Depends on type of energy discussed
Units: EES: Foot-pound force (ft lbf), British thermal unit (Btu); SI: joule (J)
Conversions: To convert Btu to J, multiply by 1.054. To convert ft lbf to J, multiply by 1.356.

A body is said to possess energy when it is capable of doing work. More generally, energy is the capacity for producing an effect. Energy may be classified as *stored energy* or *energy in transition.* Chemical energy is stored in high explosives and fuels such as coal, gas, and oil. Energy is stored in an automobile battery.

When a switch in an electric circuit is closed, electrical energy flows from a power plant to an electric motor or other electrical device. Energy flows from an automobile engine to the wheels through a gear train to drive the vehicle.

The heating engineer is concerned with the flow of heat energy in systems to provide thermal comfort. The refrigeration engineer is concerned with the removal of heat from areas where it is undesirable and the discharge of this energy to the atmosphere or to cooling water.

The complete or partial transformation of energy from one of the many forms in which it may exist into other forms of energy takes place in accordance with the *law of conservation of energy.* This law states that *energy can be neither created nor destroyed;* therefore, when energy is transformed either completely or partially from one form to another, the total amount of energy remains the same. In all types of engineering problems except those dealing with nuclear reactions, the laws of conservation of energy and conservation of matter apply.

1-13 MECHANICAL POTENTIAL ENERGY

Definition: Stored energy due to position relative to some selected datum
Symbol: PE
Units: EES: ft lbf; SI: J
Conversion: To convert ft lbf to J, multiply by 1.356.

Stored energy associated with the position of bodies is called *mechanical potential energy.* In order to establish the value of potential energy, it is necessary to define a reference datum, usually the surface of the earth. A body on or near the earth's surface is attracted to the earth by the gravitational force, which is called the weight (*w*) of the body. If the body is lifted from the earth's surface to an elevation of *Z* ft above the earth, the body has stored (potential) energy equivalent to the work required to lift the body, or

$$PE = wZ \text{ (ft lbf)} \qquad (1\text{-}15)$$

It should be noted that potential energy is a function of gravitational force or weight. If mass is known, instead of weight, we may use Eq. (1-4), $w/g = m/g_c$, and potential energy becomes

$$PE = \frac{g}{g_c}mZ \text{ (ft lbf)} \qquad (1\text{-}16)$$

Since potential energy depends upon the location of an arbitrary datum of reference, the absolute value of PE is relative. In most problems, we are interested in the *difference* in values of PE between two conditions of the system; therefore, the absolute value is of little importance in most cases.

In terms of SI units, Eq. (1-16) becomes

$$PE = mgZ \text{ (N·m or J)} \qquad (1\text{-}16a)$$

where

$$m = \text{mass (k)}$$
$$g = 9.80 \text{ m/s}^2$$
$$Z = \text{elevation (m)}$$

1-14 MECHANICAL KINETIC ENERGY

Definition: Energy possessed by a body due to its velocity

Symbol: KE

Units: EES: ft lbf; SI: J

Conversion: To convert ft lbf to J, multiply by 1.356.

Kinetic energy is the energy a body possesses as a result of its motion or velocity. The amount of kinetic energy a body possesses is a function of its mass and its velocity squared, or

$$KE = \frac{1}{2g_c}mV^2 \text{ (ft lbf)} \qquad (1\text{-}17)$$

Remembering again from Eq. (1-4) that $w/g = m/g_c$, Eq. (1-17) may be written as

$$KE = \frac{wV^2}{2g} \text{ (ft lbf)} \qquad (1\text{-}17a)$$

As with potential energy, we are normally concerned with a *change* in kinetic energy, if and when the velocity changes.

In SI units, kinetic energy becomes

$$KE = \frac{mV^2}{2} \text{ (N·m or J)} \qquad (1\text{-}17b)$$

where

$$m = \text{mass (kg)}$$
$$V = \text{velocity (m/s)}$$

1-15 INTERNAL ENERGY

Definition: Molecular energy due to molecular motion and molecular attractive forces

Symbols: EES: u (specific), U (total)

Units: EES: Btu/lbm (specific), Btu (total); SI: J/kg (specific), J (total)

Conversions: To convert Btu/lbm to J/kg, multiply by 2324. To convert Btu to J, multiply by 1054.

Matter was discussed as being composed of molecules. Molecules, like tangible bodies, have mass. In the solid state, they are held together in relatively fixed position by powerful forces. However, the atoms of which they are composed vibrate. In the liquid and gaseous states, the molecules have motion of translation and rotation and spin. Consequently, because of their mass and motion, the molecules have kinetic energy stored within them. Since the activity of the molecules is some function of temperature, any change in temperature is accompanied by a change in the kinetic energy stored in the molecules.

Also, molecules are attracted to one another by forces that are very large in the solid phase, that are much less in the liquid phase, and that tend to vanish in the gas phase where the molecules are far apart in comparison to their size. In the melting of a solid or the evaporation of a liquid, it is necessary that the powerful molecular attractive forces be overcome. The energy required to bring about this change of phase is stored in the molecules as *molecular potential energy* and will be released when the substance returns to its initial state.

1-16 DISPLACEMENT OR FLOW ENERGY

The quantity pv represents energy associated with a unit mass of fluid by virtue of the displacement of its boundaries from one position to another at constant pressure. One element of mass in transport exerts pressure on the element immediately ahead and in turn has pressure exerted on it by the immediately following elements. The distance of displacement of the elements times the pressure introduces flow work or displacement energy. We will discuss this in Chapter 2.

1-17 ENTHALPY

Definition: A property combining the properties of internal energy, pressure, and specific volume in a defined way

Symbols: EES: h (specific), H (total)

Units: EES: Btu/lbm (specific), Btu (total); SI: J/kg (specific), J (total)

Conversions: To convert Btu/lbm to J/kg, multiply by 2324. To convert Btu to J, multiply by 1054.

Pressure, temperature, specific volume, and internal energy are properties that define the state of a substance. In general, if a proper combination of two of the properties is known for a given state, the others are fixed and can be computed. Frequently, in equations in thermodynamics, the internal energy term and the product of pressure and specific volume appear in the same equation. Therefore, it has been found convenient to group these three properties into a new, single term called *enthalpy, h,* which is defined as

$$h = u + \frac{Pv}{J} \text{ (Btu/lbm)} \qquad (1\text{-}18)$$

where

u = internal energy (Btu/lbm)
P = pressure (lbf/ft^2)
v = specific volume (ft^3/lbm)
J = conversion unit (778 ft lbf/Btu)

In SI units, Eq. (1-18) becomes

$$h = u + Pv \text{ (J/kg)} \qquad (1\text{-}18a)$$

1-18   ENTROPY

Entropy, like enthalpy, is a mathematical function of observable properties of a substance, and therefore changes in entropy are not evident to the human senses. An increase in entropy indicates an increase in the degradation of energy or a decrease in the availability of energy. Mathematically, entropy changes between states 1 and 2 are defined as

$$_1ds_2 = \int_1^2 \frac{dQ}{T} \qquad (1\text{-}19)$$

where

Q = heat
T = absolute temperature
s = entropy

This expression occurs repeatedly in thermodynamic analyses, and entropy is a useful coordinate in plotting processes and making charts. Differences in entropy and not absolute values are used.

A common hydraulic analogy to entropy is found in the potential energy associated with the position of water at the top of a waterfall and the unharnessed dissipation of that energy in its drop to the foot of the falls. The potential energy of the water is first converted to kinetic energy during free-fall and is then dissipated as heat upon impact. The amount of energy remains un-

changed, but it has suffered degradation and is no longer easily converted into useful work as it might have been had a water turbine been interposed in the path of its fall. This degradation is loosely equivalent to the throttling of a gas from a high pressure to a low pressure through an orifice in that again there is a decrease in the availability of the energy, although the energy degradation is measurable as an increase in entropy and has not been accomplished by the accomplishment of useful work. The interposing of an expansion engine in lieu of the orifice would have accomplished conversion of part of the energy to useful work.

1-19   REVERSIBILITY

All naturally occurring changes or processes are irreversible. Like a clock, they tend to "run down" and cannot "rewind" themselves. Familiar examples are the transfer of heat with a finite temperature difference, the mixing of two gases, a waterfall, a chemical reaction. But, all of these changes *can be reversed.* That is, we can transfer heat from a region of low temperature to one of higher temperature; we can separate a gas into its components; we can cause water to flow uphill. The important point is that we can do these things *only at the expense of some other system,* which itself becomes run down.

A process is said to be reversible *if its direction can be reversed at any stage by an infinitesimal change in external conditions.* If we consider a connected series of equilibrium states, each representing only an infinitesimal displacement from the adjacent one but with the overall result a finite change, then we have a reversible process.

All actual processes can be made to approach more or less closely a reversible process by the suitable choice of conditions, but the strictly reversible process is purely a concept that aids in the analysis of certain problems. However, the approach of actual processes to this ideal limit can be made almost as close as we please. The closeness of approach is generally limited by economic factors rather than by purely physical ones. The truly reversible process would require an infinite time for its completion, but we are generally in more of a hurry than that. The sole reason for the invention of the concept of the reversible process is to establish a standard for the comparison of actual processes. The reversible process is one that gives the maximum accomplishment; that is, it yields the greatest amount of work or requires the least amount of work to bring about a given change. It tells us the maximum efficiency toward which we may strive but which we never expect to equal. Without such an absolute standard or yardstick, the attempts of engineers to improve processes would be merely shots in the dark with no goal at which to aim. With the reversible process as our standard, we know at once whether an actual process is already highly efficient or whether it is very inefficient and therefore capable of considerable improvement.

Since the reversible process represents a succession of equilibrium states, each only a differential step from its neighbor, the reversible process can be represented as a continuous line on a state diagram (*p–v, T–s,* etc.). On the other hand, the irreversible process cannot be so represented. One can note the terminal states and indicate the general direction of change, but it is inherent in the nature of the irreversible process that the complete path of the change is indeterminate and therefore cannot be drawn as a line of a thermodynamic diagram.

Irreversibilities always lower the efficiencies of processes. Their effect in this respect is identical with that of friction, which is one cause of irreversibility. Conversely, no process more efficient than a reversible process can even be imagined. The reversible process is an abstraction, an idealization, which is never achieved in practice. It is, however, of enormous utility because it allows calculation of work from knowledge of the system properties alone. In addition, it represents a standard of perfection that cannot be exceeded because (1) it places an upper limit on the work *output* from a given work-producing process and (2) it places a lower limit on the work *input* for a given work-requiring process.

1-20 THE PURE SUBSTANCE

A pure substance is one that has a homogeneous and an invariable chemical composition. It may exist in more than one phase, but the chemical composition is the same in all phases. Thus, liquid water, a mixture of liquid water and water vapor (steam), or a mixture of ice and liquid water are all pure substances because every phase has the same chemical composition. On the other hand, a mixture of liquid air and gaseous air is not a pure substance since the composition of the liquid phase is different from that of the vapor phase.

Sometimes, a mixture of gases, such as air, is considered a pure substance as long as there is no change in phase. Strictly speaking, this is not true; rather we should say that any mixture of gases, such as air, exhibits some of the characteristics of a pure substance as long as there is no phase change.

A change in the physical state that is familiar to all of us is the melting of ice to form water and the boiling of water to form steam. Each change of state or phase requires relatively large amounts of heat exchange.

When heat, either absorbed or rejected by a material, causes or accompanies a change in temperature of the material, the heat quantity is called *sensible heat* (q_s). The term "sensible heat" comes from the fact that the change of temperature may be detected by the sense of touch, and the temperature can be measured with a thermometer.

When heat, either absorbed or rejected by a material, brings about or accompanies a change of phase of the material, the heat quantity is sometimes called *latent heat* (q_L). Latent heat addition or removal is accompanied by *no change in temperature* of the substance. When a liquid changes to a solid, we say that a *fusion* process has taken place; when a liquid changes to a vapor, there is *vaporization;* and when a solid changes directly to a vapor, a *sublimation* process is said to have taken place. In each of these processes, energy must be exchanged between the substance and the surroundings to affect the change in phase. The temperature at which these changes will occur is dependent on the pressure exerted on the substance as well as the substance itself.

Figure 1-5 is a sketch of the pressure–temperature diagram for water. The fusion line represents the solid–liquid mixture; the vaporization line represents the liquid–vapor mixture; and the sublimation line represents the solid–vapor mixture.

The *triple point* is the condition where it is possible to maintain an equilibrium mixture of the three phases. The *critical point* (3206.2 psia, 705.4°F) is the state where the vapor phase has the same properties as the liquid phase. It is not possible to distinguish between the liquid and vapor phases at pressures and temperatures above this point.

The three equilibrium lines in Figure 1-5 (fusion, vaporization, and sublimation) may be said to designate *saturation* regions. We may say that the vaporization line represents the saturation region between liquid and

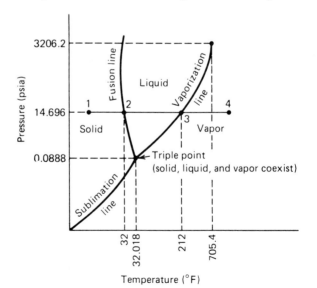

FIGURE 1-5 Pressure-temperature diagram for water (no scale).

vapor. The vapor in such a mixture is said to be *saturated vapor,* and the liquid present in the mixture is said to be *saturated liquid.*

Consider the constant pressure line 1234 of Figure 1-5 located at 14.696 psia. At point 1, the material is solid (ice). As heat is added at constant pressure to the solid, the temperature will increase (process 1–2). At point 2, any addition of heat will cause a change in phase (melting). The temperature will remain constant until the phase change has been completed. Further addition of heat will raise the temperature of the liquid (process 2–3). At point 3, any addition of heat will cause some of the liquid to vaporize. The vaporization process will continue without an increase in temperature until all the liquid has been vaporized. Any further addition of heat will produce *superheated vapor.* Processes 1–2, 2–3, and 3–4 are all sensible heat processes. The change-of-phase processes occurring at points 2 and 3 are latent heat processes.

The quantity of heat absorbed by a given mass of solid material at the fusion temperature corresponding to the pressure in melting to the liquid phase, or conversely the heat rejected by a given mass of liquid at the fusion temperature in solidifying, is called the *latent heat of fusion* and can be determined from

$$q_L = m(h_{fu}) \qquad (1\text{-}20)$$

where

q_L = quantity of heat (Btu or J)
m = mass (lbm or kg)
h_{fu} = latent heat of fusion (Btu/lbm or J/kg)

The temperature at which a liquid will change into the vapor phase is called the *saturation temperature,* sometimes called the *boiling point* or *boiling temperature.* A liquid whose temperature is at the saturation temperature corresponding to the existing pressure is called *saturated liquid,* as previously noted. The saturation temperature is different for each liquid and varies with pressure. Water boils at 180°F when the pressure above it is 7.515 psia. If the pressure above the liquid water is maintained at 100 psia, the water will not boil until its temperature is 327.86°F.

The quantity of heat absorbed by a given mass of liquid at its saturation temperature to cause it to change to the gaseous phase (vapor) at the same temperature is called the *latent heat of evaporation* and may be calculated from

$$q_L = m(h_{fg}) \qquad (1\text{-}21)$$

where

q_L = quantity of heat (Btu or J)
m = mass (lbm or kg)
h_{fg} = latent heat of evaporation (Btu/lbm or J/kg)

Once a liquid has been completely vaporized, the temperature of the vapor may be increased by further addition of heat. The heat added is called *superheat.* When the temperature of the vapor is above the saturation temperature corresponding to the pressure, the vapor is said to be *superheated vapor.*

REVIEW PROBLEMS

1.1. Convert 20°C, 40°C, and 60°C to equivalent degrees Fahrenheit.

1.2. Convert 0°F, 10°F, and 50°F to equivalent degrees Celsius.

1.3. A pressure gauge indicates 25 psi when the barometer is at a pressure equivalent to 14.50 psia. Compute the absolute pressure in psia and inches of mercury when the specific weight of mercury is 13.0 g/cm³.

1.4. A vacuum gauge reads 8 in. Hg when the atmospheric pressure is 29.00 in. Hg. If the specific weight of mercury is 13.6 g/cm³, compute the absolute pressure in psia.

1.5. A skin diver descends to a depth of 80 ft in fresh water. What is the gauge pressure on his body? The specific weight of fresh water can be taken as 62.4 lbf/ft³.

1.6. Determine the density and specific volume of the contents of a 10-ft³ tank if the contents weigh 250 lb. The local acceleration of gravity is 31.1 ft/s².

1.7. A U-tube mercury manometer, open at one end to atmospheric pressure (14.7 psia), is connected to a pressure source. If the difference in mercury levels in the tube is 7.6 in., determine the unknown pressure in psia.

1.8. Convert 500°R, 500 K, and 650°R to degrees Celsius.

BIBLIOGRAPHY

1.1. Jordan, Richard C., and Priester, Gayle B., *Refrigeration and Air Conditioning,* 2nd ed., Prentice-Hall, Inc., Englewood Cliffs, NJ, 1956.

1.2. Granet, Irving, *Thermodynamics and Heat Power,* Reston Publishing Company, Inc., Reston, VA, 1974.

2

Review of Thermodynamic Principles

2-1 INTRODUCTION

Thermodynamics is that branch of engineering science that deals with energy and its transformation. In our work in heating and ventilating, we must deal constantly with energy, its exchanges, and conversion. For example, in heating, we are concerned with the extraction of heat energy from a source such as fuel to heat a fluid medium such as water or air. The fluid medium carries the heat energy to a selected area and discharges it to that area as required. In warm-air systems, the working medium (air) approaches a perfect gas and the relatively simple thermodynamic relationships for such a gas may be applied. Hot-water heating systems also have relatively simple relationships. However, in some systems, such as steam heating, the working medium alternates between the liquid and vapor phases. Here, the thermodynamic relationships are complex, and it is necessary to use tables and charts to determine the physical and thermodynamic properties.

This chapter is concerned with the review of the more important relationships of both gas and vapor thermodynamics that apply most directly to our area of study.

Other important disciplines for our work are heat transfer and fluid flow. These will be discussed in subsequent chapters.

2-2 THERMODYNAMIC LAWS

There are two major premises upon which the science of thermodynamics is based. Because no major exceptions to these rules have ever been experienced, either accidentally or through controlled experiments, and because all attempts to disprove them have failed, they are termed thermodynamic "laws" and have been used as a foundation for further developments.

The *first law of thermodynamics* states that heat and mechanical energy are interconvertible and can be neither created nor destroyed. This is a limited form of the more general law of conservation of energy, which states that all forms of energy are interconvertible and can be neither created nor destroyed. Although these statements are all-inclusive, no inference can be drawn that the conversion of one energy form to another is necessarily complete. Only a portion of a given definite amount of heat energy can be converted to mechanical or to electrical energy, whereas all energy forms can be converted completely to heat energy. This ability for complete conversion into heat accompanied by a general tendency for all energy to be dissipated eventually in this manner has sometimes led to the description of heat as "low-grade energy."

For example, the power input to an electric motor operating a fan is converted, eventually, into heat and is absorbed by the air passing through the fan or lost to the surroundings by convection, conduction, and radiation. All electric-motor losses, such as resistance, windage, hysteresis, and friction, are transformed into heat. All electrical power converted into mechanical energy by the motor is supplied to the fan, which transfers this energy to the air passing through the fan. This mechanical energy is eventually converted to heat. Even the energy transformed into sound in the operation of a system is dissipated eventually as heat.

The *second law of thermodynamics,* according to Rudolph Clausius, a German physicist, states that it is impossible for a self-acting machine unaided by any external agency to transfer heat from one body to another at higher temperatures. Other statements of this same premise are that heat will not of itself flow from one body to another body maintained at a higher temperature and that no machine, actual or ideal, can both completely and continuously transform heat into mechanical energy. All these statements imply the same principle, but Clausius' statement is the most closely allied to air conditioning. The second law enables direct limitations to be placed upon the theoretical maximum operating efficiency that can be attained by a thermodynamic system operating under any specified set of conditions.

2-3 THE SYSTEM

In analyzing the flow of mass and energy in some apparatus, it is convenient to define the *system* as a particular region that is surrounded by real or imaginary surfaces that can be specified. The real or imaginary surfaces that enclose this space constitute the *boundaries* of the system. The region outside the boundaries that may be affected by the transfer of energy or mass is called the *surroundings.*

Systems may be divided into two types: (1) the *closed,* or fixed-mass, system and (2) the *open* system.

Perhaps the simplest type of closed system is illustrated in Figure 2-1 by a steel tank containing some fluid at some pressure and temperature. The space within the tank is the system, and the inner surface of the tank is considered to be the boundary. The tank itself and

anything outside the tank that would be affected by energy transfer constitute the surroundings. Although it is customary to consider the space within the tank as the system, a constant mass is contained within the system boundaries, and this fixed mass is sometimes referred to as the system.

Another very important type of closed system is the volume within a cylinder and frictionless, gastight movable piston as illustrated in Figure 2-2. Obviously, pistons in real machinery are neither frictionless nor gastight. However, in order to analyze many problems, it is necessary to assume idealized conditions and then modify the final results if necessary to consider imperfections. In the system illustrated in Figure 2-2, the inside surfaces of the cylinder, cylinder head, and piston are the boundaries of the system. This system has a variable volume depending on the position of the piston, whereas the system illustrated in Figure 2-1 has a constant volume. Both systems contain a constant, or fixed, mass.

In the two closed systems that have been described, the boundaries consist of the inside surfaces of metal walls. In the general case, a closed system may be considered to be a *nonflow system*—that is, as some region surrounded by stationary or moving real or imaginary boundaries across which no mass is transferred.

The open system is illustrated in Figure 2-3. It consists of a region surrounded by specified boundaries arranged so that a fluid may enter and leave the system. The system consists of the space contained within the inner surfaces of the apparatus including the volume within the inlet and outlet pipes out to some specified planes or sections in these pipes such as sections 1–1 and 2–2 in Figure 2-3. As an illustration of such a system, consider one cylinder of an automobile engine with the inlet and outlet planes in the inlet and exhaust manifolds. Then, a mixture of air and vaporized fuel passes the inlet section intermittently as the engine rotates. The products of combustion pass the outlet section intermittently. The volume of the system changes as the piston moves in the cylinder. Energy is released by combustion within the system, heat is transferred through the cylinder walls, and work is done by the moving piston. In

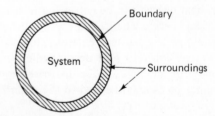

FIGURE 2-1 Constant-volume closed system.

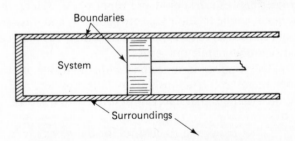

FIGURE 2-2 Variable-volume closed system.

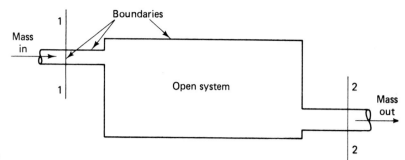

FIGURE 2-3   Open or flow system.

the general case of the open system, the mass within the system and the conditions at the inlet and outlet may vary with respect to time.

In heat exchangers, pumps, compressors, and so forth, operating under steady conditions, the flow of fluid in the inlet and outlet pipes occurs at constant velocity and the mass within the system is constant. Where factors such as velocity, pressure, temperature, and flow rate are constant with respect to time, the system is said to be a *steady-flow system*. The steady-flow system is therefore a special case of the open system. Most of our work here is concerned with the transfer and conversion of energy that occur in steady-flow systems.

2-4   WORK

When a force causes a displacement of a body, the work done on that body is the product of the displacement and the component of force acting in the direction of the displacement. If F is the constant component of the force acting in the direction of the displacement, $x_2 - x_1$, then the work done is

$$_1wk_2 = F(x_2 - x_1) \qquad (2\text{-}1)$$

where

$$F \qquad = \text{force (lbf or N)}$$
$$x_2 - x_1 = \text{distance traversed (ft or m)}$$

If F is a variable force, its magnitude when acting through a differential distance, dx, must be summed up for each variable step (Fdx) to give the total work, and this summation is represented as

$$_1wk_2 = \int_1^2 Fdx \qquad (2\text{-}1a)$$

Work is further defined as *energy that is being transferred across the boundaries of a system because of a force acting through a distance.* Work flows from the surroundings to the system, or vice versa, only when

a force produces some movement or displacement. The driving potential that produces the flow of energy is a *force*.

Referring to Figure 2-4, let it be assumed that a cylinder is closed with a frictionless, gastight movable piston. The system is the gas contained in the cylinder and has an initial pressure and volume of p_1 and V_1, respectively. On the p–V diagram of Figure 2-4, point 1 is located so as to represent to some scale that initial pressure and volume. The piston is allowed to move, and, in the general case, the pressure of the gas varies as the volume changes. A variable pressure expansion is represented in Figure 2-4 by the curve 1–2. The equation of the curve can often be expressed in the form $pV^n = C$, where C is a constant and n is some exponent between 0 and ∞. When the pressure is changing and the equation showing the relationship between p and V is known, it is possible to evaluate the work by considering a small movement of the piston during which the change in pressure is negligible. Assume that the piston

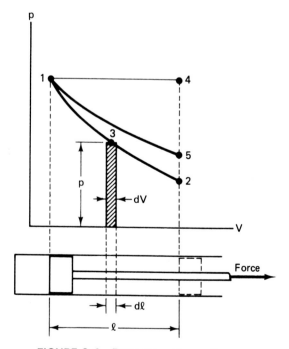

FIGURE 2-4   Pressure–volume diagram.

is at point 3 on Figure 2-4. If the piston then moves a distance $d\ell$, the work done may be written as

$$\delta wk = p\,A\,d\ell = p(A d\ell) = p(dV)$$

where $dV = Ad\ell$ = increase in volume of the system in cubic feet. The elementary cross-hatched area is pdV and represents the work done during the small movement of the piston. By the use of the calculus, the area under the curve 1–2 can be determined as follows:

$$_1wk_2 = \int_{v_1}^{v_2} pdV \text{ (ft lbf)} \qquad (2\text{-}2)$$

where

$$p = \text{pressure (lbf/ft}^2)$$
$$dV = \text{change in volume (ft}^3)$$

The units of work then are ft lbf or N·m (joule).

Equation (2-2) is the general equation for calculating work done by or on a fluid *during a frictionless process in a closed system* when the algebraic relation between pressure and volume is known.

Attention is again called to the definition of work as energy that is being transferred across the boundaries of a system because of a force acting through a distance. Referring to Figure 2-4, note that, as the piston moves to the right, the force could rotate a shaft through a suitable mechanism and drive an electric generator or lift a weight. Energy is leaving the system as work. This flow of energy stops the instant the piston ceases to move.

When work is done by the system as it expands and forces the piston to move to the right, Eq. (2-2) gives a positive result. During compression, an external force pushes the piston to the left, work is supplied from the surroundings that perform work on the system, V_2 is less than V_1, and Eq. (2-2) gives a negative result. Consequently, it is customary to consider work as positive ($+wk$) when the system is expanding and is transferring work from the system to the surroundings. Conversely, the work term is negative ($-wk$) when the system is undergoing compression and work is being transferred from the surroundings to the system.

Power is the *rate* of doing work. The horsepower (hp) is the common unit of power in EES units. One horsepower is defined as work being done at the rate of 550 ft lbf/s or 33,000 ft lbf/min. One horsepower-hour (hp-hr) is defined as the quantity of work done in one hour if the work is done continuously at an average rate of 33,000 ft lbf/min. for a period of one hour. Therefore, 1 hp-hr equals 1,980,000 ft lbf/hr.

The SI unit of power is the watt (W), which is equivalent to J/s. Also,

1 joule (J)	= 1 N·m = 0.73757 ft lbf
1 watt (W)	= 1 J/s = 0.73757 ft lbf/s
1 kilowatt (kW)	= 737.57 ft lbf/s
1 horsepower (hp)	= 0.746 kW
1 kilowatt (kW)	= 1.340 hp

The product of power and time is an energy unit. One kilowatt acting for one hour is the very common energy unit of kilowatt-hour (kW-hr). Also,

$$1 \text{ kW-hr} = 1.340 \text{ hp-hr} = 2,655,300 \text{ ft lbf/hr}$$

2-5 HEAT

Let Figure 2-5 represent a closed system containing gas and having its surroundings at a temperature higher than that of the system. It is common experience that in such a situation the temperature of the system will increase. Also, in general, the temperature of the surroundings will decrease. If conditions are allowed to proceed to equilibrium, the system and the surroundings will attain a uniform temperature. Heat is defined as *energy that is being transferred across the boundaries of a system because of a temperature difference.* The energy that is transferred was stored initially in the molecules of the surroundings as internal energy. After transfer has been completed, it is stored in the system as internal energy. The temperature of the gas in the system has increased, which means that the average velocity of translation of the molecules has increased or that the internal energy of the gas has increased.

The similarity in the definitions of heat and work should be noted. Both are forms of energy in the process of *being transferred* across the system boundaries. In the case of work, the driving potential is a *force*. In the case of heat, the driving potential is a *temperature difference*. Since both heat and work are energy in transition, these quantities may be expressed in the same units.

The EES unit of heat is the British thermal unit (Btu). The relationship between Btu and ft lbf is 1 Btu

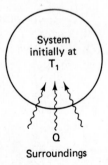

FIGURE 2-5 Heat transfer to system from surroundings across system boundary.

= 778.26 ft-lb (usually taken as 778). The SI unit of heat is the joule (J).

Since heat may be transferred to a system from the surroundings, or vice versa, a sign convention is selected such that heat transferred *to* a system is positive ($+Q$) and heat transferred *from* a system is negative ($-Q$). Note that

$$1 \text{ kw-hr} = 3413 \text{ Btu}$$
$$1 \text{ Btu} = 1055 \text{ J} = 778 \text{ ft lbf}$$

2-6 SPECIFIC HEAT

In engineering literature, the specific heat (c) is defined as the quantity of heat that must be transferred across the boundaries of a system to cause a temperature change of one degree per unit mass of the system *in the absence of friction*. Frictional effects are excluded because, in some work processes, the work can produce a change in temperature through the mechanism of fluid friction.

In general, specific heat is a variable. Therefore,

$$\delta Q = mc\, dt$$

and

$$_1Q_2 = m \int_{t_1}^{t_2} c\, dt \qquad (2\text{-}3)$$

where

$_1Q_2$ = heat transferred in Btu (J)
m = mass of substance being heated in lbm (kg)
c = specific heat in Btu/lbm-°F (J/kg-K)
t_1 = initial temperature in °F (K)
t_2 = final temperature in °F (K)

For most substances, c increases with temperature and can be expressed as some function of temperature. In a great many cases, a mean value of specific heat may be used, which permits treating the specific heat as a constant. In this case, Eq. (2-3) becomes

$$Q = mc(t_2 - t_1) \qquad (2\text{-}4)$$

If heat is added to a compressible fluid such as a gas, the volume is either constant or changing during the process of heat addition. If the volume is constant, all the heat added is used to increase the gas temperature. The heat required to change the temperature of unit mass of the gas one degree at constant volume is called the *constant-volume specific heat, c_v.* Figure

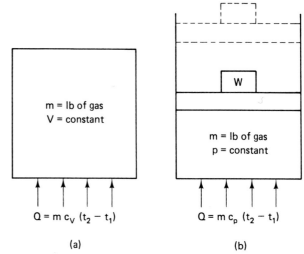

FIGURE 2-6   (a) Constant-volume and (b) constant-pressure specific heat.

2-6a illustrates a closed constant-volume system such as a sealed tank.

A gas may expand at constant pressure during the heating process, as would be the case in a vertical cylinder closed by a gastight movable piston as illustrated in Figure 2-6b. In this case, work is done by the gas as it expands, lifting the piston weight. The work so done would be stored in the mechanical potential energy of the weight in its final position shown by the dotted outline. Sufficient heat energy must be added to increase the gas temperature and also perform the work of expansion. The heat required to raise the temperature of unit mass of the gas one degree during a constant-pressure process is called *constant-pressure specific heat, c_p.* It should be obvious that c_p is greater than c_v for a particular gas because no work is done during a constant-volume heat addition process. Many of the processes involved in air conditioning may be considered constant-pressure processes.

2-7 ENERGY EQUATIONS

It is frequently necessary to evaluate the energy interchanges occurring to or from a working medium during a process. The development of the quantitative relationships necessary to accomplish such an evaluation is based upon five basic equations: (1) the *specific heat* equation, Eq. (2-3), (2) the *nonflow* and (3) the *steady-flow work* equations for an expanding substance, and (4) the *nonflow* and (5) the *steady-flow energy* equations.

Nonflow and Steady-Flow Work Equations. As noted in Section 2-3, when there is no transfer of the working substance during a process, it is termed *non-*

flow; when there is a continuous and steady flow of the working medium, it is termed *steady-flow.* Processes involving intermittent flow with rapid cycling, such as the compression of a refrigerant vapor or air in a piston compressor, are usually treated as steady-flow, although, if desired, the individual parts of the cycle may be treated as nonflow. The nonflow work equation for compressible fluid was derived in Section 2-4, Eq. (2-2), as $_1wk_2 = \int_1^2 p\,dV.$

If we consider unit mass of working substance, then we can say $_1wk_2 = \int_1^2 p\,dv,$ where dv is the change in specific volume. In the compression process illustrated in Figure 2-7, this is equivalent to the area under the curve 1–2 (area 12301). If, during compression, however, steady-flow conditions exist, this area represents only a portion of the work required for compression. The working substance enters the compression process possessing flow energy p_1v_1 (area 015601) and is discharged with flow energy p_2v_2 (area 24632). The difference in these flow energies added to the nonflow work equation results in the steady-flow work equation equivalent to the process:

$$_1wk_2 = \int_1^2 v\,dp \qquad (2-5)$$

The quantity $pv,$ you will recall, is the energy associated with a unit mass of fluid by virtue of the displacement of its boundary from one position to another at constant pressure. This was called *displacement energy* or *flow energy,* sometimes called *flow work,* as discussed in Chapter 1.

Figure 2-8 represents an elementary steady-flow device existing between planes 1 and 2. Fluid in the inlet pipe is at pressure p_1 (psfa) and has a specific volume v_1 (ft³/lbm). Pushing the fluid into the system across the system boundary requires work from the upstream fluid; otherwise, the fluid would not be flowing. If the cross-hatched volume of the inlet pipe contains 1 lbm

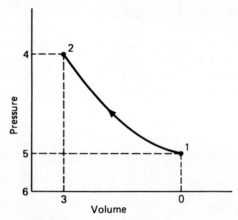

FIGURE 2-7   Compression process.

of fluid, and if A_1 is the internal cross-sectional area of the inlet pipe in square feet, we may say that $A_1\ell_1 = v_1$ (ft³). The force pushing this volume of fluid is p_1A_1 pounds force. The work done in pushing this fluid across the system boundary at section 1 is

$$wk = \text{force} \times \text{distance} = (p_1A_1)(\ell_1)$$

But $A_1\ell_1 = v_1$; therefore,

$$wk = p_1v_1 \qquad \text{(ft lbf/lbm of fluid)} \qquad (2-6)$$

Thus, when a fluid is flowing past a given section into a system, it is flowing because of work being done on it to push it past the section, and the amount of work is p_1v_1.

The same type of analysis could be made at the exit pipe section. In that case, the exit flow energy (work) would be p_2v_2. The net flow work for a given open system would be

$$\text{net flow work} = p_2v_2 - p_1v_1 \qquad (2-7)$$

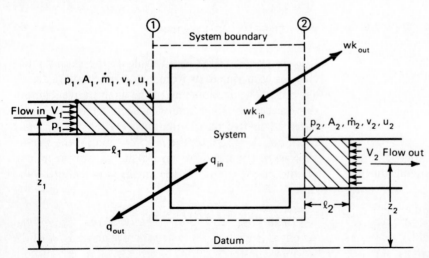

FIGURE 2-8   Flow work—steady-flow system.

Steady-Flow Energy Equation. The steady-flow, or general, energy equation is a mathematical statement of the first law of thermodynamics. It is merely an energy balance equating all energy entering a system to that leaving a system. The only restrictions on its application are the following:

1. The mass of fluid in the system must be constant. If 1 lbm of fluid enters, 1 lbm must leave at the same time.
2. Pressure, temperature, specific volume, and flow velocity are all constant with respect to time at the entrance section. Values of the same quantities at the exit section must also be constant with respect to time.
3. Energy transferred across the system boundaries as heat and work must be at a constant rate.

The mass of fluid entering the system of Figure 2-8 would be

$$\dot{m}_1 = \rho_1 A_1 V_1$$

where

$\dot{m}_1$ = mass flow rate (lbm/s or kg/s)
ρ_1 = mass density (lbm/ft^3 or kg/m^3)
A_1 = flow area (ft^2 or m^2)
V_1 = average flow velocity (ft/s or m/s)

Also, the mass flow of fluid leaving the system would be

$$\dot{m}_2 = \rho_2 A_2 V_2$$

Since mass entering and leaving must be equal for a steady-flow system,

$$\dot{m}_1 = \dot{m}_2 = \rho_1 A_1 V_1 = \rho_2 A_2 V_2 \qquad (2\text{-}8)$$

Equation (2-8) is commonly referred to as the *continuity equation.* When the fluid flowing through the system is considered *incompressible* (usually, liquids are considered incompressible at ordinary pressures), $\rho_1 = \rho_2$ and Eq. (2-8) becomes

$$\dot{Q} = A_1 V_1 = A_2 V_2 \qquad (2\text{-}9)$$

where $\dot{Q}$ = volume flow rate (ft^3/s or m^3/s).

In applying the first law of thermodynamics to the steady-flow system of Figure 2-8, it is important to include *all* energy terms identified in Chapter 1, such as potential energy, kinetic energy, internal energy, and flow work. The fluid can now do mechanical work in the amount $+ _1wk_2$, or mechanical work can be done on the fluid in the amount $- _1wk_2$. The subscripts 1 and 2 simply imply that work occurs between the fluid inlet and outlet sections, and the term is read as "work one to two." In the case of a turbine or an engine, energy leaves the system through a driven shaft. In the case of a compressor or pump, work is supplied to the system. In the engine example, some of the work generated in the device may be dissipated as friction in bearings; not all of it will appear as delivered energy from the shaft. Such a condition also exists in the opposite sense for a compressor or pump.

Heat in the amount of $+ _1q_2$ can be added to certain devices, such as a boiler, or perhaps may leave in the amount $- _1q_2$ by radiation or convection from cooling coils to colder surroundings, or the like.

The steady-flow energy equation for the device in Figure 2-8 is written by equating the energy leaving the system to that entering, or

$$Z_1\left(\frac{g}{g_c}\right) + \frac{V_1^2}{2g_c} + u_1 + p_1v_1 + _1q_2$$

$$= Z_2\left(\frac{g}{g_c}\right) + \frac{V_2^2}{2g_c} + u_2 + p_2v_2 + _1wk_2 \quad (2\text{-}10)$$

All terms shown in Eq. (2-10) must have consistent units. In the foot–pound–second (EES) system, the unit of mass is the pound mass, elevation Z is expressed in feet, V is in feet per second, and g/g_c is the lbf/lbm. Since it is customary to express the heat q and the internal energy u in Btu/lbm, the factor $J = 778$ ft lbf/Btu must be used to obtain units of Btu/lbm for all the terms. Thus,

$$\frac{Z_1}{J}\left(\frac{g}{g_c}\right) + \frac{V_1^2}{2g_cJ} + u_1 + \frac{p_1v_1}{J} + _1q_2$$

$$= \frac{Z_2}{J}\left(\frac{g}{g_c}\right) + \frac{V_2^2}{2g_cJ} + u_2 + \frac{p_2v_2}{J} + \frac{_1wk_2}{J} \quad (2\text{-}11)$$

If we recall that enthalpy h is equal to $u + pv$, Eq. (2-11) may be written as

$$\frac{Z_1}{J}\left(\frac{g}{g_c}\right) + \frac{V_1^2}{2g_cJ} + h_1 + _1q_2$$

$$= \frac{Z_2}{J}\left(\frac{g}{g_c}\right) + \frac{V_2^2}{2g_cJ} + h_2 + \frac{_1wk_2}{J} \quad (2\text{-}11a)$$

where h = enthalpy (Btu/lbm).

In SI units, the unit of mass is the kilogram and energy is expressed in terms of the newton-meter (N·m) with its name joule. The elevation Z is in meters, velocity V is in meters per second, g_c is unity, pressure p is in pascals (newtons per square meter), and specific volume is in cubic meters per kilogram. The standard weight force of 1 kilogram mass is 9.807 N, which is needed for the potential energy term of Eq. (2-10). Therefore, for joule/kilogram, we have

$$9.807Z_1 + \frac{V_1^2}{2g_c} + h_1 + {}_1q_2$$

$$= 9.807Z_2 + \frac{V_2^2}{2g_c} + h_2 + {}_1wk_2 \qquad (2\text{-}12)$$

Nonflow Energy Equation. The nonflow, or simple, energy equation is a simplified mathematical statement of the steady-flow energy equation when we eliminate all terms not involved in the nonflow process as well as those that are usually negligible. Considering unit mass of working substance, the nonflow energy equation is

$${}_1q_2 = u_2 - u_1 + {}_1wk_2 \qquad (2\text{-}13)$$

Figure 2-9 illustrates two constant-mass, nonflow systems. Figure 2-9a represents a cylinder containing a movable, frictionless piston. As the piston moves, the gas volume changes. In Figure 2-9b, the tank, considered to be rigid, represents a constant-mass, constant-volume system. In Eq. (2-13), the subscripts 1 and 2 identify the limits of the process—that is, process 1 to 2:

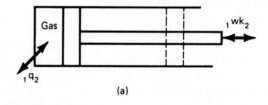

(a)

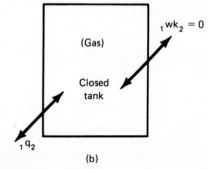

(b)

FIGURE 2-9 Nonflow systems—fixed mass.

- ${}_1q_2$ represents the heat energy transferred to or from the system during the process. A positive sign indicates *heat addition,* and a negative sign indicates *heat removal.*
- ${}_1wk_2$ represents the work transferred from or to the system during the process. A positive sign indicates *work done by the system* as gas expands against the piston, and a negative sign indicates *work done on the system* by the piston compressing the gas.
- $u_2 - u_1$ represents the change in internal energy stored in the gas within the system.

2-8 PROCESSES AND CYCLES

A *process* is a change in state. A process is described in part by a series of states passed through by a system. Often, but not always, some sort of interaction between the system and surroundings occurs during a process; the specification of this interaction completes the description of the process.

A description of a process typically involves specification of the initial and final equilibrium states, the path (if identifiable), and the interactions that take place across the boundaries of the system during the process. *Path* in thermodynamics refers to the specification of a series of states through which the system passes. Of special significance in thermodynamics is a *quasistatic* process or path. During such a process, the system internally must be infinitesimally close to a state of equilibrium at all times. That is, the path of a quasistatic process is a series of equilibrium steps. Although a quasistatic process is an idealization, many actual processes approximate quasistatic conditions closely. It is extremely helpful in engineering analysis when an actual process is a nonequilibrium one; however, it is still essential that the initial and final states be equilibrium ones. For nonequilibrium processes, certain intermediate information during the process is missing. Nevertheless, we are able to predict various overall effects, even though a detailed description is not possible.

For some processes, there are special names:

1. A process wherein the pressure does not change is termed as an *isobaric* or a *constant-pressure* process.
2. A process occurring at constant temperature is an *isothermal* process.
3. A constant-volume process is *isometric.*
4. With no heat transfer to or from the system, the process is *adiabatic.*
5. With no change in entropy, the process is *isentropic.*

A thermodynamic *cycle* is a collection of two or more processes for which the initial and final states are the same. The cycle may take place once or may be repeated over and over again.

2-9   SECOND LAW FOR A CYCLE

The first law of thermodynamics gives us a technique for energy analysis, but it does not describe how the energy will flow. The second law of thermodynamics gives direction to the energy flow. The second law was discussed briefly in Section 2-2.

The thermal efficiency, η_t, of any heat engine is the net work produced divided by the energy supplied to produce that work, or

$$\eta_t = \frac{\text{net work done}}{\text{heat supplied}} = \frac{q_a - q_r}{q_a} \qquad (2\text{-}14)$$

where q_a is heat supplied to the cycle and q_r is heat rejected from the cycle.

For the limited purpose of our thermodynamics review, the second law may be stated as follows: Given a heat engine operating in a cycle between a source of energy at a constant absolute temperature T_H and an environment at a lower absolute temperature T_L to which the engine may reject that portion of the energy received from the source which it cannot convert to work, the maximum thermal efficiency is

$$(\eta_t)_{\text{max}} = \frac{T_H - T_L}{T_H} = 1 - \frac{T_L}{T_H} \qquad (2\text{-}15)$$

Complete conversion into work of all the energy transferred to the system as heat from the source requires that the sink or receiver temperature T_L be equal to absolute zero. Since the lowest temperature at which an engine can reject heat is normally the temperature of the atmosphere or a body of water such as a river or lake, it is apparent that no machine can ever be built with a thermal efficiency approaching 100%.

The *Carnot cycle* is an ideal, thermodynamically reversible power cycle, first investigated by Sadi Carnot in 1824 as a measure of the maximum possible conversion of heat energy into mechanical energy. In its reversed form, it is used as a measure of the maximum possible performance of refrigeration equipment. Although it cannot be applied in an actual machine because of the impossibility of obtaining a completely reversible engine, it is nevertheless extremely valuable as a criterion of inherent limitations.

The direct Carnot cycle is shown graphically in Figure 2-10 on *p-v* and *T-s* coordinates. The cycle con-

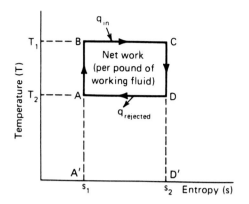

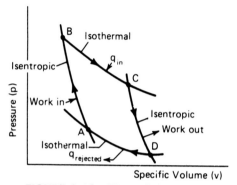

FIGURE 2-10   Direct Carnot cycle.

sists of an isothermal (constant-temperature) heat addition process BC, an adiabatic (frictionless and no-heat-loss) expansion process CD, an isothermal heat rejection process DA, and an adiabatic compression process AB to form a closed cycle. Areas on the *T-s* diagram represent the heat quantities because, from Section 1-18, $T \times \Delta s = q$.

From Figure 2-10, we may see that $T_1(s_2 - s_1)$ represents the total area A'BCD' and indicates the total heat supplied at constant temperature T_1. The quantity $T_2(s_2 - s_1)$ represents the area A'ADD' and indicates the total heat rejected at constant temperature T_2. The difference between the two areas is area ABCDA, which indicates the net cycle of work. Equation 2-14 gives a statement of thermal efficiency. Therefore, for the direct Carnot cycle,

$$\eta_{Carnot} = \frac{T_1(s_2 - s_1) - T_2(s_2 - s_1)}{T_1(s_2 - s_1)}$$

or

$$\eta_{Carnot} = \frac{T_1 - T_2}{T_1} \qquad (2\text{-}16)$$

which is the same as Eq. (2-15) with different subscripts.

2-10 THERMODYNAMIC RELATIONSHIPS FOR IDEAL GASES

In our study of air conditioning, we are involved with the behavior of liquids, gases, and vapors as various processes take place. We will now review briefly these processes and the changes in fluid properties that occur.

An *ideal* or perfect gas may be defined as one that obeys the *equation of state* and whose internal energy is a function of temperature only. The equation of state relates the properties of pressure, specific volume, and temperature and is

$$pv = RT \qquad (2\text{-}17)$$

where

p = pressure in lbf/ft^2 (N/m^2)
v = specific volume in ft^3/lbm (m^3/kg)
T = absolute temperature in °R (K)
R = gas constant in ft lbf/lbm-°R (J/kg-K)

The gas constant R is different for each gas when based on unit mass for that gas. However, for a mole (molecular weight) of a gas, the value of molal $\overline{R}$, the universal gas constant, is the same for every gas:

$$\overline{R} = 8314.41 \text{ J/kg-mol-K}$$

$$= 1545.32 \text{ ft lbf/lb-mol-°R}$$

or

$$= 1.986 \text{ Btu/lb-mol-°R}$$

The molecular weights of many gases are tabulated in Table 2-1. To find the value to use in Eq.

(2-17), divide $\overline{R}$ by the molecular weight, $\mathscr{M}W$, as follows:

$$R = \frac{\overline{R}}{\mathscr{M}} = \frac{8314.41}{\mathscr{M}W} \text{ J/kg-K} \qquad (2\text{-}18)$$

and

$$R = \frac{\overline{R}}{\mathscr{M}W} = \frac{1545.32}{\mathscr{M}W} \text{ ft lbf/lbm-°R} \qquad (2\text{-}19)$$

If Eq. (2-17) is multiplied by the quantity of mass *m,* we obtain

$$pmv = mRT$$

Since the product *mv* is equal to the total volume *V,* we have

$$pV = mRT \qquad (2\text{-}20)$$

For a fixed-mass or closed system, the relationship between pressure, volume, and temperature at the beginning and end of any process involving an ideal gas may be shown to be

$$\frac{p_1V_1}{T_1} = \frac{p_2V_2}{T_2} = \text{constant} \qquad (2\text{-}21)$$

or for unit mass,

$$\frac{p_1v_1}{T_1} = \frac{p_2v_2}{T_2} = \text{constant} \qquad (2\text{-}21a)$$

Since the internal energy of a perfect gas is dependent on absolute temperature, the change in inter-

TABLE 2-1
Properties of gases

Gas	Chemical Formula	Molecular Weight	R, ft lbf lbm-°R	c_p Btu lbm-°R	c_p kJ kg-K	c_v Btu lbm-°R	c_v kJ kg-K	k
Air	—	28.97	53.34	0.240	1.0	0.171	0.716	1.400
Argon	Ar	39.94	38.66	0.125	0.523	0.075	0.316	1.667
Carbon dioxide	CO_2	44.01	35.10	0.203	0.85	0.158	0.661	1.285
Carbon monoxide	CO	28.01	55.16	0.249	1.04	0.178	0.715	1.399
Helium	He	4.003	386.0	1.25	5.23	0.753	3.158	1.667
Hydrogen	H_2	2.016	766.4	3.43	14.36	2.44	10.22	1.404
Methane	CH_4	16.04	96.35	0.532	2.23	0.403	1.69	1.32
Nitrogen	N_2	28.016	55.15	0.248	1.04	0.177	0.741	1.400
Oxygen	O_2	32.00	48.28	0.219	0.917	0.157	0.657	1.395
Steam	H_2O	18.016	85.76	0.445	1.863	0.335	1.402	1.329

Source: Howell, R. H., and Sauer, R. J., *Environmental Control Principles* published by the American Society of Heating, Refrigerating, and Air Conditioning Engineers, Inc., Atlanta, GA, 1985. Used with permission.

TABLE 2-2

Values of polytropic exponents *n* for specific processes

Isentropic* (s = constant)	frictionless adiabatic	$n = k = C_p/C_v$
Isothermal (t = constant)	constant temperature	$n = 1$
Isobaric (p = constant)	constant pressure	$n = 0$
Isometric (v = constant)	constant volume	$n = \infty$
Polytropic	any	$n = 0$ to ∞

* If a process takes place without heat transfer and is frictionless, it is reversible and follows a path of constant entropy (*s*) and hence is called an *isentropic* process. Without heat transfer a process is called *adiabatic*.

nal energy in a nonflow constant-volume process may be given by

$$u_2 - u_1 = c_v(T_2 - T_1) \quad \text{(Btu/lbm)} \quad (2\text{-}22)$$

The enthalpy change of a perfect gas during all processes (nonflow and steady-flow) may be shown to be

$$h_2 - h_1 = c_p(T_2 - T_1) \quad \text{(Btu/lbm)} \quad (2\text{-}23)$$

The relationship between the constant-pressure specific heat c_p and the constant-volume specific heat c_v for a perfect gas may be shown to be

$$c_p = c_v + \frac{R}{J} \quad (2\text{-}24)$$

In general, the polytropic pressure–volume relationship is

$$pv^n = \text{constant} \quad (2\text{-}25)$$

where the exponent *n* varies from 0 to ∞. Values of *n* for specific processes are shown in Table 2-2.

Many processes encountered in air conditioning involve working fluids that alternate between the vapor and liquid phases and therefore cannot be treated thermodynamically as perfect gases. Some fluids, such as air, may be treated as a perfect gas, and Table 2-3 summarizes thermodynamic relationships for your reference. You should refer to any good textbook in thermodynamics for further information. Table 2-3 has been developed using Eqs. (2-17) through (2-25) and the nonflow energy equation, Eq. (2-13).

2-11 TABLES AND GRAPHS FOR PROPERTIES OF REAL FLUIDS

Many fluids used in air conditioning systems must be treated as imperfect gases or as vapors. Refrigerants and steam are the two most common. In most systems, these fluids alternate between the gaseous phase and the liquid phase and sometimes may coexist in varying pro-

portions in the two phases at the same time. Although this greatly complicates the thermodynamic relationships, the actual solution of problems is simplified by the use of tabular data and charts giving thermodynamic properties of these fluids.

Before discussing tables and charts of thermodynamic properties for real fluids, we should review briefly what is meant by *saturation* and *superheat*. If the temperature of a liquid is at a point when any small amount of heat addition will cause it to boil, the liquid is said to be "saturated" and the corresponding temperature is called the *saturation temperature*. The pressure on the liquid is called the *saturation pressure*.

Figure 2-11 illustrates the processes of saturation, superheating, and condensation using the familiar fluid, water. Figure 2-11a shows the water at the saturation temperature of 212°F (100°C) when the pressure above the liquid is 14.696 psia (101.325 kPa). At this state, if heat is added (q_{in}) in even a small amount, saturated steam (vapor) will be produced above the liquid. The liquid is said to "boil." Note that this process of evaporation (boiling) takes place at constant pressure and constant temperature. As long as any liquid remains in the container, the temperature and pressure will remain at the above temperature and pressure.

In Figure 2-11b, the outlet valve is partially closed, restricting the escape of steam. Under these conditions, the pressure above the liquid will increase above standard atmospheric pressure to a pressure of, say, 40.0 psia. At this pressure, the water will not start to boil until its temperature is 267.26°F. Again, we have the state of saturated liquid and saturated vapor. Therefore, we can say that the saturation temperature of water for 40.0 psia is 267.26°F.

If the saturated steam is heated further in a heat-exchange device, at constant pressure, the temperature of the steam is increased above the saturation temperature, and the steam is said to be "superheated." The heating device is called a *superheater*.

In Figure 2-11c, heat is added to saturated water at 14.696 psia (101.325 kPa) and 212°F (100°C) to produce saturated steam. The saturated steam enters a heat-exchange device where heat is extracted from the steam (q_{out}) to produce saturated liquid. The heat-ex-

TABLE 2-3

Thermodynamic relationships for gases (constant specific heat)

Process	Isometric or Constant Volume	Isobaric or Constant Pressure	Isothermal or Constant Temperature	Isentropic or Constant Entropy (Reversible Adiabatic)	Polytropic	Free Expansion (Irreversible Adiabatic)
Pressure-Volume Relationship pv^n = constant	$n = \infty$ $V_1 = V_2$	$n = 0$ $p_1 = p_2$	$n = 1$ $p_1 V_1 = p_2 V_2$	$n = k$ $p_1 V_1^k = p_2 V_2^k$ $\dfrac{T_2}{T_1} = \left(\dfrac{V_1}{V_2}\right)^{k-1} = \left(\dfrac{P_2}{P_1}\right)^{(k-1)/k}$	$p_1 V_1^n = p_2 V_2^n$ $\dfrac{T_2}{T_1} = \left(\dfrac{V_1}{V_2}\right)^{n-1} = \left(\dfrac{P_2}{P_1}\right)^{(n-1)/n}$	
Heat added to gas	$mC_v(T_2 - T_1)$	$mC_p(T_2 - T_1)$	$\dfrac{mRT}{J} \ln\left(\dfrac{V_2}{V_1}\right)$	0	$m\left(\dfrac{n-k}{n-1}\right) C_v(T_2 - T_1)$	0
Work done by gas (nonflow process)	0	$p(V_2 - V_1)$	$pV \ln\left(\dfrac{V_2}{V_1}\right)$	$\dfrac{p_1 V_1 - p_2 V_2}{k - 1}$	$\dfrac{p_1 V_1 - p_2 V_2}{n - 1}$	0
Work done by gas (steady-flow process)	$V(p_2 - p_1)$	0	$pV \ln\left(\dfrac{P_2}{P_1}\right)$	$\dfrac{k}{k - 1}(p_1 V_1 - p_2 V_2)$	$\dfrac{n}{n - 1}(p_1 V_1 - p_2 V_2)$	0
Gain in entropy of gas	$mC_v \ln\left(\dfrac{T_2}{T_1}\right)$	$mC_p \ln\left(\dfrac{T_2}{T_1}\right)$	$\dfrac{mR}{J} \ln\left(\dfrac{V_2}{V_1}\right)$	0	$m\left(\dfrac{n - k}{n - 1}\right) C_v \ln\left(\dfrac{T_2}{T_1}\right)$	$\dfrac{mR}{J} \ln\left(\dfrac{V_2}{V_1}\right)$
Gain in internal energy of gas	$mC_v(T_2 - T_1)$	$mC_v(T_2 - T_1)$	0	$-mC_v(T_2 - T_1)$	$mC_v(T_2 - T_1)$	0

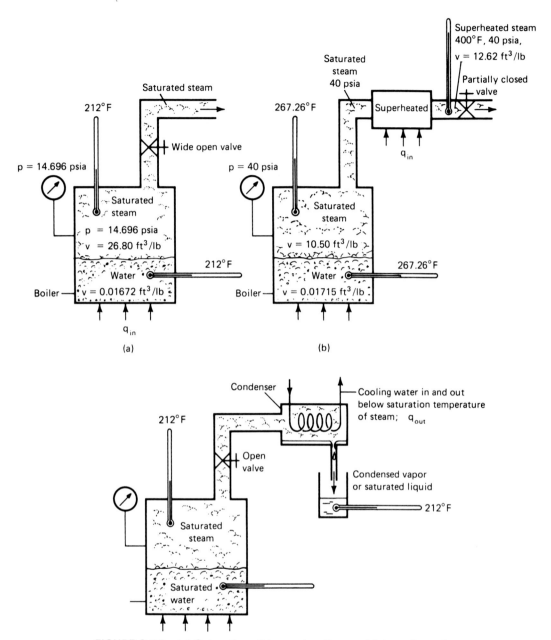

FIGURE 2-11   (a) Saturation, (b) superheating, and (c) condensation.

change device is called a *condenser,* and the process carried out at constant pressure is called *condensation*. The heat that is removed from the steam in the condenser is exactly equal to the amount of heat added to the water in the boiler ($q_{in} = q_{out}$) neglecting radiation losses.

Certain points may be noted at this time. First, for each pressure, there is a corresponding saturation temperature. Second, as the water changes to steam, there is a large increase in specific volume, approximately 1600 times at atmospheric pressure. Third, as saturated steam is superheated at constant pressure, there is an increase in specific volume.

Tables of thermodynamic properties of many substances are available, and, in general, all have the same form. Of particular interest to us in the study of heating and ventilating systems are those for water–steam. Tables A-1 and A-2 (see Appendix A) list properties of pressure, temperature, specific volume, internal energy, enthalpy, and entropy for saturated water and steam. Table A-3 lists corresponding properties for superheated steam. Subscripts used in the tables are as follows:

f, for saturated liquid.

g, for saturated vapor.

fg, for change in specific property from liquid to vapor.

It may be seen, in studying the tables, that the following relationships are true:

$$v_g = v_f + v_{fg}$$
$$h_g = h_f + h_{fg}$$
$$u_g = u_f + u_{fg}$$
$$s_g = s_f + s_{fg}$$

Similar property tables are available for refrigerants when we are working in the field of refrigeration.

As is the case with several working fluids, such as water, refrigerants, and so forth, it is possible for the liquid and vapor phases to coexist during processes. The two-phase fluid properties may be calculated from saturation properties if the *quality, x,* is known, where the quality is defined as the mass of vapor divided by the mass of mixture. The following equations are useful in the determination of two-phase mixtures where x is the percent quality expressed as a decimal:

$$h = h_f + x(h_{fg}) \qquad (2\text{-}26)$$
$$v = x(v_g) + (1 - x)v_f \qquad (2\text{-}27)$$
$$s = s_f + x(s_{fg}) \qquad (2\text{-}28)$$

It is sometimes convenient to show fluid properties in graphical form using *phase diagrams*. Figure 2-12 shows four such phase diagrams for water in its three phases of solid, liquid, and vapor. Of the several different combinations of properties for the coordinates, the ones shown are most common.

Figure 2-12a is the *pressure–volume* diagram, Figure 2-12b is the *temperature–entropy* diagram, Figure 2-12c is the *pressure–enthalpy* diagram, and Figure 2-12d is the *enthalpy–entropy* diagram (usually called a *Mollier chart*). Figures 2-12a, b, and c have boundary curves separating the phase areas and are identically numbered.

The *saturated-liquid line* (3–4) and *saturated-vapor line* (4–6), together with the *triple-point line* (2–3–5) bound the areas in which the liquid and vapor phases coexist in varying proportions. In this area, the mass proportionality of the homogeneous mixture of the two phases is called *quality,* as identified previously. A quality of 90% would indicate a mixture of 10% saturated liquid and 90% saturated vapor by weight.

In the area to the left of the saturated-liquid line and above the triple-point temperature, the substance is a *subcooled liquid.* In the area to the right of the saturated-vapor line and both above and below the triple-

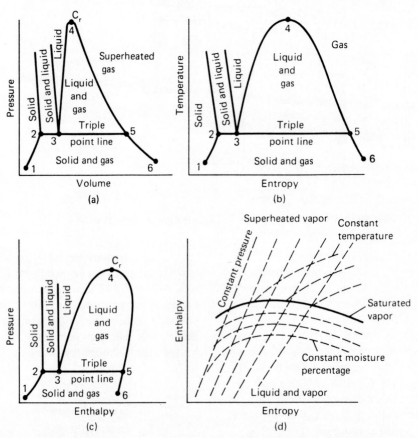

FIGURE 2-12 Phase diagrams for water (not drawn to scale).

point temperature, the substance is a *superheated vapor.*

The critical point (4) at the junction of the saturated-liquid and saturated-vapor lines defines the *critical temperature* above which liquidization of the substance cannot occur, except under extreme pressure. Above the *critical pressure,* the latent heat of vaporization becomes zero, the boundary line between the liquid and vapor phases disappears, and there is no recognizable phenomenon such as vaporization (boiling) and condensation. There is no observable evidence of a phase change when the water passes from liquid to vapor or vice versa.

The *saturation-solid line* (1–2), the lower portion of the *saturated-vapor line* (5–6), and the *triple-point isotherm* (2–3–5) bound the area in which the solid and vapor phases coexist in varying proportions. In the area to the left of the saturated-solid line and below the triple-point line (2–3) (at the triple-point temperature) is a unique series of state points at which the substance may exist in all three phases (solid, liquid, and gas) in equilibrium. Below the triple-point temperature, the heat required to change substance from a solid directly to a gas is termed the *latent heat of sublimation.* At the triple-point temperature, the heat required to change from a solid to a liquid (along 2–3) is termed the *latent heat of fusion.* Above this temperature, the heat required to change from a liquid to a vapor is termed the *latent heat of vaporization.*

Figure 2-12d is the enthalpy–entropy diagram for steam. The coordinates and steam property lines plotted on such a diagram make it especially useful in steam power work, where isentropic expansion and compression processes, the throttling process, and so forth, may be illustrated graphically. Its usefulness for our work in air conditioning is rather limited.

Figures 2-12a and b are useful in showing processes in a graphical way when we are discussing steam. Figure 2-12c is used most often in studying the processes involved in vapor compression refrigeration using refrigerants.

2-12 GAS MIXTURES

Mixtures of different gases are so commonly encountered in engineering applications that methods for computing the properties of mixtures are most necessary.

The individual gases in a gaseous mixture are called the *constituents* of the mixture. If we consider a gas mixture composed of constituent gases *a*, *b*, and *c*, then the mass *m* of the mixture will be

$$m = m_a + m_b + m_c \qquad (2\text{-}29)$$

The *mass fraction* of each constituent would be m_a/m, m_b/m, and m_c/m, respectively, with symbols Y_a, Y_b, and Y_c.

The total moles of mixture, N, is the summation of the moles of the constituents, or

$$N = N_a + N_b + N_c \qquad (2\text{-}30)$$

The *mole fraction, X,* for each constituent would be N_a/N, N_b/N, and N_c/N, represented by X_a, X_b, and X_c, respectively.

Let us consider that we have the constituents of a gas mixture separated and existing at the same temperature, T, and pressure, p. The equation of state may be written for each of the gases. For constituent a,

$$pV_a = N_a \overline{R} T \qquad (2\text{-}31)$$

The properties for the mixture of the gases would have a volume V; the properties of the mixture are denoted by the absence of a subscript. For the mixture, the equation of state is

$$pV = N \overline{R} T \qquad (2\text{-}32)$$

and we find the volume fraction of gas by dividing Eq. (2-31) by Eq. (2-32), or

$$\frac{V_a}{V} = \frac{N_a}{N} = X_a \qquad (2\text{-}33)$$

where X_a is the mole fraction of gas constituent a. Therefore, the volume fraction is equal to the mole fraction.

Amagat's Law. Amagat's law of additive volumes is as follows: *The total volume of a mixture of gases is equal to the sum of the volumes that would be occupied by each constituent gas at the mixture temperature, T, and pressure, p.*

Another approach in analyzing gas mixtures applies when the constituents occupy the total mixture volume, V, at the same temperature, T. The ideal gas equation of state may be written for each constituent and for the total mixture. For constituent a,

$$P_a V = N_a \overline{R} T \qquad (2\text{-}34)$$

and

$$PV = N \overline{R} T \qquad (2\text{-}35)$$

If Eq. (2-34) is divided by Eq. (2-35), we have

$$\frac{P_a}{P} = \frac{N_a}{N} = X_a \qquad (2\text{-}36)$$

TABLE 2-4

Solution to illustrative problem 2-1

Constituent (1)	Volumetric Percentages (2)	Moles/100 Moles (3)	Partial Pressure, $P_t \times$ Col. 3/100 (4)	Molecular Weight (5)	Col. 3 × Col. 5 (6)	Weight Percentages (7)
Carbon dioxide	0.20	0.20	0.03	44	8.8	0.525
Methane	94.40	94.40	14.16	16	1510.4	90.066
Ethane	2.90	2.90	0.435	30	87.0	5.188
Oxygen	0.20	0.20	0.03	32	6.4	0.382
Nitrogen	2.30	2.30	0.345	28	64.4	3.840
	100.0	100.0	15.0		1677	100.0

Column 6 is the weight of 100 moles of the mixture. Therefore, the molecular weight of the gas is 1677/100 = 16.77 lbm/mol. If the gas mixture is considered to be a perfect gas, then $pV = mRT$, and, for density, $V = 1$ ft³. Then,

$$144(15.0)\,(1) = m\left(\frac{1545.3}{16.77}\right)(460 + 60)$$

$$m = 0.0451 \text{ lbm/ft}^3 \text{ (density)}$$

Therefore, the ratio of the partial pressure, the pressure of constituent *a* occupying the mixture volume at the same temperature and volume to total pressure is equal to the mole fraction.

Dalton's Law.   Dalton's law of partial pressure states: *The total mixture pressure, p, is the sum of the pressure that each gas would exert if it were to occupy the vessel alone at volume V and temperature T.* In equation form for constituent gases *a, b,* and *c,* we have

$$p = p_a + p_b + p_c \qquad (2\text{-}37)$$

Mixture Properties.   The total mixture properties, such as internal energy, enthalpy, and entropy, may be determined by adding the properties of the constituents at the mixture conditions. The internal energy and enthalpy are functions of temperature only; therefore,

$$U = N\bar{u} = N_a\bar{u}_a + N_b\bar{u}_b + N_c\bar{u}_c \qquad (2\text{-}38)$$

$$H = N\bar{h} = N_a\bar{h}_a + N_b\bar{h}_b + N_c\bar{h}_c \qquad (2\text{-}39)$$

where $\bar{u}_a$, $\bar{u}_b$, $\bar{u}_c$ and $\bar{h}_a$, $\bar{h}_b$, $\bar{h}_c$ are the specific internal energy and enthalpy of the constituents on the mole basis, energy per unit mole, at the mixture temperature.

The entropy of a constituent gas is a function of the temperature and pressure. The entropy of the mixture is the sum of the constituent entropies:

$$S = N\bar{s} = N_a\bar{s}_a + N_b\bar{s}_b + N_c\bar{s}_c \qquad (2\text{-}40)$$

where $\bar{s}_a$, $\bar{s}_b$, and $\bar{s}_c$ represent the entropy per mole of the constituent gases at the mixture temperature T and at their partial pressures p_a, p_b, and p_c.

The molecular weight ($\mathcal{M}$W) is defined as *the apparent (or average) molecular weight of the mixture.* Then,

$$\mathcal{M}W = X_a(\mathcal{M}W)_a + X_b(\mathcal{M}W)_a$$
$$+ X_b(\mathcal{M}W)_b + X_c(\mathcal{M}W)_c \qquad (2\text{-}41)$$

If we consider that all the constituent gases are ideal gases, the gas constant R and specific heat values are as follows:

$$R = Y_a R_a + Y_b R_b + Y_c R_c \qquad (2\text{-}42)$$

$$c_v = Y_a(c_v)_a + Y_b(c_v)_b + Y_c(c_v)_c \qquad (2\text{-}43)$$

$$c_p = Y_a(c_p)_a + Y_b(c_p)_b + Y_c(c_p)_c \qquad (2\text{-}44)$$

ILLUSTRATIVE PROBLEM 2-1

A Texas natural gas has the following volumetric analysis: 0.20% CO_2 (carbon dioxide), 94.4% CH_4 (methane), 2.9% C_2H_6 (ethane), 0.20% O_2 (oxygen), and 2.3% N_2 (nitrogen). The gas is under a pressure of 15.0 psia and a temperature of 60°F. Calculate the partial pressure of each constituent gas in the mixture, the mole concentration, the weight analysis, the molecular weight, and the density.

Solution: In this type of problem, it is most convenient to present the partial solution in tabular form as shown in Table 2-4.

REVIEW PROBLEMS

2.1. A gas initially at 150°F and having $c_p = 0.23$ Btu/lbm-°R and $c_v = 0.170$ Btu/lbm-°R is placed within a cylinder containing a frictionless piston. If 900 Btu are added to 15 lbm of the gas in the cylinder in a nonflow constant-pressure process, determine the final gas temperature and the work done on or by the gas.

2.2. Air enters a gas turbine at 150 psia and 540°F and leaves at 15 psia and 140°F. If the average value of the constant-pressure specific heat is 0.24 Btu/lbm-°R, determine the work output of the turbine per lbm of the air.

2.3. A gas flows through a pipe. At an upstream station (1), the gas properties are $p_1 = 100$ psia, $t_1 = 950°F$, and $v_1 = 4.0$ ft³/lbm. At a downstream section (2), the gas properties are $p_2 = 76$ psia, $t_2 = 580°F$, and $v_2 = 3.86$ ft³/lbm. The assumed specific heat at constant volume is 0.32 Btu/lbm-°R. If no work is done and if velocities are small, determine the magnitude and direction of the heat transfer. Assume the pipe is horizontal.

2.4. A steam boiler is required to produce 5000 lbm/hr of steam superheated at 200 psia and 820°F when supplied with feed water at 200 psia and 100°F. How much heat must be added to the feed water to convert it to steam at the conditions stated?

2.5. Water flows through a heat exchanger and heat is transferred from the water to heat air. The entering water is at 300°F and 100 psia. The leaving water is at 200°F and 80 psia. Determine the amount of heat transferred from the water.

2.6. Dry saturated steam at 200 psia flows adiabatically in a constant-diameter pipe. At the pipe outlet, the pressure is 100 psia and the temperature is 400°F.
   (a) How much energy was transferred?
   (b) Was this into or out of the system?
   (c) Was this heat or work?

2.7. A sample of steam at 100 psia passes through a throttling calorimeter. The pressure and temperature in the calorimeter are 14.696 psia and 240°F, respectively. What is the quality and the moisture content of the 100 psia steam?

2.8. A refrigerant gas is compressed from 30 to 180 psia, and the specific volume decreased from 9 to 2 ft³/lbm. Determine the polytropic exponent of the compression process.

2.9. A refrigerant gas has a specific volume of 10 ft³/lbm and is compressed from 10 to 66 psia. If the compression exponent (n) is 1.25, determine the specific volume at the end of the compression.

2.10. Air at standard atmospheric pressure (14.7 psia) and at 70°F is compressed, through a pressure ratio of 7 to 1, to a final pressure of 102.9 psia in a steady-flow compressor.
   (a) Find the work required to compress each pound of air isentropically (reversible adiabatic) by the pressure–volume method.
   (b) Repeat part (a) using the enthalpy-difference (temperature-difference) method.
   (c) If, because of turbulence, friction, and internal leakage, it is found that 33% more work is required to compress each pound of air than is indicated by the reversible-adiabatic computation, find the work per pound of air.
   (d) Find the actual air horsepower needed to deliver the air at the rate of 50 lbm/hr.

2.11. Steam is delivered from a boiler at 85 psig pressure, and its quality is 95%. The barometric pressure is 14.7 psia. What are the enthalpy, specific volume, and temperature of the delivered steam?

2.12. By volume analysis, the dry flue gas from a furnace consists of the following: $CO_2 = 11\%$, $O_2 = 8\%$, $CO = 1.0\%$, and $N_2 = 80\%$.
   (a) Compute the weight analysis of the flue gas.
   (b) Compute its molecular weight.
   (c) Compute its density measured at 68°F and 14.7 psia.

BIBLIOGRAPHY

2.1. Jordan, R. C., and Priester, G. B., *Refrigeration and Air Conditioning,* 2nd ed., Prentice-Hall, Inc., Englewood Cliffs, NJ, 1956.

2.2. Granet, Irving, *Thermodynamics and Heat Power,* Reston Publishing Company, Inc., Reston, VA, 1974.

2.3. Burghardt, M. D., *Engineering Thermodynamics with Applications,* Harper and Row Publishers, New York, 1978.

2.4. Van Wylen, G. J., and Sontag, R. E., *Fundamentals of Classical Thermodynamics,* John Wiley and Sons, Inc., New York, 1973.

3

Psychrometrics

3-1 INTRODUCTION

It is of great significance for us to thoroughly understand and be able to determine the thermal properties of air as it is being processed in air conditioning equipment. The science involving the thermal properties of moist air, the measuring and control of the moisture content of air, and the effect of atmospheric moisture on materials and human comfort may be called *psychrometrics*. In this chapter, we discuss the properties of air, the psychrometric chart, and fundamental psychrometric processes.

Air conditioning of buildings concerns mainly the comfort of people and not the maintenance of exact conditions as may be required for industrial products. The conditions conducive to comfort (see Chapter 5) have been found to depend on air temperature, moisture content of the air, motion, and air purity. The physiological behavior of the human body demands equality between the rate of internal chemical heat production and the rate of external heat loss. The human body maintains a remarkable system of temperature control to regulate this loss, which occurs by radiation, convection, and evaporation. The relative proportion of each depends on the total heat production due to activ-

ity, the amount of clothing, the temperature of the surrounding walls, and the properties of the ambient air.

It is evident that conditions conducive to comfort are complicated by physiological and physical factors. Much research has been conducted by the medical and engineering professions for the explanations of the effect of these properties and has led to the development of optimum standards published by ASHRAE.

The earth's atmosphere, the air we breathe, and the air in most air-conditioned spaces are mainly mechanical mixtures of dry air and water vapor. While the relative weight of moisture is small (1% to 3% of the weight of atmospheric air), it is nevertheless one of the most important factors in human comfort and in its effect on most organic materials. It profoundly affects the processing of material such as textiles, paper, confections, pharmaceutical products, and tobacco. It affects both the storage and processing of many food products and even influences combustion in blast furnaces and the yield of iron. Its effect on human activities is altogether disproportional to its relative weight.

The study of psychrometry, including the properties of air–water vapor mixtures and their measurement and control, is of great importance as a preliminary to the proper design and control of an air conditioning system. In order to have a clear understanding of the process of evaporation in the humidifying of air and the process of condensation in the dehumidifying and cooling of air, the fundamental laws of gases and of partial pressures in gaseous mixtures must be understood.

3-2 COMPOSITION OF DRY AIR AND MOIST AIR

Atmospheric air is a complex mixture of several gases including nitrogen, oxygen, carbon dioxide, water vapor, and traces of other gases. Also, the air usually contains some particulate matter such as smoke, dust, pollen, and additional vapors in small concentrations.

Dry air (abbreviated da) exists when all the water vapor and contaminants have been removed from atmospheric air. Extensive measurements have shown that the composition of dry air is relatively constant, but small variations in the amounts of individual components occur with time, geographical location, and altitude. The approximate percentage composition of dry air by volume is as follows: nitrogen, 78.084; oxygen, 20.9476; argon, 0.934; carbon dioxide, 0.0314; neon, 0.001818; helium, 0.000524; methane, 0.002; sulfur dioxide, 0 to 0.0001; hydrogen, 0.00005; and very minor components such as krypton, xenon, and ozone, 0.0002. The apparent molecular weight, or weighted average molecular weight of all components, for air is 28.9645. The gas constant for dry air is $R_a = 53.352$ ft lbf/lbm-°F abs. or 287 J/kg-K.

The common designation, *air,* means moist air in air conditioning work. However, because of its nearly constant composition, dry air is used as the basis for defining the properties of air.

Moist air is a binary (or two-component) mechanical mixture of dry air and water vapor. Atmospheric air is constantly in contact with liquid or solid water and picks up moisture from these sources by the processes of evaporation and sublimation. The quantity of water thus picked up depends principally on the temperature and quantity of water available for evaporation. Conversely, when air containing large quantities of water vapor is cooled, it loses moisture through condensation. Thus, the amount of moisture vapor in moist air varies from zero (dry air) to a maximum that depends on temperature and pressure. The latter condition refers to *saturation,* a state of neutral equilibrium between moist air and the condensed water phase (liquid or solid). Unless otherwise stated, saturation refers to a flat interface surface between the moist air and the condensed phase. The molecular weight of water is 18.015. The gas constant is $R_v = 85.778$ ft lbf/lbm-°F abs. or 462 J/kg-K.

Standard air as defined by *ASHRAE Handbook 1989 Fundamentals* at sea level is air at 59°F (15°C) and a barometric pressure of 29.921 in. mercury (101.325 kPa). The temperature is assumed to decrease linearly with increasing altitude throughout the lower atmosphere. The lower atmosphere is assumed to consist of dry air that behaves as an ideal gas. Gravity is also assumed constant at a standard value of 32.174 ft/s² (9.807 m/s²).

Humidity: In a broad sense, the water-vapor content of air is referred to as its humidity, but since the water-vapor content may be expressed in terms of volumes, weights, moles, or pressure, the term *humidity* has been defined in various ways. For many years, air conditioning engineers have defined humidity as the mass of water vapor carried by unit mass of dry air. This is the definition of *humidity ratio.* Other names that have been used over the years are absolute humidity and specific humidity. For our work, we will use humidity ratio as defined above.

The mass of water vapor contained in air, when expressed in pounds per pound of dry air, is a relatively small number. Therefore, it is sometimes expressed in grains of water vapor per pound of dry air. There are 7000 grains to the pound.

Saturation: The term *saturation* denotes the maximum amount of water vapor that can exist in one cubic foot of space at a given temperature and is essentially independent of the mass and pressure of the air that may simultaneously exist in the same space. Frequently, we speak of "saturated air." However, it must be remembered that the *air* is not saturated; it is the contained water vapor that may be saturated at the air temperature.

3-3 PROPERTIES OF AIR–WATER VAPOR MIXTURES

In 1911, Dr. Willis H. Carrier (3.1) published relations for moist-air properties together with a psychrometric chart. These formulas became standards for the industry.

In the 1940s, Goff and Gratch (3.2) at the University of Pennsylvania published primary source data for moist-air properties that served for several decades.

Recently, ASHRAE Research Project RP216 conducted at the National Bureau of Standards reexamined available thermodynamic data for water and moist air. New formulations for thermodynamic properties were developed and expressed in terms of current temperature scales. Table A-4 (see Appendix A) is a portion of the moist-air property table calculated from these new formulations.

In comparison with these formulations, *perfect gas* equations can be used in a majority of air conditioning problems to substantially increase the ease, rapidity, and economy of calculations with only a slight loss in accuracy. Threlkeld (3.3) has shown that errors in calculating humidity ratio, enthalpy, and volume of saturated air at 29.921 in. mercury pressure are less than 0.7% for a temperature range of $-60°F$ to $120°F$ when perfect gas relationships are used. Furthermore, these errors decrease with decreasing pressure. In this chapter, we will deal primarily with perfect gas relationships in the determination of moist-air properties.

When moist air is considered a mixture of independent perfect gases, dry air, and water vapor, each is assumed to obey the perfect gas equation of state as follows:

1. *For dry air,*

$$p_a V = n_a \overline{R} T \quad \text{or} \quad p_a V$$
$$= m_a R_a T \quad \text{or} \quad p_a = \rho_a R_a T$$

or

$$p_a v_a = R_a T \qquad (3\text{-}1)$$

2. *For water vapor,*

$$p_v V = n_v \overline{R} T \quad \text{or} \quad p_v V$$
$$= m_v R_v T \quad \text{or} \quad p_v = \rho_v R_v T$$

or

$$p_v v_v = R_v T \qquad (3\text{-}2)$$

where

p_a = partial pressure of dry air
p_v = partial pressure of water vapor
V = total volume of mixture
n_a = number of moles of dry air
n_v = number of moles of water vapor
$\overline{R}$ = universal gas constant,
  1545.32 ft lbf/lb-mol-°R,
  8314.41 J/kg-mol-K
T = absolute temperature

The mixture also obeys the perfect gas equations:

$$p_b V = n \overline{R} T \qquad (3\text{-}3)$$

or

$$(p_a + p_v)V = (n_a + n_v)\overline{R} T \qquad (3\text{-}3\text{a})$$

where $p_b = p_a + p_v$ is the mixture total pressure and $n = n_a + n_v$ is the total number of moles in the mixture (see Section 2-12).

3-4 VAPOR DENSITY (ρ_v)

Vapor density is the mass of water vapor contained in unit volume of space. The quantity may be calculated by using Eq. (3-2).

3-5 DRY-BULB TEMPERATURE (t)

The dry-bulb (DB) temperature is the air temperature measured by an accurate thermometer or thermocouple. When measuring the dry-bulb temperature of the air, the thermometer should be shielded to reduce the effects of direct radiation.

3-6 WET-BULB TEMPERATURE (t_w)

Figure 3-1 is a schematic drawing of a device to measure the wet- and dry-bulb temperatures. The various instruments used to take these measurements are called *psychrometers*.

When unsaturated air is passed over a wetted thermometer bulb, water evaporates from the wetted surface and latent heat absorbed by the vaporizing water causes the temperature of the wetted surface and the enclosed thermometer bulb to fall. As soon as the wetted surface temperature drops below that of the surrounding atmosphere, heat begins to flow from the warmer air to the cooler surface, and the quantity of heat transferred in this manner increases with an increasing drop in temperature. On the other hand, as the temperature drops, the vapor pressure of the water becomes lower, and, hence, the rate of evaporation decreases. Eventually, a temperature is reached where the rate at which heat is transferred from the air to the wetted surface by convection and conduction is equal to the rate at which the wetted surface loses heat in the form of latent heat of vaporization. Thus, no further drop in temperature can occur. This is known as the wet-bulb (WB) temperature.

As moisture evaporates from the wetted bulb, the air surrounding the bulb becomes more humid. Therefore, in order to measure the wet-bulb temperature of the air in a given space, a continuous sample of the air

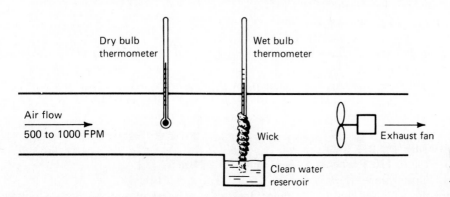

FIGURE 3-1 Steady-flow device for measuring wet- and dry-bulb temperatures.

must pass around the bulb. The purpose of the fan in Figure 3-1 is to cause the air to be drawn across the wetted bulb. Conventional air velocities used are between 500 and 1000 fpm for normal size thermometer bulbs. Soft, fine-meshed cotton tubing is recommended for the wick; it should cover the bulb plus about an inch of the thermometer stem. The wick should be watched and replaced before it becomes dirty or crusty. Using distilled water is recommended to give greater accuracy for a longer period of time.

Figure 3-2 shows a device called a *sling psychrometer.* It is a commonly used device especially for checking conditions on a job. The instrument is rotated by hand to obtain the air movement across the bulbs. The instrument is rotated until no further change is indicated on the wet bulb. The reading taken at that time is the air wet-bulb temperature.

Thermodynamic wet-bulb temperature (t^*), sometimes called *adiabatic saturation temperature,* will be discussed in Section 3-14.

3-7 PARTIAL PRESSURE OF WATER VAPOR (p_v)

The partial pressure of water vapor in an air–water vapor mixture was shown in Eq. (3-2). Several equations for calculating this partial pressure have been proposed and used. Dr. Carrier's (3.1) equation, first presented in 1911, has been frequently used with a high degree of accuracy. The equation makes use of the easily obtainable wet- and dry-bulb temperatures, and its present form is

$$p_v = p_w - \frac{(p_b - p_w)(t - t_w)}{2831 - 1.43t_w} \qquad (3-4)$$

where

p_w = partial pressure of water vapor saturated at wet-bulb temperature t_w
p_b = barometric pressure

t, t_w = dry- and wet-bulb temperatures, respectively, in °F
p_v, p_b, and p_w must have consistent units, either in. Hg or psia

At temperatures below 32°F, Eq. (3-4) applies only for temperatures of air and water vapor over supercooled water. For partial pressures of water vapor over ice, the denominator becomes $3160 - 0.09t_w$, and

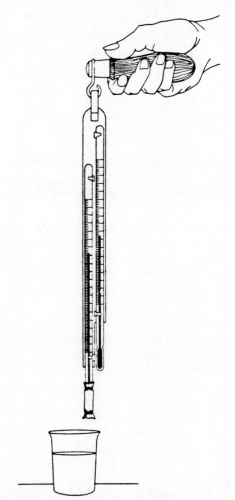

FIGURE 3-2 Sling psychrometer. (Courtesy of The Trane Company, LaCrosse, WI)

p_w must be the partial pressure of water vapor over ice at t_w, the temperature of an iced wet bulb.

ILLUSTRATIVE PROBLEM 3-1

A sample of moist air has a dry-bulb temperature of 70°F and a wet-bulb temperature of 60°F. The barometric pressure is 29.90 in. Hg. Determine the partial pressure of the water vapor and of the dry air in the sample of moist air.

Solution: From Appendix Table A-1 at 60°F WB, find p_w = 0.2563 psia. The barometric pressure of 29.90 in. Hg is converted by (29.90) (0.491) = 14.681 psia. By Eq. (3-4)

$$p_v = p_w - \frac{(p_b - p_w)(t - t_w)}{2831 - 1.43 t_w}$$

$$= 0.2563 - \frac{(14.681 - 0.2563)(70 - 60)}{2831 - (1.43)(60)}$$

$$= 0.2083 \text{ psia} \qquad (ans.)$$

Since $p_b = p_v + p_a$,

$$p_a = p_b - p_v$$
$$= 14.681 - 0.2083$$
$$= 14.477 \text{ psia} \qquad (ans.)$$

3-8 DEW-POINT TEMPERATURE (t_{dp})

During the various seasons of the year, especially during the summer months, in localities when the water supply is cool, it is a common sight to see the outside surface of bare cold-water pipes covered with moisture. Another common sight is that of a glass of ice water with its outside surface covered with a film of moisture. The term often used to describe the appearance of moisture on cold surfaces is "sweating," as though the moisture came through the walls of the pipe or the glass.

What is actually happening is that the outside of the pipe or the glass is at or below the saturation temperature corresponding to the partial pressure of the water vapor in the surrounding air. This saturation temperature is known as the dew-point (DP) temperature when condensation first starts to appear on the cold surface as the moist air is cooled at constant pressure.

In Illustrative Problem 3-1, we calculated the partial pressure of the water vapor in the air to be 0.2038 psia. Referring to Appendix Table A-1, we find that the saturation temperature corresponding to a pressure of 0.2038 psia is 53.3°F by interpolation. Therefore, 53.3°F is the dew-point temperature of the air sample. If any surface located in this air sample were at that temperature, moisture would start to condense on the surface.

In air conditioning work, the dew-point temperature is frequently called *air dew point*. However, this is a misnomer because the air does not condense nor does it have anything to do with the cooling and condensation of the water vapor. Actually, the same cooling and condensation of the water vapor would take place if there were no air present and the entire process were carried out in a closed vessel under vacuum. Since the term is in common use, however, we will use it in our discussion.

At the dew-point temperature and below, the air is said to be "saturated" because the air is mixed with the maximum possible weight of water vapor. If the mixture of air and water vapor is cooled at constant pressure, but above the dew-point temperature, there will be no condensation. However, as the mixture of air and water vapor is cooled, the volume of each component will contract in the same proportion because both are cooled through the same temperature range. In other words, if a mixture consisting of 1 pound of dry air and 0.15 pound of water vapor is cooled, the smaller volume will still contain 1 pound of dry air and 0.15 pound of water vapor as both gases will contract in the same proportion. Changes in the temperature of an air-water vapor mixture do not affect the amount of water vapor mixed with each pound of air as long as the mixture is not cooled down to the dew-point temperature. Under these conditions, the mass of water vapor per pound of dry air will remain the same regardless of the temperature changes. An air-water vapor mixture at a dry-bulb temperature higher than its dew-point temperature is said to be "unsaturated" and the water vapor in the mixture is superheated.

3-9 HUMIDITY RATIO (W)

Humidity ratio was defined in Section 3-2 as the mass of water vapor associated with unit mass of dry air. By applying Dalton's law of partial pressures and the perfect gas law, neglecting intermolecular forces, an expression for calculating humidity ratio can be derived as follows: The volume, V, occupied by unit mass of dry air at its partial pressure is from

$$p_a V = m_a R_a T$$

or

$$V = \frac{m_a R_a T}{p_a} = \frac{1(53.352)(T)}{p_b - p_v} \qquad (a)$$

where p_b is the barometric (total) pressure of the moist air. The mass of water vapor, W, in unit mass of dry air of V cubic feet is from

$$p_v V = W R_v T$$

or

$$W = \frac{p_v V}{R_v T} = \frac{p_v V}{(85.778)\,(T)} \qquad (b)$$

Substituting Eq. (a) into Eq. (b), we have

$$W = \frac{p_v \left(\dfrac{53.352 T}{p_b - p_v} \right)}{85.778 T}$$

which reduces to

$$W = 0.622 \left(\frac{p_v}{p_b - p_v} \right) \qquad \text{lbv/lbda} \qquad (3\text{-}5)$$

3-10 SATURATION RATIO (μ)

Saturation ratio is defined as the ratio of the actual humidity ratio to the humidity ratio of air saturated at the same pressure and temperature, or

$$\mu = \frac{W}{W_s} \times 100 \qquad (3\text{-}6)$$

where W_s is calculated from Eq. (3-5) using p_{vs} in place of p_v. p_{vs} is the partial pressure of the water vapor saturated at the air dry-bulb temperature. Saturation ratio is sometimes referred to as percent humidity, percent saturation, or degree of saturation.

3-11 RELATIVE HUMIDITY (ϕ)

Relative humidity (RH) is the ratio of the mole fraction of water vapor in moist air to the mole fraction of saturated air at the air dry-bulb temperature. In equation form,

$$\phi = \frac{X_v}{X_{vs}} \times 100$$

Since we are treating the dry air and water vapor as ideal gases, we may use partial pressures, or

$$\phi = \frac{p_v}{p_{vs}} \times 100 \qquad (3\text{-}7)$$

ILLUSTRATIVE PROBLEM 3-2

From the data of Illustrative Problem 3-1, determine the relative humidity, humidity ratio, and saturation ratio of the air sample.

Solution: From Illustrative Problem 3-1, $p_v = 0.2038$ psia. Referring to Appendix Table A-1 at 70°F, find $p_{vs} = 0.3632$ psia. By Eq. (3-7),

$$\phi = \frac{p_v}{p_{vs}} \times 100 = \left(\frac{0.2038}{0.3632} \right) (100) = 56\% \quad (ans.)$$

By Eq. (3-5),

$$W = 0.622 \left(\frac{p_v}{p_b - p_v} \right)$$

$$= 0.622 \left(\frac{0.2038}{14.681 - 0.2038} \right)$$

$$= 0.00876 \text{ lbv/lbda} \qquad (ans.)$$

By Eq. (3-5),

$$W_s = 0.622 \left(\frac{p_{vs}}{p_b - p_{vs}} \right)$$

$$= 0.622 \left(\frac{0.3632}{14.681 - 0.3632} \right)$$

$$= 0.01578 \text{ lbv/lbda} \qquad (ans.)$$

By Eq. (3-6),

$$\mu = \frac{W}{W_s} \times 100 = \left(\frac{0.00876}{0.01578} \right) (100) = 55.5\% \quad (ans.)$$

3-12 VOLUME (v_m)

The volume, v_m, of a moist-air mixture is expressed in terms of a unit mass of *dry air* as

$$v_m = \frac{V}{M_a} = \frac{V}{28.9645 n_a} \qquad (3\text{-}8)$$

where

V = total volume of mixture
M_a = total mass of dry air
n_a = number of moles of dry air

By Eqs. (3-1) and (3-8), with $p_b = p_a + p_v$,

$$v_m = \frac{\overline{R}T}{28.9645\,(p_b - p_v)} = \frac{R_a T}{p_b - p_v} \qquad (3\text{-}9)$$

Using Eq. (3-5),

$$v_m = \frac{R_a T}{p_b}(1 + 1.6078W) \qquad (3\text{-}10)$$

In Eqs. (3-9) and (3-10), v_m is the volume, T is absolute temperature, p_b is total pressure, p_v is the partial pressure of water vapor, and W is the humidity ratio.

3-13   ENTHALPY (h)

The enthalpy of a mixture of perfect gases equals the sum of the individual partial enthalpies of the components (see Section 2-12). Therefore, for an air–water vapor mixture with dry air as the reference, the enthalpy may be written as

$$h = h_a + Wh_g \qquad (3\text{-}11)$$

where h_a is the specific enthalpy of dry air and h_g is the specific enthalpy of saturated water vapor at the temperature of the mixture. Each term has units of energy per unit mass of dry air. If it is assumed that we have a perfect gas, the enthalpy is a function of temperature only ($h = c_p t$). In most air conditioning processes, *enthalpy changes* are important. Thus, if 0°F or 0°C is selected as a reference state where the enthalpy of dry air is zero, and if the specific heats c_{pa} and c_{pv} are assumed constant,

$$h_a = c_{pa}t$$
$$h_v = h_g + c_{pv}t$$

The enthalpy of saturated water vapor, h_g, at 0°F is 1061 Btu/lb and at 0°C is 2501 kJ/kg. The specific heats c_{pa} and c_{pv} are 0.240 Btu/lb-°F (1.0 kJ/kg-K) and 0.444 Btu/lb-°F (1.805 kJ/kg-K), respectively. Eq. (3-11) becomes, for EES units,

$$h = 0.240t + W(1061 + 0.444t) \qquad (3\text{-}12)$$

with the temperature, t, in °F and humidity ratio, W, in units of lbv/lbda. For SI units,

$$h = t + W(2501 + 1.805t) \qquad (3\text{-}12a)$$

with the temperature, t, in °C and humidity ratio, W, in units of kgv/kgda.

3-14   THERMODYNAMIC WET-BULB TEMPERATURE (t*)

Figure 3-3 represents an idealized, fully insulated flow device where unsaturated moist air enters at dry-bulb temperature t_1, enthalpy h_1, and humidity ratio W_1. When this air is brought into contact with the water at a lower temperature, the air is both cooled and humidified. If the system is fully insulated so that no heat is transferred into or out of the system, the process is adiabatic; and if the water is at a constant temperature, the latent heat of evaporation can come only from the sensible heat given up by the air in cooling. The quantity of water present is assumed to be large (large surface area and quantity) compared to the amount evaporated into the air. We assume that there is no temperature gradient in the body of water.

If the temperature reached by the air as it leaves the device where it is saturated is identical to the temperature of the water, this temperature is called the *adiabatic saturation temperature* or, more commonly, the *thermodynamic wet-bulb temperature (t*)*.

Thus, in Figure 3-3, the saturated air leaving the device will have properties t_2^*, h_2^*, and W_2^*. Liquid water must be supplied to the device having an enthalpy h_{f2} at t_2^* for the process to be steady-flow. Assuming steady-flow conditions do exist, the energy equation for the process is

$$h_1 + (W_2^* - W_1)h_{f2}^* = h_2^* \qquad (3\text{-}13)$$

The asterisk is used to denote properties at the thermodynamic wet-bulb temperature. The temperature corresponding to h_2 for the given values of h_1 and W_1 is the defined thermodynamic wet-bulb temperature.

Equation (3-13) is *exact* since it defines the thermodynamic wet-bulb temperature t^*. Substituting the approximate perfect gas relationship for h from Eq. (3-12), the corresponding expression for h_1^*, and the approximate relationship $h_{f2}^* = t^* - 32$ into Eq. (3-13), and then solving for the humidity ratio W_1 gives

$$W_1 = \frac{(1093 - 0.556t^*)W_2^* - 0.240(t_1 - t^*)}{1093 + 0.444t_1 - t^*} \qquad (3\text{-}14)$$

where t_1 and t^* are in °F.

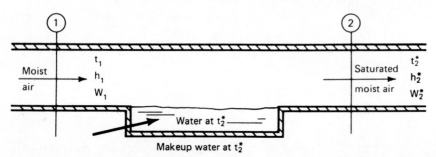

FIGURE 3-3 Adiabatic saturation of air.

The corresponding equation in SI units is

$$W_1 = \frac{(2501 - 2.381t^*)W_2^* - (t_1 - t^*)}{2501 + 1.805t_1 - 4.186t^*} \quad (3\text{-}14a)$$

where t_1 and t^* are in °C.

ILLUSTRATIVE PROBLEM 3-3

Air enters an adiabatic saturator at 80°F dry-bulb and 14.696 psia and leaves saturated at 65°F. Water added is at 65°F. Compute the humidity ratio W_1 and relative humidity ϕ_1.

Solution: Because the mixture leaving is saturated, $p_{v2} = p_{vs2}$ at 65°F. From Appendix Table A-1 at 65°F, find $p_{vs2} = 0.3093$ psia. By Eq. (3-5),

$$W_2^* = 0.622\left(\frac{p_{vs2}}{p - p_{vs2}}\right)$$

$$= 0.622\left(\frac{0.3093}{14.696 - 0.3093}\right)$$

$$= 0.01337 \text{ lbv/lbda}$$

By Eq. (3-14),

$$W_1 = \frac{(1093 - 0.556 \times 65)(0.01337) - 0.240(80 - 65)}{1093 + (0.444)(80) - 65}$$

$$= 0.0101 \text{ lbv/lbda} \qquad (ans.)$$

Using Eq. (3-5),

$$W_1 = 0.622\left(\frac{p_{v1}}{p - p_{v1}}\right)$$

$$0.0101 = 0.622\left(\frac{p_{v1}}{14.696 - p_{v1}}\right)$$

$$p_{v1} = 0.2348 \text{ psia}$$

By Eq. (3-7),

$$\phi = \frac{p_{v1}}{p_{vs1}} \times 100$$

From Table A-1 at 80°F, find $p_{vs1} = 0.5073$ psia. Then,

$$\phi_1 = \left(\frac{0.2348}{0.5073}\right)(100) = 46.3\% \qquad (ans.)$$

The process discussed in this section is called the *adiabatic saturation process* (to be discussed in more detail in the next chapter). The usefulness of the foregoing discussion lies in the fact that the temperature of the saturated air–water vapor mixture leaving the system is a function of the temperature, pressure, and relative humidity of the mixture entering and the exit pressure. Conversely, knowing the entering and exit pressures and temperatures, we may determine the relative humidity and humidity ratio of the entering mixture, as shown in Illustrative Problem 3-3.

In principle, there is a difference between the wet-bulb temperature, t_w, and the temperature of adiabatic saturation, t^*. The wet-bulb temperature is a function of both heat and mass transfer rates, while the adiabatic saturation temperature is a function of a thermodynamic equilibrium process. However, in practice, it has been found that for air–water vapor mixtures at atmospheric pressures and temperatures, the wet-bulb and adiabatic saturation temperatures are essentially equal numerically.

3-15 THERMODYNAMIC PROPERTIES OF MOIST AIR[1]

Table A-4 in Appendix A shows values of thermodynamic properties, calculated from RP216 relations, for standard atmospheric pressure 14.696 psia or 29.92 in. Hg. The properties in this table are based on the *thermodynamic temperature scale*. This ideal scale differs only slightly from the practical temperature scales used for actual physical measurements. Brief explanations on the data in each column are given next.

[1]Table A-4 in Appendix A and its description are taken with permission from *ASHRAE Handbook 1989 Fundamentals,* American Society of Heating, Refrigerating, and Air Conditioning Engineers, Atlanta, GA.

Symbols Used in Appendix Table A-4

t = Fahrenheit temperature.

W_s = humidity ratio *at saturation,* the condition at which the gaseous phase (moist air) exists in equilibrium with a condensed phase (liquid or solid) at the given temperature and pressure (standard atmospheric pressure). At given values of temperature and pressure, the humidity ratio W can have any value from zero to W_s.

v_a = specific volume of dry air, ft³/lb.

v_{as} = $v_s - v_a$, the difference between the volume of moist air *at saturation,* per lb of dry air, and the specific volume of the dry air itself, ft³/lbda, at the same pressure and temperature.

v_s = volume of moist air *at saturation* per lb of dry air, ft³/lbda.

h_a = specific enthalpy of dry air, Btu/lbda. The specific enthalpy of dry air has been assigned the value of zero at 0°F and standard atmospheric pressure.

h_{as} = $h_s - h_a$, the difference between the enthalpy of moist air *at saturation,* per lb of dry air, and the specific enthalpy of the dry air itself, Btu/lbda, at the same pressure and temperature.

s_a = specific entropy of dry air, Btu/lb-°F (abs.). The specific entropy of dry air has been assigned the value of zero at 0°F and standard atmospheric pressure.

s_s = specific entropy of moist air *at saturation* per lb of dry air, Btu/lbda-°F (abs.).

h_w = specific enthalpy of condensed water (liquid or solid) in equilibrium with saturated air at a specified temperature and pressure, Btu/lb water. Specific enthalpy of liquid water has been assigned the value of zero at its triple point (32.018°F) and saturation pressure. *Note:* h_w is greater than the steam-table enthalpy of saturated pure condensed phase by the amount of the enthalpy increase governed by the pressure increase from saturation pressure to one atmosphere, plus influence from the presence of air.

p_s = vapor pressure of water in saturated moist air, psia or in. Hg. This pressure, p_s, differs negligibly from the saturation vapor pressure of pure water, p_{vs}, at least for the conditions shown.

ILLUSTRATIVE PROBLEM 3-4

What is the relative humidity of moist air that has a dry-bulb temperature of 80°F and a wet-bulb temperature of 72°F? The barometric pressure is 29.92 in. Hg.

Solution: Refer to Appendix Table A-4. At 72°F wet-bulb, find h_s = 35.841 Btu/lbda. At 80°F dry-bulb, find h_a = 19.222 Btu/lbda. Then,

$$h_{as} = 35.841 - 19.222 = 16.619 \text{ Btu/lbda}$$

This is the heat of the vapor. From Table A-4, the value of h_{as} = 16.619 falls between 16.094 and 16.677 with corresponding p_v values of 0.69065 and 0.71479 in. Hg. By interpolation, p_v corresponding to h_{as} = 16.619 is 0.7128 psia. At 80°F dry-bulb, p_{vs} = 1.03302 in. Hg. So, by Eq. (3-7),

$$\phi = \left(\frac{p_v}{p_{vs}}\right) \times 100 = \left(\frac{0.7128}{1.03302}\right)(100) = 69.0\% \quad (ans.)$$

Note: p_v may also be calculated by use of Eq. (3-4).

ILLUSTRATIVE PROBLEM 3-5

Moist air exists at 80°F dry-bulb and 60°F dew-point when the barometric pressure is 29.92 in. Hg. What is the relative humidity of the moist air?

Solution: By definition, the 60°F dew-point temperature is the saturation temperature corresponding to the actual partial pressure of the water vapor in the air. From Appendix Table A-4 at 60°F, find $p_v = p_s$ = 0.52193 in. Hg. At 80°F, find p_{vs} = p_s = 1.03302 in. Hg. The relative humidity then, is

$$\phi = \frac{0.52193}{1.03302} = 0.505, \text{ or } 50.5\% \quad (ans.)$$

ILLUSTRATIVE PROBLEM 3-6

What is the enthalpy of moist air at 70°F dry-bulb temperature and 40% relative humidity? Barometric pressure is 29.92 in. Hg.

Solution: By Eq. (3-12), $h = 0.240t + W(1061 + 0.444t)$. From Appendix Table A-4 at 70°F, find $p_{vs} = p_s$ = 0.73966 in. Hg. Relative humidity $\phi = p_v/p_{vs}$; then, $p_v = \phi p_{vs} = 0.40(0.73966) = 0.2958$ in. Hg. By Eq. (3-5),

$$W = 0.622\left(\frac{p_v}{p_b - p_v}\right)$$

$$= 0.622\left(\frac{0.2958}{29.92 - 0.2958}\right)$$

$$= 0.00621 \text{ lbv/lbda}$$

So,

$$h = 0.240(70) + 0.00621(1061 + 0.444 \times 70)$$

$$= 23.58 \text{ Btu/lbda} \qquad (ans.)$$

ILLUSTRATIVE PROBLEM 3-7

Moist air exists at 80°F dry-bulb, 60°F dew-point, and 29.92 in. Hg barometric pressure. Determine (1) humidity ratio, (2) saturation ratio, (3) relative humidity, (4) enthalpy, and (5) specific volume of dry air.

Solution:

1. From Appendix Table A-4 at dew-point temperature of 60°F, find $p_v = p_s = 0.52193$ in. Hg. By Eq. (3-5),

$$W = 0.622 \left(\frac{0.52193}{29.92 - 0.52193} \right)$$

$$= 0.0110 \text{ lbv/lbda} \qquad (ans.)$$

2. From Table A-4 at $t = 80°F$ dry-bulb, find $W_s = 0.02234$ lbv/lbda. By Eq. (3-6),

$$\mu = \frac{W}{W_s} = \frac{0.0110}{0.02234} = 0.492, \text{ or } 49.2\% \quad (ans.)$$

3. From Table A-4 at $t = 80°F$, find $p_{vs} = p_s = 1.03302$ in. Hg. By Eq. (3-7),

$$\phi = \frac{p_v}{p_{vs}} = \frac{0.52193}{1.03302} = 0.505, \text{ or } 50.5\% \quad (ans.)$$

4. By Eq. (3-12),

$$h = 0.240t + W(1061 + 0.444t)$$
$$= 0.240(80) + 0.0110(1061 + 0.444 \times 80)$$
$$= 31.26 \text{ Btu/lbda} \qquad (ans.)$$

5. By Eq. (3-1), $p_a v_a = R_a T$, where p_a is the partial pressure of the dry air in the moist air, may be used to find v_a. By Eq. (3-3), $p_a = p_b - p_v = 29.92 - 0.52193 = 29.398$ in. Hg = 14.434 psia. Then,

$$v_a = \frac{R_a T}{p_a} = \frac{(53.352)(80 + 460)}{144(14.434)}$$

$$= 13.86 \text{ ft}^3/\text{lbda} \qquad (ans.)$$

3-16 THE PSYCHROMETRIC CHART

The illustrative problems in previous sections of this chapter indicated how properties of moist air may be calculated by the use of equations and tables. As may be noted, this method may be long and tedious, especially when the properties at more than one state point must be evaluated, as in any process or series of processes. In addition, it is usually much more revealing to us to be able to see the processes taking place by using some form of graphical representation.

The psychrometric chart is a plot of the psychrometric properties of moist air. Charts of various types have been designed and produced. Each has its usefulness, and selection of a chart for use is normally one of personal preference and the temperature range required.

Trane Psychrometric Chart. For general use in this text, we will be using the Trane Psychrometric Chart for normal temperatures (30°F to 115°F). Other charts are available for low temperatures (−40°F to 50°F) and for high temperatures (60°F to 250°F). A brief description of the Trane chart follows.

Figure 3-4 is a skeleton chart showing the various coordinates and identifies the various scales and lines for the psychrometric properties of air.

Dry-bulb temperature (line 1): The temperature of air read on a standard thermometer. Lines of constant dry-bulb temperature are straight, vertical lines on the chart. The dry-bulb temperature scale is at the bottom of the chart. Units are °F.

Wet-bulb temperature (line 2): The wet-bulb temperature above 32°F is the temperature indicated by a thermometer whose bulb is covered by a wet wick and exposed to a stream of air moving at a velocity of 1000 fpm. Wet-bulb temperatures below 32°F are obtained from a thermometer on which the water in the wick has frozen to ice. This is the reason the slopes of the wet-bulb temperature lines change below 32°F. The wet-bulb temperature scale is on the curved line at the left of the chart. Units are °F.

Humidity ratio (line 3): Lines of constant humidity ratio are straight, horizontal lines at right angles

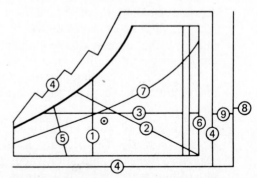

FIGURE 3-4 Skeleton psychrometric chart showing various coordinates, scales, and lines.

to the dry-bulb temperature lines. The humidity ratio scale is at the right of the chart. Units are grains of moisture per pound of dry air.

Enthalpy (line 4): The thermodynamic property that serves as a measure of the heat energy in the air above some selected datum temperature. In this case, it represents the enthalpy of 1 lb of dry air plus the enthalpy of W grains of moisture associated with it. The enthalpy scale, line 4, is shown on the upper left, and a duplication at the bottom and right side of the chart. In reading enthalpy values, it will be necessary to use a straightedge spanning the chart from side to side through the point in the chart for which the enthalpy is to be determined. Note that on this chart lines of constant enthalpy are not parallel to lines of constant wet-bulb temperature. The units of enthalpy are Btu/lbda.

Specific volume (line 5): Lines of constant specific volume. The units are ft³/lbda.

Dew-point temperature (line 6): Lines of constant dew-point temperature are horizontal, parallel to humidity ratio lines. Line 6 is the dew-point temperature scale. Units are °F.

Relative humidity (line 7): Lines of constant relative humidity curve upward from left to right.

Vapor pressure (line 8): Lines of constant partial pressure of water vapor in the air–water vapor mixture are horizontal and parallel to lines of constant humidity ratio. Line 8 is the pressure scale in units of inches of mercury absolute.

Sensible heat ratio (line 9): The ratio of sensible heat to total heat in a heating and humidifying process or a cooling and dehumidifying process. At 78°F dry-bulb temperature and 50% relative humidity is the alignment circle for the sensible heat ratio scale.

Figure 3-5 is the complete Trane Psychrometric Chart. The chart had been developed for standard sea level barometric pressure of 29.92 in. Hg but is usable with little error for barometric pressures from 29.00 to 31.00 in. Hg.

3-17 BASIC PROCESSES INVOLVING AIR–WATER VAPOR MIXTURES

Figure 3-6 represents schematically the basic process lines for air conditioning processes. Each process line shown, or combinations of process lines, may be used to show graphically the variation in the properties of an air–vapor mixture as it is heated, humidified, cooled, dehumidified, or treated in various combinations.

Chapter 4 discusses in detail the various air conditioning processes with illustrative problems commonly encountered in applied psychrometrics.

3-18 HOW TO DETERMINE PSYCHROMETRIC PROPERTIES OF MOIST AIR USING THE PSYCHROMETRIC CHART

It has been demonstrated in previous sections of this chapter that psychrometric properties of moist air may be calculated from equations and tables. We will now determine psychrometric properties of moist air using the psychrometric chart.

ILLUSTRATIVE PROBLEM 3-8

Using the psychrometric chart, determine the psychrometric properties of moist air existing at 70°F dry-bulb, 60°F wet-bulb, and 29.92 in. Hg barometric pressure.

Solution:
Step 1. Referring to Figure 3-7, locate the moist-air state on the chart where the vertical line representing 70°F DB intersects the inclined line representing 60°F WB. This point is labeled A.
Step 2. At point A by interpolating between the inclined lines of constant specific volume, find $v_A = 13.53$ ft³/lbda.
Step 3. At point A by interpolating between the curved lines of constant relative humidity, find $\phi_A = 56\%$.
Step 4. Using a straightedge, draw a straight horizontal line through point A to the right to intersect the humidity ratio, dew-point, and vapor pressure scales. Find humidity ratio, $W = 61.9$ grains/lbda; dew-point temperature, $t_{dp} = 53.6$°F; and vapor pressure, $p_v = 0.420$ in. Hg.
Step 5. Using a straightedge, span the chart from the enthalpy scale at upper left to the same enthalpy value on the lower enthalpy scale at bottom and right of chart through point A. Where the straightedge intersects the enthalpy scale, find $h_A = 26.4$ Btu/lbda. (Note that lines of constant enthalpy are *not* parallel to lines of constant wet-bulb temperature in this area of the chart.)

ILLUSTRATIVE PROBLEM 3-9

Using the psychrometric chart, determine the following psychrometric properties of moist air existing at 85°F dry-bulb and 70% relative humidity. Determine wet-bulb temperature, humidity ratio, enthalpy, dew-point temperature, specific volume, vapor pressure, and saturation ratio.

Solution:
Step 1. Referring to Figure 3-7, locate the moist-air condition on the chart where the vertical line representing 85°F DB intersects the curved line representing 70% RH. This point is labeled B.

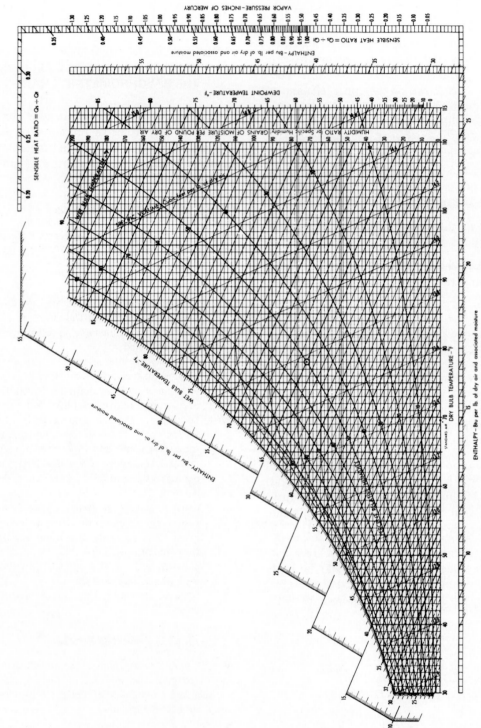

FIGURE 3-5 Psychrometric chart. (Reprinted courtesy of The Trane Company, LaCrosse, WI)

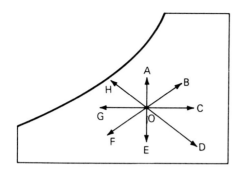

FIGURE 3-6 Schematic psychrometric chart representing various basic psychrometric processes:

OA = humidifying only
OB = heating and humidifying
OC = sensible heating only
OD = chemical dehumidifying
OE = dehumidifying only
OF = cooling and dehumidifying
OG = sensible cooling only
OH = cooling and humidifying

Step 2. Referring to point *B,* locate the sloped wet-bulb line passing through the point and read from the 100% RH curve the wet-bulb temperature of 77°F.

Step 3. Using a straightedge, span the chart through point *B* to the enthalpy scales and read $h_B = 40.4$ Btu/lbda.

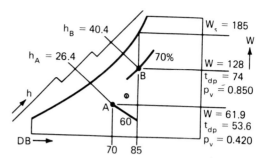

FIGURE 3-7 Solution to Illustrative Problems 3-8 and 3-9.

Step 4. Using a straightedge, draw a straight horizontal line through point *B* to the right to intersect the humidity ratio, dew-point, and vapor pressure scales. Find humidity ratio, $W = 128$ grains/lbda, dew-point temperature, $t_{dp} = 74°$F; and vapor pressure, $p_v = 0.850$ in. Hg.

Step 5. At point *B* by interpolating between the inclined lines of constant specific volume, find $v_B = 14.12$ ft³/lbda.

Step 6. By Eq. (3-6), saturation ratio $\mu = W/W_s$. W_s is the humidity ratio of air at 85°F DB and 100% RH. Extend the vertical line representing 85°F DB upward from point *B* to 100% RH. Through this point, draw a horizontal line to the right to find $W_s = 185$ grains/lbda. Therefore, $\mu = 128/185 = 0.692$.

REVIEW PROBLEMS

3.1. Calculate the humidity ratio, enthalpy, and relative humidity of moist air at 90°DB and 80°WB. Barometric pressure is 29.92 in. Hg.

3.2. Determine the same quantities as in Review Problem 3.1 using the psychrometric chart.

3.3. Moist air at a barometric pressure of 14.696 psia has a dry-bulb temperature of 80°F and a wet-bulb temperature of 60°F.
  (a) Calculate the air dew-point temperature.
  (b) Calculate the humidity ratio.
  (c) Calculate the enthalpy.
  (d) Check the answers to parts (a), (b), and (c) by using the psychrometric chart.

3.4. Moist air exists at 70°DB and 65°DP, and barometric pressure is 29.92 in. Hg.
  (a) Calculate the relative humidity, humidity ratio, saturation ratio, and enthalpy.
  (b) Determine the same properties from the psychrometric chart.

3.5. Air at 30°DB and 30°WB is heated to 80°DB without addition of moisture. Determine the following from the psychrometric chart:
  (a) Initial and final humidity ratio.
  (b) Initial and final relative humidity.
  (c) Initial and final enthalpy.
  (d) Initial and final specific volume.

  (e) Initial and final dew-point temperature.
  (f) Initial and final partial pressure of the water vapor in the air.

3.6. Moist air exists at 70°DB and 60% RH when the barometric pressure is 14.696 psia. Determine partial pressure of the water vapor in the air, wet-bulb temperature, specific volume, enthalpy, humidity ratio, and dew-point temperature.

3.7. **(a)** Calculate the humidity ratio, enthalpy, and specific volume for saturated air at 14.696 psia total pressure for $t = 70°$F.
  (b) Repeat part (a) for $t = 32°$F.
  Check calculated values with those found in Appendix Table A-4.

3.8. Moist air exists at 70°F and 50% RH in a room containing several windows. If the inside surface temperature of the windows is at 45°F, will condensation form on the windows?

3.9. Moist air exists at 32°F DB and 80% RH. How much moisture must be added to each pound of air if the final condition of the air is to be 80°F DB and 40% RH? Assume that total pressure is 14.696 psia.

3.10. Calculate the relative humidity of moist air if its wet-bulb temperature is 62°F and its dew-point temperature is 54°F. Barometric pressure is 29.92 in. Hg.

3.11. Solve Review Problem 3.10 by use of the psychrometric chart.

3.12. Calculate the dry-bulb temperature of moist air that has a relative humidity of 76% and a dew-point temperature of 66°F. Barometric pressure is 29.92 in. Hg.

3.13. Solve Review Problem 3.12 by use of the psychrometric chart.

3.14. Calculate the dew-point temperature of moist air at 80°F DB and 72°F WB. Barometric pressure is 29.92 in. Hg.

3.15. Solve Review Problem 3.14 by use of the psychrometric chart.

3.16. Calculate the dew-point temperature of moist air having a dry-bulb temperature of 80°F and a relative humidity of 50%. Barometric pressure is 29.92 in. Hg.

3.17. Solve Review Problem 3.16 by use of the psychrometric chart.

3.18. Moist air exists at 90°F DB and 40% RH. The barometric pressure is 12.56 psia.
  (a) Determine the dew-point temperature.
  (b) Determine the enthalpy of the air.

3.19. Moist air exists at 70°F DB, 60°F WB, and 29.92 in. Hg.
  (a) Calculate the relative humidity.
  (b) Calculate the vapor density.
  (c) Calculate the dew-point temperature.
  (d) Calculate the humidity ratio.
  (e) Calculate the volume occupied by the mixture associated with one pound of dry air.

3.20. Outdoor air at 35°F DB and 60% RH is to be humidified to a final state of 70°F DB and 60% RH. The barometric pressure is 29.00 in. Hg. How many pounds of water must be added to each pound of air to obtain the final state of the air? Check by using the psychrometric chart. Why do the two results differ?

3.21. Air leaves a well-insulated (adiabatic) saturator at 70°F and 29.92 in. Hg barometric pressure. It enters the saturator at 80°F, and water is supplied at 70°F. Find the humidity ratio, degree of saturation, enthalpy, and relative humidity of the entering air.

3.22. A sample of moist air has a dry-bulb temperature of 76°F and a relative humidity of 50%, and the barometric pressure is 29.8 in. Hg.
  (a) Calculate the weight of 1 ft³ of the mixture of air and moisture.
  (b) Calculate the weight of moisture per pound of dry air.

BIBLIOGRAPHY

3.1. Carrier, W. H., "Rational Psychrometric Formulas," *Transaction ASME,* vol. 33 (1911), p. 1005.

3.2. Goff, J. A., and Gratch, S., "Thermodynamics of Moist Air," *ASHVE Transactions,* 51, 1945.

3.3. Threlkeld, James L., *Thermal Environmental Engineering,* 2nd ed., Prentice-Hall, Inc., Englewood Cliffs, NJ, 1970.

3.4. *ASHRAE Handbook 1989 Fundamentals,* American Society of Heating, Refrigerating, and Air Conditioning Engineers, Atlanta, GA, 1989.

4

Applied Psychrometrics

4-1 INTRODUCTION

Let us take a brief look at the overall requirements of a year-round air conditioning system. Figure 4-1 shows a schematic layout of the space to be conditioned with its possible heat loads, the air conditioner with its load, a circulating air fan, and connecting ductwork. In Figure 4-1, we note that some of the return air (ra) exhausted from the space is mixed (mix) with fresh outdoor air (oa), which, after passing through the air conditioner, becomes supply air (sa) to the space.

The symbol $\dot{q}_s$ represents the sensible heat transfer (gain or loss) from all sources, and $\dot{m}_w$ represents moisture transfer rates. The symbol $\dot{q}_L$ designates the energy associated with the moisture transfer and is given by $\Sigma \dot{m}_w h_w$, where h_w is the specific enthalpy of the added (or removed) moisture. Solar radiation and internal loads are always heat gains to the conditioned space. Heat transmission through solid boundaries of the space due to temperature differences as well as energy transfers because of infiltration may represent a gain or a loss.

At the air conditioner, it should be noted that the energy transfer $\dot{q}_c$ and the moisture transfer $(\dot{m}_w)_c$ cannot be obtained from the space heat and moisture loads alone. The effect of the loads caused by outdoor air

ventilation loads must be included as well as the space loads. The HVAC system designer must recognize that items such as fan energy, duct transmission, roof and ceiling transmission, heat of lighting, bypass and leakage, type of return system, location of main fans, and actual versus design space conditions are all related to one another, to component sizing, and to system arrangement.

The psychrometric chart, introduced in Chapter 3, is one of the most useful tools of the practicing air conditioning engineer. The chart assists in the solution of simple as well as complex processes involved in the conditioning of moist air. Most of the complex problems consist of various combinations of simple processes, such as sensible heating or cooling, humidification, or dehumidification. The chart provides a means for showing the various processes graphically.

Also, the most powerful analytical tools for problem solving are (1) the first law of thermodynamics or energy balance and (2) the conservation of mass or mass balance.

In some air conditioning systems, moist air is taken from the conditioned space (return air), passed through the conditioner, and returned to the space as

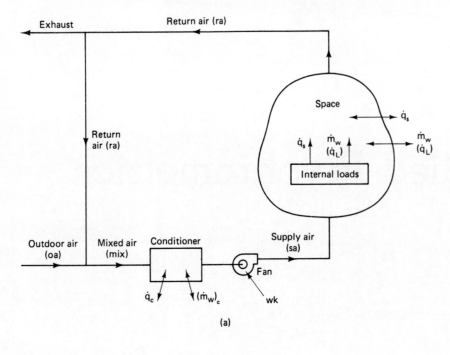

(a)

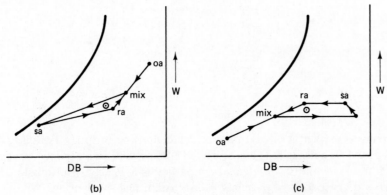

(b)   (c)

FIGURE 4-1 (a) Schematic of air conditioning system, (b) psychrometric process (cooling and dehumidification mode), and (c) psychrometric processes (heating and humidification mode).

supply air. In most systems, some of the return air is mixed with fresh outdoor air, passed through the conditioner, and returned to the space as supply air. Figure 4-1a illustrates the latter type. Figure 4-1b is a sketch showing typical process lines for the *cooling and dehumidification* mode. Figure 4-1c is a sketch showing typical process lines for the *heating and humidification* mode.

In this chapter, we will use the conservation laws and the psychrometric chart to analyze the various basic air conditioning processes and to progress into the more complex processes.

4-2 SENSIBLE HEATING OR COOLING OF MOIST AIR

When moist air is heated or cooled without the gain or loss of moisture, the process is called *sensible heating* or *sensible cooling,* respectively. Figure 4-2 is a sketch

showing these two processes as horizontal lines on a psychrometric chart. Figure 4-2a represents a heating coil, and Figure 4-2b represents a cooling coil. In both, the humidity ratio remains constant as well as the dewpoint temperature. Changes in the dry-bulb and wetbulb temperatures are evident. Figure 4-2c shows the two processes as a horizontal line. Line 1–2 is sensible heating, and line 2–1 is sensible cooling.

The steady-flow and material-balance equations that apply to the sensible heating processes are as follows:

$$\dot{m}_{a1}h_1 + {_1}\dot{q}_{s2} = \dot{m}_{a2}h_2 \qquad (4\text{-}1)$$

$$\dot{m}_{a1} = \dot{m}_{a2} \qquad (4\text{-}2)$$

$$\dot{m}_{a1}W_1 = \dot{m}_{a2}W_2 \qquad (4\text{-}3)$$

where

$\dot{m}_a$ = pounds of air flowing per unit of time (usually lb/hr)

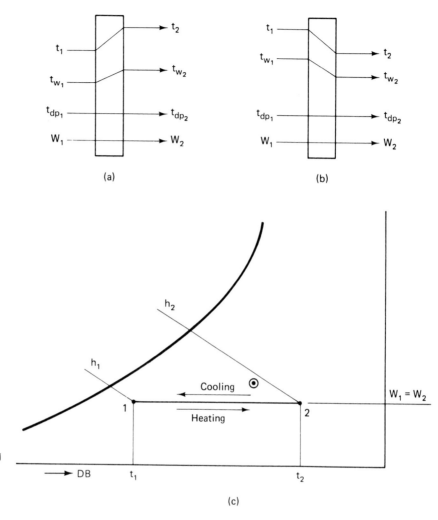

FIGURE 4-2 (a) Sensible heating process, (b) sensible cooling process, and (c) graphical representation.

$_1\dot{q}_{s2}$ = sensible heat added between state 1 and state 2 per unit of time (usually Btu/hr)

W = humidity ratio (lb water vapor/lb dry air)

h = enthalpy of moist air (Btu/lb dry air)

Therefore, for sensible heating of moist air, combining Eqs. (4-1) and (4-2), we have

$$_1\dot{q}_{s2} = \dot{m}_a(h_2 - h_1) \qquad (4\text{-}4)$$

Since the humidity ratio is constant during the process, it is closely true that the enthalpy of moist air is given by

$$h = 0.240t + W(1061 + 0.444t) \qquad (3\text{-}12)$$

If t_1 and t_2 are substituted into Eq. (3-12) to obtain h_1 and h_2 and if the results are substituted into Eq. (4-4), we would have

$$_1\dot{q}_{s2} = \dot{m}_a(0.240 + 0.45W)(t_2 - t_1) \qquad (4\text{-}5)$$

where t_1 and t_2 are the initial and final dry-bulb temperatures, respectively, for the sensible heating processes.

Equation (4-5) assumes a constant specific heat of superheated steam of 0.450. It is often convenient to combine the two specific heat values for an average value of humidity ratio, say, 0.010 lb vapor/lb dry air. This gives a single specific heat value for moist air of 0.244, and Eq. (4-5) becomes

$$_1\dot{q}_{s2} = 0.244\dot{m}_a(t_2 - t_1) \qquad (4\text{-}6)$$

For the sensible cooling process, Eq. (4-6) gives a positive value of $\dot{q}_s$ if the subscripts are reversed.

Air-handling components in HVAC systems, such as fans, ducts, and dampers, are selected on the basis of *volume* flow rate rather than *mass* flow rate. This introduces some difficulty in problem solutions because the volume flow rate *varies* with temperature and pressure (or, the specific volume), whereas the mass flow rate *is constant*. Therefore, if the volume flow rate is to be determined, it is necessary to specify the point in the system where the volume flow rate is to be determined and find the specific volume of the air at that point. With the mass flow rate, $\dot{m}_a$, known and the specific

50   *Chap. 4 / Applied Psychrometrics*

volume of the air at the point, we may calculate the volume flow rate in cubic feet per minute from

$$\text{cfm} = \frac{(\dot{m}_a)(v)}{60} \qquad (4\text{-}7)$$

where $\dot{m}_a$ is the mass flow rate in lb/hr, 60 min./hr, and v is the specific volume of the air at the point in question in ft³/lbda.

For uniformity in the manufacturing industry, air-handling equipment is normally rated on the basis of "standard air," defined as air having a density of 0.075 lbda/ft³ (1.204 kgda/m³), which corresponds to approximately 60°F (15.5°C) at saturation and 69°F (20.6°C) dry at 14.7 psia (101.4 kPa). Using the density, 0.075 lb/ft³, equal to $1/v$, in Eq. (4-7), gives

$$\dot{m}_a = (60)(0.075)(\text{cfm}) = 4.5(\text{cfm})$$

If this value of $\dot{m}_a$ is substituted into Eq. (4-4), we have

$$_1\dot{q}_{s2} = 4.5(\text{cfm})(h_2 - h_1) \qquad (4\text{-}8)$$

The corresponding relation in SI units is

$$_1\dot{q}_{s2} = 4.33(\text{L/s})(h_2 - h_1) \text{ watts} \qquad (4\text{-}9)$$

where L/s is liters per second.

If we substitute the values above for $\dot{m}_a$ into Eq. (4-6), we have

$$_1\dot{q}_{s2} = 0.244(4.5)(\text{cfm})(t_2 - t_1)$$

or

$$_1\dot{q}_{s2} = 1.10(\text{cfm})(t_2 - t_1) \qquad (4\text{-}10)$$

The corresponding relationship in SI units is

$$_1\dot{q}_{s2} = 1.232(\text{L/s})(t_2 - t_1) \text{ watts} \qquad (4\text{-}11)$$

Caution: You should recall that, of the series of equations just developed for $_1\dot{q}_{s2}$, Eq. (4-4) is the only one that may be called *exact*. The remaining equations, Eqs. (4-5) through (4-11), can be called *approximate* because, as you have seen, certain assumptions were made about the moist-air mixture in their development. Nonetheless, Eqs. (4-5) through (4-11) are used quite generally in actual work because of their simplicity, and the dry-bulb temperatures are easy to read from the psychrometric chart. Another reason for using these equations instead of Eq. (4-4) is because of the difficulty in reading accurate enthalpy values from the chart, which is quite necessary in some problems. For example, assume that a system requires an airflow rate of 50,000 lb/hr of air and an enthalpy change $(h_2 - h_1)$ equal to

15.5 Btu/lbda. The required $_1\dot{q}_{s2}$ would be 775,000 Btu/hr. However, if the enthalpy difference read from the psychrometric chart had been 15.7 Btu/lbda, the resulting $_1\dot{q}_{s2}$ would have been 785,000 Btu/hr, with an error of about 1.3%. This is well within the accuracy of Eqs. (4-5) through (4-11).

ILLUSTRATIVE PROBLEM 4-1

How much heat is required to heat 3000 cfm of moist air at 45°F DB and 35°F WB to a final temperature of 90°F DB without change in humidity ratio?

Solution: Since this problem states no change in humidity ratio, the process is a sensible heating process.

Step 1. On the psychrometric chart (see Figure 4-3), locate point 1 at the intersection of 45°F DB and 35°F WB. At this point, find $W_1 = 14$ grains/lbda, $h_1 = 13.0$ Btu/lbda, and $v_1 = 12.76$ ft³/lbda.

Step 2. On the psychrometric chart, locate point 2 by drawing a horizontal line (constant humidity ratio) through point 1 and extend to point 2 at 90°F DB. At this point, find $W_2 = 14$ grains/lbda, $h_2 = 23.9$ Btu/lbda, and $v_2 = 13.9$ ft³/lbda.

Step 3. The mass of air, found for the initial air state, is from Eq. (4-7), or $\dot{m}_a = 60(\text{cfm})/v_1 = 60(3000)/12.76 = 14,100$ lb/hr.

Step 4. The sensible heat added, by Eq. (4-4), is

$$_1\dot{q}_{s2} = \dot{m}_a(h_2 - h_1) = 14,100(23.9 - 13.0)$$
$$= 153,690 \text{ Btu/hr (exact)} \qquad (ans.)$$

or, by Eq. (4-6), is

$$_1\dot{q}_{s2} = 0.244\dot{m}_a(t_2 - t_1) = 0.244(14,100)(90 - 45)$$
$$= 154,820 \text{ Btu/hr (approximate)}$$

or, by Eq. (4-10), is

$$_1\dot{q}_{s2} = 1.10(\text{cfm})(t_2 - t_1) = 1.10(300)(90 - 45)$$
$$= 148,500 \text{ Btu/hr (approximate)}$$

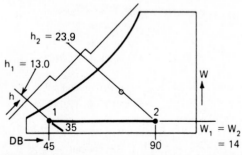

FIGURE 4-3 Solution to Illustrative Problem 4-1.

ILLUSTRATIVE PROBLEM 4-2

How much heated air must be supplied to a space within a building to offset a heat loss of 38,200 Btu/hr? The space air condition is to be maintained at 70°F DB and 50% RH, and the air to be supplied is at a temperature of 105°F DB. Assume no moisture addition within the space.

Solution: Since there is no moisture addition in the space, the supply air will have the same humidity ratio as the space air. The supply air cools to the space condition as it gives up its sensible heat to overcome the space heat loss. Therefore, the supply air undergoes a sensible cooling process from 1 to 2 as illustrated on Figure 4-4.

Step 1. On the psychrometric chart at 70°F DB and 50% RH, locate the space air condition, point 2. At this point, read $W_2 = 55$ grains/lbda, $h_2 = 25.3$ Btu/lbda, and $v_2 = 13.51$ ft³/lbda.

Step 2. On the chart, locate the supply air condition by constructing a horizontal line ($W_1 = W_2$) through point 2 and extend to supply air temperature, 105°F DB, point 1. At this point, find $W_1 = 55$ grains/lbda, $h_1 = 33.9$ Btu/lbda, and $v_1 = 14.41$ ft³/lbda.

Step 3. The mass of supply air may be found from Eq. (4-4) as

$$_1\dot{q}_{s2} = \dot{m}_a(h_1 - h_2)$$

Note that the subscripts on h have been reversed to give a positive value of $_1\dot{q}_{s_2}$. Then,

$$38,200 = \dot{m}_a(33.9 - 25.3)$$

$$\dot{m}_a = 4440 \text{ lb/hr}$$

Step 4. If the supply air volume is to be found, we make use of Eq. (4-7), or

$$\text{cfm} = \frac{\dot{m}_a v_1}{60} = \frac{(4440)(14.41)}{60}$$

$$= 1066 \text{ at supply air conditions} \qquad (ans.)$$

An approximate solution could be obtained by use of Eq. (4-10), or

$$_1\dot{q}_{s2} = 1.10(\text{cfm})(t_1 - t_2)$$

$$38,200 = 1.10(\text{cfm})(105 - 70)$$

$$\text{cfm} = 992 \text{ (approximate)}$$

4-3   ADIABATIC MIXING OF TWO AIRSTREAMS

A frequently encountered process in air conditioning is the mixing of two or more airstreams having different psychrometric properties. Figure 4-5 represents the schematic drawing of the two airstreams mixing and the psychrometric chart for the process. The fundamental equations that apply to this mixing process are as follows:

$$\dot{m}_{a1}h_1 + \dot{m}_{a2}h_2 = \dot{m}_{a3}h_3 \qquad (4\text{-}12)$$

$$\dot{m}_{a1} + \dot{m}_{a2} = \dot{m}_{a3} \qquad (4\text{-}13)$$

$$\dot{m}_{a1}W_1 + \dot{m}_{a2}W_2 = \dot{m}_{a3}W_3 \qquad (4\text{-}14)$$

If $\dot{m}_{a3}$ is eliminated from Eqs. (4-12) through (4-14), we would have

$$\frac{\dot{m}_{a1}}{\dot{m}_{a2}} = \frac{h_2 - h_3}{h_3 - h_1} = \frac{W_2 - W_3}{W_3 - W_1} \qquad (4\text{-}15)$$

This defines a straight line on the psychrometric chart between points 1 and 2. Point 3 lies on the line between

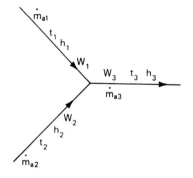

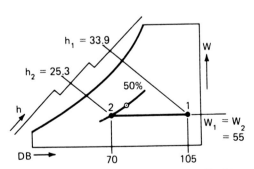

FIGURE 4-4   Solution to Illustrative Problem 4-2.

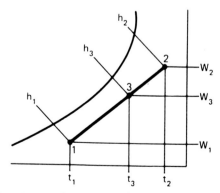

FIGURE 4-5   Schematic of the mixing process.

points 1 and 2, and the line segments produced are proportional to the masses of air mixed, the temperature differences, humidity ratio differences, and the enthalpy differences.

Making use of Eqs. (4-12) through (4-15), the following three equations become useful in the evaluation of the mixed-air properties:

$$t_3 = \frac{\dot{m}_{a1} \times t_1 + \dot{m}_{a2} \times t_2}{\dot{m}_{a1} + \dot{m}_{a2}} \qquad (4\text{-}16)$$

$$W_3 = \frac{\dot{m}_{a1} \times W_1 + \dot{m}_{a2} \times W_2}{\dot{m}_{a1} + \dot{m}_{a2}} \qquad (4\text{-}17)$$

$$h_3 = \frac{\dot{m}_{a1} \times h_1 + \dot{m}_{a2} \times h_2}{\dot{m}_{a1} + \dot{m}_{a2}} \qquad (4\text{-}18)$$

The solution of a mixed-air problem in psychrometrics normally makes use of one of Eqs. (4-16) through (4-18), and then the remaining mixed-air properties are determined by graphical means. This will be shown by Illustrative Problem 4-3 to follow.

If the specific volumes of the two airstreams do not vary appreciably from each other (about 0.5 ft³/lbda), then Eq. (4-16) could be used, substituting cfm in place of $\dot{m}_a$, as follows:

$$t_3 = \frac{\text{cfm}_1 \times t_1 + \text{cfm}_2 \times t_2}{\text{cfm}_1 + \text{cfm}_2} \qquad (4\text{-}19)$$

Also, in some problems involving the mixing of two airstreams, data is given on the percent of the total flow entering at the two conditions. This is handled by substituting the percentage flow, expressed as a decimal, in place of the mass flow quantities.

ILLUSTRATIVE PROBLEM 4-3

Three hundred cfm of air at 35°F DB and 100% RH are mixed with 600 cfm of air at 85°F DB and 50% RH. What will be the dry-bulb temperature, humidity ratio, and enthalpy of the mixture?

Solution: (See Figure 4-6.)
Step 1. Determine the psychrometric properties of the two airstreams from the psychrometric chart.

	Condition 1	Condition 2
	300 cfm at 35°F DB, 100% RH	600 cfm at 85°F DB, 50% RH
Specific volume:	$v_1 = 12.55$ ft³/lbda	$v_2 = 14.02$ ft³/lbda
Humidity ratio:	$W_1 = 30$ grains/lbda	$W_2 = 90$ grains/lbda
Enthalpy:	$h_1 = 13.1$ Btu/lbda	$h_2 = 34.5$ Btu/lbda

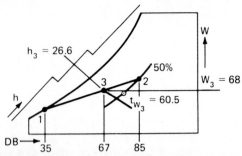

FIGURE 4-6 Solution to Illustrative Problem 4-3.

Step 2. Determine the mass of the two airstreams. Using Eq. (4-7) to determine lb/min., we have

$$\dot{m}_{a1} = \frac{\text{cfm}_1}{v_1} = \frac{300}{12.55} = 23.9 \text{ lb/min.}$$

$$\dot{m}_{a2} = \frac{\text{cfm}_2}{v_2} = \frac{600}{14.02} = 42.8 \text{ lb/min.}$$

Step 3. Determine the psychrometric properties of the mixed air. On the psychrometric chart, draw a straight line joining point 1 and point 2. Calculate the dry-bulb temperature of the mixture by using Eq. (4-16):

$$t_3 = \frac{\dot{m}_{a1} \times t_1 + \dot{m}_{a2} \times t_2}{\dot{m}_{a1} + \dot{m}_{a2}}$$

$$= \frac{23.9 \times 35 + 42.8 \times 85}{23.9 + 42.8}$$

$$= 67.08°F \text{ (say, 67°F)} \qquad (ans.)$$

Where $t_3 = 67°F$ intersects the mixture line between points 1 and 2 determines the mixed-air condition, point 3. From the chart at point 3, find $t_{w3} = 60.5°F$, $W_3 = 68$ grains/lbda, $h_3 = 26.6$ Btu/lbda, dew-point $t_{dp} = 56.4°F$, and partial pressure $p_v = 0.455$ in. Hg.

Step 4. Determine the dry-bulb temperature of the mixture by using approximate Eq. (4-19):

$$t_3 = \frac{300 \times 35 + 600 \times 85}{300 + 600}$$

$$= 68.3°F \qquad (ans.)$$

This varies 1.3°F from the previous value of $t_3 = 67°F$. The significance of the difference must be evaluated by engineers as to the accuracy with which they should be working. The exact Eq. (4-16) should be used when possible.

4-4 HUMIDIFICATION OF AIR[1]

As stated previously, the purpose of environmental control is to make people more comfortable and the labor of people and equipment more efficient. Temperature, humidity, cleanliness, air movement, and thermal radiation interrelate to create conditions in which people are more comfortable or less comfortable. In a home, business, or industry, comfort or production efficiency is greatly affected by changes in these environmental variables.

Least evident of these variables to human perception is humidity. All of us will recognize and react more quickly to temperature changes, odors, heavy dust in the air, drafts, or radiant heat from sunlight or a radiator than we will to a change in relative humidity. However, as relative humidity interrelates with temperature and others of these variables, it becomes a vital ingredient in total environmental control.

Frequently in air conditioning work, it is necessary to humidify the air by introducing moisture into the air. The moisture added may be already in the vapor state, or it may be liquid. (It may also be solid, although infrequently.) Humidification is frequently required during cold weather conditions because the cold outside air infiltrating into a heated building or being brought in by a mechanical ventilation system is normally very dry (low humidity ratio, even though its relative humidity may be high). If the infiltration or ventilation air is not humidified, a low dew-point temperature, relative humidity, and humidity ratio could exist within the heated building.

ILLUSTRATIVE PROBLEM 4-4

Assume that outside air at 30°F DB and 40% RH infiltrates into a building where the inside is being maintained at 72°F DB. What would be the theoretical relative humidity of the infiltrated air at 72°F?

Solution: Locate on the psychrometric chart the condition of the outside air at 30°F DB and 40% RH. At this point, find the humidity ratio of 9.5 grains/lbda. If there is no moisture addition in the building, the infiltration air will have the same humidity ratio at 72°F.

At a humidity ratio of 9.5 grains/lbda and 72°F, find the relative humidity to be slightly less than 9.0%.   (*ans.*)

Since one to three complete air changes occur every hour in most buildings through infiltration (and many more times with forced makeup or exhaust), cold outdoor air replaces the warm indoor air. The heating system heats this cold, moist outdoor air and it becomes warm, dry indoor air. This condition is unhealthy for people and may be troublesome for some industrial processes.

Indoor relative humidity calculated as we did in Illustrative Problem 4-4 should be called the "theoretical" indoor relative humidity. This condition very seldom actually exists in a building. Relative humidity values observed on a hygrometer will almost always exceed this theoretical value. The reason is that this dry, heated air will absorb moisture from any source it can within the building. It will absorb moisture from any hygroscopic material in the building, as well as from nasal passages and skin of human beings. This is not "free" humidification; it is the most expensive when translated into terms of human health and comfort, material deterioration, and production difficulties. Moreover, it requires the same amount of energy whether the moisture is absorbed from people and materials or added to the air by an efficient humidification system.

ILLUSTRATIVE PROBLEM 4-5

Let us examine the theoretical system of Illustrative Problem 4-4 using enthalpy as the base and determine the energy exchanges taking place.

Solution: From the psychrometric chart at 30°F DB and 40% RH, find h_{oa} = 8.8 Btu/lbda. If this air is heated to 72°F without moisture addition, the enthalpy is h_{ra} = 18.7 Btu/lbda, and the resulting relative humidity of 9.0%.

If the actual inside relative humidity were 25% by actual measurement by a hygrometer, the enthalpy at 72°F and 25% RH would be h_{ra} = 21.7 Btu/lbda, with the additional moisture coming from hygroscopic materials and people in the area. The additional energy 21.7 − 18.7 = 3.0 Btu/lbda, which is 16%, must come from the heating system.

If a humidification system is used and moisture added to achieve a comfortable 35% RH, the enthalpy at 72°F and 35% RH is 23.6 Btu/lbda, which is only (23.6 − 21.7)/21.7 = 8.0% increase over the "inevitable" energy load of 21.7 Btu/lbda. This is substantially less than (23.6 − 18.7)/18.7 = 26%, which is the theoretical increase from 9.0% RH and 35% RH.

If the inside air temperature were only 68°F at 35% RH (because people can be more comfortable at a lower temperature with higher humidity levels), the enthalpy is only 21.8 Btu/lbda, which is a very slight increase of 21.8 − 21.7 = 0.1 Btu/lbda in energy. In many cases, there will be a decrease in the required energy.

The introduction of water into the air for purposes of humidification is relatively easy and can be accomplished with rather simplified equipment. However, before we introduce this equipment, we will discuss briefly the principle of *vaporization* of a liquid.

[1]Extracted from *The Armstrong Humidification Handbook,* Armstrong Machine Works, Three Rivers, Michigan, p. 2–3, with permission.

Vaporization[2] of a liquid, such as water, may take place in two different ways: (1) by evaporation and (2) by boiling. *Evaporation* of a liquid can occur only at the free surface of the liquid and may take place at any liquid temperature between its saturation temperature and the dew-point temperature of the surrounding air. (*Free surface* means the interface between the liquid surface and the surrounding air.) *Boiling* takes place both at the liquid free surface and within the body of the liquid and can occur only at the saturation temperature corresponding to the absolute pressure on the liquid.

Water left standing in a pan open to the atmosphere as shown in Figure 4-7a will eventually evaporate completely, even though the actual water temperature is below its saturation temperature. However, if the open pan is placed inside a closed container as shown in Figure 4-7b, only a portion of the water will evaporate; the balance will remain in the pan for an indefinite length of time if the closed container remains sealed. The reason for this difference in behavior is found in the difference between the vapor pressure of the water in the pan and the pressure exerted by the vapor pressure of moisture in the air in the sealed container. *Vapor pressure difference* is the controlling factor in evaporation and also condensation of water; or we may say that the dew-point temperature of the air above the liquid is the controlling factor.

Consider an open pan of water, such as in Figure 4-7a, sitting in a room where the air temperature is 74°F and the relative humidity is 40%; the corresponding dew-point temperature is the saturation temperature of the moisture vapor at its vapor pressure, or $p_v = \phi \times p_{vs} = 0.40 \times 0.8462 = 0.3385$ in. Hg. The dew-point temperature at 0.3385 in. Hg (see Table 4-1, given later in this chapter) is 48.2°F. Although the water temperature in the pan could fall to the wet-bulb temperature

[2]Information extracted from *Trane Air Conditioning Manual,* The Trane Company, LaCrosse, Wisconsin, 1965, with permission.

($t_w = 58.8$°F) of the surrounding air, it will in a case such as this probably assume a temperature that is closer to the air dry-bulb temperature ($t = 74$°F). Assume, for purposes of discussion, that the water assumes a temperature of 65°F. The vapor pressure corresponding to the air dew point was 0.3385 in. Hg. At the same time, the vapor pressure of the water at 65°F is 0.6222 in. Hg. Since the vapor pressure of the surrounding air is less than that of the water, evaporation of the water into the air will occur.

This difference in vapor pressure is the driving force that causes the vapor to diffuse from the liquid surface into the air above. As long as the air currents above the liquid continually carry away the moisture being evaporated, evaporation will continue.

On the other hand, when the pan of water is placed inside a closed container as in Figure 4-7b, the pressure of the vapor mixed with the air in the container is continually increasing due to the water evaporating from the pan. Eventually, *vapor pressure equilibrium* will be established when the vapor pressure in the air and of the water are equal and no further evaporation takes place. When evaporation ceases, the dew-point temperature of the air inside the container will be equal to the temperature of the water in the pan. Under these conditions, the air is said to be saturated.

From this, it is apparent that if the temperature of the water is greater than the air dew point, evaporation takes place; and when the air dew point is equal to the water temperature, evaporation ceases.

The evaporation of a liquid requires that the latent heat of vaporization be supplied. As a result of the water evaporation into the air, the water remaining tends to cool as it surrenders the necessary latent heat from vaporization. In the preceding discussion of the pan of water, it has been assumed that heat would flow from the surrounding air into the water, as no insulation was indicated. As a result, even though the water tends to cool slightly, its temperature remains nearly constant because of heat flow into it from the surrounding air.

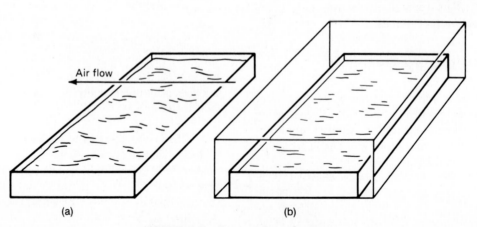

Air flow

(a)   (b)

FIGURE 4-7 Evaporation of water.

4-5 HEAT EXCHANGE BETWEEN AIR AND WATER[3]

When air is brought in contact with water at a temperature different from the air wet-bulb temperature, an exchange of heat, as well as moisture, will take place between the air and water.

1. If the water temperature is higher than the air wet-bulb temperature, the water temperature will drop and the air wet-bulb temperature will rise because the water surrenders heat to the air.
2. If the water temperature is lower than the air wet-bulb temperature, the water temperature will rise and the air wet-bulb temperature will drop.
3. In any exchange of heat between water and air, the water temperature can never fall, or rise, to the initial air wet-bulb temperature.
4. Whenever an exchange of heat occurs between air and water, the temperature of both must change.

There is one *important exception* to item 4 here. In *air washers,* to be discussed later, where the water is continually recirculated being neither heated nor cooled externally, the temperature of the spray water does not change because there is no addition or removal of heat from the system. The latent heat required for evaporation of water comes from the sensible heat in the air, which causes a reduction in air dry-bulb temperature.

In order to humidify air with water sprays, the temperature of the spray water must always be higher than the required final dew point of the air. Such an amount of water must be used that, as the water cools down from its initial temperature, its final temperature will still be above the required final air dew point. It is also necessary to remember that, besides the relationship between the final air dew-point temperature and the water temperature, the exchange of heat and moisture between air and water depends upon the air dry-bulb and wet-bulb temperatures.

4-6 METHODS OF HUMIDIFICATION[4]

If you are humidifying a hospital operating room, obviously your design criteria are different than those for humidifying a textile mill, an office building, a labora-

tory, or a home. Different types of operations have substantially different requirements for the achievement of proper relative humidity. These requirements determine what means of humidification you should use.

Three basic humidifying media are available: (1) steam, (2) evaporative pan, and (3) water spray. Each has particular advantages and limitations that determine its suitability for a particular application. A fourth medium, wetted element humidifiers, applies primarily to private residences or similar small buildings where very low humidifying capacities are required (see Figures 4-8 and 4-9).

Steam is ready-made water vapor that needs only to be mixed with the air. With evaporative pan humidification, air flows across the surface of heated water in a pan and absorbs the water vapor. Both steam and evaporative pan humidification are essentially *isothermal* processes where little change in air temperature occurs.

Water spray humidification is an *adiabatic* process if evaporation of the water is caused by the water absorbing sensible heat from the air reducing the air dry-bulb temperature. The latent heat of evaporation reduces the sensible heat of the air by about 1000 Btu for each pound of water evaporated.

Let us examine the basic requirements for effective humidification and see how each of the humidifying media satisfies these requirements.

Requirements of Effective Humidification

Capacity: Small-capacity requirements can be met most economically with evaporative pan or self-contained steam-generating unit humidifiers, installed either in the air-handling system or individually within the area(s) to be humidified. Larger-capacity needs, depending on the application, can best be met by either water spray or steam humidification systems, which may be installed either in nonducted industrial applications or in air-handling systems. As a rule, steam humidification can provide larger capacities than water spray units.

Maintenance: Steam humidification systems are largely maintenance free. Maintenance requirements for both water spray and evaporative pan humidification depend on mineral content of the water. With water spray equipment, minerals dispersed in the water mist settle out as dust when the droplets evaporate. Spray nozzles can become clogged with mineral deposits and require regular cleaning to maintain capacity. Similarly, evaporative pan humidifiers are subject to "liming up," particularly on the heating coils in the pan, although chemical additives automatically or manually added to the water in the pan can reduce this problem by as much as 50%.

[3]Information extracted from *Trane Air Conditioning Manual,* The Trane Company, LaCrosse, Wisconsin, 1965, pp. 69–76, with permission.

[4]Extracted from *The Armstrong Humidification Handbook,* Armstrong Machine Works, Three Rivers, Michigan, pp. 9–12, with permission.

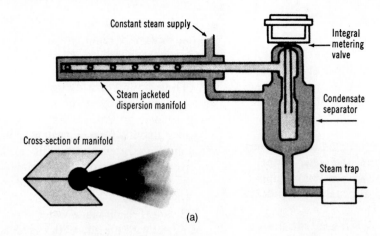

Constant steam supply

Integral metering valve

Steam jacketed dispersion manifold

Condensate separator

Cross-section of manifold

Steam trap

(a)

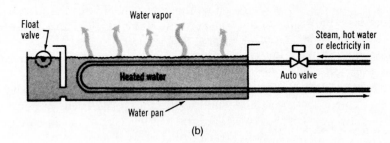

Float valve

Water vapor

Steam, hot water or electricity in

Heated water

Auto valve

Water pan

(b)

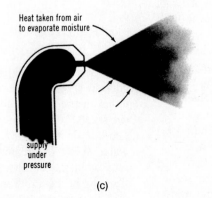

Heat taken from air to evaporate moisture

supply under pressure

(c)

FIGURE 4-8 Humidification method (a) using dry steam, (b) using evaporated water, and (c) using spray water. (Courtesy of Armstrong Machine Works, Three Rivers, MI)

Response to control: Since steam is ready-made water vapor, it needs only to be mixed with the air to satisfy the demands of the system. Response to control is much faster than with either water spray or evaporative pan, where evaporation must take place before humidified air can be circulated.

Control of output: Operation of the three basic types of systems is regulated by a humidity controller located either in the space humidified or in the return air duct of an air duct system. Water spray and evaporative pan equipment operate intermittently in response to control. Output for these types of humidifiers is de-

termined by size and number of spray nozzles or by water temperature and surface area, respectively. Steam humidifiers can meter the output by means of a modulating control valve in response to the humidity controller, by positioning the humidifier operator anywhere from closed to fully open. Steam humidifiers can thus respond more precisely to system demand than can the other two types.

Sanitation: The high temperatures inherent in steam humidification make it virtually a sterile medium. Assuming boiler makeup water is of satisfactory quality and there is no condensation, dripping, or spitting in

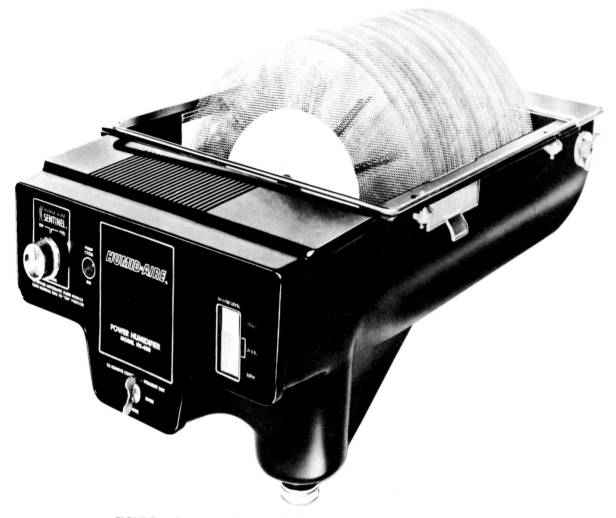

FIGURE 4-9 HUMID-AIRE FH-400, a wetted element humidifier.
(Courtesy of Adams Manufacturing Company, Cleveland, OH)

the air ducts, no bacteria or odors will be disseminated with steam humidification. Evaporative pan humidifiers can sustain bacteria colonies in the reservoir and distribute them throughout the humidified space. High water temperatures, water treatment, and regular cleaning and flushing of the humidifier help to minimize the problem, however. Water spray systems can distribute large amounts of bacteria and can present odor problems in air-handling systems. Drip from spray nozzles and unevaporated water discharge can collect in ducts, around drains and drip pans, and on eliminator plates, encouraging the growth of algae and bacteria.

Corrosion: Whenever liquid moisture is present in a humidification system, corrosion can be a problem. Since steam and evaporative pan humidifiers are designed to discharge only water vapor, corrosion in ductwork is not a problem. Evaporative pan units are subject to corrosion and deterioration, the severity of which depends on materials of construction and fre-

quency of cleaning. Steam units are largely self-cleaning; scale or sediment, whether formed in the unit or entrained in the supply steam, are drained from the humidifier through the drain trap. Water spray humidifiers and all ductwork, eliminator plates, and so forth, in proximity to them are highly susceptible to corrosion.

Costs: Evaluation of costs in selecting a humidification system should include installation, operating and maintenance costs, as well as first cost. However, total humidification costs should run far less than heating or cooling system costs.

First costs, of course, vary with the size of the units. Priced on a capacity basis, larger-capacity units are the most economical, regardless of the types of humidifier; that is, one humidifier capable of delivering 1000 lb of humidification per hour costs less than two 500-lb/hr units of the same type.

Of the three basic humidification media, steam humidifiers will provide the highest capacity per first-cost dollar; water spray is next; and evaporative pan is the least economical, assuming capacity needs are 75 lb/hr or more.

Installation costs for the three types cannot be accurately formulated, as availability of water, steam, electricity, and so forth, close to the humidifiers can vary greatly between jobs. Operating costs are similar between the three types, although they are insignificant compared to heating and cooling system costs.

Maintenance costs vary widely with water spray and to a lesser degree with evaporative pan equipment, requiring substantially more maintenance than steam humidification. Table 4-1 summarizes the capabilities of each type.

Recommended Applications

Steam: Recommended for virtually all commercial, institutional, and industrial applications. Where steam is not available, small-capacity needs up to 50 to 75 lb/hr can be best met using self-contained steam-generating units; above this capacity range, central system steam humidifiers are most effective and economical. Steam should be specified with caution where humidification is being used in small, confined areas to add large amounts of moisture to hygroscopic materials.

Evaporative pan: Recommended only as an alternative to self-contained steam-generating unit humidifiers for small commercial or institutional applications. Generally not recommended where load requirements exceed 50 to 75 lb/hr.

Water spray: Recommended for industrial applications where evaporative cooling is required. Typical application is summer time humidification of textile mills in the southern United States.

Wetted element humidifiers: These units, as mentioned previously, are recommended for private residences or similar small buildings where very low humidifying capacities are required. Figure 4-9 represents a humidifying device that exposes a large wetted surface to an airstream. The unit shown would be mounted in the warm airstream in a duct system. The unit consists of a series of rotating, alloy metal, screened disks rotated at slow speed (2 rpm) by a small electric motor drive. The disks rotate in a bath of water in the pan where water is picked up and held in the mesh of the screen by surface tension. The wetted disks rotate into

TABLE 4-1

Comparison of humidification methods

	Steam	Evaporative Pan	Water Spray
Effect on air temperature	Virtually no change	Small temperature rise	Substantial temperature drop (if no heating of spray water)
Unit capacity per unit size	Small to very large	Small	Small
Maintenance frequency	Annual	Weekly to monthly	Weekly to bimonthly
Response to control	Immediate	Slow	Slow
Control of output	On-off or full range modulation	On-off	On-off
Sanitation/Corrosion	Sterile medium; corrosion free	Pan subject to corrosion; bacteria can be present	Subject to severe corrosion and bacteria problems
Cost: Price (per unit of capacity)	Low	High	Low to medium
Installation	Varies with availability of steam, water, electricity, etc.		
Operating	Low	Low	Low
Maintenance	Low	High	Very high

Source: Reproduced from *The Armstrong Humidification Handbook,* Armstrong Machine Works, Three Rivers, Mich., with permission.

the airstream, presenting a large surface area for water evaporation. The disks rotate only when humidification is called for by the humidity control system.

If we are concerned with only the humidification of moist air, Figure 4-10a illustrates a schematic system where we assume adiabatic mixing of moist air with moisture (steam or liquid water). The equations that apply are as follows:

$$\dot{m}_{a1}h_1 + \dot{m}_w h_w = \dot{m}_{a2}h_2 \qquad (4\text{-}20)$$

$$\dot{m}_{a1} = \dot{m}_{a2} \qquad (4\text{-}21)$$

$$\dot{m}_{a1}W_1 + \dot{m}_w = \dot{m}_{a2}W_2 \qquad (4\text{-}22)$$

By solving Eq. (4-22) for $\dot{m}_w$, substituting into Eq. (4-20), and rearranging, we have

$$h_w = \frac{h_2 - h_1}{W_2 - W_1} = \frac{\Delta h}{\Delta W} \qquad (4\text{-}23)$$

Equation (4-23) is significant because the ratio $\Delta h / \Delta W$ indicates that the slope of the humidification process line on a psychrometric chart is a function of the enthalpy of the added moisture. Therefore, the air may be heated or cooled during humidification dependent on the enthalpy of the moisture added (see Section 4-5). Figure 4-10b indicates some possible paths for the humidification process.

ILLUSTRATIVE PROBLEM 4-6

An evaporative-pan type of humidifier is to be used in a duct system to humidify air supplied to a building. The supply air upstream of the humidifier has a dry-bulb temperature of 90°F and a wet-bulb temperature of 58°F. The downstream air leaving the humidifier must have a humidity ratio of 70 grains/lbda (0.010 lbv/lbda). The quantity of air flowing through the duct upstream of the humidifier is 8000 cfm. If the water in the pan is maintained at 212°F (14.7 psia) by a steam immersion coil, determine (1) the moisture addition to the air (lb/hr) and (2) the resulting psychrometric properties of the air downstream from the humidifier. Figure 4-11 shows a sketch of the equipment.

Solution:

Step 1. Determine the psychrometric properties of the air at its initial state. On a psychrometric chart, locate point 1 at $t_1 = 90°F$ and $t_{w_1} = 58°F$. From the chart, find $W_1 = 21.0$ grains/lbda, $h_1 = 24.9$ Btu/lbda, and $v_1 = 13.92$ ft³/lbda.

Step 2. Determine the final state of the air. The final humidity ratio $W_2 = 70$ grains/lbda (given). The final enthalpy of the air may be found from Eq. (4-23), $h_2 - h_1 = (W_2 - W_1)h_w$, where $h_w = h_g$ at 212°F, or 1150.5 Btu/lbv. Thus,

$$h_2 - 24.9 = \left(\frac{70 - 21}{7000}\right)(1150.5) = 33.0 \text{ Btu/lbda}$$

On the psychrometric chart, find the point of intersection where $h_2 = 33.0$ and $W_2 = 70$, and find $t_2 = 92°F$ and $t_{w_2} = 69°F$. See Figure 4-12.

Step 3. Determine the amount of moisture addition. The mass of airflow is, by Eq. (4-7),

$$\dot{m}_a = \frac{(60)(\text{cfm})}{v_1} = \frac{(60)(8000)}{13.92}$$

$$= 34{,}480 \text{ lb/hr}$$

and by Eq. (4-22),

$$\dot{m}_w = \dot{m}_a(W_2 - W_1) = 34{,}480\left(\frac{70 - 21}{7000}\right)$$

$$= 241.4 \text{ lb/hr} \qquad (ans.)$$

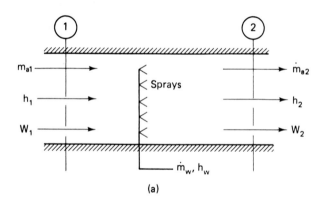

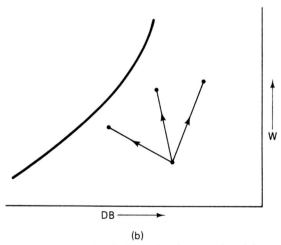

(a)

(b)

FIGURE 4-10 Schematic representing injection of moisture into moist air.

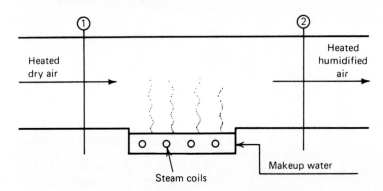

FIGURE 4-11 Schematic of open-pan humidifier (Illustrative Problem 4-6).

4-7 COMBINED HEATING AND HUMIDIFICATION OF AIR

During cold weather, it may be necessary to provide for both heating and humidification of air to be circulated in a building to maintain comfort conditions. The heated air must be introduced into the space to be conditioned at a sufficiently high dry-bulb temperature and humidity ratio to offset heat losses and to maintain a satisfactory humidity condition.

Figure 4-13 is a typical heating and ventilating air-handling unit. The unit consists of several sections: a mixing box section where return air from the conditioned space is mixed with outside air for ventilation in a fixed or variable ratio that is adjusted by damper position, a heating coil and humidifier section, and a fan section. The unit shown is called a *draw-through unit* because of the location of the fan with respect to the other components.

Figure 4-14 is a generalized sketch of an air-handling system where moist air enters at point 1 and leaves at point 2. As the air passes through the system, it may have sensible heat and/or moisture added. The

system is a steady-flow system, and the following equations apply:

$$\dot{m}_{a1} W_1 + \dot{m}_w = \dot{m}_{a2} W_2 \qquad (4\text{-}22)$$

$$\dot{m}_{a1} h_1 + {}_1q_{s2} + \dot{m}_w h_w = \dot{m}_{a2} h_2 \qquad (4\text{-}24)$$

$$\dot{m}_{a1} = \dot{m}_{a2} \qquad (4\text{-}25)$$

where $\dot{m}_w$ is the pounds of water evaporated, or the pounds of vapor (steam) added per hour, and h_w is the corresponding enthalpy of the moisture added in Btu/lb of moisture.

Since $\dot{m}_{a_1} = m_{a_2}$, we divide Eq. (4-24) by $\dot{m}_a$ and we have

$$h_1 + \frac{{}_1\dot{q}_{s2}}{\dot{m}_a} + \frac{\dot{m}_w h_w}{\dot{m}_a} = h_2 \qquad (4\text{-}26)$$

Also, from Eqs. (4-24) and (4-25), we have

$$_1\dot{q}_{s2} = \dot{m}_a(h_2 - h_1) - \dot{m}_w h_w \qquad (4\text{-}27)$$

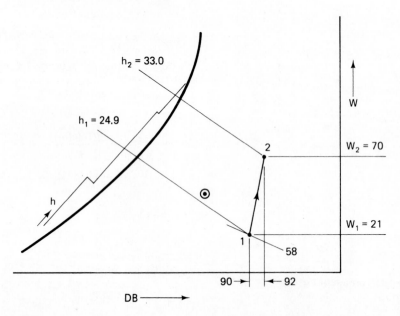

FIGURE 4-12 Solution to Illustrative Problem 4-6.

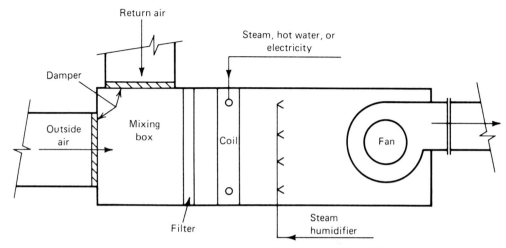

FIGURE 4-13 Schematic of typical heating and humidifying unit.

and, from Eq. (4-22), we have

$$\dot{m}_w = \dot{m}_a(W_2 - W_1) \qquad (4\text{-}28)$$

Substitution of Eq. (4-28) into Eq. (4-26) gives

$$h_1 + \frac{{}_1\dot{q}_{s2}}{\dot{m}_a} + (W_2 - W_1)h_w = h_2 \qquad (4\text{-}29)$$

Equation (4-29) is the general equation that states that the enthalpy of the air leaving, h_2, is equal to the enthalpy of the air entering, h_1, plus the sensible heat added plus the latent heat added by the moisture.

If the process were adiabatic (${}_1\dot{q}_{s_2} = 0$), then Eq. (4-29) would become

$$h_1 + (W_2 - W_1)h_w = h_2 \qquad (4\text{-}30)$$

In the process represented by Eq. (4-30), ideally the air entering at point 1 is brought to its "temperature of adiabatic saturation," commonly known as the "thermodynamic wet-bulb temperature" (see Section 3-14). The process represents cooling of the air at *constant wet bulb* as the water absorbs sensible heat from the air to provide the latent heat for vaporization. This is the process used to cool and humidify air with recirculated

spray water in an air washer with no external heating or cooling of the spray water. The process is called *evaporative cooling* (discussed later).

By solving Eq. (4-28) for $\dot{m}_a$ and substituting into Eq. (4-27) and rearranging terms, we have

$$\frac{\Delta h}{\Delta W} = \frac{h_2 - h_1}{W_2 - W_1} = \frac{{}_1\dot{q}_{s2}}{\dot{m}_w} + h_w$$
$$= \frac{{}_1\dot{q}_{s2} + \dot{m}_w h_w}{\dot{m}_w} \qquad (4\text{-}31)$$

Equation (4-31) defines the enthalpy–humidity ratio where

1. The rate of airflow has disappeared.
2. The ratio $\Delta h/\Delta W$ is the slope of a line on the psychrometric chart related only to the total space load and the moisture added.
3. If the condition of the air at point 1 or point 2 is set, the values of h and W for the other point may be allowed to vary but must always be on a straight line on the psychrometric chart having a slope of $\Delta h/\Delta W$.

Some psychrometric charts include an alignment quadrant that gives various values of $\Delta h/\Delta W$, which provides a means for determining the direction of heating and humidifying process line on the chart. The following problem illustrates the process.

ILLUSTRATIVE PROBLEM 4-7

Moist air at 40°F DB and 35°F WB enters a heating and humidifying device at the rate of 3000 cfm at a barometric pressure of 29.92 in. Hg. While passing through the chamber, the air absorbs sensible heat at the rate of 153,000 Btu/hr and picks up 83 lb/hr of saturated steam at 230°F. Determine the

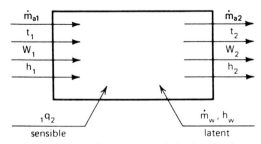

FIGURE 4-14 Schematic of device for heating and humidifying air.

dry-bulb and wet-bulb temperatures of the air leaving the device.

Solution (*Method 1*): (See Figure 4-15.)

Step 1. Determine the psychrometric properties of air at the initial condition. On the psychrometric chart (Figure 4-15), at 40°F DB and 35°F WB, find $v_1 = 12.64$ ft³/lbda, $W_1 = 22$ grains/lbda, and $h_1 = 13.0$ Btu/lbda.

Step 2. Determine the mass of air flowing. From Eq. (4-7),

$$\dot{m}_{a1} = \dot{m}_{a2} = \frac{(60)(\text{cfm})}{v_1} = \frac{(60)(3000)}{12.64}$$

$$\dot{m}_a = 14{,}240 \text{ lb/hr}, \quad \text{or} \quad \frac{14{,}240}{60} = 237.3 \text{ lb/min.}$$

Step 3. Determine the psychrometric properties of air at the final condition. From Eq. (4-22),

$$W_2 = \frac{\dot{m}_a W_1 + \dot{m}_w}{\dot{m}_a} = W_1 + \frac{\dot{m}_w}{\dot{m}_a}$$

$$= \frac{22}{7000} + \frac{83}{14{,}240}$$

$$= 0.00898 \text{ lbv/lbda}$$

or

$$W_2 = (0.00898)(7000)$$

$$= 62.86 \text{ grains/lbda} \qquad \textit{(ans.)}$$

From Eq. (4-26),

$$h_2 = h_1 + \frac{\dot{q}_{s2}}{\dot{m}_a} + \frac{\dot{m}_w h_w}{\dot{m}_a}$$

From Appendix Table A-1, h_w at 230°F = h_g at 230°F = 1157.1 Btu/lb steam. Thus,

$$h_2 = 13.00 + \frac{153{,}000}{14{,}240} + \frac{(83)(1157.1)}{14{,}240}$$

$$= 13.00 + 10.74 + 6.74$$

$$= 30.48 \text{ Btu/lbda} \qquad \textit{(ans.)}$$

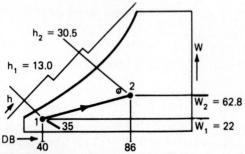

FIGURE 4-15 Solution to Illustrative Problem 4-7 (Method 1).

Note: In this solution for h_2, the value 10.74 represents the sensible heat added, and the value 6.74 represents the latent heat added per pound of air passing through the device.

Step 4. Locate the final air condition on the psychrometric chart. Draw a horizontal line on the chart for $W_2 = 62.86$ grains/lbda. Draw an inclined line on the chart for $h_2 = 30.48$ Btu/lbda to intersect the W_2 line. This point of intersection is the final air condition leaving the device, point 2. At point 2, read the dry-bulb temperature as 86°F and the wet-bulb temperature as 65.8°F.

Step 5. Connect points 1 and 2 by a straight line. This line represents the process of heating and humidification within the device.

Solution (*Method 2*): (See Figure 4-16.)

Step 1. Determine the sensible heat ratio (SHR) for the heating and humidifying process. From the solution of Method 1, it was determined that the sensible heat addition to the air was $q_s = 10.74$ Btu/lbda and the latent heat addition was $q_L = 6.74$ Btu/lbda. The SHR is defined as the sensible heat divided by the total heat for a process. For this problem,

$$\text{SHR} = \frac{q_s}{q_s + q_L} = \frac{10.74}{10.74 + 6.74} = 0.614$$

Step 2. Plot the SHR on the psychrometric chart. In Figure 4-16, at a point where the 78°F DB line intersects with the 50% RH curve, note a small circle. This is the alignment circle for the SHR scale located on the upper right of Figure 4-16. Using a straightedge, draw a construction line between the alignment circle and an SHR value of 0.614. Parallel to this construction line, draw the process line through the known point 1 at 40°F DB and 35°F WB. Point 2 must be on this process line.

Step 3. Determine the location of point 2. Refer again to Method 1 solution, step 3; the humidity ratio of the air leaving the device was $W_2 = 62.86$ grains/lbda. Determine the point of intersection of W_2 and the process line drawn in step 2 above. This determines point 2 on the psychrometric chart, and we can read the dry-bulb temperature as 84°F and the wet-bulb temperature as 65.2°F.

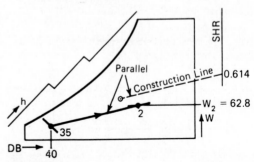

FIGURE 4-16 Solution to Illustrative Problem 4-7 (Method 2).

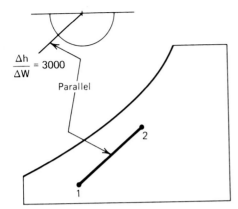

FIGURE 4-17 Alignment quadrant used with some psychrometric charts (see Illustrative Problem 4-7, Method 3).

Note: The small variation in the results obtained is quite possible with the graphical-type solution. In using the SHR scale, extreme care must be taken if a high degree of accuracy is to be expected.

Solution (Method 3):   (See Figure 4-17.) After the development of Eq. (4-31), it was indicated that some psychrometric charts provide an alignment quadrant to identify the enthalpy–humidity ratio ($\Delta h/\Delta W$) for a heating and humidifying process. Figure 4-17 is a schematic drawing of such a chart and quadrant.

Step 1. Determine the enthalpy–humidity ratio for the process. By Eq. (4-31),

$$\frac{\Delta h}{\Delta W} = \frac{\dot{q}_{s2}}{\dot{m}_w} + h_w$$

$$= \frac{153,000}{83} + 1157.1$$

$$= 3004 \quad \text{(say, 3000)} \qquad \text{(ans.)}$$

Step 2. On the quadrant, draw a construction line from its center through 3000 on the scale. Parallel to this line, draw the heating and humidifying process line on the psychrometric chart through the known conditions at point 1.

Step 3. Determine the position of point 2 on the process line by using either Eq. (4-22) or (4-29).

4-8   HEATING AND HUMIDIFICATION USING SPRAY EQUIPMENT

Frequently, spray water equipment is used with or without various arrangements of heating coils to preheat and/or reheat the air passing through the spray chamber. The spray chamber is frequently called an *air washer* because it performs the secondary task of cleaning and removing dust and other particulate from the airstream. However, for our discussion at this point, we will be concerned primarily with the spray chamber as a device to humidify the air.

In Sections 4-4 and 4-5, there was a discussion of the behavior of air and water in contact with each other. Humidification of air by spraying water into it is a direct application of the principles of evaporation and heat exchange between air and water when they are in contact with each other.

In most instances, air washers are used in industrial applications where the device performs the two objectives of humidification and cleaning of the air. Figure 4-18 shows a sectional view of a large, central station washer. It consists of a spray chamber in which a number of spray nozzles and risers are installed. The washer illustrated has one bank of spray nozzles. Sometimes two, and occasionally three, banks are installed. Nozzles used for air washers ordinarily have a capacity of approximately 1¼ to 2 gpm per nozzle. The quantity of water delivered by each bank can be varied by installing a larger or smaller number of nozzles. The water is forced through the nozzle by pumping at pressures from 20 to 40 psi to obtain good atomization of the water. The washer housing is a rectangular steel casing, closed at the top and sides and mounted on a shallow watertight tank base. Inlet baffles distribute the entering air uniformly through the chamber, and at the leaving end eliminators remove the entrained water droplets in the airstream preventing so-called carryover. The eliminators also pick up dirt particles. The eliminator plates are designed and installed so that the direction of airflow is changed a number of times while flowing through them. In this way, the air impinges against the wet surfaces of the plates and the dust particles and water droplets are deposited on the plates. The plates are washed down by a continuous stream of water from flooding nozzles. Air washers will not always remove greasy particles and soot. Tobacco smoke will ordinarily pass through an air washer.

Some odors can be removed from the air passing through an air washer. An odor is usually due to the vapor of some compound mixed with the air. Many of these vapors will dissolve in water. Of course, as the water becomes saturated with the soluble vapors, it becomes less and less able to remove odors from the air. Furthermore, the water itself acquires an odor due to the material in solution; therefore, under these conditions, it must be changed frequently or continuous overflow used.

Air washers are designed to be installed on the suction side of the fan. They should never be installed on the discharge side unless they have been specially built for this purpose. There will be no difficulty with water leaking through the joints of the washer installed on the suction side of a fan. However, if a washer is installed

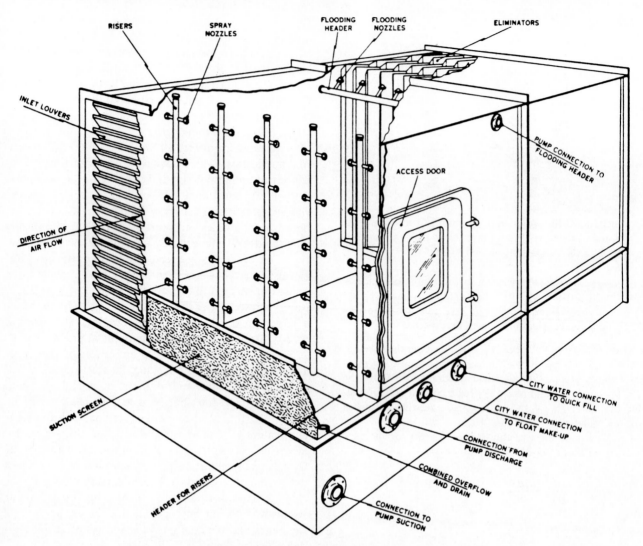

FIGURE 4-18 Industrial-type air washer. (Courtesy of The Trane Company, LaCrosse, WI)

on the discharge side of a fan, the air pressure developed by the fan is likely to cause water leakage through the joints in spite of the rubber gaskets usually installed.

If a set of eliminator plates is installed after each bank of sprays, a washer having two banks is known as a *two-stage washer*. Two-stage washers are illustrated in Figure 4-19, along with a single-bank air washer for comparison. Sometimes, a two-stage washer may have two or three banks of spray nozzles in each stage. Ordinarily, the water in air washers is recirculated. When the air is to be humidified, either the air or the water or sometimes both are heated before entering the washer. If the air is to be cooled and dehumidified in the washer, the water must first be cooled. Heating or cooling of the water can be accomplished by supplying either hot or chilled water to the heat exchanger shown in the spray water piping of Figure 4-19.

The total cross-sectional flow area of air washers

is usually based on an air velocity of 500 fpm. If higher air velocities are used, trouble may be experienced with entrained moisture being carried over from the washer.

Some washers, smaller in size than the industrial type, are used in built-up air-handling units with various arrangements of heating and tempering coils, as well as means for heating or cooling the spray water used in the washer. Figure 4-20 is a schematic drawing of such a built-up air-handling unit. The arrangement is general, and only the components actually required to perform the heating and humidifying process would be included. On the sketch is shown the variation in air properties as the air passes through the unit.

Several common arrangements are used:

1. Preheating of the air prior to its contact with recirculated spray water, followed by reheating of the air.

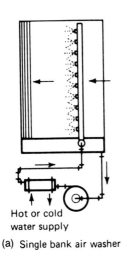 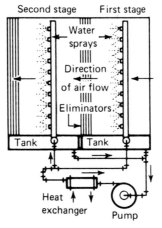 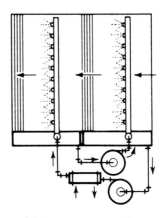

(a) Single bank air washer   (b) Two-stage air washer   (c) Two-stage counterflow air washer

FIGURE 4-19 Types of air washers.

2. Heating of the spray water only.
3. Moderately preheating the air and then passing it through heated spray water.
4. Passing the air through a spray washer using recirculated water with no preheating or reheating of the air.

Arrangement 1 normally uses recirculated spray water having no external heating and uses a reheating coil to

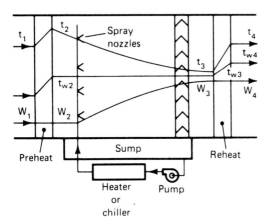

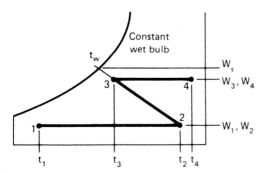

FIGURE 4-20 Schematic and psychrometric chart for heating and humidifying devices using a spray washer.

heat the humidified air to a final temperature that may be required to provide sensible heating of a building. Arrangement 2 is seldom used if the incoming air temperature is expected to be below 32°F at any time. Arrangement 3 is used for special cases when arrangement 2 cannot meet the requirements. Since it involves additional equipment and piping, it is a more expensive system. Arrangement 4 is the familiar process of adiabatic saturation of air when the air dry-bulb temperature is lowered at constant wet bulb as some of the water evaporates to humidify the air. In hot, dry climates, a device of this type becomes a very simple, useful unit for summer cooling.

Refer again to Figure 4-20 and notice that, as the air is preheated, its dry- and wet-bulb temperatures increase, but the humidity ratio remains constant. Passing through the spray chamber, the wet-bulb temperature remains constant as the dry-bulb temperature decreases and the humidity ratio increases. The reheating process increases the dry-bulb and wet-bulb temperatures at constant humidity ratio. Therefore, we have three elementary processes in series: a sensible heating process, an adiabatic saturation process, and another sensible heating process.

Humidifying efficiency, frequently called *effectiveness,* is a measure of the completeness of the humidifying process using a spray chamber. This efficiency is dependent on several factors:

- Number of spray nozzles, and banks of nozzles, and the direction of the sprays with respect to the airflow direction.
- Production of a fine water mist by the nozzles, which depends on the nozzle design, size, and water pressure.
- Air velocity through the chamber.
- Ratio of mass of air to water flow.

• Time of contact between water and air, or actually the length of the unit.

Under favorable conditions, nearly complete saturation of the air at the air wet-bulb temperature may be obtained. The amount of moisture added to the air depends on the entering air wet bulb and on the spray chamber design. Commercial wet washers are less than 100% effective. That is, complete saturation of the air is not obtained. The saturation effectiveness will normally range from 80% to 90%. Referring to Figure 4-20, the saturation efficiency is defined as

$$E_h = \frac{\text{actual change in air dry-bulb temperature}}{\text{maximum possible change in dry bulb}}$$

or

$$E_h = \frac{t_2 - t_3}{t_2 - t_w} \times 100 \qquad (4\text{-}32)$$

and also

$$E_h = \frac{W_3 - W_2}{W_s - W_2} \times 100 \qquad (4\text{-}33)$$

where t_2 and t_3 are the dry-bulb temperatures of the air entering and leaving the spray chamber, respectively; t_w is the constant wet-bulb temperature for the process; W_2 and W_3 are the humidity ratios of the entering and leaving air, respectively; and W_s is the humidity ratio of the air saturated at the wet-bulb temperature.

ILLUSTRATIVE PROBLEM 4-8

Five thousand cfm of air at 40°F DB and 30°F WB are to be heated and humidified to a final condition of 90°F DB and 30% RH. The process of heating will be done by using a preheating coil and a reheating coil arranged as shown in Figure 4-20. The humidification process is to be done by using a spray chamber with recirculated unheated water. Assume that the humidifying process will have an effectiveness of 82%. Determine the following: (1) the psychrometric properties of the air entering and leaving the spray washer, (2) the amount of preheat and reheat required in Btu/hr, and (3) the amount of water evaporated into the air in lb/hr.

Solution: (See Figure 4-21.)
Step 1. Locate initial and final air conditions on the psychrometric chart.

Initial Condition (1)	Final Condition (4)
$t_1 = 40°F$ DB	$t_4 = 90°F$ DB
$t_{w1} = 30°F$ WB	$t_{w4} = 67.2°F$ WB
$W_1 = 10$ grains/lbda	$W_4 = 63$ grains/lbda
$h_1 = 11.2$ Btu/lbda	$h_4 = 31.6$ Btu/lbda
$v_1 = 12.62$ ft³/lbda	$v_4 = 14.04$ ft³/lbda

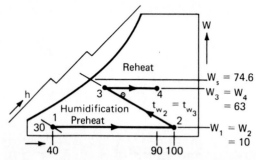

FIGURE 4-21 Solution to Illustrative Problem 4-8.

Step 2. Construct the horizontal *preheat* line through point 1 and extend to the right. The air condition entering spray chamber must be on this line.
Step 3. Construct the horizontal *reheat* line through point 4 and extend to the left. The air condition leaving spray chamber must be on this line.
Step 4. Determine the psychrometric properties of the air entering and leaving the chamber. The humidifying process is assumed to have an effectiveness of 82%; therefore, with $W_4 = W_3$ and $W_2 = W_1$, Eq. (4-33) becomes

$$E_h = \frac{W_4 - W_1}{W_s - W_1}$$

where W_4 is the humidity ratio of the air leaving the spray chamber. Solving Eq. (4-33) for W_s, we have

$$W_s = \frac{W_4 - W_1}{E_h} + W_1$$

$$= \frac{63 - 10}{0.82} + 10$$

$$= 74.6 \text{ grains/lbda}$$

In Figure 4-21, locate W_s of 74.6 on the humidity ratio scale and extend a horizontal line to the left to intersect the 100% RH line. This point locates the wet-bulb temperature for the humidifying process as 59.2°F. Where this constant wet-bulb line intersects the preheat process line and the reheat process line will be the entering and leaving air condition, respectively, for the humidifying process, points 2 and 3. From Figure 4-21, then we have

Point 2	Point 3
$t_2 = 100°F$ DB	$t_3 = 66.6°F$ DB
$t_{w2} = 59.2°F$ WB	$t_{w3} = 59.2°F$ WB
$W_2 = 10$ grains/lbda	$W_3 = 63$ grains/lbda
$h_2 = 25.6$ Btu/lbda	$h_3 = 25.8$ Btu/lbda
$v_2 = 14.13$ ft³/lbda	$v_3 = 13.44$ ft³/lbda

Step 5. Determine the quantity of preheat and reheat. From Eq. (4-7),

$$\dot{m}_a = \frac{(60)(\text{cfm})}{v_1} = \frac{(60)(5000)}{12.62}$$

$$= 23,772 \text{ lb/hr} \qquad (ans.)$$

Remember that this *mass* flow rate *is constant* throughout the flow system. The *volume* flow rate (cfm) *changes* because the specific volume is changing. By Eq. (4-4),

$$_1\dot{q}_{s2} = \dot{m}_a(h_2 - h_1) = 23,772(25.6 - 11.2)$$

$$= 342,300 \text{ Btu/hr} \quad \text{(preheat)} \qquad (ans.)$$

By Eq. (4-4),

$$_3\dot{q}_{s4} = \dot{m}_a(h_4 - h_3) = 23,772(31.6 - 25.8)$$

$$= 137,800 \text{ Btu/hr} \quad \text{(reheat)} \qquad (ans.)$$

Step 6. Determine the mass of water evaporated. Each pound of air absorbs $W_3 - W_2$ grains of moisture; therefore, by Eq. (4-28),

$$\dot{m}_w = \dot{m}_a(W_3 - W_2) = 23,772\left(\frac{63 - 10}{7000}\right)$$

$$= 179.9 \text{ lb/hr} \quad \text{(evaporated)} \qquad (ans.)$$

4-9 HEATING AND HUMIDIFYING AIR WITH HEATED SPRAY WATER

As noted in Section 4-8, air may be heated and humidified by passing it through a spray chamber where the spray water is heated externally. The initial spray water temperature must be high enough so that as it contacts the air and transfers its heat to the air and cools, the final air temperature must be above the initial air temperature. The water temperature can fall only to the wet-bulb temperature of the leaving air (ideally). The changes in the water temperature and air temperature within the spray chamber depend on the relative masses of air and water circulated.

If both the air and water temperatures are changing, the process line on the psychrometric chart is not a straight line. The path of the process is a pursuit curve of the water temperature, which is represented on the saturation curve (100% RH). To draw the pursuit curve on a psychrometric chart, it is necessary to use the following relationship:

heat absorbed by air = heat given up by water

or

$$\dot{m}_a(h_B - h_A) = \dot{m}_w c_p(t_{A'} - t_{B'}) \qquad (4-34)$$

where $\dot{m}_a$ is lb/hr of airflow through the spray chamber; $\dot{m}_w$ is lb/hr of water sprayed; h_A and h_B are the air enthalpy values from the psychrometric chart for two points, A and B; $t_{A'}$ and $t_{B'}$ are the spray water temperatures (in °F); and c_p is specific heat of water, assumed constant and equal to unity.

Figure 4-22 represents a procedure for the development of the pursuit curve on the psychrometric chart. To draw the curve, the initial air condition, point A, and the initial water temperature, point A', are plotted and then connected by a straight line. Select point B on this line a short distance from A. The water temperature at point B' may then be calculated from Eq. (4-34) with known values of $\dot{m}_a$ and $\dot{m}_w$. Connect points B and B' with a straight line. Next, select a point such as C a short distance from B and again use Eq. (4-34) to calculate the water temperature at point C'. The procedure is continued until the final water temperature, D', is approximately equal to the wet-bulb temperature of the air passing through D. A straight line drawn through point A and the final air dry bulb, point D, continued to intersect the saturation curve (shown dashed) defines the humidifying efficiency (effectiveness) of the process. For Figure 4-22, the value of E_h would be $(W_D - W_A)/(W_s - W_A)$ expressed as a decimal.

This method of plotting the pursuit curve is time consuming, and a simpler solution of such problems using heated spray water may be obtained by the use of Figure 4-23. This graph is based upon the assumption that the heat absorbed by the air is equal to the heat given up by the water. Also, the final water temperature and final air wet-bulb temperature are the same. Actually, the final water temperature will be higher than the final air wet-bulb temperature. Nevertheless, this chart enables many problems involving washers to be solved easily by first finding the theoretical solution and then

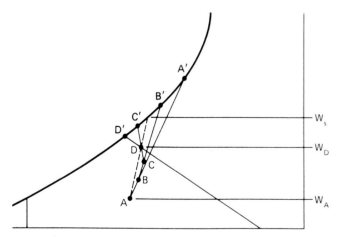

FIGURE 4-22 Construction of pursuit curve (heating and humidifying with heated spray water).

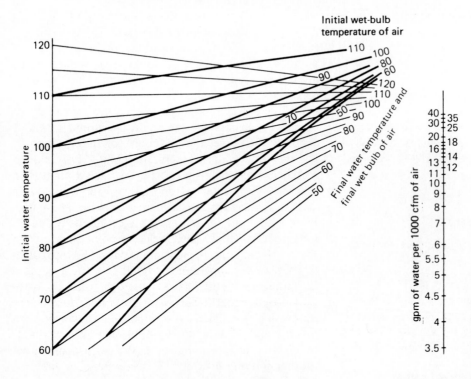

FIGURE 4-23 Spray water air-washer chart for humidification. (Courtesy of The Trane Company, LaCrosse, WI)

applying a correction factor (CF).[5] The correction factor is determined by

$$CF = \frac{\text{actual change in air enthalpy}}{\text{theoretical change in air enthalpy}}$$

or

$$CF = \frac{\Delta h \text{ actual}}{\Delta h \text{ theoretical}} \quad (4\text{-}35)$$

In effect, this correction factor corresponds to the humidifying efficiency, E_h, determined by Eqs. (4-32) and (4-33).

ILLUSTRATIVE PROBLEM 4-9[6]

An air quantity of 15,000 cfm enters a washer at 70°F DB and 50°F WB. The washer is supplied with 150 gpm of water at 90°F. For a correction factor of 0.85, determine the water temperature in the tank and the condition of the air leaving the washer.

Solution: (See Figure 4-24.)
Step 1. Determine final water temperature and theoretical final air wet-bulb temperature. The water-to-air ratio is 150/15 = 10 gpm/1000 cfm of air. Using Figure 4-23,

[5]*Trane Air Conditioning Manual,* The Trane Company, LaCrosse, Wisconsin, 1965, p. 227.

[6]*Trane Air Conditioning Manual,* The Trane Company, LaCrosse, Wisconsin, 1965, p. 232.

set the right end of the straightedge on the right scale at 10 gpm/1000 cfm; set the left end of the straightedge on the initial water temperature of 90°F on the left scale. Where a line between these two points crosses the initial air wet bulb of 50°F, read the final water temperature of 73.5°F, which is also the *theoretical* final air wet-bulb temperature.

Step 2. Determine actual leaving air condition from the washer. In Figure 4-24, at entering air condition of 70°F DB and 50°F WB, find h_1 = 20.2 Btu/lbda. Theoretically, the air could be heated and humidified to saturation at 73.5°F, and line 1–2 represents the process through the washer. The enthalpy of saturated air at 73.5°F is h_2 = 37.2 Btu/lbda. Theoretical change in enthalpy is $h_2 - h_1$ = 37.2 − 20.2 = 17.0 Btu/lbda. The actual change in enthalpy is the theoretical change times the correction factor. Actual enthalpy change = 17.0 × 0.85 = 14.45 Btu/lbda. The actual enthalpy of air leaving the washer will be h_3 = 20.2 + 14.45 = 34.65 Btu/lbda. In Figure 4-24,

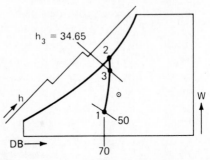

FIGURE 4-24 Solution to Illustrative Problem 4-9.

construct a constant-enthalpy line at 34.65 and extend to intersect line 1–2 at point 3. At point 3, find $t_3 = 73.5°F$, $t_{w3} = 70.7°F$. *(ans.)*

A water-to-air ratio of 5 gpm/1000 cfm is common for a single-bank washer, while approximately 10 to 15 gpm/1000 cfm are usual ratios for two- and three-bank washers. Initial water temperature will vary within the range of 60°F to 120°F depending upon the inlet and outlet air temperatures. Ordinarily, the operator has the following items to vary to control the performance of the washer:

- Inlet air temperature by changing the amount of heating medium to the preheat coil.
- Inlet water temperature by controlling the heating capacity of the heat exchanger.
- Water quantity by taking a bank of nozzles out of service.

REVIEW PROBLEMS

4.1. Moist air at 35°F DB and 32°F WB is to be conditioned to 70°F DB and 50% RH. Using the psychrometric chart and any required calculations, determine the following:
  (a) The initial and final values of humidity ratio, enthalpy, specific volume, and dew-point temperature.
  (b) If the initial air volume entering the conditioning device is 5000 cfm, determine (1) required heating capacity (Btu/hr), (2) required moisture addition (lb/hr), and (3) pounds per hour of air flowing through the conditioner.

4.2. Three thousand cfm of air at 40°F DB and 35°F WB are mixed with 5000 cfm of air at 90°F DB and 74°F WB in a steady-flow device. Determine the dry-bulb and wet-bulb temperatures, humidity ratio, and enthalpy of the resulting mixture.

4.3. Moist air at an initial condition of 40°F DB and 32°F WB is to be heated and humidified to a final condition of 90°F DB and 30% RH. The process of humidification is to be carried out in a spray chamber (air washer) using recirculated spray water with no external heating or cooling of the spray water. The spray washer has a humidifying effectiveness of 88%. Preheating and reheating of the air will be done by heat-transfer coils. If the initial air volume flow rate is 8000 cfm, determine the following:
  (a) How much preheat and reheat capacity is required (Btu/hr).
  (b) How much water must be added to the spray chamber as makeup (gpm).

4.4. Using the same initial and final air conditions of Review Problem 4.3, determine the range of preheated air temperatures that would permit the use of an open-pan type of humidifier containing water maintained at 212°F by immersed heating coils. Assume that the air pressure is 14.7 psia.

4.5. A building interior is to be heated by forced circulation warm air supplied at 105°F DB. The calculated sensible heat loss from the building is 560,000 Btu/hr with the building interior maintained at 68°F DB and 60% RH when the outdoor air is at 30°F DB and 24°F WB. Proper ventilation requires that 25% of the air circulated must be outdoor air. If the building heat loss is all sensible heat, determine the following:
  (a) The required supply air quantity (lb/hr and cfm).
  (b) If humidification is performed by spraying steam into the airstream after the mixing of outdoor and recirculated air, how much steam (lb/hr) is required, assuming the steam is at 16 psia, dry and saturated?
  (c) How much heat must be supplied by the heat-transfer coils after humidification?

4.6. Air at 60°F DB and 50°F WB is to be preheated and then humidified to a final condition of 100°F DB and 40% RH in a steady-flow device. Humidification is to be performed by spraying steam into the air. The steam pressure is 20 psia and has a quality of 80%.
  (a) To what temperature must the air be preheated before humidification?
  (b) If 4000 cfm of air at the initial condition enter the process, how much steam (lb/hr) is required to humidify the air?
  (c) What must be the heating capacity of the preheating coil (Btu/hr)?

4.7. A building has a calculated heat loss of 400,000 Btu/hr and a sensible heat ratio of 0.80. The building interior is to be maintained at 70°F DB and 50% RH when the outdoor air is at 35°F DB and 40% RH. Outdoor air required for ventilation is 3000 cfm. Air supplied to the building interior is to be at 115°F DB. Humidification of the supply air to the building interior, if required, will be done by using dry saturated steam at 18 psia after the air has passed through the heating device.
  (a) How much air must be supplied to the building interior (lb/hr and cfm)?
  (b) How much steam, if any, is required for humidification (lb/hr)?
  (c) Find the required temperature rise of the air through the heat-transfer device and heat transfer (Btu/hr).
  (d) Make a sketch of a psychrometric chart showing all process lines and points that pertain to the solution of the problem.

4.8. Moist air at 60°F DB and 30°F dew-point temperature enters a humidifying device at the rate of 4000 cfm.

Dry saturated steam at 30 psia is sprayed into the air-stream. Following humidification, the air is heated by a heat-transfer device to a final condition of 90°F DB and 55°F dew-point temperature.

(a) How much steam must be added to the air (lb/hr)?

(b) How much heat must be supplied by the heat-transfer device (Btu/hr)?

(c) What are the dry- and wet-bulb temperatures after humidification?

(d) Make a sketch of a psychrometric chart showing all process lines and points that pertain to the solution of the problem.

4.9. A building interior is to be maintained at 70°F DB and 50% RH when outdoor design conditions are 30°F DB and 26°F WB. Heat losses from the building are 250,000 Btu/hr sensible and 44,100 Btu/hr latent. Latent heat transfer is due to infiltration of cold, dry outdoor air. Ventilation requires that 1000 cfm of outdoor air be introduced into the supply air. The supply air to the building interior is to be at 100°F DB. The conditioning equipment is shown in Figure 4-25.

(a) Determine the quantity of supply air required (lb/hr and cfm).

(b) Determine the capacity of the heating device (Btu/hr) (1) if the humidifier is a spray washer using recirculated spray water, without heating or cooling of the water; (2) if the humidifier is a steam humidifier using steam at 16 psia, dry and saturated.

(c) Make a sketch of a psychrometric chart showing all process lines and points that pertain to the solution of the problem.

4.10. A high-humidity chamber is to be maintained at 75°F DB and 70% RH when outdoor design conditions are 30°F DB and 80% RH. The sensible heat loss from the chamber is 260,000 Btu/hr. Due to contamination within the chamber, it will be necessary to use 75% outdoor air and 25% recirculated air for the supply air mixture. The proposed equipment to condition the air is shown in Figure 4-26. The preheat coil will heat the outdoor air to 70°F. The spray washer has a humidifying effectiveness of 90% and will use recirculated spray water with no heating or cooling of the spray water. The steam humidifier will use dry saturated steam at 20 psia. Supply air temperature to the chamber will be 110°F DB.

(a) Determine the dry- and wet-bulb temperatures, humidity ratio, and enthalpy for the air at each numbered point in the system.

(b) Determine the required supply air quantity (lb/hr and cfm).

(c) Determine the heating capacity required for the preheat coil and the reheat coil (Btu/hr).

(d) Determine the amount of water absorbed by the air in the air washer (lb/hr).

(e) Determine the amount of steam absorbed by the air in the steam humidifier (lb/hr).

(f) Make a sketch of a psychrometric chart showing all process lines and points that pertain to the solution of the problem.

4.11. Five thousand cfm of moist air at 40°F DB and 30°F WB are to be heated and humidified to a final state of 90°F DB and 30% RH. The air will initially be heated to a dry-bulb temperature of 60°F, humidified, and then reheated to the final state. Humidification is to be done using saturated steam at 20 psia.

(a) Determine the air properties and all significant points as it passes through the equipment.

(b) Determine the amount of preheat and reheat required.

(c) Determine the amount of moisture added to the air.

4.12. Five thousand cfm of moist air at 40°F DB and 30°F WB are to be heated and humidified to a final state of 90°F DB and 30% RH. The air will be preheated and then humidified with saturated steam at 20 psia. To what dry-bulb temperature must the air be preheated?

4.13. Three thousand cfm of air at 10°F DB and 80% RH are heated to a dry-bulb temperature of 40°F without the addition of moisture. The air is then mixed with 6000 cfm of air at 70°F DB and 70% RH. The resulting mixture is heated to 100°F DB without moisture addition. Assume barometric pressure is 29.92 in. Hg.

(a) Calculate the properties of the air at the final state point.

(b) Calculate the total heat added to the air.

4.14. Five thousand cfm of moist air at 90°F DB and 60°F WB pass through a spray washer using recirculated spray water with no external heating or cooling of the water. The humidifying effectiveness is 90%.

(a) Determine the properties of the air leaving the washer.

(b) Determine the mass of water evaporated into the air.

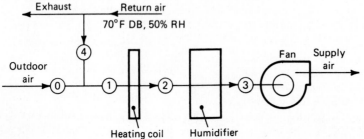

FIGURE 4-25   Sketch for Review Problem 4-9.

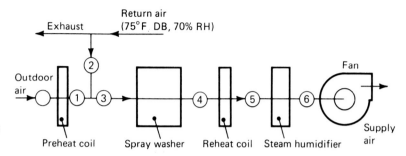

FIGURE 4-26 Sketch for Review Problem 4-10.

4.15. Moist air at 70°F DB and 51°F WB is to be humidified to a final state of 70% RH by using steam at 20 psia and 90% quality. The airflow rate at the initial state is 4000 cfm.

(a) Determine the final dry-bulb and wet-bulb temperatures of the air.

(b) Determine the amount of moisture added.

4.16. An air conditioning unit for an industrial building must provide 4500 cfm of outdoor air, which enters a preheat coil at 5°F DB and 50% RH. The preheat coil heats the air to 40°F DB. This air then mixes with 13,500 cfm of recirculated air at 70°F DB and 40% RH. The mixed air passes through a spray washer supplied with 90 gpm of water at 110°F, and the washer has a correction factor of 0.83. The humidified air is then reheated to a dry-bulb temperature of 115°F before entering the supply air fan, which delivers it to a duct system for distribution within the building. (See Figure 4-27 for equipment layout.)

(a) Determine the psychrometric properties of the air as it flows through the equipment (numbered points).

(b) If the heat supply to the spray water heat exchanger were turned off so that the water in the washer was recirculated without any heat being added, what would be the properties of the air through the equipment?

(c) Show all processes on a psychrometric chart.

4.17. Air is supplied to a certain space from the outside, where the temperature is 20°F and the relative humidity is 70%. It is desired to keep the space at 70°F and 60% relative humidity. The barometric pressure is 29.50 in. Hg. If the air supplied is ventilation air, at space conditions, how much moisture and how much heat must be supplied to each pound of air entering the space?

4.18. Moist air at 90°F DB and 60°F WB is humidified adiabatically with steam. The steam supplied contains 15% moisture saturated at 20 psia. When sufficient steam is added to humidify the air to 50% RH, what will be the dry-bulb temperature of the humidified air? The barometric pressure is 29.92 in. Hg.

4.19. Moist air at 80°F DB and 55°F WB is to be humidified with the air dry-bulb temperature remaining constant.

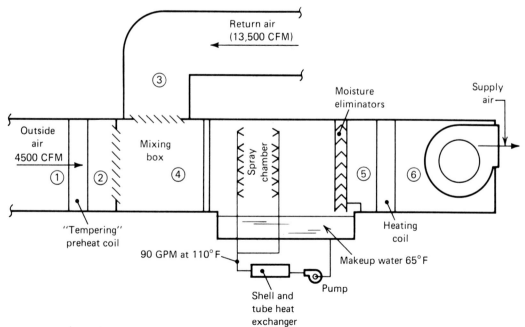

FIGURE 4-27 Schematic of equipment arrangement for Review Problem 4-16.

Saturated, wet steam is to be supplied at 15 psia. What quality must the steam have **(a)** to provide saturated air and **(b)** to provide air at 70% RH?

4.20. Moist air at 90°F DB and 65°F WB is adiabatically mixed with water supplied at 150° in such proportions that the mixture has a relative humidity of 80%. Determine the dry-bulb temperature of the mixture.

4.21. A building has a sensible heat loss of 250,000 Btu/hr and is heated by forced circulation warm air supplied at 115°F DB and having a humidity ratio of 0.006 lbv/lbda. Air returns to the furnace heater at 65°F DB with no significant change in humidity ratio. How many pounds of air must be circulated per hour for heating? This is equivalent to how many cubic feet of air per minute measured at supply condition?

4.22. Air at 40°F DB and 35°F WB, and with a barometric pressure of 29.92 in. Hg, is heated and then humidified under steady-flow conditions to 70°F DB and 40% RH. The airflow rate is 4000 cfm measured at the initial condition.
(a) What weight of water vapor must be added per hour?
(b) If the water supplied is at 50°F, how much heat must be added to the air?
(c) Determine the air dry-bulb temperature after heating and before humidification.

4.23. A building has a calculated sensible heat loss of 190,000 Btu/hr and a latent heat loss of 33,000 Btu/hr. Latent heat transfer is due to infiltration of dry, cold outside air. At design conditions, the building interior will be maintained at 72°F DB and 40% RH when the outside air is at 35°F DB and 50% RH. Ventilation requires that 2000 cfm of outside air be introduced into the supply air. Supply air to the building interior will be at 110°F DB.

(a) Determine the quantity of supply air required (lb/hr and cfm).
(b) Determine the amount of saturated, dry steam at 15 psia required by the humidifier.
(c) Determine the temperature rise of the air passing through the furnace.
(d) Determine the furnace capacity.

4.24. Saturated, dry steam at 40 psia is to be used to humidify moist air. The moist air enters the humidifier at 60°F DB and 50°F WB at the rate of 1000 cfm. Barometric pressure is 14.696 psia.
(a) Determine the mass flow rate of steam required to saturate the air.
(b) Determine the temperature of the saturated air.

4.25. Moist air at 40°F DB and 35°F WB enters a mixing chamber at the rate of 4000 cfm. The other airstream entering the mixing chamber is at 70°F DB and 60% RH at a rate of 2000 cfm. The resulting mixture passes through a heat-transfer coil, which heats the mixture to a dry-bulb temperature of 90°F. The heated mixture then passes through a spray washer that uses recirculated spray water and has a humidifying effectiveness of 90%. The humidified air passes through a final heat-transfer coil, which produces a final air temperature of 115°F.
(a) Determine the mass flow of air through the system.
(b) Determine the mass of water absorbed by the air.
(c) Determine the heat added by each of the heating coils.
(d) Determine the dry-bulb and wet-bulb temperatures for the processes.
(e) Show all processes on a psychrometric chart.

BIBLIOGRAPHY

4.1. *The Armstrong Humidification Handbook,* Armstrong Machine Works, Three Rivers, MI.

4.2. *Trane Air Conditioning Manual,* The Trane Company, LaCrosse, WI, 1965.

4.3. *Carrier System Design Manual,* Part 1, *Load Estimating,* Carrier Air Conditioning Company, Syracuse, NY, 1972.

5

Comfort in Air Conditioning

5-1 INTRODUCTION

Air conditioning has been defined as the simultaneous control of temperature, humidity, air movement, and the quality of air in a given space. The use of the conditioned space determines the temperature, humidity, air movement, and quality of the air to be maintained. Air conditioning is able to provide widely varying atmospheric conditions—from those necessary for drying processes to those necessary for high-humidity process application. Air conditioning can maintain any atmospheric condition regardless of variations in outdoor weather. The range of temperatures and humidities used in comfort air conditioning is a small band. The location of this band on the psychrometric chart depends on the season of the year.

Cleanliness and air movement must be considered. Air should be cleaned—that is, freed of dust and soot particles and odor. Air cleanliness is important from the standpoint of human health. Also, the walls and ceilings of rooms supplied with filtered air will need cleaning less frequently than those supplied with unfiltered air. Considerable dust is carried into a room by an unfil-

tered air supply. Some spaces, called "clean rooms," require special attention as to air filtration.

Air should circulate freely in the room to which it is delivered. This will allow it to absorb heat and moisture uniformly throughout the entire room during the cooling cycle and to deliver heat and moisture uniformly during the heating cycle. At the same time, the air movement should be gentle, or it will cause objectional draft.

Comfort can be defined as any condition that when changed will make a person uncomfortable. Though this sounds paradoxical, it just means that a person is not aware of the best air conditioning systems when comfortable. If a person is uncomfortably warm, then the temperature or humidity or both are too high. In a first-class air conditioning system, one is not really conscious of the temperature or humidity because one is comfortable. Neither is there any disturbance because of equipment noise or air movement.

Finding the best conditions for comfort and health has been the object of considerable research work.

ASHRAE Handbook 1989 Fundamentals is probably the most up-to-date and complete source of information relating to the physiological aspects of thermal comfort. Thus, much information is available concerning factors affecting human comfort.

5-2 HEAT BALANCE OF THE HUMAN BODY

Individual comfort depends on how fast the body is losing heat. The human body may be compared to a furnace using food as fuel. Food is largely carbon and hydrogen; energy contained in the fuel (food) is released by oxidation. The oxygen comes from the air, and the principal products of combustion are carbon dioxide and water vapor. The physician calls this process *metabolism.*

The human body is essentially a constant-temperature device. Its deep body temperature is kept at 98.6°F (37°C) by a delicate, complex temperature-regulating mechanism within the body.

The object of air conditioning is to assist the body in controlling its cooling rate. This is true for both winter and summer conditions. In summer, the job is to increase the cooling rate of the body; in winter, it is to decrease the cooling rate.

It is possible to compile a heat balance for the human body and study the ways the body's cooling rate can be controlled. As mentioned previously, metabolism is the process by which the body produces heat. This will be shown on the left side of the following equation. On the right side of the equation will be shown the ways the body loses heat. Thus,

$$\text{heat produced} = \text{heat lost}$$

$$M = \pm \dot{q}_s \pm \dot{q}_r + \dot{q}_L \qquad (5\text{-}1)$$

where

M = metabolism (Btu/hr or W)
$\dot{q}_s$ = sensible heat exchange (Btu/hr or W)
$\dot{q}_r$ = radiant heat exchange (Btu/hr or W)
$\dot{q}_L$ = latent heat exchange, evaporation (Btu/hr or W)

Two things can cause the deep body temperature to rise: (1) fever or (2) continuous heavy activity. The body temperature will fall in freezing weather if it is not properly protected. If a rise or fall of deep body temperature were considered, Eq. (5-1) would have a "stor-

age" term and a mechanical work term, and for a normal, healthy comfortable body it would be written as

$$S = M - (\pm W) \pm \dot{q}_L \pm \dot{q}_r \pm \dot{q}_s \qquad (5\text{-}2)$$

where

S = rate of heat storage; proportional to the time rate of change in intrinsic body heat
W = mechanical work accomplished

The left side of Eq. (5-1) can be the basal metabolism. This is the amount of heat generated by an unclothed human at rest in an air temperature of about 70°F (21.1°C). For an individual of average weight and body build, the basal metabolism is about 240 Btu/hr (70.3 W). About 60 Btu/hr (17.6 W) is used by the heart and lungs; the balance is used for oxidation. Except for clinics and hospitals, air conditioning engineers have only passing interest in basal metabolism. The people they must keep comfortable are clothed and have various degrees of activity. So, the left side of Eq. (5-1), *M,* becomes the actual metabolic level of the individual. This is the heat generated by the individual and may vary widely depending on activity levels. Naturally, the higher the metabolic rate, the higher the heat-transfer rate from the body to the environment.

The right side of Eq. (5-1) represents the heat loss from the body. The terms represent the ways the body is able to keep itself at a constant temperature. This is quite a remarkable feat, considering that activity may vary from sleeping to heavy work and in an air-conditioned atmosphere or in hot sun. Each of the heat loss terms will be discussed briefly in following sections.

Both the air and surfaces of the enclosure surrounding the occupant are sinks for the metabolic heat emitted by the occupant. Air circulates around the occupant and the surfaces. The occupant also exchanges radiant heat with the surrounding surfaces, such as glass and outside walls. Air is brought into motion within a space thermally or by mechanical forces.

Both nude and clothed subjects feel comfortable at the same mean skin temperature of 91.5°F (33°C). The range of skin temperature within which no discomfort is experienced is about ±2.5°F (1.5°C). The necessary criteria, indices, and standards for use where human occupancy is concerned are given in *ASHRAE Handbook 1989 Fundamentals,* Chapter 8.

In choosing optimal conditions for comfort and health, the energy expended during the course of routine physical activities must be known since body heat production increases in proportion to exercise intensity.

Table 5-1 presents probable metabolic rates for various typical activities. The values are expressed in met units and in Btu/hr·ft² of body surface area. *One met unit is defined as 18.43 Btu/hr·ft²* (58.2 W/m²; or 50 kcal/m²).

TABLE 5-1

Typical metabolic heat generation for various activities[a]

Activity	Btu/hr · ft²	met[b]
Resting:		
Sleeping	13	0.7
Reclining	15	0.8
Seated, quiet	18	1.0
Standing, relaxed	22	1.2
Walking (on the level):		
0.89 m/s	37	2.0
1.34 m/s	48	2.6
1.79 m/s	70	3.8
Office activities:		
Reading, seated	18	1.0
Writing	18	1.0
Typing	20	1.1
Filing, seated	22	1.2
Filing, standing	26	1.4
Walking about	31	1.7
Lifting/packing	39	2.1
Driving/flying:		
Car	18-37	1.0-2.0
Aircraft, routine	22	1.2
Aircraft, instrument landing	33	1.8
Aircraft, combat	44	2.4
Heavy vehicle	59	3.2
Miscellaneous occupational activities:		
Cooking	29-37	1.6-2.0
Housecleaning	37-63	2.0-3.4
Seated, heavy limb movement	41	2.2
Machine work,		
sawing (table saw)	33	1.8
light (electrical industry)	37-44	2.0-2.4
heavy	74	4.0
Handling 50-kg bags	74	4.0
Pick and shovel work	74-88	4.0-4.8
Miscellaneous leisure activities:		
Dancing, social	44-81	2.4-4.4
Calisthenics/exercise	55-74	3.0-4.0
Tennis, singles	66-74	3.6-4.0
Basketball	90-140	5.0-7.6
Wrestling, competitive	130-160	7.0-8.7

[a] Source: *ASHRAE Handbook 1989 Fundamentals.* Used with permission from the American Society of Heating, Refrigerating and Air Conditioning Engineers, Atlanta, GA.
[b] One met = 18.43 Btu/hr · ft².

The most useful measure of body surface area, proposed by Dubois (1916), is described by

$$A_D = 15.6(m^{0.425})(h^{0.725}) \qquad (5-3)$$

where

A_D = Dubois surface area (in.²)
m = body mass (lb)
h = body height (in.)

An average size man has a mass of 154 lb (70 kg) and a height of 68 in. (1.73 m); so, his body surface, A_D, = 2830 in.², or 19.6 ft² (1.8 m²). Then, for an average size man, the met unit corresponds to approximately 350 Btu/hr (100 W, or 90 kcal/hr). The met corresponds to the average heat production for a sedentary man.

In Table 5-1, all values are expressed in multiples of the met unit. Values are typical and primarily used in engineering planning. Activities chosen are those performed steadily rather than intermittently. The highest energy level a person can maintain for any continuous length of time is approximately 50% of his or her maximum capacity to utilize oxygen (maximum energy capacity). A normal, healthy man has a maximum capacity of approximately 12 mets at 20 years of age, which drops to 7 mets at 70 years of age. Women tend to have maximum levels about 30% lower. Well-trained athletes have a maximum as high as 20 mets. For an untrained person at 35 years of age, the overall average can be considered as 10 mets. Activities performed continuously above the 5-met level by an untrained 35-year-old man may prove exhausting and uncomfortable.

5-3 SENSIBLE HEAT LOSS

Note the plus/minus sign in front of the sensible heat term in Eq. (5-1). This means that the body can either gain or lose sensible heat. The key is air temperature. If the air temperature is below the skin temperature, the sensible heat term is "plus"; that is, the body is losing sensible heat to the air. If the air temperature is above the skin temperature, the sensible heat term is "minus"; that is, the body is gaining sensible heat from the air.

Some of the heat liberated in a furnace is transferred to the air surrounding the steel furnace walls. Similarly, some of the heat liberated within the body is transmitted to the air in contact with the skin. Although the deep body temperature remains at 98.6°F, the skin temperature varies. The skin temperature will vary from 40°F to 105°F (4.5°C to 40.6°C) according to the temperature, humidity, and velocity of the surrounding air. If the temperature of the surrounding air falls, the skin temperature will also fall.

The range of skin temperature, at any location on the body, is comparatively small. A change of 10 degrees in room temperature does not produce anything like a change of 10 degrees in skin temperature. For instance, one series of tests reported that the forehead temperature ranged from 93°F (33.9°C) at 64°F (17.8°C) to 100°F (37.8°C) in air at 96°F (35.6°C). So, as the temperature of the air rises, the difference in temperature between the skin and air decreases. At normal indoor air temperature, there is a steady flow of sensible heat from the skin to the surrounding air. The amount of this sensible heat depends upon the temperature difference between the skin and air. Inasmuch as this difference decreases as the temperature of the surrounding air rises, less sensible heat will be transferred to the surrounding air. When the surrounding air is at about 70°F (21.1°C), most people lose sensible heat at such a rate that they are comfortable.

5-4 RADIANT HEAT LOSS

The body gains or loses heat by radiation according to the difference between two temperatures: (1) the body surface (bare skin and clothing) temperature and (2) the mean radiant temperature. Mean radiant temperature (MRT) is a weighted average of the temperature of all the surfaces in direct line of sight of the body.

If the MRT is below the body temperature, the radiant heat loss term, $\dot{q}_r$, in Eq. (5-1) is "plus." Then, the body is losing radiant heat to surfaces it can see. If the MRT is above the body temperature, $\dot{q}_r$ is "minus" and the body is gaining radiant heat. It should be kept in mind that the body loses (or gains) sensible and radiant heat according to its surface temperature. For a comfortable adult normally dressed, the weighted average between bare skin and clothed surfaces is 80°F (26.7°C).

Surface temperatures have much to do with comfort. Suppose, during the heating season, a person is working at a desk facing the center of a room with his or her back 5 ft (1.5 m) from an outside wall. The wall surface temperature is 59°F (15°C). Assume the air distribution in the room is almost perfect (74°F, 2 in. below the ceiling; 73°F, 2 in. above the floor; and 73.5°F, all around including the space between the person's back and the wall). Will the person be comfortable? Probably not. Because radiant heat loss from the body to the wall is so high, a chill may be caused. What can be done to correct this situation? The temperature of the wall surface cannot be conveniently changed. The position of the desk might be changed so that the person's back was against an inside wall. The surface temperature of the inside wall will be near the air temper-

ature, probably 73°F. So, the radiant heat loss would be less because of the smaller temperature difference. If the desk cannot be moved, what else might be done? The temperature of the air might be increased by turning up the thermostat. Increasing the air temperature will decrease the sensible heat loss from the body. Suppose the air temperature is set at 77°F (25.0°C). This will decrease the sensible heat loss by the same amount the radiant heat loss was decreased by moving the desk from the outside wall. This balances the heat loss from the body. The trouble is that everyone else in the room is too warm.

Exactly the same thing might be true during the cooling season. One might be too warm because of the radiant heat the body gains from a warm outside wall or window. Now the sensible heat loss from the body should be *increased* by *decreasing* the air temperature. This puts the body heat loss in balance, but everyone else in the room is too cool.

There is an important point to this discussion. It was assumed that air distribution was excellent; it was also assumed that air temperature could be raised or lowered at will. Still, everyone in the room will not be comfortable all the time. This means that there is more to comfort than mechanical equipment. Good equipment properly operated will provide comfort in most instances, but it must be correctly applied. (It just will not do the job in a sheet iron shack or a tent.) The building construction and orientation and the type of occupancy all influence comfort.

5-5 LATENT HEAT LOSS (EVAPORATION)

The last term in Eq. (5-1) is latent heat loss or perspiration. Under all conceivable situations, this term will always be positive. Even at rest, the body requires about 100 Btu/hr (29.3 W) to evaporate moisture into the inhaled air to keep the lungs moist. If it were not for this fact, lung tissues would stick together and not enough air would enter the lungs to keep alive. It is possible to "see" the breath when exhaling on a frosty morning. This is evidence that the air leaving the lungs has a high moisture content.

Evaporation heat regulation is the body process that sustains life outside an air-conditioned space. Equation (5-1) explains this. Suppose one goes outside when the air temperature is 100°F (37.8°C). The sensible heat loss is "minus" because the body is *gaining* heat from the air. The MRT is much higher than the body surface temperature: The sidewalk, street, building walls, and everything the body "sees" is above body surface temperature. Thus, the second term, $\dot{q}_r$, is also "minus" because the body is *gaining* radiant heat. But, as one walks down the sidewalk, the left side of Eq. (5-1) (metabolism) is about 700 Btu/hr (205 W). So, to maintain its heat balance, the body must lose about 1025 Btu/hr (300 W) by evaporation. Nature takes care of this by automatically opening the sweat glands. Moisture on the body surface then evaporates. Most of the heat required to convert (evaporate) the moisture to a vapor comes from the body. As long as the air will carry away the water vapor, the skin temperature will remain cool. This in turn keeps the deep body temperature at very nearly 98.6°F (37°C).

Moisture loss by the body, whether in the exhaled air or from the skin, is in the form of steam at very low pressures. The body supplies the necessary latent heat to evaporate the moisture that it loses.

What bearing does the loss of moisture have on comfort when the temperature of the surrounding air rises from the comfortable levels to the uncomfortable levels? As the dry-bulb temperature of the air rises, less sensible heat is lost by the body, but the latent heat loss increases because the loss of moisture from the body increases. Thus, at 70°F (21.1°C), 290 Btu/hr (85 W) of sensible heat and 110 Btu/hr (32 W) of latent heat is lost. At 80°F (26.7°C), the sensible heat loss drops to nearly zero, but the latent heat loss increases to nearly 400 Btu/hr (117 W). Still, in both cases, the total loss is the same—approximately 400 Btu/hr (117 W). When the outside temperature rises to about 90°F (32.2°C), all the body heat is carried away as latent heat.

When people work under conditions of high temperature and high humidity, both the sensible heat loss and the evaporation of moisture from their skins are retarded. Under these conditions, the rate of sensible heat transfer and evaporation must be increased by blowing air at high velocity over the body. In this way, the evaporation of moisture from the surface of the skin is greatly accelerated, with a consequent increase in the total heat transferred from the skin to the air.

5-6 COMFORT CONDITIONS

The objective of a comfort air conditioning system is to provide a comfortable environment for an occupant or occupants of a residential, public, medical, factory, or office building. It may be for a number of transient occupants in a commercial establishment such as a store, a bowling alley, a beauty salon, or a restaurant. It may be for an assembly of occupants gathered in a large space such as a church, a theater, an auditorium, a pavilion, or a factory loft or floor.

The comfort environment is the result of simultaneous control of temperature, humidity, cleanliness, and air distribution within the occupant's vicinity. This set of factors includes mean radiant temperature as well as the air temperature, odor control, and control of the proper acoustic level within the occupant's vicinity.

There appears to be no rigid rule as to the best environmental conditions for comfort for all people. Under the same conditions of temperature, humidity, and air movements, a healthy, young man may be slightly warm while an elderly woman is cool. A customer who steps into a conditioned store from the bright sun of the street experiences a sense of relief, while the active clerk who has been in the store for several hours may be a bit too warm for perfect comfort. The dancers in a night club may feel somewhat warm, while the patrons seated at tables are comfortable or perhaps even slightly cool.

The comfort of an individual is affected by many variables. Health, age, activity level, clothing, sex, food, and acclimatization all play their part in determining the elusive "best comfort conditions" for any particular person or group of persons. Hard and fast rules that will apply to all conditions and to all people cannot be given. The best that can be done is to approximate those conditions under which a majority of the occupants of a room will feel comfortable.

In order to evaluate the sensation of comfort for the human body, three classes of environmental indices are used: (1) the *direct*, (2) the *rationally derived*, and (3) the *empirical*. These are as follows:

Direct Indices

- Dry-bulb temperature
- Wet-bulb temperature
- Dew-point temperature
- Relative humidity
- Air movement

Rationally Derived Indices

- Mean radiant temperature
- Operative temperature
- Humid operative temperature
- Heat stress index
- Index of skin wettedness

Empirical Indices

- Effective temperature
- Black globe temperature
- Corrected effective temperature
- Wet-bulb–globe temperature index
- Wind chill index

Not all of these indices are directly useful to the design engineer, and many of these are defined for research purposes. Complete definition of these indices may be found in *ASHRAE Handbook 1989 Fundamentals*.

Physiologists have recognized that sensations of comfort and of temperature may have different physiological and physical bases, and each should be considered separately. This dichotomy was recognized in ANSI/ASHRAE Standard 55–1981, *Thermal Environment Conditions for Human Occupancy,* where thermal comfort is defined as "that state of mind which expresses satisfaction with the thermal environment." Unfortunately, little research to date fulfills this definition. Most current predictive charts are based on comfort defined as "a sensation that is neither slightly warm nor slightly cool."

Since about 90% of indoor occupation and leisure time is spent at or near the sedentary activity level, ASHRAE research has been limited to sedentary subjects, lightly clothed at 0.5 to 0.6 clo. (The *clo unit* is

used to specify the insulation effect of clothing: clo = 0, for the nude body; clo = 1.0, for a typical business suit with vest.) Predictive methods available are believed accurate within stated limitations. During physical activity, a change occurs in our physiology. Physiological thermal neutrality in the older (sedentary) sense does not exist. Some form of thermal regulation always occurs within the human body during exercise. Temperature and comfort during activity have different bases than during the sedentary condition. Sedentary skin and body temperatures used as indices of comfort will likely prove false during moderate-to-heavy activity.

Figure 5-1 shows the winter and summer comfort zones specified in ANSI/ASHRAE Standard 55–1981. The temperature ranges are appropriate for current seasonal clothing habits in the United States. Summer clothing includes light slacks and short-sleeve shirt or a comparable ensemble with an insulation value of 0.5 clo. Winter clothing is heavy slacks, long-sleeve shirt, and sweater or jacket with an insulation value of 0.9

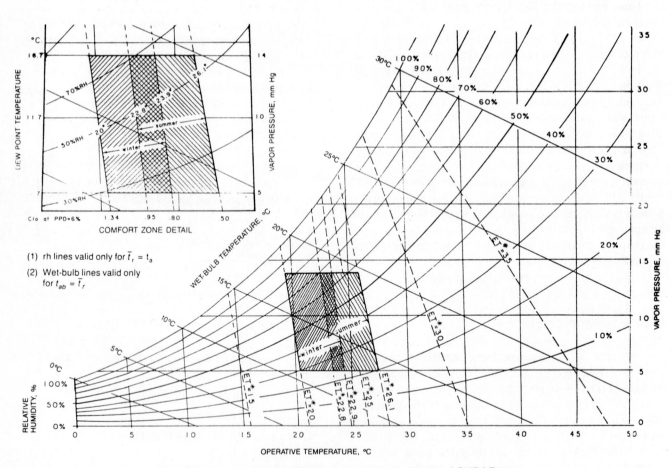

FIGURE 5-1 Standard effective temperature and the ASHRAE comfort zones. (From ANSI/ASHRAE Standard 55–1981, *Thermal Environmental Conditions for Human Occupancy.* Used with permission of American Society of Heating, Refrigerating, and Air Conditioning Engineers, Atlanta, GA)

clo. The temperature ranges are for sedentary and slightly active persons.

The winter zone is for air speeds less than 0.15 m/s (30 fpm). The standard allows the summer comfort zone to extend above 26°C (79°F) if the average air movement is increased to 0.275 m/s (55 fpm) for each degree K of temperature increase to a maximum temperature of 28°C (82°F) and air movement of 0.8 m/s (157 fpm).

The temperature boundaries of the comfort zones in Figure 5-1 can be shifted −0.6°C per 0.1 clo for clothing levels other than 0.5 and 0.9. The zones can also be shifted lower for increased activity levels.

Thermal comfort conditions must be fairly uniform over the body to prevent local discomfort. The radiant temperature asymmetry should be less than 5°C in the vertical direction and 10°C in the horizontal direction. The vertical temperature difference between head and foot should not exceed 3°C.

Figure 5-1 applies generally to altitudes from sea level to 3000 m (10,000 ft) and to the most common indoor thermal environment in which mean air velocity is less than 30 fpm in winter and 50 fpm in summer. For these cases, the thermal environment is well specified by the two variables shown—operative temperature and humidity.

The *operative temperature* is the uniform temperature of a radiantly black enclosure in which an occupant would exchange the same amount of heat by radiation plus convection as in the actual nonuniform environment. Operative temperature is numerically the average, weighted by respective heat-transfer coefficients of the air and mean radiant temperatures. At speeds of 0.40 m/s (80 fpm) or less and mean radiant temperature less than 50°C (120°F), operative temperature is approximately the simple average of the air and mean radiant temperature.

A wide range of environmental applications are covered by the ANSI/ASHRAE Standard 55–1981.

The comfort envelope defined by the Standard applies only for sedentary and slightly active, normally clothed people at low air velocities, when the MRT is equal to the air temperature. For other clothing, activities, air temperatures, and so forth, the Standard recommends the use of Fanger's General Comfort Charts, which may be found in *ASHRAE Handbook 1989 Fundamentals*. Examples of these charts are shown in Figure 5-2.

With the increasing need to conserve energy, the federal government has mandated adjustments to the allowable winter and summer thermostat settings in public buildings. A question then arises relative to temperatures as low as 68°F (20°C) in winter and as high as 80°F (27°C) in summer. For an individual to be comfortable, one has no choice but to wear substantially heavier clothing in winter and lighter clothing in summer.

The most commonly recommended inside design temperatures for comfort occur where the summer and winter zones of Figure 5-1 overlap—that is, between 73°F (22.5°C) and 76°F (24.5°C). The effect relative humidity on human comfort has not been completely established. Nevertheless, humidity extremes are assumed to be undesirable, and, for human comfort, relative humidity should be kept within a broad range between 30% and 70%.

The indoor conditions to be maintained within a building for comfort considerations are assumed to be the average conditions at the breathing line, 3 ft to 5 ft (1 m to 1.5 m) above the floor. These average conditions should not be affected by abnormal or unusual heat gains or losses from the interior or exterior of the building.

Table 5-2 gives ranges of winter indoor design dry-bulb temperatures most commonly found in practice. However, the most comfortable dry-bulb temperature to be maintained depends on the humidity level that exists.

Temperature and Humidity Fluctuations. Environmental temperature and humidity often fluctuate as the control system operates the heating and humidifying system equipment. Studies have shown that allowable fluctuating limits stated in ANSI/ASHRAE Standard 55–1981 are conservative; and, in the thermal comfort range, the rate of change of temperature should not exceed 4.0°F/hr (2.2°C/hr) if the peak-to-peak variation in dry-bulb temperature is 2.0°F (1.1°C) or greater. For MRT fluctuations, the rate of 1.8°F/hr (1.0°C/hr) should not be exceeded if the peak-to-peak variation in MRT is 1.5°F (0.8°C) or greater. Humidity limits are quite broad and usually pose no problems for modern controls; the rate of relative humidity change should not exceed 20%/hr if the peak-to-peak variation in humidity is 10% or greater.

Air Distribution Within Conditioned Spaces. In the preceding paragraph, temperature and humidity fluctuations within a given space were discussed briefly. Air introduced into a conditioned space as supply air plays an important role in maintaining equilibrium within a space. The supply air should be distributed in such a way that the occupied zone (floor level to 6 ft above the floor) has only minor horizontal or vertical temperature variations and has the proper quantity of air delivered to all sections of the space according to heating and ventilating requirements. Yet both of these requirements must be met without creating drafts.

Interactions Between Air Velocity, Temperature, and Activity

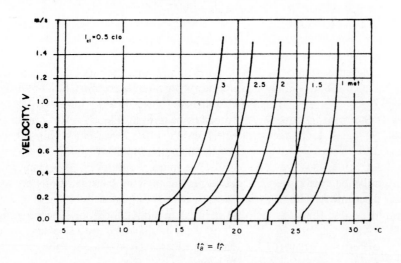

Interactions Between Air Temperature, Mean Radiant Temperature, and Velocity

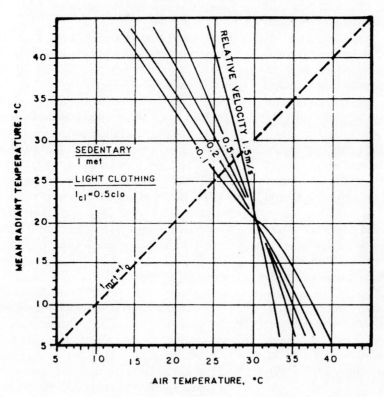

FIGURE 5-2 Examples of Fanger's Charts. (From *ASHRAE Handbook 1989 Fundamentals*. Used with permission of American Society of Heating, Refrigerating, and Air Conditioning Engineers, Atlanta, GA)

A *draft* may be defined as a noticeable air current. Drafts are objectionable; yet there must be air motion, or the occupants will feel uncomfortable. Heat and moisture must be carried away from the body as they are liberated, or a stagnant film of warm and moist air would envelop each occupant. The type of occupancy, the physical arrangement of the space, the acceptable noise level, and the degree of activity of the occupants all have a bearing on the permissible air velocity in a conditioned space. Generally, an air velocity in the

TABLE 5-2
Winter indoor dry-bulb temperatures usually specified

Type of Building	Dry Bulb °F.
SCHOOLS	
Classrooms	72–74
Assembly rooms	68–72
Gymnasiums	55–65
Toilets and baths	70
Wardrobe and locker rooms	65–68
Kitchens	66
Dining and lunch rooms	65–70
Playrooms	60–65
Natatoriums	75
HOSPITALS	
Private rooms	72–74
Private rooms (surgical)	70–80
Operating rooms	70–95
Wards	72–74
Kitchen and laundries	66
Toilets	68
Bathrooms	70–80
THEATERS	
Seating Space	68–72
Lounge rooms	68–72
Toilets	68

Type of Building	Dry Bulb °F.
HOTELS	
Bedrooms and baths	75
Dining rooms	72
Kitchens and laundries	66
Ballrooms	65–68
Toilets and service rooms	68
HOMES	73–75
STORES	65–68
PUBLIC BUILDINGS	72–74
WARM AIR BATHS	120
STEAM BATHS	110
FACTORIES AND MACHINE SHOPS	60–65
FOUNDRIES AND BOILER SHOPS	50–60
PAINT SHOPS	80

space of 15 to 25 fpm (0.08 to 0.13 m/s) is considered to be still air, while air moving at 65 fpm would constitute a draft to most people.

In order to heat a conditioned space, the supply air may be introduced into the space at 10°F to 50°F (5.6°C to 28°C) *above* the design space temperature and at a velocity considerably above 15 fpm. Likewise, for cooling a conditioned space, the air may be supplied at 12°F to 30°F (6.7°C to 16.7°C) *below* the required final temperature. The average air velocity through supply air diffusers may be 500 to 700 fpm (2.5 to 3.5 m/s). A good air distribution system for the space must do the following: (1) entrain room air into the supply air such that when the combined airstream reaches the occupied zone it will be at the correct temperature so that it will not be objectionable; (2) reduce the air velocity before reaching the occupied zone to a point that there will be freedom from drafts; (3) provide a turbulent, eddying air motion within the entire occupied zone; and (4) keep air noise from the supply outlets and return air inlets below the objectionable level.

Although it may seem that these requirements are stringent, it should be pointed out that an air conditioning system may have the best mechanical equipment, properly sized for the load, and yet, from the owner's standpoint, be unsatisfactory if the air distribution is not providing comfort for the occupants. There will be

a more complete discussion of air distribution in Chapter 13.

5-7 HUMIDITY LEVELS WITHIN A STRUCTURE

In Section 5-6, it was noted that for comfort the humidity level should be between 30% and 70%; however, much more must be said concerning moisture levels within buildings. Too often, the behavior of moisture is given insufficient attention in building design and construction. This should be carefully considered, regardless of comfort requirements. Moisture is present as a vapor in all air and as absorbed moisture in most building materials. Problems involving moisture may arise from changes in moisture content, from the presence of excessive or insufficient moisture in materials, or from the effects associated with its change of phase (condensation).

Water vapor originates from activities such as cooking, laundering, bathing, and our breathing and perspiring. And, of course, many industrial processes produce varying amounts of water vapor. Typical amounts of water vapor produced from domestic activities are shown in Table 5-3. Tables in *ASHRAE Handbook 1989 Fundamentals,* Chapter 26 may be used for estimating moisture quantities in lb/hr (g/s) emitted by

TABLE 5-3

Moisture production in residences

Operation			Moisture lb	(kg)
Floor mopping—80 ft² (7.4 m²) kitchen			2.40	(1.09)
Clothes drying* (not vented)			26.40	(11.97)
Clothes washing*			4.33	(1.96)
Cooking (not vented)*—from food and from gas				
Breakfast	0.34 (0.16)	0.56 (0.25)	0.90	(0.41)
Lunch	0.51 (0.23)	0.66 (0.33)	1.17	(0.53)
Dinner	1.17 (0.53)	1.52 (0.69)	2.69	(1.22)
Bathing—shower			0.50	(0.23)
Bathing—tub			0.12	(0.05)
Dishwashing*				
Breakfast			0.20	(0.09)
Lunch			0.15	(0.07)
Dinner			0.65	(0.29)
Human contribution—adults (per hour)				
When resting			0.2	(0.09)
Working hard			0.6	(0.27)
Average			0.4	(0.18)
Houseplants (per hour)			0.04	(0.02)

*Based on family of four.

occupants and appliances by dividing the latent heat values in Btu/hr (W) by 1100 Btu/lb (2560 kJ/kg).

Interior and exterior building materials should allow water vapor to pass 5 times more rapidly than materials inside the walls or roof. If this condition is met, any moisture that may get into the walls or roof will move through it.

Selecting and applying humidification or dehumidification equipment involves considering both the environmental requirements of the occupants or process and the limitations imposed by the thermal and permeable characteristics of the building enclosure. As these may not always be compatible, a compromise solution may be necessary, particularly in the case of existing buildings.

The extent to which a building may be humidified in winter depends on the ability of the walls, roof, and other elements of the building enclosure to prevent or tolerate moisture condensation. The formation of condensed moisture or frost on surfaces exposed to the building interior, or visible condensation, can result in deterioration of the surface finish, mold growth, subsequent indirect moisture damage and nuisance, and reduction of visibility through windows. If the walls and roof have not been specifically designed to prevent entry of moist air or vapor from inside, concealed condensation within these constructions may occur even at low interior humidities and give rise to serious deterioration.

The environmental requirements for a particular occupancy or process may dictate a specific relative humidity, a required range of relative humidity, or certain limiting maximum or minimum values. The following classifications give guidance for most applications.

Human Comfort. As stated previously, the effect of relative humidity on human comfort has not been completely established. Nonetheless, humidity extremes are assumed to be undesirable, and, for human comfort, relative humidities should be kept within the broad range of 30% to 70%.

Static Electricity. Electrostatic charges are generated when materials of high electrical resistance move against one another. Such charges may cause unpleasant sparks to people walking over carpets; difficulties in handling sheets of paper, fibers, and fabrics; objectionable clinging of dust to oppositely charged objects; or dangerous situations when explosive gases are present. Increasing the relative humidity of an environment tends to prevent the accumulation of such charges, but the optimum level of humidity depends to some extent on the materials involved. With many materials, relative humidities of 45% or more are usually required to reduce or eliminate electrostatic effects. Hospital operating rooms, where explosive mixtures of anesthetics are used, constitute a special and critical case with regard to electrostatic charges. Relative humidities of 50% or more are usually required, and other special grounding arrangements and restrictions are imposed as to types of clothing the occupants wear. From a consideration of both comfort and safety, conditions of 72°F (22°C) and 55% RH are usually recommended for hospital operating rooms.

Prevention and Treatment of Disease. Relative humidity has a significant effect on the control of airborne infection. At 50% RH, the mortality rate of certain organisms is highest; for example, influenza virus loses much of its virulence. The mortality rate decreases both above and below this value. The value of 65% RH is regarded as optimum for nurseries for premature infants, while a value of 50% RH is suitable for full-terms and observational nurseries.

Visible Condensation. Condensation occurs on any interior surface when the dew-point temperature of the air in contact with it exceeds the surface temperature (see Chapter 3). The maximum permissible relative humidity that may be maintained without condensation is thus influenced by the thermal properties of the enclosure and the interior and exterior environment. In general, windows present the lowest interior surface temperature in most buildings and provide the best guide to permissible indoor humidity levels for no condensation (see Table 5-4).

Concealed Condensation. The humidity level that a building can tolerate without serious difficulties from concealed condensation may be much lower than indicated by visible condensation criteria. The migration of water vapor through the inner building envelope by diffusion or air leakage brings it into contact with surfaces at temperatures approaching the outside temperature. Unless the building has been designed to elimi-

nate or effectively reduce this possibility, the permissible humidity level may be limited by the ability of the building enclosure to handle internal moisture rather than prevent the occurrence of moisture.

The humidification or dehumidification load depends primarily on the rate of ventilation of the space to be conditioned, but other sources of moisture gain or loss should be considered, as noted in this section. Chapters 3 and 4 of this book contain equations and examples for determining loads and energy requirements associated with humidification and dehumidification.

5-8 QUALITY AND QUANTITY OF AIR SUPPLIED

The quantity of air available remains constant though its quality does not. In a sense, we breathe today the same air used and reused by countless preceding generations. Our air is continually regenerated for us by nature.

But the demands of modern life and industry are so complex and urgent that it is no longer practical to depend upon nature to provide air of the precise quality required, at a specific time or in a particular place. A climate neither too cold nor too hot, too dry nor too humid, all year-round is not part of our heritage. To make our immediate atmosphere more suitable to our needs, we have learned to "condition" air mechanically. The process is not without cost.

Whether simple or intricate, an air conditioning system represents a capital investment in equipment—in fans, ducts, dampers, diffusers, and grilles; in boilers, compressors, tempering and cooling coils, filters, air washers, and pumps; and in thermostats and regulators of all types. It represents an operating charge for electrical energy, fuel, water, and operating and maintenance labor—all expended for the sole purpose of obtaining the precise quality and quantity of air desired and delivering it when and where required.

Once air is conditioned, contrary to being "free," it becomes a valuable commodity because of the conditioning energy that has been invested in it. The protection of that investment by conserving the conditioned air becomes important.

However, air becomes contaminated by "use," and we are as much concerned about air purity and freshness as we are about its temperature and humidity. Here we may anticipate nature by resorting to mechanical and chemical purification to restore air freshness while at the same time conserving its thermal and psychrometric value.

Air purity is relative. Except possibly on a mountain top immediately after a rainstorm, no air is entirely

TABLE 5-4
Maximum relative humidity without window condensation

Natural Convection, Indoor Air at 23.3°C (74°F)		
Outdoor Temp., °C (°F)	Single Glazing	Double Glazing
4.4 (40)	39	59
−1.1 (30)	29	50
−6.7 (20)	21	43
−12.2 (10)	15	36
−17.8 (0)	10	30
−23.3 (−10)	7	26
−28.9 (−20)	5	21
−34.4 (−30)	3	17

free of entrained impurities. Even mountain air, for example, will be found to contain a trace (though negligible) of sulfur dioxide. The extent to which impurities are present in air is termed their *concentration*. Concentration is normally expressed in one of two different ways: (1) mass per unit volume of air (g/m³) or (2) volume per number of volumes of air (parts per million, ppm).

For social and economic reasons, we have found it necessary to congregate in confined areas and consequently must meet the problem of obtaining uncontaminated air. All air that surrounds dwellings and industry accumulates foreign substances. These contribute no appreciable ill effects as long as their concentration is not excessive. There is, however, a "threshold" concentration of impurities that is considered optimal or tolerable for each contaminating substance under specified circumstances of exposure. The criteria establishing allowable concentrations are often a compromise. The concentration tolerated in a foundry, for example, would not be considered suitable in a home. On the other hand, the degree of air purity demanded in a factory devoted to the making of highly sensitive photographic films or precision instruments may be substantially higher than that which would be completely acceptable in the finest residence.

When air purification is employed to preserve the freshness of conditioned air so as to conserve its thermal and psychrometric value—its heat or lack of heat, its moisture or dryness—the process is called *air recovery*.

Regardless of the amount of outside air entering a building by natural or mechanical methods, the air must meet certain quality standards. ASHRAE Standard 62-1989, *Ventilation for Acceptable Indoor Air Quality,* lists maximum allowable pollutant concentrations in air used for the ventilation of spaces used for human occupancy. These are shown in Table 5-5.

The intrinsic worth or utility of an air conditioning system is dependent on two factors: (1) the cost of the installation and (2) the cost of operation. Both of these factors are directly related to the amount of new, outdoor air required by the space to be conditioned. The internal heat gain or loss is a constant dictated by the locale, exposure, construction, occupancy, and function of the space. This constant represents the irreducible minimum of conditioning required. The air change—that is, the volume of outdoor air that must be supplied and conditioned—constitutes the variable and, when added to the internal load, determines the size and cost of the conditioning system.

The vitalizing quality in outdoor air is its oxygen content, which is almost exactly 21% by volume. The *freshness* of the air is entirely a function of its freedom from entrained impurities.

Ventilation standards, insofar as they are related to the above, establish the volume of air that must be supplied to a building or enclosure in order to (1) replenish the oxygen consumed and (2) dilute internally generated air-contaminating impurities.

The air needed to satisfy the first requirement must invariably be new, outdoor air regardless of whether or not it is also fresh—that is, uncontaminated. The air required to serve the second purpose must obviously be fresh regardless of whether or not it is outdoor

TABLE 5-5

National primary ambient-air quality standards for outdoor air as set by the U.S. Environmental Protection Agency

Contaminant	Long-Term Concentration Averaging			Short-Term Concentration Averaging		
	µg/m³	ppm		µg/m³	ppm	
Sulfur dioxide	80	0.03	1 yr	365	0.14	24 hr
Total particulate	75[a]	—	1 yr	260	—	24 hr
Carbon monoxide				40,000	35	1 hr
Carbon monoxide				10,000	9	8 hr
Oxidants (ozone)				235[b]	0.12[b]	1 hr
Nitrogen dioxide	100	0.055	1 yr			
Lead	1.5	—	3 mon.[c]			

Source: Reprinted from ASHRAE Standard 62-1989, *Ventilation for Acceptable Indoor Air Quality,* with permission from American Society of Heating, Refrigerating, and Air Conditioning Engineers, Atlanta, GA.

[a] Arithmetic mean.

[b] Standard is attained when expected number of days per calendar year with maximum hourly average concentration above 0.12 ppm (235 µg/m³) is equal to or less than 1, as determined by Appendix H to subchapter C, 40 CFR 50.

[c] Three-month period is a calendar quarter.

air. Far more air is required for dilution of impurities than for oxygen replenishment; the former may, in fact, be 10 to 20 times the latter.

All air expelled from an air-conditioned space carries with it cooling or heating energy. All air that can be recirculated conserves this conditioning energy. Whenever stale, vitiated, or otherwise contaminated conditioned air can be converted to its original freshness at a cost less than that required to replace it, the failure to employ such conversion adds to the cost and operation of the conditioning plant.

5-9 MEETING OUTDOOR AIR REQUIREMENTS

Ventilation may be defined as the process of supplying and removing air by natural or mechanical means to and from any space. Such air may or may not be conditioned. Further, *ventilation air* is defined as that portion of supply air which is outdoor air plus any recirculated air that has been treated for the purpose of maintaining acceptable indoor air quality. Ventilation air becomes supply air after passing through the conditioning equipment. Ventilation may use 100% outdoor air. The term *makeup air* may be used synonymously with *outdoor air,* and *return air* and *recirculated air* are often used interchangeably.

Outdoor air that flows through a building either intentionally as ventilation air or unintentionally as infiltration (or exfiltration) is important for two reasons. Outdoor air is often used to dilute indoor air contaminants, and the energy associated with conditioning this outdoor air is a significant space load. The magnitude of these airflow rates should be known at maximum load to properly size equipment, as well as at average conditions to properly estimate average or seasonal energy consumption. Minimum air exchange rates need to be known to assure proper control of indoor air contaminant levels. In large buildings, the effect of infiltration and ventilation on distribution and interzone airflow patterns, which include smoke circulation patterns in the event of fire, should be determined (see Chapter 58 of the 1987 HVAC Volume published by ASHRAE).

Buildings have three different modes of air exchange: (1) forced ventilation, (2) natural ventilation, and (3) infiltration or exfiltration. These modes differ significantly in how they affect energy, air quality, and thermal comfort. They also differ in their ability to maintain a desired air exchange rate. The air exchange rate in a building at any given time generally includes all three modes, and they all must be considered even when only one is expected to dominate.

The air exchange rate associated with a *forced ventilation system* depends on the airflow rates through the system fans, the airflow resistance associated with the air distribution system, the airflow resistance between the zones of the building, and the air tightness of the building envelope. If any one of these factors is not at the design level or not properly accounted for, the building air exchange rate can be quite different from its design value.

Forced ventilation provides the greatest opportunity for control of the air exchange rate and air distribution within the building through the proper design, installation, operation, and maintenance of the ventilation system. An ideal forced ventilation system has a sufficient ventilation rate to control indoor contaminant levels and, at the same time, avoids overventilation and the associated energy penalty. In addition, it maintains good thermal comfort.

Forced ventilation is generally mandatory in larger buildings, where a minimum amount of outdoor air is required for occupant health and comfort and where a mechanical exhaust system is advisable or necessary. Forced ventilation has generally not been used in residential and other envelope-dominated structures. However, tighter, more energy-conserving buildings require ventilation systems to assure an adequate amount of outdoor air for maintaining acceptable indoor air quality.

Natural ventilation through intentional openings is caused by pressures from wind and indoor–outdoor air temperature differences. Airflow through open windows and doors and other design openings can be used to provide adequate ventilation for contaminant dilution and temperature control. Unintentional openings in the building envelope and the associated infiltration can interfere with desired natural ventilation air distribution patterns and lead to larger-than-design airflow rates. Natural ventilation is sometimes defined to include infiltration, but in this book it does not.

Infiltration is the uncontrolled flow of air through unintentional openings in the building envelope driven by wind, temperature difference, and appliance-induced pressures. Infiltration is least reliable in providing adequate ventilation and distribution because it depends on weather conditions and the location of unintentional openings. It is the main source of ventilation in envelope-dominated buildings and is also an important factor in mechanically ventilated buildings.

The amount of ventilation necessary for a given type of space for various kinds of occupancy and activity has evolved mainly through long experience. Where the findings of such experience have been sufficiently conclusive, ventilation requirements have been reduced to standardized practices, codes, and, in some instances, statutes. These generally stipulate the ventilation (fresh air) requirements in terms such as cubic feet

TABLE 5-6a

Outdoor air requirements for ventilation[a]

	Commercial facilities (offices, stores, shops, hotels, sports facilities)					
	Estimated Maximum[b] Occupancy P/1000 ft^2 or 100 m^2	Outdoor Air Requirements				
Application		cfm/ person	L/s · person	cfm/ft^2	L/s · m^2	Comments
Dry cleaners, laundries:						Dry-cleaning processes may require more air.
Commercial laundries	10	25	13			
Commercial dry cleaners	30	30	15			
Storage, pick-up	30	35	18			
Coin-operated laundries	20	15	8			
Coin-operated dry cleaners	20	15	8			
Food and beverage service:						
Dining rooms	70	20	10			
Cafeteria, fast food	100	20	10			
Bars, cocktail lounges	100	30	15			Supplementary smoke-removal equipment may be required.
Kitchens (cooking)	20	15	8			Makeup air for hood exhaust may require more ventilating air. The sum of the outdoor air and transfer air of acceptable quality from adjacent spaces shall be sufficient to provide an exhaust rate of not less than 1.5 cfm/ft^2 (7.5 L/s · m^2).
Garages, repair, service stations:						
Enclosed parking garages				1.50	7.5	Distribution among people must consider worker location and concentration of running engines; stands where engines are run must incorporate systems for positive engine exhaust withdrawal. Contaminant sensors may be used to control ventilation.
Auto repair rooms				1.50	7.5	
Hotels, motels, resorts, dormitories:		cfm/room	L/s · room			Independent of room size.
Bedrooms		30	15			
Living rooms		30	15			
Baths		35	18			Installed capacity for intermittent use.
Lobbies	30	15	8			
Conference rooms	50	20	10			
Assembly rooms	120	15	8			
Dormitory sleeping areas	20	15	8			See also food and beverage services, merchandising, barber and beauty shops, garages.
Gambling casinos	120	30	15			Supplementary smoke-removal equipment may be required.

TABLE 5-6a (Continued)

	Commercial facilities (offices, stores, shops, hotels, sports facilities)					
Application	Estimated Maximum[b] Occupancy P/1000 ft² or 100 m²	Outdoor Air Requirements				Comments
		cfm/ person	L/s · person	cfm/ft²	L/s · m²	
Offices:						
Office space	7	20	10			Some office equipment may
Reception areas	60	15	8			require local exhaust.
Telecommunication centers and data entry areas	60	20	10			
Conference rooms	50	20	10			Supplementary smoke-removal equipment may be required.
Public spaces:				cfm/ft²	L/s · m²	
Corridors and utilities				0.05	0.25	
Public restrooms, cfm/wc or urinal		50	25			Mechanical exhaust with
Locker and dressing rooms				0.5	2.5	no recirculation is recommended.
Smoking lounges	70	60	30			Normally supplied by transfer air, local mechanical exhaust, with no recirculation recommended.
Elevators				1.00	5.0	Normally supplied by transfer air.
Retail stores, sales floors, and show room floors:						
Basement and street	30			0.30	1.50	
Upper floors	20			0.20	1.00	
Storage rooms	15			0.15	0.75	
Dressing rooms				0.20	1.00	
Malls and arcades	20			0.20	1.00	
Shipping and receiving	10			0.15	0.75	
Warehouses	5			0.05	0.25	
Smoking lounges	70	60	30			Normally supplied by transfer air, local mechanical exhaust, with no recirculation recommended.
Specialty shops:						
Barbers	25	15	8			
Beauty salons	25	25	13			
Reducing salons	20	15	8			
Florists	8	15	8			Ventilation to optimize plant growth may dictate requirements.
Clothiers, furniture stores				0.30	1.50	
Hardware, drugs, fabrics	8	15	8			
Supermarkets	8	15	8			
Pet shops				1.00	5.00	
Sports and amusement:						
Spectator areas	150	15	8			When internal combustion
Game rooms	70	25	13			engines are operated for
Ice arenas (playing areas)				0.50	2.50	maintenance of playing

(continued)

TABLE 5-6a (Continued)

	Commercial facilities (offices, stores, shops, hotels, sports facilities)					
	Estimated Maximum[b] Occupancy P/1000 ft² or 100 m²	Outdoor Air Requirements				
Application		cfm/ person	L/s · person	cfm/ft²	L/s · m²	Comments
						surfaces, increased ventilation rates may be required.
Swimming pools (pool and deck area)				0.50	2.50	Higher values may be required for humidity control.
Playing floors (gymnasiums)	30	20	10			
Ballrooms and discos	100	25	13			
Bowling alleys (seating areas)	70	25	13			
Theaters:						Special ventilation will be needed to eliminate special stage effects (e.g., dry ice vapors, mists, etc.).
Ticket booths	60	20	10			
Lobbies	150	20	10			
Auditoriums	150	15	8			
Stages, studios	70	15	8			
Transportation:						Ventilation within vehicles may require special considerations.
Waiting rooms	100	15	8			
Platforms	100	15	8			
Vehicles	150	15	8			
Workrooms:						Spaces maintained at low temperatures (−10°F to +50°F, or −23°C to +10°C) are not covered by these requirements unless the occupancy is continuous. Ventilation from adjoining spaces is permissible. When the occupancy is intermittent, infiltration will normally exceed the ventilation requirement.
Meat processing	10	15	8			
Photo studios	10	15	8			
Darkrooms	10			0.50	2.50	
Pharmacies	20	15	8			
Bank vaults	5	15	8			
Duplicating, printing				0.50	2.50	Installed equipment must incorporate positive exhaust and control (as required) of undesirable contaminants (toxic or otherwise).
	Institutional Facilities					
Education:						Special contaminant control systems may be required for processes or
Classrooms	50	15	8			
Laboratories	30	20	10			
Training shops	30	20	10			

TABLE 5-6a (Continued)

		Institutional Facilities				
	Estimated Maximum[b] Occupancy P/1000 ft² or 100 m²	Outdoor Air Requirements				
Application		cfm/ person	L/s · person	cfm/ft²	L/s · m²	Comments
Music rooms	50	15	8			functions including laboratory animal occupancy.
Libraries	20	15	8			
Locker rooms				0.50	2.50	
Corridors				0.10	0.50	
Auditoriums	150	15	8			
Smoking lounges	70	60	30			Normally supplied by transfer air, local mechanical exhaust, with no recirculation recommended.
Hospitals, nursing and convalescent homes:						
Patient rooms	10	25	13			Special requirements or codes and pressure relationships may determine minimum ventilation rates and filter efficiency. Procedures generating contaminants may require higher rates.
Medical procedure	20	15	8			
Operating rooms	20	30	15			
Recovery and ICU	20	15	8			
Autopsy rooms				0.50	2.50	Air shall not be recirculated into other spaces.
Physical therapy	20	15	8			
Correctional facilities:						
Cells	20	20	10			
Dining halls	100	15	8			
Guard stations	40	15	8			

Source: Reprinted from ASHRAE Standard 62–1989, with permission from American Society of Heating, Refrigerating, and Air Conditioning Engineers, Atlanta, GA.

[a] Table 5-6a prescribes supply rates of acceptable outdoor air required for acceptable indoor air quality. These values have been chosen to control CO_2 and other contaminants with an adequate margin of safety and to account for health variations among people, varied activity levels, and a moderate amount of smoking. Rationale for CO_2 control is presented in Appendix D of Standard.

[b] Net occupiable space.

per minute (cfm) or liters per second (L/s) per occupant; cfm per ft² of floor area or per ft³ of space; air changes per hour or air velocity across the face of a ventilation exhaust hood.

Tables 5-6a and b from ASHRAE Standard 62–1989 show outdoor air requirements for commercial, institutional, and residential facilities to obtain acceptable indoor air quality by the *Ventilation Rate Procedure* presented in the Standard. Another procedure discussed in ASHRAE Standard 62–1989 is the *Indoor Air Quality Procedure,* which will not be discussed here.

Outdoor air for ventilation is acceptable if the contaminants in the air do not exceed the concentrations listed in Table 5-5.

Indoor air quality is considered acceptable if the required rates of acceptable outdoor air in Table 5-6 are provided for the occupied space. The required ventilation rates listed are in cfm (L/s) per person or cfm/ft² (L/s·m²) for a variety of indoor spaces. In most cases, the contamination produced is presumed to be in proportion to the number of persons in the space. In other cases, the contamination is presumed to be chiefly due to other factors, and the ventilating rates given are based on more approximate parameters. Where appropriate, the estimated density of people is listed for design purposes.

Where occupant density differs from that in Table 5-6, use the per-occupant ventilation rate for the antici-

TABLE 5-6b
Outdoor air requirements for ventilation of residential facilities (private dwellings, single, multiple)[a]

Application	Outdoor Air Requirements	Comments
Living areas	0.35 air changes per hour but not less than 15 cfm (7.5 L/s) per person	For calculating the air changes per hour, the volume of the living spaces shall include all areas within the conditioned space. The ventilation is normally satisfied by infiltration and natural ventilation. Dwellings with tight enclosures may require supplemental ventilation supply for fuel-burning appliances, including fireplaces and mechanically exhausted appliances. Occupant loading shall be based on the number of bedrooms as follows: first bedroom, two persons; each additional bedroom, one person. Where higher occupant loadings are known, they shall be used.
Kitchens[b]	100 cfm (50 L/s) intermittent or 25 cfm (12 L/s) continuous or openable windows	Installed mechanical exhaust capacity.[c] Climatic conditions may affect choice of ventilation system.
Baths, toilets[b]	50 cfm (25 L/s) intermittent or 20 cfm (10 L/s) continuous or openable windows	Installed mechanical exhaust capacity[c].
Garages:		
Separate for each dwelling unit	100 cfm (50 L/s) per car	Normally satisfied by infiltration or natural ventilation.
Common for several units	1.5 cfm/ft^2 (7.5 L/s · m^2)	

Source: Reprinted from ASHRAE Standard 62–1989, with permission from American Society of Heating, Refrigerating, and Air Conditioning Engineers, Atlanta, GA.

[a] In using this table, the outdoor air is assumed to be acceptable.

[b] Climatic conditions may affect choice of ventilation option chosen.

[c] The air exhausted from kitchens, bath, and toilet rooms may utilize air supplied through adjacent living areas to compensate for the air exhausted. The air supplied shall meet the requirements of exhaust systems as described in 5.8 and be of sufficient quantities to meet the requirements of this table. See ASHRAE Standard 62–1989.

pated occupant load. The ventilation rates for specified occupied spaces listed in Table 5-6 were selected to reflect the consensus that the provisions of acceptable outdoor air at these rates would achieve an acceptable level of indoor air quality by reasonably controlling CO_2, particulates, odors, and other contaminants common to these spaces.

Human occupants produce carbon dioxide, water vapor, particulates, biological aerosols, and other contaminants.

Carbon dioxide concentration has been widely used as an indicator of indoor air quality. Comfort (odor) criteria are likely to be satisfied if the ventilation rate is set so that 1000 ppm CO_2 is not exceeded.

5-10   AIR-ENTRAINED IMPURITIES

The impurities found in air may be divided into the following four general classifications:

1. *Dusts and smokes*—These terms define biologically inert solid particles suspended in air. Dust particles are light enough to be blown about by air currents yet are heavy enough to settle readily by gravity in comparatively still air. Smoke particles, being much smaller and lighter, may remain suspended for long periods even in still air.

2. *Mists and fogs*—These terms define biologically inert liquid particles (droplets) suspended in air. The migrations of such particles are determined by the same factors as those of dust and smoke.

3. *Microorganisms*—This term defines air-suspended bacteria, germs, viruses, and other living organisms that are light enough to "float" in air but more generally ride on inert, suspended solid or liquid particles.

4. *Gases and vapors*—These terms define matter consisting of dispersed molecules that are diffused rather than suspended in air. The motion of molecules is the result of their collisions and is influenced by temperature and pressure. For particles under 0.01 μm in size, the kinetic velocity becomes dominant, and this dimension is therefore sometimes conceived as the boundary between particles and gases. Gaseous impurities in air account for odor sensations, air "staleness," and, in

TABLE 5-7

Air-entrained impurities—relationship of size and characteristics

Size Scale Microns*	10,000	1,000	100	10	1	0.1	0.01	0.001	0.0001	0.00003
Impurity		Dusts & Droplets (Solid) (Liquid)	Aerosols (Solid or Liquid)			Smokes & Mists (Solid) (Liquid)		Vapors & Gases (Molecules)		
Typical Particles	Raindrops Sand	Rainmist	Natural Fog	Bacteria		Tobacco Smoke Hot Oil Fog Naphthalene Molecule → •			Water Molecule → • • ← Benzene Molecule	
Method of Removal	Settling Chambers Cyclone Separators		Air Filters Electrostatic Precipitators					Activated Carbon Sorption Catalytic Combustion		
Radiation Scale Wave L	"Short" Waves (Radar, etc.)		Infra-Red			Visible Ultra Violet			X-Rays	

* 1 Micron = 10^{-4} cm = $\frac{1}{25,400}$ in.

many instances, irritating or toxic effects. Such impurities comprise many thousands of organic and inorganic substances.

Table 5-7 illustrates the relationships of size and characteristics of various air-entrained impurities.

The sources of atmospheric impurities are so diversified that enumeration would be endless. The principal sources of air-contaminating impurities in residential and commercial buildings are the occupants, their habits, and services: body, apparel and respiratory emanations, tobacco smoke, cosmetics, liquor, edibles and their cooking or preparation, furnishings, painted surfaces, cleaning components, and so forth. In the average industrial establishment of a manufacturing nature, there are also contaminating sources, such as cutting and lubricating oils and greases, solvents, fumes and metallic oxides, internally generated dust, soot, and smoke. Processing industries contribute vapors and fumes in almost infinite variety. Recently, outdoor air as a source of odors has increased in importance because in some areas it may contain a high percentage of automotive exhaust and electrical power generating station emissions.

The air conditioning system itself may contribute odors. Cooling coils collect dirt and lint, both of which are moistened by the condensate; this helps mildew to form and leads to objectionable odors.

5-11 REMOVAL OF PARTICULATE MATTER

The means to be employed in the purification of air will depend upon the nature and concentration of the contaminants and the extent to which their elimination is necessary or desired. Dust, for example, is usually present to an extent warranting the use of *air filters*. Of these, various types are available, the selection of which will be dictated by the degree of efficiency demanded. Where extremely light and minute particles, such as those composing smoke and soot, prevail in objectionable concentration, elimination by the use of electrostatic precipitation will prove effective. Such filtration will likewise contribute to the control of pollen concentration. For those special instances where the hazardous accumulation of objectionable bacteria may be anticipated, the use of appropriate germicidal agents would be indicated.

One or more of these devices may be required to produce the desired quality of air in a conditioning system. Technical data on their respective engineering features are available from manufacturers, and *ASHRAE*

Handbook 1988 Equipment has some most valuable data. In addition, means of limiting the concentration of vaporous and gaseous impurities are necessary if air is to be recirculated indefinitely.

5-12 ELIMINATION OF VAPORS AND GASES

Various methods of eliminating air-entrained gaseous and vaporous contaminants have been employed. One persistent though futile approach to this problem has been the introduction of chemical agents either directly into the space or into the recirculated airstream to react with the entrained gaseous impurities in a manner tending to destroy, decompose, or otherwise alter their objectionable character. This method faces inescapable obstacles. On the one hand, agents that will actually destroy or alter the chemical composition of the diffused substances are often toxic. On the other hand, chemical agents that do not decompose or change the existing impurities are effective, if at all, merely as a screen; for example, masking a disagreeable odor with a stronger one of presumably more agreeable characteristics. Sometimes, the masking effect is not merely sensory but pathological. Some so-called deodorizing substances may contain ingredients such as formaldehyde that deaden or anesthetize the olfactory nerves, thus preventing the detection of either the masking or the offending odors. In either event, the result is one of adding to rather than reducing the prevailing contamination.

Of the methods to extract or separate gaseous impurities from air, two common processes cannot be applied practically to general ventilation. The first of these is *condensation* by a reduction in temperature. Most of the air-entrained gases and vapors are present in very low concentration even when highly objectionable. Hence, their vapor pressures are extremely low and their condensation temperatures are far below 0°F (−17.8°C). While the condensation method is useful in certain laboratory operations requiring the separation of gases by low-temperature distillation, it offers no possibility for large-scale air purification.

The second process is that of *air washing* or *scrubbing*. This method, if applied to general ventilation primarily for the purpose of air cleaning, is decidedly limited and generally costly. Its effectiveness in eliminating gaseous and vaporous impurities from air is confined to a relatively few gases and vapors that are water soluble. The vast majority of airborne odorants are organic substances that are insoluble in water and hence cannot be removed by water scrubbing. These insolubles include saturated and unsaturated hydrocarbons, sulfur and nitrogen compounds, esters, and most of the odorous

acids, aldehydes, and ketones. It is sometimes possible to treat the water to obtain solubility of some gases; for example, an 8% caustic water solution in a spray-type washer effectively extracts airborne sulfur dioxide. This, however, makes necessary controlled maintenance of the strength of the water solution and the constant change of water to prevent accumulation and concentration of dissolved impurities. Even with the closest control, it is seldom possible to prevent some reevaporation and escape of dissolved impurities. The effectiveness of the washer in eliminating soluble substances alone will further depend upon the degree of contact obtained between the air and water. So-called scrubbers, imposing tortuous air passages, air turbulence, and counterflow features to increase contact, are sometimes employed, but these add to flow resistance and, consequently, to the cost of the operation.

An added obstacle to air purification by washing is that the air becomes moisture saturated in the process, and, unless supplementary dehumidification is provided, its application to ventilation is impractical. In air conditioning, the air washer contributes the useful functions of cooling, humidification, or dehumidification (see Chapter 4) but is both inadequate and impractical as an air purifier.

Summarized, the preceding approaches to air purification are definitely limited, impractical, or uneconomical. There is, however, a simple and practical method of extracting nearly all odorous and objectionable gases and vapors from air, namely, the process of *adsorption*.

Adsorption (not to be confused with "absorption," which literally means the diffusion of one substance into the body of another) refers to the adhesion of fluid (gas or liquid) substances to the surfaces of solid substances. Because there are instances where the distinction between the two phenomena may not be very sharp, the term *sorption* and derivatives *sorbate* and *sorbent* are often used.

There are a number of substances that possess this unique physicochemical property of adsorption. The most common of these and the best adapted to practical air purification is activated carbon (charcoal). Other adsorbents include zeolite, silica gel, activated alumina, and mica.

Charcoal has a unique affinity for many gases and vapors, tenaciously holding them until they are removed by special means. While most charcoals in their natural state possess this adsorptive property to a degree, those intended for air purification are produced from particular raw materials and are specially processed or "activated." This purpose of the process is to create extensive surfaces on which adsorption can take place. This results in high adsorptive capacity for gases and vapors and distinguishes "activated carbon" from the more common varieties of charcoal. Activated carbons, in turn, vary in structural properties and are selected depending upon the purpose for which they are to be used, the most important distinction being between those adapted to deodorize or decolorize liquids and those suited to purify air and gas. The latter must be of much greater density and durability and have a finer pore structure.

The air-purifying effectiveness of activated carbon depends in part upon the duration of contact between the air and carbon. Since the thickness of the carbon bed used is limited by the allowable resistance to air-flow, the bed area must be large and the air velocity through the bed correspondingly low.

The amount of carbon employed must be sufficient to last a reasonable length of time before it becomes saturated. Upon saturation, the carbon can be reactivated to its original activity. Consequently, the total weight of carbon for any particular application is determined by the quantity of gaseous impurities to be adsorbed between reactivations.

Further information relating to activated carbon and information concerning the use of other adsorbents may be found in the system volume of the *ASHRAE Handbook* series.

BIBLIOGRAPHY

5.1. *ASHRAE Handbook 1989 Fundamentals,* American Society of Heating, Refrigerating, and Air Conditioning Engineers, Atlanta, GA, 1989.

5.2. ANSI/ASHRAE Standard 55-1981, *Thermal Environmental Conditions for Human Occupancy,* American Society of Heating, Refrigerating, and Air Conditioning Engineers, Atlanta, GA, 1981.

5.3. Nevins, A. G., Psychrometrics and Modern Comfort, presented at a joint ASHRAE–ASME meeting, November 28–29, 1961.

5.4. ASHRAE Standard 90-75, *Energy Conservation in New Building Design,* American Society of Heating, Refrigerating, and Air Conditioning Engineers, Atlanta, GA, 1975.

5.5. ASHRAE Standard 62-1989, *Ventilation for Acceptable Indoor Air Quality,* American Society of Heating, Refrigerating, and Air Conditioning Engineers, Atlanta, GA, 1989.

5.6. *ASHRAE Handbook 1988 Equipment,* American Society of Heating, Refrigerating, and Air Conditioning Engineers, Atlanta, GA, 1988.

5.7. *ASHRAE Handbook 1987 HVAC Systems and Applications,* American Society of Heating, Refrigerating, and Air Conditioning Engineers, Atlanta, GA, 1987.

6

Heat Transfer in Building Sections

6-1 INTRODUCTION

The design of a satisfactory heating system requires that we be able to make a satisfactory estimate of the heat-transfer losses for a building. Also, in order to design or select practical, efficient, and economical heat exchangers to perform satisfactorily in the heating of buildings, a knowledge of the basic principle of heat transfer must be acquired.

In most cases, heat is being transferred between two fluids, which may be stationary or moving. For example, heat is transferred from the warm inside air to the cold outside air through a building structure. Or, heat is transferred from a warm fluid in a heat exchanger to the cooler surrounding air. Since we are concerned with stationary fluids or fluids in motion, we will find that heat transfer is closely related to the characteristic behavior and properties of fluids.

Heat is a form of energy in transition, the transfer of which is subject to the first and second laws of thermodynamics. All the heat lost from a source or sources must equal that gained by the receiver or receivers involved. When unaided by mechanical means, the flow of heat is always from a higher to a lower temperature level.

In some cases, complete and rapid heat transfer is desired, such as in heat-exchange devices. In other cases, a minimum amount of heat transfer is desired, such as through building sections or from piping or ductwork carrying fluid to various heat-transfer devices in a heating system.

This chapter discusses concepts and procedures for determining overall heat-transmission coefficients by simplified methods commonly used in design. Factors that can affect the accuracy of these estimates of heat-transfer coefficients are also considered. Coefficients can be determined by testing or by computing from known thermal resistances tabulated for the various materials in the construction. Procedures for calculating will be illustrated by examples.

6-2 MODES OF HEAT TRANSFER

Essentially there are three methods (modes) of heat transfer, namely, *conduction, convection,* and *radiation.* In the usual situation, all three modes occur simultaneously. In some instances, the heat-transfer methods can be separated, but in others only the combined effect can be determined. The many variables involved in heat transfer have been the subject of much investigation in this field of science, and great advances have been made in the past 30 years increasing the fund of basic knowledge.

In our study of heat transfer, we will discuss separately the modes of heat transfer, and we will be concerned only with *steady-state heat transfer.* In steady-state heat transfer, the various temperatures throughout a system remain constant with respect to time during heat transmission. In unsteady heat transfer, temperature varies with time.

Thermal conduction is the term applied to the mechanism of heat transfer whereby the molecules of higher kinetic energy transmit part of their energy to adjacent molecules of lower kinetic energy by direct molecular action. Since the temperature is proportional to the average kinetic energy of molecules, thermal transfer will occur in the direction of decreasing temperature. The motion of the molecules is random; there is no net material flow associated with the conduction mechanism. In the case of flowing fluids, thermal conduction is significant in the region very close to a solid boundary or wall where the flow is laminar and parallel with the wall surface, and where practically no cross currents exist in the direction of the heat transfer across the solid–fluid boundary. In solid bodies, the significant mechanism of heat transfer is always thermal conduction.

Thermal convection involves energy transfer by eddy mixing and diffusion in addition to conduction. Convection is the transfer of heat between a moving fluid (gas or liquid) and a surface, or the transfer of heat from one point to another within a fluid. In convection, if the fluid moves because of a difference in density resulting from temperature changes, the process is called *natural convection,* or *free convection;* if the fluid is moved by mechanical means (pumps or fans), the process is called *forced convection.*

In the conduction and convection mechanisms, the transfer of heat is associated with matter. For *radiant heat transfer,* however, a change in energy form takes place, from internal energy at the source to electromagnetic energy for transmission, then back to internal energy at the receiver.

6-3 THERMAL CONDUCTION EQUATION

The theory of heat conduction was developed by a French mathematician, J. B. Fourier, and expresses steady-state conduction in one direction as

$$\dot{q} = -kA\frac{dt}{dx} \qquad (6\text{-}1)$$

where

$\dot{q}$ = heat-transfer rate (Btu/hr or W)

k = thermal conductivity (Btu/hr-ft-°F or W/m-°C)

A = surface area normal to the direction of heat flow (ft² or m²)

dt/dx = temperature gradient (°F/ft or °C/m)

Since t is decreasing as x increases, the negative sign gives a positive value of $\dot{q}$. When dt/dx is a constant negative value, Eq. (6-1) may be written as

$$\dot{q} = \frac{k}{x}A(t_1 - t_2)$$

$$= \frac{k}{x}A\Delta t \qquad (6\text{-}2)$$

where t_1 and t_2 are the hot and cold surface temperatures respectively, and $\Delta t = t_1 - t_2$, the temperature difference across the section. Figure 6-1 represents schematically the temperature variation through a homogeneous material.

It is important to know the units of thermal conductivity, k. These units may vary from one reference to another, and, for Eq. (6-2) to yield the proper results, units must be consistent. The units of k in the foot-pound–second system are

$$k = \frac{\dot{q}}{A} \times \frac{x}{t} = \frac{(\text{Btu})(\text{ft})}{(\text{hr})(\text{ft}^2)(°\text{F})}$$

$$= \text{Btu/hr-ft-°F}$$

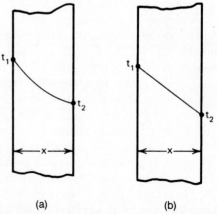

(a)   (b)

FIGURE 6-1 Temperature variation through a homogeneous material with (a) k as a variable and (b) k as a constant mean value.

or

$$k = \frac{(Btu)(in.)}{(hr)(ft^2)(°F)} = \frac{Btu}{(hr)(ft^2)(°F/in.)}$$

$$= \frac{12(Btu)(ft)}{(hr)(ft^2)(°F)}$$

At the present time, most reference data are expressed in these units.

The units of k in SI are

$$k = \frac{\dot{q}}{A} \times \frac{x}{t} = \frac{(W)(m)}{(m^2)(°C \text{ or } K)} = W/m\text{-}°C$$

For conversion, we may use the following:

$$\frac{(W)(m)}{(m^2)(°C)} = 6.938 \frac{(Btu)(in.)}{(hr)(ft^2)(°F)}$$

$$= 0.5782 \frac{(Btu)(ft)}{(hr)(ft^2)(°F)}$$

$$\frac{(Btu)(ft)}{(hr)(ft^2)(°F)} = 1.73 \frac{W}{(m)(°C)}$$

$$= 0.0173 \frac{W}{(cm)(°C)}$$

$$\frac{(Btu)(in.)}{(hr)(ft^2)(°F)} = 0.1441 \frac{W}{(m)(°C)}$$

In general, the value of k varies with temperature, density, and type of material. Most tabular data give average or mean values of k for a specific temperature or range of temperatures.

Symbols Used in Heat-Transfer Equations and Tables

U = overall heat-transmission coefficient, in units of $Btu/(hr)(ft^2)(°F)$ or $W/(m^2)(K)$, existing between the air or other fluid on the two sides of a wall, floor, ceiling, roof, or heat-transfer section under consideration.

k = thermal *conductivity,* in units of $(Btu)(in.)/(hr)(ft^2)(°F)$ or $W/(m)(°C)$.

f = film or surface coefficient (conductance) for computing heat transfer through the gas or liquid film by conduction, convection, and radiation to the adjacent medium, in units of $Btu/(hr)(ft^2)(°F)$ or $W/(m^2)(°C)$.

C = thermal *conductance,* the heat transmitted through a nonhomogeneous or composite material of the thickness and type for which C is given, in units of $Btu/(hr)(ft^2)(°F)$ or $(W/m^2)(°C)$; fluid film effects usually not considered.

R' = thermal resistance, equal to x/kA.

R = unit thermal resistance to heat flow, usually for 1 ft² of surface area. A general term for various kinds of resistance, in units of $(hr)(ft^2)(°F)/Btu$ or $(m^2)(°C)/W$.

= $1/U$ = overall resistance, fluid to fluid on each side of composite section.

= $1/C$ = units resistance of a nonhomogeneous section or material; film effects usually not considered.

= x/k = unit resistance of a homogeneous material of x units thick.

= $1/k$ = unit resistance of a homogeneous material of unit thickness.

= $1/f$ = surface film resistance.

R_a = air-space resistance.

6-4 THERMAL PROPERTIES OF TYPICAL BUILDING MATERIALS

Appendix C presents data required to calculate the heat transfer through building structural components.

Table C-1 includes the conductivity (k) or conductance (C) for many materials used in building structures, as well as the corresponding unit thermal resistance (R).

Table C-2a lists surface conduction (f) and unit resistance (R) for still and moving air films. Note that the emittance (ϵ) of the surface is considered, as well as the orientation of the surface and direction of heat flow.

Table C-2b gives values of emittance and effective emittance (E) for various combinations of reflective surfaces when considering air spaces.

Tables C-3a through C-3c list air-space unit thermal resistance for $\frac{3}{4}$-in.-, $1\frac{1}{2}$-in.-, and $3\frac{1}{2}$-in.-wide air spaces, respectively. Note that the air-space unit thermal resistance (R_a) depends not only on the air-space width but also on the position of the air space, the direction of heat flow, the mean temperature in the air space, the temperature difference across the air space, and effective emittance.

6-5 CONDUCTION THROUGH A COMPOSITE SECTION

Figure 6-2 represents a plane wall section made up of three homogeneous materials a, b, and c. The materials have mean conductivity values of k_a, k_b, and k_c, respec-

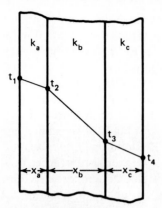

FIGURE 6-2 Temperature variation through a composite section.

tively, and thicknesses x_a, x_b, and x_c. The following assumptions are made:

1. Each of the materials is homogeneous.
2. The thermal conductivity of each material is constant with respect to temperature.
3. The surfaces of the different materials are in intimate contact so that no resistance to heat transfer exists at the interfaces.
4. The flow of heat is steady and perpendicular to the surface.

Equation (6-2) may be written as

$$\dot{q} = \frac{t_1 - t_2}{x/kA} = \frac{t_1 - t_2}{R'} \qquad (6\text{-}3)$$

where $x/kA = R'$, called *thermal resistance*.

Applying Eq. (6-3) to the composite wall section of Figure 6-2, for steady-state heat transfer we have

$$\dot{q} = \dot{q}_a = \dot{q}_b = \dot{q}_c \qquad (6\text{-}3a)$$

or

$$\dot{q} = \frac{t_1 - t_2}{x_a/k_aA_a} = \frac{t_2 - t_3}{x_b/k_bA_b} = \frac{t_3 - t_4}{x_c/k_cA_c} = \frac{t_1 - t_4}{R'_t} \quad (6\text{-}4)$$

where R'_t is the sum of all the thermal resistances in series. Considering 1 ft² of surface area, Eq. (6-4) may be written as

$$\frac{\dot{q}}{A} = \frac{t_1 - t_4}{x_a/k_a + x_b/k_b + x_c/k_c} = \frac{t_1 - t_4}{R_a + R_b + R_c}$$

$$= \frac{t_1 - t_4}{R_t} \qquad (6\text{-}5)$$

where $\dot{q}/A$ becomes the unit heat transfer in Btu/hr-ft² or W/m², and R_a, R_b, and R_c are unit thermal resistances for the materials.

From Eq. (6-2), we may write

$$\Delta t = \left(\frac{\dot{q}}{A}\right)\left(\frac{x}{k}\right) \qquad (6\text{-}6)$$

and since $\dot{q}/A$ is the same for each material in steady flow, we may say that the temperature drop, Δt, through each material of the composite wall of Figure 6-2 is

$$\Delta t_a = \left(\frac{\dot{q}}{A}\right)\left(\frac{x_a}{k_a}\right) = \left(\frac{\dot{q}}{A}\right)(R_a)$$

$$\Delta t_b = \left(\frac{\dot{q}}{A}\right)\left(\frac{x_b}{k_b}\right) = \left(\frac{\dot{q}}{A}\right)(R_b)$$

$$\Delta t_c = \left(\frac{\dot{q}}{A}\right)\left(\frac{x_c}{k_c}\right) = \left(\frac{\dot{q}}{A}\right)(R_c)$$

ILLUSTRATIVE PROBLEM 6-1

Assume that a section of wall is made up of three different materials placed as shown in Figure 6-2. Material *a* has a conductivity of 5.0 (Btu)(in.)/(hr)(ft²)(°F); for material *b*, $k = 12.0$ (Btu)(in.)/(hr)(ft²)(°F); and for material *c*, $k = 0.80$ (Btu)(in.)/(hr)(ft²)(°F). Thicknesses are as follows: $x_a = 4.0$ in., $x_b = 6.0$ in., and $x_c = 0.5$ in. If the surface temperature $t_1 = 100°F$ and the surface temperature $t_4 = 50°F$, what is the rate of heat transfer in Btu/(hr)(ft²)? What are the theoretical interface temperatures in the section?

Solution: By Eq. (6-5),

$$\frac{\dot{q}}{A} = \frac{t_1 - t_4}{x_a/k_a + x_b/k_b + x_c/k_c}$$

$$= \frac{100 - 50}{4.0/5.0 + 6.0/12.0 + 0.5/0.80}$$

$$= \frac{50}{0.80 + 0.50 + 0.625} = \frac{50}{1.925}$$

$$= 25.9 \text{ Btu/(hr)(ft²)}$$

By Eq. (6-6),

$$\Delta t_a = \left(\frac{\dot{q}}{A}\right)(R_a) = (25.9)(0.80) = 20.72°F$$

$$\Delta t_b = \left(\frac{\dot{q}}{A}\right)(R_b) = (25.9)(0.50) = 12.95°F$$

$$\Delta t_c = \left(\frac{\dot{q}}{A}\right)(R_c) = (25.9)(0.625) = 16.19°F$$

The sum $\Delta t_a + \Delta t_b + \Delta t_c$ should be equal to $t_1 - t_4 = 50°F$. The calculated sum is 49.86°F, which indicates a slight inaccuracy because of rounding-off numbers.

In the general case, Eq. (6-5) becomes

$$\frac{\dot{q}}{A} = \frac{t_1 - t_m}{x_a/k_a + x_b/k_b + x_c/k_c + \cdots + x_m/k_m}$$

$$= \frac{t_1 - t_m}{R_t} \qquad (6\text{-}7)$$

The denominator of Eq. (6-7) is the sum of all thermal resistances in series.

It is normally necessary to consider the resistance to heat flow caused by the fluid films that cling to the solid surfaces, as well as the resistance offered by air spaces within composite wall sections. The surface temperatures used in the development of Eqs. (6-5) and (6-7) are less frequently known than the bulk temperatures of the fluids on either side of the section. The method of heat transfer between a fluid (gas or liquid) and a solid surface, and through air spaces, is usually a combination of the three modes of heat transfer (conduction, convection, and radiation).

The thickness of the fluid film that clings to a solid surface varies depending on the flow conditions (laminar or turbulent), which in turn depends upon the fluid flow velocity, surface roughness, and surface orientation (horizontal, vertical, sloping). Thus, heat transfer through fluid films is influenced by the laws of fluid flow and by properties of fluids. Gas films offer much more resistance to heat transfer than do liquid films. Figure 6-3 represents the temperature profile through fluid films for both laminar and turbulent flow conditions. (The thicknesses of the films shown in the figure are greatly exaggerated.) In both types of flow conditions shown, there is a thin film of fluid at the surface that makes up what is called a *boundary layer*. For laminar flow, the boundary layer is somewhat thicker and gives a greater temperature drop from the bulk fluid temperature (t_b) to the surface temperature (t_s). Adjacent to the surface, in both laminar and turbulent flow, the fluid is essentially stationary and heat is transferred by conduction. Farther out from the surface, convection currents are set up due to changes in fluid density. Natural or free convection is an important consideration in heat transfer through building sections. Forced convection frequently occurs in heat-transfer devices. In heat transfer through the metal surfaces of heat-transfer devices, the fluid films offer the greatest resistance to heat flow and become an important consideration. In heat transfer through insulated sections, the fluid film resistance may become relatively unimportant.

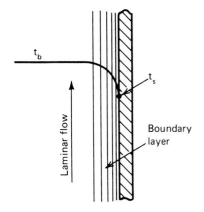

(a) Laminar fluid film

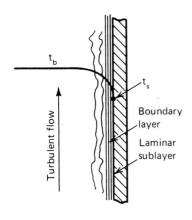

(b) Turbulent fluid film

FIGURE 6-3 Temperature change through fluid films (not drawn to scale).

Evaluation of heat transfer by convection is difficult because of the many variables involved. Mathematical expressions for convective heat transfer are best developed by an application of dimensional analysis, which is beyond the scope of this text. Textbooks on heat transfer cover this method of combining the variables to obtain practical working equations that may be used for predicting heat transfer by convection.

The usual simplified approach is to express the heat transfer by convection as

$$\dot{q}_c = h_c A(t_b - t_s) \qquad (6\text{-}8)$$

where

$\dot{q}_c$ = convective heat-transfer rate (Btu/hr or W)
A = heat-transfer area (ft^2 or m^2)

t_b = bulk or mean temperature of main stream of fluid (°F or °C)

t_s = wall surface temperature (°F or °C)

h_c = coefficient of convective heat transfer (Btu/hr-ft²-°F or W/m²-°C)

Dimensionally, h_c is equal to k/x, but since the fluid film thickness is so small, the value of h_c is usually given as a value independent of k and x for each fluid considered, but dependent on flow conditions. The coefficient of convective heat transfer h_c is sometimes called the unit surface conductance, or simply *film coefficient*. Equation (6-8) may also be expressed in terms of thermal resistance:

$$\dot{q} = \frac{t_b - t_s}{R'} \qquad (6-9)$$

where

$$R' = \frac{1}{h_c A} \text{ (hr-°F/Btu or °C/W)} \qquad (6-10)$$

or

$$R = \frac{1}{h_c} = \frac{1}{C} \text{ (hr-ft²-°F/Btu or m²-°C/W)} \qquad (6-10a)$$

where C in Eq. (6-10a) is called the *unit conductance*. The thermal resistance shown in Eq. (6-10a) may be summed with other resistances in the conduction heat-transfer Eq. (6-7).

Because of the low values of air film coefficients, especially with natural or free convection, the amount of heat transfer by thermal radiation may be equal to or greater than that by convection. This may be especially true in air spaces in building construction.

Thermal radiation is the transfer of thermal energy by electromagnetic waves and is an entirely different phenomenon from conduction and convection. In fact, thermal radiation can occur in a perfect vacuum and is actually impeded by an intervening medium between the two surfaces of unequal temperature.

The net transfer of energy by radiation from a warm to a cool body is expressed as

$$\dot{q}_r = \sigma(T_1^4 - T_2^4)(A)(e_e)(F_a) \text{ (Btu/hr or W)} \qquad (6-11)$$

where

σ = Boltzmann constant (0.1713 × 10⁻⁸ Btu/hr-ft²-°R⁴ or 5.673 × 10⁻⁸ W/m²-K⁴)

T_1, T_2 = absolute temperatures of warm and cool surfaces, respectively (°R or K)

A = surface area (ft² or m²)

e_e = effective absorptivity or emissivity factor, expressing the degree to which the two surfaces approach an ''ideal black body'' (an ideal black body is one that could absorb [or emit] all the radiant energy falling on it)

F_a = factor to account for the geometric configuration between the radiating surfaces (fortunately, $F_a = 1$ for most applications)

Values of effective emissivity, e_e, are dependent on the emissivity of each of the surfaces. For infinite parallel planes or a completely enclosed body, large compared with enclosing body,

$$e_e = \frac{1}{1/e_1 + 1/e_2 - 1}$$

For a completely enclosed body, small compared with enclosing body,

$$e_e = e_1$$

Emissivity or absorptivity values depend somewhat on the temperature of the surface and may be found in heat-transfer references.

Although Eq. (6-11) is a suitable relationship for describing radiant heat exchange, it is not convenient for computation where other modes of heat transfer are in operation. For such cases, it is convenient to define an *equivalent conductance for radiation* by the expression

$$\dot{q}_r = h_r A(t_1 - t_2) \qquad (6-12)$$

The unit conductance, h_r, thus defined is a function of the shape- or emissivity factor, as well as the temperatures of the radiator and receiver.

The calculation of heat transfer by convection and radiation may be made by use of equations and tables that apply; but because of uncertainties, theory and experiments have been combined to treat convection and radiation as a single combined process for fluid films and air spaces. The process is expressed by

$$\dot{q}_{rc} = h_{rc} A(t_1 - t_2) \qquad (6-13)$$

where

$\dot{q}_{rc}$ = heat transfer due to combined radiation and convection (Btu/hr or W)

h_{rc} = conductance of surface film or air space for combined radiation and convection (Btu/hr-ft²-°F or W/m²-°C)

In this text, as was noted in Section 6-3, the fluid film conductance will be used as *f* and the air-space conductance will be used as $1/R_a$ to conform with the usual data presented for HVAC work, and both have been presented in Tables C-2 and C-3 in the Appendix.

6-6 OVERALL HEAT-TRANSMISSION COEFFICIENT (*U*)

Figure 6-4 represents a composite wall section similar to Figure 6-2 except that fluid (air) films have been added and an air space has been included in the construction. We may write the overall heat-transfer equation now from the fluid on one side to the fluid on the other side using the same form as Eq. (6-7), and we have

$$\frac{\dot{q}}{A} = \frac{t_i - t_o}{1/f_i + x_2/k_2 + 1/C_a + x_1/k_1 + 1/C_1 + 1/f_o}$$

$$= \frac{\Delta t}{R_t} \qquad (6-14)$$

where

t_i, t_o = inside and outside air temperatures, respectively (°F or °C)

f_i, f_o = inside and outside film coefficients (Btu/hr-ft²-°F or W/m²-°C)

C_a = conductance of the air space (Btu/hr-ft²-°F or W/m²-°C)

C_1 = conductance of a nonhomogeneous material (Btu/hr-ft²-°F or W/m²-°C)

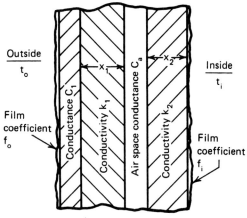

FIGURE 6-4 Composite wall section.

As with Eq. (6-7), each term in the denominator of the right side is the unit thermal resistance offered by the individual components, the sum being the total unit thermal resistance, R_t.

It is convenient in calculating heat transfer to use the *overall heat-transmission coefficient,* called the *U factor,* for composite building sections. *U* is defined as $1/R_t$, and Eq. (6-14) becomes

$$q = UA\Delta t \qquad (6-15)$$

where

$\dot{q}$ = heat transfer (Btu/hr or W)

U = overall heat-transmission coefficient (Btu/hr-ft²-°F) or W/m²-°C)

A = surface area of section considered (ft² or m²)

Δt = temperature change across the section from fluid to fluid (°F or °C)

6-7 CALCULATION OF HEAT-TRANSMISSION COEFFICIENTS

Appendix Tables C-1 through C-3 give representative values of unit thermal resistances for structural materials, air films, and air spaces. Because there are variations in commercially available materials of the same type, not all values shown will be in exact agreement with the data of individual manufacturers. The exact value for a certain manufacturer's product can be secured from unbiased tests or from guaranteed data given by the manufacturer.

The most exact method of determining the heat-transmission coefficient (*U*) for a given combination of building materials assembled as a building section is to test a representative section in a guarded hotbox. However, such a test is difficult and expensive, and it is usually not practical to test all such sections of interest in building construction.

Experience has shown that *U*-values that are calculated using data such as that presented in Appendix Tables C-1 through C-3 are in good agreement with values determined by guarded hotbox measurements.

As may be noted, the denominator of Eq. (6-14) represents the sum of the individual unit thermal resistances in series for a particular heat flow path. The reciprocal of this total thermal resistance is the heat-transmission coefficient (*U*) for that heat flow path.

ILLUSTRATIVE PROBLEM 6-2

Determine the total thermal resistance (R_t) and overall heat-transmission coefficient (*U*) for the exterior vertical wall of a building that has the structural components listed in the results

Results Table—Illustrative Problem 6-2

Component	Unit Resistance, R
Interior surface film (Appendix Table C-2a), still air, vertical surface, heat flow horizontal, nonreflective, $f_i = 1.46$	$1/f_i = 0.68$
$\frac{1}{2}$-in. cement plaster, sand aggregate (Table C-1, $k = 5.0$)	$x/k = 0.5/5.0 = 0.10$
$\frac{3}{8}$-in. gypsum plasterboard (Table C-1, $C = 3.1$)	$1/C = 1/3.1 = 0.32$
2-by-4 nominal stud space, providing a $3\frac{1}{2}$-in. vertical air space, nonreflective (Table C-3c, assume mean temp. = 50°F, temp. diff. = 30°F, and $E = 0.85$)	$= 0.91$
$\frac{5}{8}$-in. fir plywood sheathing (Table C-1, $C = 1.29$)	$1/C = 1/1.29 = 0.77$
Building paper	$= 0.00$
Siding, asbestos cement, 0.25 in., lapped (Table C-1, $C = 4.76$)	$1/C = 1/4.76 = 0.21$
Outside air film, assume 15 mph wind (Table C-2a, $f_o = 6.00$)	$1/f_o = \underline{0.17}$
	$R_t = 3.16$

Thus,

$$U = \frac{1}{R_t} = \frac{1}{3.16} = 0.316 \text{ (say, 0.32) Btu/hr-ft}^2\text{-°F}$$

table above (assume winter conditions with $t_o = 0°F$ and $t_i = 70°F$).

Solution: The results table for Illustrative Problem 6-2 appears above.

ILLUSTRATIVE PROBLEM 6-3

Assume that the wall section of Illustrative Problem 6-2 has a 2-in.-thick, foil-faced, mineral wool, batt-type insulation installed on the inside portion of the air space between the 2-by-4 wood studs. The manufacturer of the insulation quotes an R=value of 7.0 for the insulation. Determine the total thermal resistance and U for the insulated wall.

Solution: The air space between the studs has now been reduced to $1\frac{1}{2}$ in. (approximately), and one side of the air space has a reflective surface. From Appendix Table C-2b, the effective emittance, E, is 0.05. From Table C-3b, with $E = 0.05$, we will assume that the mean temperature in the new air space is 0°F and that the temperature difference is 20°F. These assumptions give a new $R_a = 2.66$. Therefore,

From Illustrative Problem 6-2	$R_t =$	3.16
Less warm air space	$=$	-0.91
Plus cold air space	$=$	$+2.66$
Plus insulation	$=$	$+7.00$
	$R_t =$	$\overline{11.91}$

Thus,

$$U = \frac{1}{R_t} = \frac{1}{11.91} = 0.0839 \text{ (say, 0.084) Btu/hr-ft}^2\text{-°F}$$

Note that, in Illustrative Problems 6-2 and 6-3, it was necessary to assume an air-space mean temperature and a temperature difference across the air space. To be accurate, these assumptions should be checked and a new R_a selected when it is judged necessary. We must know the inside and outside design temperatures and the manner in which the temperature changes through the wall. This will be shown in a later section.

6-8   SERIES AND PARALLEL HEAT CONDUCTION

In practical applications of heat conduction, the heat may flow through the materials involved by series or parallel paths or by a combination of both. An analysis of heat conduction in this respect is similar to that for the conduction of electricity through series and parallel circuits. In a series heat flow path, the heat resistances of the materials involved are additive, or

$$R_t = R_1 + R_2 + R_3 + \cdots \qquad (6\text{-}16)$$

Since unit conductance is the reciprocal of unit resistance,

$$1/C_t = 1/C_1 + 1/C_2 + 1/C_3 + \cdots \qquad (6\text{-}17)$$

or, for area A,

$$1/C_t = x_1/k_1 + x_2/k_2 + x_3/k_3 + \cdots \qquad (6\text{-}18)$$

When heat is conducted through parallel paths, the conductances are additive, or

$$C_t = C_1 + C_2 + C_3 + \cdots \qquad (6\text{-}19)$$

For an area consisting of $A_1 + A_2 + A_3 + \ldots$,

$$C_t = k_1 A_1/x_1 + k_2 A_2/x_2 + k_3 A_3/x_3 + \cdots (6\text{-}20)$$

If C_t is desired on a per-unit area basis, A_1, A_2, A_3, and so on are taken as percentages of the total typical area and expressed as decimals in Eq. (6-20).

For some situations, each parallel path may be considered to extend from inside to outside, and the U factor of each path may be calculated. The average U factor is then

$$U_{\text{ave}} = a U_a + b U_b + \cdots + n U_n \qquad (6\text{-}21)$$

where $a, b, \ldots, n$ are respective fractions of a typical basic area composed of several different paths whose U factors are U_a, U_b, and so forth.

ILLUSTRATIVE PROBLEM 6-4

Figure 6-5 shows a section through a wood stud wall of a building. The studs are nominal 2-by-6, 16 in. on center. The stud space is filled with loose-fill insulation that has a conductivity $k = 0.27$ Btu-in./hr-ft²-°F. Calculate the average U for the wall.

Solution:

Step 1. Determine the conductivity and conductance for each component of the wall section.

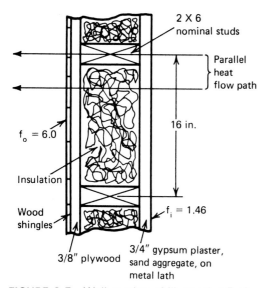

FIGURE 6-5 Wall section of Illustrative Problems 6-4 and 6-5.

Component	Conductivity, k or Conductance, C
Outside air film	$f = 6.00$
Inside air film	$f = 1.46$
Wood shingles, 16 in., 7.5-in. exposure (Appendix Table C-1, $C = 1.15$)	$C = 1.15$
Fir plywood, $\frac{3}{8}$ in. (Table C-1, $C = 2.13$)	$C = 2.13$
Insulation ($k = 0.27$ given)	$k = 0.27$
Gypsum plaster on metal lath, $\frac{3}{4}$ in., sand aggregate (Table C-1, $C = 7.70$)	$C = 7.70$
Wood studs (Table C-1, use $k = 0.80$)	$k = 0.80$

Step 2. Determine the total unit conductance of the stud space. The 16-in. center-to-center spacing of the studs gives 1.5/16 of the width occupied by one stud and 14.5/16 of the width occupied by insulation. With $k_1 = 0.80$ for the stud and $k_2 = 0.27$ for the insulation and using Eq. (6-20), we have the conductance of stud space as

$$C_{sp} = k_1 A_1/x_1 + k_2 A_2/x_2$$
$$= (0.80)(1.5/16)/5.5 + (0.27)(14.5/16)/5.5$$
$$= 0.0581 \text{ (say, 0.058)}$$

Step 3. Calculate the average U for the wall section.

Wall Component		Unit Resistance, R
Outside air film	$1/f = 1/6.00$	$= 0.17$
Wood shingles	$1/C = 1/1.15$	$= 0.87$
Fir plywood, $\frac{3}{8}$ in.	$1/C = 1/2.13$	$= 0.47$
Stud space	$1/C = 1/0.058$	$= 17.24$
Plaster, $\frac{3}{4}$ in.	$1/C = 1/7.70$	$= 0.13$
Inside air film	$1/f = 1/1.46$	$= 0.68$
		$R_t = 19.56$

Thus,

$$U = \frac{1}{R_t} = \frac{1}{19.56} = 0.051 \text{ Btu/hr-ft}^2\text{-}°F$$

If the conductance through the wood studs were neglected, the total wall resistance would be $19.56 - 17.24 + 5.5/0.27 = 22.69$, and the corresponding U would be $1/22.69 = 0.044$ Btu/hr-ft²-°F, which is approximately 14% lower when the parallel heat flow path is neglected.

ILLUSTRATIVE PROBLEM 6-5

Using Figure 6-5 and the data of Illustrative Problem 6-4, calculate U using Eq. (6-21).

Solution:

Step 1. Determine U through insulated space. This was calculated in step 3 of Illustrative Problem 6-4 and found to be 22.69, and the corresponding U was 0.044.

Step 2. Determine U through wood studs. The total unit resistance through the studs would be the total resistance found in step 3 of Illustrative Problem 6-4 minus stud space resistance plus stud resistance, or $(R_t)_s = 19.56 - 17.24 + 5.5/0.80 = 9.195$. Thus,

$$U_s = \frac{1}{9.195} = 0.108 \text{ (say, 0.11) Btu/hr-ft}^2\text{-°F}$$

Step 3. Determine the average U for the section. Using Eq. (6-21), we have

$$U = (0.044)(14.5)/16 + (0.11)(1.5)/16$$

$$= 0.0502 \text{ (say, 0.050) Btu/hr-ft}^2\text{-°F}$$

6-9 TABULATED HEAT-TRANSMISSION COEFFICIENTS OF BUILDING SECTIONS

For the convenience of the HVAC design engineer, tables have been constructed that give overall heat-transmission coefficients, U-values, for many typical building sections.

Walls, Floors, and Roofs. Appendix Table C-7 is a typical table showing various combinations of structural materials used in walls, floors, and roofs. Table C-7 is from *Load Calculation Digest, Commercial/Industrial Air Conditioning* published by General Electric Company and was constructed from procedures and data contained in *ASHRAE Handbook and Project Directory 1977 Fundamentals.* Table C-7 is intended to have a great deal of flexibility. The tabulated U-values are based on conductivity and conductance values similar to those shown in Appendix Tables C-1 through C-3. The following conditions were used:

- Equilibrium or steady-state heat transfer, eliminating effects of heat capacity.
- Surrounding surfaces at ambient air temperatures.
- Exterior wind velocity of 15 mph.
- Surface emissivity of ordinary building materials at 0.90.
- Stud space in wood frame construction not insulated.
- In construction involving air spaces, U-values shown calculated for areas between framing.
- Air spaces of $\frac{3}{4}$ in. or more in width.
- Variations in conductivity with mean temperature neglected.

It should also be noted that the effects of poor workmanship in construction and installation have an increasingly greater effect on heat transmission as the U-value becomes numerically smaller. A factor of safety may be applied as a precaution where it is judged desirable.

6-10 TEMPERATURE GRADIENT IN A WALL SECTION

The temperature in a wall section will vary from the inside to the outside surface. As indicated by Eq. (6-6), the temperature drop through the various components of a building structure are proportional to the thermal resistance. For a wall section, we may say that

$$\frac{R_x}{R_t} = \frac{t_i - t_x}{t_i - t_o}$$

or

$$t_x = t_i - \frac{R_x}{R_t}(t_i - t_o) \qquad (6\text{-}22)$$

where

t_x = temperature at selected point in construction
t_i = inside air temperature
t_o = outside air temperature
R_x = sum of unit thermal resistance from inside air to selected point
R_t = total unit thermal resistance of section

The usual point of interest in a wall section would be the inside wall surface temperature. This surface temperature would be lower than the inside air temperature because of resistance offered by the inside air film. If the inside surface temperature is at or below the inside air dew-point temperature, condensation may occur on the surface. Condensation on interior surfaces may cause damage to plaster and woodwork, as well as constituting a nuisance. In winter, insulated walls are effective in reducing condensation problems, but if walls cannot be effectively insulated, the inside air relative humidity must be reduced to lower its dewpoint temperature. Even if the condensation problem does not exist, people may feel a chill when seated next to a cold exterior wall because of radiation heat loss from their body to the cold wall surface.

ILLUSTRATIVE PROBLEM 6-6

The exterior wall of a building is constructed of 4-in. nominal face brick, $\frac{1}{2}$-in. cement mortar, 8-in. concrete block (three-holed oval-cored, sand-and-gravel aggregate), $\frac{5}{8}$-in. gypsum plaster (sand aggregate) on the inside surface. The inside air temperature is 70°F DB and 58°F WB. If the outside air dry-bulb temperature is −20°F, calculate (1) the inside surface temperature of the wall and (2) whether moisture will condense on the inside wall surface.

Wall Component	Resistance, R
Outside air film (assume 15 mph wind, Table C-2a, f_o = 6.00)	1/6.00 = 0.17
Face brick (Table C-1, k = 9.0)	4/9.00 = 0.44
Cement mortar (Table C-1, k = 5.00)	0.5/5.00 = 0.10
Concrete block (Table C-1, C = 0.90)	1/0.90 = 1.11
Gypsum plaster (Table C-1, C = 9.1)	1/9.10 = 0.11
Inside air film (assume still air, Table C-2a, f_i = 1.46)	1/1.46 = 0.68
	R_t = 2.61

Solution:

Since the only thermal resistance between the inside air and the wall surface is caused by the inside air film R_x = 0.68, Eq. (6-22) may be used to find the inside wall surface temperature:

$$t_x = t_i - \frac{R_x}{R_t}(t_i - t_o)$$

$$= 70 - \frac{0.68}{2.61}[70 - (-20)]$$

$$= 46.6°F \text{ (inside wall surface temperature)}$$

Referring to a psychrometric chart at 70°F DB and 58°F WB, we find that the dew point is 49°F, which is above the calculated wall surface temperature. Therefore, condensation will probably form on the wall surface.

6-11 SERIES HEAT FLOW THROUGH UNEQUAL AREAS

Frequently, unheated and unventilated spaces exist in a structure where the construction surrounding the space may be made up of two or more layers (flat or of small curvature) of unequal area. Heat flows through the layers in series. The most common such construction is a ceiling–roof combination where the attic space is unheated and unventilated. Figure 6-6 illustrates an attic space formed by a pitched roof and also a suspended ceiling below a flat roof. Another similar situation exists in the crawl space between the floor and ground when footing walls are used.

The air temperature of these unheated spaces would be some value between the inside and outside air temperature and may be estimated, assuming steady-state heat transfer, by

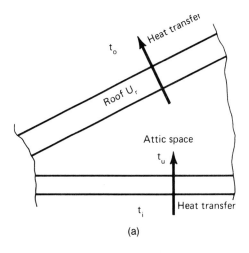

(a)

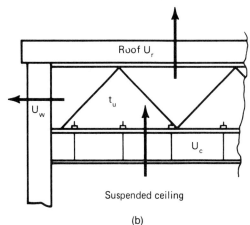

(b)

FIGURE 6-6 Typical unheated spaces in buildings: (a) pitched roof and (b) flat roof.

$$t_u = \frac{\begin{aligned}&t_i(A_1U_1 + A_2U_2 + \cdots)\\&+ t_o(A_aU_a + A_bU_b + \cdots)\end{aligned}}{\begin{aligned}&(A_1U_1 + A_2U_2 + \cdots)\\&+ (A_aU_a + A_bU_b + \cdots)\end{aligned}} \quad (6\text{-}23)$$

where

t_u = estimated air temperature in unheated space

t_i, t_o = inside and outside air temperature, respectively

Number subscripts refer to heated-to-unheated areas. Letter subscripts refer to unheated-to-outside areas.

ILLUSTRATIVE PROBLEM 6-7

Estimate the temperature in the unheated space shown in Figure 6-7. The structure is built on a slab. Exterior walls of the unheated space have a U factor of 0.25 Btu/hr-ft²-°F; roof-ceiling construction has a U factor of 0.10; interior walls have a U factor of 0.08. The inside heated spaces are maintained at 74°F, and the outside air temperature is −10°F.

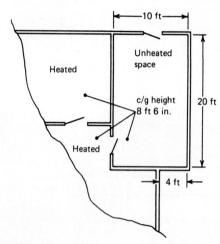

FIGURE 6-7 Sketch for Illustrative Problem 6-7.

Solution: Assume no heat loss through the slab floor because of expected low value of t_u. Neglect doors.

$$U_1 = 0.08 \text{ interior walls}$$
$$U_a = 0.25 \text{ exterior walls}$$
$$U_b = 0.10 \text{ ceiling–roof combination}$$
$$A_1 = (20 + 6)(8.5) = 221 \text{ ft}^2$$
$$A_a = (10 + 20 + 4)(8.5) = 289 \text{ ft}^2$$
$$A_b = (20)(10) = 200 \text{ ft}^2$$

By Eq. (6-23),

$$t_u = \frac{t_i(A_1U_1) + t_o(A_aU_a + A_bU_b)}{A_1U_1 + A_aU_a + A_bU_b}$$

$$= \frac{74(221)(0.08) + (-10)[(289)(0.25) + (200)(0.10)]}{(221)(0.08) + (289)(0.25) + (200)(0.10)}$$

$$= \frac{385.82}{109.93} = 3.5°F \text{ (estimated space temperature)}$$

The original assumption of no heat transfer through the floor slab is justified because of the low space temperature determined.

It is convenient in some situations to determine a combined heat-transmission coefficient for such spaces. The combined U factor would be based on the most convenient area from the air inside to the air outside and can be calculated from

$$R_t = \frac{1}{U_1} + \frac{1}{n_2U_2} + \frac{1}{n_3U_3} + \cdots + \frac{1}{n_pU_p} \quad (6-24)$$

The combined coefficient, U_t, is the reciprocal of R_t, or

$$U_t = \frac{1}{R_t}$$

where

U_t = combined coefficient to be used with A_1

R_t = total resistance of all elements in series

$U_1, U_2, \ldots, U_p$ = coefficients of transmission of A_1, $A_2, \ldots, A_p$, respectively

$n_2, n_3, \ldots, n_p$ = area ratios A_2/A_1, A_3/A_1, $\ldots$, A_p/A_1, respectively

Note that the combined coefficient U_t should be multiplied by the area A_1 and the overall temperature difference $t_i - t_o$ to determine the heat loss.

ILLUSTRATIVE PROBLEM 6-8

The top-story ceiling of a building measures 30 ft by 45 ft. An unventilated attic is formed by a pitched roof, which has an area 1.6 times the ceiling area. The ceiling U factor is 0.09 Btu/hr-ft²-°F; the roof U factor is 0.58. The air temperature just below the ceiling is 78°F, and the outside air temperature is 5°F. Find (1) the estimated attic temperature, (2) the ceiling–roof overall heat-transmission coefficient based on ceiling area, and (3) the heat loss from the building through the ceiling–roof construction in Btu/hr.

Solution:
1. Determine attic temperature. By Eq. (6-23),

$$t_u = \frac{t_i(A_1U_1) + t_o(A_aU_a)}{A_1U_1 + A_aU_a}$$

Letting

$$A_1 = A_c$$
$$A_a = A_r = 1.6A_c$$
$$U_1 = U_c \text{ (ceiling)}$$
$$U_a = U_r \text{ (roof)}$$

we have

$$t_u = \frac{t_i(A_c)(U_c) + t_o(1.6A_c)(U_r)}{A_cU_c + (1.6A_c)U_r}$$

A_c cancels and, by substitution,

$$t_u = \frac{(78)(0.09) + (5)(1.6)(0.58)}{(0.09) + (1.6)(0.58)}$$

$$= 11.45°F \text{ (estimated attic temperature)}$$

2. By Eq. (6-24),

$$R_t = \frac{1}{U_1} + \frac{1}{n_2U_2}$$

with

$$U_1 = 0.09$$
$$U_2 = 0.58$$
$$n_2 = 1.6$$

we have

$$R_t = \frac{1}{0.09} + \frac{1}{(1.6)(0.58)}$$

$$= 12.19$$

$U_t = 0.082$ Btu/hr-ft²-°F (combined U factor for ceiling–roof combination based on ceiling area)

3. Determine heat loss through ceiling–roof construction based on ceiling area A_c:

$$\dot{q} = U_t A_c(t_i - t_o)$$
$$= 0.082(30 \times 45)(78 - 5)$$
$$= 8081 \text{ Btu/hr}$$

In ASHRAE Standard 90.1, *Energy-Efficient Design for New Buildings,* requirements are stated in terms of U, where U is the combined thermal transmittance of the respective areas of gross exterior wall, roof–ceiling, and floor assemblies. The U equation for a wall could be

$$U_o = \frac{U_{\text{wall}}A_{\text{wall}} + U_{\text{window}}A_{\text{window}} + U_{\text{door}}A_{\text{door}}}{A_o} \quad (6\text{-}25)$$

where

U_o = average thermal transmittance of gross wall area (Btu/hr-ft²-°F)

A_o = gross area of exterior walls (ft²)

U_{wall} = thermal transmittance of all elements of opaque wall area (Btu/hr-ft²-°F)

A_{wall} = opaque wall area (ft²)

U_{window} = thermal transmittance of window area (Btu/hr-ft²-°F)

A_{window} = window area, including sash (ft²)

U_{door} = thermal transmittance of door area (Btu/hr-ft²-°F)

A_{door} = door area (ft²)

Where more than one type of wall, window, and/or door is used, the $U \times A$ term for that exposure shall be expanded to include its subelements as

$$U_{\text{wall}_1} \times A_{\text{wall}_1} + U_{\text{wall}_2} \times A_{\text{wall}_2} + \cdots$$

6-12 FENESTRATION

Fenestration refers to any glazed aperture in the building envelope, such as windows, skylights, and light-transmitting partitions. Fenestration components include (1) glazing materials; (2) framing, mullions, muntins, and dividers; (3) external shading devices; (4) inter-nal shading devices; and (5) integral (between-glass) shading systems.

Fenestration (1) satisfies human needs for visual communication with the outside world; (2) admits solar radiation to provide supplemental daylight and heat, and, in some cases, outside air; (3) provides egress in low-rise buildings in case of fire or other emergencies; and (4) enhances the exterior and interior appearance of a building.

The overall heat-transmission coefficient (U_w) for fenestration units may be found in Appendix Tables C-4 and C-5. *Note:* It will be necessary to study the Supplementary Notes for Tables C-4 and C-5 to be able to select the proper U-value for a given application and to understand the limitations placed on the table values.

ILLUSTRATIVE PROBLEM 6-9

Determine the overall U-value for a vertical, double-glazed window with $\frac{3}{8}$-in. air space, wood frame, and overall size of 3 ft by 4 ft.

Solution: Assuming product type R, the overall U-value from Appendix Table C-4 is 0.50 Btu/hr-ft²-°F.

ILLUSTRATIVE PROBLEM 6-10

Calculate the overall U-value for a skylight that has a double-glazed, 0.25-in. acrylic dome with a 0.75-in. air space and a thermally broken aluminum frame installed on a roof with a 5-by-12 slope.

Solution: Assuming product type R, from Appendix Table C-4, Part A, the U-value for a thermally broken aluminum frame, double-glazing, 0.25-in. acrylic, and 0.50-in. or greater air space is 0.60 Btu/hr-ft²-°F if mounted vertically. From Part B of Table C-4 for a 5-by-12 slope (23°), the resulting overall U is 0.72 Btu/hr-ft²-°F.

ILLUSTRATIVE PROBLEM 6-11

Calculate the overall U-value for a wood-framed, 30-by-80-in. swinging French door with eight 11-by-16-in. glass panes, each consisting of clear double-glazing with a 0.50-in. air space.

Solution: Refer to Figure 6-8, and, for space allowances, see Appendix Figure C-1. The areas of center-of-glass, edge-of-glass, and frame are as follows:

center-of-glass = [(11 − 5)(16 − 5)]8 = 528 in.²

edge-of-glass = [(2.5)(16)(2) + (2.5)(6)(2)]8 = 880 in.²

gross door area = (30)(80) = 2400 in.²

frame area = 2400 − (880 + 528) = 992 in.²

% center-of-glass = 528/2400 = 22%

% edge-of-glass = 880/2400 = 37%

% frame = 992/2400 = 41%

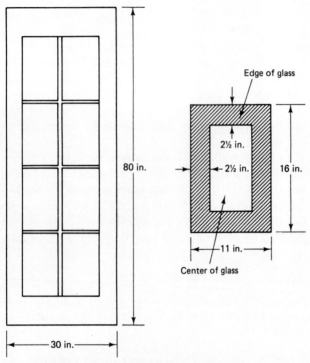

80 in.

30 in.

Edge of glass

2½ in.

2½ in.

16 in.

11 in.

Center of glass

FIGURE 6-8 Door layout for Illustrative Problem 6-11.

From Appendix Table C-4, find $U_f = 0.4$, $U_{eg} = 0.59$, and $U_{cg} = 0.49$. Using Eq. (b) in the Supplementary Notes for Table C-4, the weighted average overall U-value is

$$U = (0.49)(0.22) + (0.59)(0.37) + (0.40)(0.41)$$
$$= 0.49 \text{ Btu/hr-ft}^2\text{-}°\text{F}$$

6-13 SLAB DOORS

The U-values shown in Appendix Table C-6 are for exterior wood and steel doors. The values for wood doors were calculated and those for steel doors were taken from hotbox tests (Sabine et al. 1975; Yellott 1965) or from manufacturers' test reports. An outdoor surface conductance of 6.00 Btu/hr-ft²-°F was used, and the indoor surface conductance was taken as 1.46 Btu/hr-ft²-°R for vertical surfaces with horizontal heat flow. All values given are for exterior doors without glazing. If an exterior door contains glazing, the glazing should be analyzed as a window.

REVIEW PROBLEMS

6.1. An insulating material has a thermal conductivity (k) of 0.27 Btu-in./hr-ft²-°F. How many inches of the material should a contractor install if specifications call for an insulation resistance (R) of 14?

6.2. The section of a masonry wall is shown in Figure 6-9. Calculate the overall thermal resistance and U factor for winter conditions. The wall construction consists of 4 in. of face brick, $\frac{1}{2}$ in. of cement mortar, 8-in. sand-and-gravel concrete block (three-hole oval-cored), $1\frac{1}{2}$-by-2-in. furring strips on block surface, and metal lath and plaster (sand aggregate) attached to the furring strips. (Neglect resistance of furring strips.)

6.3. A section of a wood frame wall is shown in Figure 6-10. Calculate the overall thermal resistance and U factor for winter conditions. The wall construction consists of lapped wood siding ($\frac{1}{2}$-by-8 in.), $\frac{1}{2}$-in. fir plywood, $3\frac{1}{2}$-in. stud space with 2-in. fiberglass batt insulation between studs, and $\frac{1}{2}$-in. gypsum wall board on inside surface. (Neglect resistance of wood studs.)

6.4. For the stud wall of Review Problem 6.3, correct the U factor for the presence of the wood studs if the studs are placed 16 in. on center.

6.5. The section of a ceiling–floor combination is shown in

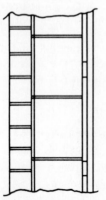

FIGURE 6-9 Sketch for Review Problem 6.2.

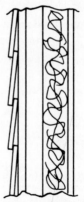

FIGURE 6-10 Sketch for Review Problems 6.3 and 6.4.

Figure 6-11. The floor is asphalt tile on 1-in. fir plywood nailed to 2-by-6-in. joists. The ceiling is $\frac{3}{4}$-in. metal lath and plaster with lightweight aggregate. Calculate the total thermal resistance and U factor on the basis of heat flow up to an unoccupied space above.

6.6. Consider the ceiling–floor combination of Review Problem 6.5, except with the ceiling above an unheated basement and with $\frac{1}{2}$-in. fir plywood nailed to the joists instead of the metal lath and plaster. Heat flow is down. Calculate the total thermal resistance and U factor for this combination.

6.7. A sloping (pitched) roof has asphalt shingles and felt laid on $\frac{5}{8}$-in plywood nailed to 2-by-6 nominal wood rafters placed 20 in. on center. Gypsum plasterboard ($\frac{1}{2}$-in.) is fastened directly to the underside of the rafters. Both sides of the air space formed between the rafters have a reflective surface. Neglecting the effect of the 2-by-6 rafters, calculate the total thermal resistance and U factor for the roof.

6.8. If the roof section of Review Problem 6.7 has ordinary building material surfaces in the air space and the effect of the rafters is included, calculate the total thermal resistance and U factor for the roof.

6.9. A masonry wall is built up of 4-in. face brick, $\frac{1}{2}$-in. cement mortar, 8-in. concrete block (sand-and-gravel aggregate, three-holed oval-cored) finished on the inside surface with metal lath and plaster on 2-in.-square furring strips. Lightweight aggregate in the plaster and the furring strips form a 2-in. air space. If the interior of the room is maintained at 70°F and the outside air is 5°F, calculate the estimated temperature on the interior wall surface. What relative humidity could be maintained in the interior space without condensation on the walls?

6.10. If the air space in the wall section of Review Problem 6.9 is filled with corkboard insulation (density of 8 lb/ft³), calculate the estimated temperature of the inside wall surface.

6.11. A pitched roof of a building has a calculated heat-transmission coefficient $U_r = 0.43$ and a horizontal ceiling heat-transmission coefficient $U_c = 0.15$. The ratio of roof area to ceiling area is 1.4. If the attic space is unvented, calculate the heat-transmission coefficient for the ceiling–roof combination based on the ceiling area.

6.12. Using Figure 6-12 and the following data, estimate the temperature in the unheated space. The exterior walls of the unheated space have a U factor of 0.25 Btu/hr-ft²-°F, the ceiling-attic structure has a U factor of 0.09

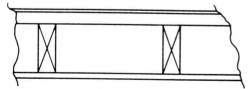

FIGURE 6-11 Sketch for Review Problems 6.5 and 6.6.

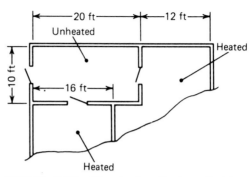

FIGURE 6-12 Sketch for Review Problem 6.12.

Btu/hr-ft²-°F, and the interior walls have a U factor of 0.06 Btu/hr-ft²-°F. Design temperatures are 70°F inside and −10°F outside. The ceiling height is 9 ft.

6.13. The exterior wall of a room in a commercial building has a U factor of 0.24 Btu/hr-ft²-°F and measures 20 ft long and 8 ft high. It contains a $1\frac{3}{4}$-in. solid-core wood door measuring 3 ft by 7 ft and two 48-by-30-in. single-glazed windows with thermally broken aluminum frames. Assuming parallel heat flow paths for the wall, door, and windows, determine the overall thermal resistance and U factor for the combination for winter conditions.

6.14. Determine the total thermal resistance and U factor for a floor section (heat flow down) constructed as follows: asphalt tile; $\frac{3}{4}$-in. fir plywood fastened to nominal 2-by-4-in. wood studs laid flat forming a $1\frac{1}{2}$-in. air space; 4-in. concrete slab, sand-and-gravel aggregate (oven-dried); 8-in. air space; and metal lath and gypsum plaster $\frac{3}{4}$-in. thick with sand aggregate.

6.15. Assume that a composite wall similar to that in Figure 6-2 is built up of 4-in. common brick, 8-in. concrete (sand-and-gravel aggregate), with $\frac{1}{2}$-in. cement plaster (sand aggregate) on the inside of the concrete. The inside air conditions are 76°F and 30% RH, and the outside air conditions are 20°F and 100% RH with 15-mph wind velocity.
  (a) Calculate the total thermal resistance of the wall section including the air film resistances.
  (b) Calculate the overall U factor for the section.
  (c) Calculate the temperature of the inside wall surface at design conditions.
  (d) Will condensation occur on the inside surface?

6.16. Determine the conduction heat loss through one exposed wall of a commercial building that measures 60 ft by 12 ft. The wall contains eight double-glazed windows with aluminum frames with no thermal break ($\frac{1}{4}$-in. air space). Each window measures 3 ft by 6 ft. The wall also contains three $2\frac{1}{4}$-in.-thick solid-core wood doors, each measuring 3 ft by 7 ft. The masonry wall has 4-in. face brick on the outside, backed with 6-in.-thick concrete (gravel aggregate), $1\frac{1}{2}$-in. air space formed by furring strips, and metal lath and gypsum plaster (lightweight aggregate) $\frac{3}{4}$-in. thick. The air temperatures inside and outside are 75°F and 10°F, respectively.

6.17. A nominal 2-by-4-in. wood stud wall is built up as follows: 4-in. face brick; $\frac{3}{4}$-in. air space (between brick and sheathing); $\frac{3}{8}$-in. fir plywood sheathing; $\frac{3}{4}$-in.-thick metal lath and gypsum plaster (sand aggregate).

(a) Determine the U factor for the stud wall, neglecting the effect of studs and assuming winter conditions.

(b) If the stud space contains batt-type insulation (paper-backed), determine the wall U factor.

(c) Correct the U factor of parts (a) and (b) for the presence of the wood studs placed 16 in. on center.

6.18. A flat masonry roof has $\frac{3}{8}$-in. built-up roofing on top of a 6-in. concrete deck (gravel aggregate, 140 lb/ft³, not dried). Inside is a suspended ceiling 18 in. below the deck. The ceiling is of $\frac{1}{2}$-in. acoustical tile (18 lb/ft³).

(a) Calculate the U factor for winter conditions.

(b) Determine the U factor if 1-in.-thick mineral fiberboard, wet-felted core insulation is added.

(c) How much additional thermal resistance, other than the 1-in. material of part (b), is needed if the overall U factor is to be 0.08 Btu/hr-ft²-°F?

6.19. One wall of a residence is 24 ft long and 8 ft high and has a U factor of 0.15 Btu/hr-ft²-°F. The wall contains a solid-core wood flush door $1\frac{1}{4}$-in. thick that measures 3 ft by 7 ft and two single-glazed wood-framed windows, each measuring 60 in. by 30 in. Assuming parallel heat flow paths for the wall, door, and windows, calculate the overall thermal resistance and U factor for the combination, assuming winter conditions.

6.20. An outside wall is constructed of 6-in. concrete (sand-and-gravel aggregate) not dried, 140 lb/ft³ density, and $\frac{1}{2}$-in.-thick gypsum board furred out 2 in. on wood furring strips.

(a) Determine the winter and summer U factors for this wall.

(b) Determine what thickness of cellular polyurethane insulation to use in the air space to produce an overall U factor of approximately 0.10 Btu/hr-ft²-°F.

6.21. A ceiling–floor combination contains the following materials: metal lath and gypsum plaster (lightweight aggregate) 0.75-in. thick; nominal 2-by-8-in. wood joists; 7.25-in. nonreflective air space; 0.75-in. wood subfloor; vapor-permeable felt; 0.625-in. fir plywood; and asphalt tile.

(a) Calculate the total thermal resistance through the ceiling–floor combination assuming heat flow upward (1) between the floor joist and (2) at the floor joist.

(b) If the floor tile, plywood, and subfloor are removed and R-19 insulation is placed between the floor joist, determine the total thermal resistance (1) between the floor joist and (2) at the floor joist.

(c) If the floor joists are placed 16 in. on center, determine the average U factor for parts (a) and (b).

6.22. A wood stud wall has the following materials in its construction: wood shingles (16 in., 7.5-in. exposure); $\frac{1}{2}$-in. fir plywood sheathing; nominal 2-by-4-in. studs; $\frac{1}{2}$-in. gypsum plasterboard; and $\frac{1}{2}$-in. gypsum plaster (lightweight aggregate). The inside and outside air temperatures are 70°F and −20°F, respectively. Calculate the total thermal resistance and U factor for the wall (uncorrected for studs).

6.23. If the wall section of Review Problem 6.22 has 2-in. foil-faced batt insulation placed on the inside surface of the stud space, determine the total thermal resistance and U factor for the wall (uncorrected for studs). The insulation material is mineral wool.

6.24. For the insulated stud wall of Review Problem 6.23, correct the U factor to account for the presence of the 2-by-4 wood studs placed 16 in. on center.

6.25. Determine the temperature on each side of the air space for the wall section of Review Problems 6.22 and 6.23. Using these values, check the values of the air-space mean temperature and the temperature drop across the air space and compare them with the assumed values used when solving those two problems. Calculate the wall inside surface temperature of the uninsulated and of the insulated wall section.

6.26. A room in a house has one wall facing an enclosed porch. The room wall measures 22 ft long by 8 ft high and contains one 2-ft-8-in. by 6-ft-8-in. wood door $1\frac{1}{4}$-in. thick and two double-hung, double-glazed wood-framed windows that measure 2-ft-6-in. by 4-ft-6-in. The wood door has a double-glazed glass panel that forms 70% of the door area. The air space in the door and windows is $\frac{1}{4}$ in. The wall is constructed of wood shingles (16 in., 7.5-in. exposure), $\frac{1}{2}$-in. medium-density particle board, 2-by-6-in. nominal wood studs, R-19 insulation in the stud space, and metal lath and $\frac{3}{4}$-in. gypsum plaster (lightweight aggregate). When the temperature in the enclosed porch is 30°F and the room temperature is 70°F, determine the heat loss from the room to the porch.

6.27. A masonry wall has the following construction materials: 4-in. face brick, backed by 8-in. precast concrete block (three-holed oval-cored, cinder aggregate); $1\frac{1}{2}$-in. air space provided by furring strips; and $\frac{3}{4}$-in. metal lath and gypsum plaster (lightweight aggregate) on the inside. Determine the total thermal resistance and U factor for the wall. Assume winter conditions.

6.28. Determine the U factor for the masonry wall of Review Problem 6.27 if the $1\frac{1}{2}$-in. air space is removed and $\frac{1}{2}$-in. gypsum plasterboard is installed using $\frac{3}{4}$-in. nailing strips for mounting instead of the metal lath and plaster. The plasterboard has a reflective surface facing the $\frac{3}{4}$-in. air space.

6.29. The outside wall of a building is constructed of 4-in. common brick, backed by 8-in. concrete (sand-and-gravel aggregate, dried), and $\frac{3}{4}$-in. cement plaster (sand aggregate) on the inside. The inside air temperature is maintained at 75°F DB and 65°F WB. How many inches of insulating slab (expanded perlite, organic-

bonded) would be required in the wall structure to prevent moisture condensation on the inside wall surface when the outside air temperature is −10°F?

BIBLIOGRAPHY

6.1. *ASHRAE Handbook 1989 Fundamentals,* American Society of Heating, Refrigerating, and Air Conditioning Engineers, Atlanta, GA, 1989.

6.2. *ASHRAE Handbook and Product Directory 1977 Fundamentals,* American Society of Heating, Refrigerating, and Air Conditioning Engineers, Atlanta, GA, 1977.

6.3. *ASHRAE Cooling and Heating Load Calculation Manual* (ASHRAE GRP 158), American Society of Heating, Refrigerating, and Air Conditioning Engineers, Atlanta, GA.

6.4. *Load Calculation Digest, Commercial/Industrial Air Conditioning,* General Electric Company, 1971.

6.5. *Trane Air Conditioning Manual,* The Trane Company, LaCrosse, WI, 1965.

6.6. Sabine, H. J., M. B. Lacher, D. R. Flynn, and T. L. Quindry. 1975. Acoustical and thermal performance of exterior residential walls, doors and windows. National Bureau of Standards, Building Science Series 77, November.

6.7. Yellott, J. I. 1965. Thermal and mechanical effects of solar radiation on steel doors. ASHRAE *Transactions* 71(2):42.

7

Heating and Ventilating Loads

7-1 INTRODUCTION

This chapter discusses accepted methods for evaluating the heating and ventilating loads for buildings. Regardless of the type of heating system selected for a particular building, the first basic requirement is an accurate determination of these loads. Rule-of-thumb methods should not be used, and the necessity of an accurate heating load calculation cannot be overemphasized. It is not just a question of the heating system having sufficient capacity to handle the design load, but of selecting the capacity required to obtain optimum performance. Also, an accurate determination of the loads may suggest to the HVAC engineer possibilities for load reduction and a better understanding of operation at partial loads.

Most heating systems operate intermittently at a fixed rate with short, infrequent cycles in mild weather and nearly continuously during the most severe weather. An ideal arrangement is to have the heating system capable of operating continuously with varying output that exactly matches the building heating load with climate changes. This ideal can be approached in careful design if an accurate calculation is made for the building heating load and if correctly sized heating equipment components are selected.

7-2 ELEMENTS OF THE HEATING AND VENTILATING LOADS

The elements that enter into the calculations of heating and ventilating loads for a building are (1) heat transmission through all surfaces exposed to unheated spaces; (2) heat required to warm the air that infiltrates through cracks around doors and windows and air that enters through doorways when people enter or leave the building; (3) heat required to warm air mechanically brought into the building as ventilation air (infiltration air should be distinguished from ventilation air); and (4) miscellaneous loads, such as humidification of the outside air from infiltration (and ventilation).

The design heating load calculation is usually based on a constant winter design temperature. These design temperatures normally occur at night, and no credit to the heating load is generally made for the heat given off by internal sources such as people, lights, and appliances. However, if an internal heat source is known to be relatively constant at all hours, credit may be taken, but good judgment must be used by the engineer. Solar heat gain during the day is not usually credited to a reduction of the design heating load because the heating equipment must have adequate capacity to meet the most adverse conditions normally expected, which generally occur at night.

Storage of heat by the building structure is a useful consideration in the selection of capacity for cooling equipment, but credit for storage in reducing the heating load for the purpose of heating equipment selection must be used with care. Such credit for storage will allow reduction in heating capacity requirements when a temperature swing is permitted for continuous operation only. Temperature swing allows the space temperature to drop a few degrees during periods at design load.

The practice of "turning down the heat" and allowing large drops in space temperature when a building is not occupied does not allow a reduction in heating capacity because of its noncontinuous nature. Although this practice may lead to energy savings based on total average heat requirements, it leads to an increase in heating equipment capacity to provide the "pickup" needed when heating operation is resumed. It may be desirable to provide additional capacity when intermittent operation is planned because of the imposed additional load from pickup.

7-3   GENERAL PROCEDURE FOR CALCULATING THE HEATING LOAD

Before a heating load can be properly estimated, a complete survey must be made of the physical data. Such information as orientation and dimension of building components; construction materials for roof, walls, ceiling, interior partitions, and fenestration; size and use of the spaces; and the surrounding conditions outdoors and in adjoining spaces are all important. The more exact such information is the more accurate the load estimation will be.

The general procedure for calculating heat losses from a structure is

1.   Select the design outdoor weather conditions: temperature, humidity, wind direction, and speed.
2.   Select the design indoor air temperature to be maintained in each heated space when outside design conditions exist.
3.   Estimate temperatures in adjacent unheated spaces.
4.   Select or compute the heat-transmission coefficient (*U* factors) for all building components through which heat losses are to be calculated.
5.   Determine all surface areas through which the heat is lost.
6.   Compute heat-transmission losses for each

kind of wall, glass, floor, ceiling, and roof in the building by multiplying the heat-transmission coefficient in each case by the area of the surface and the temperature difference between indoor and outdoor air or adjacent unheated space.

7.   Compute heat losses from basement or grade-level slab floors by methods presented in this chapter.
8.   Select infiltration air quantities and compute the heating load caused by infiltration around doors and windows.
9.   When positive ventilation using outdoor air is provided by an air-heating or air-conditioning unit, the energy required to warm the outdoor air to room temperature must be provided by the unit. The principle for calculation of this load component is identical to that for infiltration. If mechanical exhaust from the room is provided in an amount equal to the outdoor air drawn in by the unit, the unit must also provide for the natural infiltration losses. If no mechanical exhaust is used, and the outdoor ventilation air supplied equals or exceeds the natural infiltration that would occur without ventilation, natural infiltration may be neglected.
10.   The sum of the transmission losses or heat transmitted through the confining walls, floors, ceiling, glass, and other surfaces, plus the energy associated with the cold entering air by infiltration or required to replace mechanical exhaust, represents the total heating load.
11.   In buildings that have a reasonably steady internal heat release of appreciable magnitude from sources other than the heating system, a computation of this heat release under design conditions should be made and deducted from the total of the heat losses computed above.
12.   Additional heating capacity may be required for intermittently heated buildings to bring the temperature of the air, confining surfaces, and building contents to the design indoor temperature within a specified time.

7-4   DESIGN OUTDOOR WEATHER CONDITIONS

Prior to the calculation of the design heat loss from a structure, it is necessary to establish the design outside air temperature and weather condition for the area in which the structure will be located. The ideal heating system would provide enough heat to match the heat loss from the structure. Studies of weather records show that severe weather conditions do not repeat themselves each year, and weather conditions vary over the heating season. If the heating system were designed for the most severe conditions, the system would have a great excess of capacity during most of the time.

In many cases, occasional failure of a heating system to maintain a preselected indoor design temperature during brief periods of severe weather is not critical. However, the successful completion of some industrial or commercial processes may depend upon close regulation of indoor temperatures.

Table B-1 (see Appendix B) contains weather data for selected locations in the United States. The *ASHRAE Handbook 1989 Fundamentals* has a more complete tabulation of weather data.

Before selecting a design outdoor temperature from Table B-1, a heating engineer should consider each of the following questions:

1. Is the heat capacity of the structure high or low?
2. Is the structure insulated?
3. Is the structure exposed to high wind?
4. Is the ventilation or infiltration load high?
5. Is there more glass area than normal in the structure?
6. During what part of the day will the structure be used?
7. What is the nature of occupancy?
8. Will there be long periods of operation at reduced indoor temperature?
9. What is the amplitude between maximum and minimum daily temperature in the locality?
10. Are there local conditions that cause significant variation from temperatures reported by the weather bureau?
11. What auxiliary heating devices will be in the building?
12. What is the expected cost of fuel or energy?

The designer must keep in mind, before reaching a final decision on design outdoor temperature, that, if the design indoor–outdoor temperature difference is exceeded, the indoor temperature will probably fall. The question is to determine how large a drop in indoor temperature can be tolerated.

Finally, there is a factor, perhaps intangible, that should not be ignored. It is the performance expected by the owner of the system. The designer of the system should make clear to the owner the various factors considered in the design. In order to judge whether expected performance can be assured, the designer needs a full understanding of the basis on which the capacities of all the system components are derived or determined, the limits of accuracy of published performance data, and the accelerating capacity of certain types of equipment. There is no substitute for experienced engineering judgment in this type of problem.

Winter outside design dry-bulb temperatures are presented in Appendix Table B-1 for selected cities. The 97.5% values shown represent the frequency level that would be equaled or exceeded by 97.5% of the total hours in the months of December, January, and February (a total of 2160 hours) in the Northern Hemisphere. In a normal winter, there would be approximately 54 hours at or below the 97.5% value. Bibliographic reference 7.2 also lists a 99% value and design temperatures for many more locations. In general, the 97.5% temperature gives satisfactory design heat loss calculations.

7-5 DESIGN INDOOR AIR TEMPERATURES

The temperature to be maintained inside a structure is understood to be the dry-bulb temperature at the breathing line, which is 3 ft to 5 ft above the floor and at a location where the temperature-sensing device (thermostat) is not exposed to a condition of abnormal heat gain or heat loss. Design indoor air temperatures usually specified vary in accordance with the intended use of the building.

The main purpose of Chapter 5 was to define design indoor conditions that would make most of the occupants comfortable. The proper dry-bulb temperature to be maintained for comfort depends upon the relative humidity and air motion.

Table 7-1 gives representative design inside air temperatures for particular types of spaces. As may be seen, a range of temperatures is given in some cases.

In making the actual heat loss calculations for a structure, it is sometimes necessary to modify the design inside temperature so that the air temperature at the proper level within a space will be used. The air temperature within a given space will usually vary from floor to ceiling. The allowance for air temperature variation from floor to ceiling may be somewhat difficult to predict since it depends on many factors, primarily the type of heating system—that is, whether the air moves through the space by natural convection or by forced circulation. Circulation by natural convection has a tendency to cause a greater temperature differential floor-to-ceiling than does forced circulation.

It is impractical to set up rigid rules for determining temperature variations. However, normally if the floor-to-ceiling height is 12 ft or less, no variation is considered. For higher ceilings, an allowance of approximately 1.0% per foot of height above the 5-ft level may be made for ceilings up to 15 ft and approximately one-tenth of 1.0 degree per foot above this level. The temperature at the floor level may be 3 to 6 degrees lower in most cases when calculating heat loss through floors in high-ceiling spaces.

TABLE 7-1

Representative design inside air temperatures

Type of Building	(°F)	(°C)	Type of Building	(°F)	(°C)
Schools			*Theaters*		
Classrooms	70–72	21–23	Seating space	68–72	20–23
Assembly rooms	68–72	20–23	Lounge rooms	68–72	20–23
Gymnasiums	55–65	13–18	Toilets	68	20
Toilets and baths	70	21			
Locker rooms	65–68	18–20	*Hotels*		
Kitchens	66	19	Bedrooms and baths	70	21
Lunch rooms	65–70	18–21	Dining rooms	70	21
Playrooms	60–65	16–18	Kitchens	68	20
			Ballrooms	65–68	18–20
Hospitals			Public toilets	68	20
Private rooms	70–80	21–27			
Operating rooms	70–95	21–35			
Wards	68	20			
Kitchens	66	19			
Bathrooms	70–80	21–27			

The design inside air temperature may be different for various spaces within a given structure because of different requirements for these spaces. Normally, however, no distinction of 2 or 3 degrees should be made. Frequently, a single design inside temperature (say, 70°F) is used throughout the occupied areas of the structure (say, a home). Most heating systems can be readily adjusted and controlled to deliver heat to various spaces to maintain them at the individual temperature desired, even if a single design inside air temperature has been used to calculate the heat losses.

Temperatures in unheated spaces may be estimated by use of Eq. (6-23). The heat loss from heated rooms to unheated rooms or spaces must be based on this estimated or assumed temperature.

7-6 MEASUREMENTS FOR HEAT-TRANSMISSION AREAS

The basic equation for the calculation of heat transmission by conduction and convection through composite sections of a building structure was presented in Chapter 6 as

$$\dot{q} = UA\Delta t = UA(t_i - t_o) \qquad (6\text{-}15)$$

and we have discussed the methods used to determine the U factor, design inside air temperature t_i, and design outside air temperature t_o. The remaining quantity in Eq. (6-15) is the area A through which the heat is transferred. It is important to determine these areas with as much accuracy as possible.

The owner of the structure may indicate an inten-

tion of completing an unfinished space at some future time. In this case, it is necessary to obtain specific information as to how this space will be constructed and to estimate the heat losses from such a space. Rooms below a future attic room should be considered as having a cold ceiling because the attic space is unheated, and it may be some time before it is finished and heated. The heat loss of a crawl space or basement should be calculated and included in the total heat loss of the building. Garages are seldom heated with the central heating system that heats the living areas of a house, unless certain precautions are taken in the system design. If the unheated garage is attached to the house, a cold wall exists between the two building components.

When the heat loss of a new structure is to be calculated, the dimensions recorded on architectural plans of the building should be used to determine areas. If it is necessary to scale the dimensions, it is advisable to check at least one dimension for each direction on the sheet of the plans to be certain the scale is reasonably accurate. If errors are found, ask the architect whether the plans or the dimensions are correct.

It may be necessary to accurately measure spaces within an existing structure. Plans for existing structures should be used *only* after comparing them to the structure in the event that alterations or remodeling has occurred since the plans were originally drawn.

Inside dimensions of spaces are usually used in calculating areas. If measurements of an existing structure are being made, measure the distance from the surface of one wall to the surface of the opposite wall. If building plans are being used, it is sufficiently accurate to use the dimensions recorded on the plan, even though they show the distance from the center line of interior walls and to the outside of sheathing on exterior walls. Record these dimensions to the *nearest foot*. For example, if the distance is 14 ft 6 in., record 15 ft as the distance. Ceiling heights should be measured and recorded to the *nearest one-half foot*. For example, a ceiling height of 7 ft 9 in. should be recorded as 8 ft.

7-7 TYPES OF HEAT-TRANSMITTING AREAS

Heat loss calculations should begin with the *exposed walls*. An exposed wall is considered to be one that faces the outside weather or one that faces an unheated space. These areas should be determined as separate quantities. If a space has only one exposed wall, the length of the exposure is the linear feet of the exposed wall. If a space has two or more exposed walls, the total linear feet of exposed wall is the sum of the individual wall lengths.

The *gross exposed wall area* of a space is the length of exposed wall times the ceiling height. In many cases, two or more different wall constructions are used for a particular space. The gross exposed wall area is calculated separately for each construction. The actual height of the particular construction would be used instead of the ceiling height if the division between construction is horizontal. Record wall areas to the nearest square foot.

A closet is considered a part of the room into which it opens. The length of exposed wall, if any, is included in the linear feet of exposed wall of the room. If the closet has a cold ceiling or floor, it is also included in the heat loss calculations. Large closets or dressing rooms should be considered as separate rooms and should have separate heat loss calculations.

The presence of cabinets, shelves, and bookcases on exposed walls does not affect the heat loss calculation. Calculate the exposed wall area as if no shelves, bookcases, or cabinets existed.

When a stairway is located adjacent to a cold wall, the exposed area up to the ceiling of the second floor is included with that of the first floor hall or room. Unless a heating unit is to be installed in the second floor hall, the heat loss through the entire ceiling of the stairwell and the second floor hall is included with that of the first floor hall or room opening to the stairway.

Basement wall area above grade is calculated separately from basement wall area below grade. The heat loss per square foot of basement wall area above grade is considerably more than that below grade.

Windows and outside doors and other glass surfaces such as skylights should be measured inside their casings. These measurements should be recorded to the nearest 0.1 ft and areas should be recorded to the

nearest 0.1 ft^2. The *total* window, door, or skylight area should be recorded to the nearest square foot for heat loss calculations.

Most doors and windows come in stock sizes. Some architects specify the opening size for windows rather than glass size, but, in most drawings, the full opening dimensions are not specified. Instead, window sizes are specified as follows:

- Double-hung (DH), or casement, and so forth.
- Number of lights (LT) or panes.
- Dimensions in inches of each light or pane.

For example, a window specification that reads 9 × 12-12LT-DH refers to a double-hung window having 12 lights (panes), each 9 in. wide by 12 in. high. A specification that reads 26/24-2LT-DH indicates that there are two sashes in the double-hung window and the glass in each sash is 26 in. wide by 24 in. high. If there is no reference to the type of window, check with the architect or builder.

The opening size of a window is always greater than the glass size. The opening width is approximately 4 in. greater than the total glass width. The opening height is approximately 6 in. greater than the total glass height. For a 9 × 12-12LT-DH window, the total glass width would be 32 in. The glass height would be 25 in. per sash or 50 in. total. The opening height would be 56 in. A 1-in. allowance is normally made for the mullions between the panes. The recorded dimensions of the 9 × 12-12LT-DH window would be 2.7 ft × 4.7 ft (32 in. = 2 ft 8 in. or 2.7 ft; 56 in. = 4 ft 8 in. or 4.7 ft). Similarly, the opening size of the 26/24-2LT-DH window would be 30 in. (26 + 4) wide and 54 in. (24 + 24 + 6) high. The dimensions would be recorded as 2.5 ft × 4.5 ft.

When a *storm sash* is used, record the dimensions of the inside window, but indicate the presence of the storm sash. When other forms of double glazing are used, they should also be indicated.

The *fit* of a window does not affect the conducted heat loss, but the fit must be noted because it will greatly affect the calculation of infiltration, which we will discuss later in this chapter.

Net exposed wall area is the gross exposed wall area minus the exposed window and door areas. If two or more wall constructions are used in a space, care must be taken to subtract only areas of windows and doors that are actually located in each construction. Most designers distinguish between the heat loss through windows from that lost through doors. When this is done, obviously the door and window areas must be tabulated separately.

Cold partitions, floors, and ceilings separate

heated spaces and unheated spaces. A room has a cold ceiling when it is located directly beneath an attic, roof, or any unheated space. Thus, all spaces in a one-story building have cold ceilings. In a two-story building, all second-story spaces have cold ceilings. Always check elevation plans of a building to determine whether all of the ceiling should be considered cold, or note which portion of the ceiling area should be considered.

When a basement or a crawl space is heated, the heat loss from these spaces should be calculated, and the floor above it is *not* considered to be a cold floor. It is recommended that these spaces be heated, particularly in colder climates. In buildings with unheated basements, crawl spaces, or slab floors, the first-story floor is considered to be a cold floor. If a room or bay extends over a foundation wall, porch, breezeway, or other unheated space, that part of the floor exposed is considered as a cold floor exposed to outside temperatures. Floors of spaces located over an unheated garage should be considered as exposed to outside temperature. Cold-storage areas in a basement located under a first-story space also create a cold floor for the space above.

In a heated basement, the concrete floor is a cold floor. Although basement floors have a low heat loss per square foot, it is necessary to calculate this heat loss since it may be appreciable for a large area.

In areas having a mild winter climate, heat is seldom provided in or underneath floors located over crawl spaces, and adequate crawl-space ventilation must be provided. Heat loss through these floors can be reduced by insulating them, resulting in a warmer floor.

7-8 HEAT LOSS BY INFILTRATION

Outside air will leak into a building no matter how well constructed it may be, and an equal amount of conditioned air will leak out. This leakage of outside air into a building is called *infiltration*. A certain amount of fresh outdoor air is desirable and necessary, of course, but when possible it should be placed under positive control. Fresh outside air should be introduced to a building through a duct or ducts, provided specifically for that purpose, and should be conditioned before it is distributed to the interior spaces (mechanical ventilation).

Infiltration heat loss differs from other forms of heat loss. It is not conducted through the structural components of the building. The amount of heat necessary to condition the incoming infiltrated air to inside conditions is made up of the sensible heat to raise the air temperature and the latent heat to evaporate the re-

quired moisture to humidify the air. The heat required to raise the temperature of the infiltrated air is given by

$$\dot{q}_s = \dot{m}_{oa} C_p(t_i - t_o) \qquad (7\text{-}1)$$

where

$\dot{m}_{oa}$ = mass of infiltrated air (lbm/hr or kg/s)

C_p = specific heat of moist air (Btu/lbm-°F or J/kg-°C)

Infiltration is normally estimated on the basis of volume flow of air at outside conditions. Equation (7-1) then becomes

$$\dot{q}_s = \frac{\dot{Q}_{oa} C_p(t_i - t_o)}{v_{oa}} \qquad (7\text{-}1a)$$

where

$\dot{Q}_{oa}$ = volume flow rate (ft³/hr or m³/s)

v_{oa} = specific volume (ft³/lbm or m³/kg)

The latent heat required to humidify the infiltrated air is given by

$$\dot{q}_L = \dot{m}_{oa}(W_i - W_o)h_{fg} \qquad (7\text{-}2)$$

where

$W_i - W_o$ = difference in design humidity ratio (lbv/lbda or kgv/kgda)

h_{fg} = latent heat of vaporization at indoor conditions (Btu/lbv or j/kgv)

In terms of volume flow of air, Eq. (7-2) becomes

$$\dot{q}_L = \frac{\dot{Q}_{oa}}{v_{oa}} (W_i - W_o)h_{fg} \qquad (7\text{-}2a)$$

We will see that the heat loss calculated by Eqs. (7-1a) and (7-2a) may account for a large portion of the heating load.

7-9 ESTIMATION OF INFILTRATION AIR QUANTITIES

Outside air infiltration/exfiltration may account for a significant portion of the heating load for buildings. Thus, it is important to make an adequate estimate of its contribution to both the design load and the seasonal energy requirements. Air infiltration is also an important factor in determining the relative humidity that occurs in buildings or in establishing the amount of humidification required to maintain humidity levels during the heating season.

The rate of airflow into and out of a building due to infiltration, exfiltration, or natural ventilation depends on the magnitude of the pressure difference between the inside and outside of the structure and on the resistance to airflow offered by openings and interstices in the building. The pressure difference exerted on the building enclosure may be caused either by wind or by a difference in density of the inside and outside air. The effect of the difference in density, often called *chimney* or *stack effect,* is often a major factor contributing to air leakage. In winter, the warm inside air tends to rise and leak out of the upper portions of the structure, and it is replaced by colder, denser air that leaks into the lower portions of the structure. This effect is more pronounced in tall buildings than it is in one- or two-story buildings. Therefore, for residential-type buildings, only air leakage caused by wind pressure is considered.

The wind pressure may be calculated from

$$p_w = \frac{\rho V_w^2}{2g_c}$$

where ρ is the outside air density in lb/ft^3, V_w is wind velocity in ft/s, and p_w is velocity pressure in force per unit area. The velocity pressure head for a given wind speed for standard air at 0.075 lb/ft^3 density may be expressed as

$$p_w = 0.000482 V_w^2 \qquad (7\text{-}3)$$

where p_w is wind velocity pressure head in in. water gauge and V_w is wind speed in miles per hour. In SI units, Eq. (7-3) becomes

$$p_w = 0.00472 V_w^2 \qquad (7\text{-}3a)$$

where p_w is wind velocity pressure head in mm water gauge and V_w is wind speed in kilometers per hour.

Two methods are used to estimate the quantity of air infiltration in buildings. In one method, the estimate is based on measured leakage characteristics of building components and a selected pressure difference. This is known as the *crack method* since cracks around doors and windows are usually the major source of air leakage. The other method is called the *air-change method.* This method assumes that the air volume within a building is replaced by outside air a certain number of times each hour, the number of changes per hour presumably dependent on the type, use, and location of the space. The infiltration rate of an individual building depends

on weather conditions, equipment operation, and occupant activities. The rate can vary by a factor of 5 from weather effects alone.

ASHRAE states that typical infiltration values in housing in North America vary by a factor of 10, from tight housing with seasonal average air-change rates of 0.2 per hour to housing with air-change rates as great as 2.0 per hour. If the air-change rate method is used to estimate infiltration, Table 7-2 may be helpful; however, experience and good judgment are required to obtain satisfactory results.

The crack method for determining infiltration rates is most widely accepted by HVAC engineers because it is believed to be the more accurate method, providing that leakage characteristics and pressure differences can be properly evaluated. The method is based on the fact that the quantity of air that leaks through building components is roughly proportional to the length of crack, the width of crack, and the square root of the pressure difference across the crack. The pressure difference across the crack depends on the amount of pressurization within the building. Pressures inside the building due to wind action alone will depend on the resistance of cracks or openings and their location with respect to wind direction. In large buildings, the tightness of internal space separations also may be a factor. If the openings are uniformly distributed around the walls of a building, inside pressures will usually be within $\pm 0.2p_w$. If the openings on the windward side predominate, inside pressures up to $+0.8p_w$ may occur, approaching the positive values on the outside. Conversely, if openings on the leeward side predominate, inside pressures may have negative values from $-0.2p_w$ to $-0.4p_w$ or greater.

As noted previously, the crack method of estimat-

TABLE 7-2

Air changes taking place under average conditions in residences, exclusive of air provided for ventilation*

Kind of Room or Building	Air Changes per Hour
Rooms with no windows or exterior doors	0.5
Rooms with windows or exterior doors on one side only	1
Rooms with windows or exterior doors on two sides	1.5
Rooms with windows or exterior doors on three sides	2
Entrance halls	2

Source: Reprinted from *ASHRAE Handbook 1977 Fundamentals,* with permission from American Society of Heating, Refrigerating, and Air Conditioning Engineers, Atlanta, GA.

*For rooms with weatherstripped windows or storm sash, use $\frac{2}{3}$ these values.

ing infiltration air quantities has been most generally accepted in the industry. However, there are different methods of estimating crack length or effective leakage area. The current procedure presented in Chapter 23 of *ASHRAE Handbook 1989 Fundamentals* uses a simple, single-zone approach to calculate infiltration rates in houses that requires effective leakage area at 0.016 in. water pressure difference. Data is presented in terms of leakage area per component, which means either per component, per unit surface area, or per unit length of crack—whichever is appropriate. When the total building leakage area has been determined, an equation is used to calculate the infiltration rate. The equation corrects for local building shielding, stack effect, and wind velocity.

The method presented here has been used with success for a number of years.

The first step in using the crack method consists of measuring the linear feet of crack around doors and windows for a particular building space. In frame construction, measure also the crack length between foundation and sill. The second step is to determine the "fit" of the windows and doors. Windows of exactly the same type have vastly different air-leakage characteristics. The fit of the window (tight, average, or loose), the use of weatherstripping, and the use of storm sash have definite effects on the amount of air leakage. Even though it may be normal to assume that windows in a new building will be tight fitting, this is not always the case. In fact, all windows with movable sash cannot be tighter than average fit if they are to be opened or closed. Determining the fit of installed DH windows is relatively easy. On these windows, lock the window, grasp the meeting rail of the window, and shake it. If the window rattles, it is loose fitting (or poorly fitted). Otherwise, consider it as average fit.

There is a difference between "length of crack" and "infiltration rate." Although it is true that air leakage through a crack depends upon its length, it also depends upon the width and shape of the crack. Furthermore, air leakage per foot of crack for doors differs from that per foot of crack for windows. It is also possible that various windows in a space will have different air-leakage rates per foot of crack. This will usually occur when different types of windows are used in a space or when some windows are weatherstripped or equipped with storm sash while others are not. Table 7-3 gives infiltration data for double-hung wood windows.

Infiltration through door cracks: Air leakage through door cracks must be considered in relation to the type of door and room or building involved. For residences and small buildings where doors are used infrequently, infiltration can be based on the air leakage through cracks between door and frame. Such doors vary greatly in fit because of the tendency to warp. For a well-fitted door, the leakage values for a poorly fitted double-hung window may be used. If a door is poorly fitted, twice this figure should be used. If the door is weatherstripped, the values may be reduced by one-half. A single door that is frequently opened, as might be the case in a small store, should have a value applied 3 times that for a well-fitted door. This extra allowance is for opening and closing losses and is kept from being greater by the fact that doors are not used as much in the coldest and windiest weather. Large rates of leakage can occur with doors in use, and this must be taken into account in commercial establishments where high traffic rates are expected.

The amount of air entering each time the doors are opened depends on the type of door and whether there are doors in one wall or more than one wall. Tables 7-4 and 7-5 have been set up to give the amount of air per passage for single-swing doors, swing doors with vestibules, and revolving doors. Table 7-4 is for applications where the doors are in one wall only, while Table 7-5 covers applications with doors in more than one wall.

Table 7-6 represents the expected number of entrance passages per occupant per hour for various commercial-type establishments.

The use of these tables may be explained by the following problem.

ILLUSTRATIVE PROBLEM 7-1

Assume that a particular drug store has an average occupancy of 20 persons. The entrance is a single swinging door, and there are entrances in one wall only. Estimate the amount of air (ft³/hr) that infiltrates due to door openings.

Solution:

Step 1. For drug stores, Table 7-6 indicates 8 entrance passages per occupant per hour for each occupant in the store. Therefore, the total passages per hour is 8 × 20 = 160/hr.

Step 2. Since there are doors in one wall only, refer to Table 7-4 and find that, with 160 passages per hour, the infiltration rate is 110 ft³/passage for a single swinging door. In other words, each person entering or leaving would allow 110 ft³ of air into the drug store. Since there are 160 passages per hour, the total infiltration due to door openings is 160 × 110 = 17,600 ft³/hr.

Table 7-6 may not provide the necessary data for all required establishments. A relationship developed by ARI may be helpful to determine the number of entrance passages per hour. Three of the more important

TABLE 7-3

Infiltration through window and door crack (expressed in ft³/hr per ft of crack)[d]

Type of Window or Door	Pressure Difference (inches of water)			
	0.05	0.10	0.20	0.30
A. Wood, Double-hung Window (Locked) (around sash crack only)	20	25	50	75
1. Non-weatherstripped, loose fit.[a]	28	77	122	150
2. Non-weatherstripped, average fit.[b]	18	27	43	57
3. Weatherstripped, loose fit.	19	28	44	58
4. Weatherstripped, average fit.	11	14	23	30
B. Frame Wall Leakage[c] (Leakage between frame of wood, double-hung window and wall)				
1. Around frame in masonry wall (not caulked).	8	17	26	34
2. Around frame in masonry wall (caulked).	2	3	5	6
3. Around frame in wood frame wall.	6	13	21	29
C. Double-hung metal windows				
1. Non-weatherstripped (unlocked)	47	74	104	137
2. Non-weatherstripped (locked)	45	70	96	125
3. Weatherstripped (unlocked)	19	32	46	60
D. Single-sash Metal				
1. Industrial (horizontally pivoted)	108	176	244	304
2. Residential casement	32	52	76	100
3. Residential (vertically pivoted)	88	145	186	221
E. Doors				
1. Well fitted	69	110	154	199
2. Poorly fitted	138	220	308	398

Source: Data primarily from research papers in *Trans. ASHVE,* Vols. 30, 34, 36, 37, and 39; also from *ASHRAE Handbook—1981 Fundamentals,* with permission of the American Society of Heating, Refrigerating and Air-Conditioning Engineers, Atlanta, Ga.

[a] A 0.094 in. crack and clearance represent a poorly fitted window, much poorer than average.

[b] The fit of the average double-hung wood window was determined as 0.0625 in. crack and 0.047 in. clearance by measurements on approximately 600 windows under heating season conditions.

[c] The values given for frame leakage are per foot of sash perimeter, as determined for double-hung wood windows. Some of the frame leakage in masonry walls originates in the brick wall itself, and cannot be prevented by caulking. For the additional reason that caulking is not done perfectly and deteriorates with time, it is considered advisable to choose the masonry frame leakage values for caulked frames as the average determined by the caulked and no caulked tests.

[d] Multiply table values by 0.0258 to obtain *l/s* per meter of crack.

TABLE 7-4

Infiltration through entrances with doors in one wall only (summer)*

No. of Passages per Hour, Up to	Single Swing Doors		Swing Doors-Vestibule		Revolving Doors	
	Infiltration per Passage, Cu. Ft.	Heat Gain per Passage per Deg. Differ.	Infiltration per Passage, Cu. Ft.	Heat Gain per Passage per Deg. Differ.	Infiltration per Passage, Cu. Ft.	Heat Gain per Passage per Deg. Differ.
300	110	1.98	83	1.49	30	0.54
500	110	1.98	83	1.49	29	0.52
700	110	1.98	82	1.47	27	0.49
900	109	1.96	82	1.47	25	0.45
1100	109	1.96	82	1.47	23	0.41
1200	108	1.95	82	1.47	21	0.38
1300	108	1.95	82	1.47	19	0.34
1400	108	1.95	81	1.46	18	0.32
1500	108	1.95	81	1.46	17	0.31
1600	108	1.95	81	1.46	16	0.29
1700	107	1.93	81	1.46	15	0.27
1800	105	1.89	80	1.44	14	0.25
1900	104	1.87	80	1.44	13	0.23
2000	100	1.80	79	1.42	12	0.22
2100	96	1.73	79	1.42	11	0.20

*Increase "Infiltration per Passage, Cu. Ft." by 50% for winter application.

Source: Reprinted with permission. *Handbook of Air Conditioning, Heating, and Ventilating*, 3rd ed., Industrial Press, New York, 1979.

factors that determine the rate of infiltration through frequently opened doors are as follows:

- Number of door openings per hour.
- Type of door.
- Wind velocity.

The number of door openings can be determined by actual count or by the following relationship:

$$\text{door openings/hour} = \frac{P \times F}{t_o \times n} \qquad (7\text{-}4)$$

where

TABLE 7-5

Infiltration through entrances with doors in more than one wall (summer)*

No. of Passages per Hour, Up to	Single Swing Doors		Swing Doors-Vestibule		Revolving Doors	
	Infiltration per Passage Cu. Ft.	Heat Gain per Passage per Deg. Differ.	Infiltration per Passage, Cu. Ft.	Heat Gain per Passage per Deg. Differ.	Infiltration per Passage, Cu. Ft.	Heat Gain per Passage per Deg. Differ.
300	168	3.02	125	2.25	48	0.86
500	168	3.02	125	2.25	46	0.83
700	168	3.02	125	2.25	43	0.78
900	168	3.02	125	2.25	39	0.70
1100	168	3.02	125	2.25	33	0.59
1200	168	3.02	125	2.25	30	0.54
1300	168	3.02	125	2.25	28	0.50
1400	168	3.02	125	2.25	25	0.45
1500	168	3.02	125	2.25	23	0.41
1600	167	3.01	125	2.25	21	0.38
1700	163	2.93	124	2.23	19	0.34
1800	159	2.86	122	2.20	18	0.32
1900	156	2.81	120	2.16	17	0.30
2000	152	2.74	118	2.12	16	0.29
2100	147	2.64	115	2.07	15	0.27

*Increase "Infiltration per Passage, Cu. Ft." by 50% for winter application.

Source: Reprinted with permission. *Handbook of Air Conditioning, Heating, and Ventilating*, 3rd ed., Industrial Press, New York, 1979.

TABLE 7-6

Entrance passages per occupant per hour

Banks	8	Dress shops	3	Office Buildings	2
Barber shops	4	Drug stores	8	Professional offices	4
Brokers offices	8	Furriers	3	Public buildings	3
Candy and soda stores	6	Hospital rooms	4	Restaurants	3
Cigars and tobacco stores	25	Lunchrooms	6	Shoe stores	4
Department stores	8	Men's shops	4	Variety stores	12

Source: Reprinted with permission. *Handbook of Air Conditioning, Heating, and Ventilating*, 3rd ed., Industrial Press, New York, 1979.

P = number of people in the conditioned space
F = factor for arrivals and departures (use 2 for light traffic and 1.33 for heavy traffic; the factor 1.33 assumes that one-third of the arrivals and departures will occur simultaneously
t_o = average time of occupancy (hr)
n = number of doors

In using the crack method to determine infiltration rates, the basis of calculations recommended by ASHRAE is as follows: The amount of crack used for computing the infiltration heat loss should not be less than one-half the total length of crack in the outside walls of a room. For a building having no partitions, air entering through the cracks on the windward side must leave through the cracks on the leeward side. Therefore, take one-half the total crack for computing each side and end of the building. In a room with one exposed wall, take all the crack; with two exposed walls, take the wall having the most crack; and with three or four exposed walls, take the wall having the most crack; but in no case take less than half the total crack.

In small residences, the total infiltration loss of the house is generally considered to be equal to the sum of the infiltration losses of the various rooms. However, this is not necessarily accurate because at any given time infiltration will take place only on the windward side or sides and not on the leeward side. Therefore, for determining the total heat requirements of larger buildings, it is more accurate to base the total infiltration loss on the wall having the most crack; but in no case take less than half the total crack in the building.

Measurement of crackage should be done with as much care as in determining other measurements for heat loss calculations. Determine the running feet of crackage around the edges of doors by adding the width and height of the door and multiplying the sum by 2. A 3-by-7-ft door would have a crackage of 2(3 + 7) = 20 ft. Determine the running feet of crackage for *each* window by adding the lengths of cracks in each sash. Crackage should be measured, even for *fixed* sash.

Crack between the movable sections of a sash, like the meeting rail, is measured only *once*. It is impossible to give a general rule for calculating the crackage on all types of windows. For some common types, however, the following rules may be used:

1. For double-hung windows, use 2 times the height plus 3 times the width of the sash.
2. For single-sash casement windows, use twice the height plus twice the width.
3. For horizontal sliding windows, use 3 times the height plus 2 times the width.
4. For jalousies, awnings, and architectural projections, use twice the height plus the width multiplied by one more than the number of glass shutters.
5. For fixed windows, use twice the height plus twice the width.

All dimensions are to be expressed in feet. In new construction, the dimensions of double-hung windows are usually specified on architectural drawings. The dimensions specified are usually the glass size, but they may be converted to opening sizes using the method previously outlined for areas. The opening size is used to calculate crack length. For other types of windows, it is necessary to refer to the architect or builder or to the manufacturer's catalog of the particular window to obtain the crackage length. These windows are often specified on building plans by the manufacturer's model number, and reference to this published material or measurement of the installed window is the method used to obtain the necessary information.

Air leakage through walls: A large amount of cold air leakage occurs through building walls if the wall is porous and poorly constructed. In severely cold climates, buildings are usually built to prevent unwanted air leakage. Problems of excess air leakage through building construction are more common in warmer climates. As a result, the most difficult heating problem may occur in warmer climates. The necessity for good construction to reduce unwanted air leakage cannot be emphasized too strongly. Good construction practices include the following:

- Tight siding and sheathing construction.
- Using a tightly fitting building paper between sheathing and siding and between subflooring and finish flooring.
- Insulating stud spaces or placing draft stops in the spaces.
- Using a good vapor barrier beneath the plaster base.
- Making a tight seal around the floor mouldings and around window frames.

Infiltration and air for combustion: Fuel-burning appliances (gas, oil, coal, wood) require combustion air. The presence of these appliances within the building can have a pronounced effect upon the rate of infiltration. It is recommended that combustion air from the outside be introduced directly to the fuel-burning appliance to reduce the effect on infiltration. This may not be practical or desirable in the case of a fireplace.

When a fireplace is not operating and the damper is open, large quantities of air will rise through the chimney flue. This air has to leak into the building through crackage. The amount of leakage depends upon the tightness of the building construction, design of the chimney, and wind velocity, but the infiltration rate could be doubled. If the damper is closed and is tight fitting (seldom the case), the problem is eliminated.

When the fireplace is in operation, the draft through the chimney is greatly increased. The air for the fireplace, when possible, should be introduced through the central heating system and not allowed to enter by infiltration. The air required when a fireplace is in operation could be as high as 200 cfm. It is normal to assume about 50 cfm of infiltration per fireplace if the fireplace is not operating and the damper is closed.

Direct-fired warm-air furnaces are often located within the confines of the building. Most codes require that outside air for combustion be introduced directly to the space where the furnace is located. Combustion air required is 12 to 16 ft³ per 1000 Btu in furnace capacity.

7-10 HEAT TRANSMISSION THROUGH BASEMENT WALLS AND FLOORS AND SLAB FLOORS AT OR NEAR GRADE

The basement interior of a building is considered a conditioned space if a minimum temperature of 50°F is maintained over the heating season. In many instances,

the building heating plant, water heater, and heating ducts and/or pipes are in the basement so that it remains at or above 50°F.

Wall Above Grade. Heat loss through windows and basement walls above grade is calculated the same as for other above-grade building components, using an appropriate *U*-value, above-grade area, and a design temperature difference for the basement.

Wall Below Grade. The heat loss through below-grade walls and floors depends on (1) the temperature difference between air inside the basement and outside air, (2) the structural materials of the walls and floors, and (3) conductivity of the surrounding earth. Because of thermal inertia, ground temperatures vary with depth, and there is a substantial time lag between changes in outdoor temperature and corresponding changes in ground temperature. As a result, the ground-coupled heat transfer is less amenable to steady-state representation than is the case for above-grade building elements.

Heat transmission from below-grade portions of basement walls to ambient air cannot be approximated by one-dimensional heat conduction. The heat flow through this portion of the wall is not uniform with the depth below grade. Appendix Table C-8 lists heat loss values at different depths for an uninsulated and insulated concrete wall. The conductivity of the soil was assumed to be 9.6 Btu-in./hr-ft²-°F (1.4 W/m²-K). Also listed in the table are the lengths of the heat flow paths through the soil (circular path).

Basement Floors. The same steady-state design used for below-grade basement walls can be applied to the basement floor, except the length of the heat flow path is greater. Thus, heat loss through the basement floor is smaller than that from the wall. Therefore, an average value for the heat loss through the floor can be multiplied by the floor area to give total heat loss from the floor. Typical values are estimated and shown in Appendix Table C-9.

In summary, heat loss from the below-grade portion of the basement per degree Fahrenheit design temperature difference can be estimated using Appendix Tables C-8 and C-9. For the wall, the values of heat loss through each square foot are selected from Table C-8 and are added together, and the total is multiplied by the perimeter of the house. For the floor, the average heat loss per square foot is estimated from Table C-9 and multiplied by the floor area. The resulting two values may then be added and multiplied by the appropriate design temperature difference to obtain the below-grade heat loss in Btu/hr.

Design Temperature Difference. Selecting the appropriate design temperature difference can be a problem. Although inside design temperature is given by basement air temperature, none of the usual external design temperatures are applicable because of the heat capacity of the soil. However, ground surface is known to fluctuate about a mean value by an amplitude, *A*, which will vary with geographic location and surface cover. Thus, suitable external design temperatures can be obtained by subtracting amplitude *A* for the location from the mean annual air temperature, t_a. Values for t_a can be obtained from meteorological records; and *A* varies from 27°F through central Canada to 18°F through the southern United States (see Appendix Figure C-2).

If the basement is completely below grade and unheated, the air temperature will normally range between that in the rooms above and that of the outside ground. Of course, basement windows will lower basement temperature when it is cold outside, and heat given off by the heating plant will increase the basement temperature. The exact basement temperature is indeterminate if the basement is not heated. As previously noted, the presence of the heating plant, water heater, and heating ducts will normally produce a temperature of at least 50°F.

ILLUSTRATIVE PROBLEM 7-2

The full basement of a house measures 28 ft wide by 40 ft long and is 7 ft 6 in. high from basement ceiling to slab floor. The top $1\frac{1}{2}$ ft of the foundation wall is above grade, and the wall is 8 in. concrete with sand-and-gravel aggregate. The top $4\frac{1}{2}$ ft of the wall is insulated with rigid insulation ($R = 4.17$) applied to the interior surface. Estimate the total heat loss from the basement if the inside air temperature is 70°F. The external design temperature is 0°F; the mean outside air temperature from winter records is 45°F; and the ground surface temperature amplitude is 18°F.

Solution:

Step 1. Determine the *U*-value for the above-grade wall and the heat loss above grade.

Wall Component	R
Inside air film	0.68
Outside air film	0.17
Insulation	4.17
Concrete, 8 in. (Appendix Table C-1, 1/k = 0.08)	0.64
	R_t = 5.66

$$U = \frac{1}{R_t} = \frac{1}{5.66} = 0.177 \text{ (say, 0.18) Btu/hr-ft}^2\text{-}°F$$

$$\dot{q} = UA(t_i - t_o)$$
$$= (0.18)[(2)(28) + (2)(40)](1.5)(70 - 0)$$
$$= 2754 \text{ Btu/hr} \quad \text{(heat loss above grade)}$$

Step 2. Determine below-grade loss (wall and floor). Insulation on the wall down to $4\frac{1}{2}$ ft below the ceiling gives 3 ft of wall insulated below grade and 3 ft of wall uninsulated.

Wall Depth (below grade)	Appendix Table C-8 (Btu/hr-ft-°F)
1st ft	0.152
2nd ft	0.116
3rd ft	0.094
4th ft	0.119
5th ft	0.096
6th ft	0.079
	0.656 Btu/hr-ft-°F

basement perimeter = 2(28 + 40) = 136 ft
unit wall loss = (0.656)(136) = 89.2 Btu/hr-°F

From Appendix Table C-9, using shortest width (28 ft) and depth of foundation wall below grade (6 ft), find 0.025 Btu/hr-ft²-°F.

floor area = (28)(40) = 1120 ft²
unit floor loss = (0.025)(1120) = 28 Btu/hr-°F
total unit heat loss below grade = 89.2 + 28 = 117.2 Btu/hr-°F
design temperature below grade = 70 − (45 − 18) = 43°F
total heat loss below grade = (117.2)(43) = 5040 Btu/hr

Step 3. Determine basement total heat loss:

$$\dot{q} = 2754 + 5040 = 7794 \text{ Btu/hr} \quad \text{(ans.)}$$

Slab Floor At or Near Grade. Two types of concrete floors in basementless houses are (1) unheated floors, relying for warmth on heat delivered above floor level by the heating system, and (2) heated floors, containing heating ducts or pipes that constitute a radiant slab or portion thereof for complete or partial heating of the building. The perimeter insulation of a slab-on-grade floor is very important for comfort and energy conservation. In unheated slab floors, the edge must be insulated to keep the floor warm. However, an unheated slab floor should not be used in severe climates (above 4000 degree-days) because downdrafts from windows or exposed walls can create pools of chilly air over considerable areas of the floor. In heated slab floors, the floor edge *must* be insulated to prevent excessive heat loss from the heating pipe or duct embedded in the floor or from a baseboard heating unit.

Experiments have shown that heat loss from an unheated concrete slab floor is mostly through the perimeter rather than through the floor into the ground. The total heat loss is more nearly proportional to the length of the perimeter than the area of the floor, and it can be estimated by the following equation for both unheated and heated slab floors:

$$\dot{q} = F_2 P(t_i - t_o) \qquad (7\text{-}5)$$

where

$\dot{q}$ = heat loss through the perimeter (Btu/hr)
F_2 = heat loss coefficient (Btu/hr-°F per ft of perimeter; see Appendix Table C-10)
t_i = indoor temperature (°F) (for the heated slab, t_i is the weighted average heating duct or pipe temperature)
t_o = outdoor design temperature (°F)
P = perimeter or exposed edge of floor (ft)

7-11   HEAT TRANSMISSION FROM CRAWL SPACES

A crawl space can be considered a half-basement. To prevent ground moisture from evaporating and causing a condensation problem, sheets of vapor retardant—for example, polyethylene film—are used to cover the ground surface. Most codes require crawl spaces to be adequately vented all year round. However, venting the crawl space in the heating season causes substantial heat loss through the floor.

The space may be insulated in several ways. The crawl space ceiling (floor above crawl space) can be insulated, or the perimeter wall can be insulated either on the outside or on the inside of the wall. If the floor above is insulated, the crawl space vents should be kept open because the temperature in the crawl space is likely to be below the dew point of the indoor space. If the perimeter is insulated, the vents should be closed in the heating season and open the remainder of the year.

Crawl Space Temperature. The crawl space temperature depends on factors such as venting, heating ducts, and the heating plant. When the crawl space is well ventilated, its temperature is close to that of the ambient air temperature. When the crawl space vents are closed for the heating season, or when the space is used as plenum (as part of the forced-air heating system), the crawl space temperature approaches the indoor conditioned space. In the former case, the floor above the crawl space, the heating ductwork, and the utility pipes should be insulated similarly to the walls and ceiling of a house.

The following steady-state equation can be used to estimate the temperature of the crawl space:

$$\dot{q}_f = \dot{q}_p + \dot{q}_g + \dot{q}_a \qquad (7\text{-}6)$$

where

$\dot{q}_f$ = heat loss through floor to crawl space (Btu/hr)
$\dot{q}_p$ = heat loss through foundation walls and sill box from crawl space (Btu/hr)
$\dot{q}_g$ = heat loss into ground (Btu/hr)
$\dot{q}_a$ = heat loss due to ventilation of crawl space (Btu/hr)

Latta and Boileau (1969) estimated the air exchange rate for an uninsulated basement at 0.67 air changes per hour under winter conditions. In more detail, Eq. (7-6) can be stated as

$$U_f A_f(t_i - t_c) = U_p A_p(t_c - t_o)$$
$$+ U_g A_g(t_c - t_g) + 0.67 H_c V_c(t_c - t_o) \qquad (7\text{-}7)$$

where

t_i = indoor temperature, or air above ceiling of crawl space (°F)
t_o = outdoor temperature (°F)
t_g = ground temperature, constant (°F)
t_c = crawl space temperature (°F)
A_f = area of floor above (ft²)
A_p = area of perimeter, exposed foundation wall plus sill box (ft²)
A_g = area of ground below ($A_f = A_g$) (ft²)
U_f = average heat-transfer coefficient through floor (Btu/hr-ft²-°F)
U_g = average heat-transfer coefficient through ground, horizontal air film and 10 ft of soil, (Btu/hr-ft²-°F)
U_p = combined heat-transfer coefficient of sill box and foundation wall, both above and below grade, (Btu/hr-ft²-°F)
V_c = volume of crawl space (ft³)
H_c = volumetric heat capacity of air = 0.018 Btu/ft³-°F
0.67 = assumed air exchange rate (volume per hour)

ILLUSTRATIVE PROBLEM 7-3

A crawl space of 1200 ft³, with a 140-ft perimeter, is considered. The construction of the perimeter wall is shown in Figure 7-1. The indoor, outdoor, and deep-down ground temperatures are 70, 10, and 50°F, respectively. Estimate the heat loss and crawl space temperature with and without insulation.

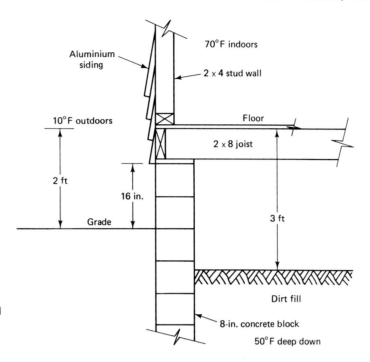

FIGURE 7-1 Uninsulated crawl space sketch for Illustrative Problem 7-3.

The heat-transmission coefficient (*U*-value) for each component is indicated in Appendix Table C-11.

Solution: Three cases are examined.

Case A. The crawl space is vented and uninsulated. The crawl space temperature approaches that of the outdoors, 10°F. The heat loss is calculated as

$$\dot{q}_f = U_f A_f (\Delta t)$$
$$= (0.25)(1200)(70 - 10)$$
$$= 18,000 \text{ Btu/hr}$$

Case B. The crawl space is vented. The floor above is insulated with *R*-11 blankets; there is no insulation on the perimeter. The temperature of the crawl space approaches that of the outdoors, 10°F. The heat loss is calculated as

$$\dot{q}_f = U_f A_f (\Delta t)$$
$$= (0.076)(1200)(70 - 10)$$
$$= 5470 \text{ Btu/hr}$$

Case C. The crawl space is not vented during the heating season. The floor above is not insulated; the perimeter wall is insulated with $R = 5.4$ down to 3 ft below grade. Then,

$$\dot{q}_f = U_f A_f (t_i - t_c) = (0.25)(1200)(70 - t_c)$$
$$\dot{q}_p = 140(0.62)(t_c - 10)$$
$$\dot{q}_g = 1200(0.077)(t_c - 50)$$
$$\dot{q}_a = 1200(3)(0.67)(0.018)(t_c - 10)$$

Substitution into Eq. (7-7) yields $t_c = 51.5°F$, the estimated crawl space temperature. The heat loss through the floor is

$$\dot{q}_f = (0.25)(1200)(70 - 51.5)$$
$$= 5550 \text{ Btu/hr}$$

In summary, we may say that the base case *A* can potentially lose the most heat. However, where the floor above is insulated, the crawl space must be vented to eliminate any condensation potential, and the heating ductwork and utility pipelines in the basement must be adequately insulated. When the perimeter is insulated, the vents must be closed during the heating season and opened for the rest of the year. The heating ductwork and utility pipelines do not need insulation.

7-12 AUXILIARY HEAT SOURCES

The heat supplied by persons, lights, motors, and machinery should always be determined in the case of theaters, assembly halls, industrial plants, and commercial buildings such as stores, office buildings, and so forth; but allowances for such heat sources must be made only after careful consideration of all local conditions. In many cases, these heat sources may materially affect the size of the heating plant and may have a marked effect on the operation and control of the system. In any evaluation, however, the night, the weekend, and any other unoccupied period must be evaluated. In general, where audiences are present, the heating system must have adequate capacity to bring the building to a stipulated indoor temperature before the audience arrives. In industrial plants, quite a different condition exists, and heat sources, if always available during occupancy, may be substituted for a portion of the heating requirements. In no case, however, should the actual heating installation (exclusive of heat sources) be reduced below that required to maintain at least 40°F in the building.

Inefficiencies in motors and machinery they drive, if both are located in the heated space, appear as room heat. This heat is retained in the room, if the product manufactured is not removed, until its temperature is the same as the room temperature. If power is transmitted to the machinery from outside, only the heat equivalent to the brake horsepower supplied is used. In some industries, this is the chief source of heating; it can frequently overheat the building even in zero weather, thus requiring year-round cooling.

7-13   INTERMITTENTLY HEATED BUILDINGS

For intermittently heated buildings, additional heat is required for raising the temperature of the air, building material, and material contents of the building to the specified indoor temperature. The rate at which this additional heat must be supplied depends on the heat capacity of the structure and its material contents and the time in which these are to be heated.

Because design outdoor temperatures generally provide a substantial margin for outdoor temperatures typically experienced during operating hours, many engineers make no allowance for this additional heat in most buildings. However, if optimum equipment sizing is used (minimum safety factor), the additional heat should be computed and allowed for as conditions require. In the case of churches, auditoriums, and other intermittently heated buildings, additional capacity of 10% should be provided.

7-14   EXPOSURE FACTORS

Some designers use empirical *exposure factors* to increase calculated heat loss from rooms or spaces on the side (sides) of the building exposed to prevailing winds.

However, use of exposure factors should be unnecessary with the method of calculating heat loss described in this chapter. They may be considered as *safety factors* for the rooms or spaces exposed to prevailing winds to allow for additional capacity for the spaces, or used to *balance the radiation,* particularly in the case of multistory buildings. Tall buildings may have severe infiltration heat losses, induced by the stack effect, that will require special consideration. Although a 15% exposure allowance frequently is assumed, the actual allowance, if any, must be largely a matter of experience and judgment; no authentic test data is available from which to develop rules for the many conditions encountered.

7-15   HEAT LOSS CALCULATIONS FOR A BUILDING

Since the results desired in evaluating heat loss from buildings is the sum of all losses, an orderly tabulation of calculations is most desirable. Such a tabulation gives a clear, orderly presentation of dimensions of components through which heat is being transmitted and also provides a means for quickly checking results. Many *heat load forms* are available, and all have their advantages and disadvantages. You may find the tabulation worksheet of Table 7-7 useful. Such a worksheet would be made up for each room or space to be heated. Adding all the total heat losses from each room or space provides the total building heat loss.

ILLUSTRATIVE PROBLEM 7-4

Figure 7-2 represents the floor plan of a medium-sized restaurant building consisting of the restaurant, a lounge, and a kitchen. Make a complete heat loss calculation for the restaurant, lounge, main entrance, and kitchen. The following information has been obtained through a preliminary survey of the building plans.

TABLE 7-7
Tabulation worksheet for heat loss

Name of Room or Space:					
Room Size: L =	; W =	; Ceiling Height =		; Volume =	
		Area × U × Δt			$\dot{q}$(Btu/hr)
Gross exposed wall				=	
Windows				=	
Doors				=	
Net exposed wall				=	
Cold ceiling				=	
Cold floor				=	
Cold partition				=	
Floor edge loss				=	
Infiltration				=	
Miscellaneous				=	
				Total loss =	————

Auxiliary calculations as required:

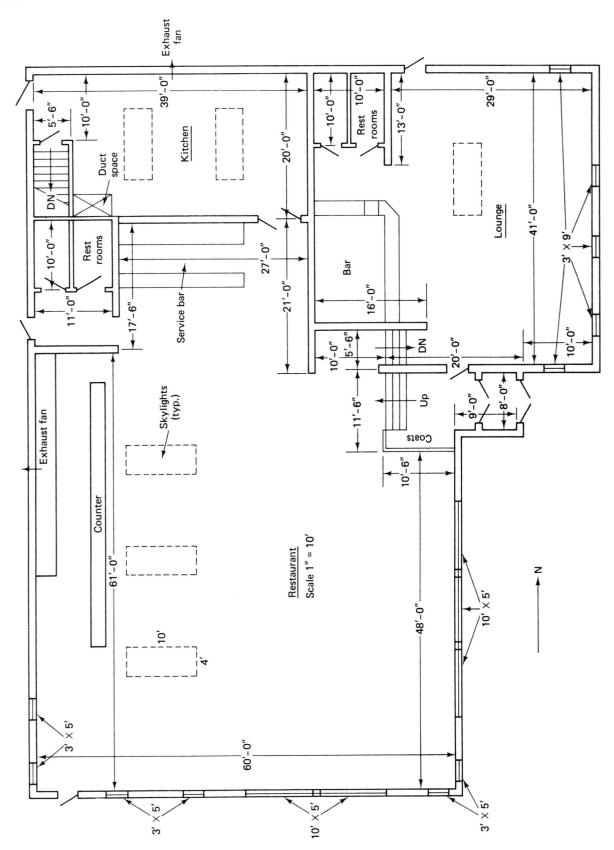

FIGURE 7-2 Floor plan used for Illustrative Problem 7-4.

Exhaust fan

39'-0"

10'-0"

5'-6"

Duct space

DN

Kitchen

Rest rooms

10'-0"

11'-0"

17'-6"

Service bar

20'-0"

27'-0"

21'-0"

Rest rooms

10'-0"

10'-0"

13'-0"

29'-0"

Lounge

41'-0"

3' × 9'

10'-0"

Bar

16'-0"

DN

10'-0"

5'-6"

20'-0"

9'-0"

8'-0"

Up

11'-6"

Coats

10'-6"

Exhaust fan

Counter

Skylights (typ.)

61'-0"

10'

4'

Restaurant
Scale 1" = 10'

N

48'-0"

10' × 5'

3' × 5'

60'-0"

3' × 5'

10' × 5'

3' × 5'

129

	Design Conditions	
	Summer	*Winter*
Outside air design	95°F/76°F	10°F, 80% RH
Inside air design	78°F, 50% RH	70°F, 40% RH
Outside air daily temperature range	20°F	
Wind velocity	7.5 mph	15 mph
Location 40° north latitude		

Structural Data

1. Full basement under restaurant and kitchen. Basement height, floor to underside of exposed bar joist, 8 ft 0 in. Foundation walls, 12-in. concrete, sand-and-gravel aggregate. Basement floor, 3-in. concrete slab. Top of foundation walls at exterior grade-level.
2. No basement under entrance and lounge. Four-inch-thick concrete slab on grade with footing walls, 1-in.-thick slab insulation ($R = 5.4$) between floor slab edge and footing walls.
3. Exterior walls of restaurant and lounge
   (a) 4-in. face brick
   (b) 3/4-in. air space
   (c) 8-in. concrete block, three-holed oval-cored, lightweight aggregate
   (d) 2-in. polyurethane, expanded, placed between 2-by-2-in. spruce furring strips that are fastened to the concrete block, placed on 1-ft centers
   (e) 1/4-in. fir plywood panels on interior surface
4. Exterior walls of kitchen
   (a) Same as restaurant and lounge, except no insulation, furring strips, and plywood
   (b) Interior surface, 5/8-in. gypsum plaster, lightweight aggregate, applied directly to the concrete blocks
5. Floors of restaurant and kitchen
   (a) 3-in.-thick concrete slab, lightweight aggregate (density 100 lb/ft), carpet and rubber pad, no ceiling below
   (b) Carpet and rubber pad on the 4-in. slab of lounge
6. Roof
   (a) Flat roof over entire building at uniform level
   (b) 2-in. concrete slab, sand-and-gravel aggregate, 2-in. cellular glass insulation, 3/8-in. built-up roofing
   (c) Suspended metal lath and plaster ceiling (gypsum plaster, lightweight aggregate, 3/4-in. thick)
7. Ceiling heights
   (a) Restaurant and kitchen: 10 ft
   (b) Lounge and entrance: 9 ft
8. Windows of restaurant and lounge: Aluminum frame, thermal break, type *C*, double-glazing with 1/2-in. air space. Frame caulked in masonry walls. Windows have indoor shading. Sizes shown on Figure 7-2.
9. Skylights: Aluminum frame, thermal break, type *C*, double-glazing with 1/2-in. air space; horizontal, mounted on curbing 4 in. above roof deck.

10. Doors
    (a) Main entrance: 3 ft by 7 ft, 3/8-in.-thick plate glass, $k = 5.5$ Btu-in./hr-ft²-°F, metal frame caulked in masonry walls
    (b) Kitchen entrance: 4 ft by 7 ft, 2 1/4-in.-thick, solid-core flush door, insulating glass, no storm door
    (c) Emergency doors (restaurant and lounge): 3 ft by 7 ft, 1 3/4-in.-thick, steel door, type *C*, weatherstripped

Miscellaneous Data

1. Equipment located in restaurant area
   (a) Two 3-gallon coffee urns, gas-fired
   (b) One electric toaster, 4 slices wide
   (c) Food warmer, water-pan, 4 burner, gas-fired
   (d) Two deep fat fryers, 32-lb capacity each
   (e) Grill, large, gas-fired
   (f) Electric range, 4 burner
   (g) One 2-cu-ft oven, electric
2. Exhaust fans
   (a) Restaurant: 1500 cfm for exhaust hood over cooking area
   (b) Lounge: 1000 cfm, ceiling over bar area
   (c) Kitchen: 2000 cfm
   (d) Restrooms: 200 cfm each
3. Electric lights
   (a) Restaurant: 3 watts per ft² of floor area, fluorescent
   (b) Lounge: 1 watt per ft² of floor area, incandescent
   (c) Kitchen: 5 watts per ft² of floor area, fluorescent
4. Restaurant hours: 6:00 A.M. to 9:00 P.M.
5. Lounge hours: 11:00 A.M. to 11:00 P.M.
6. Restaurant seating capacity: 100 people, 6 waitresses, 1 cashier, 1 cook
7. Lounge seating capacity: 40 people, 2 waitresses, 1 bartender

Solution:

1. Determine heat-transmission coefficients. For the exterior walls (restaurant and lounge),

Wall Component	Resistance, R
Outside air film ($f_o = 6.0$)	0.17
4-in. face brick ($17k = 0.11$)	0.44
3/4-in. air space	1.26
8-in. concrete block	2.00
2-in. polyurethane	12.50
1/4-in. fir plywood	0.31
Inside air film ($f_i = 1.46$)	0.68
	$R_t = \overline{17.36}$

$$U = \frac{1}{17.36} = 0.0576 \text{ Btu/hr-ft}^2\text{-°F}$$

Make correction for furring strips:

$$R_f = 17.36 - 12.5 + 2(1.25) = 7.36$$

$$U_f = \frac{1}{7.36} = 0.136$$

By Eq. (6-21),

$$U_{ave} = 2/12(0.136) + 10/12(0.0576)$$
$$= 0.07 \text{ Btu/hr-ft}^2\text{-}°F$$

For the exterior wall (kitchen),

$$R_t = 17.36 - 12.5 - 0.31 + 0.39 = 4.94$$

$$U = \frac{1}{4.94} = 0.202 \text{ (say, 0.20) Btu/hr-ft}^2\text{-}°F$$

For the roof,

Roof Component	Resistance, (R)
Outside air film ($f_o = 6.0$)	0.17
Built-up roofing	0.33
2-in. insulation	5.72
2-in. concrete	0.16
Ceiling	0.47
3 1/2-in. air space (assumed)	0.93
Inside air film	0.61
	$R_t = 8.39$

$$U = \frac{1}{8.39} = 0.119 \text{ (say, 0.12) Btu/hr-ft}^2\text{-}°F$$

For the windows,

$$U = 0.59 \text{ (Appendix Table C-4)}$$

For the skylights,

$$U = 0.71 \text{ (Appendix Table C-4)}$$

For the doors (main entrance),

$$R_t = 0.17 + 0.375/5.5 + 0.68 = 0.918$$

$$U = \frac{1}{0.918} = 1.089 \text{ (say, 1.09) Btu/hr-ft}^2\text{-}°F$$

For the door (kitchen),

$$U = 0.33 \text{ Btu/hr-ft}^2\text{-}°F$$

For the door (emergency),

$$U = 0.46 \text{ Btu/hr-ft}^2\text{-}°F$$

2. Determine heat-transfer areas.

Restaurant	Area (ft²)
Gross wall	1765
Windows	340
Doors	42
Net wall	1383
Gross roof	4285
Skylights	120
Net roof	4165

Lounge	Area (ft²)
Gross wall	720
Windows	135
Doors	21
Net wall	564
Gross roof	1439
Skylights	40
Net roof	1399

Kitchen	Area (ft²)
Gross wall	490
Doors	28
Net wall	462
Gross roof	725
Skylights	80
Net roof	645

Main Entrance	Area (ft²)
Gross wall	153
Doors	42
Net wall	111
Gross roof	156

3. Determine conduction and infiltration losses.

Restaurant	U	×	A	×	Δt	=	Heat Loss
Windows	0.59	×	340	×	60	=	12,000
Doors	0.46	×	42	×	60	=	1,160
Net wall	0.07	×	1383	×	60	=	5,800
Skylights	0.71	×	120	×	60	=	5,100
Net roof	0.12	×	4165	×	60	=	30,000
Floor (assumed negligible)						=	0
Infiltration (see calculations below)						=	2,780
						Total =	56,840 Btu/hr

Infiltration: By Eq. (7-3) wind pressure $p_w = 0.000482 V_w = 0.000482(15)^2 = 0.108$ (say, 0.10) in. water gauge. The amount of crackage to use is usually very difficult to estimate. The fixed windows would have a small amount of crackage around the frames caulked in masonry walls. Emergency doors are weatherstripped. From Table 7-3, for fixed windows, use 3 ft³/hr per ft of crack, and, for emergency doors, use 110 ft³/hr per ft of crack. Assuming half of the crack length for infiltration, we have for the windows [(30 + 5)

(2) + (20 + 5)(2) + (16 × 6)]/2 = 108 ft of crack and for the doors (20)(2)/2 = 20 ft of crack. Thus, infiltration for the windows = (3)(108)/60 = 5.4 cfm and for the doors = (110)(20)/60 = 36.7 cfm. Total infiltration for the restaurant = 42.1 cfm.

$$\text{Infiltration heat loss} = 1.10(\text{cfm})(t_i - t_o)$$
$$= 1.10(42.1)(70 - 10)$$
$$\dot{q}_{\text{inf}} = 2780 \text{ Btu/hr}$$

Lounge	U	×	A	×	Δt	=	Heat Loss
Windows	0.59	×	136	×	60	=	4,800
Doors	0.46	×	21	×	60	=	580
Net wall	0.07	×	564	×	60	=	2,370
Skylights	0.71	×	40	×	60	=	1,700
Net roof	0.12	×	1399	×	60	=	10,070
Floor edge loss (0.56)(80)(60)						=	2,700
Infiltration (see calculations below)						=	2,620
						Total =	24,800 Btu/hr

Infiltration (crackage): We shall take 50% of the window crackage plus all door crackage. Window crack length = (24)(5)/2 = 60 ft, and door crack length = 20 ft. Thus, infiltration for the windows = (3)(60)/60 = 3.0 cfm and for the doors = (110)(20)/60 = 36.7 cfm. Total infiltration for the lounge = 39.7 cfm.

$$\text{Infiltration heat loss} = 1.10(39.7)(70 - 10)$$
$$= 2620 \text{ Btu/hr}$$

Kitchen	U	×	A	×	Δt	=	Heat Loss
Door	0.33	×	28	×	60	=	550
Net wall	0.20	×	462	×	60	=	5,550
Skylights	0.71	×	80	×	60	=	3,400
Net roof	0.12	×	645	×	60	=	4,640
Infiltration (see below)						=	0
						Total =	14,140 Btu/hr

Infiltration (crackage): Will be assumed negligible because of exhaust air fan and required makeup air, which must be supplied.

Main Entrance	U	×	A	×	Δt	=	Heat Loss
Doors	1.09	×	42	×	60	=	2,750
Net wall	0.07	×	111	×	60	=	470
Net roof	0.12	×	156	×	60	=	1,120
Floor edge loss (0.56)(17)(60)						=	570
Infiltration (see calculations below)						=	45,240
						Total =	50,150 Btu/hr

Infiltration: Will be for crackage and for door openings from people entering and leaving the building. From Table 7-6, assume 3 passages per hour per occupant. Assume 100 occupants (restaurant plus lounge) at any one time. The num-

ber of passages per hour is 3 × 100 = 300 passages per hour. From Table 7-4, we find 83 ft³ per passage × 1.5 for winter, or 125 ft³/passage. Infiltration due to door openings = (300)(125) = 37,500 ft³/hr, or 625 cfm. For crackage, crack length is (6 × 2) + (3 × 7) = 33 ft. Table 7-3 gives 220 ft³/hr per foot of crack for poorly fitted doors (doors opened frequently are seldom better than poorly fitted); however, the vestibule reduces this by 50%. Therefore, infiltration due to crackage is (33)(220/2) = 3630 ft³/hr, or 60.5 cfm.

$$\text{Infiltration heat loss} = 1.10(685.5)(70 - 60)$$
$$= 45,240 \text{ Btu/hr}$$

4. Summarize heat losses.

Restaurant	56,840
Lounge	24,800
Kitchen	14,140
Main entrance	50,150
Building loss	145,930 Btu/hr

ILLUSTRATIVE PROBLEM 7-5

For the building of Illustrative Problem 7-4 determine the required amount of outside air for ventilation of the (1) restaurant, (2) lounge, and (3) kitchen.

Solution: Table 5-5 presents ventilation air quantities for various types of areas, giving minimum and recommended quantities of outside air for occupants of these areas. The presence of an exhaust fan in each of the areas considered influences the selection of outside air quantities to be used. The air entering the areas by infiltration can be considered as part of the total ventilation air requirements.

1. *Restaurant:* From the calculations of Illustrative Problem 7-4, the infiltration by crackage was 42.1 cfm. It is also safe to assume that most, if not all, of the infiltration air entering through the main entrance will also enter the restaurant. The calculated infiltration for the main entrance was 686 cfm. Therefore, the total outside air entering the restaurant by infiltration is 728 cfm. Now, the exhaust fan located in the restaurant would be exhausting at the rate of 1500 cfm when it is operating (an assumed condition when restaurant is operating). There should be a lower pressure in the kitchen than in the restaurant to assure that cooking odors from the kitchen will not enter the restaurant. We shall allow an additional 200 cfm outside air for this. We shall also allow 100 cfm additional air to account for restroom exhaust. Therefore, the total outdoor air brought in through the heating equipment would be 1500 + 200 + 100 − 728 = 1072 cfm. If we can assume that the total number of people in the restaurant at any one time is 80, then the ventilation air per person is 1800/80 = 22.5 cfm, which seems adequate when we refer to Table 5-5.

2. *Lounge:* From Illustrative Problem 7-4, the infiltration rate was 39.7 cfm. The exhaust fan is rated at 1000 cfm. We shall allow 100 cfm of exhaust for restrooms. Therefore, the total amount of outside ventilation air through the heating equipment would be $1100 - 39.7 = 1060.3$ cfm. If we can assume that a maximum of 35 people would be in the lounge at any one time, the outside air per person is $1100/35 = 31.4$ cfm, which is a bit high but reasonable, especially if we assume that smoking takes place in such areas.

3. *Kitchen:* A commercial kitchen presents a special problem in most cases. Normally, the number of people working in a kitchen does not represent a problem. It is usually better to exhaust slightly more air than air supplied as makeup. We have already allowed an additional 200 cfm of air in the restaurant; therefore, outside air as makeup in the kitchen will be 1800 cfm. With the exhaust fan operating at 2000 cfm, we should have a reasonably balanced pressure within the building.

ILLUSTRATIVE PROBLEM 7-6

For the building of Illustrative Problem 7-4, determine the heat load on the heating equipment due to introduction of outside air for ventilation and the heat load due to required humidification.

Solution: The total ventilation air through the heating equipment is 1072 cfm for the restaurant, 1060 cfm for the lounge, and 1800 cfm for the kitchen, or 3932 cfm. By approximate Eq. (4-10),

$$\dot{q}_s = 1.10(\text{cfm}) (t_i - t_o)$$
$$= 1.10(3932) (70 - 10)$$
$$= 259,500 \text{ Btu/hr}$$

For the humidification load, we must have the humidity ratios for the outside air and inside air. From the psychrometric chart, $W_i = 44$ grv/lbda, or 0.00628 lbv/lbda. For the outside air, at 10°F, $P_{vs} = 0.0629$ in. Hg. At 80% RH, $p_v = (0.80)(0.0629) = 0.0503$ in Hg. By Eq. (3-5),

$$W_o = 0.622 \left(\frac{0.0503}{29.92 - 0.0503} \right)$$
$$= 0.00105 \text{ lbv/lbda}$$

The specific volume of outside air is 12.06 ft³/lb. The quantity of outside air that must be humidified is the total of infiltration and mechanical ventilation: 1800 cfm for the restaurant, 1100 cfm for the lounge, and 1800 cfm for the kitchen, or a total of 4700 cfm. By Eq. (7-2a),

$$\dot{q}_L = \frac{\dot{Q}_{oa}}{v_{oa}}(W_i - W_o)h_{fg}$$
$$= \frac{(4700)(60)}{11.85}(0.00628 - 0.00105)(1055)$$
$$= 131,300 \text{ Btu/hr}$$

The total heating capacity of the heating and ventilating equipment would be $145,930 + 259,500 + 131,300 = 536,730$ Btu/hr.

7-16 AIR REQUIRED FOR SPACE HEATING

The computation of air supply quantities required for space heating was discussed in Chapter 4. The procedure used is always recommended when latent heat loss is involved or when a relatively large amount of outdoor ventilation air is used. In many cases, especially in residential or light commercial establishments, when latent heat loss is relatively small and may be neglected or included with the sensible heat loss, the air supply quantity may be calculated from

$$\dot{q}_t = \dot{m}_{sa}C_p(t_{sa} - t_{ra}) \tag{7-8}$$

or

$$\dot{q}_t = \frac{\dot{Q}_s}{v_{sa}} C_p(t_{sa} - t_{ra}) \tag{7-8a}$$

or (approximately)

$$\dot{q}_t = 1.10(\text{cfm}_{sa})(t_{sa} - t_{ra}) \tag{7-8b}$$

where

$\dot{q}_t$ = sensible heat loss for a space, including latent heat loss if negligible (Btu/hr)
$\dot{m}_{sa}$ = mass of supply air (lb/hr)
v_{sa} = specific volume of supply air (ft³/lb)
$\dot{Q}_s$ = volume of supply air (ft³/hr)
t_{sa} = supply air temperature
t_{ra} = room air temperature

The temperature difference $t_{sa} - t_{ra}$ is normally less than 100°F. However, some small-capacity residential-type warm-air furnaces are rated on a 100-degree rise in air temperature through the heat exchanger. These units also have a rated capacity in cfm and a limited fan static pressure rise.

Most commercial warm-air systems are built-up units where the heat exchanger and fan may be selected to give performance levels that will meet any requirements.

If a system is being designed for year-round operation, the airflow quantities required for cooling are normally greater than for heating because of the differences in calculated heat loads and the $t_{ra} - t_{sa}$ difference used for cooling and $t_{sa} - t_{ra}$ used for heating. Usually, the airflow quantity determined by the cooling loads determines the supply air quantity.

As may be seen by Eq. (7-8), the quantity of supply air, for a given space heat load, is dependent on the supply air temperature difference $t_{sa} - t_{ra}$. A large temperature difference reduces the air supply quantity required. Another factor should also be considered. The total air circulated as supply air must be large enough to promote air motion in the space that is within comfort levels (20 to 40 fpm), or to provide for a reasonable number of air changes per hour.

7-17   ENERGY ESTIMATING METHODS

It is often necessary to estimate the energy requirements and fuel consumption of HVAC systems for either short- or long-term operation. These can be much more difficult to calculate than design heat loss and gain or required system capacity since they involve the integration over the period in question of the influence of many factors that may vary with time. It is difficult to accurately foresee the way in which all the factors involved may vary throughout the time period. In addition to this, the calculations required to take all such variations into account become very involved. For these and other reasons, records of past operating experience of the building in question, when they are available, provide the most reliable and usually the most accurate basis for the prediction of future requirements.

Although the procedures used for estimating energy requirements vary considerably in degree of sophistication, they all have three common elements: the calculation of (1) space load (heat loss or gain), (2) secondary equipment load, and (3) primary equipment energy requirements. Here, *secondary* refers to equipment that distributes the heating, cooling, or ventilating medium to the conditioned spaces, while *primary* refers to central plant equipment that converts fuel or electrical energy to a heating or cooling effect.

The space load calculations determine the amount of energy that must be added to or extracted from a space to maintain thermal comfort. The simplest procedures assume that the energy required to maintain comfort is a function of a single parameter, the outdoor dry-bulb temperature. The more reliable methods include consideration of solar effects, internal gains, heat storage in the walls and interiors, and the effects of wind on both the building envelope heat transfer and infiltration. The most sophisticated procedures are based on hourly profiles for climate conditions and operational characteristics of a number of "typical" days of the year or a full 8760 hours of operation.

The second step translates the space load into a load on the secondary equipment. This can be either a simple estimate of duct or piping losses or gains or a complex hour-by-hour simulation of an air system such as a variable air volume with outdoor air cooling. This step must include the calculation of all forms of energy required by the secondary system—that is, electrical energy to operate fans and/or pumps, as well as energy in the form of heated or chilled water.

The third step calculates the fuel and energy required by the primary equipment to meet these loads. It considers equipment efficiencies and part-load characteristics. Often, it is necessary to keep track of the different forms of energy, such as electrical, natural gas, or oil. In some cases, where calculations are made to assure compliance with codes and standards, these energies must be converted to "source" energy or resources consumed, as opposed to energy delivered to the building boundary.

The sophistication of the calculation procedure can often be inferred from the number of separate ambient conditions and/or time increments used in the calculations. Thus, a simple procedure may use only one measure, such as annual degree-days, and would be appropriate only for simple systems and applications. Such methods may be referred to as *single-measure methods*. Improved accuracy may be obtained by the use of more information, such as the number of hours anticipated under particular conditions of operation. These methods, of which the "bin method" is the most well known, are referred to as *simplified multiple-measure methods*. The most elaborate methods currently in use perform energy balance calculations at each hour over some period of analysis, typically one year. These are called *detailed simulation methods,* of which there are a number of variations. These methods require hourly weather data, as well as hourly estimates of internal loads such as lighting and occupants.

A variety of energy analysis procedures are presently available. Each of these procedures has its strengths and weaknesses. In selecting the procedure to be used for a specific project, it is important that the limitations of the procedure be recognized. The available procedures may be divided into two categories: (1) those that can be implemented manually and (2) those that require the use of computers. Between these two categories are procedures that can be implemented manually but that have fairly extensive computational requirements and may be more convenient on computers.

Our discussion here will be limited to the simple procedure using the degree-day method, which generally applies only to residences and to small commercial establishments. A more complete discussion of energy estimating methods may be found in *ASHRAE Handbook 1989 Fundamentals*.

7-18 THE DEGREE-DAY METHOD

The traditional degree-day method for estimating heating energy requirements is based on the assumption that, on a long-term average, solar and internal heat gains will offset heat loss when the mean daily outdoor temperature is 65°F (18.3°C) or higher and that fuel consumption will be proportional to the difference between the mean daily temperature and 65°F (18.3°C). In other words, on a day when the mean daily temperature is 20 degrees below 65°F (11.1 degrees below 18.3°C), twice as much fuel is consumed as on days when the mean daily temperature is 10 degrees below 65°F (5.6 degree-days below 18.3°C). Thus,

$$\text{degree-day (DD)} = (65 - t_a)(1 \text{ day}) \qquad (7\text{-}9)$$

where t_a is the mean daily temperature. The number of degree-days for any longer period is the sum of all such products for as many days as the period covers.

If the daily mean temperature for the days of the year, over a period of several years, were plotted against the number of days in a year, the result would be a diagram similar to Figure 7-3. The diagram would be different for every section of the United States. The shaded area represents the annual degree-days per year. The points where the curve crosses the 65°F line indicate the dates for the beginning and the end of the heating season. If we let t_x be the average height of the shaded area and d be the number of days in the heating season, we may say that

$$\text{DD} = d \times t_x$$

or

$$t_x = \frac{\text{DD}}{d}$$

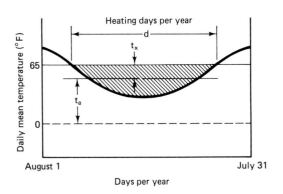

FIGURE 7-3 Degree-day diagram.

But, from Figure 7-3, we see that $t_x = 65 - t_a$; therefore,

$$65 - t_a = \frac{\text{DD}}{d}$$

or

$$t_a = 65 - \frac{\text{DD}}{d} \qquad (7\text{-}10)$$

where t_a is the average outside air temperature during the heating period of d number of days over which time DD degree-days accumulated.

Appendix Table B-2 shows average degree-days per month and yearly totals. Figure B-3 indicates the number of days in a normal heating season, and Figure B-4 indicates graphically the number of degree-days per year.

One of the first relations used to calculate annual fuel usage was developed as follows. If the heat loss calculations for a residence were accurate for the design conditions, and if the heat loss at any other outside air temperature were proportional to that at design conditions, then the heat loss from the building for the heating season could be expressed as

$$H = \frac{24\dot{q}d(t_n - t_a)}{t_i - t_o} \qquad (7\text{-}11)$$

where

H = seasonal heat loss (Btu)
$\dot{q}$ = hourly heat loss from building at design conditions (Btu/hr)
d = number of days in heating season
t_i = inside design temperature (°F)
t_o = outside design temperature (°F)
t_a = average outside temperature during period (°F) from Eq. (7-10)
t_n = weighted average inside air temperature (°F)
24 = hours per day

The weighted average inside air temperature (t_n) may be used if the temperature control system is designed to change the inside temperature during certain periods of the day, such as using night set-back. It is calculated by

$$t_n = \frac{t_1 \times \text{hours at } t_1 + t_2 \times \text{hours at } t_2}{24} \qquad (7\text{-}12)$$

The annual fuel consumption is estimated by

$$F = \frac{H}{\text{heat per sale unit} \times \text{efficiency of combustion}} \quad (7\text{-}13)$$

Equation (7-11) is very conservative, often giving estimates up to 30% high. It assumes constant temperatures for very definite hours each day throughout the entire heating season. It does not take into account factors that are difficult, if not impossible, to evaluate, such as temporary opening of windows, abnormal heating above or below the inside design temperature, poor heating system design layout, and heat gains from solar effects and interior sources. It is based on an average heating season, on–off operation of the heating system, and partial-load operation, and it does not consider wind-chill factors.

The latest degree-day method presented in *ASHRAE Handbook 1989 Fundamentals* has undergone several stages of refinements in an attempt to make it agree as closely as possible with the available measured data on an average basis. The current form is

$$F = \frac{24 \times \text{DD} \times \dot{q}}{\Delta t \times k \times \text{HHV}} (C_D) \quad (7\text{-}14)$$

where

F = fuel or energy consumption for the estimate period

DD = number of degree-days for the estimate period

$\dot{q}$ = design heat loss including infiltration and ventilation (Btu/hr or W)

Δt = design temperature difference, $t_i - t_o$ (°F or °C)

k = correction factor that includes effects of rated full-load efficiency, part-load performance, oversizing, and energy conservation devices

HHV = heating value of fuel (units consistent with q and F)

C_D = empirical correction factor for heating effect versus degree-days

Although empirical, this equation has some basis in physical fact. In an approximate sense, $\dot{q}/\Delta t$ may be thought of as an overall inside-to-outside thermal conductance (UA) for the building. In a similar sense,

the DD can be thought of as the inside-to-outside temperature difference averaged over the day so that the product yields the daily heating load. Division by the fuel heating value yields the ideal amount of fuel required for the day. The constant is for unit conversion (24 hr/day). The k and C_D factors are empirical and attempt to account for various effects noted in their definitions.

Values of C_D may be taken from Table 7-8. Correction factor k is empirical and should not be confused with any ratings for "seasonal efficiency." For electric resistance heating, $k = 1.0$. For gas heating, $k = 0.55$, and, for oil heating, $k = 0.80$, but values may vary between 0.55 and 0.80 for either.

TABLE 7-8

C_D values

Degree-Days	C_D
1000	0.81
2000	0.75
3000	0.70
4000	0.65
5000	0.60
6000	0.61
7000	0.62
8000	0.64
9000	0.66

Source: Adapted from *ASHRAE Handbook 1989 Fundamentals,* with permission from American Society of Heating, Refrigerating, and Air Conditioning Engineers, Atlanta, GA.

ILLUSTRATIVE PROBLEM 7-7

A large residence in Portland, Maine, has a calculated design heat loss of 120,000 Btu/hr based on design temperatures $t_i = 70$°F and $t_o = -5$°F. The oil-fired hot-water boiler installed has a rated net output of 150,000 Btu when fired with No. 2 fuel oil at the rate of 1.35 gallons per hour. Estimate the fuel oil consumption (gallons/season) for a typical heating season. No. 2 fuel oil has HHV = 140,000 Btu/gal.

Solution: From Appendix Table B-2, annual average degree-days is 7681. Assume $k = 0.60$ and $C_D = 0.64$. By Eq. (7-14),

$$F = \frac{24 \times \text{DD} \times \dot{q}}{\Delta t \times k \times \text{HHV}} (C_D)$$

$$= \frac{(24)(7681)(120,000)}{[70 - (-5)](0.60)(140,000)} (0.64)$$

$$= 2247 \text{ gal/season}$$

Application of this procedure is limited to residential buildings, where the building envelope transmission and infiltration are the major factors contributing to the load. In commercial buildings with highly varying internal loads, sophisticated control systems, and complex air systems or plant arrangements, the degree-day method is totally inadequate.

7-19 THE VARIABLE-BASE DEGREE-DAY METHOD

The 65°F base, which was established on the basis of studies made by the American Gas Association and the National District Heating Association, has been modified due to several factors. Residential envelope insulation levels and internal loads have increased significantly, infiltration rates have decreased through tighter construction practices, and occupants have lowered indoor temperature settings to reduce energy consumption. Consequently, use of the 65°F base results in a significant overestimation of the heating energy requirements.

Recent research at the National Bureau of Standards shows that the use of a variable-base degree-day method for residential buildings yields results that compare well with those obtained by using the DOE-2 computer program. However, the improved procedure is still not applicable for commercial buildings and has limited capabilities for analyzing time-dependent energy consumption factors that are not included in design load calculations. One should refer to *ASHRAE Handbook 1989 Fundamentals* for further discussion of the variable-base degree-day method.

7-20 THE BIN METHOD (HEATING)

Heating loads can be estimated using the bin (temperature frequency) method. The bin method involves making instantaneous energy calculations at several different outdoor dry-bulb temperatures and weighing each result by the number of hours of temperature occurrence within each bin. The bins are usually 5°F (2.78°C) in size and are often collected in three daily 8-hour shifts. Mean coincident wet-bulb temperature data, for each dry-bulb bin, are used to calculate latent loads for infiltration and ventilation. The bin method considers both occupied and unoccupied building conditions and gives credit for internal loads by adjusting the balance point. For further discussion of the bin method, one should refer to *ASHRAE Handbook 1989 Fundamentals*.

7-21 COMPUTER PROGRAMS

Numerous comprehensive building energy analysis programs that perform hourly calculations have been developed. These programs model the hourly envelope heat-transfer process as well as the detailed HVAC system and equipment simulation to estimate the building energy usage. These programs can be very useful and provide reliable results; however, they face several barriers to widespread usage. They require a mainframe computer with large memory and high speed. The cost of leasing these programs can be high. Additionally, special training in the use of particular programs due to their complexity and access to a computer terminal are required. Therefore, the average consulting firm may not have the resources or the volume of projects to justify their use.

In addition, because of the complexity of the algorithms, average practicing engineers find it difficult to evaluate the accuracy or credibility of the results obtained. The services of specially trained energy consultants are often needed to make use of these programs.

Currently, many HVAC manufacturers have developed software for the personal computer that is excellent in producing reliable load calculations for both heating and cooling. In addition, many of them perform energy analysis and produce an energy profile including the annual cost of operation.

7-22 COMPARISON OF ENERGY COSTS FOR VARIOUS TYPES OF FUELS

It is sometimes required in cost analyses to compare the relative unit cost of using alternate fuels such as gas, oil, coal, and electricity. The difficulty of such a comparison originates with the fact that fuels are sold on a different basis, such as cents per 1000 ft³ of gas, cents per gallon of oil, dollars per ton of coal, and cents per kW-hr for electricity. Since a person should be interested in the unit of heat being purchased, the Btu, all costs should be reduced to utilize heat units as a basis of comparison. A convenient common basis is the *therm* (100,000 Btu) or, in some cases, one million Btu. The cost of utilized heat in cents/therm may be computed for any fuel by

$$\text{cents/therm} = \frac{\text{price per sale unit (cents)} \times 100{,}000}{\text{Btu per sale unit} \times \text{utilization efficiency}} \quad (7\text{-}15)$$

REVIEW PROBLEMS

7.1. A room in a residence has two exposed walls, measures 20 ft long by 15 ft wide, and has a ceiling height of 8 ft 6 in. The 20-ft side contains three windows, and the 15-ft side contains two windows. The windows are double-hung, wood-sash, single-glazed, and weather-stripped; each measures 3 ft by 5 ft. The long side of the room faces the prevailing wind direction, and the wind speed is assumed to be 15 mph. The overall heat-transmission coefficient (U) for the walls is 0.28 Btu/hr-ft²-°F. Determine the heat loss from the room through the walls and windows if the inside and outside air temperatures are 70°F and 10°F, respectively.

7.2. A large room in an old building having one exposed wall is fitted with 10 poorly fitted, double-hung, wood-sash windows, 3 ft 6 in. wide by 6 ft high. Outside design conditions are −10°F with 15-mph wind speed. Estimate the reduction in heat loss from infiltration if the windows are weatherstripped. Assume that one-half of the crack length contributes to infiltration.

7.3. Outside air at 30°F and 60% RH infiltrates into a small building at the rate of 2000 ft³/hr.
 (a) How much moisture must be evaporated inside the building to maintain there a relative humidity of 40% when the temperature is 70°F?
 (b) How much heat is required for this humidification?

7.4. A workshop area in an industrial building has one exposed wall containing eight vertically pivoted, steel-sash windows, each measuring 28 in. wide by 74 in. high. The wall is on the windward side of the building and also contains two 3-ft-6-in. by 7-ft-6-in. well-fitted doors. The shop area has interior well-fitted doors leading to the remainder of the building. For a wind speed of 10 mph, inside and outside air temperatures of 65°F and 10°F, respectively, estimate the infiltration rate and associated heat loss when **(a)** the interior doors are open and **(b)** the interior doors are closed.

7.5. A small restaurant has a seating capacity of 50 people. The main entrance has two swinging double doors that open into a vestibule. There are two other exterior doors in the restaurant; however, they are for emergency use only. Estimate the amount of air that infiltrates due to door openings.

7.6. A certain building in Boston, Massachusetts, has a calculated heat loss of 150,000 Btu/hr based on an inside design air temperature of 70°F. Estimate the probable No. 2 fuel oil consumption for an average heating season.

7.7. A large office building in Philadelphia, Pennsylvania, has a calculated design heat load of 1,000,000 Btu/hr based on an inside air design temperature of 70°F. The building is heated with steam. (Assume 1000 Btu/lb of steam.)

 (a) Estimate the steam consumption for a heating season.
 (b) What is the average outside air temperature during the heating season (65°F base)?
 (c) If the inside air temperature is reduced to 65°F from 6:00 P.M. to 7:00 A.M., what is the weighted average inside air temperature?
 (d) Estimate the probable savings in steam consumption that could be saved by reducing the inside air temperature during part of the 24-hour period (night set-back), as indicated in part (c).

7.8. A controlled-temperature test room is to be held at 115°F. The test room is to be built in a storage space of an industrial plant where the air temperature is 60°F during the winter. The interior dimensions of the test room are 20 ft long by 15 ft wide, and the ceiling height is 9 ft. One of the 15-ft walls is the exterior wall of the building, which is constructed similar to wall construction No. 18a (Appendix Table C-7). The remaining three walls are constructed of 2-by-4 nominal wood studs placed 16 in. on center, faced with 3/8-in. fir plywood on each side. The stud space is filled with mineral wool batts. The test room has one 4-ft-by-7-ft, $1\frac{3}{8}$-in.-thick, solid-core, flush wood door, but no windows. If the outside design air temperature is 0°F, determine the following:
 (a) The U factor for the stud walls.
 (b) The required thickness of glass fiber, organic-bonded slabs (4 to 9 lb/ft³) to be applied to the exterior wall to give a U factor similar to the other three walls.
 (c) The design heat loss from the test room in Btu/hr, neglecting heat loss through floor and ceiling.

7.9. A partition wall 20 ft long and 8 ft high is built in an existing industrial building to close off a section of the plant. Using material on hand, the wall was constructed of 1-in. nominal pine boards (25/32-in. thick), tongue and groove, 4-in. mineral wool (resin binder) with a resistance of 3.51 (hr)(ft²)(°F)/(Btu)(in.), and an inside cover of pine boards similar to the outside layer. The boards are held in place over the insulation by 3/4-in. steel through-bolts in such a way that one bolt is used for each 1 1/2 ft² of wall area. In addition, a lattice of nominal 2-by-4 wood studs for bracing and cross-ties on 6-ft centers were attached to the wall for rigidity. Assume that steel has a conductivity of 28 (Btu)(ft)/(hr)(ft²)(°F). Disregarding the thermal effect of the wood bracing, determine **(a)** the U factor for the wall, disregarding the presence of the steel bolts, and **(b)** the same, but considering the effect of the steel bolts.

7.10. A large business office has two exposed walls. The long side has nine loose-fitting, nonweatherstripped, double-hung wood-framed windows, each measuring 3 ft wide by 6 ft high. This is the windward side of the building. There are six such windows in the other exposed wall. The window frames are caulked in the masonry walls. If the design wind velocity is 15 mph, estimate the infiltration heat loss when the inside air condition is 70°F DB and 30% RH and the outside air condition is 35°F DB and 30°F WB. What would be the reduction in heat loss if the windows were weather-stripped?

7.11. A commercial building has two sets of double-swinging doors on the windward side of the building. The doors open into a vestibule. Each swinging door is a standard 3-by-7-ft door. The average traffic rate of people on a busy day is 300 persons per hour, with a maximum rate of 500 per hour. Estimate the rate of air infiltration through the doors when the wind velocity is 10 mph normal to the building on the door side.

7.12. A room has four vertically pivoted, steel-sash windows that measure 25 in. wide by 62 in. high located on the windward side of the room. The room also has two tight-fitting interior doors that open into the central part of a relatively open building. At design conditions, the outside air is at 10°F dry-bulb temperature and 100% RH, and the inside air is at 70°F and 50% RH. The wind velocity is 10 mph normal to the windows.

    (a) Estimate the probable infiltration with both interior doors open.

    (b) Estimate the probable infiltration with both interior doors closed.

    (c) Estimate the heat loss due to infiltration at the two conditions.

7.13. A small commercial building has a calculated heat loss at design conditions of 200,000 Btu/hr sensible and 25,000 Btu/hr latent. The latent heat is due to outside air infiltration. The building is to be heated with forced-circulation warm air supplied at 45°F above the inside air state at 70°F DB and 30% RH.

    (a) Using the psychrometric chart, calculate the required supply air quantity (cfm) at design conditions.

    (b) Calculate the supply air quantity required based on sensible heat only.

    (c) If the building is located in Oklahoma City, estimate the quantity of natural gas required to heat the building during an average winter. Assume that the heating value of natural gas is 1000 Btu/ft³.

7.14. A certain building located in Boston, Massachusetts, has a design sensible heat loss of 150,000 Btu/hr (assume no latent loss) based on 70°F DB inside and 10°F outside.

    (a) For a heating season that extends from October 1 to May 31, estimate the average outside air temperature.

    (b) Also estimate the probable fuel oil consumption using No. 2 oil having a heating value of 140,000 Btu/gallon.

7.15. Determine the conduction heat loss through the four sides of an industrial building that measures 40 ft wide and 60 ft long, with a ceiling height of 12 ft. The walls are constructed of 4-in. face brick, 8-in. concrete blocks (three-holed oval-cored, sand-and-gravel aggregate), and 3/4-in. metal lath and gypsum plaster (lightweight aggregate) fastened to 2-by-2-in. furring strips placed 24 in. on center. The air space contains 1-in.-thick cellular glass slab insulation bonded to the concrete blocks. The building contains a total of 12 single-glazed, aluminum-frame (no thermal break) windows, each measuring 3 ft by 6 ft, and three 4-ft-by-7-ft steel doors, 1 3/4-in. thick with mineral wool core and no glazing. The inside and outside air temperatures are 70°F and 10°F, respectively.

7.16. A single-story house is built on a concrete floor slab laid at grade level. Insulation (*R*-5.4) is placed against the footing walls from the edge of the slab to the footer. The footing walls are 8-in. block with brick facing. The inside and outside design temperatures are 70°F and 20°F, respectively, and design degree-days are 4200. The house is rectangular and measures 30 ft by 50 ft. Determine the slab floor loss in Btu/hr.

7.17. The price of energy is continuously changing, but at one period the following price quotations were applicable: $5.00 per 1000 ft³ of gas (1000 Btu/ft³), 93 cents per gallon of No. 2 fuel oil (140,000 Btu/gallon), and 9 cents per kW-hr for electricity. Assume utilization efficiencies are 82% for gas, 75% for oil, and 100% for electricity.

    (a) Compare the energy costs per therm of utilized energy using the above energy sources.

    (b) Compute the operating costs per hour for heating a large residence with a calculated heat loss of 160,000 Btu/hr when operating at this maximum design demand.

7.18. The kitchen in a house has two exposed walls and a pitched roof–ceiling combination. The wood stud walls have a *U* factor of 0.15 Btu/hr-ft²-°F, and the roof-ceiling combination has a *U* factor of 0.09 Btu/hr-ft²-°F. The east wall contains a double-glazed (3/8-in. air space), aluminum-frame (with thermal break) window that measures 48 in. wide by 30 in. high. The south wall contains a similar window that measures 24 in. wide by 36 in. high, and a solid-core wood door that measures 3 ft by 7 ft and is 1 3/8 in. thick. The roof-ceiling has a 28-degree slope and contains two double-glazed (1/2-in. air space), aluminum-frame (with thermal break) skylights that measure 18 in. by 36 in. The east and south walls measure 15 ft and 20 ft by 8.5 ft high, respectively. The roof–ceiling measures 17 ft by 20 ft. Estimate the heat loss by conduction through the walls and roof of the kitchen if the inside and outside design temperatures are 70°F and 10°F, respectively.

BIBLIOGRAPHY

7.1. *ASHRAE Cooling and Heating Load Calculation Manual* (ASHRAE GRP 158), American Society of Heating, Refrigerating, and Air Conditioning Engineers, Atlanta, GA.

7.2. *ASHRAE Handbook 1989 Fundamentals,* American Society of Heating, Refrigerating, and Air Conditioning Engineers, Atlanta, GA, 1989.

7.3. Howell, R. H., and Sauer, Jr., H. J., *Environmental Control Principles,* a textbook supplement to the *ASHRAE Handbook 1985 Fundamentals,* American Society of Heating, Refrigerating, and Air Conditioning Engineers, Atlanta, GA.

7.4. Knebel, David E., *Simplified Energy Analysis Using the Modified Bin Method,* American Society of Heating, Refrigerating, and Air Conditioning Engineers, Atlanta, GA.

7.5. Strock, C., and Koral, R. L., *Handbook of Air Conditioning, Heating, and Ventilating,* 3rd ed., Industrial Press, New York, 1979.

7.6. Latta, J. K., and G. G. Boileau. 1969. Heat losses from house basements. *Canadian Building* 19(10):39.

8

Heat-Transfer Devices

8-1 INTRODUCTION

After the calculations have been made to determine the heating and ventilating requirements for a building, we are ready to select the type of system that will meet these requirements. The type of system selected will determine the type of heat-transfer devices to be used in the various areas to be conditioned.

The number and types of heat-transfer devices are many. This chapter will identify the most common types, present some performance data of these units, and describe some of the rating procedures used to establish standards for the equipment.

The numerous heat-transfer devices may be classified generally as (1) *energy conversion units* or (2) *terminal units*. Some devices encountered may be classified as combination units. One such type is the space heater located within the space to be heated. The space heater burns a fuel such as coal, gas, oil, or wood and some of the released heat energy is delivered by radiation and convection directly to the space to be heated. Another type is the electric resistance heater, which, when located within the space, converts electrical energy into heat energy, which is delivered directly to the space by radiation and convection. Basically, however, the energy conversion units convert some of the energy in

fuels into heat energy that is transferred to a working fluid. The working fluid circulates to the various terminal units where the heat is transferred to the space. The most common energy conversion units are boilers and the hot-air furnaces.

8-2 BOILER TYPES

Steam and hot-water boilers for heating and process are built of steel or cast iron in a wide variety of types and sizes for low- or high-pressure operation. The nationally recognized code governing the construction of low-pressure steel and cast-iron heating boilers is the ASME Code for Heating Boilers. Some states and municipalities have their own codes, which apply locally, but they are usually patterned after the ASME Code.

The maximum allowable working pressures for heating boilers are limited by the ASME Code for Heating Boilers to 15 psig for steam and 160 psig for hot-water boilers, with a maximum temperature limitation of 250°F (121°C). Hot-water boilers are generally designed for a working pressure of 30 psig and may be equipped for higher working pressures for either heating purposes or for hot water supply only when designed, tested, and stamped for higher pressures.

The nationally recommended code governing the design and construction of high-pressure steel boilers is the ASME Code for Power Boilers. This code is used almost universally in the design of steam boilers for operating pressures over 15 psig. High-pressure boilers used for combination heating and process loads, or process loads alone, are normally designed for 150 psig and higher, and most local codes require that boilers operating over 15 psig must have constant attendance by a licensed operator. Unusually large or specialized heating systems are sometimes equipped with boilers designed for pressures above the limits of the code for hot-water heating boilers. These boilers fall under the jurisdiction of ASME Code for Power Boilers.

Boilers may be classified in a number of different ways, such as the following:

1. *Materials of construction*—Most low-pressure boilers are constructed of cast iron or steel, but nonferrous metals are used for some.
2. *Design pressures*—As previously noted, heating boiler pressures are 15 psig for steam and 160 psig for water. Boilers designed for higher pressures are power boilers and are normally constructed of steel.
3. *Raw energy used*—Coal (hand-fired or stoker-fired), oil, gas, or electricity. Most boilers are designed for just one type of fuel, but some may be designed for dual-fuel operation.
4. *Type of application*—Space heating or domestic hot water service.
5. *Construction*—Sectional, round, firetube, or watertube.
6. *Type of combustion air-handling system*—Natural draft, induced draft, or forced draft.

Cast-iron boilers are usually constructed with vertical sections and are square or rectangular. Large boilers are usually shipped in sections and assembled at the place of installation. However, some boilers are shipped factory-assembled as *boiler–burner units,* or *packaged-boilers,* having all components in an assembled unit ready for connections to the piping, fuel, and electric power. In the majority of cast-iron boilers, the sections are assembled with push nipples and tie rods, but external headers and screw nipples are used in some. Many sectional boilers are provided with large push nipples to permit the circulation of water between adjacent sec-

tions at both the waterline and bottom of the boiler. This is also necessary to permit the use of an indirect water heater with the boiler for summer–winter domestic hot water supply. Many sectional-type boilers may be enlarged by the addition of sections and corresponding plate work.

Figure 8-1 shows one of the larger-capacity sectional-type cast-iron boilers. The No. 88 shown, manufactured by Weil-McLain Company, Inc., a division of the Marley Company, is available as a boiler–burner unit complete with light oil, gas, or combination gas–light oil burner. It is also available as a boiler only, without burner.

The No. 88 is designed for efficient trouble-free heating in apartment buildings, schools, churches, offices, and other commercial and institutional buildings. The boiler is available for water or steam service with net I-B-R ratings from 690,400 to 4,034,800 Btu/hr.

The unit features forced-draft firing at over 82% operating efficiency. It is available in individual sections, with factory-assembled sections, or as a fire-tested package unit. Outstanding design and construction features include provision for multiple tankless water heaters, patented section sealing method, Hydro-Wall design, large 9-in. top port opening, insulated steel jacket, built-in air eliminator in water boilers, simplified piping, no refractory combustion chamber, no separate base, and Weil-McLain cast-iron construction.

The No. 88 is pressurized for forced-draft firing and therefore does not require a chimney for draft; only a 3-ft vent above the building roof is necessary. This feature is particularly valuable in replacement installations since an existing chimney with insufficient draft because of low height or poor construction is not a problem. Other advantages of a forced-draft boiler are that no mechanical draft equipment is required, boiler room space requirements are reduced, and a pressurized boiler is more efficient.

Capacities of cast-iron heating boilers generally range from capacities required for small residences up to 13,000 MBh (MBh means thousands of Btu/hr) gross output. Table 8-1 presents the engineering performance data for the No. 88 boiler shown in Figure 8-1.

Figure 8-2 shows two high-efficiency, cast-iron, residential package-type boiler–burner units. Figure 8-2a is the Burnham V1 Series boiler designed specifically for hot-water heating systems. Figure 8-2b is the Burnham V7 Series boiler designed specifically to meet the increasing residential demand for quality oil-fired steam heat, especially in metropolitan areas. Both the V1 and V7 boilers are fully factory-packaged cast-iron boilers given a capacity range suitable for most residences. Annual efficiency of up to 85% is claimed, and instantaneous heat coils may be provided for domestic

FIGURE 8-1   Sectional cast-iron boiler for hot water or steam: No. 88 multifuel boiler. (Courtesy of Weil-McLain, Michigan City, IN. A Marley Company)

hot water. Tables 8-2 and 8-3 show ratings for the V1 and V7 boilers, respectively, along with optional tankless heaters.

Steel boilers may be of the *firetube* type, in which the combustion gases pass through the tubes as boiler water circulates around them, or of the *watertube* type, in which the combustion gases circulate around the tubes and the water passes through the tubes. Either the firetube or watertube type may be designed with integral water-jacketed furnaces or arranged for refractory-lined brick or refractory-lined jacketed furnaces. (Those with integral water-jacketed furnaces are called *firebox boilers* and are the most common type.) They are usually shipped in one piece, ready for piping connections. A typical firetube boiler is shown in Figure 8-3, with performance data presented in Table 8-4. Capacities for steel heating boilers range from those required by small residences up to about 23,500 MBh gross output.

TABLE 8-1

Performance ratings of No. 88 boiler shown in Figure 8-1

Boiler Unit Number (Steam or Water)	I-B-R Burner Capacity△		Gross I-B-R Output Btu/hr +	Net I-B-R Ratings‡			Net Sq. Ft Water ***	Boiler hp	Net Firebox Volume Cu. Ft	Stack Gas Volume cfm ****	Positive Pressure in Firebox □	Approx. Shipping Wt. (Lbs.) Boiler Only	Boiler Water Content Gals.	I-B-R Vent Dia. Inches
	Light Oil GPh **	Gas MBh ○		Steam Sq. Ft	Steam Btu/hr	Water Btu/hr								
▲ 488R*F•	6.9	996	794,000	2,480	595,600	690,400	4,605	23.7	11.02	370	.42	2,500	109	10
▲ 488*F•	7.0	1,010	810,000	2,530	607,700	704,300	4,695	24.2	11.02	376	.43	2,500	109	10
▲ 588*F•	9.4	1,357	1,084,000	3,390	813,200	942,600	6,285	32.4	14.45	507	.44	3,035	132	10
▲ 688*F•	11.8	1,703	1,358,000	4,275	1,026,500	1,180,900	7,875	40.6	18.08	639	.46	3,570	155	10
▲ 788*F•	14.2	2,049	1,632,000	5,230	1,255,400	1,419,100	9,460	48.8	21.61	772	.47	4,105	178	12
▲ 888*F•	16.6	2,396	1,904,000	6,155	1,477,100	1,655,700	11,040	56.9	25.14	906	.49	4,640	201	12
▲ 988R*F•	17.2	2,497	1,992,000	6,445	1,546,600	1,732,200	11,550	59.9	28.67	954	.45	5,175	224	14
▲ 988*F•	18.8	2,713	2,176,000	7,040	1,689,400	1,892,200	12,615	65.0	28.67	1,031	.50	5,175	224	14
▲ 1088R*F•	20.0	2,886	2,304,000	7,455	1,788,900	2,003,500	13,355	68.8	32.20	1,101	.49	5,710	247	14
▲ 1088*F•	21.5	3,103	2,452,000	7,930	1,903,700	2,132,200	14,215	73.2	32.20	1,184	.52	5,710	247	14
▲ 1188*F•	23.5	3,392	2,724,000	8,810	2,114,900	2,368,700	15,790	81.4	35.73	1,299	.53	6,245	270	14
▲ 1288*F•	26.0	3,753	3,000,000	9,705	2,329,200	2,608,700	17,390	89.6	39.26	1,443	.55	6,780	293	14
▲ 1388*F•	28.5	4,113	3,270,000	10,580	2,538,800	2,843,500	18,955	97.7	42.79	1,588	.56	7,315	316	14
▲ 1488*F•	31.0	4,474	3,550,000	11,485	2,756,200	3,087,000	20,580	106.0	46.32	1,735	.58	7,850	339	16
▲ 1588*F•	33.0	4,763	3,820,000	12,360	2,965,800	3,321,700	22,145	114.1	49.85	1,854	.59	8,385	362	16
▲ 1688R*F•	34.5	4,979	3,980,000	12,875	3,090,100	3,460,900	23,070	118.9	53.38	1,945	.59	8,920	385	16
▲ 1688*F•	35.5	5,124	4,090,000	13,230	3,175,500	3,556,600	23,710	122.2	53.38	2,002	.61	8,920	385	16
▲ 1788*F•	38.0	5,485	4,370,000	14,135	3,392,900	3,800,000	25,335	130.5	56.91	2,152	.62	9,455	408	18
▲ 1888*F•	40.5	5,845	4,640,000	15,010	3,602,500	4,034,800	26,900	138.6	60.44	2,303	.64	9,990	431	18

Source: Courtesy of Weil-McLain, Michigan City, IN. A Marley Company.

▲ Substitute "BL" for light oil, "BG" for gas, "BGL" for gas-light oil, or "H" for boiler only for use with approved burners. Add prefix "A" to designator for factory-assembled No. 88 (example ABL-488). Substitute "P" for "B" for fire-tested package unit (example PL-488). Boilers with "R" in model number are furnished with reduced ratings.

* Substitute "S" for steam, "W" for water.

• For T-Intermediate section(s) and tankless heater(s) add suffix "(number required) TIH"; for T-Intermediate section(s) with cover plate(s) only add suffix "(number required) TIP"; for T-intermediate section(s) with storage heater(s) add suffix "(number required) TISH."

△ Burner input based on maximum of 2,000 ft. altitude—for higher altitudes consult Weil-McLain Application Engineering Department.

** No. 2 fuel oil—Commercial Standard Spec. CS75-56. Heat value of oil—140,000 Btu/G.

○ Consult Weil-McLain Application Engineering Department for gas pressure required.

+ Gross I-B-R ratings have been determined under the I-B-R provision governing forced draft boiler-burner units.

‡ Net I-B-R ratings are based on net installed radiation of sufficient quantity for the requirements of the building and nothing need be added for normal piping and pick-up. Water ratings are based on a piping and pick-up allowance of 1.15. Steam ratings are based on the following allowances: 488 and 588—1.333; 688—1.323; 788—1.300; 888—1.289; and 988 through 1888—1.288. An additional allowance should be made for gravity hot water systems or for unusual piping and pick-up loads. Consult Application Engineering Department.

*** Based on average water temperature of 170°F in heat distributing units.

**** Stack gas volume at outlet temperature.

□ With 0.10" WC positive pressure at flue collar.

Note: Water boilers available upon special request at 80 PSI working pressure.

(a)

(b)

FIGURE 8-2   Residential cast-iron boilers: (a) V1 water boiler and (b) V7 steam boiler. (Courtesy of Burnham Corporation, Hydronics Division, Lancaster, PA)

TABLE 8-2

V1 water boiler ratings

Boiler Number (1)	Burner Capacity GPh (2)	DOE Heating Capacity MBh	I-B-R Net Rating Water MBh (3)	Tankless Heater Capacity V1-1 Heater GPm (4)	DOE Annual Efficiency %
V-13A	.75	89	77.4	$3\frac{1}{4}$*	83.5
V-13A RO/FO	.75	89	77.4	$3\frac{1}{4}$*	83.5
V-14A	1.05	124	107.8	$3\frac{1}{2}$	83.4
V-14A RO/FO	1.05	124	107.8	$3\frac{1}{2}$	83.4
V-15A	1.35	160	139.1	4	83.3
V-16A	1.65	195	169.6	4	83.2
V-17A	1.90	225	195.7	4	83.2
V-18A	2.10	248	215.7	4	83.2
V1-72	.60	73	63.5	$3\frac{1}{4}$*	85.4
V1-96	.80	97	84.3	$3\frac{1}{4}$*	85.4

TABLE 8-2 *(Continued)*

| Boiler Number | *Optional Tankless Heaters* | |
	Tankless Heater Number	Tankless Heater Capacity (GPm) (4)
V-14AT	V1-2	4
V-15AT	V1-2	$4\frac{1}{2}$
V-16AT	V1-2	$4\frac{1}{2}$
V-17AT	V1-2	5
V-18AT	V1-2	5

Source: Courtesy of Burnham Corporation, Hydronics Division, Lancaster, PA.

DOE heating capacity and annual efficiency are based on U.S. Government Standard tests.

*Uses tankless heater V1-2.

Notes:
1. Add suffix "T" to denote boiler with tankless heater back section and with tankless heater shown.
2. The I-B-R burner capacity is based on oil having a heat value of 140,000 Btu/gal.
3. Net I-B-R ratings shown are based on piping and pickup allowance of 1.15.
4. Tankless heater ratings are based on 40°–140°F rise with boiler temperature at 200°F—intermittent draw.

Electric boilers generally can be classified in one of the following categories:

1. Electric steam boiler with (a) immersion heater-type consisting of resistance heaters immersed below the water line or (b) electrode-type consisting of bare electrodes immersed in the water. Heat is generated by resistance of the water to the flow of electricity between the electrodes.

FIGURE 8-3 Firetube steel boiler for steam or hot water.

2. Electric hot-water boilers, which are (a) conventional boilers (large internal volume) with immersed electric resistance heaters, (b) instantaneous-type (minimum internal volume) with immersed electric resistance heaters, or (c) instantaneous-type with wraparound electric heaters. The heater is bonded to the outside of the water shell.

Residential and small commercial model electric boilers are generally quite small and compact and are often mounted on the wall to save floor space. Some of

TABLE 8-3

V7 steam boiler ratings

| Boiler Model (1) | Burner Capacity GPh (2) | DOE Heating Capacity MBh (3) | I-B-R Net Rating | | | AFUE | |
			Water MBh (4)	Steam MBh (5)	Steam Sq. Ft	Steam	Water
V-73	0.75	89	77	67	278	83.0	84.1
V-74	1.05	125	109	94	391	82.9	83.9
V-75	1.35	160	139	120	500	82.9	83.8
V-76	1.65	195	170	146	610	82.8	83.6
V-77	1.90	225	196	169	703	82.9	83.7
V-78	2.10	249	217	187	778	82.9	83.8

Notes:
1. Add suffix "T" to denote boiler with tankless heater back section and with tankless heater shown.
2. The I-B-R burner capacity is based on oil having a heat value of 140,000 Btu/gal.
3. DOE heating capacity and annual efficiency are based on U.S. Government Standard tests at 13.0% CO_2 with No. 1 maximum smoke.
4. Net I-B-R ratings are based on piping and pickup allowance of 1.15.
5. Net I-B-R ratings are based on piping and pickup allowance of 1.33.

TABLE 8-4

Performance and rating of firetube boiler shown in Figure 8-3

Unit Number		3R1-F	3R2-F	3R3-F	3R4-F	3R5-F	3R6-F	3R7-F	3R8-F	3R9-F	3R10-F	3R11-F	3R12-F
SBI gross output	MBh.	324	396	468	540	648	792	936	1080	1260	1440	1620	1800
horsepower		10	12	14	16	19	24	28	32	38	43	48	54
steam per hour 212°F.	lbs.	334	408	482	557	668	816	965	1113	1299	1484	1670	1855
steam	sq. ft.	1350	1650	1950	2250	2700	3300	3900	4500	5250	6000	6750	7500
SBI net rating—steam	sq. ft.	1010	1240	1460	1690	2020	2480	2930	3380	3940	4500	5060	5620
Firing rate—gas 1000 Btu/cu. ft.		405	495	585	675	810	990	1170	1350	1575	1800	2025	2250
oil 140,000 Btu/gal.		2.9	3.5	4.2	4.8	5.8	7.1	8.4	9.6	11.2	12.9	14.5	16.1
Heating surface—Total	sq. ft.	53	65	77	88	106	129	153	177	206	236	265	294
primary	sq. ft.	20.5	22.8	25.2	27.5	33.0	36.6	40.4	44.5	49.5	54.1	58.4	62.6
Furnace vol. net	cu. ft.	7.8	9.1	10.5	11.7	15.4	17.9	20.4	23.3	29.5	33.4	37.1	40.8
Safety valve capacity steam per hour	lbs.	424	520	616	704	848	1032	1224	1416	1648	1888	2120	2352
Approximate dry weight	lbs.	1650	1800	1950	2100	2350	2700	3000	3300	3700	4150	4500	4900
Vent diameter—forced draft	in.	6	6	6	6	8	8	8	8	10	10	10	10
Stack diameter—natural draft	in.	12	12	12	12	15	15	15	15	18	18	18	18
Stack height—natural draft	ft.	30	35	35	40	35	40	40	45	35	40	45	50

Source: Courtesy of Kewanee Boiler Corporation.

147

TABLE 8-3 *(Continued)*

Boiler Model	Coil No.	Tankless Heater Rating—GPm	
		Steam Blr.	Water Blr.
V-73	V1-2	$2\frac{3}{4}$	3
V-74	V1-2	3	$3\frac{1}{4}$
V-75	V1-2	$3\frac{1}{4}$	$3\frac{1}{2}$
V-76	V1-2	$3\frac{3}{4}$	$3\frac{3}{4}$
V-77	V1-2	$3\frac{3}{4}$	4
V-78	V1-2	4	$4\frac{1}{2}$

Optional Tankless Heaters

Source: Courtesy of Burnham Corporation, Hydronics Division, Lancaster, PA.

Note: Tankless heater rating GPm based on 100°F rise with 190°F boiler water temperature with 40°F inlet water (intermittent draw).

the smaller models are designed to be mounted behind a finned-tube enclosure.

Large commercial electric boilers of the electrode type are available for floor mounting with steam capacities from 1070 to 9500 lb steam per hour (200 to 300 kw) and hot water capacities from 1440 to 6520 MBh (400 to 20,000 kW). Pressure selection of these boilers is from low pressure to 2000 psig.

8-3 BOILERS FOR SPECIAL APPLICATIONS

Boilers for *domestic hot water supply* are classified as *direct* if the water heated passes through the boiler and as *indirect* if the water heated does not come in contact with the water or steam in the boiler.

Direct heaters are built to operate at the pressures found in city water mains and are tested at pressures from 200 to 300 psig. The life of direct water heaters depends almost entirely on the scale-forming properties of the water supplied and the temperatures maintained. If low water temperatures are maintained, the life of the heater will be much longer due to decreased scale formation and minimized corrosion. Direct water heaters may be designed to burn refuse and garbage.

Indirect heaters generally consist of steam boilers in connection with heat exchangers of the coil or tube types, which transfer the heat from the steam to the water. This type of installation has the following advantages:

1. A steam boiler can operate at low pressure.
2. The boiler is protected from scale and corrosion.
3. The scale formed on the heat exchanger parts may be removed by cleaning or by replacing the heat exchanger components. The accumulation of scale does not affect efficiency,

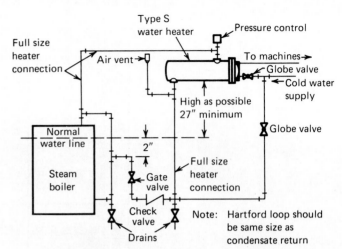

FIGURE 8-4 Indirect hot-water heater showing connections to steam boiler.

although it will affect the capacity of the heat exchanger.
4. Discoloration of water may be prevented if the water supply comes in contact with only nonferrous metal.

Where a steam or forced-circulation hot-water heating boiler system is installed, the domestic hot water may be heated by an indirect heater attached to the boiler. For most satisfactory performance in a steam system, the heater is placed just below the water line of the boiler. In a forced-circulation hot-water system, it should be located as high as practical with respect to the boiler; see Figure 8-4.

8-4 SELECTION OF BOILERS

In addition to the obvious choices of materials of construction, steam or hot water, type of fuel, and so forth, the selection of a boiler for a particular application depends on the load the boiler must handle and the boiler operational efficiency.

The *maximum* or *gross load* on the boiler is the sum of the following four load components:

1. *Radiation load*—The estimated heat emission in Btu/hr of the connected radiation. The connected radiation is determined by calculating the heat losses from each room. The sum of these heat losses is the total required heat emission of connected radiation, expressed in Btu/hr.
2. *Domestic hot water load*—The estimated maximum heat in Btu/hr required to heat water for domestic use. (Obviously, if the

domestic hot water is supplied by a separate heating device, this boiler load does not exist.)

I-B-R recommends that (a) no allowance is necessary for domestic hot water service unless there are more than two bathrooms to be served, or if the estimated use of hot water will not exceed 75 gal per day; and (b) if the estimated use of domestic hot water exceeds these limits, the following allowances should be made: (1)For a storage-type heater, allow 120 Btu/hr for each gallon of water storage tank capacity, and (2) for a tankless heater, allow 12,000 Btu/hr for each bathroom in excess of two.

3. *Piping tax*—The estimated heat loss from all sections of piping in the system.
4. *Warm-up* or *pickup allowance.*

The *net boiler load* is the sum of items 1 and 2.

Boiler operational efficiency is an important factor in selecting a space heating boiler, but the issue is not as simple as it appears because there are three different kinds of boiler efficiency:

1. *Combustion efficiency* expresses how efficiently the boiler burns fuel. Combustion efficiency is input minus stack (chimney) loss, divided by the input, and ranges from 75% to 86% for most heating boilers.
2. *Overall efficiency* (steady-state efficiency) expresses how efficiently the boiler uses the heat from combustion while operating. Overall efficiency is gross output divided by input. Gross output is measured in the steam or water leaving the boiler and depends on individual installation characteristics. Overall efficiency is lower than combustion efficiency by the percentage of heat lost from the outside surface of the boiler (this loss is usually called *radiation loss*). Overall efficiency can be precisely determined only by laboratory tests under fixed test conditions.
3. *Seasonal efficiency* expresses how efficiently the boiler uses fuel over the entire heating season.

While each type of efficiency is important, *seasonal* efficiency is the one that really counts; it determines how much the building owner will pay for fuel over the course of a heating season. For good seasonal efficiency, a boiler must have good steady-state efficiency *and* good combustion efficiency. But, a boiler with high steady-state efficiency (say, 80%) can easily

have a seasonal efficiency of only 65%. It all depends on downtime losses, losses that occur when the boiler is not operating. Downtime losses depend on boiler construction, type of application, and design of the system. When a boiler shuts off, the heat in the boiler will continue to radiate through the jacket. This is lost energy. In addition, boiler room air continues to flow through many boilers after the burner shuts off. This air is drawn up by the chimney, thus cooling off the boiler. When the burner starts again, it must reheat the boiler water back to operating temperature. In addition, it must reheat the chimney to produce proper draft for efficient combustion. The more often the burner cycles, the greater the losses and the lower the seasonal efficiency.

If a boiler is selected that is much smaller than required for the design load, it will more closely match the heating load of the building for a large part of the season, meaning fewer on-and-off cycles. When it can no longer keep up with the heat load, a second boiler picks up the extra load, and then a third boiler, if necessary.

Therefore, instead of just using one boiler for a job, two boilers, each having half the required capacity, or three boilers with one-third the required capacity, should be specified. The individual boilers will cycle one-half to one-third less than a single boiler, thus increasing seasonal efficiency. However, the key to maximizing this efficiency is to make sure that each boiler is completely isolated from the others so that nonoperating boilers are not hot with system water. This eliminates most of the jacket heat losses of nonoperating boilers. Prewired control panels are available to assure proper sequencing and temperature control.

Energy-saving multiple-boiler systems are available featuring two or more residential or commercial boilers that work together to provide the same total output as one large boiler but with greater energy efficiency and performance.

Figure 8-5 shows a multiple-boiler system (MBS) utilizing Weil-McLain high-efficiency residential gas boilers. The system sequences each boiler in the system so that only those boilers needed to handle the heating load are fired. This reduces standby losses and increases seasonal operating efficiency.

Manual isolation valves in the MBS installation allow individual boilers to be serviced without shutting down the entire system. The Weil-McLain Easy-Fit Piping System, featuring an exclusive easy-fit manifold, and prewired control panels make the Weil-McLain MBS systems easier and faster to install than a large commercial boiler. Additional boilers may be added to the system should heating requirements increase.

Multiple-boiler systems are controlled by one of

FIGURE 8-5 Multiple-boiler system utilizing Weil-McLain high-efficiency residential gas boilers. (Courtesy of Weil-McLain, Michigan City, IN. A Marley Company)

13 Weil-McLain energy management control systems, which assure proper function and control boiler cycling for increased operating efficiency and reduced fuel usage. The control systems automatically sequence individual boilers to energize the necessary units to meet load demands. Control panels are also available for multiple zoning with either zone valves or circulators and for combined space heating and domestic hot water service.

8-5 BOILER RATING AND TESTING CODES

Heating boilers are usually rated according to standards developed by (1) the Hydronics Institute, formerly the Institute of Boiler and Radiator Manufacturers (I-B-R), and the Steel Boiler Institute (SBI); (2) the American Gas Association (AGA); and (3) the American Boiler Manufacturers Association (ABMA).

The Hydronics Institute has adopted a standard for rating cast-iron and steel heating boilers ("Testing and Rating Standard for Cast-Iron and Steel Heating Boilers") based on performance obtained under controlled test conditions. The gross output obtained by testing is limited by certain factors, such as flue gas temperature, draft, CO_2 in the flue gas, and minimum overall efficiency. This standard applies primarily to oil-fired equipment but is also used for power gas ratings for dual-fueled units.

Boilers designed for burning gas are design-certified by AGA based on tests conducted in accordance with the American National Standard Z21.13 ("Gas Fired Low Pressure Steam and Hot Water Heating Boilers").

ABMA has adopted test procedures for commercial-industrial and packaged firetube boilers based on the ASME Performance Test Code PTC 4.1 ("Steam-Generating Equipment"). The units are tested for performance under controlled test conditions with minimum levels of efficiency required.

In 1978, the U.S. Department of Energy issued a test procedure applying to all gas- and oil-fired boilers up to 300,000 Btu/hr (90 kW) input. This procedure supersedes industry rating methods used previously. The DOE test procedure determined both on-cycle and off-cycle losses based on a laboratory test involving cyclic conditions. The test results are applied to a computer program, which simulates an actual installation and results in an Annual Fuel Utilization Efficiency (AFUE). The steady-state efficiency developed during the test is similar to combustion efficiency and is the basis for determining heating capacity, a term equivalent to gross output.

The I-B-R emblem and the SBI emblem are the property of the Hydronics Institute, registered in its name in the United States Patent Office and with the Registrar of Trademarks of Canada. The emblems were designed and are to be used only to indicate I-B-R and SBI approval of ratings in specific units. Therefore, the use of the emblems by manufacturers is restricted to those portions of their literature where I-B-R and SBI ratings are shown. No one has the right to state that their boilers have been tested under the Institute Boiler Standard or to publish I-B-R or SBI ratings or to use the I-B-R or SBI emblem in connection therewith unless (1) the manufacturer has executed a license with the institute and (2) the institute has advised the manufacturer in writing the exact I-B-R or SBI ratings that are applicable to the specific unit.

All I-B-R or SBI rated boilers have been manufactured in accordance with the latest edition of the ASME Boiler and Pressure Vessel Code.

I-B-R and SBI ratings are expressed in terms of:

1. "Gross I-B-R or SBI Output," which is the total quantity of Btu's that the boiler will deliver per hour and at the same time meet all the limitations of the code.
2. "Net I-B-R or SBI Rating," which is the amount of installed radiation that will be served by the boiler based on the normal allowance for piping and pickup.

Note: Net ratings equal gross output divided by the piping and pickup factor. All net I-B-R and SBI ratings are based on piping and pickup factors of 1.15 for water, 1.333 for steam up to 1254 MBh gross output, decreas-

ing regularly from 1.333 to 1.288 between gross outputs of 1256 to 1936 MBh, and 1.288 for larger steam sizes. The manufacturer should be consulted before selecting a boiler for installation on a job having unusual piping and pickup requirements. It is essential that advice of the manufacturer of the boiler be solicited before selecting a boiler for replacement on an existing installation that has a larger amount of piping than would normally be installed or where other unusual conditions exist, such as intermittent operation, temperature conditions other than normal, and unbalanced systems.

The net I-B-R or SBI ratings usually show steam ratings in square feet of equivalent direct radiation, EDR, and Btu/hr, and water ratings in Btu/hr.

8-6 WARM-AIR FURNACES

If a warm-air furnace is defined as a heat-transfer device in which heat is released on one side of a heat exchanger surface and heat is absorbed by circulating air on the other side, a large number of devices can be included in this category, some of which are illustrated in schematic form in Figure 8-6.

Early Types

1. *Stove*—The earliest ancestor of the present-day warm-air furnace is the parlor stove, sometimes referred to as the "pot-bellied" stove. Aside from the cheery aspects of the flame that could be viewed through the isinglass openings, the stove was a utility that was tolerated but not aesthetically admired.

2. *Pipeless furnace*—Warmed air rose by gravity action and was delivered into the room above through a large grilled opening in the floor. Return air from the room passed through the same floor grille but into the cooler down-flow concentric passage.

3. *Gravity hot-air furnaces (pre-1918)*—These early systems were installed without benefit of engineering knowledge. Airflow was restricted, and the furnaces were aptly described as "hot-air" furnaces since the air temperatures were in excess of 200°F.

4. *Gravity warm-air furnaces (pre-1930)*—The early experimental work at the University of Illinois by Professors A. C. Willard and A. P. Kratz in the period following 1918 showed that, for example, markedly improved results could be obtained by use of streamlined, amply sized return-air ducts, large warm-air stacks, and registers with ample free area. In these improved systems, the actual operating register-air temperature was much less than the assumed design value of 175°F, even under design heater loads.

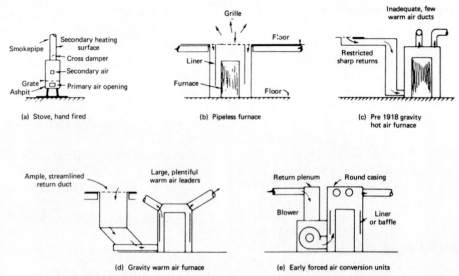

(a) Stove, hand fired

(b) Pipeless furnace

(c) Pre 1918 gravity hot air furnace

(d) Gravity warm air furnace

(e) Early forced air conversion units

FIGURE 8-6   Development stages of warm-air furnaces.

5. *Early forced-air conversion models (pre-1940)*—Even in the greatly improved gravity warm-air furnace system, the surfaces had to be extensive and the furnaces were large in size. The obvious step to increase the heat-transfer rate was to use some mechanical means for circulating the air. Propeller fans located in the return-air boot or in the supply bonnet were found to be lacking in pressure and capacity. When centrifugal fans were placed in a compartment attached to the casing, it was found that furnace casings had to be made smaller to prevent bypassing the heating surface. Later, casings became rectangular and smaller; fans, filters, burners, and controls were integrally built into the casing, and a whole new line of products were introduced. In the trade, the centrifugal fans were referred to as "blowers."

Current Forced-Air Furnace Types.   The current types of furnaces, some of which are illustrated in Figure 8-7, show a great diversity of shapes and arrangements. Among the principal types used for residences and small buildings, the following arrangements are common:

1. Low-boy arrangement with furnace and blower in separate compartments and with air discharged upward. Used for basement installation, but also usable in furnace closets located on the first story.
2. High-boy arrangement with up-flow of air in which the blower is located at the bottom of the casing and the heat exchanger is located in the top. Used for either basement installation or for first-story installation.
3. High-boy arrangement with down-flow of air

in which the blower is located at the top of the casing and the heat exchanger is located below the blower. Used mainly for first-story installation in which warm air is discharged downward.

4. Horizontal arrangement in which the blower and heat exchanger are located side by side and the air is discharged horizontally. Used for basement, attic, or crawl space installation.

The current types are integrally designed and coordinated units in which the blower and control equipment are selected and arranged for the specific unit. In fact, in the smaller-capacity units, the furnace package consists of a factory-wired and factory-assembled heat exchanger, blower, blower motor, filter, controls, and humidifier.

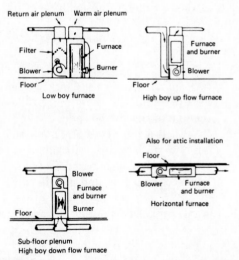

Low boy furnace

High boy up flow furnace

Sub-floor plenum
High boy down flow furnace

Horizontal furnace

FIGURE 8-7   Current types of warm-air furnaces.

Special Types of Warm-Air Furnaces. The dividing line between a warm-air furnace and a device that is not a warm-air furnace is difficult to define sharply. The following are included here as special types of warm-air furnaces:

1. *Space heater*—This reminder of the early parlor stove still persists. Most space heaters are gravity circulation systems, but some are provided with fans. No ducts are attached. Oil-fired and gas-fired space heaters are available. Those equipped to burn wood or coal are again becoming popular because of the high cost and short supplies of gas and fuel oil. Modern versions of the space heaters may include thermostatic temperature control equipment and operate at very reasonable efficiencies.

2. *Floor furnace*—This device is reminiscent of the earlier pipeless furnace but on a much smaller scale and with refinements of automatic firing and controls. The unit is usually attached to the ceiling joists of the basement and is ductless. The circulating air temperature is extremely high with gravity circulating units.

3. *Wall furnace*—This device is a space-saving unit that is installed in the wall of a room and is usually a gravity circulation, pipeless unit. Some are tall, and some are low enough to be installed just above the baseboard.

4. *Direct-fired unit heater*—Unit heaters can be direct-fired, either with or without a circulating fan behind the heat exchanger surface. Used mainly for industrial or commercial installation.

5. *Direct-fired floor model unit heaters*—Extremely large unit heaters are available either with or without duct connections. Used for commercial and industrial buildings. Capacities in excess of 1000 MBh are available.

6. *Industrial warm-air furnaces*—Industrial warm-air furnaces, or heavy-duty furnaces, are available with capacities in excess of 1000 MBh. Used for schools, churches, commercial buildings, and industrial buildings.

The styles, models, and types of warm-air furnaces are extremely diverse. Models are available for burning wood, coal, coke, fuel oil, kerosene, natural gas, bottled gas, and manufactured gas. The diversity was the result of demand—the demand of special building types and the demands of both the builder and the public for small, efficient, and economical units.

Furnace Performance Terminology. A few terms appear in catalog descriptions of warm-air furnaces that require definition (see also Figure 8-8).

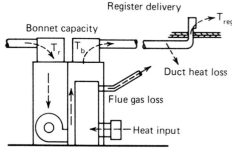

FIGURE 8-8 Terminology related to furnace performance.

1. *Input* or *heat input* is the rate at which heat is released inside the furnace, expressed in units of Btu/hr.

(a) *For gas:*

$$\frac{\text{input,}}{\text{Btu/hr}} = \frac{\text{cu. ft gas}}{\text{per hr}} \times \frac{\text{heating value,}}{\text{Btu/ft}^3}$$

(b) *For fuel oil:*

$$\frac{\text{input,}}{\text{Btu/hr}} = \frac{\text{gal oil}}{\text{per hr}} \times \frac{\text{heating value,}}{\text{Btu/gal}}$$

(c) *For coal or solid fuel:*

$$\frac{\text{input,}}{\text{Btu/hr}} = \frac{\text{lb fuel}}{\text{per hr}} \times \frac{\text{heating value,}}{\text{Btu/lb}}$$

2. *Bonnet capacity* is the heat available in the air at the furnace supply air bonnet, expressed in units of Btu/hr.

bonnet capacity

$$= (\text{cfm})_b(60)(0.24)(\rho_b)(t_b - t_r) \qquad (8\text{-}1)$$

where

$(cfm)_b$ = airflow rate in cfm measured at bonnet temperature, t_b

ρ_b = air density at atmospheric pressure and bonnet temperature, t_b

t_b = bonnet supply air temperature (°F)

t_r = bonnet return air temperature (°F)

3. *Duct heat loss* is the heat loss from the warm-air supply ducts to the spaces surrounding the ducts, expressed in units of Btu/hr (sometimes expressed as a percentage, say, 5% to 10%, of the bonnet capacity).

4. *Register delivery* is the heat available in the supply air that is delivered from the registers into the spaces to be heated, expressed in units of Btu/hr.

register delivery =

$$(\text{cfm})_{\text{reg}}(60)(0.24)(\rho_{\text{reg}})(t_{\text{reg}} - t_r) \qquad (8\text{-}2)$$

where

$(cfm)_{reg}$ = airflow rate in cfm measured at register air temperature, t_{reg}

ρ_{reg} = air density at atmospheric pressure and register air temperature, t_{reg}

t_{reg} = register supply air temperature (°F)

5. *Bonnet efficiency* is the ratio of the bonnet capacity to the input.

6. *Duct transmission efficiency* is the ratio of register delivery to the bonnet capacity.

7. *Flue gas loss* is the heat loss of the flue gases, expressed as a percentage of the heat input, and includes both sensible and latent heat losses.

8. *Combustion efficiency* is 100 minus the flue gas loss, expressed as a percentage.

Testing and Rating of Furnaces. A given furnace in combination with a given blower can operate in a number of given ways and provide almost an indefinitely large number of different capacities and efficiencies. For example, the rate of heat input to the furnace could be successively increased with a fixed speed of the blower and a set of performance data taken. After this, the whole set of tests could be repeated with another blower speed. From the standpoint of a commercial rating, it becomes necessary, therefore, to hold some of the variables at a constant value and vary only one item. This is shown by a typical performance curve for a furnace such as that illustrated in Figure 8-8. For this furnace, the rate of fuel input was successively increased after each test, but each time the blower speed was adjusted so that the air temperature rise through the furnace was maintained at a constant 100°F. The typical curves in Figure 8-9 show some interesting trends common to all furnace performance tests:

1. In common with most heat-transfer devices, the maximum efficiency occurs at a low value of input. In fact, the efficiency gradually decreases as the input increases.

2. The capacity increases with the increase in input, both not in a linear relationship. Theoretically, if the input could be made large enough, the capacity could be extended indefinitely.

3. The flue gas temperatures constantly increase with an increase in input since the heat exchanger surface is fixed in amount.

4. Since the tests were conducted with a constant rise in temperature of the air passing through the furnace (100°F), the airflow rate also increased. (Note that it would be possi-

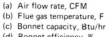
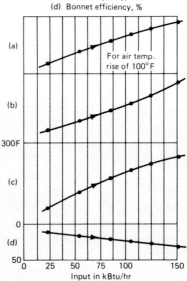

FIGURE 8-9 Typical trends of furnace performance tests.

ble to conduct tests with a constant airflow rate and a constantly increasing air temperature rise.)

From the standpoint of the engineer, a complete set of performance curves of the type shown in Figure 8-9 is much more informative and useful than a single-point rating; but, in general, a single-point rating is the only kind of rating that is acceptable to the trade. Every manufacturer is interested in a rating of the single-point type. The user is also interested in a single-point rating so that a furnace can be selected for a given application with reasonable assurance that it is adequate but not too large.

Acceptable Limits. Theoretically, the furnace whose performance curves are shown in Figure 8-9 could be rated at any value between, say, 10,000 Btu/hr up to 160,000 Btu/hr input. As a matter of fact, if no limits of any kind were imposed, the furnace could have been tested with higher and higher inputs until finally the furnace disintegrated. It becomes obvious that in order for a manufacturer to specify a single-point rating, it is necessary for the testing laboratory to impose limits, or boundaries, as to what is acceptable. The single-point rating that is finally chosen would not necessarily be a point of maximum efficiency since the maximum efficiency was obtained at an abnormally low input and capacity. On the other hand, the single-point rating would not be one for which the equipment was strained to the last notch. The arbitrary limits that happen to be selected by industry are not important for this discussion. What is important is that industry has sensibly and wisely agreed upon certain limits not only for the safeguard of the consumer but primarily for protection of the manufacturer who hopes to stay in business. Industry could mutually agree upon a value of only 75% as a minimum bonnet efficiency that would be acceptable. Or, a flue-gas temperature of, say, 800°F could be considered a maximum, or the surface temperature of the heat exchanger surface could be limited to, say, 900°F, or the draft could be limited, or all of these limits could be imposed simultaneously. Whatever capacity that corresponded with the most stringent limit would then be the capacity for rating purposes. In practice, an extremely stringent set of limits have been imposed by the American National Standard Approval Requirements for Gas-Fired Gravity and Forced-Air Central Furnaces (ANSI Z21.47), which is used universally in testing and rating. All gravity gas furnaces certified by the American Gas Association (AGA) under these requirements are assigned a rating based on 75% efficiency for gravity furnaces and 80% efficiency for forced-air furnaces.

Oil-fired furnaces equipped with pressure-type or rotary burners should bear the Underwriters' Laboratory (UL) label showing compliance with UL 296, which is basically a safety standard, and the Commercial Standard label (CS 75), which is a performance standard. In addition, the complete furnace should bear the UL 727 label and be so listed.

Selection of Furnace for Given Installation. The selection of a furnace for a given installation is simple in principle since it consists of selecting a furnace whose capacity is adequate to take care of the heat losses for the building under design winter weather conditions. In practice, a number of minor points arise that require clarification. For example, if the register delivery is to be made equal to the design heat loss, the manufacturer's catalog should either state the delivery or give some means of estimating the duct transmission efficiency. In this connection, the design manuals of the National Environmental Systems Contractors Association (NESCA) arbitrarily assume that register delivery will be 0.85 times the bonnet capacity or, in other words, that the duct transmission efficiency will be 85%. This assumption may be greatly in error in large installations. Another question that arises in connection with design heat loss is that of determining what portion of the house would be included in the heat loss calculations. For example, should the basement or crawl space heat loss be considered?

For the sake of uniformity and consistency, one set rule is offered for the selection of furnaces as follows (Strock and Koral, see the Bibliography):

Calculate the design heat loss for the entire structure, including the basement or crawl space, and select a furnace whose bonnet capacity is equal to this heat loss.

Another very important consideration in furnace selection is to realize that a great many installations are for year-round application where cooling equipment is installed in the ductwork for summer air conditioning. The blower must have adequate capacity to handle the required flow rate of air for summer cooling and to overcome the additional resistance of the cooling coil located in the air duct.

8-7 HEAT TRANSFER THROUGH HOLLOW CYLINDERS

One of the simplest terminal heat-transfer devices could be considered to be a circular hollow cylinder (say, a pipe), shown schematically in Figure 8-10. If a fluid is flowing inside the cylinder at temperature t_i and the fluid surrounding the outside of the cylinder was at temperature t_o, there would be a flow of heat from the higher to the lower temperature. If the two temperatures t_i and t_o were considered constant throughout the length L, we may express the steady-flow heat transfer by

$$\dot{q} = UA(t_i - t_o) \qquad (8\text{-}3)$$

where U is the overall heat-transmission coefficient in Btu/hr-ft²-°F or W/m²-°C, A is the surface area of the cylinder in ft² or m², and $t_i - t_o$ is the overall temperature difference (fluid to fluid) in °F or °C.

Since we are dealing here with a surface area that is different on the outside than on the inside, we must refer to this area as A_o (outside surface area) or A_i (in-

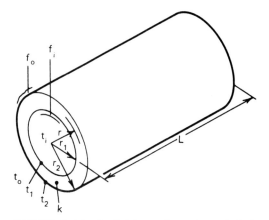

FIGURE 8-10 Radial heat flow through hollow cylinder.

side surface area); also, the U factor is dependent on which surface area we are referring to.

The U factor of Eq. (8-3) requires further explanation. This is the reciprocal of the sum of all the thermal resistances in the heat flow path. These thermal resistances are the inside fluid film resistance, f_i, the resistance of the metal wall, and the resistance of the outside fluid film, f_o. The fluid film resistances are dependent on the type and rate of fluid flow, the properties of the fluids, the size of fluid flow path, the nature of state change (evaporation, condensation), if any, that is produced by the heat transfer, and whether convection or radiation, or both, are involved. Heat transfer books and *ASHRAE Handbook 1989 Fundamentals* present many relationships and much data concerned with evaluating film coefficients.

The thermal resistance of the cylinder wall itself must take into account the fact that the inner and outer surface areas are not equal. At the inner surface, the area is $2\pi r_1 L$, and, at the outer surface, the area is $2\pi r_2 L$. Under steady-flow conditions, the heat transfer through the two areas must be equal. It can be shown that the heat transferred in Btu/hr is given by

$$\dot{q} = \frac{2\pi kL}{\ln(r_2/r_1)}(t_1 - t_2)$$

$$= \frac{2\pi kL}{2.3 \log_{10}(r_2/r_1)}(t_1 - t_2) \qquad (8\text{-}4)$$

where t_1 and t_2 are the inner and outer surface temperatures, L is the cylinder length measured along the cylinder axis, k is the thermal conductivity of the material, and r_1 and r_2 are the inner and outer radii, respectively. The thermal current is $\dot{q}/A$ and may be referred to either the inside or outside cylinder surface. Based on the inside surface area,

$$\frac{\dot{q}}{A_1} = \frac{\dot{q}}{2\pi r_1 L} = \frac{k}{r_1 \ln(r_2/r_1)}(t_1 - t_2) \qquad (8\text{-}5)$$

From Eq. (8-5), we may see that the thermal resistance of the cylinder wall must be

$$R = \frac{r_1 \ln(r_2/r_1)}{k} \qquad (8\text{-}6)$$

or, based on the outside surface area, the thermal resistance would be expressed as

$$R = \frac{r_2 \ln(r_2/r_1)}{k} \qquad (8\text{-}7)$$

Now, we may combine the resistance of fluid films and the cylinder wall to obtain the total thermal resistance. Based upon the outside surface area, the total resistance is

$$R_{to} = \frac{1}{f_i}\left(\frac{r_2}{r_1}\right) + \frac{r_2 \ln(r_2/r_1)}{k} + \frac{1}{f_o} \qquad (8\text{-}8)$$

Based upon the inner surface area, the total resistance is

$$R_{ti} = \frac{1}{f_i} + \frac{r_1 \ln(r_2/r_1)}{k} + \frac{1}{f_o}\left(\frac{r_1}{r_2}\right) \qquad (8\text{-}9)$$

The U factor now for Eq. (8-3) is simply the reciprocal of either Eq. (8-8) or Eq. (8-9), depending upon which surface area is being considered; usually, we are concerned with the outside surface area.

For thin wall cylinders such as pipe, the thermal resistance of the pipe wall, center term of Eq. (8-8), is very small in comparison to the resistances offered by the fluid films, especially if the fluids are gases. Therefore, frequently it is assumed that the inner and outer surface temperatures of the pipe wall are equal ($t_1 = t_2$). This is the assumption made for Table D-3 (see Appendix D) and discussed in the following sections.

8-8 HEAT LOSSES FROM BARE PIPE[1]

In the preceding section, we referred to the outside fluid film heat-transmission coefficient, f_o, as being dependent on both radiation and convection conditions. Heat is lost or gained by pipes by both radiation and convection. Radiation was discussed briefly in Chapter 6 where the Stephan-Boltzmann equation was introduced [Eq. (6-11)]. It is presented here in a slightly different form as

$$\frac{\dot{q}_r}{A} = 0.1713 e_e \left[\left(\frac{T_1}{100}\right)^4 - \left(\frac{T_2}{100}\right)^4 \right] \qquad (8\text{-}10)$$

where

$\dot{q}_r/A$ = Btu transferred by radiation per hour per square foot of pipe surface

e_e = effective emissivity of pipe

T_1 = absolute temperature of pipe (°F + 460)

T_2 = absolute temperature of surroundings (°F + 460)

---

[1]This section is extracted with permission from *Handbook of Air Conditioning, Heating, and Ventilating,* 3rd ed., Industrial Press, New York, 1979.

Heat transferred by convection can be determined by the Rice-Heilman formula, which resulted from work at Mellon Institute, as follows:

$$\frac{\dot{q}_c}{A} = C\left(\frac{1}{d}\right)^{0.2} \left(\frac{1}{T_{ave}}\right)^{0.181} (t_1 - t_2)^{1.266} \quad (8\text{-}11)$$

where

$\dot{q}_c/A$ = Btu transferred by convection per hour per square foot of pipe surface
C = constant with a value of 1.016 for horizontal pipe and 1.394 for vertical plates
d = outside diameter of pipe (in.)
T_{ave} = absolute average temperature of hot body and surrounding air (°F + 460)
t_1 = temperature of pipe surface (°F)
t_2 = temperature of surrounding air (°F)

The emissivity in the radiation formula is the "effective emissivity," taking into account the absorbtivity of the bodies receiving the radiation. An emissivity of 0.94 was used for oxidized steel pipe to determine the values in Appendix Table D-1; a value of 0.44 is used for tarnished copper tube to determine the values of Appendix Table D-2.

No convection formula is directly available for vertical pipes. However, the emission can be assumed to be closely approximating that of a steel plate. In this connection, note that C in Eq. (8-11) becomes 1.394. In the case of vertical pipes, d in the convection formula is not the diameter of the pipe but the height of the plate (pipe) in inches, in which the value of $(1/d)^{0.20}$ becomes constant when $d = 24$ in. Therefore, the convection formula for vertical pipes or vertical surfaces becomes

$$\frac{\dot{q}_c}{A} = 1.39 \times 0.53 \left(\frac{1}{T_{ave}}\right)^{0.181} (t_1 - t_2)^{1.266} \quad (8\text{-}12)$$

where the terms are as defined previously. For vertical pipe, the radiation is the same as for horizontal pipe.

The tables that follow in this section are based on the assumption that the outside surface of the pipe is at the same temperature as the fluid flowing in the pipe; this is not strictly true but is close enough for all practical estimating purposes.

Also, the tables assume an air temperature of 70°F and, in the case of radiation, a temperature of 70°F for the surrounding walls, machinery, and so forth. It can be seen, then, that if the surroundings are at temperatures appreciably above or below 70°F, the heat transferred by both radiation and convection would be ap-

preciably lower than or above, respectively, the values given in the tables. Consequently, the tables should not be used for cases where the air temperatures and surrounding bodies' temperatures are lower than 60°F or higher than 80°F.

The convection formulas are both based on free convection with no appreciable air motion from fans or open doors; in other words, the tables apply to "still" air condition.

In the formula for radiation, values for e_e other than those given for steel and iron pipe and copper tube are as follows:

Surface	Emissivity, e_e
Aluminum, polished	0.08
Aluminum paint	0.40
Brass	0.05
Cast iron	0.20
Lead	0.08
Nickel	0.06
Paint	0.94
Tin	0.08
Nonmetallic surfaces	0.90

Small iron pipe of 1/2-in. size, frequently left uninsulated, can be profitably painted with aluminum paint, which, with little effort, serves to reduce the emissivity of the pipe and, consequently, the radiation.

Although the emissivity of copper pipe is substantially lower than that for steel, so that the radiation loss is less, the convection loss is the same for copper as for iron where the conditions are the same. Therefore, it is not true that there is no reason to insulate hot lines because they are made of copper.

Many tables are available on heat losses from bare pipe; most of them, however, are on the basis of either Btu per square foot of pipe surface per hour or "Btu per linear foot of pipe per degree temperature difference (between pipe surface and air) per hour." Note, then, that Appendix Table D-1 is in Btu per linear foot per hour, with no additional calculation necessary.

ILLUSTRATIVE PROBLEM 8-1

A steel pipe, nominal size 3 in., carries hot water at 210°F under 10 psig of pressure. If the air temperature surrounding the pipe is 70°F, what is the heat loss per foot of pipe? If the pipe is 100 ft long, what would be the hourly heat loss?

Solution: From Appendix Table D-1, if the pipe temperature is assumed equal to the water temperature, 210°F, and the air temperature is 70°F, then the heat loss per linear foot is 303 Btu/hr-ft. Therefore, a 100-ft section would have a heat loss of 303 x 100 = 30,300 Btu/hr.

TABLE 8-5

Heat emission of pipe coils placed vertically on a wall (pipes horizontal) containing steam at 215°F (101.7°C) and surrounded with air at 70°F (21.1°C) (Btu/linear foot (W/m) of coil per hour [not linear feet (meter) of pipe])

Size of Pipe	1 in.	(25 mm)	$1\frac{1}{4}$ in.	(32 mm)	$1\frac{1}{2}$ in.	(40 mm)
Single row	132	(127)	162	(156)	185	(178)
Two	252	(242)	312	(300)	348	(335)
Four	440	(423)	545	(524)	616	(592)
Six	567	(545)	702	(675)	793	(762)
Eight	651	(626)	796	(765)	907	(872)
Ten	732	(704)	907	(872)	1020	(981)
Twelve	812	(781)	1005	(966)	1135	(1091)

Source: Reprinted from *ASHRAE Handbook 1988 Equipment,* with permission of American Society of Heating, Refrigerating, and Air Conditioning Engineers, Atlanta, GA.

Heating units made up of bare pipe coils were used some number of years ago mainly in industrial applications. The coils were built-up in a serpentine arrangement and hung vertically against a wall or horizontally at the ceiling. They are seldom, if ever, installed at the present time. However, for estimating replacement of any existing units, it is desirable to have capacity values of such old pipe coils. (See Table 8-5).

8-9   HEAT LOSS FROM INSULATED PIPE

In most cases for heating and cooling piping systems, we are concerned with reducing the heat loss or gain in the piping. Insulation installed on pipes and fittings is an effective method of saving energy.

Figure 8-11 illustrates a pipe section covered with a single thickness of insulation. Again, if we concern ourselves with only the heat transfer from the inner surface of the pipe to the outer surface of the insulation, we will be able to write a heat-transfer conduction equation similar in form to Eq. (8-4), and it would appear as

$$\dot{q} = \frac{t_1 - t_3}{R_t} = \frac{2\pi L(t_1 - t_3)}{\ln(r_2/r_1)/k_1 + \ln(r_3/r_2)/k_2} \quad (8\text{-}13)$$

where

$\dot{q}$ = Btu/hr heat loss through the composite structure for L feet of length

t_1, t_3 = inner surface temperature of pipe and outer surface temperature of insulation, respectively

k_1 = conductivity of metal pipe

k_2 = conductivity of insulation (see Table 8-6)

Again, to simplify our calculations, we will assume the resistance offered by the metal pipe is very small and concern ourselves with only the insulation.

For pipe covering in single thicknesses, the heat loss from hot pipes can be expressed as

$$\dot{q}_1 = \frac{k(t_2 - t_3)}{r_3\ln(r_3/r_2)} \quad (8\text{-}14)$$

where

$\dot{q}_1$ = heat loss (Btu/hr-ft² of outer surface of insulation)

k = conductivity of insulation

t_2 = temperature of inner surface of insulation (assumed equal to pipe temperature)

t_3 = temperature at outer surface of insulation

r_2 = inner radius of insulation

r_3 = outer radius of insulation

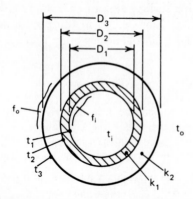

FIGURE 8-11   Radial heat flow through composite cylinder.

TABLE 8-6

Heat conductivity of pipe insulating materials (value to be used in the absence of specific data for the exact material and brand name used)

Insulation	Approx. Use Range (°F)	Approx. Density lb/ft³	Mean Temperature (°F) 100	200	300	400	500	600	700	800
			Conductivity, k, (Btu in.)/(hr ft² °F)							
85% Magnesia	600	11	0.35	0.38	0.42	0.46	—	—	—	—
Laminated asbestos	700	30	0.40	0.45	0.50	0.55	—	—	—	—
4-ply corrugated asbestos	300	12	0.57	0.62	—	—	—	—	—	—
Molded asbestos	1000	16	0.33	0.38	0.43	0.48	0.53	0.58	—	
Mineral fiber[a] wire-reinforced	1000	10	0.29	0.35	0.42	0.49	0.56	0.63	—	—
Diatomaceous silica	1600	22	—	—	—	0.64	0.66	0.68	0.70	0.72
Calcium silicate	1200	11	0.32	0.37	0.42	0.46	0.51	0.56	—	—
Mineral fiber,[a] molded	350	9	0.26	0.31	0.39	—	—	—	—	—
Mineral fiber,[a] fine fiber, molded	350	3	0.23	0.27	0.31	—	—	—	—	—
Wood felt	225	20	0.33	0.37	—	—	—	—	—	—

[a]Includes glass, mineral wool, rock wool, etc.

Source: Reprinted with permission. *Handbook of Air Conditioning, Heating, and Ventilating*, 3rd ed., Industrial Press, New York, 1979.

The objection to the practical application of Eq. (8-14) is that the temperature of the outer surface of the insulation, t_3, must be known.[2] A more useful formula is

$$\dot{q}_2 = \frac{t_p - t_a}{\dfrac{r_2 \ln(r_2/r_1)}{k} + \dfrac{1}{f_o}} \tag{8-15}$$

where

$\dot{q}_2$ = heat loss (Btu/hr-ft² of outer surface of pipe insulation)
t_p = temperature of pipe surface (°F)
t_a = temperature of ambient air (°F)
r_2 = radius of outer surface of insulation (in.)
r_1 = radius of outer surface of pipe (in.)
k = conductivity of insulation (Btu/ft²-hr-°F-in. of thickness)
f_o = surface conductance of outer surface of insulation (Btu/ft²-hr-°F)

Since, in most cases, one knows or can assume the temperature of the ambient air, Eq. (8-15) is more readily applicable than Eq. (8-14).

For very low air movements and low rates of heat transmission, the value of $1/f_o$ can be taken as 0.6.

The heat loss figure is usually, in actual problems, desired in Btu per linear foot of pipe rather than in terms of Btu per square foot of outer surface. The outer surface of insulation per linear foot of pipe is $(2\pi r_2)/12$; so Eq. (8-15) becomes

$$\dot{q} = \frac{0.523 r_2 (t_p - t_a)}{\dfrac{r_2 \ln(r_2/r_1)}{k} + 0.6} \tag{8-16}$$

where

$\dot{q}$ = Btu loss per hr per linear ft of pipe
r_2 = outer radius of insulation (in.)
r_1 = outer radius of pipe (in.)
t_p = temperature of pipe (°F)
t_a = temperature of ambient air (°F)

Equation (8-16) was used to calculate Appendix Table D-3 and shows the heat loss for a range of values of k and a range of values of $t_p - t_a$.

The problem becomes more complicated when two or more separate layers of pipe insulation are applied. The equation for this condition would be

$$\dot{q}_2 = \frac{t_p - t_a}{\dfrac{r_n \ln(r_1/r_p)}{k_1} + \dfrac{r_n \ln(r_2/r_1)}{k_2} + \dfrac{r_n \ln(r_n/r_{n-1})}{k_n} + \dfrac{1}{f_o}} \tag{8-17}$$

where q_2, t_p, t_a, and f_o are as given in Eq. (8-15) and r_1, r_2, . . . ,r_n are the outer radius, in inches, of the first, second, and nth layer of insulation, respectively; r_p is the outer radius of pipe, in inches; k_1, k_2, . . . , k_n are conductivities of the first, second, and nth layer of insulation, respectively.

[2]This section is extracted with permission from *Handbook of Air Conditioning, Heating, and Ventilating*, 3rd ed., Industrial Press, New York, pp. 8-177 to 8-178, 1979.

TABLE 8-7

Effect of air velocity on surface resistance

Heat Transmitted Btu/(hr ft²)	Velocity of Air, fpm			
	0	**100**	**200**	**400**
	Value of $1/f_o$			
0	—	0.56	0.50	0.41
50	0.60	0.52	0.45	0.39
100	0.55	0.48	0.41	0.36
150	0.50	0.45	0.39	0.34
200	0.48	0.42	0.37	0.32
300	0.43	0.38	0.33	0.29
500	0.36	0.33	0.28	0.25

Source: Reprinted with permission. *Handbook of Air Conditioning, Heating, and Ventilating,* 3rd ed., Industrial Press, New York, 1979.

The problem of two or more layers of insulation occurs more frequently in high-temperature work where, for example, the temperature may be too high for 85% magnesia, so the first layer may be calcium silicate or diatomaceous silica.

The value of thermal conductivity k for specific insulation is available from the manufacturer, and it is suggested that calculations be based on data that apply to the material and brand name of the insulation to be used. In the absence of such data, Table 8-6 has been included as a rough guide. Mean temperatures in Table 8-6 are the arithmetic mean between the inside surface and the outside surface of the insulation.

Appendix Table D-3 was calculated on the basis of still air with a $1/f_o = 0.60$. Actually, the surface resistance decreases as the air movement increases and as the rate of heat transfer is increased. Table 8-7 gives the effect of air movement on surface resistance.

When insulated piping is located where people or animals may come in contact with it, the criterion of economical heat loss in selecting insulation thickness may be outweighed by the need to maintain the outside surface temperature of the insulation at a safe level. The outside surface temperature may be estimated from

$$t_o = t_p - \left(\frac{x/k}{x/k + 1/f_o} \right) (t_p - t_a) \qquad (8\text{-}18)$$

where t_o is the outside surface temperature of the insulation in terms of t_p, t_a, k, and f_o as previously defined and x is the insulation thickness in inches.

ILLUSTRATIVE PROBLEM 8-2

A 4-in. nominal steel pipe carries water at 300°F and is insulated with 2 in. of 85% magnesia whose density is 12 lb/ft³. What will be the estimated outside surface temperature of the insulation if $t_a = 70°$F?

Solution: Assume $t_p = 300°$F and the approximate mean temperature in the insulation is 200°F. From Table 8-6, we find $k = 0.38$, and for still air conditions, $1/f_o = 0.60$. By Eq. (8-18),

$$t_o = 300 - \left(\frac{2/0.38}{2/0.38 + 0.6} \right) (300 - 70) = 94°F$$

This result of 94°F for the outside surface temperature is close to the 100°F used to estimate the arithmetic mean temperature in the insulation of 200°F. If there had been a significant difference between the calculated t_o and the estimated t_o, then a new value should be assumed to get a new arithmetic mean temperature, a new k, and the calculation repeated.

8-10 RADIATORS, CONVECTORS, BASEBOARDS, AND FINNED-TUBE UNITS

Radiators, convectors, baseboards, and finned-tubes are types of terminal heat-transfer units commonly used in hot-water heating and steam heating systems. They transfer heat from the system fluid (water or steam) to the surroundings by a combination of radiation and natural convection, and their function is to maintain the desired air temperature in the spaces where they are located. In general, these types of heat-transfer devices should be placed at points of greatest heat loss from the space in which they are located. For example, such units are located under windows, along exposed walls, and at door openings.

In general, those heat-transfer terminal units that have a large portion of their heated surface exposed to the space (radiators and cast-iron baseboards) emit a larger portion of heat by radiation than do units having completely or partially concealed heating surfaces (convectors, finned-tubes, finned-pipes). Also, finned-pipe heating elements constructed of steel emit a larger portion of heat by radiation than do finned-tube elements constructed of nonferrous materials. The heat output ratings of these various units are normally expressed in units of Btu/hr (Btuh), 1000 Btu/hr (MBh), or in square feet equivalent direct radiation (EDR).

Note: 240 Btuh = 1 ft² EDR with 1 psig steam condensing in the heat-transfer unit.

Ratings in EDR are largely being abandoned by many manufacturers in the industry.

Radiators.   The small-tube cast-iron radiators, with a length of $1\frac{3}{4}$ in. per section, occupy less space than the older column-type and the large-tube-type units and are particularly suited to installation in recesses.

The Institute of Boiler and Radiator Manufacturers (I-B-R), now the Hydronics Institute, in cooperation with the National Bureau of Standards, established the Simplified Practice Recommendation R174-65 ("Cast-Iron Radiators") for small-tube radiators. This data is no longer available, but Table 8-8 shows the dimensions and ratings of the units currently manufactured. Column, wall-type, and large-tube radiators are no longer manufactured, but many of these units are still in use. Refer to *ASHRAE Handbook 1988 Equipment,* Chapter 28, for ratings of such units.

Convectors. Convectors consist of a finned cast-iron, finned steel, or finned copper heating element installed within a cabinet-type enclosure. The same finned heating element will produce different capacities depending on the height of the enclosure and on the location of the inlet and outlet air openings in the enclosure. Air enters the enclosure below the heating element and leaves the enclosure through the outlet grille located above the heating element. The heat output from convectors is largely due to free convection heating of the air as it passes over the heating element. Since convectors are manufactured in a wide variety of enclosure types, heights, widths, and depths of enclosure, a wide range of capacities are available. The units may be floor mounted (free standing), wall hung, or recessed and may have outlet grilles, and arched inlets or inlet grilles are desired.

Convector water ratings are quoted for different cabinet enclosure types, heights, widths, and depths, an average water temperature in the heating element, and water temperature drop in the unit and are normally expressed in MBh (thousands of Btu/hr) or watts (W). Steam ratings are expressed in sq. ft EDR and Btu/hr or watts.

Figure 8-12 illustrates one type of convector, and Table 8-9 quotes only a partial listing of capacities available for this type of unit. Capacity ratings are also available for other average water temperatures and water temperature drops, as well as ratings using low-pressure steam.

To illustrate the use of Table 8-9, assume that a heating capacity of 6200 Btu/hr is needed for a particular building space and a convector similar to the type illustrated in Figure 8-12 is to be selected. The average water temperature in the convector will be 190°F. The space for the convector limits its height to 32 in. and a maximum depth of 6 in. How long should the unit be to deliver the required heating capacity? Referring to Table 8-9, we find that a 32-in.-long unit delivers 6.3 MBh with a 10-degree temperature drop, 5.6 MBh with a 20-degree drop, and 5.4 MBh with a 30-degree drop. Decision: Select the 32-in.-long unit and a 10-degree design water temperature drop in the unit.

Baseboard Units. Baseboard terminal heat-transfer units are readily available and are manufac-

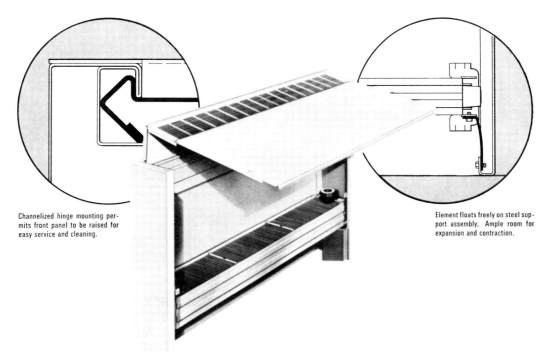

Channelized hinge mounting permits front panel to be raised for easy service and cleaning.

Element floats freely on steel support assembly. Ample room for expansion and contraction.

FIGURE 8-12 Front discharge, floor-mounted convector with front panel in raised position to show heating element (steam or hot water). (Courtesy of The Trane Company, LaCrosse, WI)

TABLE 8-8

Small-tube cast-iron radiators

Number of Tubes per Section	Catalog Rating per Section[a]		Section Dimensions (A) Height[c]	Section Dimensions (B) Width		(C) Spacing[b]	(D) Leg Height[c]
	ft²	Btu/hr (W)	in. (mm)	in. (mm)	in. (mm)	in. (mm)	in. (mm)
3	1.6	384 (113)	25 (635)	3.25 (83)	3.50 (89)	1.75 (44)	2.50 (64)
4	1.6	384 (113)	19 (483)	4.44 (113)	4.81 (122)	1.75 (44)	2.50 (64)
	1.8	432 (127)	22 (559)	4.44 (113)	4.81 (122)	1.75 (44)	2.50 (64)
	2.0	480 (141)	25 (635)	4.44 (113)	4.81 (122)	1.75 (44)	2.50 (64)
5	2.1	504 (148)	22 (559)	5.63 (143)	6.31 (160)	1.75 (44)	2.50 (64)
	2.4	576 (169)	25 (635)	5.63 (143)	6.31 (160)	1.75 (44)	2.50 (64)
6	2.3	552 (162)	19 (483)	6.81 (173)	8 (203)	1.75 (44)	2.50 (64)
	3.0	720 (211)	25 (635)	6.81 (173)	8 (203)	1.75 (44)	2.50 (64)
	3.7	888 (260)	32 (813)	6.81 (173)	8 (203)	1.75 (44)	2.50 (64)

Source: *ASHRAE Handbook 1988 Equipment*, American Society of Heating, Refrigerating, and Air Conditioning Engineers, Atlanta, GA.

[a] These ratings are based on steam at 215°F (101.7°C) and air at 70°F (21.1°C). They apply only to installed radiators exposed in a normal manner, not to radiators installed behind enclosures, grilles, or under shelves. For Btu/hr (W) ratings at other temperatures, refer to manufacturers' catalogs.

[b] Length equals number of sections multiplied by 1.75 in. (44 mm).

[c] Overall height and leg height, as produced by some manufacturers, are 1 in. (25 mm) greater than shown in columns A and D. Radiators may be furnished without legs. Where greater than standard leg heights are required, this dimension shall be 4.5 in. (114 mm).

TABLE 8-9

Capacity table for convectors (Trane models SKF and AFG) for various AWT and 65°F EAT[a,b]

180 F Average Water

HEIGHT	DEPTH	26			32			38			44			50			56			62		
14	4	2.5	2.2	2.1	3.2	2.8	2.7	3.9	3.4	3.3	4.6	4.0	3.9	5.3	4.6	4.5	6.0	5.2	5.1	6.7	5.8	5.7
16	6	3.4	3.0	2.9	4.5	3.9	3.8	5.8	5.0	4.9	7.0	6.2	6.0	8.2	7.4	7.1	9.4	8.6	8.2	10.6	9.8	9.3
18	4	2.6	2.3	2.2	3.2	2.8	2.7	4.1	3.6	3.4	5.0	4.4	4.1	5.9	5.2	4.8	6.8	6.0	5.5	7.7	6.8	6.2
20	6	3.9	3.4	3.3	4.9	4.3	4.2	6.2	5.4	5.2	7.4	6.5	6.2	8.6	7.6	7.2	9.8	8.7	8.2	11.0	9.8	9.2
	8	4.8	4.2	4.0	6.0	5.3	5.1	7.2	6.4	6.2	8.6	7.5	7.3	9.9	8.6	8.4	11.2	9.8	9.5	12.5	10.9	10.5
24	4	2.7	2.3	2.2	3.4	3.0	2.9	4.2	3.7	3.6	5.0	4.4	4.3	5.8	5.1	5.0	6.6	5.8	5.7	7.4	6.5	6.4
26	6	4.3	3.8	3.6	5.4	4.8	4.6	6.6	5.8	5.5	7.7	6.7	6.5	8.8	7.7	7.5	9.9	8.7	8.5	11.0	9.7	9.5
	8	5.1	4.5	4.3	6.5	5.7	5.5	7.8	6.8	6.6	9.2	8.0	7.7	10.5	9.2	8.9	11.9	10.4	10.1	13.2	11.6	11.2
30	4	2.8	2.5	2.4	3.6	3.2	3.1	4.4	3.8	3.7	5.2	4.5	4.4	6.0	5.2	5.1	6.8	5.9	5.8	7.6	6.6	6.5
32	6	4.5	4.0	3.8	5.7	5.0	4.8	6.9	6.0	5.8	8.0	7.0	6.8	9.2	8.0	7.8	10.4	9.0	8.8	11.6	10.0	9.8
	8	5.4	4.8	4.6	6.7	5.9	5.7	8.0	7.0	6.8	9.4	8.2	7.9	10.8	9.4	9.1	12.2	10.7	10.3	13.5	11.9	11.4
36	4	3.1	2.6	2.5	3.8	3.3	3.2	4.5	4.0	3.8	5.3	4.7	4.5	6.2	5.4	5.2	7.0	6.1	5.9	7.8	6.8	6.6
38	6	4.8	4.2	4.0	5.8	5.0	4.9	7.0	6.2	6.0	8.3	7.3	7.1	9.6	8.3	8.1	10.8	9.4	9.1	12.0	10.5	10.1
	8	5.7	5.1	4.9	7.2	6.3	6.1	8.7	7.6	7.3	10.1	8.9	8.6	11.7	10.2	9.9	13.1	11.5	11.1	14.6	12.8	12.3

190 F Average Water

HEIGHT	DEPTH	26			32			38			44			50			56			62		
14	4	2.8	2.5	2.4	3.5	3.1	3.0	4.3	3.9	3.7	5.1	4.7	4.4	5.9	5.5	5.1	6.7	6.3	5.8	7.5	7.1	6.5
16	6	3.8	3.4	3.2	5.0	4.4	4.2	6.4	5.7	5.5	7.9	7.0	6.7	9.4	8.3	7.9	10.9	9.6	9.1	12.4	10.9	10.3
18	4	2.9	2.6	2.5	3.6	3.2	3.1	4.5	4.0	3.9	5.4	4.8	4.7	6.3	5.6	5.5	7.2	6.4	6.3	8.1	7.2	7.1
20	6	4.3	3.9	3.7	5.5	4.9	4.7	6.9	6.1	5.9	8.2	7.3	7.0	9.5	8.5	8.1	10.8	9.7	9.2	12.1	10.9	10.3
	8	5.3	4.7	4.5	6.7	6.0	5.7	8.1	7.2	6.9	9.6	8.5	8.2	11.0	9.8	9.4	12.5	11.1	10.6	13.9	12.4	11.9
24	4	3.0	2.7	2.5	3.8	3.4	3.2	4.7	4.2	4.0	5.6	5.0	4.8	6.5	5.8	5.6	7.4	6.6	6.4	8.3	7.4	7.2
26	6	4.8	4.3	4.1	6.1	5.4	5.2	7.3	6.5	6.2	8.6	7.6	7.3	9.9	8.7	8.4	11.2	9.8	9.5	12.5	10.9	10.6
	8	5.7	5.1	4.9	7.2	6.4	6.2	8.7	7.7	7.4	10.2	9.1	8.7	11.8	10.5	10.0	13.3	11.8	11.3	14.8	13.1	12.6
30	4	3.2	2.8	2.7	4.1	3.6	3.5	4.9	4.3	4.2	5.8	5.2	4.9	6.7	6.1	5.6	7.6	7.0	6.3	8.5	7.9	7.0
32	6	5.1	4.5	4.3	6.3	5.6	5.4	7.7	6.8	6.5	9.0	8.0	7.6	10.3	9.1	8.8	11.6	10.3	9.9	12.9	11.5	11.1
	8	6.1	5.4	5.2	7.5	6.7	6.4	9.0	8.0	7.6	10.5	9.3	8.9	12.0	10.7	10.2	13.6	12.0	11.6	15.1	13.4	12.9
36	4	3.5	3.1	2.9	4.3	3.8	3.6	5.1	4.5	4.3	6.0	5.3	5.1	6.9	6.1	5.9	7.8	6.9	6.6	8.7	7.7	7.4
38	6	5.3	4.7	4.5	6.4	5.7	5.5	7.9	7.0	6.7	9.3	8.3	7.9	10.7	9.5	9.1	12.0	10.7	10.2	13.3	11.9	11.4
	8	6.5	5.7	5.5	8.1	7.2	6.9	9.7	8.6	8.2	11.3	10.1	9.6	13.0	11.6	11.1	14.7	13.0	12.5	16.3	14.5	13.9

200 F Average Water

HEIGHT	DEPTH	26			32			38			44			50			56			62		
14	4	3.1	2.8	2.6	3.9	3.5	3.3	4.8	4.3	4.1	5.7	5.1	4.9	6.6	5.9	5.7	7.5	6.7	6.5	8.4	7.5	7.3
16	6	4.2	3.8	3.6	5.5	5.0	4.7	7.1	6.4	6.0	8.7	7.8	7.4	10.3	9.2	8.8	11.9	10.6	10.2	13.5	12.0	11.6
18	4	3.2	2.9	2.7	4.0	3.6	3.4	5.0	4.5	4.3	6.0	5.4	5.2	7.0	6.3	6.1	8.0	7.2	7.0	9.0	8.1	7.9
20	6	4.8	4.3	4.1	6.1	5.5	5.2	7.6	6.8	6.5	9.1	8.2	7.7	10.6	9.6	9.0	12.1	11.0	10.3	13.6	12.4	11.6
	8	5.9	5.3	5.0	7.4	6.7	6.3	9.0	8.1	7.7	10.6	9.5	9.0	12.2	11.0	10.4	13.8	12.4	11.7	15.4	13.9	13.1
24	4	3.3	2.9	2.8	4.2	3.8	3.6	5.2	4.7	4.4	6.2	5.6	5.2	7.2	6.5	6.0	8.2	7.4	6.8	9.2	8.3	7.6
26	6	5.3	4.8	4.5	6.7	6.0	5.7	8.1	7.3	6.9	9.5	8.6	8.1	10.9	9.9	9.3	12.3	11.2	10.5	13.7	12.5	11.7
	8	6.3	5.7	5.4	8.0	7.2	6.8	9.6	8.6	8.2	11.3	10.2	9.6	13.0	11.7	11.1	14.7	13.2	12.5	16.3	14.7	13.9
30	4	3.5	3.2	3.0	4.5	4.1	3.8	5.4	4.9	4.6	6.4	5.8	5.4	7.4	6.7	6.2	8.4	7.6	7.0	9.4	8.5	7.8
32	6	5.6	5.0	4.8	7.0	6.3	6.0	8.5	7.7	7.2	9.9	8.9	8.4	11.4	10.3	9.7	12.8	11.5	10.9	14.2	12.7	12.1
	8	6.7	6.0	5.7	8.3	7.5	7.1	9.9	8.9	8.4	11.6	10.4	9.9	13.3	12.0	11.3	15.0	13.5	12.7	16.7	15.0	14.2
36	4	3.8	3.4	3.2	4.7	4.2	4.0	5.6	5.0	4.8	6.6	5.9	5.6	7.6	6.8	6.5	8.6	7.7	7.3	9.6	8.6	8.2
38	6	5.8	5.2	5.0	7.1	6.4	6.0	8.7	7.8	7.4	10.3	9.3	8.8	11.8	10.6	10.0	13.3	12.0	11.3	14.8	13.3	12.6
	8	7.0	6.4	6.0	8.9	8.0	7.6	10.7	9.6	9.1	12.5	11.3	10.6	14.4	13.0	12.2	16.2	14.6	13.8	18.0	16.2	15.3

210 F Average Water

HEIGHT	DEPTH	26			32			38			44			50			56			62		
14	4	3.3	3.1	2.9	4.2	3.9	3.6	5.2	4.8	4.4	6.2	5.7	5.2	7.2	6.6	6.0	8.2	7.5	6.8	9.2	8.4	7.6
16	6	4.5	4.2	3.9	5.9	5.5	5.1	7.7	7.1	6.6	9.4	8.7	8.0	11.1	10.3	9.4	12.8	11.9	10.8	14.5	13.5	12.2
18	4	3.5	3.2	3.0	4.3	4.0	3.7	5.4	5.0	4.6	6.5	6.0	5.5	7.6	7.0	6.4	8.7	8.0	7.3	9.8	9.0	8.2
20	6	5.2	4.8	4.4	6.6	6.1	5.6	8.2	7.6	7.0	9.8	9.1	8.4	11.4	10.6	9.8	13.0	12.1	11.2	14.6	13.6	12.6
	8	6.4	5.9	5.5	8.0	7.4	6.8	9.7	9.0	8.3	11.4	10.6	9.8	13.1	12.2	11.3	14.9	13.8	12.8	16.6	15.4	14.2
24	4	3.6	3.3	3.1	4.5	4.2	3.9	5.6	5.2	4.8	6.7	6.2	5.7	7.8	7.2	6.6	8.9	8.2	7.5	10.0	9.2	8.4
26	6	5.7	5.3	4.9	7.2	6.7	6.2	8.7	8.1	7.5	10.3	9.5	8.8	11.9	10.9	10.1	13.5	12.3	11.4	15.1	13.7	12.7
	8	6.8	6.3	5.8	8.6	8.0	7.4	10.3	9.6	8.9	12.2	11.3	10.5	14.0	13.0	12.0	15.9	14.7	13.6	17.6	16.3	15.1
30	4	3.8	3.5	3.2	4.9	4.5	4.2	5.8	5.4	5.0	6.9	6.4	5.9	8.0	7.4	6.8	9.1	8.4	7.7	10.2	9.4	8.6
32	6	6.0	5.6	5.2	7.6	7.0	6.5	9.2	8.5	7.9	10.7	9.9	9.2	12.3	11.4	10.5	13.8	12.8	11.8	15.3	14.2	13.1
	8	7.2	6.7	6.2	9.0	8.3	7.7	10.7	9.9	9.2	12.5	11.6	10.7	14.4	13.3	12.3	16.2	15.0	13.9	18.0	16.7	15.4
36	4	4.2	3.8	3.4	5.1	4.7	4.3	6.0	5.6	5.2	7.1	6.6	6.1	8.2	7.6	7.0	9.3	8.6	8.0	10.4	9.6	8.9
38	6	6.3	5.8	5.5	7.7	7.1	6.6	9.4	8.7	8.0	11.1	10.3	9.5	12.7	11.8	10.9	14.4	13.3	12.3	16.0	14.8	13.7
	8	7.7	7.0	6.6	9.6	8.9	8.2	11.6	10.7	9.9	13.5	12.5	11.6	15.5	14.4	13.3	17.5	16.2	15.0	19.4	18.0	16.7

[a]Capacities expressed in MBh for 10, 20, and 30°F. Temperature drops are located below each cabinet length.
[b]See manufacturer's catalog for complete data.

Source: Reproduced in part, courtesy of The Trane Company, La Crosse, Wis.

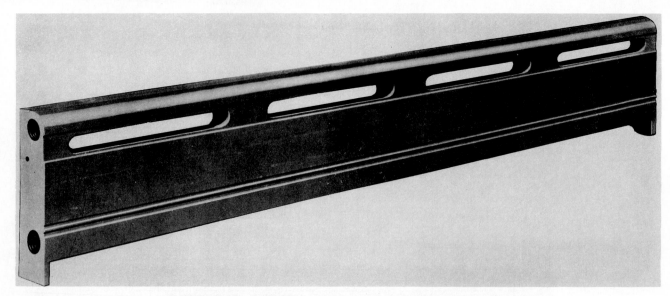

FIGURE 8-13 BASERAY cast-iron radiant baseboard. (Courtesy of Burnham Corporation, Hydronics Division, Lancaster, PA)

tured in two types: (1) radiant-convector and (2) finned-tube.

The *radiant-convector* type of baseboard is made of cast iron or steel. The units have air openings at the top and bottom to permit circulation of room air over the wall side of the unit, which has extended surface to provide increased heat output. A large portion of the heat emitted is transferred by convection. Figure 8-13 shows a type of radiant-convector baseboard from one manufacturer. BASERAY assemblies are available in lengths from $1\frac{1}{2}$ to 6 linear ft shipped in one piece. Longer assemblies are shipped in two or more pieces or subassemblies, none of which exceeds 6 linear ft. The BASERAY cross section indicates a thickness of $2\frac{1}{2}$ in. and a height of $9\frac{7}{8}$ in. Table 8-10 gives BASERAY ratings for steam and hot water.

The *finned-tube* baseboard has a finned-pipe or finned-tube heating element concealed by a long, low, sheet metal enclosure or cover that is open at the top and bottom for air circulation. Nearly all the heat output from such units is by natural convection. The finned heating elements may be steel pipe with steel fins or copper tubes with aluminum fins.

The capacity of baseboard heating units varies over a wide range depending on the physical dimensions and materials used. Ratings are usually expressed in Btuh per linear foot of unit at a specific average water temperature or a standard steam pressure.

Figure 8-14 shows a type of baseboard heating unit, and Table 8-11 gives the ratings for this unit. In selecting this type of baseboard, the designer should avoid using a unit with too high an output per linear

TABLE 8-10

BASERAY ratings (steam and hot water)

Model No.	Flow Rate (lb/hr)	Steam Rating		Water Ratings (Btu/hr per linear foot at average water temperature indicated)						
		sq. ft	Btu/hr at 215°F	170°F	180°F	190°F	200°F	210°F	220°F	230°F
9A	2000	3.40	820	550	620	690	750	810	880	940
	500	3.40	820	520	590	650	710	770	830	890

Source: Courtesy of Burnham Corporation, Hydronics Division, Lancaster, PA.

Where water flow rate through the BASERAY is not known, the rating at the standard flow rate of 500 lb/hr (1 gpm) must be used.

I-B-R ratings are determined from tests made in accordance with the I-B-R Testing and Rating Code for baseboard-type radiation, including an allowance of 15% for heating effect permitted by this code.

Maximum allowable working pressure is 30 psig.

TABLE 8-11

Model 75WL-3 (3/4" tube) Therma-Trim I-B-R approved water ratings
(capacities in Btu/hr per linear foot with 65°F entering air)

Number of Linear Feet	Water Flow Rate—500 lb/hr or 1 gpm Average Water Temperature (°F)					
	170	180	190	200	210	220
1	510	570	630	690	750	810
2	1,020	1,140	1,260	1,380	1,500	1,620
3	1,530	1,710	1,890	2,070	2,250	2,430
4	2,040	2,280	2,520	2,760	3,000	3,240
5	2,550	2,850	3,150	3,450	3,750	4,050
6	3,060	3,420	3,780	4,140	4,500	4,860
7	3,570	3,990	4,410	4,830	5,250	5,670
8	4,080	4,560	5,040	5,520	6,000	6,480
9	4,590	5,130	5,670	6,210	6,750	7,290
10	5,100	5,700	6,300	6,900	7,500	8,100
11	5,610	6,270	6,930	7,590	8,250	8,910
12	6,120	6,840	7,560	8,280	9,000	9,720
13	6,630	7,410	8,190	8,970	9,750	10,530
14	7,140	7,980	8,820	9,660	10,500	11,340
15	7,650	8,550	9,450	10,350	11,250	12,150
16	8,160	9,120	10,080	11,040	12,000	12,960
17	8,670	9,690	10,710	11,730	12,750	13,770
18	9,180	10,260	11,340	12,420	13,500	14,580
19	9,690	10,830	11,970	13,110	14,250	15,390
20	10,200	11,400	12,600	13,800	15,000	16,200

Number of Linear Feet	Water Flow Rate—2000 lb/hr or 4 gpm Average Water Temperature (°F)					
	170	180	190	200	210	220
1	540	600	670	730	790	860

Source: Courtesy of Weil-McLain, Michigan City, IN. A Marley Company.

I-B-R approved water ratings are based on the active (finned) length and include the 15% addition for heating effect allowed by the I-B-R Testing and Rating Code for Baseboard Type of Radiation. The active length is 3" less than the enclosure length. Ratings apply to the assembly with the damper installed and adjusted to the normally open position.

The heating elements are constructed of $\frac{3}{4}$" nominal copper tubing expanded into $2\frac{1}{8}$" × 2 $\frac{5}{16}$" flanged aluminum fins, which are spaced 51 to the foot; fin thickness is .007". Elements are unpainted.

Use of I-B-R ratings at the 4 gpm flow rate is limited to installations in which the water flow rate through the baseboard unit is equal to or greater than 4 gpm. Where the water flow rate is not known, the I-B-R ratings at the standard flow rate of 1 gpm must be used. Flow rates exceeding 6 gpm should be avoided because of possible noise.

Pressure drop through Therma-Trim baseboard is 0.047" of water per linear foot at the 1 gpm flow rate. At the 4 gpm flow rate, pressure drop is 0.525" of water per linear foot.

The following steam ratings are certified by Weil-McLain:

Model 75WL-3 steam rating: 1 lb steam—65°F entering air—3.7 sq. ft EDR or 820 Btu/hr per linear foot.

foot. Optimum comfort for room occupants is usually obtained when units are selected so that they are installed along as much of the exposed wall of a space as possible.

The basic advantage of the baseboard unit is that its normal placement is along cold walls and below areas where the greatest heat loss occurs (below windows). Other advantages claimed are that it is inconspicuous, that it offers a minimum of interference with furniture placement, and that it distributes the heat near the floor. This last characteristic reduces the floor-to-ceiling temperature gradient at least 2°F to 4°F and tends to produce a more uniform temperature throughout the room.

Commercial Finned-Tube or Finned-Pipe.   These heating units are similar, as to component parts, to baseboard units. The main difference is in the size and enclosure design, resulting in a higher capacity per unit length than baseboard units.

Finned-tube heating elements are available in several pipe sizes, either steel or copper—1 to 2 in. I.P.S. for steel and $\frac{3}{4}$ to $1\frac{1}{4}$ nominal tube size for copper. Fins are usually mechanically bonded to the pipe and tube and are steel or aluminum, either square or rectangular depending on the manufacturer.

The enclosures are sheet-metal and have an open or grilled air inlet below the heating element and normally a grilled air outlet at the top or top front above the heating element.

The ratings of commercial finned-pipe units, like baseboard units, are quoted in Btuh per linear foot at a specific average water temperature or specific steam pressure. Figure 8-15 illustrates typical shapes of finned-tube enclosures, and Table 8-12 gives ratings of one enclosure type with different finned-tube (or pipe) installed.

To standardize rating procedures for convectors, baseboards, and commercial finned-pipe, various codes have been established and used by industry. The generally accepted method of testing and rating of both ferrous and nonferrous convectors is given in Commercial Standard CS140-47 ("Testing and Rating Convectors"), which has been developed cooperatively by the Convector Manufacturer Association, the Institute of Boiler and Radiator Manufacturers, other members of the trade, and the National Bureau of Standards.

Baseboard testing and rating is based upon I-B-R Testing and Rating Code for Baseboard Radiation, and finned-pipe testing and rating is based upon I-B-R Test-

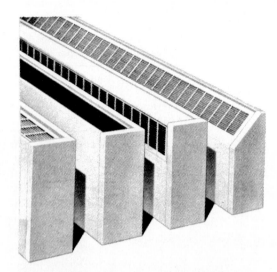

FIGURE 8-15   Trane wall fin convection–radiation unit enclosure types, heating element not shown (steam or hot water). (Courtesy of The Trane Company, LaCrosse, WI)

ing and Rating Code for Finned Tube (Commercial) Radiation.

8-11   UNIT HEATERS, UNIT VENTILATORS, AND MAKEUP AIR HEATERS

These units are grouped together because they all have the same basic components, which are a finned heating element, a motor-driven fan, and an enclosure. Filters, dampers, directional outlets, direct collars, inlet or outlet grilles or diffusers, combustion chambers, and flues may also be included depending on the required function and application.

Unit Heaters.   These units have the primary function of heating. Ventilation is sometimes provided

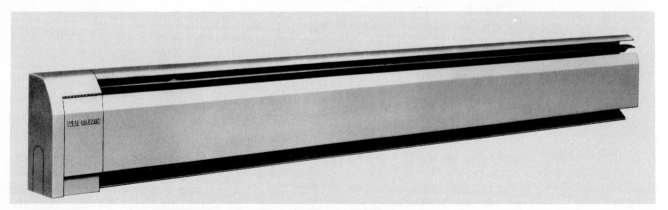

FIGURE 8-14   Model 75WL-3 Therma-Trim baseboard. (Courtesy of Weil-McLain, Michigan City, IN. A Marley Company)

TABLE 8-12

Ratings of Trane wall fin elements with type S enclosure*

Ratings of Wall Fin Elements With Type S Enclosure

Element	Rows	Enclosure	Installed Height	Steam Capacity/Ft. 1 PSI at 65 F Air: EDR Sq. Ft.	Steam: Btu/Hr.	Hot Water Capacity—Btu/Hr./Ft. at 65 F Air, Average Water Temp. (IBR Factor—Steam to Hot Water): 220F (1.05)	210F (0.95)	200F (0.86)	190F (0.78)	180F (0.69)	170F (0.61)
Steel—1¼" Series—40 Fins—2½" x 5¼" x .027" Finish—Black	1	12S	15⁷⁄₃₂	5.45	1310	1380	1240	1130	1020	900	800
		16S	18⁷⁄₃₂	5.60	1340	1410	1270	1150	1050	920	820
		20S	21⁷⁄₃₂	5.70	1370	1440	1300	1180	1070	950	840
	2*	16S	18⁷⁄₃₂	8.75	2100	2210	2000	1810	1640	1450	1280
		20S	21⁷⁄₃₂	9.00	2160	2270	2050	1860	1680	1490	1320
	2**	20S	21⁷⁄₃₂	9.40	2260	2370	2150	1940	1760	1560	1380
Steel—1¼" Series—52 Fins—2½" x 5¼" x .027" Finish—Black	1	12S	15⁷⁄₃₂	6.40	1530	1610	1450	1320	1190	1060	930
		16S	18⁷⁄₃₂	6.90	1650	1730	1570	1420	1290	1140	1010
		20S	21⁷⁄₃₂	7.25	1740	1830	1650	1500	1360	1200	1060
	2*	16S	18⁷⁄₃₂	9.15	2200	2310	2090	1890	1720	1520	1340
		20S	21⁷⁄₃₂	9.90	2380	2500	2260	2050	1860	1640	1450
	2**	20S	21⁷⁄₃₂	10.25	2460	2580	2340	2120	1920	1700	1500
Copper-Aluminum—1¼" Series—60 Fins—2½" x 5¼" x .015" Finish—Black	1	12S	15⁷⁄₃₂	7.65	1840	1930	1750	1580	1440	1270	1120
		16S	18⁷⁄₃₂	8.55	2050	2150	1950	1760	1600	1410	1250
		20S	21⁷⁄₃₂	9.35	2240	2350	2130	1930	1750	1550	1370
	2*	16S	18⁷⁄₃₂	10.15	2440	2560	2320	2100	1900	1680	1490
		20S	21⁷⁄₃₂	11.25	2700	2840	2570	2320	2110	1860	1650
	2**	20S	21⁷⁄₃₂	11.45	2750	2890	2610	2370	2150	1900	1680
Copper-Aluminum—1" Series—68 Fins—2" x 3¼" x .011" Finish—Black	1	10S	14³⁄₃₂	4.70	1130	1190	1070	970	880	780	690
		14S	17³⁄₃₂	5.30	1290	1330	1210	1090	990	880	790
		18S	20⅜	5.90	1410	1480	1340	1210	1100	970	860
	2*	14S	17³⁄₃₂	6.15	1470	1540	1400	1260	1150	1010	900
		18S	20⅜	6.75	1620	1700	1540	1390	1260	1120	990
	2**	18S	20⅜	7.25	1740	1830	1650	1500	1360	1200	1060

*4" CENTERS **8" CENTERS

SLOPING TOP Type S

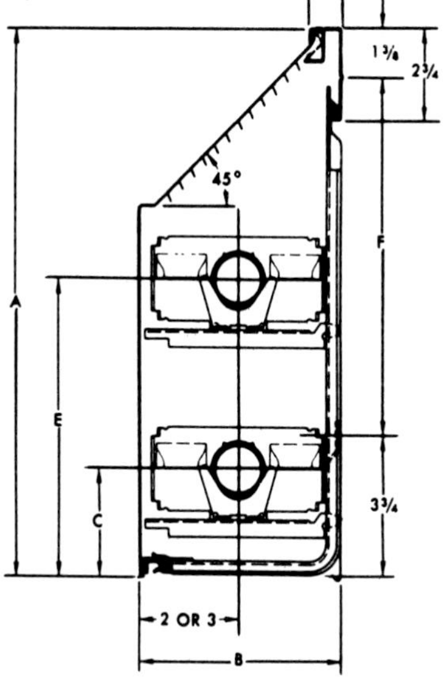

Types S and F Enclosure Dimensions

Dimensions	3¼" Fin			5¼" Fin		
A	10	14	18	12	16	20
B	4	4	4	6	6	6
C Min.	2⅛	2⅛	2⅛	2½	2½	2½
C Max.	3⅝	3⅝	3⅝	3⅛	3⅛	3⅛
D*	1⅝	1⅝	1⅝	1⅛	1⅛	1⅛
E Min.	—	6⅛	10⅛	—	6½	10½
E Max.	—	7⅝	11⅝	—	7⅛	11⅛
F	4⅛	8⅛	12⅛	6⅛	10⅛	14⅛

*See complete catalog data for other enclosure types.

by introducing outside air, but the majority of these units have 100% air recirculation.

Unit heaters can be classified according to (1) the heating medium used, (2) the type of fan, or (3) the arrangement of the unit components.

The *heating medium* may be steam, hot water, gas indirect-fired, oil indirect-fired, or electric. Three types of fans, *propeller, centrifugal,* and *remote air mover,* can be considered. Propeller fan units may be *horizontal-blow* or *down-blow.* Centrifugal fan units may be of the smaller *cabinet* type or larger *industrial* type. Units with remote air movers are known as *duct unit heaters.* The arrangement of unit components identifies the *draw-through,* in which the fan draws air through, and the *blow-through,* in which the fan blows the air through the heating element. The indirect-fired units are always of the blow-through type.

Unit heaters have relatively large heating capacities in compact casings, the ability to project heated air in a controlled manner over a considerable distance, and a relatively low installed cost. They are, therefore, usually applied where the required heating capacity, the volume of the heated space, or both, are too large to be handled adequately or economically by other means. They eliminate extensive ductwork.

Unit heaters are used for spot or intermittent heating, such as blanketing outside doors. They are used where filtration of heated air is required.

Unit heaters are used to heat garages, factories, warehouses, showrooms, stores, and laboratories, as well as corridors, lobbies, vestibules, and similar auxiliary spaces in buildings.

In this section, we discuss three types of unit heaters: (1) floor-mounted centrifugal fan unit heater, cabinet type; (2) propeller fan unit heater, horizontal-blow type; and (3) propeller gas-fired propeller fan unit heater.

Figure 8-16 shows a floor-mounted *cabinet heater* manufactured by The Trane Company. The external appearance is similar to a convector. A centrifugal fan and filters have been added as well as a different heating coil design. The fan is employed to force air across the heating element, which greatly increases the capacity of the unit, and, in addition, provides desirable air circulation within the space to be heated. These units are available in a wide variety of cabinet styles and may be floor mounted, wall hung, ceiling mounted, or recessed. The airflow may be upward (normal) or downward through the unit (called *inverted* airflow). The low-level fan noise makes these units suitable for building entrances, below windows in occupied areas, and in any location where high heating capacity is required. Table 8-13 gives a typical capacity table using hot water for a few of the sizes available from one manufacturer.

Figures 8-17 and 8-18 show two types of propeller fan unit heaters. These types are normally employed in factories and warehouses where the units are suspended from the ceiling or mounted on wall brackets. The propeller fan forces air across the heat-transfer coil, and the air is directed outward or downward by means of adjustable louvers (Figure 8-17). Figure 8-18 shows a vertical projection unit where the fan, mounted vertically, draws the air horizontally across the heat-transfer

coil and blows the air downward. These unit heaters are available in a wide range of sizes, varying in capacity from 5 to 500 MBh, and use either steam or hot water as the heating medium.

Table 8-14 gives the performance ratings of the horizontal delivery unit heater (Figure 8-17), the vertical delivery unit heater (Figure 8-18), and the "power-throw" unit (not shown) using steam as the heating medium. Standard rating of steam unit heaters has been based on using saturated steam at 2 psig (13.8 kPa) pressure at the heater coil, air at 60°F (16°C), 29.92 in. Hg (101.13 kPa) barometric pressure entering the heater, and the heater operating free of external resistance to airflow. The capacity of the heater increases as the steam pressure increases and decreases as the entering air temperature increases. Refer to manufacturers' catalogs for correction factors.

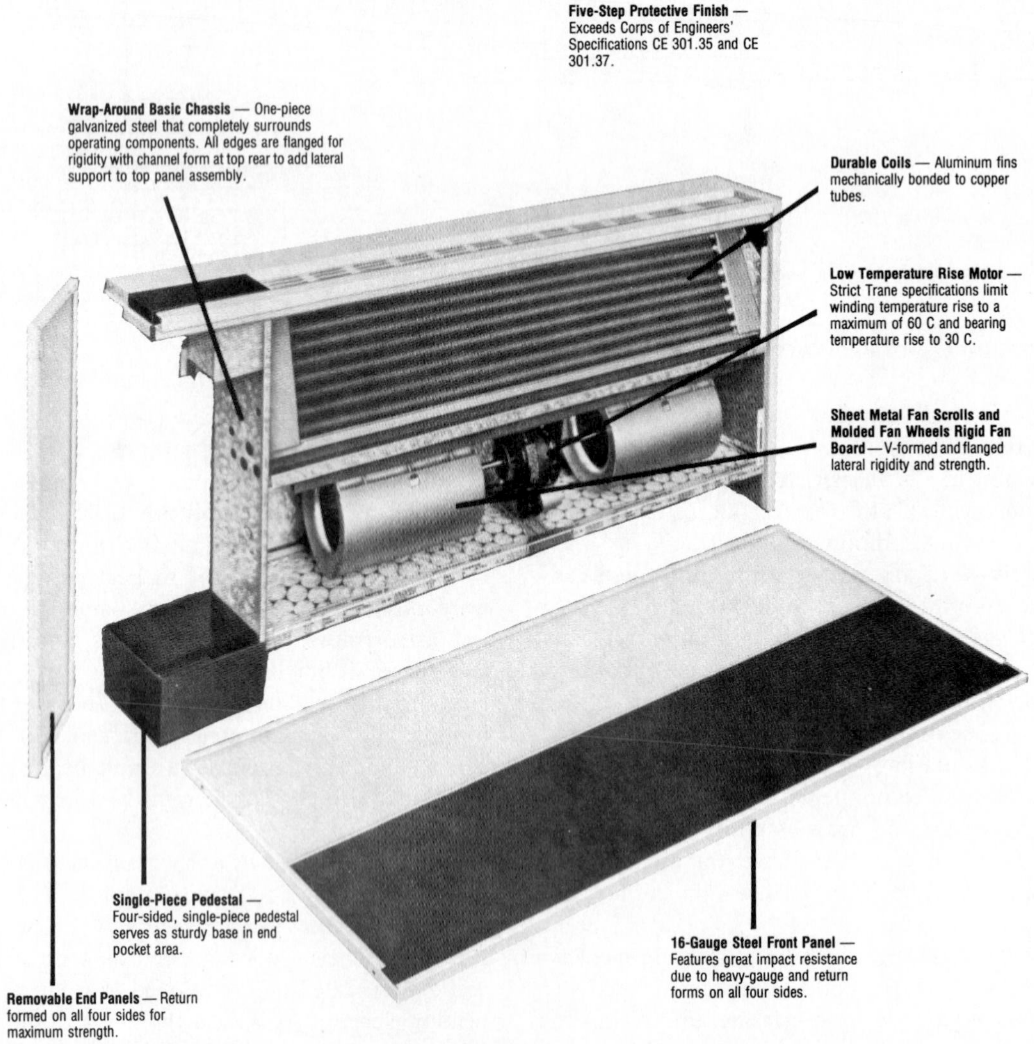

Five-Step Protective Finish — Exceeds Corps of Engineers' Specifications CE 301.35 and CE 301.37.

Wrap-Around Basic Chassis — One-piece galvanized steel that completely surrounds operating components. All edges are flanged for rigidity with channel form at top rear to add lateral support to top panel assembly.

Durable Coils — Aluminum fins mechanically bonded to copper tubes.

Low Temperature Rise Motor — Strict Trane specifications limit winding temperature rise to a maximum of 60 C and bearing temperature rise to 30 C.

Sheet Metal Fan Scrolls and Molded Fan Wheels Rigid Fan Board — V-formed and flanged lateral rigidity and strength.

Single-Piece Pedestal — Four-sided, single-piece pedestal serves as sturdy base in end pocket area.

16-Gauge Steel Front Panel — Features great impact resistance due to heavy-gauge and return forms on all four sides.

Removable End Panels — Return formed on all four sides for maximum strength.

FIGURE 8-16 Trane force-flow cabinet heater, front and end panel removed to show fans and heating element (steam or hot water). (Courtesy of The Trane Company, LaCrosse, WI)

FIGURE 8-17 Horizontal delivery unit heater. (Courtesy of Modine Manufacturing Company, Racine, WI)

FIGURE 8-18 Vertical delivery unit heater. (Courtesy of Modine Manufacturing Company, Racine, WI)

Table 8-15 gives the performance ratings for the unit heaters of Figures 8-17 and 8-18 using hot water as the heating medium. Rating of hot-water unit heaters is usually based on water at 200°F, water temperature drop of 20°F, entering air at 60°F, 29.92 in. Hg barometric pressure (93°C, 11°C drop, 16°C entering air at 101.325 kPa), and the heater operating free of external resistance to airflow. Variations in entering water tem-

TABLE 8-13

A-coil hot water capacity, 60°F EAT and 180°F EWT*[a]

UNIT SIZE	HIGH SPEED (1100 RPM)					LOW SPEED (700 RPM)				
	GPM	WPD	MBH	WTD	FAT	GPM	WPD	MBH	WTD	FAT
02	0.50	0.07	11.1	44.4	102.8	0.50	0.07	9.6	38.3	118.2
	0.75	0.15	13.6	36.5	112.5	0.75	0.15	11.3	30.2	128.6
	1.00	0.24	15.3	30.9	119.2	1.00	0.24	12.4	24.9	135.3
	1.50	0.50	17.6	23.8	128.0	1.50	0.50	13.7	18.5	143.5
	2.00	0.83	19.0	19.3	133.5	2.00	0.83	14.5	14.7	148.3
	3.00	1.70	20.7	14.0	140.0	3.00	1.70	15.4	10.4	153.7
	5.00	4.20	22.3	9.0	146.0	5.00	4.20	16.2	6.6	158.6
	1.90	0.76	18.8	20.0	132.6	1.35	0.41	13.4	20.0	141.5
	0.63	0.11	12.5	40.0	108.1	0.46	0.06	9.2	40.0	116.1
03	0.50	0.08	14.9	54.9	100.2	0.50	0.08	13.0	51.5	115.1
	0.75	0.17	18.4	49.2	109.7	0.75	0.17	15.4	40.9	125.3
	1.00	0.28	20.9	42.0	116.4	1.00	0.28	17.0	34.0	132.1
	1.50	0.58	24.1	32.5	125.2	1.50	0.58	18.9	25.4	140.5
	2.00	0.97	26.2	26.5	130.8	2.00	0.97	20.1	20.2	145.5
	3.00	1.99	28.6	19.3	137.4	3.00	1.99	21.5	14.4	151.2
	5.00	4.90	31.0	12.6	143.7	5.00	4.90	22.7	9.1	156.3
	2.87	1.84	28.4	20.0	136.8	2.03	0.99	20.2	20.0	145.7
	1.08	0.33	21.5	40.0	118.2	0.78	0.18	15.6	40.0	126.2
04	0.50	0.11	17.0	68.6	95.5	0.50	0.11	15.2	61.1	108.7
	0.75	0.23	21.4	57.8	104.7	0.75	0.23	18.4	49.5	119.1
	1.00	0.38	24.6	49.9	111.4	1.00	0.38	20.6	41.7	126.1
	1.50	0.77	28.9	39.2	120.5	1.50	0.77	23.4	31.6	135.1
	2.00	1.28	31.7	32.3	126.4	2.00	1.28	25.1	25.5	140.7
	3.00	2.63	35.2	23.9	133.5	3.00	2.63	27.1	18.4	147.1
	5.00	6.49	38.5	15.7	140.5	5.00	6.49	28.9	11.8	153.0
	3.75	3.90	36.7	20.0	136.9	2.71	2.19	26.6	20.0	145.6
	1.46	0.73	28.6	40.0	119.9	1.07	0.42	21.0	40.0	127.6
06	0.50	0.13	21.3	85.4	91.9	0.50	0.13	19.2	77.0	104.3
	0.75	0.27	27.2	73.1	100.8	0.75	0.27	23.7	63.6	114.6
	1.00	0.45	31.6	64.0	107.5	1.00	0.45	26.8	54.1	121.9
	1.50	0.93	37.8	51.1	116.8	1.50	0.93	30.9	41.7	131.4
	2.00	1.54	41.9	42.6	123.0	2.00	1.54	33.5	34.0	137.3
	3.00	3.16	47.1	31.9	130.7	3.00	3.16	36.6	24.8	144.4
	5.00	7.80	52.2	21.3	138.4	5.00	7.80	39.5	16.0	151.1
	5.38	8.89	52.8	20.0	139.4	3.88	4.97	38.2	20.0	148.1
	2.19	1.82	43.2	40.0	124.8	1.60	1.03	31.5	40.0	132.7

WPD = WATER PRESSURE DROP.
WTD = WATER TEMPERATURE DIFFERENCE.
FAT = FINAL AIR TEMPERATURE.

[a]This is only a partial listing of capacity and sizes. See the complete data for other sizes in manufacturer's catalog.
Source: Reprinted courtesy of The Trane Company, La Crosse, Wis.

TABLE 8-14

Steam performance data, 2 psig steam, 60°F entering air

Type	Model No.	Btu/hr	Sq. Ft EDR	Sound Class	Max. Mounting Height (ft)[a]	High Motor Speeds — Air Data: Heat Throw or Spread @ Max. Height[a]	Lower Mtg. Height (ft)	Heat Throw @ Lower Mtg. Height	Cfm[c]	Outlet Velocity (fpm)	Final Air Temp. (°F)	Condensate Lb/hr	Motor Data: Hp[b]	Approx. rpm
Horizontal Delivery	HS-18	18,000	75	II	8	17	—	—	340	625	107	18	16 Mhp	1550
	HS-24	24,000	100	II	9	18	—	—	370	695	119	25	$\frac{1}{25}$	1550
	HS-33	33,000	138	II	10	21	—	—	630	690	108	35	$\frac{1}{25}$	1550
	HS-47	47,000	196	III	12	28	8	33	730	810	119	49	$\frac{1}{12}$	1550
	HS-63	63,000	263	II	14	29	8	39	1,120	690	111	66	$\frac{1}{12}$	1550
	HS-86	86,000	358	III	15	31	10	43	1,340	835	118	89	$\frac{1}{8}$	1625
	HS-108	108,000	450	III	17	31	10	43	2,010	790	109	111	$\frac{1}{8}$	1625
	HS-121	121,000	504	III	16	25	10	38	1,775	715	122	126	$\frac{1}{6}$	1075
	HS-165	165,000	688	IV	19	40	10	56	3,240	880	106	170	$\frac{1}{3}$	1075
	HS-193	193,000	804	IV	18	38	10	53	2,900	810	121	200	$\frac{1}{3}$	1075
	HS-258	258,000	1075	V	19	44	12	60	4,560	750	111	267	$\frac{1}{2}$	1075
	HS-290	290,000	1208	V	20	46	12	60	4,590	765	117	300	$\frac{1}{2}$	1075
	HS-340	340,000	1417	V	20	**46**	12	56	5,130	735	120	352	$\frac{1}{2}$	1075
"Power-Throw"	PT-279	279,000	1163	V	16	**100**	—	—	5,460	2165	111	288	$\frac{1}{2}$	1075
	PT-333	333,000	1388	VI	17	**110**	—	—	5,980	2165	116	345	$\frac{3}{4}$	1140
	PT-385	385,000	1604	VI	17	**115**	—	—	7,900	1860	110	398	1	1140
	PT-500	500,000	2083	VI	18	**130**	—	—	10,390	2520	108	518	$1\frac{1}{2}$	1140
	PT-610	610,000	2542	VI	20	**140**	—	—	11,750	2315	112	631	$1\frac{1}{2}$	1140
Vertical Delivery	V-42	42,000	175	II	15	11	—	—	950	825	103	43	$\frac{1}{30}$	1050
	V-59	59,000	246	II	19	14	—	—	1,155	1005	111	61	$\frac{1}{30}$	1050
	V-78	78,000	325	II	20	15	—	—	1,590	1065	109	81	$\frac{1}{15}$	1050

Model No.	Btu/hr	Sq. Ft EDR	Sound Class	Max. Mounting Height (ft)	Heat Throw or Spread (ft)	Cfm	Outlet Velocity (fpm)	Final Air Temp. (°F)	Condensate Lb/hr	Hp	Approx. rpm
V-95	95,000	396	II	20	15	—	1,665	118	99	$\frac{1}{15}$	1050
V-139	139,000	579	III	24	18	—	2,660	112	144	$\frac{1}{6}$	1075
V-161	161,000	671	IV	27	20	—	3,200	115	167	$\frac{1}{3}$	1075
V-193	193,000	804	IV	30	22	—	3,500	116	200	$\frac{1}{3}$	1075
V-212	212,000	883	IV	30	22	—	3,610	120	219	$\frac{1}{3}$	1075
V-247	247,000	1029	V	34	26	—	4,820	111	256	$\frac{1}{2}$	1075
V-279	279,000	1163	V	37	30	—	5,460	111	288	$\frac{1}{2}$	1075
V-333	333,000	1388	V	37	30	—	5,980	116	345	$\frac{3}{4}$	1140
V-385	385,000	1604	VI	36	30	—	7,900	110	398	1	1140
V-500	500,000	2083	VI	44	37	—	10,390	108	518	$1\frac{1}{2}$	1140
V-610	610,000	2542	VI	43	36	—	11,750	112	631	$1\frac{1}{2}$	1140

Reduced Motor Speeds
(requires solid-state motor speed controller)

						Air Data				Motor Data	
Type	Model No.	Btu/hr	Sq. Ft EDR	Sound Class	Max. Mounting Height (ft) [a,d]	Heat Throw or Spread (ft) [a]	Cfm [c]	Outlet Velocity (fpm)	Final Air Temp. (°F)	Condensate Lb/hr	Hp [b]	Approx. rpm
Horizontal Delivery	HS-18	14,000	58	II	8	10	220	415	118	14	16 Mhp	1000
	HS-24	18,000	75	II	9	11	230	440	131	18	$\frac{1}{25}$	1000
	HS-33	25,000	104	II	10	13	395	440	118	26	$\frac{1}{25}$	1000
	HS-47	38,000	158	III	12	17	450	515	137	36	$\frac{1}{12}$	1000
	HS-63	47,000	195	II	14	17	685	430	122	49	$\frac{1}{12}$	1000
	HS-86	64,000	265	III	15	19	825	525	131	66	$\frac{1}{8}$	1000
	HS-108	81,000	340	III	17	19	1,255	500	119	84	$\frac{1}{8}$	1000

Source: Courtesy of Modine Manufacturing Company, Racine, WI.

[a] Horizontal units with horizontal louvers opened 30° from the vertical plane; vertical types equipped with cone jet deflector, blades in full open position.
[b] For most popular motor used on these models.
[c] Cfm for horizontal types is entering cfm; cfm for vertical and "power-throw" types is leaving cfm.
[d] High motor speed.

TABLE 8-15

Hot water performance data, 200°F entering water temperature, 60°F entering air temperature, 20°F water temperature drop

Type	Model No.	Btu/hr	Gpm	Pressure Drop (ft of water)	Sound Class	Air Data — High Motor Speeds							Motor Data	
						Max. Mounting Height (ft)[a]	Heat Throw or Spread @ Max. Height[a]	Lower Mounting Height (ft)	Heat Throw @ Lower Mtg. Height	Cfm[c]	Outlet Velocity (fpm)	Final Air Temp. (°F)	Hp[b]	Approx. rpm
Horizontal Delivery	HS-18	12,800	1.3	0.5	II	9	18	—	—	340	615	94	16 Mhp	1550
	HS-24	15,700	1.6	0.8	II	10	19	—	—	370	675	98	$\frac{1}{25}$	1550
	HS-33	24,500	2.5	0.2	II	11	23	—	—	630	675	95	$\frac{1}{25}$	1550
	HS-47	29,000	2.9	0.3	III	13	30	8	36	730	785	96	$\frac{1}{12}$	1550
	HS-63	47,000	4.7	0.8	II	15	31	8	42	1,120	680	98	$\frac{1}{12}$	1550
	HS-86	63,000	6.3	1.4	III	16	33	10	46	1,340	820	102	$\frac{1}{8}$	1625
	HS-108	81,000	8.1	3.2	III	18	33	10	46	2,010	775	96	$\frac{1}{8}$	1625
	HS-121	90,000	9.0	4.0	III	17	27	10	41	1,775	700	106	$\frac{1}{6}$	1075
	HS-165	133,000	13.5	7.9	IV	20	43	10	60	3,240	870	97	$\frac{1}{3}$	1075
	HS-193	139,000	14.0	2.0	IV	19	41	10	57	2,900	790	103	$\frac{1}{3}$	1075
	HS-258	198,000	20.0	5.0	V	20	47	12	65	4,560	740	99	$\frac{1}{2}$	1075
	HS-290	224,000	22.0	5.8	V	22	50	12	65	4,590	750	104	$\frac{1}{2}$	1075
	HS-340	273,000	27.0	11.0	V	22	50	12	60	5,130	720	108	$\frac{1}{2}$	1075
"Power-Throw"	PT-279	180,000	18.5	0.3	V	17	108	—	—	5,460	2165	92	$\frac{1}{2}$	1075
	PT-333	230,000	23.0	0.4	VI	18	117	—	—	5,980	2165	97	$\frac{3}{4}$	1140
	PT-385	270,000	27.3	0.5	VI	18	124	—	—	7,900	1860	94	1	1140
	PT-500	340,000	34.0	0.6	VI	19	138	—	—	10,390	2520	91	$1\frac{1}{2}$	1140
	PT-610	430,000	43.0	0.8	VI	22	151	—	—	11,750	2315	95	$1\frac{1}{2}$	1140
Vertical Delivery	V-42	30,000	3.0	0.6	II	16	12	—	—	950	825	90	$\frac{1}{30}$	1050
	V-59	43,000	4.3	0.6	II	20	15	—	—	1,155	1005	96	$\frac{1}{30}$	1050
	V-78	57,000	5.7	0.6	II	22	16	—	—	1,590	1065	95	$\frac{1}{15}$	1050

Vertical Delivery (continued)

Model No.	Btu/hr	Gpm	Pressure Drop (ft of water)	Sound Class	Max. Mounting Height (ft)	Heat Throw or Spread (ft)		Cfm	Outlet Velocity (fpm)	Final Air Temp. (°F)	Hp	Approx. rpm
V-95	68,000	6.8	0.5	II	22	16	—	1,665	1120	100	$\frac{1}{15}$	1050
V-139	105,000	10.5	2.4	III	26	19	—	2,660	1285	98	$\frac{1}{6}$	1075
V-161	123,000	12.5	2.2	IV	29	22	—	3,200	1420	101	$\frac{1}{3}$	1075
V-193	140,000	14.0	1.8	IV	32	24	—	3,500	1690	99	$\frac{1}{3}$	1075
V-212	156,000	15.5	1.4	IV	32	24	—	3,610	1740	102	$\frac{1}{3}$	1075
V-247	184,000	18.5	1.9	V	37	28	—	4,820	1910	97	$\frac{1}{2}$	1075
V-279	210,000	21.0	2.0	V	40	32	—	5,460	2165	97	$\frac{1}{2}$	1075
V-333	257,000	26.0	3.8	V	40	32	—	5,980	2165	102	$\frac{3}{4}$	1140
V-385	300,000	30.0	2.8	VI	39	32	—	7,900	1860	98	1	1140
V-500	383,000	38.0	4.2	VI	47	40	—	10,390	2520	96	$1\frac{1}{2}$	1140
V-610	430,000	43.0	0.8	VI	46	39	—	11,750	2315	95	$1\frac{1}{2}$	1140
V-680	680,000	68.0	17.5	VI	41	34	—	10,700	2110	125	$1\frac{1}{2}$	1140

Reduced Motor Speeds
(requires solid-state motor speed controller)

Type	Model No.	Btu/hr	Gpm	Pressure Drop (ft of water)	Sound Class	Air Data					Motor Data	
						Max. Mounting Height (ft)[a,d]	Heat Throw or Spread (ft)[a]	Cfm[c]	Outlet Velocity (fpm)	Final Air Temp. (°F)	Hp[b]	Approx. rpm
Horizontal Delivery	HS-18	9,800	1.3	0.5	II	9	11	220	400	100	$\frac{1}{6}$ Mhp	1000
	HS-24	12,000	1.6	0.8	II	10	12	230	425	105	$\frac{1}{25}$	1000
	HS-33	19,000	2.5	0.2	II	11	14	395	430	105	$\frac{1}{25}$	1000
	HS-47	22,000	2.9	0.3	III	13	18	450	490	105	$\frac{1}{12}$	1000
	HS-63	36,000	4.7	0.8	II	15	18	685	420	105	$\frac{1}{12}$	1000
	HS-86	48,000	6.3	1.4	III	16	20	825	515	115	$\frac{1}{8}$	1000
	HS-108	62,000	8.1	3.2	III	18	20	1,255	490	105	$\frac{1}{8}$	1000

Source: Courtesy of Modine Manufacturing Company, Racine, WI.

[a] Horizontal units with horizontal louvers opened 30° from the vertical plane; vertical types equipped with cone jet deflector, blades in full open position.

[b] For most popular motor used on these models.

[c] Cfm for horizontal types is entering cfm; cfm for vertical and "power-throw" types is leaving cfm.

[d] High motor speed.

perature, entering air temperature, and water flow rate affect capacity. Refer to manufacturers' catalogs for correction factors.

Figure 8-19 shows a gas indirect-fired unit heater. This type of unit, or one using oil, may be preferred when the number of units does not justify the expense or space requirements of a new boiler system or where individual metering of the fuel supply is required. Gas indirect-fired units are usually either of the horizontal propeller type (Figure 8-19) or of the industrial centrifugal type. Some codes limit the use of indirect-fired unit heaters in some applications.

Table 8-16 gives the manufacturer's performance table for the Model PA unit shown in Figure 8-19. Gas-fired unit heaters are rated in both input and output, in accordance with approval requirements of the American Gas Association. Ratings of oil-fired unit heaters are based on heat delivered at the heater outlet.

Unit heaters are customarily rated at free-delivery. If outdoor air intakes, air filters, or inlet or outlet ducts are used, a reduction in air and heating capacity will result due to the added resistance to airflow, unless fan speed is increased. Since this reduction in capacity depends on the characteristics of the heater and on the

FIGURE 8-19 Gas indirect-fired unit heater. (Courtesy of Modine Manufacturing Company, Racine, WI)

TABLE 8-16

Performance rating table for gas indirect-fired unit heater

Performance														
Model Number														
		PA30	PA50	PA75	PA105	PA130	PA150	PA170	PA200	PA225	PA250	PA300	PA350	PA400
Total Input Btu/hr		30,000	50,000	75,000	105,000	130,000	150,000	170,000	200,000	225,000	250,000	300,000	350,000	400,000
Total Output Btu/hr		22,500	38,500	57,750	80,850	100,100	112,500	127,500	152,000	173,250	192,500	231,000	269,500	304,000
Cfm @ 70°F		400	675	855	1400	1725	1900	2540	2900	2900	4275	4275	4400	5300
Fpm		470	555	585	825	740	815	955	825	830	1200	1075	975	995
Air Temp. Rise (°F)		52	53	63	53	54	55	46	49	55	42	50	57	53
Mounting Height (Max. Ft)		8	8	10	10	12	12	12	16	16	20	20	20	20
Heat Throw (ft)		18	23	25	40	45	48	50	50	50	65	65	65	65
Motor Data	Hp	8 Mph	25 Mph	25 Mph	1/15	1/15	1/6	1/6	1/6	1/6	1/3	1/3	1/2	1/2
	Amps (@ 115 V)	.51	1.0	1.0	2.5	2.5	2.8	2.8	2.8	2.8	5.4	5.4	6.8	6.8
	Rpm	1550	1550	1550	1050	1050	1075	1075	1075	1075	1075	1075	1075	1075
	Shaded Type	shaded pole	shaded pole	shaded pole	shaded pole	perm. pole	perm. split cap.	perm. split cap.	perm. split cap.	perm. split cap.	perm. split cap.	perm. split cap.	perm. split cap.	split cap.

Source: Courtesy of Modine Manufacturing Company, Racine, WI.

Note: Ratings shown are for elevations up to 2000 feet. For elevations above 2000 feet, ratings should be reduced at the rate of 4% for each 1000 feet above sea level.

type, design, and speed of the fans, no specific percentage reduction can be assigned at a given added resistance. The engineer should always consult with the manufacturer when other than free-delivery conditions are to be met.

Unit Ventilators. Unit ventilators (sometimes called "classroom" unit ventilators because of their frequent use in school classrooms) are a specifically designed fan-coil unit that provides for a variety of applications. *ASHRAE Handbook 1988 Equipment* identifies two general types of unit ventilators.

The *heating unit ventilator* denotes an assembly of components, the principal functions of which are to heat, ventilate, and cool a given space by the introduction of outside air in quantities up to 100%. The heating medium may be steam, hot water, gas, or electricity. The essential components of a heating unit ventilator are the electric motor-driven centrifugal fans, heating element, dampers, filters, inlet grilles, and outlet grilles (diffusers), all of which are encased in a housing (see Figure 8-20). Some or all of the automatic controls are also included in the unit housing.

The *air conditioning unit ventilator* is similar to the heating unit ventilator, but, in addition to the normal winter-season function of heating, ventilating, and cooling with outdoor air, it is equipped to cool and dehumidify the air during the summer season. It is usually arranged and controlled to introduce a fixed quantity of outdoor air for ventilation during warm periods in mild weather. The air conditioning unit ventilator may be provided with a variety of combinations of heating and cooling elements. Some of the more common arrangements include (1) combination hot and chilled water coil (two-pipe), (2) separate hot and chilled water coils (four-pipe), (3) hot water or steam coil and direct expansion refrigerant coil, and (4) electric heating coil and chilled water or direct expansion coil.

Unit ventilators are available for floor mounting, recessed applications, wall or ceiling mounting, and with various airflow and capacity ratings. They can be arranged for either blow-through or draw-through operation.

Unit ventilators are used primarily for schools, meeting rooms, offices, and other areas where density of occupancy indicates the need for controlled ventilation. The typical unit is equipped with a system of control that permits the heating, ventilating, and cooling effect to be varied while the fans are operating continuously. In normal operation, the discharge air temperature from a unit is varied in accordance with room requirements. When heating is required, the delivered air is above room temperature. When heat is generated within the room from people, lights, solar radiation, and so forth, in sufficient quantity to exceed heat losses, the temperature of the delivered air is below that of the room. With the heating unit ventilator, this can be done whenever the outdoor air temperature is below the room

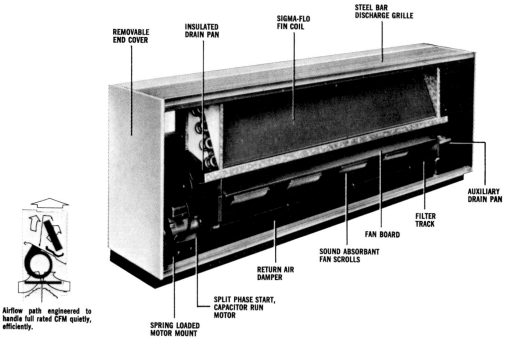

FIGURE 8-20 Classroom unit ventilator. (Courtesy of The Trane Company, LaCrosse, WI)

temperature by bringing in more outdoor air, thereby providing what is called *ventilation cooling*. Air conditioning unit ventilators can provide mechanical cooling in cases where the outdoor air temperature is too high to be used effectively for ventilation cooling.

Since the unit ventilator has a dual function of introducing outdoor air for ventilation and maintaining a specified room condition, the required heating capacity is the sum of heat required to bring outdoor air up to room temperature and the heat required to offset room losses. The ventilation cooling capacity of a unit ventilator is determined by the air volume delivered by the unit and the room temperature.

ILLUSTRATIVE PROBLEM 8-3

A room has a heat loss of 23,000 Btu/hr at a winter outdoor air design temperature of 0°F and an inside design temperature of 70°F. To obtain the required number of air changes, a 1250-cfm unit ventilator is to be used, and 20% (250-cfm) outdoor air will be used. Minimum discharge air temperature from the unit is 60°F. Determine the ventilation heat requirement, the total heating requirement, and the ventilation cooling capacity of the unit with outdoor air temperature below 60°F.

Solution: The heat required for the ventilation air may be found from the equation of Chapter 4:

$$q_v = 1.10(\text{cfm}_{oa})\,(t_i - t_o) = 1.10(250)\,(70 - 0)$$
$$= 19,250 \text{ Btu/hr}$$

The total heating capacity is $19,250 + 23,000 = 42,250$ Btu/hr. The total ventilation cooling capacity may be found from

$$q_{vc} = 1.10(\text{cfm}_{hc})\,(t_i - t_f)$$

where

$$q_{vc} = \text{ventilation cooling capacity (Btu/hr)}$$
$$\text{cfm}_{hc} = \text{volume of air handled by unit}$$
$$t_i = \text{required room temperature (°F)}$$
$$t_f = \text{unit discharge air temperature (°F)}$$

So,

$$q_{vc} = 1.10(1250)\,(70 - 60) = 13,750 \text{ Btu/hr}$$

In general, when a unit ventilator is to be selected, we must consider the following: (1) unit air capacity, cfm; (2) minimum percentage outdoor air; (3) heating and/or cooling requirements; (4) control cycle required; and (5) location of unit.

Of primary consideration in selecting the unit air capacity are the ventilation cooling capacity and the number of occupants in the space. Other factors to be considered are legal requirements, volume of the space, density of occupancy, and the usage of the space. The number of air changes required for a specific application also depends on window area, orientation, and the maximum outdoor temperature at which the unit will be able to cool by ventilation.

For classrooms, air changes per hour to prevent overheating at 55°F (13°C) outdoor air varies from 6 (for rooms with small, north windows) to 12 (for rooms with large, south windows). Factories and kitchens require 30 to 60 air changes or more. Office areas may need from 10 to 15 air changes per hour.

The minimum amount of outdoor air for ventilation is determined after the total air capacity has been established. It may be governed by codes or calculated by the engineer to meet the ventilation air needs of the particular application (see ASHRAE Standard 62-1989, *Ventilation for Acceptable Indoor Air Quality*).

In the absence of other criteria, 5 cfm (2.36 L/s) per occupant is used frequently as minimum outdoor air for classroom applications (see Chapter 5). Heating capacity should always be selected after selecting the unit air capacity for mild-weather cooling.

Air Makeup Units.   *ASHRAE Handbook 1988 Equipment* identifies these units as an assembly of elements, the principal function of which is to properly condition makeup air being introduced into a space, customarily to make up for air being exhausted from that space. The air exhausted may be from a particular process or may be general exhaust from the space or building.

Whatever the method of exhaust, a volume of air is removed from a structure and must be replaced with new outdoor air. If the temperature, humidity, or both, within a structure are controlled, the environmental control system must provide capacity to condition the replacement air if the space condition is to be maintained. The most practical way to condition the replacement air is through the use of a makeup air unit.

Certain symptoms of an "air-starved" building may be recognized: For example, (1) gravity stacks will back-vent; (2) exhaust systems do not perform at rated volume, resulting in poor control of contaminants; (3) the perimeter of the building will be cold due to a high rate of infiltration; (4) there will be severe indraft at exterior doors; (5) exterior doors will be difficult to open; and (6) heating systems will not be able to maintain uniform comfort conditions throughout the building. The center core will become overheated.

Makeup air units may be required to do heating, cooling, humidifying, dehumidifying, and filtering. They may be required to replace the air at space conditions, or they may be used for part or all of the heating and cooling load.

Various types of air makeup units may be identified, such as *blow-through* or *draw-through* referring to the relative location of the fan with respect to the heat-transfer device. The fans may be either centrifugal (single or double width), forward or backward inclined (flat plate or airfoil blades), axial flow (propeller, ducted propeller, vaneaxial, tubeaxial, or axial centrifugal). *Heating sources* include gas (direct- or indirect-fired), oil (direct- or indirect-fired), steam, hot water, electric blast coils, heat-transfer fluids, partial recirculation (heat conservation), and summer supply only. *Cooling sources* include direct expansion (DX) refrigerant or chilled water coils, evaporative cooling, sprayed coils, or chilled water air washers. *Filters* may be automatic or manual roll type, throw away or cleanable panels, bag type, electrostatic, or a combination of these types.

8-12 AIR-HEATING COILS

Heat-transfer coils are widely used for forced-convection heating of air. The total coil surface may consist of a single coil section or several coil sections assembled in a bank. In this section, we are concerned with those heat-transfer coils used for comfort heating and air conditioning using steam or hot water.

Extended-surface coils consist of a primary and a secondary heat-transfer surface. The primary surface is the surface of the round tubes or pipes that are arranged in a repetitive pattern with respect to the airflow. The secondary surface (fins) consists of thin metal plates or a spiral ribbon uniformly spaced or wound along the length of the primary surface. It is in intimate contact with the primary surface for good heat transfer. The

bond must be maintained permanently to ensure continuation of rated performance.

Copper and aluminum are the materials most commonly used in the fabrication of extended-surface coils. Tubing made of steel or various copper alloys is used in applications where corrosive forces might attack the coils from either inside or outside. The most common combination for low-pressure applications is aluminum fins on copper tubes. Low-pressure steam coils are usually designed to operate up to 150 to 200 psi (1.0 to 1.4 MPa) gauge. Above that point, tube materials such as red brass, Admiralty, or Cupro-Nickel are selected.

Finned-tube coils owe their popularity to their relative compactness and to the ease with which they can be installed. These coils are used extensively as the heat-transfer device in central station air-handling units; factory-assembled, self-contained air conditioners; and room air terminals. The application of each type of coil is limited to the field within which it is rated. Other limitations are imposed by code regulations, choice of materials, fluids used, and condition of the air handled, or by an economic analysis of possible alternatives for each installation.

Water Coils. Figure 8-21 illustrates two types of hot-water heating coils. The performance of water coils for heating depends on the elimination of air from the system and the proper distribution of the water in the individual tube circuits. Air elimination in the system piping is discussed in a later chapter. Figure 8-22 illustrates some methods used for connecting water coils to a piping system.

To produce the most efficient capacity without ex-

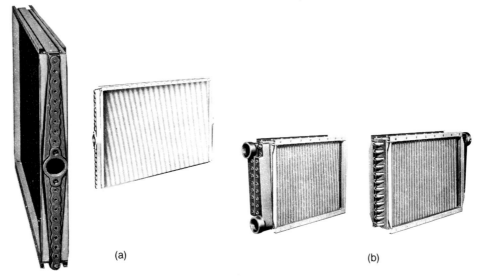

(a)   (b)

FIGURE 8-21 Typical hot-water heating coils: (a) single-row and (b) double-row. (Courtesy of The Trane Company, LaCrosse, WI)

cessive water-pressure drop through the coil, various circuit arrangements are used. A single-tube serpentine circuit can be used on small booster heaters requiring small water quantities up to a maximum of 5 gpm (0.32 L/s). With this arrangement, a single tube handles the entire water quantity provided the tube makes a number of passes across the airstream.

The most common circuiting arrangement is often called *single-row serpentine* or *standard* circuiting. With this arrangement, all the tubes in each coil row are supplied with an equal amount of water through a manifold, commonly called the *coil header*. When the water volume is small, so that water flow is laminar, *turbulators* are sometimes in each tube circuit to produce turbulent water flow.

When hot-water coils are used with entering air temperature below freezing (an antifreeze brine would be preferable to water), piping the coil for *parallel flow* rather than *counterflow* should be considered. This arrangement places the highest water temperature on the entering air side. Coils piped for counterflow have the water enter the tube row on the exit air side of the coil.

Water coils are usually designed to be self-venting by supplying water to the coil so that the water flows upward in the coil, forcing air out through the return water connection. This design also ensures that the coil is always completely filled with water, regardless of the water volume supplied. Water circuits must be designed to ensure complete coil drainage.

Steam Coils. Heating coils using steam as the heating medium are made in two general types. First are those used for general heating applications where no attempt is made to obtain uniform air temperature over the length of the coil, or which are *not* used in airstreams that drop below 32°F temperature. Second are those called *steam-distributing* heating coils, made with perforated inner tubes to distribute the steam along the full length of the inner surface of the outer tubes.

The first type, illustrated in Figure 8-23, is most

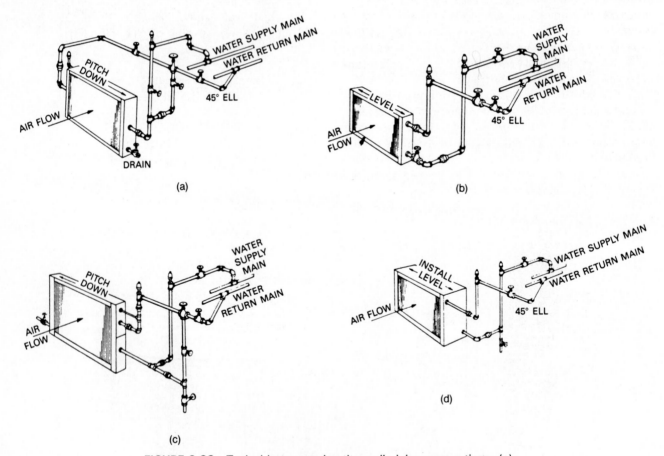

FIGURE 8-22 Typical hot-water heating coil piping connections: (a) heating coil pitched down toward supply end—vent is used in top plug of return header if water velocity through coil is below 1.5 fps; (b) heating coil installed with tubes level; (c) heating coil pitched away from supply end—vent is used if water velocity is below 1.5 fps; and (d) two-row heating coil installed for hot water.

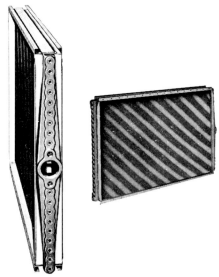

FIGURE 8-23 Single-row steam heating coil with opposite end connections. (Courtesy of The Trane Company, LaCrosse, WI)

often used in air-handling systems where face and by-pass dampers may be used or where comparatively short tubes can be used.

The so-called steam-distributing type of heating coils are normally used where the steam flow to the coil must be throttled in order to obtain temperature control, where uniform temperatures are required along the length of the coil, or where freezing temperatures may be encountered with the entering air. This type of coil is illustrated in Figure 8-24, which shows the inner steam-distributing tubes. Steam condenses in the outer tubes where heat transfer takes place and drains to the return header.

For proper performance of steam heating coils,

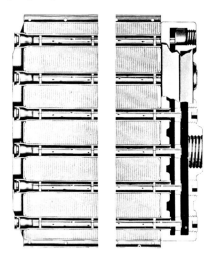

FIGURE 8-24 Section view of steam heating coil with supply and return headers and steam-distributing tubes. (Courtesy of The Trane Company, LaCrosse, WI)

condensate and air or other noncondensables must be eliminated rapidly and the steam must be distributed uniformly to the individual tubes. Noncondensable gas (such as carbon dioxide) remaining in the coil causes chemical corrosion and results in early coil failure.

Particular care in steam piping, controls, and installation is necessary to provide optimum performance from steam heating coils. Figure 8-25 illustrates several methods frequently used to connect steam coils into the steam piping system.

Application of Heating Coils.   Airflow in heating coils may be vertical or horizontal although the latter arrangement is more common.

For steam heating, the coils may be installed with the tubes in a vertical or horizontal position. Horizontal-tube coils should be pitched or inclined toward the return connection to drain the condensate. Water heating coils generally have horizontal tubes to avoid air and water pockets. Where water coils may be exposed to below-freezing temperatures, drainability must be considered, or a fluid such as glycol must be put in the coil. If a coil is to be drained and exposed to below-freezing temperatures, it should first be flushed with an antifreeze solution.

When the leaving air is controlled by modulation of the steam to the coil, steam-distributing tube-type coils are optimum for providing uniform exit air temperatures, as noted previously. Correctly designed steam-distributing tube coils limit the exit air temperature stratification to a maximum of 5 to 6°F (about 3°C) over the entire length of the coil even when the steam supply is modulated to a small fraction of the full-load capacity.

As an added precaution, in the application of both steam and water heating coils, the outdoor-air inlet dampers usually close automatically when the fan is stopped (system shut down). This arrangement minimizes the danger of freezing. In steam systems with very low-temperature outdoor conditions, such as −20°F (−29°C) or below, it is desirable to have the steam valve go to full open position when the system is shut down. If the outside air is used for proportioning building makeup air, the damper used should be an opposed-blade design.

Heating coils are designed to allow for expansion and contraction resulting from the temperature ranges within which they operate. Care must be taken to prevent imposing strains from the piping on the piping connections. Expansion loops, expansion or swing joints, or flexible connections usually provide this protection.

It is good practice to support banked coils individually in an angle-iron frame or a similar supporting structure. With this arrangement, the lowest coil is not

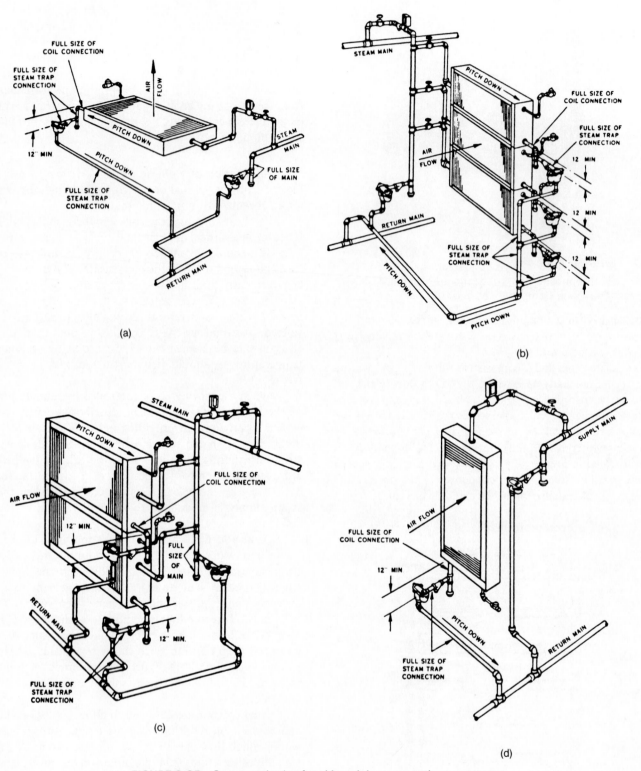

FIGURE 8-25 Some methods of making piping connections to steam heating coils.

required to support the weight of the coils stacked above. This design also facilitates the removal of individual coils in a multiple-coil bank for repair or replacement.

Low-pressure steam systems, coils controlled by modulating the steam supply, or both, should have a vacuum breaker or be drained through a vacuum-return system to ensure proper condensate drainage (see Chapters 11 and 12).

Heating Coil Selection. ASHRAE Handbook 1988 Equipment suggests that the following factors should be considered in coil selection:

- The required duty or capacity considering the other system components.
- Temperature of the air entering the coil.
- Available heating media.
- Space and dimensional limitations.
- Air quality.
- Permissible resistances for both the air and heating media.
- Characteristics of individual coil designs.
- Individual installation requirements, such as type of control to be used.
- Coil face velocity.

The duty required may be determined from the calculated heating and/or ventilation load. There may be a choice of heating media, as well as operating temperature, depending on whether the installation is new or is being modified. The air handled may be limited by the use of ventilating ducts for air distribution, or it may be determined by requirements for satisfactory air distribution or ventilation.

The resistance of the airflow through the coil influences the required fan power and speed. This resistance may be limited to allow the use of a given size of fan motor or to keep the operating expense low. It may be limited because of sound-level requirements.

The permissible water flow resistance of a hot-water coil may be dictated by the available pump head from a given size of pump and pump motor.

The performance of a heating coil depends on the correct choice of the original equipment and on proper application and maintenance. For steam coils, performance and selection of the type and size of steam trap are of utmost importance (see Chapter 12).

Coil ratings are based on uniform face velocity. Nonuniform airflow through the coil will affect its performance. Nonuniform airflow may be caused by air entrance at odd angles or by inadvertent blocking of a portion of the coil face. Complete mixing of return air

and outside air is essential to the proper operation of a coil. Mixing damper design is critical to the operation of the system. Systems in which the air passes through a fan before flowing through a coil do not ensure proper air mixing. To obtain rated performance, air quantity in the field must correspond with design quantity and must always be maintained.

The selection of heating coils is relatively simple because it involves dry-bulb temperature and sensible heat only, without the complication of simultaneous latent heat loads, as in cooling coils. Heating coils are usually selected from charts or tables giving final air temperatures, steam saturation pressures, and water temperatures. Oversizing of modulating steam valves and steam coils in the system should be avoided because such oversizing makes proper control more difficult (see Chapter 16).

The selection of hot-water coils is more complicated because of the added variable of water velocity and the fact that the water temperature decreases as the water flows through the coil circuits. Therefore, in selecting hot-water heating coils, the water velocity and the mean temperature difference (between the hot water flowing in the tubes and the air passing over the fins) must be considered. Most coil manufacturers have computer programs to assist in the coil selection process.

Coil Ratings. Steam and hot-water coils are usually rated within the following limits, which may be exceeded for special applications:

1. *Air face velocity*—Between 200 to 1500 fpm (1 to 8 m/s), based on air at standard density of 0.075 lb/ft³ (1.2 kg/m³).
2. *Entering air temperature*—From −20°F to 100°F (−29°C to 38°C) for steam coils; 0°F to 100°F (−18°C to 38°C) for hot-water coils.
3. *Steam pressure*—From 2 to 250 psi (14 to 1720 kPa) gauge at the coil steam supply connection (pressure drop through the steam control valve must be considered).
4. *Hot-water temperature*—Between 120°F and 250°F (50°C and 120°C).
5. *Water velocities*—From 0.5 to 8 fps (0.2 to 2.5 m/s).

Individual installations vary widely, but the following values can be used as a guide.

The most common air face velocities used are between 500 and 1000 fpm (2.5 and 5 m/s). Delivered air temperatures vary from about 72°F (22°C) for ventilation only to about 150°F (66°C) for complete heating. Steam pressures vary from 2 to 15 psi (14 to 103 kPa) gauge, with 5 psi (35 kPa) gauge being the most common. A minimum steam pressure of 5 psi (35 kPa) gauge is recommended for systems with entering air temperatures below freezing. Water temperatures for comfort heating are commonly between 180°F and 200°F (80°C to 95°C), with water velocities between 4 and 6 fps (1.2 and 1.8 m/s).

Water quantity is usually based on about 20°F (11°C) temperature drop through the coil. Air resistance is usually limited to $\frac{3}{8}$ to $\frac{5}{8}$ in. (90 to 155 Pa) of water for commercial buildings and to about 1 in. (250 Pa) for industrial buildings. High-temperature water systems have water temperatures commonly between 300°F and 400°F (150°C and 200°C), with up to 100°F (56°C) drop through the coil.

Most coil manufacturers have their own methods of producing performance rating tables from a suitable number of coil performance tests. A method of testing air-heating coils is given in ASHRAE Standard 33-78, *Method of Testing for Rating Forced-Circulation Air Cooling and Heating Coils*. A basic method of rating to provide a fundamental means for establishing thermal performance of air-heating coils by extension of test data, as determined from laboratory tests on prototypes, to other operating conditions, coil sizes, and row depths of a particular surface design and arrangement is given in ARI Standard 410-81, *Forced Circulation Air-Cooling and Air-Heating Coils.*

8-13 SHELL-AND-TUBE HEAT EXCHANGERS (HEATING)

Shell-and-tube heat exchangers are frequently used in heating systems. One type of heat exchanger is illustrated in Figure 8-26. This type of unit is designed to heat water with steam. The water flows through the U-bent tubes, while the steam condenses inside the shell and on the tubes. The shell, tube sheets, spacer plates, and tie rods are normally made of steel. The U-tubes are normally of $\frac{3}{4}$-in. OD seamless copper and may serpentine through the length of the shell two, four, or six times (called two-pass, four-pass, and six-pass).

Another type of shell-and-tube heat exchanger used when there is a source of high-temperature water is a water-to-water unit. These units are constructed in the same general way as shown in Figure 8-26, except the inside of the shell has baffles spaced along its length to force the water flow in a definite path through the shell.

The usual installation arrangement is to have the medium to be heated passing through the tubes and

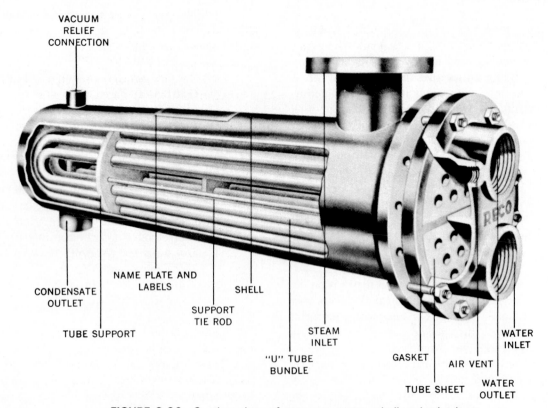

FIGURE 8-26 Section view of steam-to-water, shell-and-tube heat exchanger.

the medium being cooled passing through the shell. The steam-to-water heat exchanger is an instantaneous-type heater (water enters the tubes at one temperature and leaves at the outlet at the desired final temperature). Most manufacturers' catalogs list ratings for the most commonly required temperature rises for the water through a wide range of steam pressures. They are available in two-, four-, or six-pass construction and usually are cataloged for lengths up to 10 ft and shell diameters from 4 in. to 30 in.

The most common application for steam-to-water, shell-and-tube units is to use existing steam from a process or existing steam heating system to heat water for use in an auxiliary heating system, to heat domestic hot water, and so forth.

For heating water with steam, the following conditions must be known: (1) water flow rate through the tubes, gpm; (2) water temperature in and out, °F; and (3) the steam condition in the heater shell. In selecting a steam-to-water heat exchanger, a few miscellaneous hints may be in order, such as the following:

1. Water flow velocity through the tubes should be limited to less than 7.5 ft/s because:

(a) Water velocities of 7.5 ft/s and above can become erosive. Rapid wear of the tubing is the result.

(b) Any small accumulation of scale in the unit that has been rated at high velocity causes a very sharp drop-off in heating capacity.

(c) The high pressure drop resulting from very high velocities can make pump selections difficult and costly.

2. Fouling factors as set forth by TEMA (Tubular Exchanger Manufacturers Association) should always be considered. Water from different localities varies in mineral content. In the process of being heated, the minerals are precipitated in the form of lime, scale, and so forth. They then collect on the tube walls, and the ability of the unit to transfer heat is reduced. To offset a loss in heater capacity from fouling, the size of the heater should be increased so that, after scale has collected, the unit will still operate at its rated capacity.

3. When a steam-to-water heat exchanger is used where human contact with any part of the system is possible, insulation of the heater shell and piping should always be used. Also, a low steam pressure (15 psig or less) should be used.

REVIEW PROBLEMS

8.1. A hot-water heating system in a building has an installed radiation capacity of 119,000 Btu/hr. Domestic hot water requirements are 3 gpm of 140°F water. Select a boiler for the system.
(a) What is the piping and pickup factor?
(b) What is the overall efficiency using No. 2 fuel oil (heating value = 140,000 Btu/gal)?

8.2. An oil-fired boiler uses 2.1 gph of No. 2 fuel oil (heating value = 140,000 Btu/gal) to give a gross output of 220,000 Btu/hr.
(a) What efficiency of operation is obtained?
(b) If the net I-B-R rating is 145,000 Btu/hr, what total piping and pickup factor was allowed?

8.3. A commercial building, which is to be heated continuously, has a calculated design heat loss of 1,950,000 Btu/hr. The piping will be typical of such buildings. Select, by number, the boiler best suited for this installation using data from Table 8-1 and assuming forced draft.
(a) What reserve capacity will the unit have in percent of design heating load?
(b) What overall efficiency is indicated by the tabulated data when the unit is fired with (1) gas and (2) No. 2 fuel oil?

8.4. Estimate the hourly heat loss per linear foot of 2 in. schedule 40 steel pipe covered with canvas-wrapped 85% magnesia insulation $1\frac{1}{2}$ in. thick when 200°F water

flows in the pipe and the surrounding air temperature is 70°F.

8.5. A 4-in. schedule 40 steel pipe carries saturated steam at 400°F and is insulated with $2\frac{1}{2}$ in. of 85% magnesia insulation, density 11 lb/ft³. What will be the estimated outside surface temperature of the insulation if the surrounding air is at 80°F?

8.6. A space within a building has a heat loss of 8200 Btu/hr at design conditions. The heating system is forced-circulation hot water with an average water temperature of 200°F. Select a convector to heat the space using Table 8-9. The convector height is limited to a maximum of 26 in.

8.7. For the same data of Review Problem 8.6, how many linear ft of Model 75WL-3 baseboard should be used? Assume (a) 1.0 gpm and (b) 4.0 gpm.

8.8. A space in a building has an exterior wall that is 32 ft long. The heat loss from the space has been calculated and was found to be 48,000 Btu/hr. It has been determined that Trane type *S* wall fin will be used to provide the heating. Forced-circulation hot water at an average temperature of 190°F is to be used in the system. Specify the size and type of heating element, enclosure height, and length to be used.

8.9. A cabinet heater is to be installed in the entrance of a building where the calculated heat loss is 30,000 Btu/hr at design conditions. Using Table 8-13, determine the unit size required, water flow rate, water pressure drop, and final air temperature, assuming 180°F entering water temperature and 60°F entering air temperature.

8.10. A machine shop in an industrial building has a calculated heat loss of 120,000 Btu/hr. The decision of the heating engineer is to use Modine Model HS unit heaters in the space. The forced-circulation hot-water system circulates 200°F water with a system water temperature drop of 20°F. Select the unit size and number of heaters to use. What is the capacity, gpm, final air temperature, and water pressure head drop for each unit?

8.11. A room in a residence has a calculated design heat loss of 9200 Btu/hr. The exposed wall of the room is 18 ft long and contains two windows, each having a sill height of 30 in. above the floor. The heating system is

forced-circulation hot water with a supply water temperature of 200°F and a system design water temperature drop of 20°F.
(a) What size convector(s) should be used?
(b) How many linear ft of 75WL-3 baseboard are needed?

8.12. The two exposed walls of a corner room of a house are 18 ft and 14 ft in length and are 8.5 ft high. The walls have several windows with a sill height of 36 in. The design heat loss for the room is 15,200 Btu/hr. Consider the use of cast-iron baseboard on the outside walls, or two convectors. A hydronic hot-water system is to be used with an average water temperature of 190°F. Select units from the tables in this chapter and show Btu/hr excess capacity, or show Btu/hr inadequacy for unsuitable equipment.

BIBLIOGRAPHY

8.1. *ASHRAE Handbook 1988 Equipment,* American Society of Heating, Refrigerating, and Air Conditioning Engineers, Atlanta, GA, 1988.

8.2. *SBI Testing and Rating Code for Steel Boilers,* Hydronics Institute, Berkeley Heights, NJ, 10th ed., November 1967.

8.3. *I-B-R Testing and Rating Code for Low Pressure Cast-Iron Heating Boilers,* Hydronics Institute, Berkeley Heights, NJ, November 1967.

8.4. *I-B-R Ratings for Cast-Iron Boilers,* Hydronics Institute, Berkeley Heights, NJ, May 1978.

8.5. *Steel Boiler Ratings,* Hydronics Institute, Berkeley Heights, NJ, May 1978.

8.6. *ASHRAE Standard Method of Testing for Rating Unit Ventilators* (ASHRAE Standard 71-73, 1973), American Society of Heating, Refrigerating, and Air Conditioning Engineers, Atlanta, GA.

8.7. *Standard Code for Testing and Rating Steam Unit Heaters,* Air Moving and Conditioning Association, AMCA Bulletin No. 20, 1950.

8.8. *Standard Code for Testing and Rating Hot Water Unit Heaters,* adopted jointly by American Society of Heating and Ventilating Engineers and Industrial Unit Heater Association, IUHA Bulletin No. 11, April 1953, 2nd ed.

8.9. Strock, C., and Koral, R. L., *Handbook of Air Conditioning, Heating, and Ventilating,* 3rd ed., Industrial Press, New York, 1979.

8.10. *Trane Air Conditioning Manual,* The Trane Company, LaCrosse, WI, 1965.

8.11. *Cooling and Heating Coils, Catalog D-Coil 1,* The Trane Company, LaCrosse, WI, 1973.

9

Fluid Flow Fundamentals and Piping Systems

9-1 INTRODUCTION

This chapter deals with basic fluid mechanics, and our discussion will be limited primarily to the flow of fluids. Design of heating and ventilating systems normally involves fluids that carry the heat to or from various heat-exchange devices. These fluids must be transported from one place to another in pipes and ducts. The proper design and layout of a piping system or duct system involve consideration of the behavior of fluids as they flow through the conduits. Much of the sizing of pipes and ducts is done by use of tables and charts, which are convenient and save much valuable time. However, before such tables and charts are used, it is desirable to study and understand the basic principles of fluid flow through closed conduits. This should lead us to a more complete understanding of the tables and charts, and we should be able to use them with a clearer understanding of their worth and limitations.

9-2 FLUID MECHANICS

Fluid mechanics is the study of the physical behavior of fluids and fluid systems and the laws describing this behavior. In general, the behavior of fluids follows laws similar to those that describe the behavior of solid bodies, which are familiar from studies in statics and dynamics. The major difference lies in the interpretation that must be placed on the laws due to the nature of the substance with which we are dealing.

A *fluid* is a substance that has a definite mass and volume but that has no definite shape. A fluid cannot sustain a shear force under equilibrium conditions. To be more precise, there is one additional restriction that must be made to the foregoing statement to cover all possible cases. That is, a fluid has a definite mass and volume at constant pressure and temperature. There are two basic classes of fluids: (1) *liquids* and (2) *gases.*

Liquids are fluids having definite volumes independent of the size of the container. Under conditions of constant pressure and temperature, liquids will assume the shape of the container and fill the part of it equal in volume to the volume of the liquid. Liquids are generally considered to be incompressible in air conditioning work; that is, their volumes do not change with a change in pressure. The specific volume of a liquid *does* change appreciably with temperature, however. Therefore, we must be careful to account for this in our calculations. For instance, in a closed hot-water heating system, if the system is initially filled with cold water and this water is then heated to the system operating temperature, the volume of the water may increase by 5% or 6%. Since the liquid is confined, excessive pressures would be encountered that could easily burst pipes or connections in the system. Liquids, when exposed to atmospheric pressure, are said to have a "free surface."

Gases are fluids that are compressible. Unlike liquids, which have a definite volume for a given mass, the volume of a given mass of gas will change to fill the container. The behavior of gases was described generally by the gas laws presented previously. Gases cannot have the "free surface" that liquids can display. Although gases are compressible, they may be treated as incompressible in most problems because pressure and temperature changes may be small during a gas flow process.

9-3   FLUID PROPERTIES

Most of the fluid properties have been discussed in Chapter 1. These properties were specific volume, density, specific weight, internal energy, enthalpy, entropy, temperature, pressure, and specific gravity. A remaining fluid property necessary in our study of fluid flow is viscosity.

Viscosity is that property of a fluid that determines its resistance to a shearing force. Viscosity is due to the interaction between fluid molecules (cohesion and molecular momentum). Viscosity expresses the readiness with which fluid flows when it is acted upon by an external force. The coefficient of absolute viscosity, or, simply, *absolute viscosity,* is a measure of its resistance to internal deformation or shear. Molasses is a highly viscous fluid; water is comparatively much less viscous; and the viscosity of gases is quite small compared with water.

Fluids used in air conditioning systems have viscosities that are predictable. Some fluids not commonly encountered have viscosities that depend upon previous working of the fluid. Printer's ink, wood pulp slurries, and catsup are examples of fluids possessing such thixotropic properties of viscosity.

Considerable confusion exists concerning the units used to express viscosity. Proper units *must* be used whenever substituting values of viscosity into formulas.

Absolute Viscosity, μ.   In the English Engineering System of units (pound force, foot, second), the units of absolute viscosity are

$$\frac{\text{lbf·s}}{\text{ft}^2} = \frac{\text{slugs}}{\text{ft·s}} \quad \text{or} \quad \frac{\text{lbm}}{\text{ft·s}}$$

It should be recalled that 1 slug is equal to 32.174 lbm; therefore, 1 slug/ft·s is equal to 32.174 lbm/ft·s.

The most commonly used unit of absolute viscosity is in the cgs (centimeter, gram, second) system. The unit is called the *poise*. The units would be

$$\mu \text{ (poise)} = \frac{\text{dyn·s}}{\text{cm}^2} = \frac{\text{gm}}{\text{cm·s}}$$

The poise (P) is a relatively large number; therefore, *centipoise* (cP) is frequently quoted in the literature. One centipoise is equal to 0.01 poise.

Conversion of the unit poise to SI units gives an equivalent term called the *pascal-second* (Pa·s). One poise is equal to 0.1 Pa·s. The conversion can be made with the following SI unit equivalents:

$$1 \text{ pascal (Pa)} = 1 \text{ newton/square meter (N/m}^2)$$

Therefore,

$$1 \text{ Pa·s} = 1 \text{ N·s/m}^2$$

A summary of units and conversions for absolute viscosity follows.

Symbol: μ (mu)
Units: EES: lbf·s/ft², slugs/ft·s, and lbm/ft·s; cgs: gm/cm·s (poise); SI: N·s/m² = Pa·s
Conversion: To convert lbf·s/ft² to poise, multiply by 478.8. To convert lbf·s/ft² to N·s/m², multiply by 47.88. To convert lbm/ft·s to N·s/m², multiply by 1.488. To convert poise to N·s/m², multiply by 0.1.

Kinematic Viscosity, ν.   Kinematic viscosity is the ratio of the absolute viscosity to the mass density of the fluid.

In the English Engineering System of units, we found that absolute viscosity μ may have units of lbf·s/ft², slugs/ft·s, or lbm/ft·s. Therefore, the kinematic viscosity units would be defined as

$$\nu = \frac{\mu}{\rho} = \frac{\mu}{\gamma/g} = \frac{\mu \times g}{\gamma}$$

$$= \frac{(\text{lbf·s/ft}^2)(\text{ft/s}^2)}{\text{lbf/ft}^3} = \text{ft}^2/\text{s}$$

where γ, the specific weight, is equal to $\rho \times g$. Also,

$$\nu = \frac{\mu}{\rho} = \frac{\text{lbm/ft·s}}{\text{slugs/ft}^3} = \text{ft}^2/\text{s}$$

And,

$$v = \frac{\mu}{\rho} = \frac{\text{slugs/ft·s}}{\text{lbm/ft}^3} = \text{ft}^2/\text{s}$$

In the cgs system of units, the unit of kinematic viscosity is called the *stoke* (S), which has dimensions of cm²/s. *Centistoke* (cS) is frequently quoted, and 1 cS is equal to 0.01 S. Occasionally, it is desirable to convert stoke to poise. It should be recognized that stoke (cm²/s) multiplied by the mass density (gm/cm³) will yield the dimensions of gm/cm·s, which is poise. Mass density in gm/cm³ is equal numerically to the fluid specific gravity (SG). Therefore,

$$\mu \text{ (poise)} = v \text{ (stokes)} \times SG \qquad (9\text{-}1)$$

In SI units, kinematic viscosity has units of m²/s.

A summary of units and conversions for kinematic viscosity follows.

Symbol: v (nu)

Units: EES: ft²/s; cgs: cm²/s (stoke); SI: m²/s

Conversion: To convert ft²/s to stoke, multiply by (30.48)². To convert ft²/s to m²/s, multiply by 0.0929. To convert stoke to m²/s, multiply by 0.0001.

Saybolt Universal Viscosity. The measurement of the absolute viscosity of fluids (especially gases and vapors) requires elaborate equipment and considerable experimental skill. On the other hand, a rather simple instrument can be used for measuring the kinematic viscosity of oils and other viscous liquids. The instrument adopted as a standard in this country is the Saybolt Universal Viscosimeter. In measuring kinematic viscosity with this instrument, the time required for a small volume of liquid (60 cm³) to pass through a calibrated orifice at a specified temperature is determined. (Viscosity is a measure of the "flowability" at a given temperature.) The elapsed time is measured in seconds with the fluid sample maintained at a constant temperature, usually at 100°F or 210°F. The elapsed time in seconds is the Saybolt seconds universal (SSU) viscosity at the particular temperature.

If the elapsed time, t, is between 32 and 100 s for a control temperature of 100°F, conversion to other viscosity units is as follows:

$$v \text{ (centistokes)} = 0.226t - \frac{195}{t} \qquad (9\text{-}2)$$

If the elapsed time, t, is greater than 100 s for a control temperature of 100°F,

$$v \text{ (centistokes)} = 0.220t - \frac{135}{t} \qquad (9\text{-}3)$$

In Eqs. (9-2) and (9-3), t is the Saybolt seconds universal (SSU) obtained by test.

ILLUSTRATIVE PROBLEM 9-1

The viscosity of a fluid measured in a Saybolt Viscosimeter is 140 s at 100°F, and the specific gravity measured at 60°F is 0.875. Determine the fluid viscosity in (1) centistokes, (2) centipoise, (3) ft²/s, (4) lbf·s/ft², and (5) Pa·s, at a temperature of 100°F.

Solution

1. Since elapsed time (SSU) is greater than 100 s, use Eq. (9-3):

$$
\begin{aligned}
v \text{ (centistokes)} &= 0.220t - \frac{135}{t} \\
&= (0.220)(140) - \frac{135}{140} \\
&= 30.8 - 0.964 \\
&= 29.84 \text{ cS} \qquad (ans.)
\end{aligned}
$$

2. Specific gravity of the fluid at 100°F may be found from Eq. (1-14):

$$
\begin{aligned}
(SG)_t &= SG_{60/60} - 0.00035(t - 60) \\
SG_{100} &= 0.875 - 0.00035(100 - 60) \\
&= 0.861
\end{aligned}
$$

From Eq. (9-1),

$$
\begin{aligned}
\mu \text{ (centipoise)} &= v \text{ (centistokes)} \times SG \\
&= (29.84)(0.861) \\
&= 25.69 \text{ cP} \qquad (ans.)
\end{aligned}
$$

3. (ft²/s)(30.48)² = stokes

or

$$
\begin{aligned}
v \text{ (ft}^2\text{/s)} &= \frac{29.84/100}{(30.48)^2} \\
&= 0.000321 \\
&= 3.21 \times 10^{-4} \text{ ft}^2/\text{s} \qquad (ans.)
\end{aligned}
$$

4. (lbf·s/ft²)(478.8) = poise

or

$$
\begin{aligned}
\mu \text{ (lbf·s/ft}^2\text{)} &= \frac{\text{poise}}{478.8} = \frac{25.69/100}{478.8} \\
&= 0.000537 \\
&= 5.37 \times 10^{-4} \text{ lbf·s/ft}^2 \qquad (ans.)
\end{aligned}
$$

5. $1 \text{ Pa·s} = 1 \text{ N·s/m}^2$

$$(\text{lbf·s/ft}^2)(47.88) = \text{N·s/m}^2$$
$$\text{N·s/m}^2 = (5.37 \times 10^{-4})(47.88)$$
$$= 2.57 \times 10^{-2} \text{ Pa·s} \qquad (ans.)$$

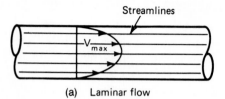

(a) Laminar flow

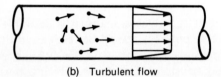

(b) Turbulent flow

FIGURE 9-1 (a) Laminar and (b) turbulent flow in a pipe.

9-4 FLUID FLOW

In general, there are two types of fluid flow: (1) free, natural, or gravity flow and (2) forced flow. Fluid flow is further subdivided into three classifications: (a) viscous, laminar, or streamlined flow, (b) turbulent flow, and (c) supersonic flow. Supersonic flow does not occur as yet in air conditioning systems.

Gravity flow is the type of flow caused by an unbalanced difference in density of portions of a fluid that are attempting to find their equilibrium level. This unbalance may result from an increase in density of an upper portion of a fluid due to cooling or a decrease in density of the lower portion of a fluid due to heating. Cooling or heating of the fluid occurs when the fluid is in contact with a cool or warm surface.

A common example of free flow occurs when air, in contact with a heat pipe and fins of a heating element, is heated with a corresponding change in air density. The heated air, less dense than the surrounding air, rises and is replaced by the more dense air in the surroundings. The resulting "draft" can be appreciable under good conditions. Gravity flow may be either laminar or turbulent.

Forced flow occurs when a fluid is forced to flow by some mechanical means, such as a fan or a pump, rather than by thermal means. Forced flow may be either laminar or turbulent. Turbulent flow is most common in air conditioning systems.

Laminar flow (viscous, streamlined) implies that all portions of the fluid move in paths parallel to one another and parallel to the walls of confining surfaces, if any. When heat is being transferred to or from the fluid flowing, natural convection currents within the fluid have a tendency to distort the streamline pattern. In such cases, the flow is classified as nonisothermal or modified laminar flow.

Figure 9-1a represents the "velocity profile" for laminar flow of a fluid through a pipe. It may be seen that the velocity varies from maximum (V_{max}) at the center of the pipe to zero velocity at the pipe walls. In fluid flow work, we are concerned with *average flow velocity* (V_{ave}), which is used in the continuity equation. For laminar flow in a pipe, $V_{\text{ave}} = 1/2 V_{\text{max}}$. Laminar flow may occur in some air conditioning equipment and also may be encountered when handling viscous fluids such as fuel oil flowing in a pipe.

Turbulent flow is the most common type of flow encountered in practice. In turbulent flow, there is random motion of the fluid particles in directions transverse to the direction of the main flow. The velocity distribution in turbulent flow is more uniform across a pipe diameter than in laminar flow. This may be observed in Figure 9-1b. Even though a turbulent motion of the fluid exists throughout the greater portion of the pipe diameter, there is always a thin layer of fluid at the pipe wall, known as the "boundary layer" or "laminar sublayer," which is moving in laminar flow.

The average velocity in turbulent flow cannot be found by a simple relationship as for laminar flow. It is normally determined from the continuity equation first presented in Chapter 2:

$$\dot{m} = \rho A V \qquad (2\text{-}8)$$

where

$\dot{m}$ = mass flow rate (lbm/s or kg/s)
A = cross-sectional flow area (ft² or m²)
V = average flow velocity (ft/s or m/s)

When the fluid is incompressible, Eq (2-8) becomes

$$\dot{Q} = A V \qquad (9\text{-}4)$$

where $\dot{Q}$ = volume flow rate (ft³/s or m³/s).

Frequently, the flow rate of a liquid is given in gallons per minute (gpm), the flow cross-sectional area in in.², or the diameter of flow path in inches. It is convenient to recognize that, using appropriate conversion units, Eq. (9-4) may be rearranged as

$$V = \frac{0.408 \times \text{gpm}}{d^2}$$

or

$$V = \frac{0.3208 \times \text{gpm}}{a} \qquad (9\text{-}5)$$

where

V = average flow velocity (ft/s)
gpm = gallons per minute
d = internal pipe diameter (in.)
a = internal pipe flow area (in.^{2})

9-5 REYNOLDS NUMBER

The work of Osborne Reynolds (1842–1912) has shown that the nature of fluid flow—that is, whether it is laminar or turbulent—depends upon the ratio of the inertia forces to the viscous forces and may be defined as the so-called Reynolds number:

$$N_R = \frac{\text{velocity} \times \text{diameter} \times \text{density}}{\text{absolute viscosity}}$$

which is a dimensionless combination of the four variables. In symbol form, Reynolds number becomes

$$N_R = \frac{VD\rho}{\mu} = \frac{VD}{\mu/\rho}$$
$$= \frac{VD}{\nu} \text{ (dimensionless ratio)} \qquad (9\text{-}6)$$

where

V = average flow velocity (ft/s or m/s)
D = characteristic dimension of system (length of surface, diameter of pipe, hydraulic radius) (ft or m)
ρ = fluid mass density (lbm/ft^3 or kg/m^3)
μ = absolute fluid viscosity [lbm/ft·s or Pa·s (N·s/m^2) or, if ρ is in slugs/ft^3, then μ must be in slugs/ft·s]
ν = kinematic viscosity (ft^2/s or m^2/s)

Caution must be used when substituting into Eq. (9-6) to use units to produce a dimensionless ratio.

For engineering work, flow in conduits is usually considered to be laminar if N_R is less than 2000 and turbulent if N_R is greater than 4000. Between the two values lies the "critical zone" where the flow may be laminar, turbulent, or a combination of both.

Reynolds number may also be calculated from the following relationships using different, and sometimes more convenient, units in the ratio:

$$N_R = \frac{123.9dV\rho}{\mu} = \frac{7740dV}{\nu} = \frac{50.6(\text{gpm})(\rho)}{d\mu} \qquad (9\text{-}7)$$

where

d = inside diameter of pipe (in.)
V = flow velocity (ft/s)
ρ = density (lbm/ft^3)
μ = absolute viscosity (centipoise)
ν = kinematic viscosity (centistokes)
gpm = gallons per minute flow rate

Occasionally, a conduit of noncircular cross section is encountered or a conduit partially filled with flowing fluid. To calculate Reynolds number for these conditions, an equivalent diameter (four times the hydraulic radius) is substituted for the dimension D. The hydraulic radius is defined as

$$R_H = \frac{\text{cross-sectional flow area in ft}^2 \text{ or m}^2}{\text{wetted perimeter in ft or m}}$$

This applies to any ordinary conduit (circular conduit not flowing full, oval, square, or rectangular) but not to extremely narrow shapes, such as annular or elongated openings, where the width is small relative to the length. In such cases, the hydraulic radius is approximately equal to one-half the width of the passage.

9-6 THE STEADY-FLOW ENERGY EQUATION

In Chapter 2, the steady-flow energy equation was discussed and is repeated here with the heat and work terms omitted:

$$\left(\frac{g}{g_c}\right)Z_1 + P_1v_1 + \frac{V_1^2}{2g_c} + JU_1$$
$$= \left(\frac{g}{g_c}\right)Z_2 + P_2v_2 + \frac{V_2^2}{2g_c} + JU_2$$

The change in internal energy $J(U_2 - U_1)$ is usually very small in fluid flow work, and letting $v = 1/\rho = $ a constant (incompressible flow), we would have

$$\left(\frac{g}{g_c}\right)Z_1 + \frac{P_1}{\rho} + \frac{V_1^2}{2g_c}$$
$$= \left(\frac{g}{g_c}\right)Z_2 + \frac{P_2}{\rho} + \frac{V_2^2}{2g_c} \qquad (9\text{-}8)$$

Each term in Eq. (9-8) has units of ft·lbf/lbm or J/kg, which is *energy per unit mass*.

If each term of Eq. (9-8) is multiplied by the ratio g_c/g (lbf/lbm), we would have

$$Z_1 + \frac{g_c P_1}{\rho g} + \frac{V_1^2}{2g} = Z_2 + \frac{g_c P_2}{\rho g} + \frac{V_2^2}{2g} \quad (9\text{-}9)$$

Since $\rho(g/g_c) = \gamma$, the specific weight, Eq. (9-9) frequently appears as

$$Z_1 + \frac{P_1}{\gamma_1} + \frac{V_1^2}{2g} = Z_2 + \frac{P_2}{\gamma_2} + \frac{V_2^2}{2g} \quad (9\text{-}9a)$$

Each term in Eqs. (9-9) and (9-9a) has units of ft·lbf/ lbf, or head (ft), and is *energy per unit weight*. This form of the steady-flow equation is Bernoulli's equation, frequently used in the solution of many problems involving incompressible fluid flow. The three terms on each side of the equation are known, respectively, as static elevation head, static pressure head, and velocity head.

With the flow of real fluids, losses due to fluid friction must be accounted for. Therefore, Eq. (9-9a) must include a term to represent these losses, as follows:

$$Z_1 + \frac{P_1}{\gamma_1} + \frac{V_1^2}{2g} = Z_2 + \frac{P_2}{\gamma_2} + \frac{V_2^2}{2g} + h_L \quad (9\text{-}10)$$

where h_L is the *lost head*.

Figure 9-2 represents a section of pipe through which a fluid is flowing under steady-flow conditions. Since Eqs. (9-9) and (9-10) represent energy levels at points 1 and 2, in units of feet (or meters) of fluid, it is convenient to represent graphically the change in energy levels as fluid flows through the pipe. A vertical scale may be established to represent these energy levels above a conveniently selected datum plane, and values

of Z, P/γ, $V^2/2g$, and h_L may be plotted above the datum plane.

The sum $(Z + P/\gamma)$ plots the *hydraulic grade line,* and the sum $(Z + P/\gamma + V^2/2g)$ plots the *energy grade line.* The energy grade line represents graphically the available energy in the flow stream at various points between section 1 and 2 and always slopes downward in the direction of fluid flow. The loss term, h_L, represents the loss of energy due to friction.

The various terms in Eq. (9-10) are frequently grouped and referred to by particular names:

$$\text{elevation head change} = Z_2 - Z_1$$
$$\text{pressure head change} = \frac{P_2 - P_1}{\gamma}$$
$$(\gamma_1 = \gamma_2 \text{ for incompressible flow})$$
$$\text{velocity head change} = \frac{V_2^2 - V_1^2}{2g}$$
$$\text{head loss} = h_L$$

ILLUSTRATIVE PROBLEM 9-2

If water is flowing through the section of pipe of Figure 9-3 at the rate of 100 gpm and the static pressure at point 1 is 100 psi, what is the static pressure at point 2? The inside diameter at point 1 is 2.5 in. and at point 2 is 1.5 in. Assume lost head h_L is 3 ft of water.
Solution: By Eq. (9-5),

$$V_1 = \frac{0.408(\text{gpm})}{d_1^2} = \frac{0.408(100)}{(2.5)^2} = 6.528 \text{ ft/s}$$

$$V_2 = \frac{0.408(\text{gpm})}{d_2^2} = \frac{0.408(100)}{(1.5)^2} = 18.133 \text{ ft/s}$$

By Eq. (9-10),

$$Z_1 + \frac{P_1}{\gamma} + \frac{V_1^2}{2g} = Z_2 + \frac{P_2}{\gamma} + \frac{V_2^2}{2g} + h_L$$

Since the datum plane is selected at the pipe centerline and the pipe is horizontal, $Z_1 = Z_2$ and they are canceled from the equation. By substitution,

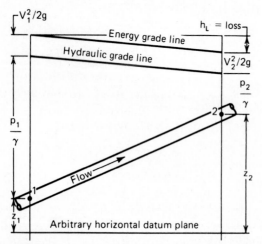

FIGURE 9-2 Steady flow of fluid through pipe section.

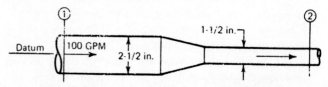

FIGURE 9-3 Sketch for Illustrative Problem 9-2.

$$\frac{144(100)}{62.4} + \frac{(6.528)^2}{2(32.2)} = \frac{144P_2}{62.4} + \frac{(18.133)^2}{2(32.2)} + 3$$

$$230.77 + 0.66 = 2.31P_2 + 5.11 + 3$$

$$P_2 = 96.7 \text{ psi} \qquad \text{(ans.)}$$

Note in the preceding calculations the magnitude of the $V^2/2g$ terms in comparison to the P/γ terms. In some cases, velocity head terms may be omitted in calculations because of their low value. However, it is recommended that they always be included in calculations to determine whether they are of importance.

9-7 FRICTION LOSSES IN A CLOSED CONDUIT

Flow of real fluids in a conduit is always accompanied by friction, which represents a loss of available energy; or we may say that there is a loss in pressure in the direction of fluid flow. This loss is expressed in Eq. (9-10) as h_L. The general equation for calculating this loss is known as the *Darcy equation*. It is an empirical equation, meaning that it is based on the results of tests rather than theory. The equation may be expressed in feet (or meters) of fluid flowing or in lbf/ft² (or N/m²) by use of Eqs. (9-11) and (9-12), respectively. For incompressible flow in closed conduits,

$$h_L = f\left(\frac{L}{D}\right)\left(\frac{V^2}{2g}\right) \qquad \text{(ft or m)} \qquad \text{(9-11)}$$

$$\Delta p_L = f\left(\frac{L}{D}\right)\left(\frac{\gamma V^2}{2g}\right) \text{ (lbf/ft}^2 \text{ or N/m}^2\text{)} \qquad \text{(9-12)}$$

where

h_L = head loss due to fluid friction (ft or m)
Δp_L = pressure loss due to fluid friction (lbf/ft² or N/m²)
f = friction factor (dimensionless)
L = length of conduit (ft or m)
D = conduit inside diameter (ft or m)
V = mean flow velocity (ft/s or m/s)
γ = specific weight of the fluid (lbf/ft³ or N/m³)
g = acceleration of gravity (ft/s² or m/s²)

The Darcy equation is valid for laminar or turbulent flow of any liquid in a closed conduit of cross-sectional diameter D. However, at high velocities with warm liquids flowing in a pipe, the downstream pressure may drop to the vapor pressure of the liquid. This may cause a phenomenon called *cavitation* to occur and calculated flow rates will be inaccurate. Cavitation may be described as the rapid formation and collapse of vapor

bubbles in the flowing stream. It will be discussed in more detail in the chapter on pumps.

The Darcy equation may be used, with suitable restrictions, for the flow of gases and vapors (compressible fluids) in closed conduits. Equations (9-11) and (9-12) give the head loss and pressure loss, respectively, due to fluid friction in a closed conduit of constant diameter carrying fluids of reasonably constant specific weight in straight pipe, whether horizontal, vertical, or sloping. For inclined conduits, vertical conduits, or conduits of varying diameter, the change in pressure due to changes in elevation, velocity, and specific weight of the fluid must be accounted for with Eq. (9-10).

The Darcy equation may be developed from the methods of dimensional analysis, with the exception of the *friction factor, f*. The friction factor must be determined experimentally. Numerous investigators over past years have determined that, for laminar flow (N_R less than 2000), the friction factor is a function of Reynolds number only; but, for turbulent flow (N_R greater than 4000), the friction factor depends upon Reynolds number and also the surface roughness of the conduit. In the region known as the "critical zone" (N_R between 2000 and 4000), the friction factor is indeterminate. It depends on various factors such as changes in pipe section or direction of flow and obstructions such as valves and fittings in the upstream piping.

For laminar flow, the friction factor f is

$$f = \frac{64}{N_R} \qquad \text{(9-13)}$$

If appropriate units are used for N_R and the value of f resulting from Eq. (9-13) is substituted into Eq. (9-12), we would have

$$\Delta p_L = 0.000668 \frac{\mu L V}{d^2} \text{ (laminar flow)} \qquad \text{(9-14)}$$

where

Δp_L = pressure loss (lbf/in.²)
μ = absolute viscosity (centipoise)
L = conduit length (ft)
V = average flow velocity (ft/s)
d = internal pipe diameter (in.)

When flow is turbulent (N_R greater than 4000), flow conditions become quite stable and the friction factor depends upon Reynolds number and the *relative roughness* (ϵ/D) of the pipe surface. The term ϵ/D is the absolute surface roughness divided by the internal pipe diameter, both expressed in feet. For very smooth pipes, such as drawn tubing, the friction factor decreases more rapidly with increasing Reynolds number than for pipe with a comparatively rough surface. Since the character of the internal surface of commercial pipe is practically independent of the diameter, the roughness of the pipe walls has a greater effect on the friction factor in the smaller pipe sizes. Consequently, pipe of small diameter will approach the very rough condition, and, in general, will have higher friction factors than large pipe of the same material.

The most useful and widely accepted data of friction factors for use with the Darcy equation has been presented by L. F. Moody and are reproduced here as Figures 9-4 and 9-5. Figure 9-4 shows the relative roughness of pipe of various materials and inside diameters. It also indicates the friction factor f for complete turbulence.

Figure 9-5 shows a plot of the friction factor against Reynolds number for laminar flow and for turbulent flow for pipe of various relative roughness condi-tions. The friction factor is normally determined by the following procedure:

1. Calculate Reynolds number (N_R).
2. Calculate the relative roughness (ϵ/D).
3. Using ϵ/D, enter Figure 9-5 at the right, and approximate a curve of relative roughness parallel to an existing curve.
4. Follow the curve of ϵ/D to the left to inter-sect with the vertical line for the calculated N_R.
5. From the point of intersection of ϵ/D and N_R, move horizontally to the left and read the friction factor f.

Use of Figures 9-4 and 9-5 will be illustrated by the solution of three common *classes* of problems involving flow in pipes.

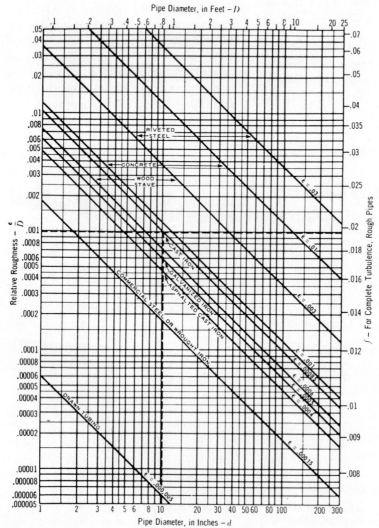

FIGURE 9-4 Relative roughness of pipe materials and friction fac-tors for complete turbulence.

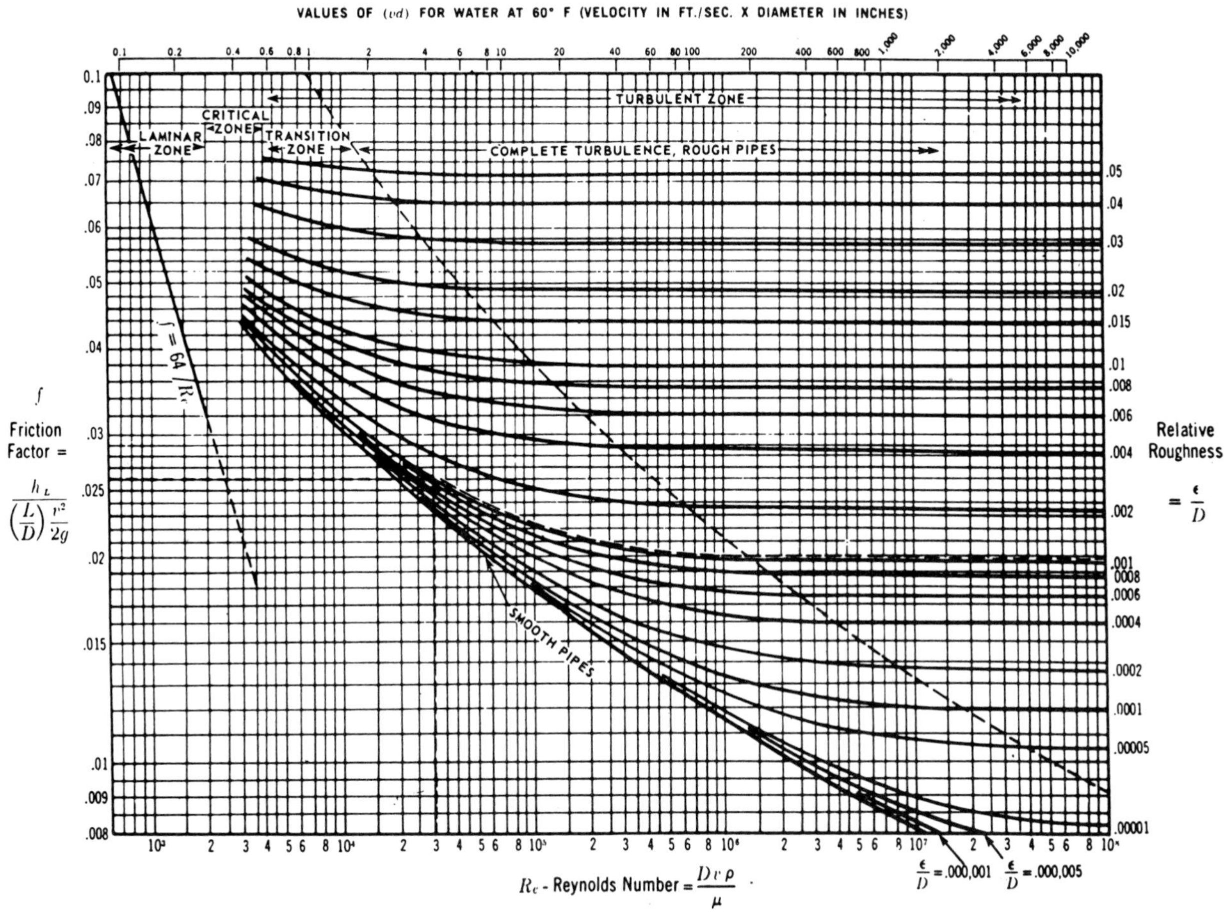

VALUES OF (vd) FOR WATER AT 60° F (VELOCITY IN FT./SEC. X DIAMETER IN INCHES)

FIGURE 9-5 Friction factors for any type of commercial pipe.

ILLUSTRATIVE PROBLEM 9-3 (CLASS I)

In this type of problem, the head loss or pressure loss is to be determined when the quantity of flow, length of pipe, size of pipe, pipe material, and fluid viscosity are known.

Problem statement: A section of 2-in. schedule 40 commercial steel pipe is 50 ft long and carries 20 gpm of fluid having a specific gravity of 0.85 and a viscosity of 150 SSU at 100°F. Determine the head loss and pressure loss in the section of pipe.

Solution:

Step 1. Determine Reynolds number for the flow. By Eq. (9-5),

$$V = \frac{0.408(\text{gpm})}{d^2} = \frac{0.408(20)}{(2.067)^2} = 1.910 \text{ ft/s}$$

By Eq. (9-3),

$$v \text{ (centistokes)} = 0.220t - \frac{135}{t}$$
$$= 0.220(150) - \frac{135}{150}$$
$$= 32.1 \text{ cS}$$

By Eq. (9-7),

$$N_R = \frac{7740 dV}{v} = \frac{7740(2.067)(1.910)}{32.1}$$

$$= 952 \text{ (less than 2000; therefore, flow is laminar)}$$

Step 2. Determine pressure loss. By Eq. (9-13), for laminar flow,

$$f = \frac{64}{N_R} = \frac{64}{952} = 0.0672$$

By Eq. (9-12),

$$\Delta p_L = f\left(\frac{L}{D}\right)\left(\frac{\gamma V^2}{2g}\right)$$

$$= 0.0672\left(\frac{50}{2.067/12}\right)\left(\frac{(62.4)(0.85)(1.91)^2}{2(32.2)}\right)$$

$$= 58.61 \text{ lbf/ft}^2 = 0.407 \text{ lbf/in.}^2 \qquad (ans.)$$

Also, by Eq. (9-14),

$$\Delta p_L = 0.000668 \frac{\mu L V}{d^2}$$

$$= 0.000668 \frac{(32.1)(0.85)(50)(1.91)}{(2.067)^2}$$

$$= 0.407 \text{ lbf/in.}^2$$

Step 3. Determine head loss. By Eq. (9-11),

$$h_L = f\left(\frac{L}{D}\right)\left(\frac{V^2}{2g}\right)$$

or note that

$$h_L = \frac{\Delta p_L}{\gamma}$$

Therefore,

$$h_L = \frac{(0.407)(144)}{(62.4)(0.85)} = 1.1 \text{ ft of fluid} \qquad (ans.)$$

ILLUSTRATIVE PROBLEM 9-4 (CLASS I)

If the fluid flowing in the pipe of Illustrative Problem 9-3 was water at 60° F with a viscosity of 1.15 cP, determine the pressure loss and head loss.

Solution:
Step 1. Determine Reynolds number.

$$V = 1.910 \text{ ft/s (from before)}$$

By Eq. (9-7),

$$N_R = \frac{123.9dV\rho}{\mu} = \frac{123.9(2.067)(1.910)(62.4)}{1.15}$$
$$= 26,540 \text{ (greater than 4000; therefore,}$$
$$\text{flow is turbulent)}$$

Step 2. Determine friction factor. From Figure 9-4, find ϵ/D for 2-in. commercial steel pipe to be 0.0009. On Figure 9-5, find point of intersection of $\epsilon/D = 0.0009$ and $N_R = 2.654 \times 10^4$ and read $f = 0.026$.

Step 3. Determine head loss and pressure loss. By Eq. (9-11),

$$h_L = f\left(\frac{L}{D}\right)\left(\frac{V^2}{2g}\right)$$

$$= 0.026\left(\frac{50}{2.067/12}\right)\left[\frac{(1.910)^2}{2(32.2)}\right]$$

$$= 0.428 \text{ ft of water} \qquad (ans.)$$

$$\Delta p_L = \gamma \times h_L = (62.4)(0.428)$$
$$= 26.71 \text{ lbf/ft}^2 = 0.185 \text{ lbf/in.}^2 \qquad (ans.)$$

or use Eq. (9-12) directly.

ILLUSTRATIVE PROBLEM 9-5 (CLASS II)

In this type of problem, the flow quantity is to be determined when the head loss or pressure loss, pipe size and material, and fluid properties are known. This type of problem requires a trial-and-error type of solution because the flow velocity and friction factor are unknown.

Problem statement: Water of 60°F flows through a 12-in. inside diameter riveted steel pipe ($\epsilon = 0.01$) with a measured head loss of 40 ft of water in a section of pipe 1000 ft long. Estimate the flow rate through the pipe in ft³/s if the viscosity is assumed to be 2.35×10^{-5} lbf·s/ft².

Solution:
Procedure:
Step 1. Solve for relative roughness ϵ/D and assume a trial value of f from Figure 9-4, assuming complete turbulence.
Step 2. Substitute trial value of f in Eq. (9-11) and solve for velocity V.
Step 3. Determine Reynolds number.
Step 4. With this Reynolds number and ϵ/D, find f from Figure 9-5.
Step 5. If f does not agree with the value assumed in step 1, use new value of f, and repeat steps 1 through 4. Repeat until f does not change.
Step 6. Calculate Q using Eq. (9-4).

1. $\epsilon/D = 0.01/1 = 0.01$. From Figure 9-4, assume f for complete turbulence is 0.038.
2. By Eq. (9-11),

$$h_L = f\left(\frac{L}{D}\right)\left(\frac{V^2}{2g}\right)$$

$$40 = 0.038\left(\frac{1000}{1}\right)\left[\frac{V^2}{2(32.2)}\right]$$

$$V = 8.23 \text{ ft/s}$$

3. By Eq. (9-6),

$$N_R = \frac{VD\rho}{\mu} = \frac{8.23(1)(62.4/32.2)}{2.35 \times 10^{-5}}$$

$$= 6.79 \times 10^5 \text{ (turbulent)}$$

4. With $N_R = 6.79 \times 10^5$ and $\epsilon/D = 0.01$, enter Figure 9-5 for an improved value of f. As may be seen, f is in the zone of complete turbulence; therefore, $f = 0.038$ and the flow velocity is 8.23 ft/s found in step 2.
5. Not required.
6. By Eq. (9-4),

$$\dot{Q} = AV = \frac{\pi D^2 V}{4}$$

$$= \frac{\pi (1)^2 (8.23)}{4} = 6.46 \text{ ft}^3/\text{s} \qquad (ans.)$$

ILLUSTRATIVE PROBLEM 9-6 (CLASS III)

In this type of problem, the pipe inside diameter is to be determined if the head loss or pressure loss, pipe material, flow quantity, and fluid properties are known. This type of problem requires a trial-and-error solution because D, V, and f are all unknown. The solution of such a problem is sometimes made easier by recognizing the following relationships: (1) By Eq. (9-11),

$$h_L = f \left(\frac{L}{D} \right) \left(\frac{V^2}{2g} \right)$$

and (2) by Eq. (9-4),

$$V^2 = \frac{\dot{Q}^2}{A^2} = \frac{\dot{Q}^2}{\left(\frac{\pi D^2}{4} \right)^2}$$

Substituting into Eq. (9-11), we have

$$h_L = f \left(\frac{L}{D} \right) \left[\frac{\dot{Q}^2}{\left(\frac{\pi D^2}{4} \right)^2 (2g)} \right]$$

Rearranging and simplifying, we obtain

$$D^5 = \left[\frac{8L\dot{Q}^2}{h_L g \pi^2} \right] f$$

All terms within the brackets are constant for a given problem; therefore, we may say that

$$D^5 = C_1 \times f \qquad (a)$$

By Eq. (9-6),

$$N_R = \frac{VD}{v}$$

From Eq. (9-4),

$$\dot{Q} = \frac{\pi D^2 V}{4}$$

or

$$V = \frac{4\dot{Q}}{\pi D^2}$$

Substituting into Eq. (9-6), we have

$$N_R = \frac{4\dot{Q} \times D}{\pi D^2 \times v}$$

or

$$N_R = \left[\frac{4\dot{Q}}{\pi v} \right] \left(\frac{1}{D} \right)$$

Again, all terms within the brackets are constant; therefore, we may say that

$$N_R = \frac{C_2}{D} \qquad (b)$$

Problem statement: What nominal size commercial steel schedule 40 pipe should be used to carry 40 gpm of water with a pressure loss of not more than 5 lbf/in.^{2} per 100 ft of length? Assume that the kinematic viscosity is 7.61×10^{-6} ft^2/s.

Solution:
Procedure:

Step 1. Assume a value of f and solve Eq. (a) for D.

Step 2. Using this value of D, solve Eq. (b) for N_R, and determine ϵ/D for the pipe.

Step 3. With N_R and ϵ/D, enter Figure 9-5 and determine f. If this f is the same as assumed value in step 1, the problem is solved.

Step 4. If f found in step 3 is different from assumed value in step 1, use the new value of f and repeat steps 1, 2, and 3 until f does not change.

1. Assume $f = 0.02$. By Eq. (a),

$$D^5 = \left[\frac{8L\dot{Q}^2}{h_L g \pi^2} \right] f = C_1 \times f$$

$$\dot{Q} = \frac{40 \text{ gpm}}{(60 \text{ s/min})(7.48 \text{ gal/ft}^3)} = 0.0892 \text{ ft}^3/\text{s}$$

$$h_L = \frac{(5 \text{ lbf/in.}^2)(144 \text{ in.}^2/\text{ft}^2)}{62.4 \text{ lbf/ft}^3} = 11.54 \text{ ft of water}$$

With substitution,

$$D^5 = \left[\frac{(8)(100)(0.0892)^2}{(11.54)(32.2)(\pi^2)} \right] f = 0.00173f$$

For assumed value of $f = 0.02$,

$$D^5 = 0.00173(0.02)$$
$$D = 0.128 \text{ ft}$$

2. By Eq. (b),

$$N_R = \frac{C_2}{D} = \left(\frac{4\dot{Q}}{\pi v}\right)\left(\frac{1}{D}\right)$$

$$= \left[\frac{(4)(0.0892)}{\pi(7.61 \times 10^{-6})}\right]\left(\frac{1}{D}\right)$$

$$= \frac{1.49 \times 10^4}{D}$$

By substituting $D = 0.128$ ft,

$$N_R = \frac{1.49 \times 10^4}{0.128} = 1.164 \times 10^5 \text{ (turbulent)}$$

From Figure 9-4, find $\epsilon = 0.00015$ for commercial steel pipe. Therefore, $\epsilon/D = 0.00015/0.128 = 0.0012$.

3. With $\epsilon/D = 0.0012$ and $N_R = 1.164 \times 10^5$, enter Figure 9-5 and find $f = 0.023$.

4. This value of f is different from the assumed value of 0.02. Use $f = 0.023$ and repeat steps 1, 2, and 3, as follows:

1.
$$D^5 = 0.00173f = 0.00173(0.023)$$
$$D = 0.132 \text{ ft}$$

2.
$$N_R = \frac{1.49 \times 10^4}{D} = \frac{1.49 \times 10^4}{0.132}$$
$$= 1.129 \times 10^5 \text{ (turbulent)}$$
$$\epsilon/D = 0.00015/0.132 = 0.0011$$

3. With $\epsilon/D = 0.0011$ and $N_R = 1.129 \times 10^5$, enter Figure 9-5 and find f approximates the value of 0.023. Therefore, since f does not change, the pipe inside diameter should be 0.132 ft or 1.584 in. From pipe size table, the nominal size pipe having an inside diameter closest to this value is 1 1/2 in. schedule 40, which has an actual inside diameter of 1.610 in.

Most of the pipe flow problems encountered in heating and cooling systems are of the class III type. The trial-and-error solution is very cumbersome as may be noted by studying Illustrative Problem 9-6. Pipe-sizing methods commonly used in system design usually involve charts and tables that greatly simplify the task. These methods will be demonstrated in following chapters.

9-8 LOSSES IN FITTINGS AND VALVES

The preceding sections have considered fluid flow through closed conduits (pipes and ducts). The friction loss calculated by Eqs. (9-8) and (9-9) assumes the loss occurred in a straight section of conduit L units long. Any conduit system also contains devices, such as fittings and valves, that cause losses (sometimes called "minor losses" in fluid mechanics textbooks). Piping designed to carry fluids in heating systems normally contains a relatively large number of fittings, valves, and other flow-restricting devices. The losses due to these fittings may be quite high in comparison to the loss in straight pipe. Therefore, it is necessary to determine the types of fittings in the system and evaluate the losses associated with them in order to accurately determine the total system loss.

Extensive work has been done by Crane Company to develop comprehensive methods to evaluate flow resistances of valves and fittings. The results of that company's work have been published in Technical Paper No. 410, *Flow of Fluids.*

The great variety of valve designs available renders it impossible to make a thorough classification. If valves were classified according to the resistance they offer to flow, those exhibiting a straight-through flow path (such as gate ball, plug, and butterfly valves) would fall in the low-resistance class and those having a change in flow path direction (such as globe valves and angle valves) would fall in the high-resistance class.

Fittings may be classified as through-flow, branching, reducing, expanding, or deflecting. Such fittings as tees, crosses, side-outlet elbows, and so forth, may be regarded as branching fittings.

Reducing or expanding fittings are those that change the area of the flow passageway. In this class are reducers and bushings. Deflecting fittings such as bends, elbows, return bends, and so forth, are those that change the direction of flow.

Some fittings, of course, may be combinations of any of the foregoing general classifications. In addition, there are types of fittings, such as couplings and unions, that offer no appreciable resistance to flow and, therefore, need not be considered.

When a fluid is flowing steadily in a long, straight pipe of uniform diameter, the flow pattern, as indicated by the velocity distribution across the pipe diameter (see Figure 9-1), will assume a certain characteristic form. Any obstruction in the pipe that changes the direction or velocity of the whole stream, or even part of it, will alter the characteristic flow pattern and create turbulence accompanying flow in a straight pipe. Because valves and fittings in a pipe line disturb the flow pattern, they produce an additional pressure drop. This loss of pressure consists of the following:

- Pressure drop within the fitting itself.
- Pressure drop in the upstream piping in excess

of that which would normally occur if there were no fittings in the line (this effect is small).

- Pressure drop in the downstream piping in excess of that which would normally occur if there were no fittings in the line (this effect may be comparatively large).

From an experimental point of view, it is difficult to measure the three items separately. Their combined effect is the desired quantity, however, and this can be accurately measured by well-known methods.

Figure 9-6 shows two sections of a pipe line of the same diameter and length. The upper section contains a globe valve. If the pressure drops, Δp_1 and Δp_2, were measured between the points indicated, it would be found that Δp_1 is greater than Δp_2. Actually, the loss chargeable to the globe valve of length c is Δp_1 minus the loss in a section of pipe of length $a + b$. The losses, expressed in terms of resistance coefficient K of various valves and fittings as given in the K factor table (Table 9-1), include the loss due to the length of the valve or fitting.

Many experiments have shown that the pressure loss due to valves and fittings is mainly dynamic and is proportional to some power of the flow velocity. When pressure drop or head loss is plotted against velocity on logarithmic coordinates, the resulting curve is therefore a straight line. In the turbulent flow range, the value of the power of V has been found to vary from about 1.8 to 2.1 for different designs of valves and fittings. However, for all practical purposes, it can be assumed that the pressure drop or head loss due to the flow of fluids in the turbulent range through valves and fittings varies as the square of the velocity.

A special situation must be considered for check valves because of their construction. The relationship of pressure drop to velocity of flow is valid only if there is sufficient flow to hold the disc in a wide-open position. Most of the difficulties encountered with check valves, both lift and swing types, have been found to be due to oversizing, which results in noisy operation and premature wear of the moving parts. The minimum velocity required to lift the disc to the full open and stable posi-

tion has been determined by tests of numerous types of check and foot valves and is given in Table 9-1. It is expressed in terms of a constant times the square root of the specific volume of the fluid being handled, making it applicable for use with any fluid.

Sizing of check valves in accordance with the specified minimum velocity for full disc lift will often result in valves smaller in size than the pipe in which they are installed; however, the actual pressure drop will be small, if any, and higher than that of a full-size valve that is used in other than the wide-open positions. The advantages are longer valve life and quieter operation. The losses due to sudden or gradual contraction and enlargement, which will occur in such installations with bushings, reducing flanges, or tapered reducers, can be readily calculated from the data given in Table 9-1.

As previously stated, pressure drop or head loss through fittings is due primarily to dynamic effects; that is, it is due to changes in flow direction and velocity (change in flow area). To a lesser extent, it is frictional, which depends upon surface roughness and Reynolds number.

Pressure drop test data for a wide variety of valves and fittings are available from numerous investigators. However, due to the time-consuming and costly nature of the tests, it is virtually impossible to obtain test data for every size and type of valve and fitting. It is therefore desirable to provide a means of reliably extrapolating available test data to envelop those items that have not been or cannot readily be tested. Manufacturers' tests are usually made under fully turbulent flow conditions and sometimes include entrance and exit losses. Commonly used methods of reporting resistances of valves and fittings are as follows:

- Equivalent length, L, of straight pipe of the same nominal pipe size having the same loss as the fitting.
- Equivalent length in pipe diameters, L/D, where D equals the pipe inside diameter in feet.
- Resistance coefficient, K.
- Flow coefficient, C_v.

The relationship between L and L/D may be readily seen by stating that

$$L = \left(\frac{L}{D}\right)\left(\frac{d}{12}\right) \qquad (9\text{-}15)$$

where d is the internal pipe diameter in inches. The relationship of the resistance coefficient K, equivalent length L/D, and flow coefficient C_v requires special attention.

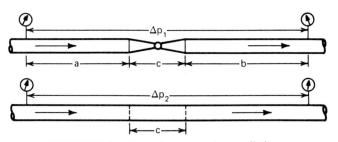

FIGURE 9-6 Flow through valve or fitting.

TABLE 9-1

K factor table of representative resistance coefficients (*K*) for valves and fittings

**PIPE FRICTION DATA FOR CLEAN COMMERCIAL STEEL PIPE
WITH FLOW IN ZONE OF COMPLETE TURBULENCE**

Nominal Size	½"	¾"	1"	1¼"	1½"	2"	2½, 3"	4"	5"	6"	8-10"	12-16"	18-24"
Friction Factor (f_T)	.027	.025	.023	.022	.021	.019	.018	.017	.016	.015	.014	.013	.012

**FORMULAS FOR CALCULATING "K" FACTORS
FOR VALVES AND FITTINGS WITH REDUCED PORT**

- **Formula 1**

$$K_2 = \frac{0.8 \sin\frac{\theta}{2}(1 - \beta^2)}{\beta^4}$$

- **Formula 2**

$$K_2 = \frac{0.5(1 - \beta^2)\sqrt{\sin\frac{\theta}{2}}}{\beta^4}$$

- **Formula 3**

$$K_2 = \frac{2.6 \sin\frac{\theta}{2}(1 - \beta^2)^2}{\beta^4}$$

- **Formula 4**

$$K_2 = \frac{(1 - \beta^2)^2}{\beta^4}$$

- **Formula 5**

$$K_2 = \frac{K_1}{\beta^4} + \text{Formula 1} + \text{Formula 3}$$

$$K_2 = \frac{K_1 + \sin\frac{\theta}{2}[0.8(1 - \beta^2) + 2.6(1 - \beta^2)^2]}{\beta^4}$$

- **Formula 6**

$$K_2 = \frac{K_1}{\beta^4} + \text{Formula 2} + \text{Formula 4}$$

$$K_2 = \frac{K_1 + 0.5\sqrt{\sin\frac{\theta}{2}}(1 - \beta^2) + (1 - \beta^2)^2}{\beta^4}$$

- **Formula 7**

$$K_2 = \frac{K_1}{\beta^4} + \beta(\text{Formula 2} + \text{Formula 4}) \text{ when } \theta = 180°$$

$$K_2 = \frac{K_1 + \beta\left[0.5(1 - \beta^2) + (1 - \beta^2)^2\right]}{\beta^4}$$

$$\beta = \frac{d_1}{d_2}$$

$$\beta^2 = \left(\frac{d_1}{d_2}\right)^2 = \frac{a_1}{a_2}$$

Subscript 1 defines dimensions and coefficients with reference to the smaller diameter.
Subscript 2 refers to the larger diameter.

SUDDEN AND GRADUAL CONTRACTION

If: $\theta \lesssim 45°$ . . . . K_2 = Formula 1
$\theta > 45° \lesssim 180°$ . . . K_2 = Formula 2

SUDDEN AND GRADUAL ENLARGEMENT

If: $\theta \lesssim 45°$ . . . . K_2 = Formula 3
$\theta > 45° \lesssim 180°$ . . . K_2 = Formula 4

Source: Reprinted with permission from Technical Paper 410, *Flow of Fluids*, Crane Company, Chicago.

TABLE 9-1

K (continued)

GATE VALVES
Wedge Disc, Double Disc, or Plug Type

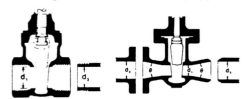

If: $\beta = 1, \theta = 0$............$K_1 = 8 f_T$
$\beta < 1$ and $\theta \lessgtr 45°$.........$K_2 =$ Formula 5
$\beta < 1$ and $\theta > 45° \lessgtr 180°$...$K_2 =$ Formula 6

GLOBE AND ANGLE VALVES

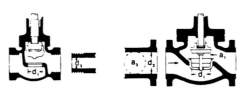

If: $\beta = 1$...$K_1 = 340 f_T$

If: $\beta = 1$...$K_1 = 55 f_T$

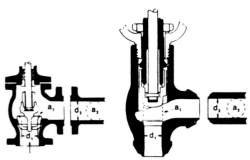

If: $\beta = 1$...$K_1 = 150 f_T$ If: $\beta = 1$...$K_1 = 55 f_T$

All globe and angle valves,
whether reduced seat or throttled,

If: $\beta < 1$...$K_2 =$ Formula 7

SWING CHECK VALVES

$K = 100 f_T$ \qquad $K = 50 f_T$

Minimum pipe velocity
(fps) for full disc lift
$= 35 \sqrt{v}$

Minimum pipe velocity
(fps) for full disc lift
$= 48 \sqrt{v}$

LIFT CHECK VALVES

If: $\beta = 1$...$K_1 = 600 f_T$
$\beta < 1$...$K_2 =$ Formula 7
Minimum pipe velocity (fps) for full disc lift
$= 40 \beta^2 \sqrt{v}$

If: $\beta = 1$...$K_1 = 55 f_T$
$\beta < 1$...$K_2 =$ Formula 7
Minimum pipe velocity (fps) for full disc lift
$= 140 \beta^2 \sqrt{v}$

TILTING DISC CHECK VALVES

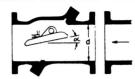

	$\alpha = 5°$	$\alpha = 15°$
Sizes 2 to 8"...$K =$	$40 f_T$	$120 f_T$
Sizes 10 to 14"...$K =$	$30 f_T$	$90 f_T$
Sizes 16 to 48"...$K =$	$20 f_T$	$60 f_T$
Minimum pipe velocity (fps) for full disc lift $=$	$80 \sqrt{v}$	$30 \sqrt{v}$

TABLE 9-1

K *(continued)*

STOP-CHECK VALVES
(Globe and Angle Types)

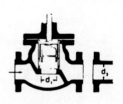

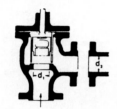

If:
$\beta = 1 \ldots K_1 = 400\, f_T$
$\beta < 1 \ldots : K_2 =$ Formula 7

Minimum pipe velocity
for full disc lift
$= 55\, \beta^2 \sqrt{v}$

If:
$\beta = 1 \ldots K_1 = 200\, f_T$
$\beta < 1 \ldots K_2 =$ Formula 7

Minimum pipe velocity
for full disc lift
$= 75\, \beta^2 \sqrt{v}$

FOOT VALVES WITH STRAINER

Poppet Disc   **Hinged Disc**

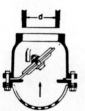

$K = 420\, f_T$

Minimum pipe velocity
(fps) for full disc lift
$= 15 \sqrt{v}$

$K = 75\, f_T$

Minimum pipe velocity
(fps) for full disc lift
$= 35 \sqrt{v}$

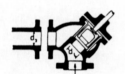

If:
$\beta = 1 \ldots K_1 = 350\, f_T$
$\beta < 1 \ldots K_2 =$ Formula 7

If:
$\beta = 1 \ldots K_1 = 300\, f_T$
$\beta < 1 \ldots K_2 =$ Formula 7

Minimum pipe velocity (fps) for full disc lift
$= 60\, \beta^2 \sqrt{v}$

BALL VALVES

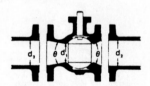

If: $\beta = 1, \theta = 0 \ldots \ldots \ldots \ldots K_1 = 3\, f_T$
$\beta < 1$ and $\theta \lessapprox 45° \ldots \ldots K_2 =$ Formula 5
$\beta < 1$ and $\theta > 45° \lessapprox 180° \ldots K_2 =$ Formula 6

If:
$\beta = 1 \ldots K_1 = 55\, f_T$
$\beta < 1 \ldots K_2 =$ Formula 7

If:
$\beta = 1 \ldots K_1 = 55\, f_T$
$\beta < 1 \ldots K_2 =$ Formula 7

Minimum pipe velocity (fps) for full disc lift
$= 140\, \beta^2 \sqrt{v}$

BUTTERFLY VALVES

Sizes   2 to 8" ... $K = 45\, f_T$
Sizes 10 to 14" ... $K = 35\, f_T$
Sizes 16 to 24" ... $K = 25\, f_T$

TABLE 9-1

K (continued)

PLUG VALVES AND COCKS

Straight-Way

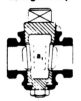

3-Way

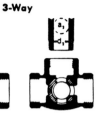

View X—X

If: $\beta = 1$, $K_1 = 18 f_T$

If: $\beta = 1$, $K_1 = 30 f_T$

If: $\beta = 1$, $K_1 = 90 f_T$

If: $\beta < 1 \ldots K_2 =$ Formula 6

MITRE BENDS

∝	K
0°	2 f_T
15°	4 f_T
30°	8 f_T
45°	15 f_T
60°	25 f_T
75°	40 f_T
90°	60 f_T

90° PIPE BENDS AND FLANGED OR BUTT-WELDING 90° ELBOWS

r/d	K	r/d	K
1	20 f_T	10	30 f_T
2	12 f_T	12	34 f_T
3	12 f_T	14	38 f_T
4	14 f_T	16	42 f_T
6	17 f_T	18	46 f_T
8	24 f_T	20	50 f_T

The resistance coefficient, K_B, for pipe bends other than 90° may be determined as follows:

$$K_B = (n - 1) \left(0.25 \, \pi \, f_T \frac{r}{d} + 0.5 \, K \right) + K$$

n = number of 90° bends
K = resistance coefficient for one 90° bend (per table)

CLOSE PATTERN RETURN BENDS

$K = 50 f_T$

STANDARD ELBOWS

90°

45°

$K = 30 f_T$

$K = 16 f_T$

STANDARD TEES

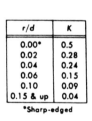

Flow thru run . . . . $K = 20 f_T$
Flow thru branch . . . . $K = 60 f_T$

PIPE ENTRANCE

Inward Projecting

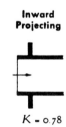

$K = 0.78$

r/d	K
0.00*	0.5
0.02	0.28
0.04	0.24
0.06	0.15
0.10	0.09
0.15 & up	0.04

*Sharp-edged

Flush

For K, see table

PIPE EXIT

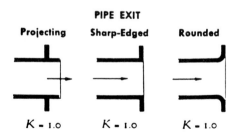

Projecting   **Sharp-Edged**   **Rounded**

$K = 1.0$ $K = 1.0$ $K = 1.0$

9-9 RESISTANCE COEFFICIENT *K*, EQUIVALENT LENGTH *L/D*, AND FLOW COEFFICIENT *C$_v$*

Velocity of fluid flow in a pipe is obtained at the expense of static head, and decrease in static head due to velocity is

$$h_v = \frac{V^2}{2g} \qquad (9\text{-}16)$$

which is defined as "velocity head" in feet (meters) of fluid. Flow of fluid through a valve or fitting in a pipe line also causes a reduction in static head, which may be expressed in terms of velocity head. The resistance coefficient *K* in the equation

$$h_L = K\left(\frac{V^2}{2g}\right) \qquad (9\text{-}17)$$

is defined as the number of velocity heads lost due to a valve or fitting. *It is always associated with the diameter in which the velocity occurs.* In most valves and fittings, the losses due to friction resulting from the length of flow path are minor in comparison to the dynamic loss. The resistance coefficient *K* is therefore considered as being independent of friction factor or Reynolds number and may be considered constant for any given obstruction (valve or fitting) in a piping system under all conditions of flow.

Equation (9-11) may be arranged as

$$h_L = \left(f\frac{L}{D}\right)\left(\frac{V^2}{2g}\right)$$

Comparing Eqs. (9-11) and (9-17), it follows that

$$K = f\left(\frac{L}{D}\right) \qquad (9\text{-}18)$$

The ratio *L/D* is the equivalent length, in pipe diameters of straight pipe, that will cause the same head loss as the obstruction under the same flow conditions. Since the resistance coefficient *K* is constant for all conditions of flow, the value of *L/D* for any given valve or fitting must necessarily vary inversely with the change in friction factor *f* for different flow conditions.

The resistance coefficient *K* would theoretically be a constant for all sizes of a given design or line of valves and fittings if all sizes were geometrically similar. However, geometric similarity is seldom, if ever, achieved because the design of valves and fittings is dictated by manufacturing economies, standards, structural strength, and other considerations. Based upon evidence from numerous investigations, it can be said that the resistance coefficient *K*, for a given line of valves or fittings, tends to vary with size as does the friction factor *f*, for straight clean commercial steel pipe at flow conditions resulting in a constant friction factor, and that the equivalent length *L/D* tends toward a constant for the various sizes of a given line of valves or fittings at the same flow conditions.

On the basis of this relationship, the resistance coefficient *K* for each illustrated type of valve and fitting is presented in Table 9-1. These coefficients are given as the product of the friction factor *f$_t$* for the desired size of clean commercial steel pipe with flow in the zone of complete turbulence and a constant, which represents the equivalent length *L/D*, for the valve or fitting in pipe diameters for the same flow conditions, on the basis of test data. This equivalent length, or constant, is valid for all sizes of the valve or fitting type with which it is identified.

The friction factors *f$_t$* for clean commercial steel pipe with flow in the zone of complete turbulence, for nominal sizes from 1/2 in. to 24 in., are tabulated at the beginning of Table 9-1.

There are some resistances to flow in piping, such as sudden and gradual contractions and enlargements, and pipe entrances and exits, that have geometric similarity between sizes. The resistance coefficients *K* for these items are therefore independent of size, as indicated by the absence of a friction factor in their values given in Table 9-1.

As previously stated, the resistance coefficient *K* is always associated with the diameter in which the velocity in the term *V²/2g* occurs. The values in the *K* factor table (Table 9-1) are associated with the internal diameter of the following pipe schedule numbers for the various ANSI classes of valves and fittings:

Class 300 and lower	Schedule 40
Class 400 to 600	Schedule 80
Class 900	Schedule 120
Class 1500	Schedule 160

When the resistance coefficient *K* is used in flow equations such as Eqs. (9-17) or (9-18), the velocity and internal diameter dimensions used in the equations must be based on the dimensions of these schedule numbers regardless of the pipe in which the valve may be installed.

An alternate procedure, which yields identical results for Eq. (9-17), is to adjust *K* in proportion to the fourth power of the diameter ratio and to base values of velocity or diameter on the internal diameter of the connecting pipe:

$$K_a = K_b \left(\frac{d_a}{d_b} \right)^4 \qquad (9\text{-}19)$$

where subscript *a* defines *K* and *d* with reference to the internal diameter of the connecting pipe. Subscript *b* defines *K* and *d* with reference to the internal diameter of the pipe for which the values of *K* were established, as given in the foregoing list of pipe schedule numbers.

When a piping system contains more than one size of pipe, valves, or fittings, Eq. (9-19) may be used to express all resistances in terms of one size. For this, subscript *a* relates to the size with reference to which all resistances are to be expressed, and subscript *b* relates to any other size in the system.

Flow Coefficient, C$_v$.   It has been found convenient in some branches of the valve industry, particularly in connection with control valves, to express the valve capacity and valve flow characteristics in terms of a flow coefficient, C$_v$. The coefficient is numerically equal to the flow rate in gpm of 60°F water, which will give a pressure drop of 1 lb/in.² (2.31 ft of water) across the device.

By substitution of appropriate EES units into the Darcy equation, it can be shown that

$$C_v = \frac{29.9 d^2}{\sqrt{K}} \qquad (9\text{-}20)$$

where *d* is the internal diameter in inches. In SI units, a flow coefficient C_{vSI} is defined as the flow rate in m³/s of water at 15°C with a pressure loss of 1 kPa given by

$$C_{vSI} = \frac{1.11 D^2}{\sqrt{K}} \qquad (9\text{-}20a)$$

where *D* is in meters.

The quantity in gallons per minute (gpm) of any liquid that will flow through the device can be determined from

$$\text{gpm} = Cv \sqrt{\Delta p_L \left(\frac{62.4}{\rho} \right)} \qquad (9\text{-}21)$$

and the pressure drop can be computed from the same relationship arranged as

$$\Delta p_L = \frac{\rho}{62.4} \left(\frac{\text{gpm}}{C_v} \right)^2 \qquad (9\text{-}22)$$

In Eqs. (9-21) and (9-22), Δp_L is the pressure drop across the valve in lbf/in.² and ρ is the fluid density in lbm/ft³.

Since the pressure loss is proportional to the square of the flow velocity, the pressure drop or lost head may be calculated at other flow rates:

$$\frac{h_{L1}}{h_{L2}} = \left(\frac{\text{gpm}_1}{\text{gpm}_2} \right)^2 \qquad (9\text{-}23)$$

In terms of the flow coefficient C_v,

$$\frac{h_L}{2.31} = \left(\frac{\text{gpm}}{C_v} \right)^2$$

or

$$h_L = 2.31 \left(\frac{\text{gpm}}{C_v} \right)^2 \qquad (9\text{-}24)$$

where C_v is in gpm and h_L is in feet of water.

Heating and cooling devices usually have pressure or head loss information furnished by the manufacturer. In fact, the flow rate may be adjusted by measurement of the pressure drop across the device. This is frequently done to balance the flows in the system. Equation (9-23) may be used to estimate the flow rate at other than specified conditions.

Laminar Flow Conditions.   Equation (9-17) is valid for computing the head loss due to valves and fittings for all conditions of flow, including laminar flow, using the resistance coefficient *K* as given in Table 9-1. When this equation is used to determine the losses in straight pipe, it is necessary to compute the Reynolds number in order to establish the friction factor *f* to be used to determine the resistance coefficient *K* for the pipe in accordance with Eq. (9-18) [*K = f(L/D)*].

Conversions between *K*, *L/D*, and *L* can be obtained for various pipe sizes by use of Figure 9-7. When using SI units, it is suggested that the *L/D* ratio be determined from Figure 9-7 using the nominal pipe size. The equivalent length in meters may then be determined using the inside diameter *D* in meters.

The relationship between resistance coefficient *K* and flow coefficient C$_v$ is shown in Figure 9-8.

ILLUSTRATIVE PROBLEM 9-7

A 6-in. class 125 iron Y-pattern globe valve has a flow coefficient C$_v$ of 600. Determine the resistance coefficient *K* and equivalent lengths *L/D* and *L* for flow in the zone of complete turbulence.

Solution: *K*, *L/D*, and *L* should be calculated in terms of 6-in. schedule 40 pipe. Nominal 6-in. schedule 40 pipe has an internal diameter (ID) of 6.065 in. By Eq. (9-20),

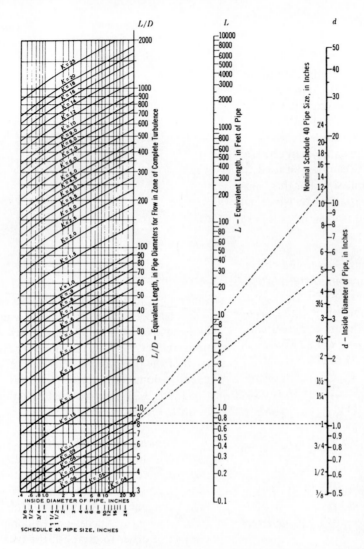

FIGURE 9-7 Equivalent lengths L and L/D and resistance coefficient K. (Reprinted from Technical Paper No. 410, *Flow of Fluids*, with permission of the publisher, Crane Company, 1976)

$$C_v = \frac{29.9d^2}{\sqrt{K}}$$

$$K = \frac{(29.9)^2(d^2)^2}{(C_v)^2} = \frac{894d^4}{(C_v)^2} = \frac{894(6.065)^4}{(600)^2}$$

$$= 3.36 \qquad \qquad (ans.)$$

Also, determine K from Figure 9-8. And, by Eq. (9-18),

$$K = f\left(\frac{L}{D}\right)$$

From Table 9-1, $f = 0.015$; therefore,

$$3.36 = 0.015\left(\frac{L}{D}\right)$$

$$\text{ratio } L/D = 224 \qquad \qquad (ans.)$$

Also, determine L/D from Figure 9-7. And, by Eq. (9-15),

$$L = \left(\frac{L}{D}\right)\left(\frac{d}{12}\right) = 224\left(\frac{6.065}{12}\right)$$

$$= 113.2 \text{ ft} \qquad \qquad (ans.)$$

Also, from Figure 9-7, determine L.

ILLUSTRATIVE PROBLEM 9-8

For 4-in. class 600 steel conventional angle valve with full seat area, find the resistance coefficient K, flow coefficient C_v, and equivalent lengths L/D and L for flow in the zone of complete turbulence.

Solution: K, L/D, and L should be given in terms of 4-in. schedule 80 pipe. Nominal 4-in. schedule 80 pipe has an ID of 3.826 in. From Table 9-1, $K = 150 f_t$, and $f_t = 0.017$. By Eq. (9-20),

$$C_v = \frac{29.9d^2}{\sqrt{K}}$$

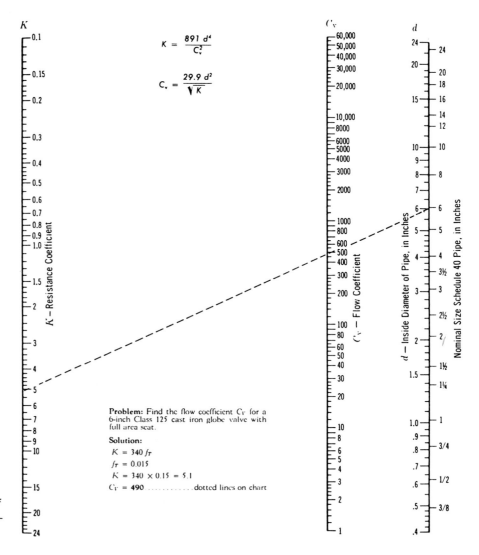

$$K = \frac{891 \, d^4}{C_v^2}$$

$$C_v = \frac{29.9 \, d^2}{\sqrt{K}}$$

Problem: Find the flow coefficient C_v for a 6-inch Class 125 cast iron globe valve with full area seat.

Solution:
$K = 340 \, f_T$
$f_T = 0.015$
$K = 340 \times 0.15 = 5.1$
$C_v = 490$ . . . . dotted lines on chart

FIGURE 9-8 Equivalents of resistance coefficient K and flow coefficient C_v. (Reprinted from Technical Paper No. 410, *Flow of Fluids,* with permission of the publisher, Crane Company, 1976)

By Eq. (9-18),

$$K = f_t \left(\frac{L}{D} \right)$$

or

$$\frac{L}{D} = \frac{K}{f_t} \text{ (for complete turbulence)}$$

By Eq. (9-15),

$$L = \left(\frac{L}{D} \right) \left(\frac{d}{12} \right)$$

Substituting gives

$$K = 150(0.017) = 2.55$$

$$C_v = \frac{29.9(3.826)^2}{\sqrt{2.55}} = 274$$

$$\frac{L}{D} = \frac{2.55}{0.017} = 150$$

$$L = 150 \left(\frac{3.826}{12} \right) = 47.8 \text{ ft}$$

9-10 SUDDEN CONTRACTIONS AND ENLARGEMENTS

The resistance to flow due to sudden enlargements may be expressed by

$$K_1 = \left(1 - \frac{d_1^2}{d_2^2} \right)^2 \qquad (9\text{-}25)$$

and the resistance due to sudden contractions by

$$K_1 = 0.5\left(1 - \frac{d_1^2}{d_2^2}\right) \qquad (9\text{-}26)$$

where subscripts 1 and 2 define the internal diameters of the small and large pipes, respectively. It is convenient to identify the ratio of diameters of the small to large pipes by the Greek letter β (beta). Using this notation, we would have

$$K_1 = (1 - \beta^2)^2 \text{ (sudden enlargements)} \qquad (9\text{-}25a)$$

and

$$K_1 = 0.5(1 - \beta^2) \text{ (sudden contractions)} \qquad (9\text{-}26a)$$

Equation (9-25) is derived from the momentum equation together with the Bernoulli equation. Equation (9-26) uses the derivation of Eq. (9-25) together with the continuity equation and a close approximation of the contraction coefficients determined by Julius Weisbach.

The value of the resistance coefficient in terms of the larger pipe is determined by dividing Eqs. (9-25) and (9-26) by β^4:

$$K_2 = \frac{K_1}{\beta^4} \qquad (9\text{-}27)$$

The losses due to gradual enlargements in pipes were investigated by A. H. Gibson and may be expressed as a coefficient, C_e, applied to Eq. (9-25). Approximate averages of Gibson's coefficients for different included angles of divergence θ are defined by the following equations:

For $\theta \leq 45°$ $\qquad C_e = 2.6 \sin \dfrac{\theta}{2}$ $\qquad (9\text{-}28)$

For $\theta > 45° \leq 180°$ $\qquad C_e = 1$ $\qquad (9\text{-}28a)$

The losses due to gradual contractions in pipes were established by analysis of Crane test data, using the same basis as that of Gibson for gradual enlargements, to provide a contraction coefficient, C_c, to be applied to Eq. (9-26). The approximate averages of these coefficients for different included angles of convergence θ are defined by the following equations:

For $\theta \leq 45°$ $\qquad C_c = 0.8 \sin \dfrac{\theta}{2}$ $\qquad (9\text{-}29)$

For $\theta > 45° \leq 180°$ $\qquad C_c = 0.5 \sqrt{\sin \dfrac{\theta}{2}}$ $\qquad (9\text{-}29a)$

The resistance coefficient K for sudden and gradual enlargements and contractions, expressed in terms

of the large pipe, is established by combining Eqs. (9-25) through (9-29). The formulas are summarized in Table 9-1.

Valves with reduced seats: Valves are often designed with reduced seats, and the transition from seat to valve ends may be either abrupt or gradual. Straight-through types, such as gate and ball valves, so designed with gradual transition, are sometimes referred to as venturi valves. Analysis of tests on such straight-through valves indicates an excellent correlation between test results and calculated values of K. See summation of formulas in Table 9-1.

The procedure for determining K for reduced-seat globe and angle valves is also applicable to *throttled* globe and angle valves. For this case, the value of β must be based on the square root of the ratio of areas: $\beta = \sqrt{a_1/a_2}$, where a_1 defines the area at the most restricted point in the flow path and a_2 defines the internal area of the connecting pipe. See formula summation in Table 9-1.

ILLUSTRATIVE PROBLEM 9-9

Given: a 6-by-4-in. class 600 steel gate valve with inlet and outlet ports conically tapered from back to body rings to valve ends. Face-to-face dimension is 22 in. and back of seat ring to back of seat ring is about 6 in. Determine K_2 for any flow condition, and L/D and L for flow in the zone of complete turbulence.

Solution: For class 600, K_2, L/D, and L should be given in terms of 6-in. schedule 80 pipe. From Table 9-1, $K_1 = 8f_t$, and

$$K_2 \text{ (in Formula 5)} =$$

$$\frac{K_1 + \sin \dfrac{\theta}{2}[0.8(1 - \beta^2) + 2.6(1 - \beta^2)^2]}{\beta^4}$$

By Eq. (9-18),

$$K = f\left(\frac{L}{D}\right) \qquad \text{or} \qquad \frac{L}{D} = \frac{K}{f_t}$$

and

$$\beta = \frac{d_1}{d_2}$$

For 4-in. schedule 80 pipe, ID = 3.826 in. For 6-in. schedule 80 pipe, ID = 5.761 in. From Table 9-1, $f_t = 0.015$ for 6-in. size. With substitution,

$$\beta = \frac{3.826}{5.761} = 0.664$$

$$\beta^2 = 0.441 \qquad \text{and} \qquad \beta^4 = 0.194$$

$$\tan \frac{\theta}{2} = \frac{0.5(5.761 - 3.826)}{0.5(22 - 6)} = 0.121$$

$$= \sin \frac{\theta}{2} \text{ (approx.)}$$

$$K_2 = \frac{(8)(0.015) + (0.121)[0.8(1 - 0.441)}{0.194}$$

$$= 1.40$$

$$\frac{L}{D} = \frac{K_2}{f_t} = \frac{1.40}{0.015}$$

$$= 93.3 \text{ diameters of 6-in.}$$
schedule 80 pipe

$$L = 93.3 \left(\frac{5.761}{12}\right)$$

$$= 48.8 \text{ ft of 6 in.}$$
schedule 80 pipe

ILLUSTRATIVE PROBLEM 9-10

A globe-type check valve with a swing-guided disc is required in a 3-in. schedule 40 horizontal pipe carrying 70°F water at the rate of 80 gpm. Determine the proper size check valve to use and its pressure drop. The valve is to be sized so that the disc is fully lifted at the specified flow.

Solution:

1. From Table 9-1, we find that $V_{min} = 40\sqrt{v}$, that $K_1 = 600$ f_t, and that $f_t = 0.018$ for either $2\frac{1}{2}$ or 3-in. pipe.
2. From Table 9-3, ID = 2.469 in. for a $2\frac{1}{2}$-in. schedule 40 pipe, and ID = 3.068 in. for a 3-in. pipe.
3. From Appendix Table A-1 at 70°F, $v = 0.016051$ ft^3/lb. Also, $\rho = 1/v = 62.30$ lb/ft^3.
4. Assume that the check valve is the same size as the pipe, or $\beta = 1$. Then,

$$V_{min} = 40\sqrt{v} = 40\sqrt{0.016051}$$
$$= 5.07 \text{ fps for full disc lift}$$

By Eq. (9-5),

$$V_{actual} = \frac{0.408(gpm)}{d^2} = \frac{0.408(80)}{(3.068)^2} = 3.47 \text{ fps}$$

5. Since V_{actual} is less than V_{min} required, try $2\frac{1}{2}$-in. check valve size. For $2\frac{1}{2}$-in. size,

$$V_{actual} = \frac{0.408(80)}{(2.469)^2} = 5.35 \text{ fps}$$

6. For the $2\frac{1}{2}$-in. size, V_{actual} is greater than V_{min}; therefore, use $2\frac{1}{2}$-in. valve size with reducers in the 3-in. pipe. With substitution,

$$\beta = \frac{d_1}{d_2} = \frac{2.469}{3.068} = 0.805$$
$$\beta^2 = 0.648 \quad \text{and} \quad \beta^4 = 0.420$$

7. From Table 9-1, Formula 7,

$$K_2 = \frac{K_1 + \beta[0.5(1 - \beta^2) + (1 - \beta^2)^2]}{\beta^4}$$

$$= \frac{(600)(0.018) + (0.80)[0.5(1 - 0.648)}{0.420}$$

$$= 26.3$$

8. From Eqs. (9-11) and (9-17),

$$h_L = f\left(\frac{L}{D}\right)\left(\frac{V^2}{2g}\right) = K_2\left(\frac{V^2}{2g}\right)$$

$$= 26.3 \left[\frac{(3.47)^2}{2(32.2)}\right]$$

$$= 4.9 \text{ ft of water}$$

So,

$$\Delta p = h_L \times \gamma = (4.9)(62.3)$$
$$= 305 \text{ psf or 2.1 psi}$$

ILLUSTRATIVE PROBLEM 9-11

Water at 60°F discharges from a tank with an average head of 22 ft of water to atmospheric pressure through 200 ft of 3-in. schedule 40 pipe, six 3-in. standard 90-degree threaded elbows, one 3-in. flanged ball valve having a 2 3/8-in. diameter seat and 16-degree conical inlet and 30-degree conical outlet, and a sharp-edged pipe entrance flush with the inside of the tank. Determine the flow velocity in the pipe and the rate of discharge through the pipe in gallons per minute.

Solution:

Step 1. Determine K values for fittings and pipe. By Eq. (9-17),

$$h_L = K\left(\frac{V^2}{2g}\right) \quad \text{or} \quad V = \sqrt{\frac{2g(h_L)}{K}}$$

By Eq. (9-5),

$$V = \frac{0.408(gpm)}{d^2} \quad \text{or} \quad gpm = \frac{Vd^2}{0.408}$$

From Table 9-1, $f_t = 0.018$. For pipe entrance, $K = 0.5$. For pipe exit, $K = 1.0$. For 90-degree elbows, $K = 30f_t$. For ball valve, $K_1 = 3f_t$. Because of reduced seat, K_2 must be found from Formula 5 of Table 9-1. However, when inlet and outlet angles (θ) differ, Formula 5 must be expanded to

$$K_2 = \frac{K_1 + 0.8 \sin \frac{\theta_1}{2} (1 - \beta^2) + 2.6 \sin \frac{\theta_2}{2} (1 - \beta^2)^2}{\beta^4}$$

Also,

$$d_1 = 2.375 \text{ in.} \quad \text{and} \quad d_2 = 3.068 \text{ in.}$$

$$\beta = \frac{d_1}{d_2} = \frac{2.375}{3.068} = 0.774$$

$$\beta^2 = 0.599 \text{ and } \beta^4 = 0.359$$

At valve inlet, $\sin 16°/2 = \sin 8° = 0.139$. At valve outlet, $\sin 30°/2 = \sin 15° = 0.259$.

$$K_2 = \frac{(3)(0.018) + (0.8)(0.139)}{0.359}$$
$$\frac{(1 - 0.599) + 2.6(0.259)(1 - 0.599)^2}{0.359}$$
$$= 0.577$$

For six pipe elbows,

$$K = 6(30f_t) = (6)(30)(0.018) = 3.24$$

For straight pipe,

$$K = f_t \left(\frac{L}{D}\right) = \frac{(0.018)(200)}{3.068/12} = 14.08$$

For entire system,

$$K = 0.5 + 14.08 + 0.577 + 3.24 + 1 = 19.4$$

Step 2. Calculate velocity and flow rate. From step 1,

$$V = \sqrt{\frac{2g(h_L)}{K}} = \sqrt{\frac{(2)(32.2)(22)}{19.4}} = 8.54 \text{ fps}$$

$$\text{gpm} = \frac{Vd^2}{0.408} = \frac{(8.54)(3.068)^2}{0.408} = 197 \text{ gpm}$$

ILLUSTRATIVE PROBLEM 9-12

Determine the total equivalent length (TEL) and head loss for 100 ft of 2-in. schedule 40 commercial steel pipe carrying 100 gpm of 60°F water. The pipe section contains three standard 90° elbows, one globe valve, and a swing check valve ($K = 100f_t$). Assume that the head loss due to friction is 7.1×10^{-2} ft of water per ft of pipe length.

Solution:

Table 9-1: $f_t = 0.019$

Elbow: $K = 30f_t = (30)(0.019) = 0.57$

Figure 9-7: $L = 4.8$ ft of pipe

Globe valve: $K = 340f_t = (340)(0.019) = 6.45$

Figure 9-7: $L = 30$ ft of pipe

Check valve: $K = 100f_t = (100)(0.019) = 1.9$

Figure 9-7: $L = 12$ ft of pipe

TEL: straight length of pipe plus equivalent length of the fittings, or

$$\text{TEL} = 100 + (3)(4.8) + 30 + 12 = 156.4 \text{ ft}$$

h_L: friction rate × TEL, or

$$h_L = (7.1 \times 10^{-2})(156.4) = 11.1 \text{ ft of water}$$

Illustrative Problems 9-9 through 9-12 illustrate the use of Table 9-1 and Figure 9-7 to determine losses in pipe fittings and valves. The values obtained may be considered to be as accurate as experimentation has given to us. Such calculations are sometimes cumbersome. In the next two chapters, we shall find tables that provide fitting equivalent lengths that are normally used in HVAC system design.

9-11 PIPING SYSTEMS

Heating, ventilating, and cooling systems may use one or more fluids circulating through closed conduits to and from various locations in the system. Typical fluids are water, steam, refrigerants, water–glycol mixtures, air, and so on. The flow of fluids in the system and in the system components must be completely controlled and coordinated. This leads to a variety of integrated piping and duct systems that must be installed throughout a building to connect all the heat-transfer devices. Some of the systems with which we are most concerned in this book are (1) hot-water piping systems, (2) steam supply and return piping systems, and (3) air duct systems. We shall discuss these systems in detail in later chapters.

Piping systems in the HVAC industry for the distribution of water are either *open* systems or *closed* systems. An open system refers to that type of system where one or both ends of the piping system is exposed to atmospheric pressure. For example, the piping system used in a spray washer for conditioning air is an open system (see Chapter 4).

Most hot-water heating systems installed at the present time are referred to as closed systems. The water is sealed in the system pipes and in the heat-transfer devices under some pressure above atmospheric pressure. The pressures used make it possible to have the temperature of the circulating water higher (above 212°F at 14.7 psia) without boiling the water in the piping.

Another reason is that the elimination of air from the system is made easier, which reduces the tendency of air to leak into the system. Air trapped in the system causes a restriction to water flow and noisy operation.

When water is used in a closed piping system, the piping arrangement takes the form of supply and return pipes along with the necessary pump, or pumps, to circulate the water through the piping and heat-transfer devices. The characteristics of the building to be heated determine the pipe routing. The system design depends on load, control and installation requirements, and cost. To meet these requirements, basic piping arrangements are selected, such as (1) series loop, (2) one-pipe, (3) two-pipe reverse-return, or (4) two-pipe direct-return. These four basic piping systems can also be combined in various ways to use the advantages of each. These basic piping systems will be discussed in detail in Chapter 10.

9-12 PIPE EXPANSION AND FLEXIBILITY

Changes in temperature cause dimensional changes in all materials. It is important in the design of any piping system to give careful consideration to the expansion and contraction of the piping caused by temperature changes in the system. For systems operating at high temperature, such as steam and hot water, the rate of expansion is high and significant movements can occur even in short runs of piping. The change in length of a pipe may be calculated using the following expression:

$$\Delta L = L_o \alpha (t - t_o) \qquad (9\text{-}30)$$

where

ΔL = change in length of pipe for the temperature change, $t - t_o$
L_o = original length of pipe at t_o
α = coefficient of linear expansion for the pipe material

Table 9-2 lists the linear expansion of pipe for various temperature changes based on Eq. (9-30) using an average coefficient of linear expansion and an original pipe length of 100 feet for steel, brass, or copper. The linear expansion of any length may be found by dividing the actual length by 100 and multiplying by the linear expansion for the temperature change under consideration from Table 9-2. For example, consider a 2-in. steel pipe, 150 ft long, that is subjected to a temperature change from 70°F to 200°F. From Table 9-2, for a temperature increase of 200° − 70° = 130°F, the expansion per 100 ft is 0.99 in. Therefore, the linear expansion for 150 ft would be (150/100) (0.99) = 1.485 in.

TABLE 9-2

Linear expansion of pipe in inches per 100 feet

Temperature Change,* (°F)	Steel	Brass and Copper
50	0.38	0.57
60	0.45	0.64
70	0.53	0.74
80	0.61	0.85
90	0.68	0.95
100	0.76	1.06
110	0.84	1.17
120	0.91	1.28
130	0.99	1.38
140	1.07	1.49
150	1.15	1.60
160	1.22	1.70
170	1.30	1.81
180	1.37	1.91
190	1.45	2.02
200	1.57	2.13
250	1.99	2.66
300	2.47	3.19
350	2.94	3.72
400	3.46	4.25

* Between fluid in pipe and surrounding air.

Even though rates of expansion may be low for systems operating in the range of 40°F to 100°F (5°C to 40°C), they can cause significant movements in long runs of piping that commonly occur in distribution systems in high-rise buildings. Therefore, in addition to design requirements for pressure, weight, and other loadings, piping systems must accommodate thermal and other movements to prevent (1) failure of pipe and supports from overstress and fatigue, (2) leakage of joints, and (3) detrimental forces and stresses on connected equipment.

Since the anchor forces and bowing of pipe anchored at both ends is generally too high, general practice is to *never anchor a straight run of pipe at both ends.* Piping systems must be allowed to expand or contract due to thermal changes. Ample flexibility can be obtained by using pipe bends and loops or supplemental devices, such as expansion joints, in the system. Figure 9-9 shows a variety of pipe bends commonly used in piping systems. Also, packless expansion joints and slip joints in a large variety of forms are commercially available.

9-13 PIPE ANCHORS AND SUPPORTS

Pipe supporting elements consist of hangers, which support from above; supports, which bear loads from below; and restraints, such as anchors and guides, which limit or direct movement, as well as support loads. Pipe-supporting elements withstand all static and dynamic

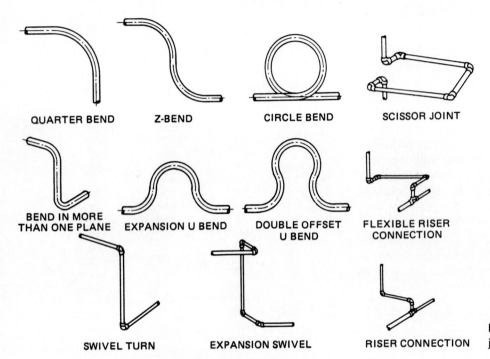

QUARTER BEND   Z-BEND   CIRCLE BEND   SCISSOR JOINT

BEND IN MORE THAN ONE PLANE   EXPANSION U BEND   DOUBLE OFFSET U BEND   FLEXIBLE RISER CONNECTION

SWIVEL TURN   EXPANSION SWIVEL   RISER CONNECTION

FIGURE 9-9   Common expansion joints and expansion bends.

conditions including (1) weight of pipe, valves, fittings, insulation, and fluid contents, including test fluids if heavier-than-normal flow media; (2) occasional loads such as ice, wind, and seismic forces; (3) forces imposed by thermal expansion and contraction of pipe bends and loops; (4) frictional, spring, and pressure thrust forces imposed by expansion joints in the system; (5) frictional forces of guides and supports; (6) other loads as might be imposed, such as water hammer, vibration, and reactive forces of relief valves; and (7) test loads and forces.

In addition, pipe-supporting elements must be evaluated in terms of stress at the point of connection to the pipe and the building structure. Stress at the point of connection to the pipe is especially important for base elbow and trunnion supports, as the limiting and controlling parameter is usually not the strength of the structural member, but the localized stress at the point of attachment to the pipe. All loads on the structure caused by piping elements should be given to and coordinated with the structural engineers.

The Code for Pressure Piping (ASME B-31) establishes criteria for the design of pipe-supporting elements and the Manufacturers Standardization Society of the Valve and Fitting Industry (MSS) has established standards for the design, fabrication, selection, and installation of pipe hangers and supports based on these codes. MSS Standard SP-69 and the catalogs of many manufacturers illustrate the various hangers and components and provide information on the types to use for various pipe systems.

The loads on most pipe-supporting elements are moderate and can be selected safely in accordance with preceding information and manufacturers' catalog data; however, some loads and forces can be very high, especially in multistory buildings and large diameter pipe when expansion joints are used at high operating pressures. Consequently, a qualified engineer should design or review the design of all anchors and pipe-supporting elements when high-load conditions exist.

9-14   STEEL PIPE AND COPPER TUBE DATA

The most common pipe material for industrial use is wrought steel. Cast iron is also used, but much less frequently. Table 9-3 gives dimensional data for schedule 40 (sometimes called standard) wrought steel pipe, as well as schedule 80 (sometimes called extra-strong) pipe. For steam and condensate lines and some hot water systems, steel pipe is used with threaded fittings. In some cases, especially for the larger pipe sizes, pipes are connected by welding. When threaded fittings are used, it is necessary to ream the ends of the pipe to eliminate burrs and to avoid the possibility of a partial closure of the internal bore. If this is not done, design computations for friction loss are uncertain. When steel pipe is used for hot water systems, it may be desirable to use galvanized pipe, which is steel pipe covered with a corrosion-resistant zinc coating. Such coatings can prevent rusting for long periods of time.

Copper tubing has been used extensively for hot water and chilled water systems (less frequently for steam systems) and for refrigeration systems. Table 9-4 gives the characteristic data for commercial copper tub-

TABLE 9-3

Dimensions and properties of steel pipe

Nom. Size and Pipe O.D., D, in.	Schedule Number or Weight[a]	Wall Thickness, in. t	Inside Diameter in. d	Surface Area Outside ft²/ft	Surface Area Inside ft²/ft	Cross-Sectional Metal Area in²	Cross-Sectional Flow Area in²	Weight of Pipe lb/ft	Weight of Water lb/ft	Mfr. Process	Joint Type	Working Pressure[b] ASTM A53 B to 400°F psig
1/4 D = 0.540	40 ST	0.088	0.364	0.141	0.095	0.125	0.104	0.424	0.045	CW	Thrd	188
	80 XS	0.119	0.302	0.141	0.079	0.157	0.072	0.535	0.031	CW	Thrd	871
3/8 D = 0.675	40 ST	0.091	0.493	0.177	0.129	0.167	0.191	0.567	0.083	CW	Thrd	203
	80 XS	0.126	0.423	0.177	0.111	0.217	0.141	0.738	0.061	CW	Thrd	820
1/2 D = 0.840	40 ST	0.109	0.622	0.220	0.163	0.250	0.304	0.850	0.131	CW	Thrd	214
	80 XS	0.147	0.546	0.220	0.143	0.320	0.234	1.087	0.101	CW	Thrd	753
3/4 D = 1.050	40 ST	0.113	0.824	0.275	0.216	0.333	0.533	1.13	0.231	CW	Thrd	217
	80 XS	0.154	0.742	0.275	0.194	0.433	0.432	1.47	0.187	CW	Thrd	681
1 D = 1.315	40 ST	0.133	1.049	0.344	0.275	0.494	0.864	1.68	0.374	CW	Thrd	226
	80 XS	0.179	0.957	0.344	0.251	0.639	0.719	2.17	0.311	CW	Thrd	642
1 1/4 D = 1.660	40 ST	0.140	1.380	0.435	0.361	0.669	1.50	2.27	0.647	CW	Thrd	229
	80 XS	0.191	1.278	0.435	0.335	0.881	1.28	2.99	0.555	CW	Thrd	594
1 1/2 D = 1.900	40 ST	0.145	1.610	0.497	0.421	0.799	2.04	2.72	0.881	CW	Thrd	231
	80 XS	0.200	1.500	0.497	0.393	1.068	1.77	3.63	0.765	CW	Thrd	576
2 D = 2.375	40 ST	0.154	2.067	0.622	0.541	1.07	3.36	3.65	1.45	CW	Thrd	230
	80 XS	0.218	1.939	0.622	0.508	1.48	2.95	5.02	1.28	CW	Thrd	551
2 1/2 D = 2.875	40 ST	0.203	2.469	0.753	0.646	1.70	4.79	5.79	2.07	CW	Weld	533
	80 XS	0.276	2.323	0.753	0.608	2.25	4.24	7.66	1.83	CW	Weld	835
3 D = 3.500	40 ST	0.216	3.068	0.916	0.803	2.23	7.39	7.57	3.20	CW	Weld	482
	80 XS	0.300	2.900	0.916	0.759	3.02	6.60	10.25	2.86	CW	Weld	767
4 D = 4.500	40 ST	0.237	4.026	1.178	1.054	3.17	12.73	10.78	5.51	CW	Weld	430
	80 XS	0.337	3.826	1.178	1.002	4.41	11.50	14.97	4.98	CW	Weld	695
6 D = 6.625	40 ST	0.280	6.065	1.734	1.588	5.58	28.89	18.96	12.50	ERW	Weld	696
	80 XS	0.432	5.761	1.734	1.508	8.40	26.07	28.55	11.28	ERW	Weld	1209
8 D = 8.625	30	0.277	8.071	2.258	2.113	7.26	51.16	24.68	22.14	ERW	Weld	526
	40 ST	0.322	7.981	2.258	2.089	8.40	50.03	28.53	21.65	ERW	Weld	643
	80 XS	0.500	7.625	2.258	1.996	12.76	45.66	43.35	19.76	ERW	Weld	1106
10 D = 10.75	30	0.307	10.136	2.814	2.654	10.07	80.69	34.21	34.92	ERW	Weld	485
	40 ST	0.365	10.020	2.814	2.623	11.91	78.85	40.45	34.12	ERW	Weld	606
	XS	0.500	9.750	2.814	2.552	16.10	74.66	54.69	32.31	ERW	Weld	887
	80	0.593	9.564	2.814	2.504	18.92	71.84	64.28	31.09	ERW	Weld	1081
12 D = 12.75	30	0.330	12.090	3.338	3.165	12.88	114.8	43.74	49.68	ERW	Weld	449
	ST	0.375	12.000	3.338	3.141	14.58	113.1	49.52	48.94	ERW	Weld	528
	40	0.406	11.938	3.338	3.125	15.74	111.9	53.48	48.44	ERW	Weld	583
	XS	0.500	11.750	3.338	3.076	19.24	108.4	65.37	46.92	ERW	Weld	748
	80	0.687	11.376	3.338	2.978	26.03	101.6	88.44	43.98	ERW	Weld	1076
14 D = 14.00	30 ST	0.375	13.250	3.665	3.469	16.05	137.9	54.53	59.67	ERW	Weld	481
	40	0.437	13.126	3.665	3.436	18.62	135.3	63.25	58.56	ERW	Weld	580
	XS	0.500	13.000	3.665	3.403	21.21	132.7	72.04	57.44	ERW	Weld	681
	80	0.750	12.500	3.665	3.272	31.22	122.7	106.05	53.11	ERW	Weld	1081
16 D = 16.00	30 ST	0.375	15.250	4.189	3.992	18.41	182.6	62.53	79.04	ERW	Weld	421
	40 XS	0.500	15.000	4.189	3.927	24.35	176.7	82.71	76.47	ERW	Weld	596
18 D = 18.00	ST	0.375	17.250	4.712	4.516	20.76	233.7	70.54	101.13	ERW	Weld	374
	30	0.437	17.126	4.712	4.483	24.11	230.3	81.91	99.68	ERW	Weld	451
	XS	0.500	17.000	4.712	4.450	27.49	227.0	93.38	98.22	ERW	Weld	530
	40	0.562	16.876	4.712	4.418	30.79	223.7	104.59	96.80	ERW	Weld	607
20 D = 20.00	ST	0.375	19.250	5.236	5.039	23.12	291.0	78.54	125.94	ERW	Weld	337
	30 XS	0.500	19.000	5.236	4.974	30.63	283.5	104.05	122.69	ERW	Weld	477
	40	0.593	18.814	5.236	4.925	36.15	278.0	122.82	120.30	ERW	Weld	581

[a] Numbers are schedule numbers per ASTM B36.10; ST = Standard Weight; XS = Extra Strong.

[b] Working pressures have been calculated per ASME/ANSI B31.9 using furnace butt weld (continuous weld, CW) pipe through 4 in. and electric resistance weld (ERW) thereafter. The allowance, A, has been taken as:
  (a) 12.5% of t for mill tolerance on pipe wall thickness, plus
  (b) An arbitrary corrosion allowance of 0.025 in. for pipe sizes through NPS 2 and 0.065 in. from NPS 2½ through 20, plus
  (c) A thread cutting allowance for sizes through NPS 2.

Because the pipe wall thickness of threaded standard weight pipe is so small after deducting the allowance, A, the mechanical strength of the pipe is impaired. It is good practice to limit standard weight threaded pipe pressures to 90 psig for steam and 125 psig for water.

Conversion Factors:
  mm = 25.4 × in.
  m²/m = 0.3048 × ft²/ft
  mm² = 645 × in.²
  kg/m = 1.49 × lb/ft
  kPa gauge = 6.89 × psig

Source: ASHRAE Handbook—1988 Equipment.

TABLE 9-4

Dimensions and properties of copper tube

Nominal Diameter	Type	Wall Thickness, t, in.	Diameter Outside, D, in.	Diameter Inside, d, in.	Surface Area Outside ft²/ft	Surface Area Inside ft²/ft	Cross-Sectional Metal Area in²	Cross-Sectional Flow Area in²	Weight of Tube lb/ft	Weight of Water lb/ft	Working Pressure[a,b,c] ASTM B88 to 250°F Annealed psig	Drawn psig
1/4	K	0.035	0.375	0.305	0.098	0.080	0.037	0.073	0.145	0.032	851	1596
	L	0.030	0.375	0.315	0.098	0.082	0.033	0.078	0.126	0.034	730	1368
3/8	K	0.049	0.500	0.402	0.131	0.105	0.069	0.127	0.269	0.055	894	1676
	L	0.035	0.500	0.430	0.131	0.113	0.051	0.145	0.198	0.063	638	1197
	M	0.025	0.500	0.450	0.131	0.008	0.037	0.159	0.145	0.069	456	855
1/2	K	0.049	0.625	0.527	0.164	0.138	0.089	0.218	0.344	0.094	715	1341
	L	0.040	0.625	0.545	0.164	0.143	0.074	0.233	0.285	0.101	584	1094
	M	0.028	0.625	0.569	0.164	0.149	0.053	0.254	0.203	0.110	409	766
5/8	K	0.049	0.750	0.652	0.196	0.171	0.108	0.334	0.418	0.144	596	1117
	L	0.042	0.750	0.666	0.196	0.174	0.093	0.348	0.362	0.151	511	958
3/4	K	0.065	0.875	0.745	0.229	0.195	0.165	0.436	0.641	0.189	677	1270
	L	0.045	0.875	0.785	0.229	0.206	0.117	0.484	0.455	0.209	469	879
	M	0.032	0.875	0.811	0.229	0.212	0.085	0.517	0.328	0.224	334	625
1	K	0.065	1.125	0.995	0.295	0.260	0.216	0.778	0.839	0.336	527	988
	L	0.050	1.125	1.025	0.295	0.268	0.169	0.825	0.654	0.357	405	760
	M	0.035	1.125	1.055	0.295	0.276	0.120	0.874	0.464	0.378	284	532
1 1/4	K	0.065	1.375	1.245	0.360	0.326	0.268	1.217	1.037	0.527	431	808
	L	0.055	1.375	1.265	0.360	0.331	0.228	1.257	0.884	0.544	365	684
	M	0.042	1.375	1.291	0.360	0.338	0.176	1.309	0.682	0.566	279	522
	DWV	0.040	1.375	1.295	0.360	0.339	0.168	1.317	0.650	0.570	265	497
1 1/2	K	0.072	1.625	1.481	0.425	0.388	0.351	1.723	1.361	0.745	404	758
	L	0.060	1.625	1.505	0.425	0.394	0.295	1.779	1.143	0.770	337	631
	M	0.049	1.625	1.527	0.425	0.400	0.243	1.831	0.940	0.792	275	516
	DWV	0.042	1.625	1.541	0.425	0.403	0.209	1.865	0.809	0.807	236	442
2	K	0.083	2.125	1.959	0.556	0.513	0.532	3.014	2.063	1.304	356	668
	L	0.070	2.125	1.985	0.556	0.520	0.452	3.095	1.751	1.339	300	573
	M	0.058	2.125	2.009	0.556	0.526	0.377	3.170	1.459	1.372	249	467
	DWV	0.042	2.125	2.041	0.556	0.534	0.275	3.272	1.065	1.416	180	338
2 1/2	K	0.095	2.625	2.435	0.687	0.637	0.755	4.657	2.926	2.015	330	619
	L	0.080	2.625	2.465	0.687	0.645	0.640	4.772	2.479	2.065	278	521
	M	0.065	2.625	2.495	0.687	0.653	0.523	4.889	2.026	2.116	226	423

Size	Type										
3	K	0.109	3.125	2.907	0.818	0.761	1.033	6.637	4.002	2.872	318	596
	L	0.090	3.125	2.945	0.818	0.771	0.858	6.812	3.325	2.947	263	492
	M	0.072	3.125	2.981	0.818	0.780	0.691	6.979	2.676	3.020	210	394
	DWV	0.045	3.125	3.035	0.818	0.795	0.435	7.234	1.687	3.130	131	246
3 1/2	K	0.120	3.625	3.385	0.949	0.886	1.321	8.999	5.120	3.894	302	566
	L	0.100	3.625	3.425	0.949	0.897	1.107	9.213	4.291	3.987	252	472
	M	0.083	3.625	3.459	0.949	0.906	0.924	9.397	3.579	4.066	209	392
4	K	0.134	4.125	3.857	1.080	1.010	1.680	11.684	6.510	5.056	296	555
	L	0.110	4.125	3.905	1.080	1.022	1.387	11.977	5.377	5.182	243	456
	M	0.095	4.125	3.935	1.080	1.030	1.203	12.161	4.661	5.262	210	394
	DWV	0.058	4.125	4.009	1.080	1.050	0.741	12.623	2.872	5.462	128	240
5	K	0.160	5.125	4.805	1.342	1.258	2.496	18.133	9.671	7.846	285	534
	L	0.125	5.125	4.875	1.342	1.276	1.963	18.665	7.609	8.077	222	417
	M	0.109	5.125	4.907	1.342	1.285	1.718	18.911	6.656	8.183	194	364
	DWV	0.072	5.125	4.981	1.342	1.304	1.143	19.486	4.429	8.432	128	240
6	K	0.192	6.125	5.741	1.603	1.503	3.579	25.886	13.867	11.201	286	536
	L	0.140	6.125	5.845	1.603	1.530	2.632	26.832	10.200	11.610	208	391
	M	0.122	6.125	5.881	1.603	1.540	2.301	27.164	8.916	11.754	182	341
	DWV	0.083	6.125	5.959	1.603	1.560	1.575	27.889	6.105	12.068	124	232
8	K	0.271	8.125	7.583	2.127	1.985	6.687	45.162	25.911	19.542	304	570
	L	0.200	8.125	7.725	2.127	2.022	4.979	46.869	19.295	20.280	224	421
	M	0.170	8.125	7.785	2.127	2.038	4.249	47.600	16.463	20.597	191	358
	DWV	0.109	8.125	7.907	2.127	2.070	2.745	49.104	10.637	21.247	122	229
10	K	0.338	10.125	9.449	2.651	2.474	10.392	70.123	40.271	30.342	304	571
	L	0.250	10.125	9.625	2.651	2.520	7.756	72.760	30.054	31.483	225	422
	M	0.212	10.125	9.701	2.651	2.540	6.602	73.913	25.584	31.982	191	358
12	K	0.405	12.125	11.315	3.174	2.962	14.912	100.554	57.784	43.510	305	571
	L	0.280	12.125	11.565	3.174	3.028	10.419	105.046	40.375	45.454	211	395
	M	0.254	12.125	11.617	3.174	3.041	9.473	105.993	36.706	45.863	191	358

Source: ASHRAE Handbook 1988 Equipment.

aWhen using soldered or brazed fittings, the joint determines the limiting pressure.

bWorking pressures calculated using ASME B31.9 allowable stresses. A 5% mill tolerance has been used on the wall thickness. Higher tube ratings can be calculated using the allowable stress for lower temperatures.

cIf soldered or brazed fittings are used on hard drawn tubing, use the annealed ratings. Full-tube allowable pressures can be used with suitably rated flare or compression-type fittings.

Conversion Factors: mm = 25.4 × in.
  $m^2/m = 0.3048 \times ft^2/ft$
  $mm^2 = 645 \times in.^2$
  kg/m = 1.49 × lb/ft
  kPa gauge = 6.89 × psig

ing. Copper tubing is seamless, deoxidized copper tube supplied in three wall thicknesses.

Type *K*, the heaviest, is usually supplied in coils for the smaller sizes and in 12- and 20-ft lengths for the larger sizes. It is furnished in either hard or soft temper and is usually specified for underground service.

Type *L* has a lesser wall thickness. It is supplied in coils up to and including 1 1/2 in. nominal and in 12- and 20-ft lengths for sizes above 1 1/2 in. It is supplied in either hard or soft temper.

Type *M* tubing, which has the thinnest wall, is furnished in hard temper in 12- and 20-ft lengths. Hard temper tubing, which is rigid, is pleasing in appearance and is most often used where runs of pipe are exposed.

Copper water tubing, such as types *K, L,* and *M,* is not intended to be joined by threaded fittings. A common type of connection is the solder joint, or sweated joint. In assembly, the tube ends are cut square to length, the end mating surfaces are cleaned with steel wool, and a flux is applied to the surfaces. The tube is inserted into the fitting, and the joint is then heated. Solder, usually of lead–tin or tin–antimony, melts on the hot surface and is drawn into the space between the fitting and tube by capillary action. After cooling and solidification, a tight and permanent joint results. Where temperatures may exceed 250°F for lead–tin solder or 312°F for tin–antimony, it is necessary to use a brazing alloy or silver solder for the joints. Type *K* and type *L* copper water tubes are also joined by compression fittings.

REVIEW PROBLEMS

9.1.   A fluid tested in a Saybolt Viscosimeter has a viscosity of 93 s at 100°F and has a specific gravity measured at 60°F of 0.850. Calculate the viscosity of the fluid at 100°F in centistokes, centipoise, ft²/s, and lb·s/ft².

9.2.   The fluid of Review Problem 9.1 is flowing through a nominal 2-in. schedule 40 steel pipe at the rate of 40 gpm. What is the flow velocity and Reynolds number of the flow? Is the flow laminar or turbulent?

9.3.   A 30% solution of ethylene glycol and water has a specific gravity of 1.049 and a viscosity of 5.6 cP at 20°F. The solution is flowing through a nominal $2\frac{1}{2}$-in. schedule 40 steel pipe at the rate of 100 gpm. What is the flow velocity and Reynolds number of the flow?

9.4.   Water is flowing in a 2-in. copper tube (type *L*) at the rate of 120 gpm. The 2-in. tube reduces to a $1\frac{1}{2}$-in. tube at a point in the flow. A pressure gauge attached to the 2-in. tube indicates a pressure of 50 psig. Downstream, in the $1\frac{1}{2}$-in. tube, a pressure gauge reads 46 psig. What is the lost pressure head in feet of water between the two gauges? Assume the system is horizontal.

9.5.   No. 3 fuel oil at 60°F flows through a 2-in. schedule 40 steel pipe at the rate of 40,000 lb/hr. Determine the flow rate in gallons per minute and the average flow velocity. Assume the specific gravity at 60°F is 0.898.

9.6.   The maximum flow rate of a liquid will be 300 gpm with a maximum average velocity of 12 ft/s through a schedule 40 steel pipe. Determine the smallest suitable nominal pipe size required and the actual average flow velocity in the pipe selected.

9.7.   Water at 150°F flows steadily in a 4-in. schedule 40 steel pipe at the rate of 400 gpm. Determine the flow rate in lb/hr, the Reynolds number, and the friction factor for this flow. Assume that the density of the water is 61.19 lb/ft³ and the viscosity is 0.43 cP at 150°F.

9.8.   Calculate the head loss and pressure loss for the conditions of Review Problem 9.7 if the pipe length is 125 ft.

9.9.   Water at 100°F flows at the rate of 100 gpm through a 2-in. type *L* copper tube. Determine the head loss and pressure loss per 100 ft of tube. Assume that the water density and viscosity are 61.99 lb/ft³ and 0.69 cP, respectively.

9.10.   Water at 80°F flows through a 2-in. schedule 40 steel pipe with a measured pressure loss of 4.5 psi in a section that is 250 ft long. Estimate the flow rate through the pipe in ft³/s and in gpm if the absolute viscosity is 0.85 cP and the water density is 62.22 lb/ft³.

9.11.   What nominal size steel pipe (schedule 40) should be used for a water flow rate of 150 gpm with a head loss of not more than 20 ft of water per 100 ft of pipe length? Assume that the water has a density of 62.4 lb/ft³ and a viscosity of 7.6×10^{-6} ft²/s.

9.12.   Calculate the head loss for 260 ft of 2-in. nominal schedule 40 steel pipe. The pipe contains four standard 90-degree elbows, one gate valve, one globe valve, and one swing check valve ($K = 50f_t$). Water is flowing in the pipe at the rate of 75 gpm and has a density of 62.4 lb/ft³ and a viscosity of 1.3 cP.

9.13.   Water at 60°F is flowing through the piping system shown in Figure 9-10 at the rate of 400 gpm. Determine the average flow velocity in the 4- and 5-in. pipe sections and the pressure difference between gauges p_1 and p_2. Assume a water density of 62.37 lb/ft³ and a viscosity of 1.1 cP.

9.14.   Water at 80°F is flowing in the 3-in. schedule 40 steel pipe shown in Figure 9-11 at the rate of 200 gpm. The valve shown is a gate valve, the check valve is a swing check valve ($K = 100f_t$ and $\beta = 1$), and all elbows are standard, threaded 90-degree elbows. Determine the pressure difference between gauges p_1 and p_2. Assume

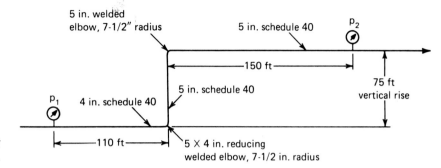

FIGURE 9-10 Piping diagram for Review Problem 9.13 (elevation).

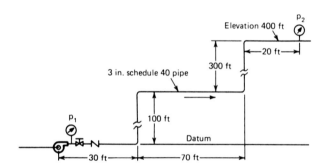

FIGURE 9-11 Piping diagram for Review Problem 9.14 (elevation).

a water density of 62.22 lb/ft^3 and a viscosity of 0.85 cP.

9.15. Figure 9-12 shows an elevation view of a section of 3-in. nominal schedule 40 steel pipe. The fluid in the pipe has a viscosity of 1.076×10^{-5} ft^2/s and a density of 56.0 lb/ft^3. The pressure gauges at points 1 and 2 read 25 psig and 20 psig, respectively. Determine the flow rate through the pipe in cfs. In which direction, 1 to 2 or 2 to 1, is the fluid flowing?

9.16. Figure 9-13 shows an elevation sketch of a pipe line connected to a large reservoir. The pipe line is 6-in. schedule 40 steel with fittings as shown. Estimate the discharge flow rate in cfs through the pipe when the reservoir level remains constant if **(a)** no losses are considered and **(b)** all losses are considered. Fluid viscosity is 2.35×10^{-5} lb·s/ft^2, and density is 62.4 lb/ft^3.

9.17. Figure 9-14 is an elevation sketch of a pipe line connected to a large reservoir that contains a fluid that has a viscosity of 1.217×10^{-5} ft^2/s. In the figure, there are 500 ft of 12-in. and 2500 ft of 8-in. schedule 40 steel pipe. Fittings are as shown.

(a) Determine the flow rate in cfs, ignoring all losses.

(b) Determine the flow rate in cfs, with only pipe friction considered.

(c) Determine the flow rate in cfs, with all losses considered.

(d) Determine the gauge pressure in the 8-in. pipe at point 3, with losses considered. (The pipe lengths from the reservoir to point 3 are 500 ft of 12-in. and 1000 ft of 8-in.)

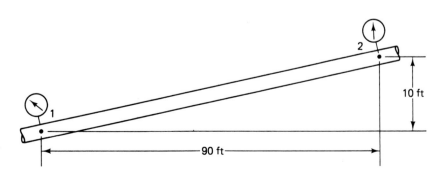

FIGURE 9-12 Sketch for Review Problem 9.15 (elevation).

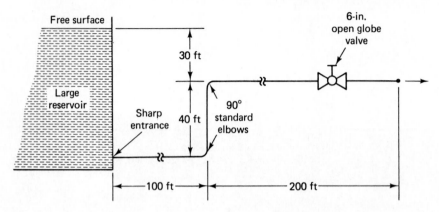

FIGURE 9-13 Sketch for Review Problem 9.16 (elevation).

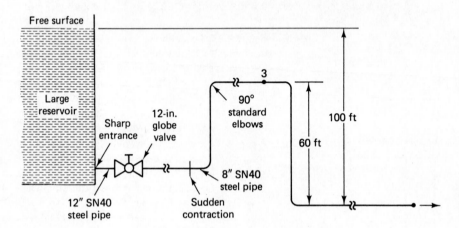

FIGURE 9-14 Sketch for Review Problem 9.17 (elevation).

BIBLIOGRAPHY

9.1.   *ASHRAE Handbook 1988 Equipment,* American Society of Heating, Refrigerating, and Air Conditioning Engineers, Atlanta, GA, 1988.

9.2.   *ASHRAE Handbook 1987 HVAC Systems and Applications,* American Society of Heating, Refrigerating, and Air Conditioning Engineers, Atlanta, GA, 1987.

9.3.   *ASHRAE Handbook 1989 Fundamentals,* American Society of Heating, Refrigerating, and Air Conditioning Engineers, Atlanta, GA, 1989.

9.4.   *Flow of Fluids Through Valves, Fittings, and Pipes,* (Technical Paper No. 410), Crane Company, Chicago, IL, 1976.

9.5.   *Pipe Friction Manual,* Hydraulic Institute, Cleveland, OH, 1961.

10

Pumps and Hydronic Heating Systems

10-1 INTRODUCTION

Chapter 9 was concerned with resistance to fluid flow in closed conduits and pipes and with a discussion of basic types of piping systems. When the fluid flowing in pipes is water, hot or cold, the systems are usually referred to as *hydronic systems*. In this chapter, we will concern ourselves with the actual design of hydronic systems and will determine the characteristics of the device that moves the water through the piping systems against system resistance. This device is called a *pump*. Generally, pumps are classified as *positive displacement* or *nonpositive displacement*.

The *reciprocating* piston pump is a positive-acting type, which means that it is a displacement pump and creates lift and pressure by displacing liquid with a piston moving in a cylinder. The chamber or cylinder is alternately filled and emptied by forcing and drawing the liquid by mechanical action. This type is called *positive* inasmuch as the only limitation on pressure that may be developed is the strength of the structural parts. Volume or capacity delivered is constant (except for slippage) regardless of pressure and is varied only by speed changes. This type of pump has limited use in the HVAC industry; however, one may find such pumps used for pumping hot and cold water, pneumatic pres-

sure systems, feeding small boilers, steam condensate return, gasoline and light-oil pumping, and small sprinkler irrigation jobs.

Another type of positive displacement pump is a *rotary* pump. The rotary pump is simple in design, has few moving parts, and, like a reciprocating pump, is positive acting. It consists primarily of two cams or gears, spur or herringbone, in mesh, the so-called idler gear driven by the "driving" gear, which is driven from an outside source of power. A close-fitting casing surrounds the cams and contains the suction and discharge connections. Liquid fills the spaces between the cam teeth and casing. The cams rotate carrying the trapped liquid, which is discharged (squeezed) out the discharge. Such operation produces a very even and continuous flow. As in the case of reciprocating pumps, the capacity delivered is constant (except for slippage) regardless of pressure. Due to the necessary close clearance and metal-to-metal contact, such units naturally work best and last longest when pumping liquids that have good lubricating qualities. These pumps are generally used where the capacity required is small but the pressure required may be medium to high. One particular application is the oil pump installed in an oil burner.

10-2 CENTRIFUGAL PUMPS

The *centrifugal pump* is by far the most frequently used pump in industry and, in particular, the HVAC industry. It is a nonpositive displacement pump, which means that its capacity is dependent on the fluid flow resistance offered by the system in which the pump is operating.

The name of this type of pump comes from the type of force exerted by a body moving in a circular path—centrifugal force. In such a pump, the fluid is forced to revolve, and it exerts a centrifugal force on the fluid in the case surrounding the revolving wheel (impeller). This force is a function of the impeller's peripheral velocity. Impeller rotation adds energy to the fluid after it enters the "eye" of the impeller. The casing collects the fluid as it leaves the impeller, reduces the fluid's absolute velocity, and guides the fluid out of the pump discharge. The fluid forced out of the casing creates a partial vacuum at the pump inlet, permitting atmospheric pressure to force more fluid into the pump suction, and the operation is continuous. There are numerous designs of impellers and casings; however, the principle of operation is the same.

The pressure energy added by the pump overcomes fluid friction caused by flow through heating equipment, piping, and valves and also raises the fluid to higher elevations in open piping systems.

Centrifugal pumps used in hydronic systems are single stage with a single- or double-entry impeller. Double-entry pumps are generally used for high-flow applications. Centrifugal pumps have either volute- or diffuser-type casings surrounding the impeller. The volute types include all pumps that collect the fluid from the impeller and discharge it perpendicular to the pump shaft. Diffuser-type casings collect the fluid from the impeller and discharge it parallel to the pump shaft. All centrifugal pumps that we will discuss here are of the volute type.

The features of the centrifugal pump that make it desirable for so many applications are simple construction, no close clearances, small size, low first cost, ease and low cost of maintenance, no excessive pressures even with the discharge valve closed, maximum practical suction lift (15 ft), smooth nonpulsating flow, quietness, no valves or reciprocating parts, high-speed operation, and capability of being belt driven, direct-motor driven, or turbine/engine driven.

The essential parts of a centrifugal pump are the rotating member (impeller) and the surrounding volute casing. Pumps are constructed of various metals, alloys, and other materials, depending on the fluid being pumped and its temperature. For our purposes here, we will limit our discussion to the materials commonly used in heating applications. The two general types are (1) bronze-fitted pumps and (2) all-bronze pumps.

Bronze-fitted pumps have a cast-iron pump body or volute and are equipped with a brass impeller, and the metal parts of the seal assembly are fabricated of brass or some other nonferrous material. These are used for circulating closed heating systems where little or no makeup water is supplied. These pumps hold up well for this type of service since the water in a closed system soon becomes chemically inert as far as corrosive action on the ferrous pump parts is concerned. These pumps are also used for pumping fresh water at relatively low temperatures since corrosive action on ferrous pump parts drops off sharply with reduction in temperature.

All-bronze pumps have the volute, impeller, and all other wetted parts made of bronze. The pump shaft is equipped with a bronze sleeve at the impeller end to prevent corrosion from affecting it at this point. The pump is used on higher-temperature fresh water where the bronze-fitted pumps would have their working parts and the cast-iron pump body affected by corrosion. Typical examples are for pumping recirculation lines for domestic hot water systems and pumping hot process water.

Any prime mover can be used to drive centrifugal pumps. Here, we are interested in the application of electric motors for this purpose. Electric motors with either sleeve or ball bearings are commonly used. Where quiet operation is essential, such as in a heating system, the sleeve bearing is preferable due to its quiet operation. Where operational noise is not a factor, the ball bearing is frequently used. Lubrication of these bearings in the proper manner is a necessity if normal operating life is to be expected. Normally, constant-speed motors are used at 1750 or 3450 rpm.

Mechanical Arrangement of Pump and Motor. The basic centrifugal pump and motor that we are discussing may be arranged as *(1) inline, (2) close-coupled,* or *(3) base-mounted.*

1. *Inline pumps* are normally used with fractional-horsepower electric motors up to about 3 hp operating at 1750 rpm. They are relatively lightweight and, because they are supported by the piping, are designed for extremely quiet operation. For this reason, sleeve bearings are used in both the motor and pumps. This type of pump is generally equipped with mechanical seals, and the motor is connected to the impeller shaft with a flexible coupling. Because of their quiet operation, these pumps are used in residences, small commercial buildings, and apartment buildings. Figure 10-1 shows an inline pump from one manufacturer.

2. *Close-coupled pumps,* like the one shown in

FIGURE 10-1   Iron and bronze booster pump. (Courtesy of ITT Fluid Transfer Division, Morton Grove, IL)

Figure 10-2a, are normally used with motors from 1/4 to 40 hp and at speeds of 1750 or 3450 rpm. Motors are furnished with sleeve or ball bearings, and the impeller is connected directly to the motor shaft. Seals may be either mechanical or packed. These pumps are somewhat less expensive than the base-mounted type (Figure 10-2b) and are used in larger systems where noise is not a factor. However, when quiet operation is required, an extra-quiet sleeve-bearing motor can be supplied. For high-temperature operation, cooling jackets are available.

   3.   *Base-mounted pumps,* like the one shown in Figure 10-2b, are normally used with motors from 1/4 hp on up and at speeds of 1750 or 3450 rpm. (Different speeds are obtained by using pulleys and a belt drive.) Motors are furnished with sleeve or ball bearings and connected to the impeller shaft by a flexible coupling.

10-3   HEAD DEVELOPED BY A CENTRIFUGAL PUMP

The pressure at any point in a liquid can be thought of as being caused by a vertical column of liquid that, due to its weight, exerts a pressure equal to the pressure at the point in question. The height of this column is called *static head* and is expressed in feet or meters. The static head corresponding to any specific pressure is dependent upon the weight of liquid according to the following relationship:

$$\text{head in feet} = \frac{\text{pressure in psi} \times 2.31}{\text{specific gravity}} \quad (10\text{-}1)$$

A centrifugal pump imparts velocity to a liquid (adds kinetic energy). This velocity energy is transformed largely into pressure energy in the diffuser as the liquid leaves the pump. Therefore, the head developed is approximately equal to the velocity energy at the periphery of the impeller. This relationship is expressed by the following well-known relationship:

$$\text{pump head } (h_p) = \frac{V_p^2}{2g} \quad (10\text{-}2)$$

where

$$h_p = \text{total developed pump head (ft)}$$
$$V_p = \text{linear velocity at periphery}$$
$$\text{of impeller (ft/s)}$$
$$g = 32.2 \text{ ft/s}^2$$

We may predict the approximate head developed by a centrifugal pump by calculating the peripheral velocity of the impeller and substituting into Eq. (10-2). A convenient formula for peripheral velocity is

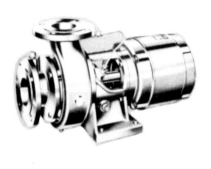

(a)

(b)

FIGURE 10-2   Centrifugal water pumps: (a) close-coupled and (b) base-mounted. (Courtesy of Taco, Inc., Cranston, RI)

$$V = \frac{\text{rpm} \times d}{229} \qquad (10\text{-}3)$$

where

V = peripheral velocity (ft/s)
d = impeller diameter (in.)

The preceding demonstrates why we must always think in terms of feet of liquid rather than pressure when working with centrifugal pumps. A given pump with a given impeller diameter and speed will raise a liquid to a certain height regardless of the weight of liquid, as shown in Figure 10-3.

10-4 SYSTEM HEAD ON A PUMP

Selection of the proper pump for a given application is of primary importance. To select a pump, it is necessary to know the required flow rate through the pump and the head or pressure against which the pump is to operate.

Equation (9-8) was written to represent the energy levels in a steady-flow fluid stream at two different points in the system. The equation may now be written to include the energy added to the fluid by the device we have been calling a pump and also the lost energy term that was discussed in Chapter 9:

$$\left(\frac{g}{g_c}\right)Z_1 + \frac{P_1}{\rho} + \frac{V_1^2}{2g_c}$$
$$= \left(\frac{g}{g_c}\right)Z_2 + \frac{P_2}{\rho} + \frac{V_2^2}{2g_c} + {}_1wk_2 + h_L \qquad (10\text{-}4)$$

Again, recall that each term in Eq. (10-4) has units of ft lbf/lbm or J/kg, which are energy units. The work term, ${}_1wk_2$, represents the energy added to the fluid by the pump, and the term, h_L, is the lost energy between points 1 and 2 of the system. Figure 10-4 shows a sketch of a steady-flow system where Eq. (10-4) applies. If each term in Eq. (10-4) is multiplied by the ratio g_c/g, we would have

$$\left(\frac{g}{g_c}\right)(Z_1)\left(\frac{g_c}{g}\right) + \frac{P_1}{\rho}\left(\frac{g_c}{g}\right) + \frac{V_1^2}{2g_c}\left(\frac{g_c}{g}\right)$$
$$= \left(\frac{g}{g_c}\right)(Z_2)\left(\frac{g_c}{g}\right) + \frac{P_2}{\rho}\left(\frac{g_c}{g}\right)$$
$$+ \frac{V_2^2}{2g_c}\left(\frac{g_c}{g}\right) + {}_1wk_2\left(\frac{g_c}{g}\right) + h_L\left(\frac{g_c}{g}\right)$$

Recalling that $\gamma = \rho(g/g_c)$ and that for incompressible flow $\gamma_1 = \gamma_2$, we would have

$$Z_1 + \frac{P_1}{\gamma} + \frac{V_1^2}{2g}$$
$$= Z_2 + \frac{P_2}{\gamma} + \frac{V_2^2}{2g} + h_p + h_L \qquad (10\text{-}5)$$

where $h_p = (g_c/g)({}_1wk_2)$ is called *pump head* and $h_L = (g_c/g)(h_L)$ is *lost head*. All terms in Eq. (10-5) now have units of ft lbf/lbf, or feet. Rearranging, combining terms, and changing signs, we have

$$h_P = (Z_2 - Z_1) + \left(\frac{P_2 - P_1}{\gamma}\right)$$
$$+ \left(\frac{V_2^2 - V_1^2}{2g}\right) + h_L \qquad (10\text{-}6)$$

This arrangement gives a positive value for pump head for convenience. The various combined terms of Eq. (10-6) are frequently referred to as follows:

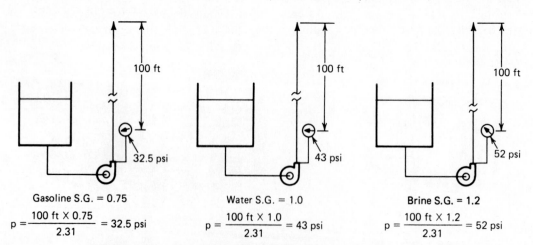

FIGURE 10-3 Identical pumps handling liquids of different specific gravities. (Adapted from *Goulds Pump Manual,* Goulds Pumps, Inc., Seneca, NY, by permission)

$(Z_2 - Z_1)$ = net static elevation head

$\left(\dfrac{P_2 - P_1}{\gamma}\right)$ = net static pressure head

$\left(\dfrac{V_2^2 - V_2^1}{2g}\right)$ = net velocity head

In Figure 10-4, an arbitrary datum was selected in such a way that both Z_1 and Z_2 were positive (above datum). It would also have been completely satisfactory to have the datum passing through the centerline of the pump. This is the usual practice to evaluate pump head. To determine the *total dynamic head* (TDH) against which a pump must operate, certain terms and definitions are used. Referring to Figure 10-5, these definitions are as follows.

Suction lift: Exists when the source of supply is below the centerline of the pump. Thus, the *static suction lift* is the vertical distance in feet from the centerline of the pump to the free level of the liquid to be pumped.

Suction head: Exists when the source of supply is above the centerline of the pump. Thus, the *static suction head* is the vertical distance in feet from the centerline of the pump to the free level of the liquid to be pumped.

Static discharge head: The vertical distance in feet between the pump centerline and the point of free discharge or the surface of the liquid in the discharge tank.

Total static head: The vertical distance in feet between the free level of the source of supply and the point of free discharge or the free surface of the discharge liquid.

Friction head (h_L): The head required to overcome the resistance to flow in the pipe and fittings. It is

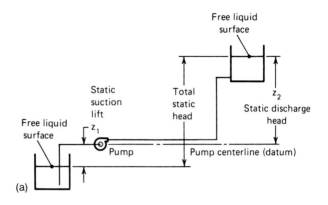

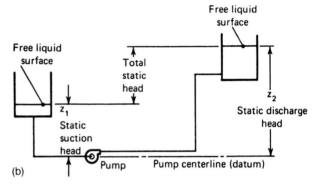

FIGURE 10-5 General pumping system (a) showing suction lift and (b) showing suction head (elevation; not drawn to scale).

dependent upon the size and type of pipe, flow rate, and nature of the liquid (see Chapter 9).

Velocity head (h_v): The energy of the liquid as a result of its motion at some velocity V. It is the equivalent head in feet through which the fluid would have to fall to acquire the same velocity; or, in other words, the head necessary to accelerate the water ($h_v = V^2/2g$) (see Chapter 9). The velocity head is usually insignificant and can be ignored in most high-head systems. However, it can be a large factor and must be considered in low-head systems.

Pressure head: Must be considered when a pumping system either begins or terminates in a tank that is under some pressure other than atmospheric. The pressure in such a tank must first be converted into feet of liquid, Eq. (10-1). A vacuum in the suction tank or a positive pressure in the discharge tank must be added to the system head, whereas a positive pressure in the suction tank or vacuum in the discharge tank would be subtracted from the system head. Conversion of inches of mercury vacuum to feet of liquid may be accomplished by

$$\text{feet of liquid} = \frac{\text{vacuum, in. Hg} \times 1.13}{\text{SG}}$$

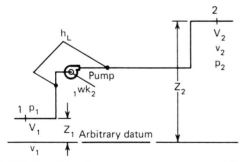

FIGURE 10-4 General pumping system (elevation; not drawn to scale).

The preceding forms of head, namely, static, friction, velocity, and pressure, are combined to make up the total system head at any particular flow rate. Following are definitions of these combined or dynamic head terms as they apply to the pump.

Total dynamic suction lift (TDSL): The static suction lift plus the velocity head at the pump suction flange plus the total friction head in the suction line. The TDSL, as determined on a pump test, is the reading of a gauge on the suction flange, converted to feet of liquid and corrected to the pump centerline, minus the velocity head at the point of gauge attachment.

Total dynamic suction head (TDSH): The static suction head minus the velocity head at the pump suction flange minus the total friction head in the suction line. The total dynamic suction head, as determined on a pump test, is the reading of a gauge on the suction flange, converted to feet of liquid and corrected to the pump centerline, plus the velocity head at the point of gauge attachment.

Total dynamic discharge head (TDDH): The static discharge head plus the velocity head at the pump discharge flange plus the total friction head in the discharge line. The total dynamic discharge head, as determined on a pump test, is the reading of a gauge at the discharge flange, converted to feet of liquid and corrected to the pump centerline, plus the velocity head at the point of gauge attachment.

Total head or total dynamic head (TDH): The total dynamic discharge head minus the total dynamic suction head or plus the total dynamic suction lift:

$$h_p = \text{TDH} = \text{TDDH} - \text{TDSH}$$
$$\text{(with a suction head)}$$

or

$$h_p = \text{TDH} = \text{TDDH} + \text{TDSL}$$
$$\text{(with a suction lift)}$$

10-5   NET POSITIVE SUCTION HEAD (NPSH) AND CAVITATION

The Hydraulic Institute defines *net positive suction head* (NPSH) as the total suction head in feet of fluid absolute, determined at the suction nozzle of the pump and corrected to datum, less the vapor pressure of the liquid in feet absolute. Simply stated, it is an analysis of energy conditions on the suction side of a pump to determine whether the liquid will vaporize at the lowest pressure point in the pump.

The pressure that a liquid exerts on its surroundings is dependent upon its temperature. This pressure, called vapor pressure, is a unique characteristic of every fluid and increases with increasing temperature. When the vapor pressure within the fluid reaches the pressure of the surrounding medium, the fluid begins to vaporize or boil. The temperature at which this vaporization occurs will decrease as the pressure of the surrounding medium decreases.

A liquid increases greatly in volume when it vaporizes. One cubic foot of water at room temperature becomes 1700 ft^3 of vapor at the same temperature. It is obvious from this that if we are to pump a liquid effectively, we must keep it in liquid form. NPSH is simply a measure of the amount of suction head present to prevent this vaporization at the lowest pressure point in the pump.

NPSH required (NPSH_R) is a function of the pump design. As the liquid passes from the pump suction to the eye of the impeller, the velocity increases and the pressure decreases. There are also pressure losses due to shock and turbulence as the liquid strikes the impeller. The centrifugal force of the impeller vanes further increases the velocity and decreases the pressure of the liquid. The NPSH_R is the positive head in feet absolute required at the pump suction to overcome these pressure drops in the pump and maintain the liquid above its vapor pressure. The NPSH_R varies with speed and capacity within any given pump. Pump manufacturers' performance curves normally provide this information.

NPSH available (NPSH_A) is a function of the system in which the pump operates. It is the excess pressure of the liquid in feet absolute over its vapor pressure as it arrives at the pump suction. The NPSH_A may be calculated for any system by

$$\text{NPSH}_A = \frac{P_B - P_v}{\gamma} + (S - h_L) \qquad (10\text{-}7)$$

where

P_B = pressure on the liquid surface in the suction tank (lbf/ft² absolute)

P_v = vapor pressure of the liquid at the liquid temperature (lbf/ft² absolute)

γ = specific weight of the liquid (lbf/ft³)

S = vertical distance (ft) from pump centerline to level of liquid in the suction tank. (S is positive if level is above pump centerline; negative if below)

h_L = total friction loss in pump suction pipe and fittings (ft of fluid)

In an existing system, the NPSH_A can be determined by a gauge reading on the pump suction. The following relationship may be used:

$$\text{NPSH}_A = \frac{P_B - P_v}{\gamma} \pm \text{GR} + h_v \qquad (10\text{-}8)$$

where

GR = gauge reading at pump suction expressed in feet (plus if above atmospheric; minus if below atmospheric) corrected to the pump centerline

h_v = velocity head in the suction pipe at the gauge connection (ft)

Cavitation is a term used to describe the phenomenon that occurs in a pump when there is insufficient NPSH_A. The pressure of the liquid is reduced to a value equal to or below its vapor pressure, and small vapor bubbles or pockets begin to form. As these vapor bubbles move along the impeller vanes to a higher-pressure area, they rapidly collapse. The collapse, or implosion, is so rapid that it may be heard as a rumbling noise. The forces during the collapse are generally high enough to cause minute pockets of fatigue failure on the impeller vane surfaces. This action may be progressive and, under severe conditions, can cause serious pitting damage to the impeller. The accompanying noise is the easiest way to recognize cavitation. Besides impeller damage, cavitation normally results in reduced capacity due to vapor present in the pump. Also, the head may be reduced and unstable, and the power consumption may be erratic. Vibration and mechanical damage such as bearing failure can also occur as a result of operating in cavitation. The only way to prevent the undesirable effects of cavitation is to ensure that NPSH_A in the system is greater than NPSH_R by the pump.

Suction Limitations. The importance of keeping within the suction limitations of *any* pump (centrifugal, rotary, piston) cannot be emphasized too greatly. A pump, by creating a vacuum at the suction (impeller eye on a centrifugal), utilizes atmospheric pressure to push the liquid into the pump. Because of this, the suction lift is limited theoretically to 33.9 ft of water maximum [(14.7)(2.31)/1.0]. Internal pump losses reduce this limitation to a practical value of approximately 15 ft. The dynamic suction lift should be calculated carefully at the required capacity to make sure that it is within the pump's capabilities. Even systems taking suction from a source above the pump can cause trouble when friction losses are too great. Always keep the pump as close to the liquid source as possible. Many pump performance curves will show the maximum practical dynamic suction lifts for a given pump or for given capacities from that same pump. Since the limitation is based on internal pump losses also, it can be seen that in any given pump the recommended suction lift is reduced as flow increases.

NPSH is a pump term that has limited usage for closed hydronic heating systems. It is of great importance in open-circuit steam condensate return systems and in open-circuit industrial applications where low suction pressures may be encountered. Pump manufacturers provide NPSH_R curves for specific pumps.

ILLUSTRATIVE PROBLEM 10-1

Figure 10-6 represents an elevation drawing of a pumping system. The pump receives water at 100°F from a storage reservoir and transfers the water through a shell-and-tube heat exchanger to the inlet of a closed storage tank where the pressure is 20 psig. The flow rate required is 99,400 lb/hr. Determine (1) pump capacity in gpm, (2) pump total dynamic head, and (3) available net positive suction head for the pump. (Assume commercial steel schedule 40 pipe.)

Solution:
1. Determine flow rate gpm. Since the flow rate is given in lb/hr, the specific gravity must be calculated at 100°F and at 180°F to determine flows in gpm:

$$\text{SG}_{100} = \frac{\rho_{100}}{\rho_{60}} = \frac{v_{60}}{v_{100}} = \frac{0.016035}{0.016130} = 0.994$$

$$\text{SG}_{180} = \frac{v_{60}}{v_{180}} = \frac{0.016035}{0.016509} = 0.971$$

By Eq. (10-11),

$$\text{gpm}_{100} = \frac{\text{lb/hr}}{500 \times \text{SG}}$$

$$= \frac{99,400}{(500)(0.994)} = 200 \text{ gpm}$$

$$\text{gpm}_{180} = \frac{99,400}{(500)(0.971)} = 204.7 \text{ gpm}$$

This shows a relatively small change in flow rate with the temperatures involved. Therefore, we will assume a flow rate constant throughout the system of 200 gpm. As far as the pump is concerned, it handles 200 gpm of 100°F water.

2. Determine pump total dynamic head.
   (a) Determine velocity of flow in the pump suction pipe and pump discharge pipe, and velocity heads.

$$3\text{-in. NPS, ID} = 3.068 \text{ in.}$$
$$2\tfrac{1}{2}\text{-in. NPS, ID} = 2.469 \text{ in.}$$

By Eq. (9-2),

$$V_{3\text{-in.}} = \frac{\text{gpm} \times 0.408}{d^2}$$

$$= \frac{(200)(0.408)}{(3.068)^2} = 8.69 \text{ ft/s}$$

$$V_{2\frac{1}{2}\text{-in.}} = \frac{(200)(0.408)}{(2.469)^2} = 13.39 \text{ ft/s}$$

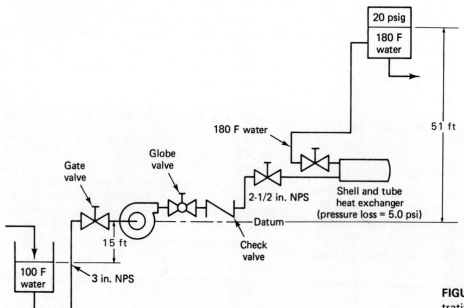

FIGURE 10-6 Diagram for Illustrative Problem 10-1.

suction velocity head $= \dfrac{V^2}{2g}$

$\qquad = \dfrac{(8.69)^2}{2 \times 32.2} = 1.2$ ft

discharge velocity head $= \dfrac{(13.39)^2}{2 \times 32.2} = 2.8$ ft

(b) Determine pressure heads. By Eq. (10-1),

suction pressure head $= \dfrac{\text{psi} \times 2.31}{\text{SG}}$

$\qquad = \dfrac{(0)(2.31)}{0.944} = 0$ ft gauge

discharge pressure head $= \dfrac{(20)(2.31)}{0.971} = 47.6$ ft gauge

(c) Determine static elevation heads.

suction = 15 ft below pump centerline to lowest level in tank (lift)
discharge = 51 ft above pump centerline to water level in discharge tank

(d) Determine losses in pipe and fittings. By Eq. (10-1),

heater head loss $= \dfrac{\text{psi} \times 2.31}{\text{SG}}$

average specific gravity of water in heater $= \dfrac{0.994 + 0.971}{2} = 0.983$

heater head loss $= \dfrac{(5.0)(2.31)}{0.983} = 11.75$ ft

Suction pipe and fitting loss (refer to Chapter 9): From Table 9-1, $f_t = 0.018$ for 2 1/2- or 3-in. pipe, assuming complete turbulence.

gate valve $K_1 = 8f_t = 8(0.018) = 0.144$

90° elbow $K_1 = 30f_t = 30(0.018) = 0.54$

pipe entrance $K = 0.5$

From Figure 9-7,

gate valve = 2 ft
3 90° elbows = 24 ft
pipe entrance = 7 ft
length of pipe = 38 ft (approximate)
total equivalent length = 71 ft (approximate)

By Eq. (9-8),

pipe loss suction $(h_L) = f_t \left(\dfrac{L}{D}\right)\left(\dfrac{V^2}{2g}\right)$

$h_L = (0.018)\left(\dfrac{71}{3.068/12}\right)\left[\dfrac{(8.69)^2}{2 \times 32.2}\right]$

$\qquad = 5.86$ (say, 6) ft

Discharge pipe and fitting loss (from Table 9-1):

globe valve $K_1 = 340f_t = 340(0.018) = 6.12$
check valve $K_1 = 100f_t = 100(0.018) = 1.80$
gate valve $K_1 = 8f_t = 8(0.018) = 0.144$
pipe exit $K = 1.0$
elbow $K = 30f_t = 30(0.018) = 0.54$

From Figure 9-7,

$$\begin{aligned}
\text{globe valve} &= 70 \text{ ft} \\
\text{check valve} &= 21 \text{ ft} \\
2 \text{ gate valves} &= 1.8 \text{ ft} \\
\text{pipe exit} &= 11.0 \\
6 \text{ elbows} &= 36.0 \\
\text{length of pipe} &= 119.0 \text{ (approximate)} \\
\text{total equivalent length} &= 258.8 \text{ (say, 259) ft}
\end{aligned}$$

By Eq. (9-8),

$$\text{pipe loss discharge } (h_L) = f_t \left(\frac{L}{D}\right)\left(\frac{V^2}{2g}\right)$$

$$h_L = (0.018)\left(\frac{259}{2.469/12}\right)\left[\frac{(13.39)^2}{2 \times 32.2}\right]$$

$$= 63.08 \text{ (say, 63) ft}$$

$$\begin{aligned}
\text{TDSL} &= \text{static suction lift} + \text{pressure head} \\
&\quad + \text{velocity head} + \text{losses} \\
&= 15 + 0 + 1.2 + 6 \\
&= 22.2 \text{ ft gauge}
\end{aligned}$$

$$\begin{aligned}
\text{TDDH} &= \text{static discharge head} + \text{pressure} \\
&\quad \text{head} + \text{velocity head} + \text{losses} \\
&= 51 + 47.6 + 2.8 + 11.75 + 63 \\
&= 176.2 \text{ ft}
\end{aligned}$$

$$\begin{aligned}
\text{pump TDH} &= \text{TDDH} + \text{TDSL} \\
&= 176.2 + 22.2 \\
&= 198.4 \text{ (say, 198) ft} \qquad (ans.)
\end{aligned}$$

Or by Eq. (10-6),

$$h_p = (Z_2 - Z_1) + \left(\frac{P_2 - P_1}{\gamma}\right)$$

$$\qquad + \frac{V_2^2 - V_1^2}{2g} + h_L$$

$$= [51 - (-15)] + \left[\frac{144(20 - 0)}{1/0.016509}\right]$$

$$+ \left[\frac{(13.39)^2 - (8.69)^2}{2 \times 32.2}\right] + (6 + 11.75 + 63)$$

$$= 66 + 47.6 + 1.6 + 80.75$$

$$= 195.9 \text{ (say, 196) ft}$$

3. Determine net positive suction head available. By Eq. (10-7),

$$\text{NPSH}_A = \frac{P_B - P_v}{\gamma} + (S - h_L)$$

Assume the following:

$$\begin{aligned}
P_B &= 14.7 \text{ psia} \\
P_v &= \text{vapor pressure at } 100°F \\
&= 0.9503 \text{ psia}
\end{aligned}$$

$$\begin{aligned}
\gamma &= \text{specific weight at } 100°F \\
&= 1/v = 1/0.01613 = 62 \text{ lb/ft}^3 \\
S &= -15 \text{ ft (lift)} \\
h_L &= 6 \text{ ft (suction loss)}
\end{aligned}$$

Then,

$$\text{NPSH}_A = \frac{(14.7 - 0.9503)(144)}{62.0} - 15 - 6$$

$$= 10.9 \text{ ft (this must be equal to or greater than NPSH}_R \text{ by pump)}$$

10-6   SPECIFIC SPEED AND PUMP TYPE

The concept of specific speed (N_s) has been developed for design purposes to show the relationship of head, capacity, and speed. Specific speed of an impeller may be defined as *the speed in rpm at which a geometrically similar impeller would operate to develop 1 ft of head when displacing 1 gpm.* It is customary to give the specific speed of an impeller for design conditions. Since specific speed is a function of impeller proportion, it is constant for any group of impellers having similar angles and dimensions. All impellers with the same specific speed would have the same efficiency except for variations due to viscosity of the fluids.

There are two steps in the derivation of the specific speed formula: (1) The speed of the impeller must be changed at constant diameter to reduce the head to 1 ft, and (2) the dimensions must be changed at constant head or velocity to get a capacity of 1 gpm.

For a constant impeller diameter, flow (Q) will vary as the speed (N), and the head (h) will vary as the square of the speed:

$$\frac{N_2}{N_1} = \frac{Q_2}{Q_1} = \sqrt{\frac{h_2}{h_1}} \qquad (a)$$

Taking the subscript 1 to indicate the original impeller and subscript 2 to indicate the conditions after securing a head of 1 ft,

$$N_2 = N_1 \sqrt{\frac{h_2}{h_1}} \qquad \text{and} \qquad Q_2 = Q_1 \sqrt{\frac{h_2}{h_1}} \qquad (b)$$

For the second step, we must change the physical size of the impeller to reduce Q_2 to a flow of 1 gpm (Q_3) but at the same time maintain the 1-ft head. This is the same as maintaining the same peripheral velocity and the same velocity at the impeller inlet. The volume of flow will vary as area and the square of the diameter. However, if the diameters are reduced, the speed must be increased to maintain an impeller velocity that will give the desired 1 ft of head. Therefore, at constant head,

$$\frac{Q_2}{Q_3} = \left(\frac{d_3}{d_3}\right)^2 = \left(\frac{N_3}{N_2}\right)^2 \qquad (c)$$

where subscript 2 has the same significance as in Eq. (b) and the subscript 3 indicates conditions after conversion to 1 gpm. Substituting Eq. (b) into Eq. (c), we have

$$N_3 = N_2 \sqrt{\frac{Q_2}{Q_3}} = N_1 \sqrt{\frac{h_2}{h_1}} \sqrt{Q_1} \sqrt{\frac{h_2}{h_1}}$$

Remember that $Q_3 = 1$ gpm and $h_2 = 1$-ft head. N_3 is the specific speed, Q is measured in gpm, and h is the head in feet for *one stage*. Q for a double-suction impeller is taken as one-half the total flow for the impeller. Therefore,

$$N_s = \frac{N\sqrt{Q}}{(h)^{3/4}} \qquad (10\text{-}9)$$

where

N = impeller speed (rpm)
Q = gpm for pump at best efficiency point
h = total head per stage at best efficiency point

The specific speed determines the general shape or class of the impeller, as shown in Figure 10-7. As the specific speed increases, the ratio of the impeller outlet diameter, D_2, to the inlet or eye diameter, D_1, decreases. This ratio becomes 1.0 for a true axial-flow impeller.

Radial-flow impellers develop head principally through centrifugal force. Pumps of higher specific speeds develop head partly by centrifugal force and partly by axial force. A higher specific speed indicates a pump design with head generation more by axial forces and less by centrifugal forces. An axial-flow or propeller pump with a specific speed of 10,000 or greater generates its head exclusively through axial forces.

Generally, impellers with low specific speeds are used where head requirements are high and capacity requirements are low. Higher specific speed impellers are for low head requirements and higher capacities.

10-7 CENTRIFUGAL PUMP OPERATING CHARACTERISTICS AND PERFORMANCE

We are concerned primarily with capacity, head, power, and efficiency of centrifugal pumps. The flow rate through a pump (capacity) may be expressed as mass per unit time (lbm/hr) or volume flow rate (cfs); however, in practice, it is normally volume flow rate in gallons per minute (gpm).

Equation (10-4) expressed the energy added (work) to a unit mass of fluid. The total work done per unit of time on the fluid would be found by multiplying by the total mass flow per unit of time. Work per unit of time is power. Therefore, we may say that power added to the fluid passing through a pump is a function of the mass flow rate per unit of time and the total head developed, or

$$\text{WHP} = \frac{\dot{m}_w \times \text{TDH}}{33{,}000 \times 60} \qquad (10\text{-}10)$$

where

WHP = horsepower added to the fluid by the pump (frequently called water horsepower, or hydraulic horsepower)
$\dot{m}_w$ = mass flow of fluid (lb/hr)

Values of Specific Speed, N_s

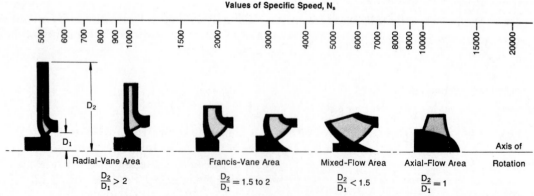

FIGURE 10-7 Impeller design versus specific speed. (From *Goulds Pump Manual*, Goulds Pumps, Inc., Seneca, NY, reproduced with permission)

TDH = total dynamic head (ft of fluid)

33,000 = ft-lb$_f$/min. per horsepower

Pump capacity in gpm and liquid specific gravity are normally used in formulas rather than the mass of fluid. Pounds per hour is converted to gpm by

$$\text{gpm} = \frac{\dot{m}_w}{500 \times \text{SG}} \qquad (10\text{-}11)$$

where 500 is the product of 60 min./hr and 8.33 lb/gal of water at 60°F. Substituting Eq. (10-11) into Eq. (10-10) gives

$$\text{WHP} = \frac{\text{gpm} \times \text{TDH} \times \text{SG}}{3960} \qquad (10\text{-}12)$$

The brake horsepower (BHP) or input power to a pump is greater than the water horsepower due to mechanical and fluid losses incurred in the pump. The pump efficiency is the ratio of WHP/BHP, or

$$\eta_p = \frac{\text{gpm} \times \text{TDH} \times \text{SG}}{3960 \times \text{BHP}} \qquad (10\text{-}13)$$

Although centrifugal pumps can be selected from rating charts (tables), performance curves give a much clearer picture of the characteristics at a given speed. Various forms of curves are used; however, the typical characteristic curve shows the total dynamic head, brake horsepower, efficiency, and net positive suction head all plotted over the capacity range of the pump. Figures 10-8 through 10-10 are nondimensional curves that indicate the general shape of the characteristic curves for the various types of pumps. They show the head, brake horsepower, and efficiency plotted as a percent of their values at the design or best efficiency point of the pump.

Figure 10-8 shows that the head curve for a radial-flow pump is relatively flat, and the head decreases gradually as the flow increases. Note that the brake

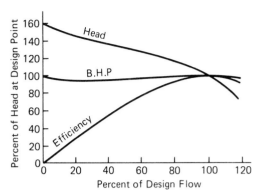

FIGURE 10-9 Mixed-flow pump.

horsepower increases gradually over the flow range with the maximum normally at the point of maximum flow.

Mixed-flow centrifugal pumps and axial-flow or propeller pumps have considerably different characteristics as shown in Figures 10-9 and 10-10. The head curve for a mixed-flow pump is steeper than for a radial-flow pump. The shutoff head is usually 150% to 200% of the design head. The brake horsepower remains fairly constant over the flow range. For a typical axial-flow pump, the head and brake horsepower both increase sharply near the shutoff as shown in Figure 10-10.

The distinction between the above-mentioned three classes of pumps is not absolute, and there are many pumps with characteristics falling somewhere between the three shown. For instance, the Francis vane impeller would have characteristics between the radial- and mixed-flow classes. Most turbine pumps are also in this range depending upon their specific speeds.

Figure 10-11 shows typical characteristic curves resulting from actual test on a radial-flow centrifugal pump operating at constant speed and for a specific impeller diameter. On the test at 200 gpm the gauges indicated a total head of 90 ft and 6.6 BHP was required. At 120 gpm the gauges indicated a total head of 120 ft and 6.1 BHP was required, and so on, for the complete test. Drawing curves through the various plotted points from test data gives the head curve and BHP curve. The

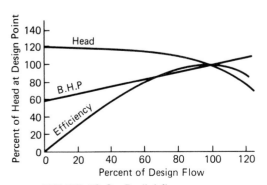

FIGURE 10-8 Radial-flow pump.

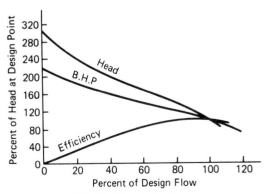

FIGURE 10-10 Axial-flow pump.

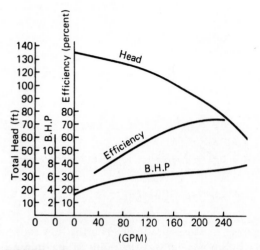

FIGURE 10-11 Typical performance curves for a centrifugal pump.

efficiency curve is next determined by calculating the efficiency using Eq. (10-13), at various flow rates. The efficiency curve results when the calculated efficiencies are plotted and joined by a curved line.

The curves of Figure 10-11 show what the pump will do with one specific impeller diameter. However, centrifugal pumps are flexible in that one casing or volute can be used with various impeller diameters cut down from the maximum. To make separate curves of several impeller diameters of each pump would result in multiplicity of sheets, which would be awkward and confusing. Therefore, a curve as shown in Figure 10-12 has been evolved and tells at a glance what the pump will do at a specified speed with various impeller diameters, from the maximum to the minimum. Each impeller is actually tested, and separate test curves are made up similar to Figure 10-11. All the head curves are then correlated on a single sheet. It can readily be seen, however, that if an attempt was made to include also all the horsepower and efficiency curves, there would be such a conglomeration of lines that it would be impossible to pick out the individual curves. This is greatly simplified by taking several points of efficiency on each of the head curves and connecting them to make a composite efficiency curve; the same is done with the horsepower curves. For instance, in Figure 10-12, 70% efficiency was obtained with a 5 7/8-in. impeller at 161 gpm and 119-ft head and at 302 gpm and 82-ft head. The same

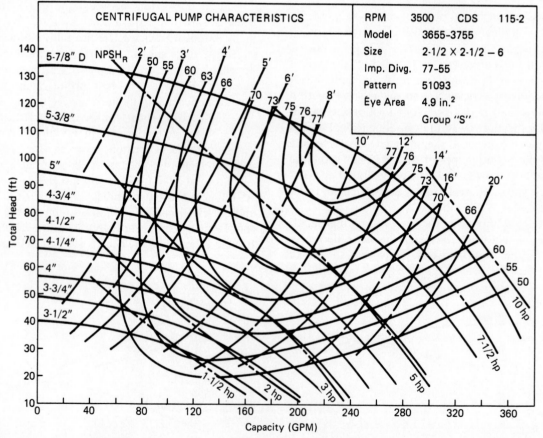

FIGURE 10-12 Centrifugal pump characteristics. (Courtesy of Goulds Pumps, Inc., Seneca, NY)

efficiency was obtained with a 5 3/8-in. impeller at 150 gpm and 98-ft head and at 264 gpm and 68 ft-head. Connecting all the points of the same efficiency results in the various constant efficiency curves.

This is done similarly to make the various horsepower curves. Note that, arbitrarily, specific diameter impellers have been tested. This is done as a check but does not mean that intermediate diameter impellers cannot be furnished. For instance, if a particular application should require 180 gpm at 105-ft head, referring to Figure 10-12, the impeller would be cut to approximately 5 1/2-in. in diameter, the efficiency would be approximately 74%, and a 7 1/2-hp motor would be required.

Referring to the horsepower curves of Figure 10-12, whenever a point of operation falls above a certain horsepower curve, use the next larger horsepower motor. Note that in a centrifugal pump the capacity increases as the head falls, with the resultant greater horsepower requirement. If on a particular application the head in feet should drop for any reason, use the next larger horsepower motor to be safe if the rating point is too close to the motor rating curve. As an example, 120 gpm at 103-ft head is required. At the rating point on Figure 10-12, it shows a 5.0-hp motor is adequate. However, it is possible the head may drop to 96 ft. At this point, the capacity goes to 160 gpm; also this causes an increase in power to above 5.0 hp. Therefore, a 7 1/2-hp motor should be selected.

Figure 10-11 shows the performance curves for a specific impeller diameter of a centrifugal pump operating at constant speed. Figure 10-12 gives the performance of a particular pump at constant speed but with various impeller diameters. However, a particular design of pump frequently consists of a multiplicity of pump sizes, and, for selecting a pump for a specific application, a composite curve of the entire line in question is the quickest and simplest method.

The practical usable limit of a centrifugal pump with its various impeller diameters is approximately inside the boundary efficiency curves (center area of Figure 10-12). This does not cover the entire range of the pump because the area outside this boundary is better covered by other sizes of the same line of pumps. If such an area of ratings were taken for each size of a pump line, a composite of the entire line could be made. Such a composite is shown in Figure 10-13, with the heavy line denoting the pump size of Figure 10-12. (Vertical scale in Figure 10-13 is reduced, which accounts for the difference in appearance of these identical curves.) On such a composite, it is impractical to try to show efficiencies, but, for the purpose of ease of selection, such a composite gives information concerning gpm, pump head, and impeller size. Curves such as shown in Figures

10-11 and 10-12 are always available for each pump size. Such curves may be referred to after a pump has been selected in order to obtain pump efficiency, impeller sizes, and so forth.

Several factors govern the rating of a centrifugal pump. Generally, these factors include the following:

- *Speed* (rpm)—The higher the speed, the higher the capacity and head.
- *Impeller diameter*—The larger the diameter, the higher the head in feet.
- *Impeller width*—The wider the impeller, the higher the capacity in gpm.

In a multistage pump, the head builds up in each stage, taking advantage of the head developed by each preceding impeller. In other words, the total head developed by a three-stage pump, where each impeller is rated at 100 ft, would be 100 × 3 or 300 ft total, the capacity remaining the same, regardless of the number of stages.

Within limits (assuming constant efficiency), it is possible to calculate pump characteristics at different operating speeds or different impeller diameters. These relationships are commonly referred to as *pump laws*.

The fundamental formulas for calculation of *variable speed performance* were developed in the section on specific speed (Section 10-6); that is,

$$\frac{N_2}{N_1} = \frac{Q_2}{Q_1} = \left(\frac{h_2}{h_1}\right)^{1/2} = \left(\frac{HP_2}{HP_1}\right)^{1/3} \quad (10\text{--}14)$$

for *constant fluid temperature and constant pump efficiency*. Any convenient units, such as gallons per minute or pounds per hour may be used for Q and feet or pounds per square inch may be used for head h, providing the ratios are dimensionless. Horsepower, HP, has been added to the formula previously given. The fact that the horsepower varies as the cube of the speed can be readily understood when it is remembered that horsepower is the product of quantity and head. Quantity varies as the speed, and head varies as the square of the speed.

Note that there is a variation in *both head and capacity* for a variation in speed. For the typical coordinates of a centrifugal pump performance curve (*h–Q* curve), the curve of the relation of these two would be a family of square curves all passing through zero–zero. Any of the points on one of these square curves theoretically would have the same efficiency and are sometimes referred to as "corresponding points."

To demonstrate the procedure to follow in making variable-speed calculations, the *h–Q* curve for the pump with a $5\frac{7}{8}$-in. impeller shown in Figure 10-12 for a speed

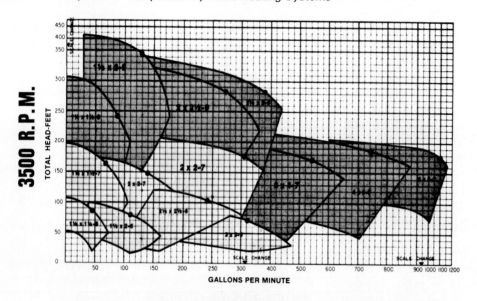

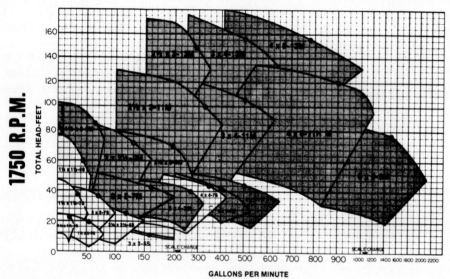

FIGURE 10-13 Typical composite centrifugal pump curves for various pump sizes. (From *Goulds Pump Manual,* reproduced courtesy of Goulds Pumps, Inc., Seneca, NY)

of 3000 rpm will be calculated. From the figure, find rpm = 3500, and select h_1 = 100 ft with Q_1 = 258 gpm; rpm$_2$ = 3000. Substitution into Eq. (10-14) yields Q_2 = 221 gpm and h_2 = 73.5 ft. The corresponding point on the 3000-rpm curve is at 221 gpm at 73.5 ft. By a similar procedure, other points on the 3000-rpm curve may be calculated.

Instead of calculating all the possible head curves at all speeds, a procedure similar to that above may be used to determine the operating conditions for one point.

ILLUSTRATIVE PROBLEM 10-2

The head-capacity curves for a centrifugal, constant-speed pump are shown in Figure 10-12. Assume that with the $5\frac{7}{8}$-in. impeller unit a variable-speed drive may be used. We wish to

determine the speed of the impeller to produce a flow of 240 gpm against a head of 95 ft. What should be the speed of this pump?

Solution: From Eq. (10-14), we may see that h = constant $\times Q^2$, or constant = h/Q^2. For this problem, the desired constant is found by using h = 95 ft and Q = 240 gpm, or constant = $95/(240)^2$ = 1.65×10^{-3}.

Next, we may select, as a trial, a value of head and corresponding capacity on the 3500-rpm curve, and, if it produces the same constant as above, it is a corresponding point.

Trial 1. On the 3500-rpm curve, select 250 gpm at 102-ft head. Then, constant = $102/(250)^2$ = 1.63×10^{-3}. This is slightly low.

Trial 2. Select 245 gpm at 103 ft. Then, constant = $103/(245)^2$ = 1.72×10^{-3}. This is too high.

The number of trials taken depends on accuracy desired and the accuracy of the performance graph. We will use 250 gpm at 102-ft head as the corresponding point. By Eq. (10-14),

$$N_2 = N_1\left(\frac{Q_2}{Q_1}\right) = 3500\left(\frac{240}{250}\right) = 3360 \text{ rpm}$$

Therefore, the pump speed should be 3360 rpm to cause a flow of 240 gpm against a head of 95 feet.

Head-capacity curves for small changes in the diameter of a given impeller may be predicted at constant speed and constant temperature from the formula

$$\frac{d_1}{d_2} = \frac{\text{gpm}_1}{\text{gpm}_2} = \left(\frac{h_1}{h_2}\right)^{1/2} = \left(\frac{HP_1}{HP_2}\right)^{1/3} \quad (10\text{-}15)$$

which is derived from the specific speed relations.

Chapter 9 and the preceding sections have presented flow of fluids in piping, general piping systems, and pumps. Much work has been done in past years to simplify the required calculations for sizing pipe and for selecting pumps in the design of forced-circulation water heating systems. Although the simplifications introduce minor inaccuracies in our calculations, much time is saved, which reduces the engineering costs. We will now proceed with accepted procedures for the design of forced-circulation water heating systems.

10-8 WATER SYSTEM DESIGN

Heating systems using hot water as a means of carrying heat to points of utilization are called *hot-water heating systems,* or *hydronic systems.* In hydronic systems, water, usually under static pressure, is heated in a heat exchanger (boiler or shell-and-tube unit) and distributed through a system of pipes either to the various terminal heat-transfer units (THTU) located in the conditioned spaces or to a process. Then, the water, reduced in temperature, is returned to the heat exchanger for reheating.

Water systems can be classified by the (1) flow generation, (2) temperature, (3) pressurization, (4) piping arrangement, and (5) pumping arrangement.

Flow Generation. There are two types of hot-water heating systems classified by flow generation: (1) the *gravity* system, which uses the difference in water density between the supply and return water columns of a circuit or system to circulate water and (2) the *forced* system, in which a pump, usually driven by an electric motor, maintains the flow. Gravity systems are seldom used at present.

Water systems are either once-through or recirculating systems. We are concerned in this chapter with *forced-recirculating systems,* or, simply, *forced-circulation systems.*

Forced-circulation hot-water heating systems have become very popular. Some of their advantages are that (1) small size tubing or pipe may be used and easily concealed in a building structure; (2) they have control versatility such that the temperature of the water being circulated may be automatically reduced or increased to change the output of the THTU's to more closely match the demand variation as outdoor climate changes; (3) there are small power requirements for the pump that circulates the water through the system; and (4) various piping arrangements are possible to meet any system demand.

Temperature Classification. ASHRAE classifies hot-water heating systems by operating temperature as follows:

1. *Low-temperature water system* (LTW) is a hot-water heating system operating within the pressure and temperature limits of the ASME boiler construction code for low-pressure heating boilers. The maximum working pressure for low-pressure heating boilers is 160 psi (1100 kPa) with a maximum temperature limitation of 250°F (121°C). The usual maximum working pressure for boilers for LTW systems is 30 psi (200 kPa), although boilers specifically designed, tested, and stamped for higher pressures may frequently be used with working pressures to 160 psi (1100 kPa). Steam-to-water or water-to-water shell-and-tube heat exchangers are also often used.

2. *Medium-temperature water system* (MTW) is a hot-water heating system operating at temperatures of 350°F (175°C) or less, with pressures not exceeding 150 psi (1030 kPa). The usual design supply water temperature is approximately 250°F to 325°F (120°C to 160°C), with a usual pressure rating for boilers and equipment of 150 psi (1030 kPa).

3. *High-temperature water system* (HTW) is a hot-water heating system operating at a temperature over 350°F (175°C) and a usual pressure of about 300 psi (2100 kPa). The maximum design supply water temperature is 400°F to 450°F (200°C to 230°C), with a pressure rating for boilers and equipment of about 300 psi (2100 kPa). The pressure–temperature rating of each component must be checked against the system's design characteristics.

This chapter covers procedures for designing and selecting the piping system and system components that apply particularly to LTW systems. These systems are used in buildings ranging in size from small dwellings to very large and complex structures. Terminal heat-transfer units include convectors, cast-iron radiators, baseboard and commercial finned-tube, fan-coil units, unit heaters, unit ventilators, multizone air-handling units, and radiant heating panels. A heat-transfer coil inside or outside the boiler is often used to supply hot water directly to the domestic water system or to a storage tank. A large storage tank may be included in the system to store energy to use when such heat input devices as the boiler or a solar energy collector are not supplying energy.

10-9 DESIGN WATER TEMPERATURE

The design water temperature is the maximum water temperature supplied to the system at design operating conditions. This water temperature should be determined as a result of economic analysis of system requirements unless it is controlled by process load requirements.

LTW systems are most widely used where loads consist primarily of space heating and domestic water heating and do not exceed 5000 MBh (1.5 MW) total (1 MBh = 1000 Btu/hr). Design water temperature normally ranges from 180°F to 250°F (82°C to 121°C), with 200°F (93°C) being very common.

MTW systems are most commonly used for space heating in large commercial and institutional buildings or in industrial applications with process loads, and where total loads range from 5000 to 20,000 MBh (1.5 to 6 MW). Design supply water temperature is approximately 250°F to 325°F (120°C to 160°C), as noted previously.

HTW systems are generally limited to campus-type district heating installations and to applications requiring process heating temperatures of 325°F (163°C) or higher. Such systems usually have loads greater than 10,000 to 20,000 MBh (3 to 6 MW). Design supply water temperatures range from 400°F to 450°F (200°C to 230°C), as noted previously.

10-10 WATER FLOW RATES IN HYDRONIC SYSTEMS

The quantity of heat transferred from the water that circulates through a hydronic system is a function of the flow rate, specific heat, and the temperature change of the water as it passes through the system. The heat transferred from the water is expressed by

$$\dot{q} = \dot{m}_w C_p (t_1 - t_2) \qquad (10\text{-}16)$$

where

$\dot{q}$ = the total system heat load (Btu/hr or W)
$\dot{m}_w$ = mass flow rate of water (lb/hr or kg/s)

C_p = specific heat of water (usually taken as 1.0 Btu/lb-°F or 4.18 kJ/kg-°C for LTW systems)
t_1 = temperature of water entering (°F or °C)
t_2 = temperature of water leaving (°F or °C)

Then, $(t_1 - t_2)$ is the design water temperature drop (TD) for the system. If the flow rate through a terminal heat-transfer unit is needed, $\dot{q}$ is the unit's heating capacity and $(t_1 - t_2)$ becomes the TD through the unit.

Equation (10-16) may be rearranged to yield flow rate in lb/hr (kg/s). For LTW systems, the flow rate is usually required in gpm (L/s). From Eq. (10-11), $\dot{m}_w = 500(\text{gpm})(\text{SG})$. Substituting into Eq. (10-16) gives

$$\dot{q} = 500(\text{gpm})(\text{SG})(C_p)(t_1 - t_2) \qquad (10\text{-}17)$$

when the average water temperature is 60°F. If the average water temperature were 180°F, the specific gravity would be 0.978, with $C_p = 1.0$, and we would have

$$\dot{q} = 490(\text{gpm})(t_1 - t_2) \qquad (10\text{-}17a)$$

For SI units, the flow rate in L/s is obtained from

$$\dot{q} = (\text{L/s})(\rho_w)(C_p)(t_1 - t_2)/1000 \qquad (10\text{-}17b)$$

where ρ_w = fluid density in kg/m³. Remember that the values of SG, ρ_w, and C_p in Eqs. (10-17) should be evaluated at the average water temperature.

10-11 DESIGN WATER FLOW RATES

The required water flow rate is the total of the flow rates required by the individual terminal units to provide the rated heating capacity at design conditions. Flow rates are determined basically by (1) assigning an arbitrary system temperature difference or (2) finding the minimum system flow rate by summing flow rates to individual terminals or zones.

Using an arbitrary system temperature difference (TD) has been common and is well suited to small system design. Using a minimum system flow rate is recommended for large systems. In large systems, the design TD should be the result, not the base point, of the design.

In small systems, flow rates to all terminal units are frequently based on an overall system TD, such as 20°F (11°C) for heating. This simplified design method is widely used and generates satisfactory system performance when applied properly; however, this method does not determine the system operating point. In larger

LTW systems, the resulting pipe sizes are often uneconomically large, and the actual system flow rate is likely to be much higher than intended. Such a design method seldom considers temperature drops higher than 20°F (11°C), which results in overdesign.

Minimum System Flow Rates. For large systems, a minimum system flow rate should provide the lowest cost and maximum control. Several methods are used to determine minimum flow rate to the individual THTU's and, by summation, the total system flow rate.

1. Various basic terminal units are capable of producing full capacity at specific temperature drops. For example, baseboard, finned-tube units, and convectors are selected frequently for 20°F to 50°F (11°C to 28°C) drop, unit heaters for 50°F (28°C), and extended surface coils in air systems for 100°F (50°C).

2. Manufacturers' catalogs list capacity ratings at various flow rates and entering temperatures (see Chapter 8). Engineers can select units for minimum flow rate (or maximum TD) to produce desired capacities.

3. Another method is to assume a constant temperature approach difference (leaving water temperature minus final heated medium temperature) for each class of equipment. For example, this approach temperature difference might be 20°F (11°C). Thus, for an air handler with 120°F (49°C) final air temperature, the leaving water temperature would be at 140°F (60°C). Similarly, a domestic water heater might have a final temperature of 180°F (82°C), which would indicate a leaving system water temperature of 200°F (93°C). Assuming a constant supply water temperature, the TD and flow rate for each terminal is determined. Flow rates determined in this manner are usually stated in gpm (L/s) or mass flow rate in lb/hr (kg/s). These flow rates are established by use of Eqs. (10-16) and (10-17). To illustrate, assume that we wish to determine the water flow rate for a LTW system for a capacity 100 MBh (29.3 kW) at 100°F TD. Using the specific heat of water equal to unity, Eq. (10-16) gives 100,000/100 = 1000 lb/hr. The mass flow rate for 30 kW at 50°C TD with a specific heat of 4.18 kJ/(kg-°C) is 30/(4.18)(50) = 0.144 kg/s.

Mass flow rate is often used, especially in large LTW systems and in MTW and HTW systems, since it simplifies calculations. Specific heat and specific gravity (or density) should be taken into account when calculating flow rate, especially for HTW system design where the change is significant. For instance, the flow rate for 100 MBh at 100°F TD and for 350°F mean water temperature ($C_p = 1.050$) is 100,000/(100)(1.050) = 952 lb/hr.

10-12 PIPING SYSTEM DESIGN AND PIPING DETAILS

After the comfort or process heating loads have been determined and all terminal heat-transfer devices have been selected and located, a tentative piping layout for the system can be made. An engineering analysis of the spaces or processes to be served, zoning, available space, and relative cost is required to determine the piping arrangement, or combination of piping arrangements, that will best meet the system needs. After the arrangement is selected, the piping system can be designed to connect all terminal units with the boiler or heat generator.

In general, the most economical distribution system layout results from running the supply and return mains by the shortest and most convenient route to terminal units and then laying out the branch or secondary circuits to connect with the mains. The piping layout is not intended to be a detailed drawing of the entire system; however, the drawing should be made to scale, showing connections to the boiler and all terminal units. The drawing is usually a single line drawing representing the pipe centerline. It should be drawn so that pipes follow the structural components and using 90-degree or 45-degree changes in direction. Water distribution mains are frequently located in corridor ceilings, above hung ceilings, wall-hung along a perimeter wall, in pipe trenches, in crawl spaces, or in basements. Water piping need not be run at a definite level or pitch but may change up or down as architectural or structural needs require.

Detail piping drawings should be made at the boiler, at terminal heat-transfer units (as required), and where special fittings and devices are employed. Special fittings and devices should be located, symbolized, and identified on all drawings. Some of these special fittings offer considerable flow resistance and must be identified to properly determine system flow resistance.

Water system piping can be divided into the following two arbitrary classifications:

1. Piping arrangements suitable for complete small systems or for terminal or branch circuits on large systems.
   (a) Series loop
   (b) One-pipe
   (c) Two-pipe *direct*-return
   (d) Two-pipe *reverse*-return
2. Main distribution piping arrangements to

carry water to and from the terminal units or circuits in a large system.

(a) Two-pipe *direct*-return
(b) Two-pipe *reverse*-return
(c) Three-pipe
(d) Four-pipe

The application usually dictates the distribution system. In small buildings, a specific piping system arrangement can normally be used throughout. In large buildings, combinations of several arrangements are usually best.

10-13 SERIES LOOP PIPING ARRANGEMENT

Perhaps the simplest of all piping arrangements is the *series loop*. It is a continuous run of constant-sized pipe or tube from a supply connection to a return connection. Terminal units are a part of the loop. Figure 10-14 shows a series of two loops on a supply and return main (split series loop). One or many series loops can be used in a complete system. Loops may connect to mains, or all loops may run directly to and from the boiler. It may be noted that the circulating hot water passes through each consecutive heat-transfer unit (frequently baseboard heating panels) in series and the cooled water returns to the boiler. The main advantages of the series loop system are its simplicity and low cost. However, certain conditions exist that limit its use. If one of the heating units is shut off, it prevents flow in the remainder of the loop. The pipe or tube size is usually limited to 1 in., with 3/4 in. being more common, thus limiting the heating capacity in the loop. Also, as the water flows through each heater, its temperature drops in proportion to the heater capacity. This means that each heater downstream of the first heater receives successively cooler supply water and may require larger heating units downstream to obtain a given heat output.

10-14 CONVENTIONAL ONE-PIPE ARRANGEMENT

Figure 10-15 illustrates the most common type of *one-pipe* system arrangement consisting of one pipe that is looped around the building and is both the supply and return main. This pipe is maintained the same size throughout its length because all the water circulated passes through this one main. At each heat-transfer unit, a supply and a return tee are installed on the main. One of the two tees is a special *diverting tee,* which creates a pressure drop in the main flow to divert part of the flow through the terminal unit and its branch circuit. The pressure drop created must be equal to, or greater than, the pressure loss through the terminal unit and its circuit. (This special diverting fitting will be discussed in detail in a later section.)

One diverting tee, located on the return side of the terminal unit circuit, is usually sufficient for upfeed (units above the main) systems. Two special tees (supply and return tees) may be required to overcome the thermal head in downfeed units. The terminal unit and its circuit must be selected for low pressure loss because the special diverting tees are designed to create a small pressure drop in order to keep the system pumping head within reasonable limits. Consult manufacturers' literature for flow rates and pressure drop data for these special tee fittings. Unit selection can be only approximate without these data.

The one-pipe system arrangement has been used extensively in residences and small commercial establishments. For larger capacity systems, the one-pipe arrangement may be designed in two or more circuits or combined with two-pipe systems (usually called primary–secondary pumping systems).

One-pipe circuits allow manual or automatic control of individual connected units. On–off rather than flow modulation control is advisable because of the relatively low pressure and low diverted flow. Cost is likely

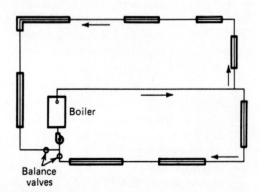

FIGURE 10-14 Two-circuit series loop system (plan).

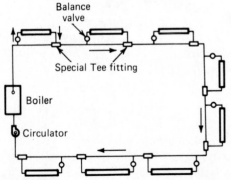

FIGURE 10-15 One-pipe heating system, single circuit (plan).

to be greater than for the series loop because of extra branch pipe and fittings, including special tees. A manual air vent in each heating unit is usually required because of the low water velocity through the unit.

Since only a fraction of the main flow is diverted in a one-pipe circuit, the flow rate and pressure drop as unit flow is controlled and are less variable than in some other circuits. One-pipe circuits can therefore be useful in large systems that cannot tolerate large uncontrolled changes in pump head and flow. (Primary–secondary connections used on any type of circuit provide superior isolation of pump pressure change.) When two or more one-pipe circuits are connected to the same two-pipe mains, the circuit flow may need to be mechanically balanced as in other types of circuits. After balancing, sufficient flow must be maintained in each one-pipe circuit to ensure adequate flow diversion to the heating units.

10-15 TWO-PIPE DIRECT-RETURN AND REVERSE-RETURN ARRANGEMENTS

Figure 10-16 represents the basic *direct-return* piping arrangement. In direct-return systems, the return main flow direction is opposite the supply main flow, and the return water from each unit takes the shortest path back to the boiler. Therefore, the distance along the piping through a heater and back to the boiler is different for each unit. This means that balancing flows may be a problem. Usually, the balancing is done by using adjustable valves, orifices, or changes in pipe sizes.

Figure 10-17 illustrates the basic *reverse-return* piping arrangement. In reverse-return circuits, the return main flow is in the same direction as the supply main flow, and the return main returns all water to the boiler after the last heater is fed. The reverse-return systems seldom need balancing valves because the water flow distance to and from the boiler is virtually the same through any heating unit. However, in practice, it is al-

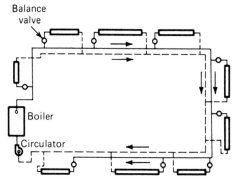

FIGURE 10-17 Two-pipe reverse-return system (plan).

ways good engineering *to install balance valves in all the circuits* to make it possible to fine-tune the system's balance.

The cost of installing a direct-return compared to a reverse-return system must be evaluated for each design. Direct-return systems use less pipe length but usually have a higher operating (pumping) cost because of the added balancing fitting losses at the same flow rate.

The two-pipe arrangements require more pipe for a given system than the one-pipe system; therefore, they are more expensive. Pipe size along the supply main decreases (increases in return main) because less water is flowing in the main after each successive heating unit, requiring more fittings and more labor. However, the flexibility of water distribution and temperature control, permissible system capacity, and simplicity make the two-pipe arrangements well suited to larger systems.

10-16 COMBINATION PIPING ARRANGEMENTS

Frequently, in system design, advantage is taken of the simplicity of the four basic piping arrangements, and their respective advantages are combined without much complication. One arrangement can grade into another, and a piping system can contain from one to all four types of piping arrangements.

10-17 SPECIAL FITTINGS AND DEVICES IN HYDRONIC SYSTEMS

Special fittings and devices are required for the proper functioning of hydronic systems. These include air-elimination devices, flow-control devices, pressure-relief valves, pressure-reducing valves, compression or expansion tanks, and special diverting tees for one-pipe systems. Provision must be made for pipe expansion and, when necessary, for elimination of vibration. These devices are briefly described in the following paragraphs.

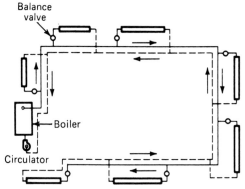

FIGURE 10-16 Two-pipe direct-return system (plan).

Drain and shutoff valves: All system low points should have drains. Separate shutoff and draining of individual equipment and piping circuits should be possible so that the entire system does not have to be drained to service a particular device.

Balance fittings: Balance fittings should be applied as needed to permit balancing of individual terminal units or major piping subcircuits. These fittings should be installed on the return side when possible.

Pitch of pipe: Piping need not be pitched but may be run level, providing the flow velocities are maintained above 1.5 ft/s (0.5 m/s).

Strainers: Strainers should be used where necessary to protect the system components. Strainers in the pump suction need to be analyzed carefully to avoid pump cavitation. Large separating chambers can serve as main air-venting points and dirt strainers ahead of pumps. Automatic control valves or spray nozzles operating with small clearances require protection from pipe scale, gravel, and welding slag, which may readily pass through its protective separator. Individual fine-mesh strainers may therefore be required ahead of each control valve.

Thermometers: Thermometers or thermometer wells should be installed to assist the system operator on routine operation and troubleshooting. Permanent thermometers with the correct scale range and separable sockets should be at all points where temperature readings are needed regularly. Thermometer wells should be installed where readings will be needed only during start-up and infrequent troubleshooting.

Flexible connectors: These devices are installed at pumps to reduce pipe vibration. Vibration may be transmitted through the water column across a flexible connection and reduce the effectiveness of the connector. Flexible connectors, however, prevent damage caused by misalignment of equipment piping flanges.

Pressure gauge cocks: Pressure gauge cocks should be installed at points requiring pressure readings. Remember that pressure gauges permanently installed in the system will deteriorate because of vibration and pulsation and will be unreliable. It is good practice to install gauge cocks and provide the operator with several quality gauges for troubleshooting.

Automatic fill valve: An automatic fill valve should be connected to the system pipe between the expansion tank and the system, which is the point of no pressure change, regardless of whether or not the system pump is operating.

FIGURE 10-18 Monoflow pipe fitting. (Courtesy of ITT Fluid Transfer Division, Morton Grove, IL)

Special tee (venturi) fittings: Figure 10-18 shows a *venturi* tee fitting manufactured by ITT Fluid Transfer Division and called a *monoflo* fitting. This specially constructed tee is engineered to produce a conversion of pressure head to velocity head in the throat of the built-in venturi. The reduction of pressure head at the throat causes the higher static pressure at the upstream standard tee to force a portion of the main flow through the branch supply riser, the heating unit, and the return riser. The flow through the return riser, reduced in temperature, joins the main flow in the venturi tee (see Figure 10-19).

The operating principle of the venturi fitting may be summarized by referring to Figure 10-20. At point 1, water enters from the upstream main. At 2, the restriction of the venturi nozzle causes an increase in flow velocity with a resultant lowering of system pressure at this point. At 3, as a pressure differential will always cause a flow from the high-pressure to low-pressure region, water continually enters at this point as long as flow exists in the main. At point 4, velocity pressure is converted back to static pressure in the main with minimum overall shock loss. This action results in (1) a constant percentage of water diversion to the heating unit at any rate of water flow through the main and fitting and (2) a jet action with its high efficiency (maximum diversion with minimum overall loss) within the flow rates recommended.

Taco, Inc., was the originator of the application

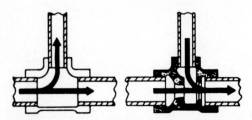

FIGURE 10-19 Typical water flow through main and riser connections using Taco venturi fittings. (Courtesy of Taco, Inc., Cranston, RI)

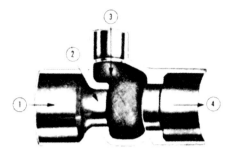

FIGURE 10-20 Operating principle of Taco venturi fittings. (Courtesy of Taco, Inc., Cranston, RI)

FIGURE 10-22 Automatic hot-water air vent. (Courtesy of Taco, Inc., Cranston, RI)

of the Bernoulli principle to forced-flow one-pipe hydronic systems. Taco *super venturi* fittings create a higher available pressure change and cause a substantially higher percentage of water diversion capacity than the *standard* fittings. These super venturis may be used with standard venturis in any one-pipe hydronic system. They are particularly well suited for downfeed heating units.

Figure 10-21 shows a sketch of typical connections for an upfeed and a downfeed heating unit using venturi tees. Generally, only one venturi tee fitting is required per heating unit whether the unit is above or below the main. The same type of venturi may be used for upfeed or downfeed.

Air eliminators: All water contains air. As the cold water is heated, the air is driven out of solution and has a tendency to collect in heating units and at high points in the system piping. This may cause uneven heating from the heating unit, noisy operation, and may completely block water circulation. The air must be

eliminated from the system to provide satisfactory performance.

Figures 10-22 through 10-24 illustrate typical devices used in hydronic systems to control the air in the system. The *air vent* shown in Figure 10-22 normally would be located on all heating units. The *high vent* shown in Figure 10-23 would be located in the piping system at all system high points. The operating principle of the air vent is as follows: Air passes through the special fiber discs and vent holes to the atmosphere. Water, following the air, wets the fiber discs, causing them to swell and completely seal the valve. As more air accumulates, the fiber discs dry and shrink, thus permitting the valve to repeat the cycle.

The operation of the high vent is as follows: When the valve shell is full of water, the valve is closed. When sufficient air accumulates, the mechanical float drops and the valve opens. As air passes out, water again fills the shell, closing the valve. As fast as air accumulates, the action is repeated.

Figure 10-24 shows one type of air eliminator. The *air scoop* shown is manufactured by Taco, Inc., is located in the supply outlet pipe from the boiler, and operates as follows:

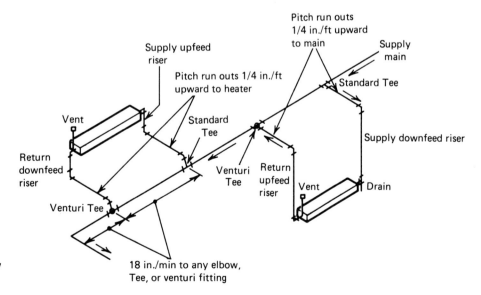

FIGURE 10-21 Venturi connections for heaters above and below main (one-pipe system).

FIGURE 10-23 High vent. (Courtesy of Taco, Inc., Cranston, RI)

1. Air, being less dense than water, will generally tend to travel along the upper portion of a horizontal pipe at velocities commonly used in hydronic systems.
2. As the air and water enter the air scoop, the air bubbles are scooped up by the first baffle and rise into the upper chambers. Any bubbles that get through the first baffle are scooped up by the second and third.
3. Air that accumulates in the first chamber is removed from the system by an air-vent valve. Air that accumulates in the second chamber passes into the expansion tank to act as a system air cushion.
4. Should the air completely fill the expansion tank and back down into the air scoop, the excess will be removed by the air vent without disturbing system operation.

Flow-control device: When the circulating pump in a hydronic system is not operating, water flow in the system should stop. However, since there is always a temperature difference in the various parts of the system, gravity circulation of the water may be expected. This could cause overheating of the building during periods when no heat is required. Also, during the summer

season, if domestic hot water is supplied by a heating coil in the boiler, the boiler water must be maintained at 160°F. This may cause gravity circulation in the piping of the heating system.

A specially designed check valve, like the one shown in Figure 10-25, is installed in the supply main, close to the boiler outlet, which closes when the pump is not in operation to prevent gravity circulation. These flow-control devices may be installed in each zoned circuit of a system that is not controlled by a motorized valve.

Relief valve: Relief valves are designed to prevent the boiler from rupturing, should the pressure exceed its working pressure. These valves operate on thermal expansion pressures and/or steam pressures.

Relief valves that carry the ASME symbol must be tested and approved under the ASME Boiler and Pressure Vessel Code and rated by their discharge capacity in Btu per hour or pounds per hour. They are installed close to or directly into the top of the boiler without any shutoff valve between the relief valve and the boiler. The drain line from the discharge of the relief valve must be full size, and a shutoff valve must never be installed in this drain. Figure 10-26 shows a typical ASME pressure-relief valve.

Reducing valve: The pressure-reducing valve for a hydronic system is designed to automatically feed makeup water into the system when the system water pressure drops below the setting of the reducing valve. The pressure-reducing valve must be located at a point in the piping system where there is no pressure change, as explained in a later section. Figure 10-27 shows a typical pressure-reducing valve.

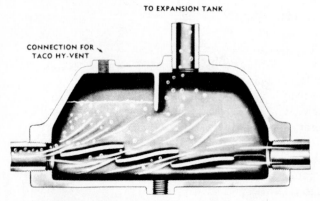

FIGURE 10-24 Air scoop. (Courtesy of Taco, Inc., Cranston, RI)

FIGURE 10-25 Flow-control valve. (Courtesy of ITT Fluid Transfer Division, Morton Grove, IL)

FIGURE 10-26 Pressure-relief valve. (Courtesy of ITT Fluid Transfer Division, Morton Grove, IL)

Figure 10-28 shows the normal placement of the various special fittings required in hydronic systems discussed in preceding sections. The locations shown are typical of a residential installation. These locations may be different in other installations, as will be noted later.

FIGURE 10-27 Pressure-reducing valve. (Courtesy of ITT Fluid Transfer Division, Morton Grove, IL)

10-18 SYSTEM PRESSURIZATION

The objectives of system pressure control are (1) to limit the pressure at all system equipment to its allowable working pressure, (2) to maintain minimum pressure at all normal operating temperatures to vent air and prevent cavitation at the pump suction and boiling of system water, and (3) to accomplish these objectives with a minimal addition of new water.

An *expansion tank* is the primary device used for system pressure control. The expansion tank, sometimes called *compression tank* or *air cushion tank,* for closed hydronic systems, is an airtight cylindrical tank located in the vicinity of the boiler, connected in such a way that, when the system is initially filled with water, air is trapped within the tank. When the system water temperature changes, it expands or contracts (depending on whether it is a heating or cooling system). The air trapped in the tank acts as a cushion providing space for the water expansion and contraction. The excess water volume in the system, resulting from increased temperature, is stored in the expansion tank during periods of high operating temperatures and is returned to the system when the system water temperature is lower. The expansion tank must be able to store the required volume of water during maximum design operating temperatures, without exceeding the maximum allowable operating pressure, and to maintain the required mini-

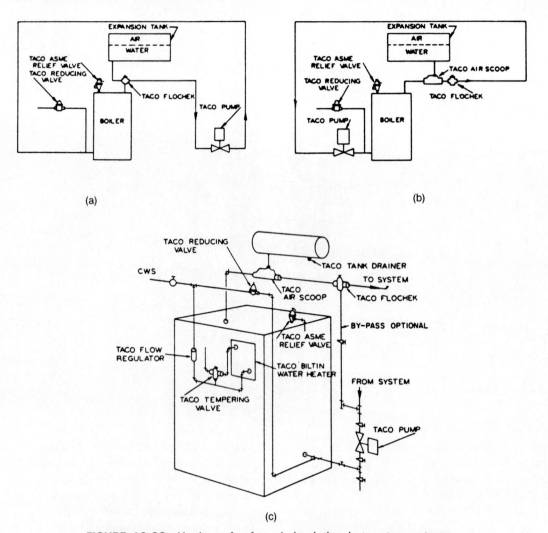

(a)

(b)

(c)

FIGURE 10-28 Hookups for forced-circulation hot-water systems: (a) recommended hookup when expansion tank is connected to suction side of pump, (b) recommended hookup when expansion tank is connected to discharge side of pump, and (c) typical boiler layout for residential installation. (Courtesy of Taco, Inc., Cranston, RI)

mum pressure when the system is cold. Some designers provide an automatic fill valve to maintain the minimum system pressure by supplying water to make up for leakage. Others supply additional system water by using a manually operated valve.

The safety relief valve installed in a hot water boiler in accordance with the ASME Boiler and Pressure Vessel Code suggests the following requirements and methods. In choosing a method to maintain the pressure above saturation, consider the following factors:

1. The pressurization system should be relatively simple and reliable. Loss of pressurization causes flashing within the system, which

can lead to problems within pumps, heat-exchange devices, and the piping system.

2. Oxygen should be excluded from the system to maintain corrosion-free characteristics, both in pressurization and in operating cycle. This precludes the use of air in direct contact with the water as a pressurizing means. Do not use cycles that periodically withdraw and reintroduce water.

3. The control of the pressurization system should maintain the inherently stable conditions of the hot water cycle. Fluctuations in system pressure or temperature should be minimized by taking advantage of the thermal storage of the cycle. This will im-

prove combustion efficiency through relatively steady firing rates.

Available fundamental methods for keeping pressure in a hydronic system at or above the desired minimum level include use of the following:

1. An elevated storage tank.
2. A pump. However, the large pressure changes encountered with small changes in volume make this method relatively impractical.
3. An expansion tank sized to accommodate the volume of water expansion, with sufficient gas space to keep the pressure range within the design limits of the system. Where air is used as the initial gas, the system should be kept tight because every recharging cycle introduces additional oxygen with the air and thus promotes system corrosion. The initial oxygen during start-up will react with the system components, leaving a basically nitrogen atmosphere in the expansion tank. The initial fill pressure sets the minimum level, while the compression of the gas space caused by water expansion determines the maximum system pressure.
4. A diaphragm expansion tank precharged to the system fill pressure and sized to accept the expansion of the water. The air charge and the water are permanently separated by a diaphragm that eliminates corrosion and noise caused by air in the system.
5. Steam pressurization and inert gas pressurization for medium-temperature and high-temperature water (MTW and HTW) systems, which are not discussed in this book.

Systems with expansion tanks are either open or closed. The *open-system expansion tank,* as shown in Figure 10-29, is vented to the atmosphere and is generally limited to installations having operating temperatures of 180°F (82°C) or less because of system water boiling and tank water evaporation problems. The tank should be at least 3 ft (1 m) above the high point of the system and connected to the suction side of the pump to prevent subatmospheric system pressures caused by pump operation. The system should have an internal overflow drain. As required by the ASME Boiler and Pressure Vessel Code, water must be prevented from freezing in the tank, the tank vent, and the pipe leading to the tank. The minimum tank volume should equal 6% of the total system water volume. Note that contin-

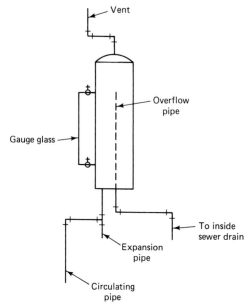

FIGURE 10-29 An open-system expansion tank.

uous oxygen contamination (absorption) occurs at the exposed water surface, making the system less corrosion-free than the closed design.

The *closed-system expansion tank* is airtight to pressurize the system for operation over a wide range of conditions (see Figure 10-28). If the tank or the amount of air in it is too small, the pressure on the system will exceed the maximum allowable and cause the safety relief valve to waste water from the system. When the system cools, the pressure will drop to a value less than minimum, making air venting impossible or drawing air into the system if automatic air vent valves are located at a high point of the piping. If the tank is too large, it will cost more and require more space. Figure 10-30 shows a diaphragm expansion tank and a sketch of its location in a closed system.

In small systems having small water volumes and low pressure from the pump and elevation, expansion tanks are small and problems are not critical. In large systems, the size of the expansion tank and pressure-control problems can be critical. Some ways of reducing expansion tank size and satisfactorily controlling pressure are as follows:

1. A compressed gas, if available, can be admitted to the expansion tank to obtain the desired pressure.
2. The additive pressure of the pump can be reduced by the use of larger pipe sizes or greater design temperature drop for the system.

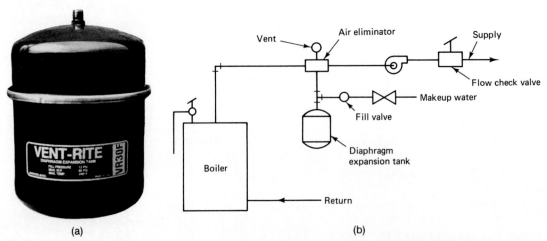

FIGURE 10-30 (a) Diaphragm expansion tank and (b) system hookup for diaphragm expansion tank. (Courtesy of Flexcon Industries, Norwood, MA)

3. The boiler can be constructed for pressures higher than 30 psig (200 kPa).
4. A steam boiler can be used with a steam-to-water heat exchanger to heat the water circulated. This is often used in multistory buildings with a steam-to-water heat exchanger located on each of the upper stories.

The point where the expansion tank is connected should be carefully selected to avoid the possibility that normal operation of automatic, check, or manual valves will isolate the tank from a hot boiler or any other part of the system.

Effect of Pump Location. The proper location of a pump in forced-flow hydronic systems, as well as the expansion tank, relief valve, and pressure-reducing valve, has an important bearing on system operation. Improper location of these system components can often lead to problems of pump cavitation and the production of negative pressures (below atmospheric) in parts of the system.

Point of No Pressure Change. Figure 10-31a illustrates the main of a closed hydronic system with a pump and compression tank shown and without means of feeding additional water. The expansion tank contains a certain volume of air, which is compressible, and the balance is water, which is essentially incompressible. To increase the pressure in the expansion tank, the air must be compressed. To compress the air requires more water. Since this is a closed system without means to receive more water, the pressure at the expansion tank cannot change, regardless of whether the pump is operating or not. Therefore, the point in the system where

the expansion tank is connected will be known as the "point of no pressure change."

Figure 10-31b illustrates the main of a closed hydronic system with the expansion tank connected *on the suction side of the pump* with the pump not operating. At the initial filling of the system with cold water, assume the pressure gauges read 12 psig, as reducing valves are normally set to fill the system at that pressure. After the water in the system is heated to operating temperature, the gauges might read 18 psig because of the water expansion. Note that all points in the system indicate the same pressure (18 psig).

Figure 10-31c indicates the same system except the pump is now operating, and it is assumed that it develops a differential pressure of 8.0 psi. Since the expansion tank is located on the suction side of the pump and this is the "point of no pressure change," the pressure at this point will remain at 18 psig. The pressure differential developed by the pump will therefore appear as a positive pressure of 26 psig (18 + 8) at the discharge of the pump. The 8.0 psi will be dissipated by system friction as indicated by the reduced gauge readings around the system.

Figure 10-31d illustrates the same system as in 10-31b and all conditions are the same, except the expansion tank is now connected *on the discharge side of the pump.* Note that the pump is not running and all pressure gauges read the same.

Figure 10-31e is the same system as in 10-31d except the pump is now running. Since the expansion tank is located on the discharge side of the pump, this point now becomes the "point of no pressure change," and the pressure at this point remains at 18 psig. The pressure differential developed by the pump now appears as a negative pressure, and the pressure on the suction side

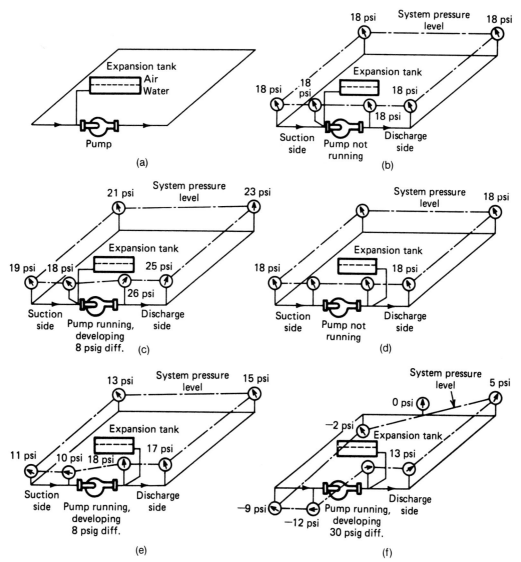

FIGURE 10-31 "Point of no pressure change" in hydronic system.
(Reproduced with permission from Taco, Inc., Cranston, RI)

of the pump will be reduced by 8.0 psi to 10 psig (18 − 8). As the friction of the system produces a drop of 8.0 psi, the pressure at various points in the system will become progressively lower.

Figure 10-31f illustrates the same system as in 10-31d and e with the expansion tank connected on the discharge side of the pump, but with a pump that develops a pressure differential of 30 psi when running. This pressure differential produces a negative gauge pressure of −12 psig (18 − 30) on the suction side of the pump. Water at −12 psig (2.7 psia) will boil and produce vapor (steam) if the water temperature is above 137°F. This vapor formed on the suction side of the pump will cause the pump to cavitate. Also, due to part of the system being subjected to vacuum, air may leak into the system creating additional problems.

Both the pressure-relief valve and pressure-reducing valves should be connected to the boiler on the expansion tank side, which is the "point of no pressure change." Figure 10-28a illustrates the hookup when the expansion tank is on the suction side of the pump. Figure 10-28b illustrates the hookup when the expansion tank is connected to the discharge side of the pump.

Water Expansion. A hydronic system's water volume is determined from heating-unit water capacities and boiler water capacity obtained from manufacturers, and pipe and tube water volume may be obtained from Table 10-1.

When the system water temperature changes, it expands or contracts (depending on whether it is being heated or cooled). The air trapped in the expansion tank

TABLE 10-1

Volume of water in standard pipe and tube

Nominal Pipe Size		Schedule No.	Standard Steel Pipe				Type L Copper Tube			
			Inside Diameter		Volume		Inside Diameter		Volume	
in.	(mm)		in.	(mm)	gal/ft	(L/m)	in.	(mm)	gal/ft	(L/m)
$\frac{3}{8}$	(10)	—	—	—	—	—	0.430	(1.09)	0.0075	(0.09)
$\frac{1}{2}$	(15)	40	0.622	(1.58)	0.0157	(0.19)	0.545	(1.38)	0.0121	(0.15)
$\frac{5}{8}$	(16)	—	—	—	—	—	0.666	(1.69)	0.0181	(0.22)
$\frac{3}{4}$	(20)	40	0.824	(2.09)	0.0277	(0.34)	0.785	(1.99)	0.0251	(0.31)
1	(25)	40	1.049	(2.66)	0.0449	(0.56)	1.025	(2.60)	0.0429	(0.53)
$1\frac{1}{4}$	(32)	40	1.380	(3.50)	0.0779	(0.97)	1.265	(3.21)	0.0653	(0.81)
$1\frac{1}{2}$	(40)	40	1.610	(4.09)	0.106	(1.32)	1.505	(3.82)	0.0924	(1.15)
2	(50)	40	2.067	(5.25)	0.174	(2.16)	1.985	(5.04)	0.161	(2.00)
$2\frac{1}{2}$	(65)	40	2.469	(6.27)	0.249	(3.09)	2.465	(6.26)	0.248	(3.08)
3	(80)	40	3.068	(7.79)	0.384	(4.77)	2.945	(7.48)	0.354	(4.40)
$3\frac{1}{2}$	(90)	40	3.548	(9.01)	0.514	(6.38)	3.425	(8.70)	0.479	(5.95)
4	(100)	40	4.026	(10.23)	0.661	(8.21)	3.905	(9.92)	0.622	(7.73)
5	(125)	40	5.047	(12.82)	1.04	(12.92)	4.875	(12.38)	0.970	(12.05)
6	(150)	40	6.065	(15.41)	1.50	(18.63)	5.845	(14.85)	1.39	(17.26)
8	(200)	30	8.071	(20.50)	2.66	(33.03)	7.725	(19.62)	2.43	(30.18)
10	(250)	30	10.136	(25.75)	4.19	(52.04)	9.625	(24.45)	3.78	(46.95)
12	(300)	30	12.090	(30.71)	5.96	(74.02)	11.565	(29.38)	5.46	(67.81)

Source: ASHRAE Handbook 1987 HVAC Systems and Applications, with permission from American Society of Heating, Refrigerating, and Air Conditioning Engineers, Atlanta, GA.

(closed system) acts as a cushion providing space for this expansion or contraction and maintenance of reasonably stable pressure in the system.

If a hydronic system is initially filled with 50°F water ($v_f = 0.016024$ ft³/lb) and is then brought up to a temperature of 200°F ($v_f = 0.016634$ ft³/lb), the system water volume increase may be found from

$$\Delta V_s = V_{sc}\left(\frac{v_{fh}}{v_{fc}} - 1\right) \qquad (10\text{-}18)$$

where

$$\Delta V_s = \text{change in system water volume}$$
$$V_{sc} = \text{volume of cold water in system}$$
$$v_{fc} \text{ and } v_{fh} = \text{specific volumes of cold water and hot water, respectively}$$

Substitution of the specific volume values for water at 50°F and 200°F gives

$$\Delta V_s = V_{sc}\left(\frac{0.016634}{0.016024} - 1\right) = V_{sc}(0.038)$$

or the system volume increases by 3.8% when heated from 50°F to 200°F.

At any given average water temperature for a hydronic system, an increase in the design temperature drop (from the common 20°F drop) can result in a smaller expansion tank since the lower circulation rate reduces pipe sizes. For example, if the average water temperature for the system were to be 180°F (82°C), the supply and return water temperature could be 190°F and 170°F (88°C and 77°C), respectively, for a 20°F (11°C) drop or 210°F and 150°F (99°C and 66°C), respectively, for a 60°F (33°C) drop. For the latter, the water circulation rate would be one-third of the former design drop. The actual reduction in water quantity, and therefore in water expansion, for any given change in water temperature difference varies with the amount of piping in the system. The water content of the hot water generator, as well as of radiation and other heat-exchange components, may or may not change in size. On larger systems, an economic study of the effects of various operating temperature ranges is often warranted.

Sizing of Expansion Tanks. Satisfactory operation of a hydronic system requires a properly sized expansion tank. Commercial tank sizes may range from 2 to 300 gallons or more, depending on the manufacturer. Frequently, in medium and large systems, it is more convenient to use two or more smaller tanks than to use one large tank. All expansion tanks should bear the ASME label signifying that the design has been approved by the ASME Construction Code.

The size of a closed-system expansion tank is determined by (1) the volume of water in the system, (2) the range of water temperatures normal to system operation, (3) the air pressure in the expansion tank when the fill water first enters it, (4) the relationship of the height of the boiler, which is usually the component with the lowest working pressure, to the expansion tank and the high point in the piping system, (5) the pressure caused by the circulating pump, (6) the location of the circulating pump with respect to the expansion tank location, and (7) the boiler.

For *residential and small commercial installations,* the expansion tank size in gallons may be estimated by dividing the total system radiation load in Btu/hr by a factor of 5000.

A more accurate method for sizing expansion tanks is to use equations presented by various authorities such as ASME and ASHRAE. The following ASME formula is applicable to system operating temperatures between 160°F and 280°F:

$$V_t = \frac{(0.00041t - 0.0466)V_s}{\dfrac{P_a}{P_f} - \dfrac{P_a}{P_o}} \quad (10\text{-}19)$$

or for SI,

$$V_t = \frac{(0.000738t - 0.03348)V_s}{\dfrac{P_a}{P_f} - \dfrac{P_a}{P_o}} \quad (10\text{-}19a)$$

where

V_t = minimum volume of expansion tank (gal or m³)
V_s = system volume (gal or m³)
t = maximum average operating temperature (°F or °C)
P_a = pressure, usually atmospheric, in expansion tank when water first enters (ft of water or kPa, absolute)
P_f = initial fill or minimum pressure in tank (ft of water or kPa, absolute) (normally the static system head)
P_o = maximum operating pressure at tank (ft of water or kPa, absolute) (atmospheric pressure head plus relief valve setting, or working pressure)

A widely used formula recommended for temperatures below 160°F (70°C) is

$$V_t = \frac{E}{\dfrac{P_a}{P_f} - \dfrac{P_a}{P_o}} \quad (10\text{-}20)$$

where E is the net expansion of water in the system when heated from minimum temperature to maximum temperature (gal or m³). The net expansion of the water is represented by the term $(0.00041t - 0.0466)V_s$ from Eq. (10-19).

Static Pressure at the Highest Point in a Hydronic System. For proper operation of a hydronic system, a positive pressure should be maintained at the highest point in the piping or heating device at all times. This pressure should be at least 1.5 psig or about 4 ft of water head to obtain positive venting in LTW systems. On boilers equipped with altitude gauges, the red hand should be set manually to the minimum desired altitude in feet. This would be the distance from the top of the boiler to the highest point in the system plus 4 ft. The black hand, which indicates the actual system pressure, should never be permitted to fall below the position of the red hand. The following relationships will be helpful in determining the minimum fill pressure to maintain the required positive pressure at the system high point.

For systems with expansion tank connected *on suction side of pump* (Figure 10-31a),

$$P_f = h + 4 \quad (10\text{-}21)$$

For systems with expansion tank connected *on discharge side of pump* (Figure 10-31d),

$$P_f = h + 4 + h_p \quad (10\text{-}22)$$

where

h = distance from top of boiler to highest point in system (ft)
h_p = operating head of pumps (ft of water)

For *diaphragm expansion tanks,* the ASME formula for minimum tank volume can be stated as

$$V_t = \frac{(0.00041t - 0.0466)V_s}{1 - \dfrac{P_f}{P_o}} \quad (10\text{-}23)$$

or for SI,

$$V_t = \frac{(0.000738t - 0.03348)V_s}{1 - \dfrac{P_f}{P_o}} \quad (10\text{-}23a)$$

where P_f is the fill pressure, which is equal to the precharge pressure.

The system for Illustrative Problems 10-3 through

10-5 has (1) a water volume of 1000 gal, (2) a high point of the system at the top of a return riser 25 ft from the top of the boiler directly above the boiler room, (3) a circulating pump having a head of 20 ft, and (4) a 30 psi boiler with an ASME-rated safety relief valve as the item of equipment having the lowest pressure rating. The P_a is atmospheric pressure, 34 ft of water absolute. The design average water temperature is 200°F. Friction losses between the high point of the system and the boiler are assumed to be negligible.

ILLUSTRATIVE PROBLEM 10-3

Determine the size of expansion tank required when the tank is connected on the suction side of the pump as shown in Figure 10-31a. Assume that the expansion tank is at the same level as the safety relief valve.

Solution: The net expansion of the water in the system is represented by the numerator of Eq. (10-19), or $(0.00041t - 0.0466)V_s$:

$$[(0.00041)(200) - 0.0466](1000) = 35.4 \text{ gal}$$

As stated, P_a is 34 ft of water absolute. The minimum fill pressure (P_f) by Eq. (10-21) is $(25 + 4) = 29$ ft of water gauge, or $(29 + 34) = 63$ ft of water absolute. P_o is equal to the safety relief valve setting minus 5 psi or 10%, whichever is larger (and which gives the lowest P_o). Thus, $(30 \text{ psi} - 5 \text{ psi})(2.31) = 57.8$ ft of water gauge, or $(57.8 + 34) = 91.8$ ft of water absolute. Substitution into Eq. (10-19) gives

$$V_t = \frac{35.4}{(34/63) - (34/91.8)} = 209.1 \text{ gal}$$

ILLUSTRATIVE PROBLEM 10-4

Determine the size of the expansion tank required when the tank is connected on the discharge side of the pump as shown in Figure 10-31d. Assume that the expansion tank is at the same level as the safety relief valve.

Solution: When the pump operates, the pressure at the top of the return riser is reduced by an amount equal to the pump head. Therefore, to maintain a positive pressure at the top of the system, the minimum pressure must be increased an amount equal to the head of the pump, 20 ft of water (given). Therefore, by Eq. (10-22), $P_f = (25 + 4 + 20) = 49$ ft of water gauge, or $(49 + 34) = 83$ ft of water absolute. The required tank volume from Eq. (10-19) gives

$$V_t = \frac{35.4}{(34/83) - (34/91.8)} = 901.5 \text{ gal}$$

ILLUSTRATIVE PROBLEM 10-5

Determine the required diaphragm tank size required for the data of Illustrative Problems 10-3 and 10-4. Using Eq. (10-23)

and $P_f = 63$ ft of water absolute and $P_o = 91.8$ ft of water absolute, we have

$$V_t = \frac{35.4}{1 - (63/91.8)} = 112.8 \text{ gal}$$

From Illustrative Problem 10-4, $P_f = 83$ ft of water absolute, and $P_o = 91.8$ ft of water absolute. Then,

$$V_t = \frac{35.4}{1 - (83/91.8)} = 369.3 \text{ gal}$$

The following two important factors in tank sizing have not been shown in Illustrative Problems 10-3 through 10-5:

1. *Using compressed air to increase the amount of air in the tank.* This factor affects the value of P_a by increasing it above atmospheric pressure. Using compressed air to *charge* the expansion tank reduces the required tank volume.

2. *The effect of a design maximum temperature substantially higher than 212°F (100°C).* It is possible for the expansion tank to operate properly without adding the pressure required to maintain liquid water at temperatures exceeding 212°F (100°C). This basis of design requires (a) that pressure increase in the system as the temperature increases from 40°F (4.4°C) to the final design temperature be sufficient to keep the pressure above that corresponding to the temperature of the water to prevent boiling and (b) that system operating personnel hold the proper pressure for various system temperatures so that they can maintain proper system pressure control. An automatic fill valve cannot be used to maintain the minimum pressure for such a system.

For further data relating to pressurization and tank sizing, refer to Chapter 15 of *ASHRAE Handbook 1987 HVAC Systems and Applications* where MTW and HTW systems are discussed.

10-19   BOILERS

An essential component of any hydronic heating system is an efficient, correctly sized boiler or heat exchanger. Boilers were discussed in some detail in Chapter 8. If the boiler is undersized, it is quite obvious that it will not supply sufficient heat for the system. Although not quite so obvious is that an oversized boiler may prove to be inefficient in its operation as well as being more costly. Research conducted by I-B-R indicates that there is no appreciable reduction in the warmup period when an oversized boiler is installed.

The boiler capacity required for a hydronic system is normally the *net rating,* or the boiler must have a net rating equal to the calculated heat loss from a building at design conditions and, also, the domestic hot water service if such water is heated by the boiler.

At times, it may be desirable to use two or more boilers instead of one in order to provide greater flexibility and higher boiler efficiency. A boiler operates most efficiently if it can operate steadily at its rated capacity. Splitting the system load between boilers will depend upon the judgment of the design engineer. An accepted practice is to size each boiler at two-thirds of the total system load so that, in the event of breakdown, the remaining boiler will carry the total load, except under extremely cold weather conditions.

Boiler Location. The location of the boiler in the hydronic system is normally dictated by the chimney location if the boiler is gas, coal, wood, or oil fired. An electric boiler, requiring no exhaust flue, may be located anywhere. If the building has a basement, the boiler would normally be located there. If the building has a slab floor without a basement or crawl space, the boiler is normally placed on a reinforced section of the first floor. Multistory buildings may have the boiler located on any floor if the piping arrangement is proper, but most often it is located in the basement or on the first floor.

Careful selection of the boiler location will result in better operation and usually lower installation costs. I-B-R recommends the following as a guide in selecting a boiler location:

1. Locate the boiler as near to a chimney as possible. Keep smoke pipe as short and direct as possible, using the least number of turns. The cross-section area of the smoke pipe and flue should not be smaller than recommended by the manufacturer.

2. Avoid connecting boiler to a fireplace flue or to a flue used for other appliances.

3. If a choice exists between using an inside or an outside chimney, make connections to the inside one. This provides better draft conditions as well as fuel economy.

4. Provide ample space around boiler and domestic hot water heater for proper installation and servicing of equipment.

5. Provide air supply for combustion in accordance to local ordinances or other regulations applicable to method of firing the boiler. When installation of firing devices is not governed by local codes, it is recommended that the American Gas Association's requirements be followed for gas-fired units and the National

Board of Fire Underwriters' requirements be followed for other fuels.

6. Provide a substantial and level foundation. If it is necessary to use waterproof concrete for the foundation, take necessary precautions to prevent cracking of the concrete due to heat from the combustion chamber.

7. If the boiler is to be located on a combustible floor, provide adequate insulation in accordance with the boiler manufacturer's recommendations and any local codes or ordinances.

8. The boiler location should conform to all local codes or ordinances governing the installation of heating equipment and firing devices.

Boiler Connections. In connecting the supply and return mains to the boiler, connections should be selected that will permit the longest flow path of water through the boiler. This will provide better balanced water temperature conditions in the boiler and improve the overall efficiency of the boiler and system. Short-circuiting of the water from return connection to outlet leads to poor operation and danger of heat stress in the boiler.

10-20 PIPE SIZING AND PRESSURE LOSSES

Chapter 9 presented means for accurately determining pipe sizes and pressure (head) losses in pipes. We will now discuss the methods used by many engineers in the design of hydronic piping systems.

The selected water flow velocity through a pipe determines the pipe size for a given gpm flow rate and also establishes the pressure loss in the pipe. At a given flow rate, higher flow velocities result in smaller pipe sizes but will also mean greater pressure loss, which will increase the power cost for pumping. It is therefore a matter of an economic balance between the cost of pipe and installation and the operating cost of the system that must be determined.

It has been found that water flow velocities in excess of 4 ft per second in 1 1/4-in. pipe or smaller are likely to cause noise in residential systems. In industrial applications, which normally have larger pipes, velocities as high as 8 ft per second are not uncommon. Table 10-2 shows recommended flow velocities through pipes of sizes normally encountered in hydronic systems.

Water flowing through a pipe produces friction, with a resulting loss in available energy as was discussed in Chapter 9. In hydronic systems, this loss of available energy is called "pressure loss" or "head loss" and is normally expressed in *feet of water per linear foot of pipe,* or *millinches of water per linear foot of pipe* (1000

TABLE 10-2

Quick pipe sizing table (recommended flow rates in gpm for all piping except branch connections on Taco venturi systems)

Nominal Pipe or Tubing Size In.	STEEL PIPE				COPPER TUBING			
	Minimum GPM	Velocity Ft./Sec.	Maximum GPM	Velocity Ft./Sec.	Minimum GPM	Velocity Ft./Sec.	Maximum GPM	Velocity Ft./Sec.
½"	–	–	1.8	1.9	–	–	1.5	2.1
¾"	1.8	1.1	4.0	2.4	1.5	1.0	3.5	2.3
1"	4.0	1.5	7.2	2.7	3.5	1.4	7.5	3.0
1¼"	7.2	1.6	16.	3.5	7.	1.9	13.	3.5
1½"	14.	2.2	23.	3.7	12.	2.1	20.	3.6
2"	23.	2.3	45.	4.6	20.	2.1	40.	4.1
2½"	40.	2.7	70.	4.7	40.	2.7	75.	5.1
3"	70.	3.0	120.	5.2	65.	3.0	110.	5.2
3½"	100.	3.2	170.	5.4	90.	3.1	150.	5.2
4"	140.	3.5	230.	5.8	130.	3.5	210.	5.6
5"	230.	3.7	400.	6.4	–	–	–	–
6"	350.	3.8	610.	6.6	–	–	–	–
8"	600.	3.8	1200.	7.6	–	–	–	–
10"	1000.	4.1	1800.	7.4	–	–	–	–
12"	1500.	4.3	2800.	8.1	–	–	–	–

Source: Copyright Industrial Press. Reproduced from *Handbook of Air Conditioning, Heating, and Ventilating,* 3rd ed., with permission of the publisher, 1979.

millinches of water head = 1 in. of water head, or 12,000 millinches of water head = 1 ft of water head). The pressure loss or head loss per foot of pipe will be called a *friction rate* (*f*).

The relationships between pipe size, flow velocity, and friction rate in LTW systems are shown in Figures 10-32 and 10-33. Such charts greatly simplify the determination of friction rates at various flow rates in different sizes of pipe.

Figure 10-32 shows head loss in feet of water per

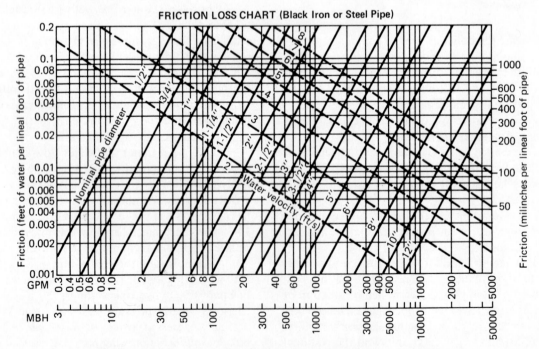

FRICTION LOSS CHART (Black Iron or Steel Pipe)

FIGURE 10-32   Pipe sizing chart for water flow through black iron or steel pipe. (Courtesy of Taco, Inc., Cranston, RI)

foot of pipe and millinches per foot of pipe plotted against gpm flow rate in various nominal sizes of black iron or steel pipe. The flow velocity in feet per second is also shown. Figure 10-33 shows the same quantities for water flow in type L copper tubing. Type L copper tube is the most common type used in hydronic systems.

In order to calculate the head loss through any section of piping, it is necessary to know the friction rate, f, and the total equivalent length (TEL) of pipe and fittings in the section of pipe. The head loss, h_L, will be

$$h_L = f \times \text{TEL} \qquad (10\text{-}24)$$

where f is friction rate (ft of water/ft of pipe).

The TEL of pipe is the measured length of straight pipe plus an allowance for the straight-pipe length equivalents of all the fittings in the particular section of pipe. In Chapter 9, it was shown that the head loss through a pipe fitting could be represented by the head loss through an equivalent length of straight pipe. Tables 10-3 and 10-4 show typical equivalent lengths of straight pipe for various fittings commonly found in hydronic systems.

Note: As may be noted in Tables 10-3 and 10-4, there are some conflicting data concerning the equivalent lengths of some fittings. This is often encountered when data are taken from different sources. The differences are relatively minor, and values from either table would be considered satisfactory in our design work. The following equation may be helpful in making conversions.

To convert head loss in feet of water to feet of pipe equivalents (EL):

$$EL = \frac{h_L \text{ (ft of water)}}{f \text{ (ft of water/ft of pipe)}} \qquad (10\text{-}25\text{a})$$

To convert head loss in inches of water to feet of pipe equivalents:

$$EL = \frac{h_L \text{ (in. of water)}}{12 \times f \text{ (ft of water/ft of pipe)}} \qquad (10\text{-}25\text{b})$$

To convert pressure loss in psi to feet of pipe equivalents:

$$EL = \frac{\text{psi} \times 2.31}{f \text{ (ft of water/ft of pipe)}} \qquad (10\text{-}25\text{c})$$

ILLUSTRATIVE PROBLEM 10-6

A section of pipe in a hydronic heating system carries 30 gpm of water. The pipe is 2-in. nominal steel pipe and scales 50 ft in length. Contained in this section of pipe are two gate valves, two 45-degree elbows, four 90-degree elbows, six flow-through tees, and one square head cock. Determine the head loss in the section of pipe.

Solution:

Step 1. Determine the friction rate. Using Figure 10-32, enter the bottom of the graph at 30 gpm and proceed vertically upward until a point of intersection is found with the 2-in. pipe size. From this point of intersection, proceed horizontally to the left scale and read $f = 0.02$ ft of water/ft of pipe, or horizontally to the right and read $f = 230$ millinches/ft of pipe.

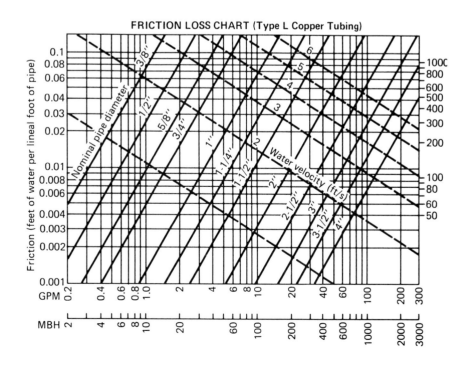

FRICTION LOSS CHART (Type L Copper Tubing)

FIGURE 10-33 Tube sizing chart for water flow through type L copper tubing. (Courtesy of Taco, Inc., Cranston, RI)

TABLE 10-3

Equivalent resistance of valves and fittings

Nominal Pipe Diameter, Inches	Value of Fitting										
	Globe Valve, Open	Gate Valve				Angle Valve, Open	Close Return Bend	Tee		Ordinary Entrance‡
		¾ Closed	½ Closed	¼ Closed	Open			Through Run	Through Side†	
	Equivalent Resistance, Feet of Pipe									
½	16	40	10	2	0.3	9	4	1.0	4	0.9
¾	22	55	14	3	0.5	12	5	1.4	5	1.2
1	27	70	17	4	0.6	15	6	1.7	6	1.5
1¼	37	90	22	4	0.8	18	8	2.3	8	2.0
1½	44	110	28	6	0.9	21	10	2.7	9	2.4
2	55	140	35	7	1.2	28	13	3.5	12	3.0
2½	65	160	40	8	1.4	32	15	4.2	14	3.3
3	80	200	50	10	1.6	41	18	5.0	17	4.5
3½	100	240	60	12	2.0	50	21	6.0	19	5.0
4	120	275	69	14	2.2	55	25	7.0	21	6.0
5	140	325	81	16	2.9	70	30	8.5	27	7.5
6	160	400	100	20	3.5	80	36	10.1	34	9.0
8	220	525	131	26	4.5	110	50	14.0	44	12.0
10	—	700	175	35	5.5	140	60	17.0	55	15.0
12	—	800	200	40	6.5	160	72	19.0	65	16.5
14	—	950	238	48	8.0	190	85	23.0	75	20.0
16	—	1100	272	52	9.0	220	100	26.0	88	22.0
18	—	1300	325	65	10.0	250	115	30.0	110	25.0

Nominal Pipe Diameter, Inches	Value of Fitting									
	Sudden Contraction§			Borda Entrance	Reducing Tee*		90° Elbow			45° Elbow
	D/d = 4	D/d = 2	D/d = 4/3		D/d = 2	D/d = 4/3	Standard	Medium Sweep	Long Sweep	
	Equivalent Resistance, Feet of Pipe									
½	0.8	0.6	0.3	1.4	1.5	1.3	1.5	1.3	1.0	0.8
¾	1.0	0.8	0.5	1.9	2.0	1.8	2.0	1.8	1.4	1.0
1	1.3	1.0	0.6	2.5	2.6	2.4	2.6	2.4	1.7	1.3
1¼	1.6	1.3	0.8	3.5	3.5	3.1	3.5	3.1	2.3	1.6
1½	2.0	1.5	0.9	4.0	4.5	3.7	4.5	3.7	2.7	2.0
2	2.5	1.9	1.2	5.0	5.3	4.5	5.3	4.5	3.5	2.5
2½	3.0	2.2	1.4	6.0	6.3	5.5	6.3	5.5	4.2	3.0
3	3.7	2.8	1.6	7.5	8.0	6.9	8.0	6.9	5.0	3.7
3½	4.4	3.3	2.0	9.0	9.5	8.0	9.5	8.0	6.0	4.4
4	5.0	3.7	2.2	11.0	11.5	9.5	11.0	9.5	7.0	5.0
5	6.0	4.7	2.9	12.0	12.5	12.0	13.0	12.0	8.5	6.0
6	7.5	5.6	3.5	15.0	16.0	14.0	16.0	14.0	10.1	7.5
8	11.0	7.2	4.5	19.0	20.0	18.0	20.0	18.0	14.0	10.0
10	13.0	9.5	5.5	24.0	25.0	22.0	25.0	22.0	17.0	13.0
12	15.0	11.0	6.5	29.0	31.0	26.0	31.0	26.0	19.0	—
14	17.0	12.5	8.0	34.0	36.0	30.0	36.0	30.0	23.0	—
16	19.0	15.0	9.0	38.0	40.0	35.0	40.0	35.0	26.0	—
18	21.0	16.5	10.0	43.0	45.0	40.0	45.0	40.0	30.0	—

Source: Copyright Industrial Press. Reproduced from *Handbook of Air Conditioning, Heating, and Ventilating*, 3rd ed., with permission of the publisher, 1979.

TABLE 10-4

Feet of pipe equivalents for various fittings used in main or trunk connections

	½ C.I.	½ COP.	¾ C.I.	¾ COP.	1 C.I.	1 COP.	1¼ C.I.	1¼ COP.	1½ C.I.	1½ COP	2 C.I.	2 COP.	2½ C.I.	2½ COP.	3 C.I.	3 COP.	4 C.I.	4 COP.	5 C.I	6 C.I	8 C.I.	10 C.I.	12 C.I.
45° Elbow	<				1.0			>	2.2	2.2	2.8	2.8	3.3	3.3	4.0	4.0	5.5	5.5	6.6	8.0	11.0	13.2	16.0
90° Elbow or Union Elbow	<				2.5			>	4.3	4.3	5.5	5.5	6.5	6.5	8.0	8.0	11.0	11.0	13.0	16.0	22.0	26.0	32.0
90° Elbow Long Turn	<				1.5			>	2.7	2.7	3.5	3.5	4.2	4.2	5.2	5.2	7.0	7.0	8.4	10.4	14.0	16.8	20.8
Gate Valve Open	<				.5			>	1.0	1.0	1.2	1.2	1.4	1.4	1.7	1.7	2.3	2.3	2.8	3.4	4.6	5.6	6.8
Globe Valve Open	17	17	22	22	27	27	36	36	43	43	55	55	67	67	82	82	110	110	134	164	220	268	328
Radiator Valve Angle	<				5.0			>															
Tee	<				5.0			>	9	9	12	12	14	14	17	17	22	22	28	34	44	56	68
Boiler	4.0	5.0	5.0	7.0	6.5	9.0	9.0	12.0	9.0	12.0	12.6	16.8	15.6	20.8	18.6	24.8	25.2	33.6	31.2	37.2	50.4	62.4	74.4
Square Head Cock Open	<				1.5			>	2.5	2.5	3.2	3.2	3.9	3.9	4.7	4.7	6.1	6.1					
Radiator C.I.	<				7.5			>															
Std. Venturi (1)			22.5	18	13.5	14	13.5	9.5	13.5	10.5	14.5		14		14								
Super Venturi (1)			37.0	31.5	27.5	28.5	33	21	32.5	26	30		29.5		28.5								
Air Scoop			2.0	2.0	2.7	2.7	4.0	4.0	4.8	4.8	6.8	6.8	8.0	8.0	13.3	13.3	15.0	15.0					
Flow Check			27	27	42	42	60	60	63	63	83	83	104	104	125	125	166	166					
3 Way Mix Valve	Refer To Manufacturers Catalog																						
Convector	Refer To Manufacturers Catalog																						
Baseboard	One Lineal Foot is equal to One Lineal Foot of Pipe																						

(1) Figures shown include Venturi Fitting and Standard Tee

Source: Courtesy of Taco, Inc., Cranston, R.I.

Step 2. Determine total equivalent length of pipe and fittings.

Type of Fitting	Equivalent Length (ft) (Table 10-4)		
2 gate valves	2 × 1.2	=	2.4
2 45° elbows	2 × 2.8	=	5.6
4 90° elbows	4 × 5.5	=	22.0
6 flow-through tees	6 × 3.5	=	21.0 (Table 10-3)
1 square head cock	1 × 3.2	=	3.2
	EL fittings	=	54.2 ft
	SL pipe	=	50.0 ft
	TEL	=	104.2 ft

Step 3. Determine head loss. By Eq. (10-24),

$$h_L = f \times \text{TEL} = (0.02)(104.2)$$
$$= 2.08 \text{ (say, 2.1) ft of water}$$

ILLUSTRATIVE PROBLEM 10-7

Determine a tentative size of type *L* copper tube to carry a flow of 15 gpm of water within recommended limits of velocity. What will be the corresponding friction rate?

Solution:

Step 1. Determine recommended flow velocity range. Refer to Table 10-2 and find, for a 1 1/2-in. tube, a mini-mum flow of 12 gpm at a velocity of 2.1 ft/s or a maximum flow of 20 gpm at a velocity of 3.6 ft/s. Also, note that a 1 1/4-in. tube would carry 13 gpm at a velocity of 3.5 ft/s. (Recall that 4 ft/s is a maximum recommended velocity in residential applications.)

Step 2. Determine tube size. Refer to Figure 10-33 and locate a flow rate of 15 gpm at the bottom of the figure. Proceed upward to locate a pipe size where the velocity would be between approximately 2.1 to 3.5 ft/s. A choice is evident between a 1 1/2- and 1 1/4-in. tube.

For a 1 1/2-in. tube at 15 gpm, the velocity is approximately 2.7 ft/s with a corresponding $f = 0.022$ ft/ft of pipe. For a 1 1/4-in. tube at 15 gpm, the velocity is approximately 4 ft/s with a corresponding $f = 0.053$ ft/ft of pipe.

For the limited data presented in the problem statement, a 1 1/4-in. tube size would be selected. However, since the friction rate is relatively high, the total head loss for a system in which the pipe is located may be too high to be reasonable.

10-21 PUMP SELECTION FOR HYDRONIC SYSTEMS

Circulating pumps used in hydronic systems can vary in size from small, inline circulators delivering up to 5 gpm at 6 or 7 ft head (4 L/s at 18 to 20 kPa above atm.) to large, base-mounted pumps handling hundreds of gpm with heads limited only by pressure characteristics of the system. *Pump operating characteristics should be carefully matched to the system operating requirements.*

The large majority of centrifugal pumps used in residential and small commercial and industrial hydronic systems are inline pumps (see Figure 10-1), with their relatively low heads and capacities, light weight, and ease of installation. Due to their availability in a large number of sizes, a selection can be made very close to the actual system requirements. However, since this type of pump is furnished with one diameter impeller and one operating speed, it has only one performance curve. Therefore, it is usually recommended that a valve be installed in the pump discharge line to make final minor adjustments of the flow to the required gpm. Inline pumps are generally designed with available heads up to 40 ft and flows from very low up to 170 gpm. Figure 10-34 shows a family of pump curves for the inline pumps shown in Figure 10-1. Each curve of Figure 10-34 shows the head-capacity curve for the pump designated. These pumps are mass-produced, are immediately available from stock, and are generally, but not always, powered for nonoverloading characteristics

throughout their entire pumping range. Table 10-5 shows the power ratings and electrical characteristics of the motors supplied for the pumps of Figure 10-34.

Pump curves are arbitrarily classified as "flat" or "steep." This refers to the general shape of the head-capacity curve. Referring to Figure 10-34, we may note that pumps in Series 100 to PR have a steep characteristic, whereas those designated LD3 to PD40 have much flatter curves. Also, the pump of Figure 10-12 has a flat curve. Flat-curve pumps are generally preferred for large, closed-circuit hydronic systems because of the influence of the pump curve on the system operating components. Large changes in capacity can be achieved with small change in head. This is advantageous in balancing multicircuit systems. Flat-curve pumps should also be selected for systems using valve control because they offer a more nearly stable pressure drop across the valves as they go to closed position and they decrease control valve "force open" possibilities.

The pump curve permits use of system curves as a method of pump selection analysis. The system curve can be used as a tool to provide better understanding of pump selection and operation as it affects overall system operation. This will be discussed in a later section.

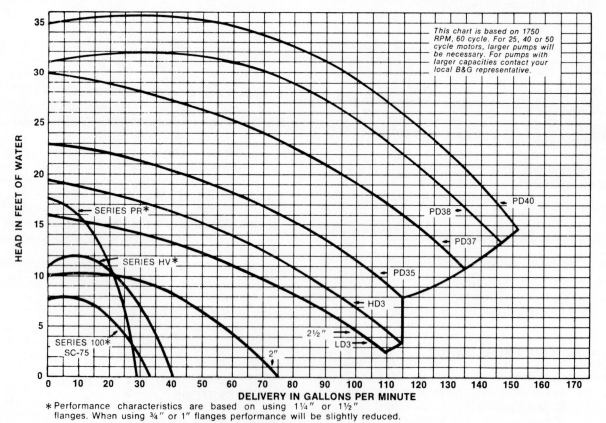

FIGURE 10-34 Performance curves for iron and bronze booster pumps like the one shown in Figure 10-1. (Courtesy of ITT Fluid Transfer Division, Morton Grove, IL)

TABLE 10-5

Power and electrical characteristics for iron and bronze booster pumps of Figures 10-1 and 10-34

Model No.	Flange Size—NPT Inches (specify size)	Standard 60-Cycle Motor Characteristics[a]		
		hp	φ	Voltage
Series 100	¾, 1, 1¼, 1½	1/12		
SC-75	Sweat			
Series PR	¾, 1, 1¼, 1½			
Series HV	1, 1¼, 1½	1/6		115—with
			1	built-in
2	2			overload
2½	2½	1/4		protection
LD3				
HD3		1/3		
PD35-S		1/2		115/230
PD35-T[b]			3	208-230/460
PD37-S	3	3/4	1	115/230
PD37-T[b]			3	208-230/460
PD38-S		1	1	115/230
PD38-T[b]			3	208-230/460
PD40-S		1½	1	115/230
PD40-T[b]			3	208-230/460

Source: ITT Fluid Transfer Division, Morton Grove, IL.
[a]Motors with special electrical characteristics are available on request at additional cost. Refer to motor data sheet HS-615 or contact your local Bell & Gossett representative for details.
[b]External overload must be provided.

Steep-curve pumps may be used in small hydronic systems where head and flow requirements are reasonably stable, such as in residential and small commercial systems.

If flow rates deviate from design flow, the operation of the system will be affected as follows.

1. *For flows below design gpm:*
   (a) Higher temperature drop (2°F higher at 10% lower flow with a 20°F design system temperature drop).
   (b) Power consumption will be less, which may in some cases result in a lower-horsepower motor.
   (c) Lower velocity in main.
   (d) Lower gpm side diversion in venturi fittings in one-pipe systems.
   (e) In some cases, a smaller size pump may be used.
2. *For flows above design gpm:*
   (a) Lower temperature drop (2°F lower at 10% higher flow with a 20°F design system temperature drop).
   (b) Power consumption may be higher, which may in some cases result in a higher-horsepower motor.

(c) Higher main velocity with possibility of creating noise in the system. The flow should not exceed the maximum recommended values shown in Table 10-2.
(d) Higher gpm diversion of venturi fittings in one-pipe systems.
(e) In some cases, a larger size pump may be required.

Figure 10-12 illustrates typical performance curves for close-coupled and base-mounted pumps.

10-22 HYDRONIC SYSTEMS DESIGN EXAMPLES

Series Loop Baseboard Heating Systems. In designing a series loop baseboard system, the maximum load carried by a loop or circuit is dependent upon the size of the baseboard connections. A baseboard unit with 3/4-in. connections will naturally carry a smaller load than one with 1-in. connections. It is the selection of the baseboard size that will determine in all cases whether one or more circuits are necessary.

Because all the water flowing through the main also flows through each baseboard, care must be taken to limit the amount of baseboard connected in series;

TABLE 10-6

Maximum recommended loop or circuit capacities using baseboards

Baseboard Size	Total Load Btu/Hr.	Velocity Based on 20F. Temp. Drop
Cast Iron Baseboard—¾″	40,000	2.4
Steel Pipe Baseboard—½″	18,000	1.9
Steel Pipe Baseboard—¾″	40,000	2.4
Steel Pipe Baseboard—1″	72,000	2.7
Steel Pipe Baseboard—1¼″	160,000	3.5
Steel Pipe Baseboard—1½″	230,000	3.7
Steel Pipe Baseboard—2″	450,000	4.6
Copper Tube Baseboard—½″	15,000	2.1
Copper Tube Baseboard—¾″	35,000	2.3
Copper Tube Baseboard—1″	75,000	3.0
Copper Tube Baseboard—1¼″	130,000	3.5

Source: Courtesy of Taco, Inc., Cranston, R.I.

otherwise, velocity noises may occur. Table 10-6 shows the maximum recommended loop or circuit capacities to be used with various sizes of baseboard units. These capacities will ensure satisfactory velocities, as indicated in Table 10-6, and also will result in reasonable friction losses.

If it is planned to use 1-in. steel pipe baseboard, the total circuit capacity should not exceed 72,000 Btu/hr. A circuit of 1-in. copper tube baseboard should not exceed a capacity of 75,000 Btu/hr. If the system load were to exceed the values, two or more circuits should be used.

It should be recognized at this point that, as the water flows through each heater in series, the water temperature decreases in proportion to the heat output from the individual units. This decrease in water temperature in the main may be readily calculated by

$$\Delta t_{\text{main}} = \Delta t_{\text{circuit}} \left(\frac{\dot{q}_{\text{heater}}}{\dot{q}_{\text{circuit}}} \right) \qquad (10\text{-}26)$$

where

Δt_{main} = temperature drop in the main caused by heat output of an installed heater
$\Delta t_{\text{circuit}}$ = design temperature drop for the circuit
$\dot{q}_{\text{heater}}$ = heat output of a particular heater
$\dot{q}_{\text{circuit}}$ = total heat output of circuit

ILLUSTRATIVE PROBLEM 10-8

Heat loss calculations for a residence give the following room-by-room heat losses:

Room	Heat Loss (Btu/hr)
1	17,400
2	6,700
3	13,400
4	6,700
5	10,000
6	5,400
7	10,700
	Total = 70,300 Btu/hr

Design a series loop hydronic system for the residence using 1-in. steel baseboard units and steel pipe in the main. Assume average water temperature in first heating unit is 190°F and a system water temperature drop of 20 degrees.

Solution:

Step 1. Determine number of circuits. Refer to Table 10-6 and find that a 1-in. steel baseboard circuit should not have a load greater than 72,000 Btu/hr. Since the system load is 70,300, the system will consist of one circuit.

Step 2. Determine average water temperatures to use in evaluating baseboard lengths. Manufacturers rate baseboard heating elements in Btu/hr per linear foot at an average water temperature in the unit. These average water temperature values are normally quoted in 10°F steps; that is, 190°F, 180°F, 170°F, and so forth. Interpolation can be used to obtain ratings at other average temperatures. To simplify the process, we will assume that the capacity of the baseboard will be evaluated at 5-degree intervals starting at 192.5°F leaving the boiler. This means that the circuit load will be divided into essentially four equal sections.

Using Eq. (10-26), the heater load to cause a 5-degree drop in temperature would be

$$\Delta t_{\text{main}} = \Delta t_{\text{circuit}} \left(\frac{\dot{q}_{\text{heater}}}{\dot{q}_{\text{circuit}}} \right)$$

$$5 = 20 \left(\frac{\dot{q}_{\text{heater}}}{70,300} \right)$$

$$\dot{q}_{\text{heater}} = 17,600 \text{ Btu/hr}$$

Therefore, we can assume that a load of 17,600 Btu/hr causes a 5-degree drop in water temperature in the main.

The approximate temperature of the water leaving the boiler then should be 192.5°F. The sum of heat loads for rooms 2 and 3 is 20,100 Btu/hr (approximately 17,600), and we will assume another 5-degree drop in temperature, giving an average water temperature of 185°F for rooms 2 and 3. In a similar manner, rooms 4 and 5 would have an average water temperature of 180°F, and rooms 6 and 7 an average water temperature of 175°F. With these average temperatures, we may now determine the unit capacity for each heating unit and the linear feet of baseboard required for each room. See Worksheet for Illustrative Problem 10-8.

Worksheet for Illustrative Problem 10-8 (series loop baseboard system)

Room	Btu/hr Load	Average Water Temperature	Baseboard Rating Btu/hr/lin. ft	Lin. ft Baseboard Required
1	17,400	190	670	26
2	6,700	185	635	11
3	13,400	185	635	21
4	6,700	180	600	11
5	10,000	180	600	17
6	5,400	175	560	10
7	10,700	175	560	19
Total =	70,300			Total = 115 ft

Step 3. Draw the sketch for the piping system. Figure 10-35 shows a scaled layout of the piping and baseboard units. Pipe will be 1-in. NPS, same as baseboard.

Step 4. Calculate flow rate, TEL, and system resistance. The flow rate may be calculated by use of Eq. (10-17a) and is

$$\text{gpm} = \frac{\dot{q}}{490(t_1 - t_2)} = \frac{70,300}{490(20)} = 7.17$$

Fittings	Equivalent Length (ft) (Table 10-4)
Length, straight pipe	= 131 (scaled from sketch)
Length, baseboard	= 115 (worksheet)
Boiler	= 6.5 (Table 10-4)
Air scoop	= 2.7 (Table 10-4)
Flow-control valve (flow check)	= 42.0 (Table 10-4)
90° elbows (14 est. at 2.5)	= 35.0 (Table 10-4)
Side-flow tee	= 6.0 (Table 10-3)
Gate valves (2 at 0.5)	= 1.0 (Table 10-4)
TEL	= 339.2 ft of pipe

From Figure 10-32, $f = 0.04$ ft of water per ft of pipe length. By Eq. (10-24),

$$h_L = f \times \text{TEL} = (0.04)(339.2)$$
$$= 13.568 \text{ (say, 13.6) ft}$$

Step 5. Select pump and expansion tank. From a pump manufacturer's catalog, select a pump having a capacity of 7.17 gpm against a head of 13.6 ft. Referring to Figure 10-34, one selection would be Series PR, 1-in. size. However, the discharge would have to be throttled to produce the required flow. Expansion tank minimum capacity would be 70,300/5000 = 14 gal.

ILLUSTRATIVE PROBLEM 10-9

Heat loss calculations for a residence give the following room-by-room heat losses:

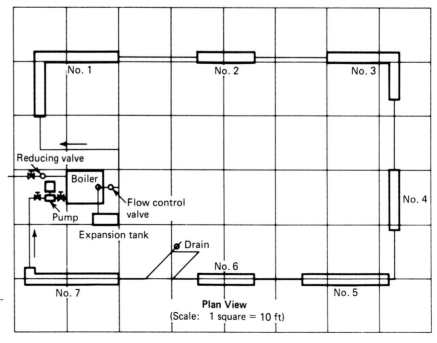

FIGURE 10-35 Series loop baseboard system (Illustrative Problem 10-8).

Room	Heat Loss (Btu/hr)
1	18,500
2	4,000
3	12,300
4	10,000
5	28,000
6	6,500
7	15,000
	Total = 94,300 Btu/hr

Design a series loop heating system for the residence using 1-in. copper tube baseboard units with copper tube main. Assume an average water temperature in the first heating unit is 190°F and a system water temperature drop of 20 degrees.

Solution:

Step 1. Determine number of circuits. Refer to Table 10-6 and find that a 1-in. copper tube baseboard circuit should not have a load exceeding 75,000 Btu/hr. Since the system load is 94,300 Btu/hr, we will use two circuits.

　　As nearly as possible, split the system load equally between the two circuits (approximately 47,000 Btu/hr per circuit). Combining Nos. 1, 2, 3, and 4 gives a circuit load of 44,800 Btu/hr (circuit 1). Combining Nos. 5, 6, and 7 gives a circuit load of 49,500 Btu/hr (circuit 2).

Step 2. Determine average water temperatures to use in evaluating baseboard unit capacity and required length. Each circuit will have a total temperature drop of 20 degrees. Assume that evaluation of average water temperature will be based on a 10-degree drop. Therefore, the baseboard capacity for a 10-degree drop would be, by Eq. (10-26),

$$\Delta t_{main} = \Delta t_{circuit} \left(\frac{\dot{q}_{heater}}{\dot{q}_{circuit}} \right)$$

For circuit 1,

$$10 = 20 \left(\frac{\dot{q}_{heater}}{44,800} \right)$$

$$\dot{q}_{heater} = 22,400 \text{ Btu/hr}$$

For circuit 2,

$$10 = 20 \left(\frac{\dot{q}_{heater}}{49,500} \right)$$

$$\dot{q}_{heater} = 24,750 \text{ Btu/hr}$$

　　The average water temperatures for evaluating baseboard capacities, then, are shown in Worksheet A for Illustrative Problem 10-9.

Step 3. Draw the sketch for the piping system. Figure 10-36 shows a scaled layout of the piping and baseboard units. Pipe will be 1-in. copper tubes for each circuit.

Step 4. Determine flows in each part of the system. By Eq. (10-17a),

$$\text{gpm} = \frac{\dot{q}}{490(t_1 - t_2)}$$

For system main,

$$\text{gpm} = \frac{94,300}{490(20)} = 9.6$$

For circuit 1,

$$\text{gpm} = \frac{44,800}{490(20)} = 4.57$$

For circuit 2,

$$\text{gpm} = \frac{49,500}{490(20)} = 5.03$$

Worksheet A for Illustrative Problem 10-9 (series loop baseboard system with two circuits)

Room	Btu/hr Load	Average Water Temperature	Baseboard Rating Btu/hr/lin. ft	Lin. ft Baseboard Required
Circuit 1				
1	18,500	190	830	22
2	4,000	190	830	5
3	12,300	180	730	18
4	10,000	180	730	14
Total =	44,800			Total = 58 ft
Circuit 2				
5	28,000	190	830	34
6	6,500	180	730	9
7	15,000	180	730	21
Total =	49,500			Total = 64 ft

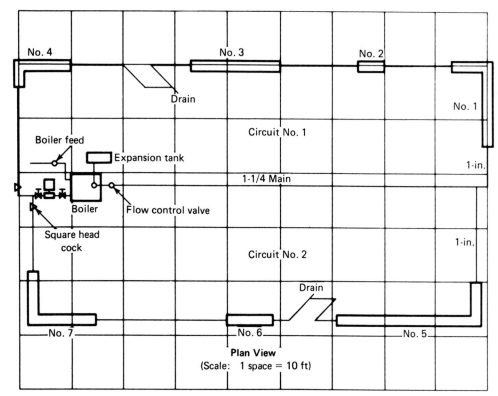

FIGURE 10-36 Series loop baseboard system with two circuits (Illustrative Problem 10-9).

Step 5. Determine system resistance and select pump. See Worksheet B for Illustrative Problem 10-9. The pump would be selected for a capacity of 9.6 gpm against a head of 7.32 ft of water.

One-Pipe Hydronic Systems Using Venturi Fittings. These systems have become very popular in recent years. They overcome some of the difficulties of the series loop systems and are limited in capacity only

Worksheet B for Illustrative Problem 10-9

	Main	*Circuit 1*	*Circuit 2*
Scaled pipe length	73.0 ft	85.0 ft	73.0 ft
Baseboard length		58.0	64.0
Boiler (1¼ in.)	12.0		
Air scoop (1¼ in.)	4.0		
Flow-control valve (flow check, 1¼ in.)	60.0		
Flow-through tees (1 in.)	2 × 1.7 = 3.4		
Gate valves (1¼ in.)	2 × 0.8 = 1.6		
90° elbows	3 × 3.5 = 10.5	7 × 2.6 = 18.2	5 × 2.6 = 13.0
Side-flow tee (1 in.)		1 × 6 = 6.0	3 × 6 = 1.8
Square head cock		1 × 1.5 = 1.5	1 × 1.5 = 1.5
Total equivalent length (TEL)	164.5 ft	168.7 ft	169.5 ft
gpm flow	9.6	4.57	5.03
Friction rate (Figure 10-33)	0.025 ft/lin. ft	0.0175	0.019
Resistance ($h_L = f \times$ TEL)	4.1 ft water	2.95	3.22
Circuit having greatest resistance (2)	3.22		
Pump head required	7.32 ft of water		

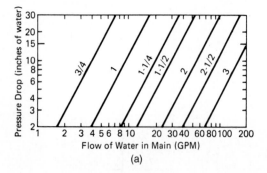

(a)

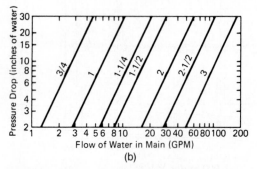

(b)

FIGURE 10-37 Pressure drop through Taco venturi fittings: (a) standard venturi, bronze and cast iron; and (b) super venturi, bronze and cast iron. (From *Taco Guide for Hydronic Engineers*, used with permission from Taco, Inc., Cranston, RI)

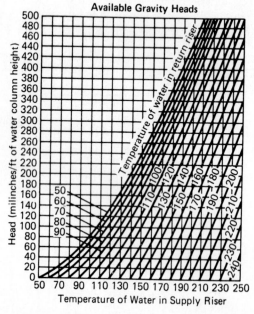

FIGURE 10-38 Available gravity heads in hydronic systems. (From *Taco Guide for Hydronic Engineers*, used with permission from Taco, Inc., Cranston, RI)

by the available size and diverting capacity of the special venturi tees used on each terminal unit.

The design of one-pipe venturi systems will involve the use of tables and figures already presented and also Figures 10-37 and 10-38. Figure 10-37 presents performance data of various sizes of venturi tees. For a given flow in the main in gpm, each size venturi within its operating range will cause a pressure drop that causes a small portion of the main flow to be diverted through a heater circuit. For example, if the main size is 1 1/4 in. carrying a flow of 15 gpm, a *standard venturi* will cause a pressure head drop of 6.0 in. of water. A *super venturi* would cause a pressure head drop of 15.0 in. of water under the same flow conditions. This pressure head drop caused by the venturi fittings is available to overcome flow resistance in the heater branch circuits.

Figure 10-38 is a graph showing the available gravity heads to cause flow in risers. This is a positive assistance when considering upflow risers but is often not considered during system design unless the risers are long. For heaters located *below* the main, the gravity effect is negative in the downfeed risers and must be considered in the proper selection of the venturi tee fittings.

ILLUSTRATIVE PROBLEM 10-10

Assume that a section of a circuit main has three heaters connected as shown in Figure 10-39. The first heater is a second-floor heater having a capacity of 10,000 Btu/hr, the second heater is a first-floor heater with a capacity of 8000 Btu/hr, and the third heater is a basement heater located below the main having a capacity of 6000 Btu/hr. Assume that manufacturers' data on the heaters states that the pressure head loss through the heaters is 2.0 in. water/gpm flow through heater. Also assume that the circuit main is 1 1/4-in. steel pipe with water flow at 16 gpm, 200°F water exists upstream of heater 1, and a 20-degree water temperature drop occurs in each heater. Determine steel pipe size for the risers on each heater, required venturi tee size, and the water temperature in the main after each heater.

Solution (Heater 1):
Step 1. Determine flow rate in heater circuit. By Eq. (10-17a),

$$\text{gpm}_{\text{heater}} = \frac{\dot{q}_{\text{heater}}}{490(t_1 - t_2)} = \frac{10,000}{490(20)} = 1.02$$

Step 2. Estimate pipe size for heater circuit and calculate circuit resistance. From Table 10-2, a 1/2-in. steel pipe should carry this flow within given velocity limits. Therefore, we shall assume that 1/2-in. pipe will be used for risers.

Equivalent length of pipe and fittings in the heater circuit must be determined in order to calculate circuit resistance, using Worksheet A for Illustrative Problem 10-10.

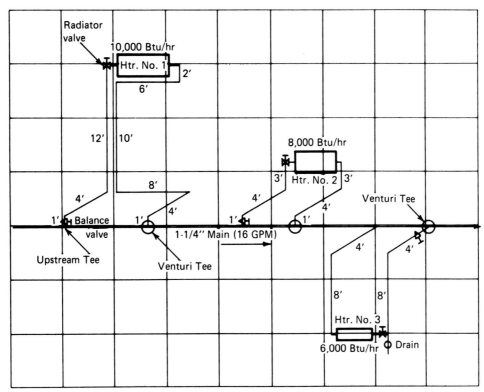

FIGURE 10-39 Section of a one-pipe venturi system (Illustrative Problem 10-10).

From Figure 10-32, for a flow of 1.02 gpm in a 1/2-in. pipe, the friction rate f is 0.015 ft of water/ft of pipe. Therefore, the equivalent length of the heater is, by Eq. (10-25b),

$$EL_{heater} = \frac{2.0(1.02)}{12(0.015)} = 11.3 \text{ ft}$$

Total equivalent length of heater circuit = 79.5 + 11.3 = 90.8 ft. By Eq. (10-24), heater circuit resistance (h_L) = $f \times$ TEL, or

$$h_L = (0.015)(90.8)$$
$$= 1.36 \text{ ft} = 16.3 \text{ in. water}$$

Step 3. Select venturi tee size. From Figure 10-37, a 1 1/4 *standard* venturi tee with a flow of 16 gpm in the main

produces a head drop of 6.7 in. water. For a 1 1/4 *super* venturi tee, the head drop is 17.0 in. water. The required drop is 16.3 in. water (step 2); therefore, a super venturi is required when 1/2-in. risers are used. At this point, we could check to see the effect of using 3/4-in. risers.

Step 4. Assume 3/4-in. risers and recalculate steps 2 and 3. By referring to Table 10-4, we find that the equivalent length of pipe and fittings does not change (79.5). However, the equivalent length of the heater does change because the friction rate changes.

Referring to Figure 10-32, 1.02 gpm flowing in 3/4-in. pipe has a friction rate of 0.0035 ft water/ft of pipe. Therefore,

$$EL_{heater\ 1} = \frac{2.0(1.02)}{12(0.0035)} = 48.6 \text{ ft}$$

Worksheet A for Illustrative Problem 10-10

Fittings		Equivalent Length (Table 10-4)
Riser length (total)		= 48.0 ft
Square head cock (balance valve)	(1 at 1.5)	= 1.5
Angle radiator valve	(1 at 5)	= 5.0
90° elbows	(8 at 2.5)	= 20.0
Side-flow tee (upstream tee used as elbow)		= 5.0
Equivalent length pipe and fittings		= 79.5 ft

Total equivalent length of heater circuit = 79.5 + 48.6 = 128.1 ft. By Eq. (10-24),

$$h_L = (0.0035)(128.1)$$
$$= 0.448 \text{ ft} = 5.38 \text{ in. water}$$

From before (step 3), a 1 1/4-in. standard venturi produces a head drop of 6.7 in. water, which is satisfactory being greater than 5.38 in. water required.

We may now conclude that a 1 1/4 by 1/2 super venturi could be used with 1/2-in. risers or a 1 1/4 by 3/4 standard venturi could be used with 3/4-in. risers.

Step 5. Determine temperature drop of water in the main caused by heater 1. If the flow of water in the main is 16 gpm and a 20-degree drop in temperature is assumed for the main, the total system heat load is, by Eq. (10-17a),

$$\text{gpm} = \frac{\dot{q}_{\text{system}}}{490(t_1 - t_2)}$$

$$16 = \frac{\dot{q}_{\text{system}}}{490(20)}$$

$$\dot{q}_{\text{system}} = 156,800 \text{ Btu/hr}$$

The water temperature drop in the main caused by heater 1 is, by Eq. (10-26),

$$\Delta t_{\text{main}} = \Delta t_{\text{circuit}}\left(\frac{\dot{q}_{\text{heater}}}{\dot{q}_{\text{system}}}\right)$$

$$= 20\left(\frac{10,000}{156,800}\right)$$

$$= 1.27 \text{ degrees}$$

Therefore, the water temperature in the main following heater 1 and entering heater 2 would be 200 − 1.27 = 198.7°F.

Solution (Heater 2):

Step 1. Determine flow and assume pipe size for heater circuit:

$$\text{gpm} = \frac{8000}{490(20)} = 0.82$$

From Table 10-2, a 1/2-in. steel pipe should carry this flow within given velocity limits.

Step 2. Determine resistance of heater circuit. From Figure 10-32, for a flow of 0.82 gpm, $f = 0.0095$ ft/ft of length. See Worksheet B for Illustrative Problem 10-10.

Step 3. Determine venturi size. From before, a 1 1/4 by 1/2 standard venturi produces a head drop of 6.7 in., which is adequate.

Step 4. Determine temperature drop in main:

$$\Delta t_{\text{main}} = 20\left(\frac{8000}{156,800}\right) = 1.02 \text{ degrees}$$

Therefore, water in main after heater 2 is 198.7 − 1.02 = 197.6°F.

Solution (Heater 3):

Step 1. Determine gravity head to be overcome. When the heating unit is below the main, it is necessary to provide sufficient head by the venturi to force water down the risers against the gravity head. This gravity head is caused by the higher-temperature water in the main and the lower-temperature water in the heater (usually assumed to be 60°F, or room temperature).

From preceding calculations, the water temperature in the main at the downfeed supply riser is 197.6°F, and the water in the return riser will be assumed at 60°F. Referring to Figure 10-38, the gravity head to be overcome is approximately 430 millinches per foot of height. Since the riser height is 8 ft, the gravity head is 8 × 430 = 3440 millinches, or 3.44 in. water.

Step 2. Select a tentative venturi size. From Figure 10-37, we have found previously that a 1 1/4-in. standard venturi tee produces a head drop of 6.7 in. and a super venturi tee produces a head drop of 17.0 in. water. Step 1 determined that 3.44-in. gravity head must be overcome; therefore, the remaining available head to overcome friction would be

---

Worksheet B for Illustrative Problem 10-10

Fittings		Equivalent Length (Table 10-4)
Riser length		= 16.0 ft
Square head cock	(1 at 1.5)	= 1.5
Angle radiator valve	(1 at 5.0)	= 5.0
90° elbows	(5 at 2.5)	= 12.5
Side-flow tee	(1 at 5)	= 5.0
Equivalent length pipe and fittings		= 40.0 ft

Equivalent length of heater $= \dfrac{2.0(0.82)}{12(0.0095)} = 14.4$ ft

Total equivalent length of heater circuit = 40 + 14.4 = 54.4 ft

Heater circuit resistance, $h_L = (0.0095)(54.4) = 0.51$ ft = 6.2 in. water

standard venturi $= 6.7 - 3.44$
$$= 3.26 \text{ in. water (very low)}$$

super venturi $= 17.0 - 3.44$
$$= 13.56 \text{ in. water}$$

At this point, we may select a 1 1/4-in. super venturi, which gives us a head of 13.56 in. water to overcome friction losses in risers and heater 3.

Step 3. Determine riser size. From Figure 10-32, for a flow rate of 0.612 gpm, the friction rate f is 0.0056 for a 1/2-in. pipe and 0.0014 for a 3/4-in. pipe. For 1/2-in. pipe risers, see Worksheet C for Illustrative Problem 10-10.

$$\text{EL}_{\text{heater 3}} = \frac{2.0(0.612)}{12(0.0056)} = 18.2 \text{ ft}$$

Total equivalent length $= 45.5 + 18.2 = 63.7$ ft. By Eq. (10-24), head loss $= f \times \text{TEL}$, or

$$h_L = (0.0056)(63.7)$$
$$= 0.357 \text{ ft} = 4.28 \text{ in.}$$

For 3/4-in. pipe risers,

$$\text{EL}_{\text{heater 3}} = \frac{2.0(0.612)}{12(0.0014)} = 72.8 \text{ ft}$$

Total equivalent length $= 45.5 + 72.8 = 118.3$ ft. By Eq. (10-24),

$$h_L = (0.0014)(118.3)$$
$$= 0.166 \text{ ft} = 1.98 \text{ in.}$$

Step 4. Determine venturi size. We may select a 1 1/4 by 1/2 super venturi with 1/2-in. pipe risers, or we may select a 1 1/4 by 3/4 standard venturi with 3/4-in. pipe risers.

Step 5. Determine temperature in main after heater 3:

$$\Delta t_{\text{main}} = \Delta t_{\text{circuit}} \left(\frac{\dot{q}_{\text{heater}}}{\dot{q}_{\text{circuit}}} \right)$$

$$= 20 \left(\frac{6000}{156,800} \right)$$

$$= 0.76 \text{ degrees}$$

Therefore, temperature in main after heater 3 is $197.6 - 0.76 = 196.8°F$.

ILLUSTRATIVE PROBLEM 10-11

Heat loss calculations for a residence give the following room-by-room heat losses:

Room	Heat Loss (Btu/hr)	Room	Heat Loss (Btu/hr)
1	6,000	6	9,000
2	10,000	7	14,000
3	12,000	8	7,000
4	8,000	9	6,000
5	15,000	10	9,000

Design a one-pipe venturi-fitting hydronic system for the residence using a 20-degree design water temperature drop for the system and a 20-degree water temperature drop through each heater (Figure 10-40). All heaters shall be considered convectors having an internal resistance to flow of 2.0 in. water per gpm water flow. Assume the water temperature leaving the boiler is at 200°F. Use copper tube for the main and all heater risers.

Solution: Assume that the risers to each of heaters 1, 2, 3, 4, 5, 8, 9, and 10 contain 10 linear ft of tubing, two 45-degree elbows, three 90-degree elbows, one angle radiator valve, and one square head cock (used as balance valve).

For heater 6 (basement, below main), assume risers contain 18 linear ft of tubing (8-ft risers), three 90-degree elbows, one angle radiator valve, and one square head cock.

For heater 7, assume risers contain 14 linear ft of tube, two 45-degree elbows, three 90-degree elbows, one angle radiator valve, and one square head cock.

Step 1. Numerous calculations are involved in problems of this type; therefore, it is convenient to tabulate results

Worksheet C for Illustrative Problem 10-10

Fittings		Equivalent Length (Table 10-4)
Riser runouts	(2 at 4)	= 8.0 ft
Risers	(2 at 8)	= 16.0
Square head cock	(1 at 1.5)	= 1.5
90° elbows	(4 at 2.5)	= 10.0
Radiator valve	(1 at 5)	= 5.0
Side-flow tee	(1 at 5)	= 5.0
Equivalent length pipe and fittings		= 45.5 ft

Flow through heater 3, (gpm) $= \dfrac{6000}{490(20)} = 0.612$ gpm

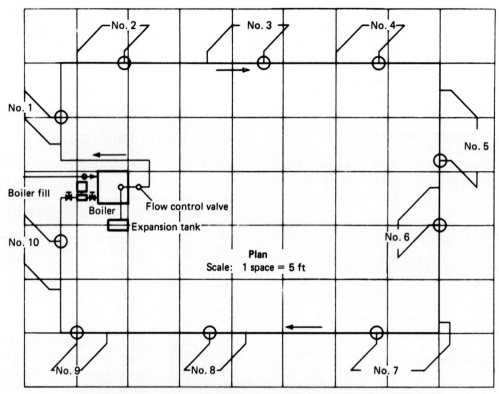

FIGURE 10-40   One-pipe venturi system (Illustrative Problem 10-11).

on a results sheet. See the results sheet for Illustrative Problem 10-11.

Step 2. Calculate flows in various parts of the system. By Eq. (10-17a),

$$\text{gpm} = \frac{\dot{q}}{490(t_1 - t_2)} = \frac{\dot{q}}{490(20)} = \frac{\dot{q}}{9800}$$

For heater 1 (typical for all heaters and system),

$$\text{gpm} = \frac{6000}{9800} = 0.612$$

Step 3. Determine equivalent length of heater circuits and select venturi sizes. Heater circuits, except for 6 and 7, are all the same. The equivalent length of pipe and fittings will be equal for each; however, the equivalent length of the heaters depends upon flow rate, tube size, and friction rate.

Results sheet for Illustrative Problem 10-11

Room or Heater	Heat Loss (Btu/hr)	Required Flow (gpm)	Branch Size (in.)	Figure 10-33 Friction Rate (ft/ft pipe)	Equivalent Length TEL (ft)	Venturi Size	Std. or Super
1	6,000	0.612	1/2	0.011	40.3	1 × 1/2	Std.
2	10,000	1.020	1/2	0.027	37.3	1 × 1/2	Super
3	12,000	1.220	1/2	0.035	36.8	1 × 1/2	Super
4	8,000	0.816	1/2	0.017	39.0	1 × 1/2	Std.
5	15,000	1.531	1/2	0.052	35.9	1 × 1/2	Super
6	9,000	0.918	1/2	0.021	44.3	1 × 1/2	Super
7	14,000	1.428	3/4	0.009	61.4	1 × 3/4	Std.
8	7,000	0.714	1/2	0.014	39.5	1 × 1/2	Std.
9	6,000	0.612	1/2	0.011	40.3	1 × 1/2	Std.
10	9,000	0.918	1/2	0.021	38.3	1 × 1/2	Std.
Total =	96,000	9.789 (9.8)					

Use 1-in. copper tube for main; f = 0.065 ft/ft tube.

Table A for Illustrative Problem 10-11

Fittings		Equivalent Length (Table 10-4)
Risers and runouts		10.0 ft
45° elbows	(2 at 1.0)	= 2.0
90° elbows	(3 at 3.5)	= 7.5
Angle radiator valve	(1 at 5.0)	= 5.0
Square head cock	(1 at 1.5)	= 1.5
Side-flow tee	(1 at 5.0)	= 5.0
Equivalent length pipe and fittings		= 31.0 ft
By Eq. (10-25b), EL $= \dfrac{\text{in. water}}{12 \times f}$		

Assume that all risers for heaters, except 6 and 7, will be 1/2-in. tube. The equivalent length of tube and fittings will be as shown in Table A for Illustrative Problem 10-11.

Heater 1

From Figure 10-33, with 0.612 gpm in 1/2-in. copper tube, $f = 0.011$ ft/ft of tube.

$$EL = \frac{2.0(0.612)}{12(0.011)} = 9.3 \text{ ft}$$

Total equivalent length of heater circuit 1 = 31.0 + 9.3 = 40.3 ft. By Eq. (10-24),

$$\begin{aligned} h_L &= f \times TEL = (0.011)(40.3) \\ &= 0.443 \text{ ft} \\ &= 5.3 \text{ in. water} \end{aligned}$$

Referring to Figure 10-37, for a flow of 9.8 gpm in the 1-in. main, a 1-in. standard venturi produces a head drop of 11.0 in. water, and a 1-in. super venturi produces a head drop of 26.0 in. water.

A 1- by 1/2-in. standard venturi may be used on heater circuit 1. For the remaining similar heater circuits, calculations would produce the sizes shown on the results sheet.

Heater 6

From Figure 10-38, with an estimated water temperature in the main of 190°F and 60°F in heater,

the gravity head to be overcome is 390 millinches/ft of riser height. If the riser height is 8 ft, the total gravity head is (8 × 390) = 3120 millinches, or 3.12 in. water.

From before, a 1-in. standard venturi produces a head drop of 11.0 in. water. Therefore, 11.0 − 3.12 = 7.88 in. water available to overcome flow resistance. (Assume riser will be 1/2-in. tube.) Fitting equivalent lengths are shown in Table B for Illustrative Problem 10-11.

From Figure 10-33, with 0.918 gpm flowing in 1/2-in. tube, $f = 0.021$ ft/ft of tube.

$$EL_{\text{heater 6}} = \frac{2.0(0.918)}{12(0.021)} = 7.3 \text{ ft}$$

Total equivalent length of heater circuit 6 = 37.0 + 7.3 = 44.3 ft. By Eq. (10-24),

$$\begin{aligned} h_L &= (0.021)(44.3) = 0.93 \text{ ft} \\ &= 11.2 \text{ in. water} \end{aligned}$$

This is higher than the 7.88 in. available. Therefore, we may use a super venturi, or we may increase the riser tube size to 3/4-in. We will use a super venturi.

Heater 7

Assume riser size to be 3/4-in. tube. Fitting equivalent lengths are shown in Table C for Illustrative Problem 10-11.

From Figure 10-33, with 1.428 gpm flowing in 3/4-in. tube, $f = 0.009$ ft/ft of tube.

Table B for Illustrative Problem 10-11

Fittings		Equivalent Length (Table 10-4)
Downfeed risers and runouts		18.0 ft
90° elbows	(3 at 2.5)	= 7.5
Angle radiator valve	(1 at 5.0)	= 5.0
Square head cock	(1 at 1.5)	= 1.5
Side-flow tee	(1 at 5.0)	= 5.0
Equivalent length tube and fittings		= 37.0 ft

Table C for Illustrative Problem 10-11

Fittings		Equivalent Length (Table 10-4)
Risers and runouts		14.0 ft
45° elbows	(2 at 1.0)	= 2.0
90° elbows	(3 at 2.5)	= 7.5
Angle radiator valve	(1 at 5.0)	= 5.0
Square head cock	(1 at 1.5)	= 1.5
Side-flow tee	(1 at 5.0)	= 5.0
Equivalent length tube and fittings		= 35.0 ft

$$EL_{heater\ 7} = \frac{2.0(1.428)}{12(0.009)} = 26.4 \text{ ft}$$

Total equivalent length of heater circuit 7 = 35 + 26.4 = 61.4 ft. By Eq. (10-24),

$$h_L = (0.009)(61.4) = 0.55 \text{ ft}$$
$$= 6.63 \text{ in. water}$$

Available head drop from a standard venturi is 11.0 in. water. Use a 1- by 3/4-in. standard venturi.

Step 4. Determine the head loss in the main. The flow in the main, as recorded on the results sheet, is 9.789 gpm, or, say, 9.8 gpm. From Figure 10-33, with 9.8 gpm flowing in the 1-in. tubing, $f = 0.065$ ft/ft of tube.

Table D for Illustrative Problem 10-11 shows the calculations required to determine the resistance in the main. The required pump head (26.2 ft) for this rather small system may seem excessive. Many hydronic systems are designed with the required pump head between 10 and 15 ft. The reason for the high head loss for the system in this example is that the 1-in. main was selected to obtain higher diversion capacities for the venturi fittings. If the main size was 1 1/4 in. (recommended in Table 10-2) to reduce the flow velocity, 3/4-in. risers would be required on all heater circuits. These changes would increase the installed cost of the system but would slightly decrease the operating cost because of possibly using a smaller

pump motor. Normally, it is a decision made by the design engineer as to which procedure to follow.

ILLUSTRATIVE PROBLEM 10-12

The two-pipe, reverse-return hydronic system is used most frequently for medium- and large-capacity systems. Figure 10-41 represents a typical system layout. The circled numbers represent the heating units having the following capacities:

Heater	Capacity (Btu/hr)	Heater	Capacity (Btu/hr)
1	6,000		
2	8,000	11	10,000
3	8,000	12	16,000
4	12,000	13	14,000
5	4,000	14	8,000
6	6,000	15	10,000
7	10,000	16	9,000
8	8,000	17	8,000
9	12,000	18	12,000
10	9,000		

The solid lines on the piping layout represent the supply main and risers; the dashed lines represent the return main and risers. The system has been designed for two circuits. For this illustrative problem, we will be concerned with circuit 1

Table D for Illustrative Problem 10-11

Fittings (1-in.)		Equivalent Length (Table 10-4)
Length of main		125.0 ft
Flow-control valve	(1 at 42)	= 42.0
Boiler	(1 at 9.0)	= 9.0
Gate valves	(2 at 0.5)	= 1.0
90° elbows	(11 at 2.5)	= 27.5
Venturis (standard)	(6 at 14.0)	= 84.0
Venturis (super)	(4 at 28.5)	= 114.0
Total equivalent length pipe and fittings		= 402.5 ft

Main $h_L = (0.065)(402.5) = 26.2$ ft water
Pump must pump 9.8 gpm against a head of 26.2 ft water.

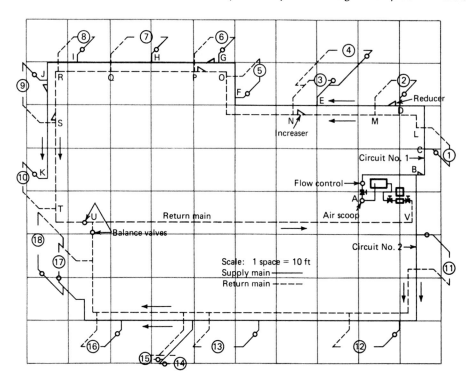

FIGURE 10-41 Two-pipe reverse-return system (Illustrative
Problem 10-12).

only. The same solution procedure would be used for circuit
2.

Problem statement: From Figure 10-41, determine the required steel pipe size for mains and risers for circuit 1 and the
required pump capacity and head. Assume that each first-
floor heater circuit contains five 90-degree elbows, one angle
radiator valve, and 8 linear ft of pipe. Second-floor heater 4
has an additional two side-flow tees, and three 90-degree elbows and 16-ft pipe. Friction loss through any heater will be
considered as 2.0 in. water per 1000 Btu/hr capacity. Assume
a system water temperature drop of 20 degrees.

Solution:

Step 1. Construct a results sheet for tabulation of data and
results. Refer to the results sheet for Illustrative Problem 10-12.

Step 2. Determine equivalent lengths of various pipe sections.
See Table A for Illustrative Problem 10-12.

Step 3. Identify circuit having greatest resistance. A study of
the piping layout will show that all heaters are connected in parallel. Therefore, if we are able to identify
the heater circuit (boiler to heater and back to boiler)
having the greatest resistance, this is the resistance
that must be overcome by the pump.

It would seem, by inspection, that the circuit to
heater 4 would have the greatest resistance, and we
will select it for illustration (see Table B for Illustrative Problem 10-12). This heater circuit starts at *A*,
continues through *B, C, D, E,* up supply riser, down
return riser to *N*, then continues through *O, P, Q, R,
S, T, U, V,* and *A*.

From Figure 10-32, average friction rate is approximately 150 millinches per foot. Pipe loss (h_L) =
72 × 150 = 10,800 millinches. Heater loss = 2.0 ×

12 = 24 in. or 24,000 millinches. Head loss (point *E*
through heater 4 to *N*) is 10,800 + 24,000 = 34,800
millinches.

On the results sheet, identify the section of pipe
and the calculated values of friction. List these friction values in column 10 and add. This sum is then

Column	Description
1	Pipe section under consideration.
2	Heat load in Btu/hr carried by each section of pipe. Note that section *AB* carries the total system load, section *BC* carries the load for circuit 1, section *CD* carries circuit 1 load minus heater 1 capacity, and so on.
3	Flow rate in gpm calculated from Eq. (10-17a), assuming gpm = $\dot{q}/9800$, or approximately $\dot{q}/10,000$.
4	Pipe size using Table 10-2 as a guide to maintain reasonable velocities.
5	Friction rate obtained from Figure 10-32 for the flow rate and pipe size of columns 3 and 4.
6	Scaled length of straight pipe from the piping layout.
7	Calculated equivalent length of the fittings in each individual section of pipe. See step 2 for typical calculations.
8	Total equivalent length; the sum of columns 6 and 7.
9	Product of the friction rate (column 5) and TEL (column 8).
10	Identifies the friction loss in the highest resistance run.

Results sheet for Illustrative Problem 10-12

Supply Pipe Section (1)	Heat Load (Btu/hr) (2)	Flow Rate (gpm) (3)	Pipe Size (4)	Friction Rate (millinches/ft) (5)	Str. Length of Pipe (ft) (6)	EL Fittings (ft) (7)	TEL (ft) (8)	Friction Loss (millinches) (9)	Resistance Longest Circuit (millinches) (10)
AB	170,000	17.0	1 1/2	275	22	77.4	99.4	27,330	27,300
BC	83,000	8.3	1 1/4	160	8	8.9	16.9	2,700	2,700
CD	77,000	7.7	1 1/4	130	16	5.9	21.9	2,800	2,800
DE	69,000	6.9	1	490	20	1.7	21.7	10,600	10,600
EF	49,000	4.9	1	250	22	4.3	26.3	6,600	
FG	45,000	4.5	1	200	12	4.4	16.4	3,300	
GH	39,000	3.9	3/4	500	16	1.4	17.4	8,700	
HI	29,000	2.9	3/4	300	18	1.4	19.4	5,800	
IJ	21,000	2.1	3/4	150	11	3.3	14.3	2,100	
JK	9,000	0.9	1/2	140	23	1.5	24.5	3,400	
Return									
LM	6,000	0.6	1/2	65	13	4.0	17.0	1,100	
MN	14,000	1.4	1/2	300	20	1.3	21.3	6,400	
NO	34,000	3.4	3/4	350	25	3.4	28.4	9,900	9,900
OP	38,000	3.8	3/4	490	9	3.3	12.3	6,000	6,000
PQ	44,000	4.4	1	180	20	1.7	21.7	3,900	3,900
QR	54,000	5.4	1	300	12	1.7	13.7	4,100	4,100
RS	62,000	6.2	1	350	13	5.0	18.0	6,300	6,300
ST	74,000	7.4	1 1/4	140	20	2.3	22.3	3,100	3,100
TU	83,000	8.3	1 1/4	180	12	7.3	19.3	3,500	3,500
UV	170,000	17.0	1 1/2	275	77	4.5	81.5	22,400	22,400
VA	170,000	17.0	1 1/2	275	20	19.8	39.8	10,900	10,900
								Total =	113,530

Table A for Illustrative Problem 10-12

Fittings—Section AB (1½ in.)		EL	*(Table 10-4)*
Air scoop	(1 at 4.8)	= 4.8	
Flow control	(1 at 63.0)	= 63.0	
Gate valve	(1 at 1.0)	= 1.0	
90° elbows	(2 at 4.3)	= 8.6	
EL fittings		= 77.4 ft	

Fittings—Section BC (1¼ in.)		EL	*(Table 10-3)*
Reducing tee	(1 at 3.1)	= 3.1	
90° elbow	(1 at 3.5)	= 3.5	
Flow-through tee	(1 at 2.3)	= 2.3	
EL fittings		= 8.9 ft	

Fittings—Section CD (1¼ in.)		EL	*(Table 10-3)*
90° elbow	(1 at 3.5)	= 3.5	
Reducing tee	(1 at 2.4)	= 2.4	
EL fittings		= 5.9 ft	

Fittings—Section EF (1 in.)		EL	*(Table 10-3)*
90° elbow	(1 at 2.6)	= 2.6	
Flow-through tee	(1 at 1.7)	= 1.7	
EL fittings		= 4.3 ft	

Fittings—Section FG *(1 in.)*		EL	*(Table 10-3)*
90° elbow	(1 at 2.6)	= 2.6	
Reducing tee	(1 at 1.8)	= 1.8	
EL fittings		= 4.4 ft	

Fittings—Section IJ *(¾ in.)*		EL	*(Table 10-3)*
90° elbow	(1 at 2.0)	= 2.0	
Reducing tee	(1 at 1.3)	= 1.3	
EL fittings		= 3.3 ft	

Fittings—Section NO *(¾ in.)*		EL	*(Table 10-3)*
90° elbow	(1 at 2.0)	= 2.0	
Flow-through tee	(1 at 1.4)	= 1.4	
EL fittings		= 3.4 ft	

Fittings—Section OP *(¾ in.)*		EL	*(Table 10-3)*
90° elbow	(1 at 2.0)	= 2.0	
Reducing tee (increaser)	(1 at 1.3)	= 1.3	
EL fittings		= 3.3 ft	

Fittings—RS *(1 in.)*		EL	*(Table 10-3)*
90° elbow	(1 at 2.6)	= 2.6	
Reducing tee (increaser)	(1 at 2.4)	= 2.4	
EL fittings		= 5.0 ft	

Fittings—TU *(1¼ in.)*		EL	*(Table 10-3 or 10-4)*
90° elbow	(1 at 3.5)	= 3.5	
Square head cock	(1 at 1.5)	= 1.5	
Flow-through tee	(1 at 2.3)	= 2.3	
EL fittings		= 7.3 ft	

Fittings—Section VA *(1½ in.)*		EL	*(Table 10-3 or 10-4)*
90° elbows	(2 at 4.5)	= 9.0	
Gate valves	(2 at 0.9)	= 1.8	
Boiler	(1 at 9.0)	= 9.0	
EL fittings		= 19.8 ft	

Table B for Illustrative Problem 10-12

Fittings—Risers to Heater 4 *(¾ in.)*		EL	*(Table 10-3)*
Side-flow tee	(4 at 5.0)	= 20	
90° elbows (¾-in.)	(8 at 2.0)	= 16	
Angle radiator valve	(1 at 12)	= 12	
EL fittings		= 48 ft	
Straight length of pipe		= 24 ft	
TEL		= 72 ft	

combined with the riser loss to give the total resistance that must be overcome by the pump. This would be

results sheet (column 10)	= 113,530
riser loss (heater 4)	= 10,800
heater loss	= 24,000
total resistance longest circuit	= 148,330 millinches
	= 12.36 ft (say, 12.4 ft)

Step 4. Select the pump capacity and head. The pump required must have a flow capacity of 17.0 gpm at a head of 12.4 ft water.

Note: The selection of the highest resistance circuit (heater 4) seemed to be the most obvious choice for evaluation. However, depending on pipe sizes selected, heater 18 on circuit 2 could also be a high-resistance system and would be checked for a complete system design. In most cases, the resistance from the boiler to each heater and back to the boiler is approximately the same because the distance traveled by the water is approximately the same. However, due to the limited number of pipe sizes available, the calculated resistance of each heater circuit may be different. Therefore, a balance valve (square head cock or equivalent) should be installed in each heater riser for final balancing of flow.

ILLUSTRATIVE PROBLEM 10-13

Another popular two-pipe system is the direct-return piping system. Figure 10-42 shows a typical layout of such a system. As may be seen, the distance through the piping to each heater and back to the boiler is different in each case. The pump head required is controlled by the circuit having the greatest resistance, which is usually the longest circuit. When the pump is selected for this longest circuit, it would cause too much flow through the shorter, or low-resistance, circuits. To balance this flow, higher resistance must be built into the shorter circuits by reducing pipe sizes or by use of balance valves or orifice devices. In many cases, a square head cock valve is used in each heater supply riser to control the flow.

Problem statement: Figure 10-42 illustrates a two-pipe direct-return hydronic heating system. The various heater capacities are as follows:

Circuit 1 Heater	Capacity (Btu/hr)	Circuit 2 Heater	Capacity (Btu/hr)
1	20,000		
2	15,000	9	15,000
3	10,000	10	18,000
4	12,000	11	8,000
5	14,000	12	12,000
6	8,000	13	20,000
7	10,000	14	16,000
8	22,000		

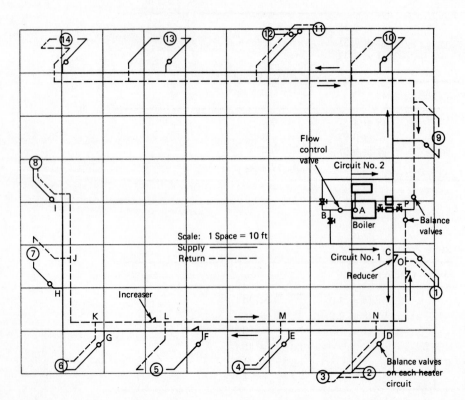

FIGURE 10-42 Two-pipe direct-return system (Illustrative Problem 10-13).

Results Sheet for Illustrative Problem 10-13

Supply Pipe Section (1)	Circuit 1 Heat Load (Btu/hr) (2)	Flow Rate (gpm) (3)	Pipe Size (4)	Friction Rate (millinches/ft) (5)	Str. Length of Pipe (ft) (6)	EL Fittings (ft) (7)	TEL (ft) (8)	Friction Loss (millinches) (9)
AB	200,000	13.6	1 1/2	180	10	81.3	91.3	16,400
BC	111,000	7.6	1 1/4	140	25	10.2	35.1	4,900
CD	91,000	6.2	1	350	20	4.3	24.3	8,500
DE	66,000	4.5	1	200	23	1.7	24.7	4,900
EF	54,000	3.7	1	150	21	1.8	22.8	3,400
FG	40,000	2.7	3/4	250	24	1.4	25.4	6,400
GH	32,000	2.2	3/4	180	20	3.4	23.4	4,200
HI	22,000	1.5	3/4	80	20	2.0	22.0	1,800
Return								
IJ	22,000	1.5	3/4	80	37	124.0	161.0	12,900
JK	32,000	2.2	3/4	180	21	2.0	23.0	4,100
KL	40,000	2.7	1 3/4	250	17	0.5	17.5	4,400
LM	54,000	3.7	1	150	28	1.7	29.7	4,500
MN	66,000	4.5	1	200	23	1.7	24.7	5,000
NO	91,000	6.2	1 1/4	350	21	5.1	26.1	9,100
OP	111,000	7.6	1 1/2	140	12	9.5	21.5	3,000
PA	200,000	13.6		180	18	19.8	37.8	6,800

Total loss longest circuit = 100,300
= 8.36 ft water

Assume that all the heaters are small, fan-coil unit heaters. Manufacturers' data indicate the head loss through the heater coils is 0.30 ft of water per gpm flow rate. The design temperature drop for the system and for each of the heaters will be 30°F with the water temperature leaving the boiler at 215°F. Assume that each heater riser system contains 22 ft of pipe, one square head cock, two globe valves, one union elbow, and six 90-degree elbows. Size the steel pipe on circuit 1 and determine the capacity and head required for the pump.

Solution:

Step 1. Construct a results sheet for data tabulation. See the results sheet for Illustrative Problem 10-13. Tabulate data and results as determined from the following steps.

Step 2. Determine flow rates in each section of piping. The design water temperature drop has been selected at 30 degrees; therefore, from Eq. (10-17a),

$$\text{gpm} = \frac{\dot{q}}{490(t_1 - t_2)} = \frac{\dot{q}}{490(30)} = \frac{\dot{q}}{14,700}$$

Calculate gpm for each section of pipe and enter values on the results sheet in column 3.

Step 3. Determine pipe sizes and friction rate for each section of pipe. Using Figure 10-32 and Table 10-2, determine the pipe size and friction rate maintaining velocities within acceptable limits. Enter values on the results sheet in columns 4 and 5.

Step 4. Determine equivalent lengths of fittings in the various sections of pipe using Tables 10-3 and 10-4. See Table A for Illustrative Problem 10-13. Enter values in column 7 of the results sheet.
Heater 8 loss (h_L) = 0.30 × 1.5 = 0.45 ft water, or, by Eq. (10-25a),

$$\text{EL}_{\text{heater}} = \frac{0.45}{0.007} = 64 \text{ ft of pipe}$$

Total equivalent length of (fittings) section *IJ* including heater 8 = 60 + 64 = 124 ft.

Step 5. Determine total equivalent length of pipe sections and calculate friction loss in each section. The total equivalent length (TEL) is the sum of columns 6 and 7 of the results sheet. Enter values in column 8.

Fittings — Section PA (1½ in.)		EL	(Table 10-3 or 10-4)
Gate valves	(2 at 0.9)	= 1.8	
90° elbows	(2 at 4.5)	= 9.0	
Boiler	(1 at 9.0)	= 9.0	
EL fittings		= 19.8 ft	

Table A for Illustrative Problem 10-13

Fittings—Section AB (1½ in.)		EL	(Table 10-3 or 10-4)
Air scoop	(1 at 4.8)	= 4.8	
Flow-control valve	(1 at 63)	= 63.0	
Side-flow tee	(1 at 9.0)	= 9.0	
90° elbow	(1 at 4.5)	= 4.5	
EL fittings		= 81.3 ft	

Fittings—Section BC (1¼ in.)		EL	(Table 10-3)
Gate valve	(1 at 0.8)	= 0.8	
90° elbows	(2 at 3.5)	= 7.0	
Reducing tee (1-in.)	(1 at 2.4)	= 2.4	
EL fittings		= 10.2 ft	

Fittings—Section CD (1 in.)		EL	(Table 10-3)
90° elbow	(1 at 2.6)	= 2.6	
Run-through tee	(1 at 1.7)	= 1.7	
EL fittings		= 4.3 ft	

Fittings—Section IJ (¾ in.)		EL	(Table 10-3 or 10-4)
Square head cock	(1 at 1.5)	= 1.5	
Globe valves	(2 at 22)	= 44.0	
Union elbow	(1 at 2.5)	= 2.5	
90° elbows	(6 at 2.0)	= 12.0	
EL fittings		= 60.0 ft	

The resistance of each pipe section is the product of the friction rate (column 5) and the TEL (column 8). Enter these products in column 9 of the results sheet.

Step 6. Determine pump head and capacity. The pump must have sufficient capacity to flow the entire gpm for the system, which is 13.6 gpm. The required pump head is the head required to overcome the losses in the longest circuit, which is the sum of column 9 of the results sheet. The loss is 100,300 millinches, or 8.36 ft of water.

Step 7. Determine riser sizes for other heaters on circuit 1. The distance from the boiler to a heater and back to the boiler is different for each heater, which means the resistance is different and the system is not in balance. Some balancing of the individual heater circuits may be accomplished by reducing the size of the risers as we work back toward the boiler. This method is limited because of the standard pipe sizes available, and a 1/2-in. size is about the minimum recommended pipe size.

As an illustration of this method, we will use heater 5 circuit. The resistance of flow from boiler to supply riser takeoff and from return riser back to boiler is the sum of the resistances in the various pipe sections already determined and shown on the results sheet. They are:

Section	Resistance	Section	Resistance
AB	16,400	LM	4,500
BC	4,900	MN	5,000
CD	8,500	NO	9,100
DE	4,900	OP	3,000
EF	3,400	PA	6,800

The sum is 66,500 millinches. The pump has a head of 100,300 millinches. This leaves 100,300 − 66,500 = 33,800 millinches to be used in the risers and heater 5.

The flow rate through heater 5 is 14,000/14,700 = 0.95 gpm. The head loss through heater is 0.30 × 0.95 = 0.285 ft of water = 3400 millinches. The available head to overcome friction in the risers is then 33,800 − 3400 = 30,400 millinches. From step 3, we found that the EL of the riser fittings was 60 ft. The straight length of pipe in the risers was given as 22 ft, and the TEL would be 60 + 22 = 82 ft, exclusive of the heater loss.

The friction rate in the risers should be approximately 30,400/82 = 370 millinches/ft. Referring to Figure 10-32, with f = 370 and 0.95 gpm, a pipe size less than 1/2 in. would be required. Use 1/2 in.; the remaining friction head would be obtained by a balancing valve, or an orifice could be used.

10-23   SYSTEM RESISTANCE CURVES

For a given impeller size and speed, a centrifugal pump has a fixed and predictable performance curve, as was noted in Section 10-7. The point on this performance curve where the pump will operate is dependent upon the resistance characteristics of the system in which the pump is operating.

Hydronic systems in HVAC are of the closed-loop type, the water being circulated through the system and returned to the pumps. No appreciable amount of water is lost from the system.

Hydronic systems are either of the full-flow or the throttling-flow type. Full-flow systems are found usually on residential or small commercial systems where pump motors are small and the energy waste caused by constant flow is not appreciable. Larger systems utilize flow-control valves, which control the flow in the system in accordance with the heating load imposed in the system. Hydronic systems can also be of continuous or intermittent operation. Most hot-water systems are in continuous operation as long as a heating load exists on the system. Condensate and boiler feed pumps, in steam systems, are often of the intermittent type, starting and stopping as the water level changes in condensate tanks or boilers.

Pumps are used to move water through hydronic systems, overcoming the friction caused by liquid flow through equipment, piping, fittings, and valves. The system resistance curve, with flow plotted horizontally and the head plotted vertically, graphically describes the head–flow relationship for a hydronic system from minimum to maximum flow. Inspection of Eqs. (10-5) and

(10-6) shows that the head required to force the liquid through any system of conduits may be divided into two categories: (1) independent heads that are constant regardless of flow rate and (2) heads that vary approximately as the square of the flow rate. In the first category are static elevation difference between reservoir and receiver, and pressure difference between reservoir and receiver (this latter item may vary as a function of flow but is usually constant). Typical of such heads are (a) static rise to the top of a cooling tower in refrigerant condenser water systems and (b) boiler pressure in steam condensate return systems and boiler return systems. The second category includes velocity head, pipe friction, shock losses, and losses through control valves and heat-transfer devices.

Losses involving control valves and heat-transfer coils may need some explanation. The head loss of a control valve and its heating coil is independent of the total flow in a system because these head losses can occur at any time regardless of the total flow in the entire system; that is, a particular coil may require its maximum flow and, therefore, maximum head loss for the coil and its valve even though the total flow in the entire system is at a minimum with minimum system friction head.

Plotting system heads against corresponding system flow rates produces a system head curve such as is shown in Figure 10-43 (solid curve). Most hydronic systems are without independent head, and the system friction head curve becomes the system resistance curve (see Illustrative Problem 10-14). Domestic and small commercial heating systems without control valves are typical of systems without independent head. In opera-

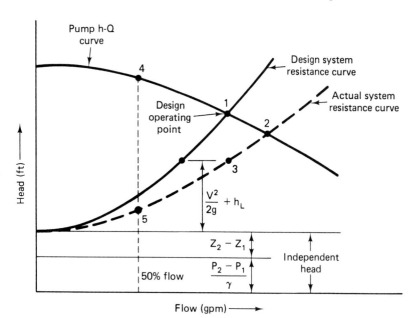

FIGURE 10-43 System resistance curves (design and actual) and pump characteristic curve.

tion, since flow is not throttled, flow occurs at only one point in the system, point 1 on Figure 10-43, at the intersection of the pump head–capacity curve and the system resistance curve. If design contingencies have included friction heads that do not actually exist, then the actual system resistance curve (dashed curve, Figure 10-43) would result when the system is put into operation.

In Figure 10-43, if the system is of the free-flowing type without control valves and with an actual system resistance curve as shown, the pump will operate at point 2, not point 1; the pump will produce a higher flow rate than the design flow rate. If the system is of the controlled-flow type with two-way valves on all heating or cooling coils, at design flow the pump will operate at point 1 and it will create an overpressure on the coils and control valves equal to the head difference between points 1 and 3. If the system flow is reduced to 50% of design on such a system, the overpressure will increase to the amount between points 4 and 5. Pump operation can be summed up as follows: (1) On a system without control valves, the pump will always operate at the point of intersection of the pump h–Q curve and the system resistance curve; (2) on controlled-flow systems, the pump will follow its h–Q curve, the difference between pump head and system head being converted into overpressure, consumed by the control valves.

ILLUSTRATIVE PROBLEM 10-14

Assume that a hydronic system has a total equivalent length of 857 ft and the pipe size is 1 1/2-in. nominal steel. Determine the system resistance curve for flow rates between 10 and 40 gpm.

Solution: Referring to Figure 10-44, determine the friction rate at flows of 10, 20, 30, and 40 gpm and calculate system resistance at these flows.

Figure 10-44 shows the plot of the system resistance curve and the h–Q curve for a constant-speed pump that may have been selected for the system. Where these two curves intersect

Flow Rate (gpm)	Friction Rate f (ft/ft of pipe)	Total System Resistance (ft) (f × TEL)
10	0.0085	7.4
20	0.030	26.3
30	0.065	56.9
40	0.120	105.0

would be the design operating point, at 32 gpm and 64 ft of head. If the resistance of the system, as installed, produced the actual system resistance curve (dashed curve), then the pump and system would operate at point 2. If the desired flow in the system were 32 gpm, the pump would operate at point 1, causing an overpressure equal to the head difference 1 to 3. This head difference would be absorbed by the control valve.

In piping systems where a number of different pipe sizes in series are involved, such as in two-pipe systems, determination of the system resistance curve becomes a complicated procedure. The procedure can be greatly simplified if it is assumed that the resistance through all the various sized piping will be increased or decreased proportionally depending on the flow volume. As the resistance through a pipe is proportional to the square of the velocity and the velocity is proportional to the flow volume, we may use the following relationship to determine the system resistance at different flow rates to closely approximate the system resistance curve:

$$\frac{h_1}{h_2} = \left(\frac{Q_1}{Q_2}\right)^2 \qquad (10\text{-}27)$$

where

h_1 = known (or calculated) head at design flow (ft)
Q_1 = system design flow rate (gpm)
h_2 = system resistance curve head point at flow rate Q_2 (ft)
Q_2 = system flow rate at condition 2 (gpm)

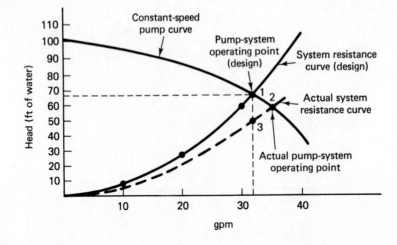

FIGURE 10-44 System resistance curve and constant-speed pump curve (Illustrative Problem 10-14).

In SI units, h_1 and h_2 are the fluid pressure divided by the density. This arrangement permits the pressure drops of selected flow rates to be determined and a typical system resistance curve to be plotted.

Overpressure Control.   Recognizing that overpressure can occur in controlled-flow systems where heating units are equipped with two-way control valves, the selection of pumps for such systems must include methods for limiting overpressure to an economical level. These methods include the following:

- Multiple pumps operating in parallel.
- Multiple pumps operating in series.
- Multispeed pump motors.
- Variable-speed pump drives.

The actual method used on a specific hydronic system depends on the economics of that system. The effects of these methods on overpressure are briefly described in the following paragraphs. The actual results secured on a particular system can be determined by developing the system resistance curve and plotting the pump h–Q curves on the same graph.

Parallel pumping: Using multiple pumps in parallel is the most common method for reducing overpressure. Figure 10-45 shows two pumps connected in parallel, the resulting h–Q pump curves for single-pump and two-pump operation, and a system resistance curve. In parallel, each pump operates at the same head and provides its share of the system flow at that head. Generally, pumps of equal size and capacity are used, and the parallel pump curve is established by doubling the flow of the single-pump curve.

Plotting the system resistance curve across the parallel pump curve shows the operating points for both single- (point 2) and parallel- (point 1) pump operation. Note that single-pump operation does not yield 50% system flow. The system curve crosses the single-pump curve (point 2) considerably to the right of its operating point (point 3) when both pumps are running. This leads to two important considerations: (1) The pumps must be powered to prevent overloading during single-pump operation, and (2) single-pump operation permits standby service up to about 80% of design flow.

It is obvious from Figure 10-45 that one pump at 50% system flow will reduce the overpressure caused by two-pump operation or for one pump selected to handle maximum design flow and head.

Series pumping: Figure 10-46 shows two identical pumps piped in series with bypasses for single-pump operation and indicates on the performance graph the effect of series pumping on a hydronic system with a system resistance curve having a large amount of system resistance. When two pumps are operating in series, each pump has the same flow but their heads are additive. For high-resistance systems, series pumping can greatly reduce the overpressure on controlled-flow systems. Series pumping should *not* be used with *flat* sys-

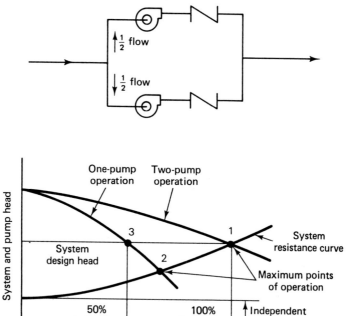

FIGURE 10-45   Pump and system curves for parallel pumping.

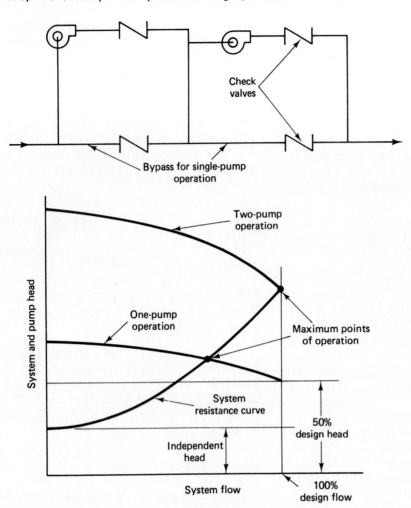

FIGURE 10-46 Pump and system curves for series pumping.

tem resistance curves similar to the one shown in Figure 10-43. For such a system, one-pump operation with series connection would result in the pump's running at shutoff head and producing no flow in the system.

Speed control: Two-speed motors, which are standard production units, are available at speeds of 1750/1150 rpm (29 and 19 rps), 1750/850 rpm (19 and 14 rps), and 3500/1750 rpm (58 and 29 rps). These motors can reduce overpressure at reduced system flow. This was discussed in Section 10-7.

Variable-speed drives have a similar effect on head–capacity curves as do two-speed motors. These drives normally have an infinitely variable speed range so that the pump, with proper controls, can follow the system resistance curve without any overpressure.

REVIEW PROBLEMS

10.1. Referring to Illustrative Problem 10-1, what is the pump horsepower for the system if the pump has an efficiency of 70%?

10.2. A piping system handling 60°F water requires a pump to flow 200 gpm against a head of 86 ft. Select a pump impeller size from Figure 10-12 and state the pump efficiency, pump motor horsepower required, and required NPSH.

10.3. A pump operating at 1750 rpm at its best efficiency point produces a flow rate of 30 gpm against a head of 25 ft and requires a power input of 0.30 BHP. A drive arrangement is available that will increase the pump speed to 2000 rpm. What would be the flow rate, head, and BHP at the speed of 2000 rpm, assuming constant efficiency?

10.4. Figure 10-47 shows a pump installed above a reservoir containing water at 80°F. Normally, the 3-in. suction pipe extends 6 ft below the reservoir water surface. Assume barometric pressure is 14.7 psia, head loss in suction pipe 10 ft of water, and the water temperature is 80°F.

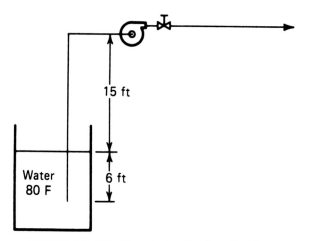

FIGURE 10-47 Sketch for Review Problem 10.4.

(a) If the water level in the reservoir is 15 ft below the pump centerline, what is the available net positive suction head?

(b) Determine available net positive suction head if the reservoir water level dropped to 20 ft below the pump centerline.

10.5. A section of pipe in a hydronic heating system carries 20 gpm of water. The pipe is 1 1/2-in. nominal steel pipe and has a scaled length of 50 ft. Contained in this section of pipe are two gate valves, 3 flow-through tees, four 90-degree elbows, two 45-degree elbows, and one square head cock valve. Determine the head loss in the section of pipe (a) using Tables 10-3 and 10-4 and (b) using methods of Chapter 9.

10.6. Determine a tentative size of steel pipe to carry a flow of 20 gpm of water within recommended limits of velocity. What will be the corresponding friction rate?

10.7. Figure 10-48 shows a plan layout for a one-pipe, two-circuit series loop hydronic heating system using baseboard terminal heating units. The baseboard units have been selected from Table 8-11 of Chapter 8. The baseboard units have the following capacities and lengths:

Baseboard Unit No.	Capacity (Btu/hr)	Length (ft)
1	18,000	22
2	8,000	10
3	10,000	16
4	14,000	20
5	6,000	8
6	9,000	16

If the supply water leaving the boiler is at 200°F and a system water temperature drop of 20 degrees is allowed, determine the following:

(a) The average water temperature in each heating unit.

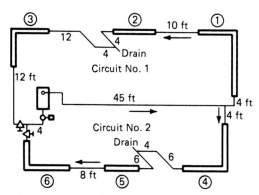

FIGURE 10-48 Sketch for Review Problem 10.7.

(b) The required copper tube size for the supply main and each loop.

(c) The required pump capacity and head to flow the system. (Include resistance of all required fittings and accessories that would normally be installed in the system.)

10.8. Figure 10-49 represents part of a one-pipe hot-water heating system. The longer circuit shown carries a total heating load of 80,000 Btu/hr and has a measured length (AB) of 150 ft. The total system heating load is 150,000 Btu/hr supplied by the main (BA), which has a measured length of 75 ft. One terminal heat-transfer unit on the longer circuit is shown. Assume that there are eight such terminal units, each having a capacity of 10,000 Btu/hr and an internal pressure head loss of 2000 millinches of water. Also, assume each THTU circuit contains seven 90-degree standard elbows, one angle valve, one balance valve, and 12 ft of straight pipe. The system main (BA) contains one boiler, one flow-control valve, four 90-degree standard elbows, one air scoop, and two gate valves. Determine the following:

(a) Steel pipe size required for the longer circuit (AB) and supply main (BA) for a system water temperature drop of 20°F.

(b) Venturi fitting size required for the THTU shown and the pipe size for the heater circuit.

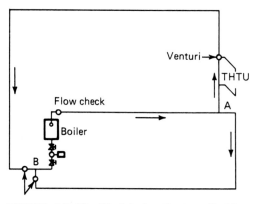

FIGURE 10-49 Sketch for Review Problem 10.8.

(c) Average water temperature in the THTU shown if the supply water at point *A* is 200°F.

(d) Pump capacity and head required for the system.

10.9. Figure 10-50 illustrates a two-pipe hot-water heating system with four horizontal-delivery unit heaters manufactured by Modine (see Table 8-15). Heater *A* is model HS-86, heaters *B* and *C* are model HS-63, and heater *D* is model HS-121. All heaters are operating on high motor speed. For a design water temperature drop of 20°F and entering water temperature of 200°F, determine the following:

(a) Heating capacity, water flow rate, water pressure drop, and final air temperature for each heater.

(b) Steel pipe size for each section of piping.

(c) Required pump capacity and head. (Assume that the supply and return mains contain one air scoop, one flow check valve, one boiler, two gate valves, one balance valve, and 90-degree standard elbows as shown. Assume the fittings at each heater include one balance valve, two globe valves, and four 90-degree standard elbows.)

10.10. Figure 10-51 represents a two-pipe reverse-return hydronic system. The heater loads in Btu/hr are as follows:

Circuit No. 1		Circuit No. 2	
Heater No. 1	12,000	Heater No. 10	10,000
2	8,000	11	12,000
3	10,000	12	8,000
4	12,000	13	9,000
5	9,000	14	15,000
6	10,000	15	10,000
7	8,000	16	8,000
8	15,000	17	6,000
9	9,000		

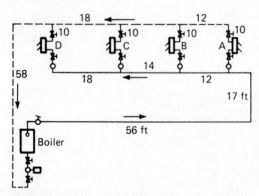

FIGURE 10-50 Sketch for Review Problem 10.9.

Supply water temperature is 200°F and the system design temperature drop is 20 degrees. Each first-floor heater branch circuit contains 8 ft of pipe, one balance valve, two side-outlet tees, six 90-degree standard elbows, and one angle valve. Each second-floor heater branch circuit contains 20 ft of pipe, one balance valve, two side-outlet tees, eight 90-degree standard elbows, and one angle valve. The system main contains one air scoop, one flow check, three gate valves, six 90-degree elbows, and one boiler. Head loss through each heater is 1 ft of water per 5000 Btu/hr capacity. Consider that the greatest head loss occurs in circuit no. 1 and determine the following:

(a) Steel pipe size for each section of piping in circuit no. 1.

(b) Required pump capacity and head.

10.11. Figure 10-52 represents an isometric view of two first-floor heaters and one second-floor heater connected to the supply main of a one-pipe hot-water heating system. Heating unit capacities are as follows: No. 1 = 8000 Btu/hr, No. 2 = 6000 Btu/hr, and No. 3 = 10,000 Btu/hr. The entire system load is 120,000 Btu/hr. Copper tube lengths are shown on each section of the diagram, and fittings are identified. Friction loss through each heater is estimated at 200 mill-inches per 1000 Btu/hr capacity. System design water temperature drop is 20 degrees.

(a) Select copper tube sizes for each section of piping.

(b) Determine the venturi sizes required.

10.12. Figure 10-53 shows the plan layout of a one-pipe, two-circuit hydronic system. The circled numbers identify the heaters in the system, which have the following capacities in Btu/hr:

First-Floor Heaters		Second-Floor Heaters		Basement Heaters	
No. 1	8,000	No. 6	11,000	No. 17	8,000
2	10,000	8	10,000	18	8,000
3	6,000	11	8,000	19	10,000
4	6,000			20	12,000
5	7,000				
7	8,000				
9	15,000				
10	9,000				
12	10,000				
13	11,000				
14	8,000				
15	11,000				
16	12,000				

Friction loss through the individual heaters will be assumed at 250 millinches per 1000 Btu/hr capacity at a water temperature drop of 20°F through the heaters.

Figure 10-53a shows a riser diagram typical of all first-floor heaters, except heaters 5, 7, and 10. Figure 10-53b shows the riser diagram typical for each

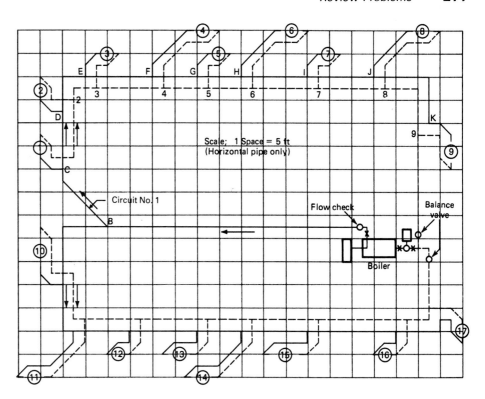

FIGURE 10-51 Sketch for Review Problem 10.10.

basement heater. Figure 10-53c shows the riser diagram for heater combinations 5 and 6, 7 and 8, and 10 and 11.

Determine the venturi size and type (standard or super) and riser sizes required for the heater circuits illustrated in Figures 10-53a through 10-53c. Assume a water temperature drop through each heater of 20°F using type *L* copper tube. Also assume a system design water temperature drop of 20°F. Tabulate the results.

10.13. Determine the size of the supply main for the system of Review Problem 10.12. Calculate the required pump head and capacity for the system.

10.14. Figure 10-12 shows the characteristic curves for a centrifugal pump using different impeller diameters. Two of these pumps having a $5\frac{3}{8}$-in. impeller have been se-

lected for a hydronic system. The pumps will be piped together in such a way that they may be operated either in parallel or in series. Determine the following:

(a) Brake horsepower for each pump when it is operated at the point of maximum efficiency but using a 1750 rpm motor.

(b) Total discharge rate for both pumps operating at rated speed, connected in parallel, when the total head is 80 ft.

(c) Total discharge when both pumps are operating in series at rated speed when the total head is 90 ft. (Assume water specific gravity is 1.0.)

10.15. Figure 10-54 shows an elevation sketch of a piping system used for pumping 60°F water from a reservoir to a storage tank. The pump suction line is 3-in. nominal schedule 40 steel pipe and is 54 ft long from

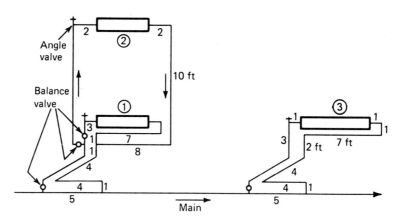

FIGURE 10-52 Sketch for Review Problem 10.11.

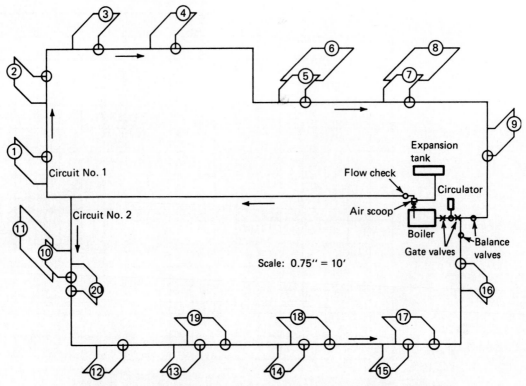

FIGURE 10-53 One-pipe hydronic system for Review Problem 10.12.

strainer to pump inlet. The pump discharge line is $2\frac{1}{2}$-in. nominal schedule 40 steel pipe and is 175 ft long from pump to inlet of tank. The design flow rate through the system is 200 gpm. For the system as shown, determine the following:

(a) Required pump head.
(b) Water horsepower.
(c) Available net positive suction head.
(d) Four points on the system resistance curve at 50, 75, 100, and 125% of the design flow rate.

10.16. Figure 10-55 shows the plan layout sketch of a one-pipe hot-water hydronic system arranged in two circuits. The total heat load on circuit no. 1 is 90,000 Btu/hr and on circuit no. 2 is 70,000 Btu/hr. The first terminal heat-transfer unit (THTU) is shown con-

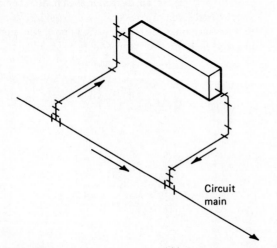

FIGURE 10-53a Review Problem 10.12: riser diagram for first-floor heaters except heaters 5, 7, 10; straight length of tube is 18 ft.

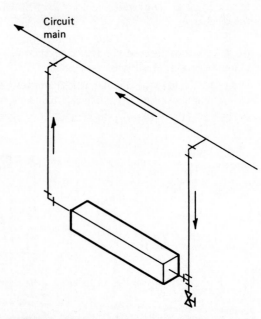

FIGURE 10-53b Review Problem 10.12: riser diagram, typical for each basement heater; straight length of tube is 24 ft.

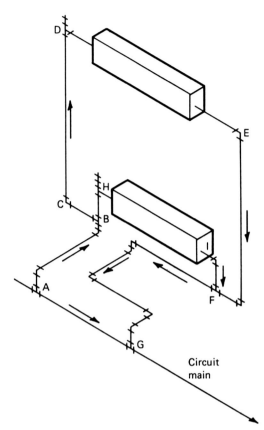

FIGURE 10-53c Review Problem 10.12: riser diagram for heater combinations 5 and 6, 7 and 8, 10 and 11; estimated tube length *ABCDEFG* is 35 ft.

nected to circuit no. 1, along with its riser diagram detail, and the unit capacity is 10,000 Btu/hr. The friction head loss through the THTU is 2000 millinches. There are eight other THTU's on circuit no. 1 that have similar capacities and riser diagrams. Circuit no. 1 has a scaled length of 95 ft (*A* to *B*). The system main (*B* to *A*) has a scaled length of 60 ft. For a system design water temperature drop of 20°F and 20°F through each THTU, determine the following:

(a) Steel pipe size for the main (*BA*), circuit no. 1 (*AB*), and risers to THTU.

(b) System head loss and required pump head and capacity.

10.17. Figure 10-56 represents the plan layout of a two-pipe hot-water piping system supplying six model HS unit heaters (see Chapter 8). Heaters *A* and *F* are model HS-86, heater *B* is model HS-33, heater *C* is model HS-63, and heaters *D* and *E* are model HS-108. All heaters are operating on high motor speed. The design water temperature drop is 20°F, and the entering water temperature is 200°F. The scaled lengths of the pipe sections in the supply and return mains are as follows:

Section	Length (ft)	Section	Length (ft)	Section	Length (ft)
AB	70	EF	36	3–4	21
BC	40	FG	35	4–5	35
CD	50	1–2	40	5–6	15
DE	30	2–3	50	6–7	227
				7–A	15

Make up and complete a results sheet similar to that for Illustrative Problem 10-12.

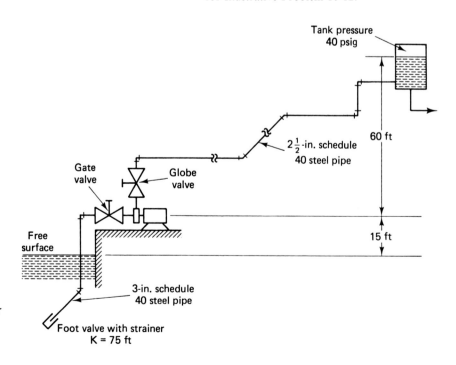

FIGURE 10-54 Piping system for Review Problem 10.15 (elevation).

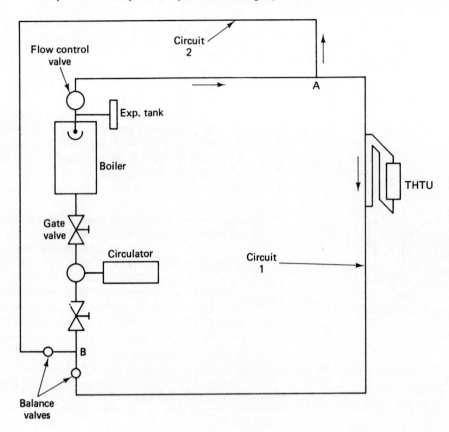

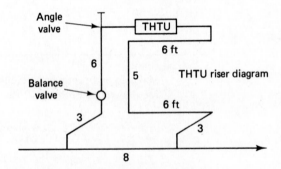

FIGURE 10-55 Diagram for Review Problem 10.16.

Assume that the supply and return mains contain one air scoop, one flow check valve, one boiler, two gate valves, one balance valve, and 90-degree standard elbows as shown. Assume that at each heater the fittings are one balance valve, two globe valves, four 90-degree standard elbows, two side-flow tees, and 12 ft of straight pipe. Determine the following:

(a) Heating capacity, water flow rate, water pressure drop, and final air temperature for each of the unit heaters.

(b) Steel pipe sizes for each section of piping.

(c) Required pump head and capacity.

10.18. A pumping system has the pump centerline located 18 ft above the free surface of the suction reservoir. The head loss in the suction pipe was calculated as 6.5 ft of water. The pump installed is the 3500-rpm model with a $5\frac{7}{8}$-in. diameter impeller with performance as shown in Figure 10-12. The flow rate of 60°F water through the pump is 200 gpm against a head of 110 ft. Determine the NPSH$_R$ from Figure 10-12, and calculate the NPSH$_A$ for the system. If the water temperature was 160°F and other factors remain constant, calculate the NPSH$_A$. Assume barometric pressure is 14.7 psia.

10.19. A water piping system has been designed to distribute 160 gpm against a head 90 ft. Using Figure 10-12, select a pump and specify the power rating of the electric drive motor.

10.20. A 3500-rpm pump of Figure 10-12 has a $5\frac{3}{8}$-in. impeller and is to be used to pump 180 gpm of lake water to a water tank on the shore above the lake. What is the maximum height that the pump can be located above the lake surface to prevent cavitation? Assume the lake water maximum temperature is 70°F, the head loss in the suction pipe is 3 ft of water, and barometric pressure is 28.5 in. Hg.

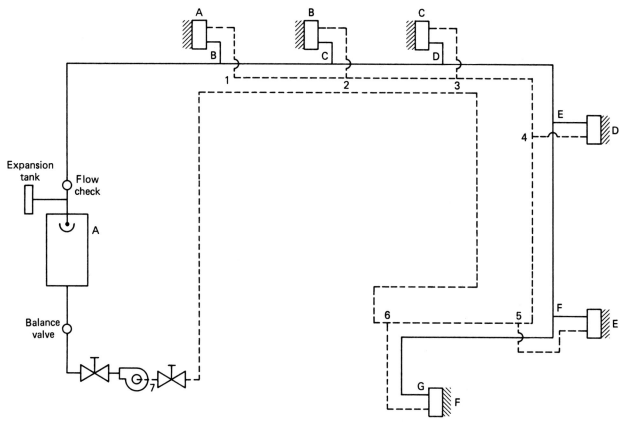

FIGURE 10-56 Piping layout for Review Problem 10.17 (plan).

BIBLIOGRAPHY

10.1. *ASHRAE Handbook 1989 Fundamentals,* American Society of Heating, Refrigerating, and Air Conditioning Engineers, Atlanta, GA, 1989.

10.2. *ASHRAE Handbook 1988 Equipment,* American Society of Heating, Refrigerating, and Air Conditioning Engineers, Atlanta, GA, 1988.

10.3. *ASHRAE Handbook 1987 HVAC Systems and Applications,* American Society of Heating, Refrigerating, and Air Conditioning Engineers, Atlanta, GA, 1987.

10.4. *Goulds Pump Manual,* Goulds Pumps, Inc., Seneca, NY.

10.5. Strock, C., and Koral, R. L. *Handbook of Air Conditioning, Heating, and Ventilating,* 3rd ed., Industrial Press, New York, 1979.

10.6. *Taco Guide for Hydronic Engineers,* Taco, Inc., Cranston, RI.

11

Steam Heating Systems

11-1   INTRODUCTION

This chapter describes the practical design and layout techniques for steam heating systems. Steam piping differs from other systems because it usually carries four fluids: steam, water (condensate), air, and carbon dioxide. For this reason, steam piping design and layout require special consideration.

Advancement in equipment design and application techniques during the past fifty years has done much toward simplifying heating system design. ASHRAE, in particular, has served to change steam heating practice from a rule-of-thumb basis to a more rational one, with a resulting standardization of types of systems, which we shall consider in the following sections.

ASHRAE Handbook 1987 HVAC Systems and Applications suggests that steam heating systems offer the following advantages:

1. Steam flows through the system unaided by external energy sources such as pumps.
2. Because of its low density, steam can be used in tall buildings where water systems create excessive pressure.
3. Terminal heat-transfer units (THTU) can be added to or removed from the system with-

out making basic changes to the system design.
4. Steam system components can be repaired or replaced by closing the steam supply to these units without the difficulties associated with draining and refilling a water system.
5. Steam is pressure–temperature dependent; therefore, the system temperature can be controlled by varying either the steam pressure or the temperature.
6. Steam can be distributed throughout a heating system with little change in temperature.

In view of these advantages, steam is applicable to the following facilities:

1. Where heat is required for process or comfort heating, such as in industrial plants, hospitals, restaurants, dry cleaning plants, laundries, and commercial buildings.
2. Where the heating medium must travel great distances, such as in facilities with scattered building locations, or where the building

height would result in excessive pressures in water systems.

3. Where intermittent changes in heat load occur.

4. Where outdoor air for ventilation is heated (especially in cold climates), such as in factories, school classrooms, auditoriums, and gymnasiums.

5. Where surplus steam from processes can be used for space heating.

6. Where there is a chance of freezing in cold climates or where subfreezing air is handled.

7. Where there may be additions or alterations of spaces or change in occupancy in a building.

8. Where extra heat is needed in buildings with large or frequently used doors, such as in department stores, garages, shipping departments, warehouses, or airplane hangers.

11-2 TERMS AND DEFINITIONS USED IN STEAM HEATING SYSTEMS

Steam system components are referred to by certain well-defined terms. It should be helpful for us to consider some of these terms before continuing our discussion of steam systems.

Air vents: A device for permitting air to be forced out of THTU's or piping and closed against water or steam. Steam cannot circulate in the system or heat the THTU surfaces until air is vented from the system.

Boiler horsepower: Heat in Btu/hr equivalent to that which would evaporate 34.5 lb of water per hour at 212°F to steam at 212°F all at 14.7 psia. This is equal to a heat output of 33,475 Btu.

Condensate: In steam heating systems, the water formed by condensation of steam in THTU's, as well as in the piping system.

Converter: A piece of equipment for heating water with steam without mixing the two. It may be used for supplying hot water for domestic purposes or for a hydronic heating system.

Cooling leg: A length of uninsulated pipe through which the condensate flows to a trap and that has sufficient cooling surface to permit the condensate to dissipate enough heat to prevent flashing when the trap opens. In the case of a thermostatic steam trap, a cooling leg may be necessary to permit the condensate temperature to drop a sufficient amount to allow the trap to open.

Drip connection: When steam system piping carries both steam and condensate, it is often desirable to drain off the condensate at various points to expedite steam flow and reduce the possibility of water hammer. The condensate is drained off to a return pipe through a "drip trap."

Dry return: The dry return is that portion of the return main located above the boiler water level and that carries condensate, air, and water vapor.

Equivalent direct radiation (EDR) (alternately, square feet of heating surfaces): By definition, the amount of heat-transfer surface area that will give off 240 Btu/hr when the heating medium is at 215°F and the ambient air is at 70°F. The equivalent square feet of direct radiation may have no direct relation to the *actual* surface area. *Note:* One pound of steam condensing per hour at 215°F equals approximately 4 EDR.

Flash steam: The rapid passing into steam of water at high temperature when the pressure it is under is reduced so that its temperature is above that of saturation for the reduced pressure. For example, if hot condensate is discharged by a trap into a low-pressure return or into the atmosphere, a certain percentage of water will immediately transform into steam. It is also called *reevaporation*.

Hartford loop: A piping arrangement at the boiler designed to prevent complete drainage of the boiler should a leak develop in the wet return. The wet return is connected to an equalizer line between the supply header and boiler return opening. The return connection is made about 2 in. below the normal water level in the boiler.

Header: Boilers, depending upon their size, have one or more outlet tappings. The vertical steam piping from the tapped outlets enters a horizontal pipe, or steam header. The steam supply mains are connected to this header.

One-pipe steam system: A steam piping arrangement consisting of a main in which the steam and condensate flow in the same pipe. There is but one connection to each THTU, and it must serve as both the supply and return.

Return main: The pipes that carry the condensate from the THTU's back to the boiler.

Riser: The vertical pipe carrying steam to the THTU's from the supply main. In a one-pipe upfeed system, the riser carries steam up to the THTU's and carries condensate downward to the supply main. In one-pipe systems, the horizontal runouts connecting the main to the riser must be pitched up to make this drainage possible.

Steam trap: A device for allowing the passage of condensate and air but preventing the passage of steam. Various steam traps are in use, each having its specific application and construction. They are thermostatic traps, bucket traps, float and thermostatic traps, and thermodynamic traps (disc traps).

Two-pipe steam system: A piping system in which one pipe serves as the steam supply and another serves as the condensate return.

Wet return: That portion of the return main located below the water level in the boiler. It is always completely filled with condensate and does not carry air or steam.

11-3 BASIC OPERATING PRINCIPLES

When steam is used for space heating, one source of the required steam is a boiler. The boiler generates steam, which is delivered to the THTU's through an appropriate piping system. There are a number of piping arrangements in use, and the simplest of these, a one-pipe system, is represented schematically in Figure 11-1. This simple system will be used to describe the principles involved in the operation of steam heating systems.

Before being placed in operation, the system is filled with water to the boiler water level. This water level will be the same in the boiler and in the vertical leg

of the return line before the boiler starts to produce steam. The steam space of the boiler, the piping system, and the THTU's will be filled with air, and additional air will be driven out of the boiler water when it is heated. This air interferes with the flow of steam through the pipes and into the THTU's.

Steam flows through the pipes because of a pressure difference. In the boiler, heat is added to the water to change it into vapor (steam) with a resulting large increase in volume. One pound of water at 212°F and atmospheric pressure has a volume of 0.016716 ft³. One pound of saturated steam at the same temperature and pressure has a volume of 26.80 ft³, or 26.80/0.016716 = 1603 times as great. When the steam is produced in the boiler at a pressure greater than atmospheric, it expands into the heating system displacing and driving out the air through the thermostatic air-vent valves placed at each THTU and at the end of the supply main. These vent valves are normally open and close off as the hot steam reaches them.

Heat is transferred from the piping and the THTU's as the steam condenses back to water with a large reduction in volume. This reduction in volume causes a lower pressure in the system, in turn, causing an increase in steam output from the boiler. Two forces, then, influence the pressure difference necessary for steam to flow: (1) application of heat in the boiler, with evaporation and an accompanying increase in volume; and (2) heat removal in the THTU's and piping, with an accompanying decrease in volume and reduction of pressure as condensation occurs. If the heat removal

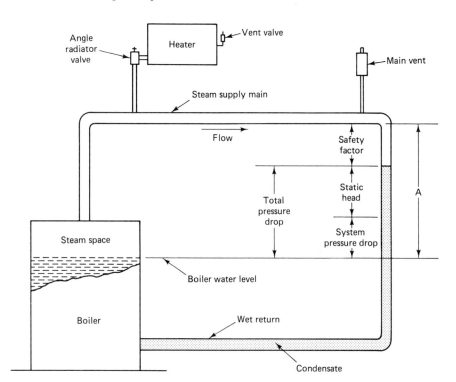

FIGURE 11-1 Sketch of one-pipe steam system illustrating system pressure drop.

rate is greater than the heat supply at the boiler, the pressure in the system falls, and if the system is airtight, transmission of heat in a decreasing amount (because of lower steam temperature) will continue while the steam pressure falls into the vacuum region.

Steam will flow through the system at a rate depending upon the pressure difference existing between the boiler and the end of the steam piping. This available pressure difference will be used up in overcoming the friction loss caused by the steam flowing through the piping. This is called the *system pressure drop*. Condensate moving through the return piping toward the boiler also encounters piping pressure drop, and energy must be made available to move this liquid.

The energy required to overcome these pressure drops is provided by the steam pressure generated by the boiler. The pressure in the steam space of the boiler will be higher than that at the end of steam piping by an amount equal to the *total* system pressure drop.

Reference to Figure 11-1 will show how this affects the relative water levels in the boiler and in the vertical leg of the system return piping. The higher pressure in the boiler causes the water to rise in this pipe, which is exposed to the lower pressure existing at the end of the system. The water column will rise until its height above boiler water level just offsets the existing pressure difference between the two points.

As Figure 11-1 shows, the head provided by the height of this column consists of the system pressure drop and the static head needed to overcome the pressure drop in the condensate return line.

Standard practices have been established in steam system design that assign certain practical values to system pressure drop. These have the effect of establishing minimum values for dimension A as shown in Figure 11-1. This dimension reflects the height of the end of the supply main above the boiler water line. This height includes the total system pressure drop plus a safety factor to ensure that flooding of the steam supply main will not take place.

For small systems having a total heat loss of not more than 100,000 Btuh, piping may be sized on the basis of 1/8 psi pressure drop each for the steam and condensate piping. The height of dimension A is calculated as follows:

system pressure drop ($\frac{1}{8}$ psi)	= 3.5 in. water
static head ($\frac{1}{8}$ psi)	= 3.5 in. water
safety factor (twice the static head)	= 7.0 in. water
total height A	= 14.0 in. water

For systems of this size, it is common practice to make the dimension A not less than 18.0 in.

For systems with heat losses greater than 100,000

Btuh, assume that the steam piping was sized for a pressure drop of 1/2 psi and the return for 4.0 in. water column. The calculation of dimension A is as follows:

system pressure drop ($\frac{1}{2}$ psi)	= 14.0 in. water
static head	= 4.0 in. water
safety factor (twice the static head)	= 8.0 in. water
total height A	= 26.0 in. water

It is standard practice for a system based on 1/2 psi pressure drop to make the dimension A not less than 28 in.

11-4 CLASSIFICATION OF STEAM HEATING SYSTEMS

Because of the various codes and regulations governing the design and operating of boilers, pressure vessels, and systems, steam systems are normally classified according to the steam operating pressure. Low-pressure systems operate at 15 psig (100 kPa above atmospheric) and under, and high-pressure systems operate at 15 psig (100 kPa above atmospheric) and over. There are many subclassifications within these broad classifications, especially for heating systems such as one- and two-pipe, gravity-return, mechanical-return, vacuum, or variable vacuum-return systems. However, these subclassifications relate to the steam and condensate piping systems or the temperature control method. Regardless of classification, all systems include a source of steam, a distribution system, and terminal heat-transfer equipment, where steam is used as the source of heat.

Steam Source. Steam can be generated directly by boilers using oil, gas, coal, wood, or waste as a fuel source or solar, nuclear, and electrical energy as a heat source, and indirectly by recovering heat from processes or equipment such as gas turbines and diesel or gas engines.

There is much interest in the cogeneration of electricity and steam for facilities that have steam requirements throughout the year. When steam is used as a power source (such as in turbine-driven equipment), extracted steam from the turbine or turbine exhaust steam may be used in heat-transfer equipment for space heating.

Steam can be provided by a facility's own boiler or cogeneration plant or can be purchased from a central utility serving a city or specific area. This distinction can be very important. A facility with its own boiler almost always has a *closed-loop* system and requires that the condensate be as hot as possible when it returns to the boiler.

Because central utilities often do not take back condensate, it is discharged by the steam-using facility and results in an *open-loop* system. If a utility does take back condensate, it rarely gives credit for its heat content. Steam is almost always purchased by the pound cost based on 1000 lb (454 kg) of steam; any heat recovered from the condensate in once-through or open-loop systems reduces the net cost per thermal unit. The heat remaining in the condensate represents 10% to 15% of the heat purchased from the utility. Using this heat effectively in a heat recovery system can reduce steam and heating costs by 10% or more.

Distribution Systems. A one- or two-pipe arrangement is standard for steam supply pipes and condensate return pipes. The one-pipe system uses a single pipe to supply steam and to return the condensate. Ordinarily, there is one connection to the THTU, and it serves as both the supply and return. One-pipe systems are normally used for low-capacity systems such as residential or small commercial establishments.

Two-pipe systems are more commonly used in heating and ventilating applications. This type of system has one pipe to carry the steam supply and another to return the condensate. The THTU's have separate connections for the supply and the return.

Distribution systems are sometimes classified as *upfeed* or *downfeed*—that is, by the direction of steam flow in the risers to the THTU's. Condensate return systems, consisting of dry return or wet return or a combination of both, may be identified as a *gravity-return* system or as a *mechanical-return* system.

When all THTU's are located above the boiler or condensate receiver water line, the return piping system is described as a gravity-return system since the condensate returns to the boiler or receiver by gravity.

If mechanical traps or pumps are used to aid the return of condensate to the boiler, the system is a mechanical-return system. The vacuum return pump, the condensate return pump, and the boiler return trap are devices used for mechanically returning condensate to the boiler.

Pressure Conditions. Steam heating systems may be divided into five classifications according to steam supply pressure:

1. High-pressure—100 psig and above.
2. Medium-pressure—15 to 100 psig.
3. Low-pressure—0 to 15 psig.
4. Vapor—vacuum to 15 psig.
5. Vacuum—vacuum to 15 psig.

Vapor and vacuum systems are identical except that a vapor system does not have a vacuum pump, but a vac-

uum system does. The most common systems are the low-pressure, vapor, and vacuum systems with steam supply pressure below 15 psig.

11-5 PIPING SYSTEM DESIGN

A steam piping system operating for air conditioning comfort conditions must distribute steam and return condensate at all operating loads. These loads can be in excess of the design load, such as for early morning warm-up, and at extreme partial loads when only a minimum of heat is required. The steam supply pipe and condensate return pipe, to operate at design load, depend on the following:

1. The initial operating steam pressure and the allowable pressure drop through the system.
2. The total equivalent length of pipe in the longest pipe run.
3. Whether the condensate flows in the same direction as the steam or in the opposite direction.

As noted previously, the most common steam heating systems are classified by a combination of piping arrangement and supply pressure conditions as follows:

1. One-pipe low-pressure
2. Two-pipe vapor
3. Two-pipe low-pressure
4. Two-pipe vacuum
5. Two-pipe medium-pressure
6. Two-pipe high-pressure

We shall discuss in detail systems 1 through 4 and briefly discuss systems 5 and 6.

11-6 ONE-PIPE LOW-PRESSURE SYSTEMS

This type of system is best adapted to small commercial establishments and to residential systems where low initial cost and ease of operation are important considerations. It is not suited to accurate automatic control. If it is applied to larger installations, it will not necessarily result in low first cost because of the necessity of using larger pipes than would be required for a two-pipe system. A diagrammatic layout of a one-pipe upfeed system is shown in Figure 11-2. As the name implies, one pipe is connected to each THTU. Steam and condensate are flowing in opposite directions in the risers and flowing in the same direction in the main, requiring a large

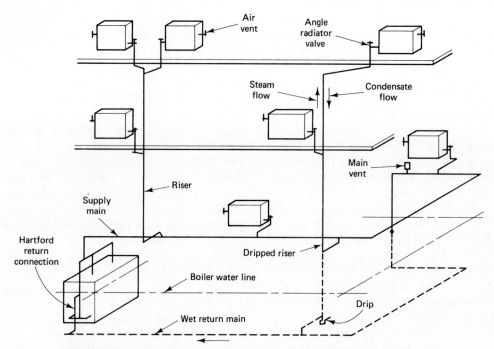

FIGURE 11-2 Diagrammatic layout of typical one-pipe upfeed gravity system.

pipe to keep the steam velocity below the level at which it would carry the condensate along with it. In the upfeed system layout shown in Figure 11-2, we should note that the steam main rises at the boiler (preferably to the ceiling) and then pitches downward from this point. The far end of the steam main is dripped into the wet return. Also, the risers lead upward to the THTU's where steam and condensate flow in opposite directions. When the condensate reaches the supply main, it flows with the steam to the drip at the far end of the main. For more extensive systems, it is desirable to drip each riser separately. System details include the following:

1. An air-vent valve, preferably automatic, should be provided for each THTU.
2. Long supply mains may be raised and dripped at intermediate points to allow elevating the main (see Chapter 12).
3. Heavy-duty air-vent valves should be installed in the main as shown in Figure 11-2.

Such valves should be located about 1 ft (0.3 m) ahead of the drip and preferably elevated above the main.

4. Horizontal branches should be pitched at least 1/2 in. per 10 ft (4 mm/m); mains, at least 1/4 in. per 10 ft (2 mm/m).
5. Wet return may be run without pitch or may be pitched in either direction. A certain amount of pitch is desirable to allow draining, if necessary, during a shutdown period.
6. THTU inlet valves of angle or gate types should be installed. Straight globe valves are not suitable unless they are installed on the side since the valve seat interferes with condensate return.
7. The lowest point in the supply main, all THTU's, and drip points must be sufficiently above the water level in the boiler to prevent flooding. The required difference in elevation is found as shown in Section 11-3.

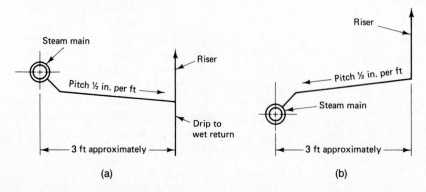

(a)　　　　　　　　　　　　　(b)

FIGURE 11-3 Runouts to risers in one-pipe upfeed system: (a) for dripped risers and (b) for undripped risers.

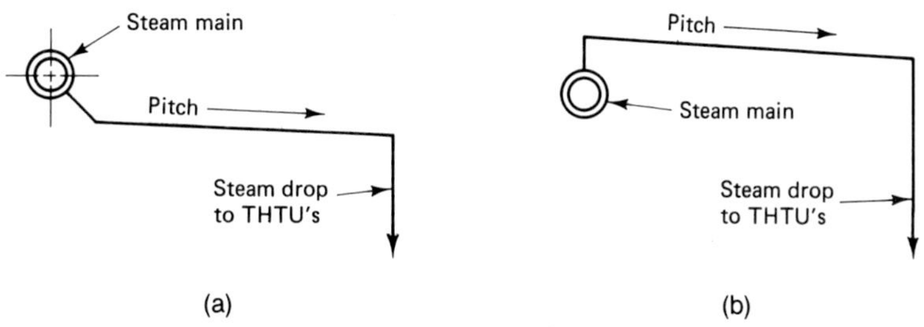

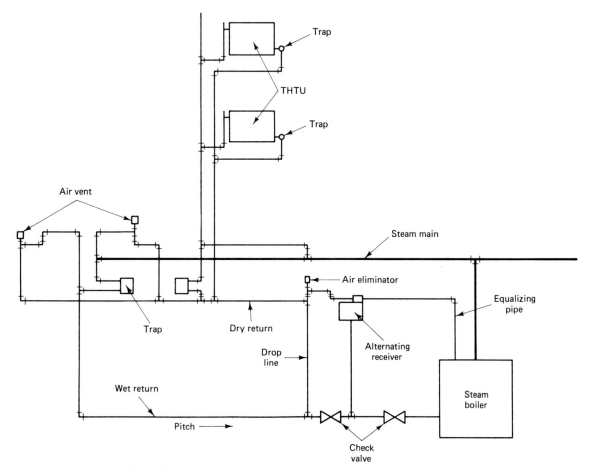

FIGURE 11-4 Runouts to risers in one-pipe downfeed system: (a) main dripped through riser and (b) runout with main dripped at end only.

8. One-pipe systems are not suitable for throttling-type control. Valves must be full open or tightly closed.
9. When risers are dripped, the riser runouts are taken from the mains as shown in Figure 11-3a. When risers are not dripped, the riser runouts are taken from the main as shown in Figure 11-3b.

One-pipe downfeed systems differ from upfeed systems in that the steam is taken from the boiler, carried to the top of the building, and distributed to the down risers, from which it is fed to the THTU's. The horizontal header connecting the downfeed risers is pitched downward to allow condensate to flow in the same direction as the steam. Thus, each downfeed riser

will take some share of the condensate. An alternate method is to take the connection to all risers, except the last one, from the top of the horizontal distributing header.

Runout connections from the main supply header differ from those used with upfeed systems (see Figure 11-4). Other elements, such as THTU's, air-vent valves, and THTU valves are the same for both systems.

11-7 TWO-PIPE VAPOR SYSTEMS

The designation ''vapor system'' is applied to layouts similar to the one shown in Figure 11-5. This type of system operates at steam boiler pressures ranging from a few ounces above atmospheric pressure as a maximum

FIGURE 11-5 Diagrammatic layout of a two-pipe vapor system.

to pressures less than atmospheric. The subatmospheric pressures are obtained by condensation of steam in the THTU's when the boiler or steam source is declining in activity. Figure 11-5 illustrates a *closed-type* upfeed system. System details include the following:

1. Packless radiator valves on each THTU.
2. Thermostatic traps on each THTU.
3. Low operating pressures—0.5 to 1.0 psig or below atmospheric.
4. Condensate returned by gravity to a receiver. This device, commonly called a *boiler-return trap* or *alternating receiver,* discharges the condensate into the boiler against boiler pressure.
5. Any air that enters the system is discharged through an air eliminator installed just above the condensate receiver. The air eliminator acts as a check valve that allows air to escape from the system but will not allow air to re-enter. Several types are available, selection depending on the capacity required and other steam specialties used (see Chapter 12). Individual air-vent valves on THTU's or mains are not advisable.
6. Each supply riser supplying THTU's above the first floor of a building (and the extreme

end of the supply main as well) is dripped to the condensate return line through combination float and thermostatic traps. These traps are not necessary if the drips are water-sealed.

7. The steam main is pitched in the direction of steam flow, and the condensate flows in the same direction as the steam. As is true for any system, trouble with water hammer is thus minimized.
8. Condensate return lines pitch toward the boiler.

11-8   TWO-PIPE LOW-PRESSURE SYSTEMS

The two-pipe low-pressure (vapor) system is similar to the vapor system except that it is usually an *open,* or *atmospheric,* system. Such a system would have its return line open to atmospheric pressure and would operate above atmospheric pressure at all times. The condensate drains by gravity to the vented condensate receiver where a condensate return pump forces the water into the boiler. The condensate return pump *cannot* cause a vacuum in the system. The source of steam is seldom above a pressure of 5 psig. The steam supply

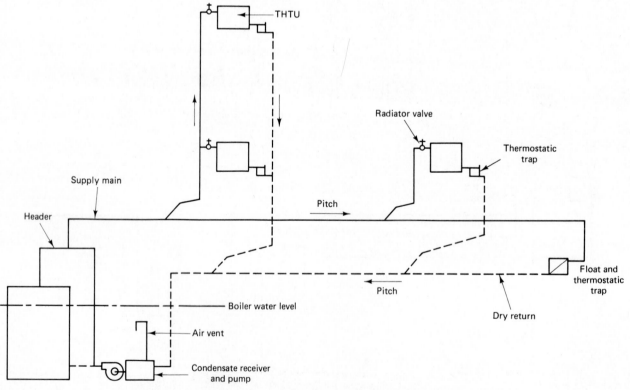

FIGURE 11-6 Diagrammatic layout of a two-pipe vapor system using a condensate return pump.

may be upfeed or downfeed. An upfeed system is shown in Figure 11-6.

A downfeed system has the main riser extended to the top of the building, and a horizontal supply main connects with the downfeed risers. The takeoffs from the main may be made as shown in Figure 11-4a or connected from the top of the main as shown in Figure 11-4b, providing the far end of the supply main is dripped. Each steam riser should have a drip trap and dirt pocket.

11-9 VACUUM SYSTEMS

In an ordinary vacuum system, the steam is supplied to the THTU's at some pressure above atmospheric, and a vacuum is maintained on the return line by use of a vacuum pump. This type of system has been applied extensively to larger installations, including many of the largest buildings in this country. Details for an upfeed system (see Figure 11-7) are as follows:

1. Throttling-type valves on steam supply to each THTU.
2. Thermostatic return traps on each THTU.
3. Systems are usually designed to operate from atmospheric to 5 psig pressure. This may be obtained from a high-pressure steam source by use of pressure-reducing valves.
4. Condensate returns to the vacuum pump where air is expelled from the system and the condensate is returned directly to the boiler or to an open-type feedwater heater. Air-vent valves are not used at the THTU's.
5. Steam supply mains are pitched in the direc-

tion of steam flow. THTU runouts are pitched downward toward the riser if the riser is dripped; otherwise, they are pitched toward the THTU.

6. All long risers should have drip connections through thermostatic traps to the return lines. Risers less than 30 to 40 ft in length are not generally dripped, although such a division point is arbitrary and the designer should use judgment in this regard.
7. The return header used to connect the vertical risers in the building basement is pitched toward the vacuum pump. If the building conditions are such that a continuous slope is impractical, *lift fittings* may be used on the vacuum system. A drain should be provided at the system low point to permit complete drainage.

11-10 PIPE SIZING FOR STEAM HEATING SYSTEMS

Successful operation of a steam heating system is dependent on proper sizing of pipe lines and attention to details of installation such as applying the correct steam specialties. (Steam specialties will be discussed in Chapter 12.)

To determine the pipe sizes with absolute accuracy on a theoretical basis from technical books is to incur an engineering expense out of proportion to the savings in first cost and operating efficiency. Also, practical limitations such as using standard pipe sizes, variations in manufacture affecting flow resistance, and installation difficulties make the use of simplified selection ta-

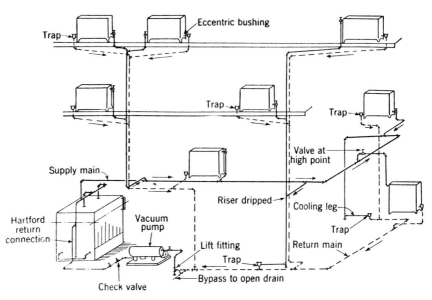

FIGURE 11-7 Diagrammatic layout of a vacuum steam system.

bles and charts advisable. On the other hand, to determine pipe sizes by ordinary rule-of-thumb methods is to set up factors of safety that are wasteful in first cost as well as operating efficiency.

Determining a pipe size for a given load in a steam heating system depends on the following principal factors: (1) steam pressure and total pressure drop that can be allowed between the source of supply and at the end of return system, (2) the maximum velocity of steam allowable for quiet and dependable operation of the system, taking into consideration the direction of condensate flow, and (3) the total equivalent length of the pipe run from the boiler or source of steam supply to the farthest THTU.

Initial Pressure and Pressure Drop.   Table 11-1 lists pressure drops commonly used with corresponding initial steam pressures for sizing steam pipes.

Theoretically, there are several factors to be considered such as initial pressure and pressure required at the end of the line, but it is most important that (1) the total pressure drop does not exceed the initial gauge pressure of the system, and in actual practice it should never exceed one-half of the initial gauge pressure; (2) the pressure drop is not great enough to cause excessive velocities; (3) there is a constant initial pressure except on systems specifically designed for varying initial pressures, such as subatmospheric, which normally operates under controlled partial vacuums; and (4) the rise in water from the pressure drop does not exceed the difference in level (for gravity-return systems) between the

TABLE 11-1

Pressure drops in common use for sizing steam pipe* (for corresponding initial steam pressures)

Initial Steam Pressure (psig)	Pressure Drop (per 100 ft)	Total Pressure Drop in Steam Supply Piping
Subatmos. or vacuum return	2–4 oz	1–2 psi
0	$\frac{1}{2}$ oz	1 oz
1	2 oz	1–4 oz
2	2 oz	8 oz
5	4 oz	$1\frac{1}{2}$ oz
10	8 oz	3 psi
15	1 oz	4 psi
30	2 psi	5–10 psi
50	2–5 psi	10–15 psi
100	2–5 psi	15–25 psi
150	2–10 psi	25–30 psi

Source: ASHRAE Handbook 1989 Fundamentals, American Society of Heating, Refrigerating, and Air Conditioning Engineers, Atlanta, GA.

*Equipment, control valves, and so forth must be selected based on delivered pressures.

lowest point on the steam main, the THTU's, or the dry return and the boiler water line (see Section 11-3).

Maximum Velocity.   For quiet operation, steam velocity should be 8000 to 12,000 fpm with a maximum of 15,000 fpm. The lower the velocity, the quieter the system. When the condensate must flow against the steam, even in limited quantity, the velocity of the steam must not exceed these limits, above which the disturbance between the steam and the counterflowing water (1) may produce objectionable noise such as water hammer or (2) may result in the retention of water in certain parts of the system until the steam flow is reduced sufficiently to permit the water to pass. The velocity at which these disturbances take place is a function of (1) pipe size, whether the pipe runs horizontally or vertically; (2) pitch of the pipe if it runs horizontally; (3) the quantity of condensate flowing against the steam; and (4) freedom of the piping from water pockets that, under certain conditions, act as a restriction in pipe size.

Equivalent Length of Run.   All tables for the flow of steam in pipes, based on pressure drop, must allow for pipe friction, as well as for the resistance of fittings and valves. These resistances are generally stated in terms of equivalent length of straight pipe. (Table 10-3 gives typical values that may be used in steam piping systems.) The total equivalent length (TEL) is the sum of the actual straight length of pipe *plus* the equivalent length of all the fittings in that section of piping.

Note: A common and practical sizing method is to assume a TEL of run and to check the assumption after pipes are sized. For this purpose, the TEL is usually assumed to be 1.5 to 2.0 times the measured length of the pipe run.

Basic Chart for Steam Pipe Sizing.   Figure 11-8 is the basic chart for determining the flow rate and velocity of steam in schedule 40 pipe for various values of pressure drop per 100 ft, based on 0 psig saturated steam and for a steam flow rate from 5 to 100,000 lb/hr. However, the velocity as read from Figure 11-8 is based on a steam pressure of 0 psig and must be corrected for the desired pressure from Figure 11-9. The complete chart is based on the Moody friction factor and *is valid where condensate and steam flow in the same direction.*

ILLUSTRATIVE PROBLEM 11-1

Determine the size of the supply main and the velocity of steam flow for a steam system for a design friction rate of 2.0 psi per 100 ft of equivalent length. The initial pressure is 10 psig, and the steam flow rate is 4000 lb/hr.

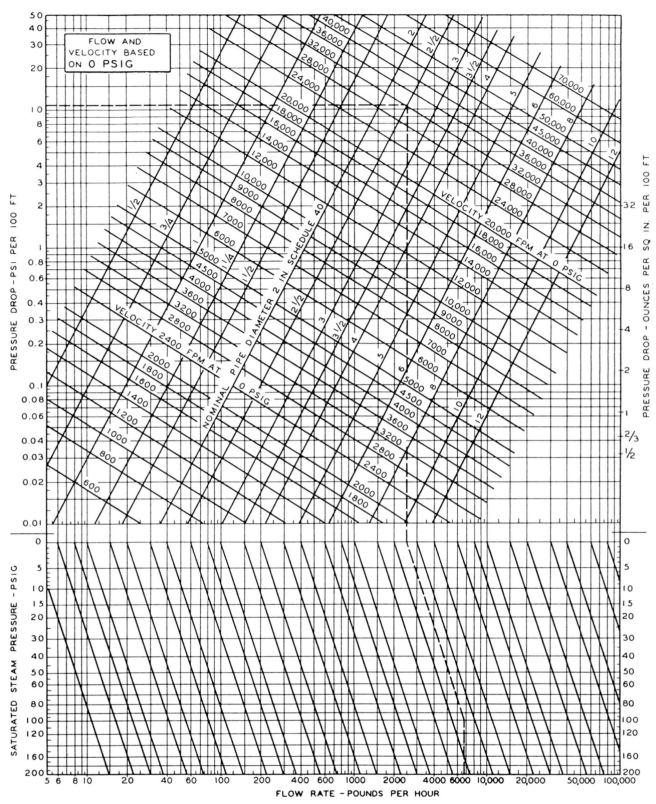

Based on Moody Friction Factor where flow of condensate does not inhibit the flow of steam.

FIGURE 11-8 Basic chart for flow rate and velocity of steam in schedule 40 pipe (based on 0 psig saturation pressure). *Note:* See Figure 11-9 for obtaining flow rates and velocities at all saturation pressures between 0 and 200 psig. (From *ASHRAE Handbook 1989 Fundamentals.* American Society of Heating, Refrigerating, and Air Conditioning Engineers, Atlanta, GA)

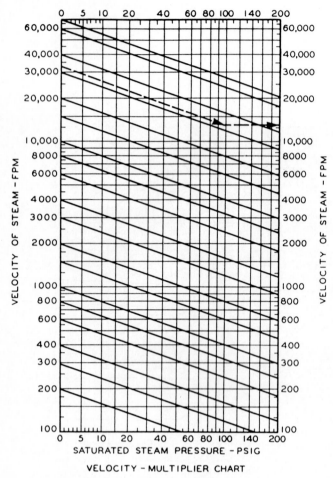

FIGURE 11-9 Velocity multiplier chart for Figure 11-8.

3.5 to 12.0 psig initial pressure. The flow rates shown for 3.5 psig can be used for saturation pressures from 1 to 6 psig, and those shown for 12.0 psig can be used for saturation pressures from 8 to 16 psig with an error not exceeding 8%. Both Figure 11-8 and Table 11-2 can be used where the flow of condensate does *not* inhibit the flow of steam.

Columns B and C of Table 11-3 are for cases where steam and condensate flow in opposite directions as in risers and runouts that are not dripped. Columns D, E, and F are for one-pipe systems and include risers, THTU valves and vertical connections, and radiator and riser runout sizes, all of which are based on the critical velocities of steam to permit the counterflow of condensate without noise.

Sizing Return Lines for Two-Pipe Systems.
Condensate from steam heating devices is generally saturated liquid or slightly subcooled liquid. The amount of subcooling is a function of the size of the heating device in comparison to the heating load and the control system for the heating device. In properly designed systems, the subcooling is usually no greater than a few degrees Fahrenheit. The condensate is then passed through a steam trap where the pressure is reduced and some of the liquid *may* flash into vapor. The amount of flashing depends on the overall pressure drop through the trap, the amount of subcooling in the heating device and the drip leg preceding the trap, and the type and condition of the trap.

The process of the fluid passing through a steam trap is a throttling or constant enthalpy process (see Chapter 2). The resulting fluid on the downstream side of the trap can be a mixture of saturated liquid and saturated vapor. The overall pressure drop and resulting flash vapor are the physical characteristics that differentiate condensate pipe sizing from other types of pipe sizing where only a single phase exists. For this reason, it is important to understand the condition of the condensate when it enters the return line from the trap.

For low-pressure steam heating systems (below 5 psig initial pressure), Table 11-4 may be used to size the return mains and branches. It should be noted that dry-return and vacuum-return pipes are only partially filled with liquid condensate; the remaining space is filled with vapor, air, and noncondensable gases. Wet return on gravity systems runs full of condensate.

For high-pressure systems (above 5 psig initial pressure), careful consideration of the condensate downstream of the trap is desirable. An excellent reference to use would be *ASHRAE Handbook 1989 Fundamentals,* Chapter 33, published by the American Society of Heating, Refrigerating, and Air Conditioning Engineers, Atlanta, GA.

Solution: Refer to Figure 11-8. Find 4000 lb/hr on the bottom scale and move vertically to the horizontal line at 10 psig. Follow along the inclined multiplier line (upward and to the left) to the horizontal 0 psig line. The equivalent mass flow rate at 0 psig is 3100 lb/hr.

Move vertically upward to intersect with the horizontal friction rate line of 2.0 psi per 100 ft. The point of intersection is between nominal pipe size of $3\frac{1}{2}$ and 4 in. Select 4-in. pipe with a corresponding steam velocity of 16,000 fpm at 0 psig.

To find the steam velocity at 10 psig, locate the value of 16,000 fpm on the ordinate of Figure 11-9. Move along the inclined multiplier line (downward and to the right) until it intersects the vertical 10 psig line. The velocity as read from the right (or left) scale is approximately 13,000 fpm.

11-11 TABLES FOR PIPE SIZING FOR LOW-PRESSURE SYSTEMS

The basic charts, Figures 11-8 and 11-9, can be used on low-pressure systems. The values in Table 11-2 provide a more rapid means of selecting pipe sizes for the various pressure drops listed and for systems operating at

TABLE 11-2

Flow rate of steam in schedule 40 pipe at initial saturation pressure of 3.5 and 12.0 psig (flow rate expressed in pounds per hour)

Nominal Pipe Size (in.)	1/16 psi (1 oz)		1/8 psi (2 oz)		1/4 psi (4 oz)		1/2 psi (8 oz)		3/4 psi (12 oz)		1 psi		2 psi	
	3.5	12	3.5	12	3.5	12	3.5	12	3.5	12	3.5	12	3.5	12
3/4	9	11	14	16	20	24	29	35	36	43	42	50	60	73
1	17	21	26	31	37	46	54	66	68	82	81	95	114	137
1¼	36	45	53	66	78	96	111	138	140	170	162	200	232	280
1½	56	70	84	100	120	147	174	210	218	260	246	304	360	430
2	108	134	162	194	234	285	336	410	420	510	480	590	710	850
2½	174	215	258	310	378	460	540	660	680	820	780	950	1,150	1,370
3	318	380	465	550	660	810	960	1,160	1,190	1,430	1,380	1,670	1,950	2,400
3½	462	550	670	800	990	1,218	1,410	1,700	1,740	2,100	2,000	2,420	2,950	3,450
4	640	800	950	1,160	1,410	1,690	1,980	2,400	2,450	3,000	2,880	3,460	4,200	4,900
5	1,200	1,430	1,680	2,100	2,440	3,000	3,570	4,250	4,380	5,250	5,100	6,100	7,500	8,600
6	1,920	2,300	2,820	3,350	3,960	4,850	5,700	7,000	7,000	8,600	8,400	10,000	11,900	14,200
8	3,900	4,800	5,570	7,000	8,100	10,000	11,400	14,300	14,500	17,700	16,500	20,500	24,000	29,500
10	7,200	8,200	10,200	12,600	15,000	18,200	21,000	26,000	26,200	32,000	30,000	37,000	42,700	52,000
12	11,400	13,700	16,500	19,500	23,400	28,400	33,000	40,000	41,000	49,500	48,000	57,500	67,800	81,000

Pressure Drop (psi per 100 ft of length) — Saturation Pressure (psig)

Source: *ASHRAE Handbook 1989 Fundamentals*, American Society of Heating, Refrigerating, and Air Conditioning Engineers, Atlanta, GA.

Notes: 1. Based on Moody friction factor, where flow of condensate does not inhibit the flow of steam.

2. Flow rates at 3.5 psig can be used to cover saturated pressures 1 to 6 psig, and rates of 12.0 psig can be used to cover saturation pressures 8 to 16 psig with an error not exceeding 8%.

3. The steam velocities corresponding to the flow rates given in this table can be found from basic chart and velocity multiplier chart, Figures 11-8 and 11-9.

TABLE 11-3

Steam pipe capacities for low-pressure systems
(for use on one-pipe systems or two-pipe systems
in which condensate flows against the steam flow)

	Capacity in Pounds per Hour				
	Two-Pipe Systems			One-Pipe Systems	
Nominal Pipe Size (in.)	Condensate Flowing Against Steam		Supply Risers Upfeed	Radiator Valves and Vertical Connections	Radiator and Riser Runouts
	Vertical	Horizontal			
(A)	(B)ᵃ	(C)ᶜ	(D)ᵇ	(E)	(F)ᶜ
$\frac{3}{4}$	8	7	6	—	7
1	14	14	11	7	7
1$\frac{1}{4}$	31	27	20	16	16
1$\frac{1}{2}$	48	42	38	23	16
2	97	93	72	42	23
2$\frac{1}{2}$	159	132	116	—	42
3	282	200	200	—	65
3$\frac{1}{2}$	387	288	286	—	119
4	511	425	380	—	186
5	1,050	788	—	—	287
6	1,800	1,400	—	—	545
8	3,750	3,000	—	—	—
10	7,000	5,700	—	—	—
12	11,500	9,500	—	—	—
16	22,000	19,000	—	—	—

Source: ASHRAE Handbook 1989 Fundamentals, American Society of Heating, Refrigerating, and Air Conditioning Engineers, Atlanta, GA.

Note: Steam at an average pressure of 1 psig is used as a basis of calculating capacities.

ᵃDo not use column B for pressure drops of less than 1/16 psig per 100 ft of equivalent length. Use Figure 11-8 instead.

ᵇDo not use column D for pressure drops of less than 1/24 psi per 100 ft of equivalent length, except on sizes 3 in. or over. Use Figure 11-8 instead.

ᶜPitch of horizontal runouts to risers and radiators should be not less than 1/2 in. per ft. Where this pitch cannot be obtained, runouts over 8 ft in length should be one pipe size larger than called for in table.

Recommendations. The following recommendations are suggested for use when sizing pipe for the various systems listed.

One-pipe low-pressure system: This type of system is used for small commercial and residential systems.

1. Size the supply main and risers for a maximum pressure drop of 1/4 psi.
2. If the total equivalent length (TEL) of the supply main to the most remote THTU is less than 200 ft, use a design friction rate of 1/16 psi per 100 ft of pipe. If the TEL is greater than 200 ft, use a design friction rate of 1/24 psi per 100 ft.

3. Size the return main and risers for a maximum pressure drop of 1/4 psi.
4. Use the same design friction rate for the return piping that is used for the supply.
5. Pitch the supply main 1/4 in. per 10 ft away from the boiler.
6. Pitch the return main 1/4 in. per 10 ft toward the boiler.
7. Size the supply main and dripped runouts using Table 11-2 or Figure 11-8.
8. Size the undripped runouts using Table 11-3, column F.
9. Size the upfeed risers using Table 11-3, column D.
10. Size the downfeed risers using Table 11-2 or Figure 11-8.
11. Where horizontal piping must be reduced in size, use eccentric reducers that permit the continuance of uniform pitch along the bottom of piping (in downward-pitched systems). Concentric reducers must be avoided on horizontal piping, since they can cause water hammer.

ILLUSTRATIVE PROBLEM 11-2

Figure 11-10 shows a sketch of a one-pipe steam system. The THTU's are shown on one of the two upfeed risers for the system. The capacity of each THTU is indicated in EDR. The initial steam pressure is 1.0 psig. Determine the pipe sizes to use for the various pipe sections indicated.

Solution: The scaled length of the supply main and riser is 250 ft. The estimated TEL will be taken as (1.5)(250) = 375 ft (using 50% allowance for fittings). Since the TEL is greater than 200 ft, the design friction rate will be taken as 1/24 psi (2/3 oz) per 100 ft of pipe.

To obtain steam flow in lb/hr, divide EDR by 4. Our design friction rate is 1/24 psi (0.04) per 100 ft. Footnote "b" of Table 11-3 indicates that we should use Figure 11-8 or Table 11-2 to find pipe size for the supply riser. See Table 11-5 for results.

Two-pipe vapor system: This type of system is used for commercial and residential applications.

1. Size the supply main and risers for a maximum pressure drop of 1/16 to 1/8 psi.
2. Size the supply main and risers for a maximum friction rate of 1/16 to 1/8 psi per 100 ft of equivalent pipe length.
3. Size the return main and risers for a maximum pressure drop of 1/16 to 1/8 psi.
4. Size the return main and risers for a maximum friction rate of 1/16 to 1/8 psi per 100 ft of equivalent pipe length.
5. Pitch the supply main 1/4 in. per 10 ft away from the boiler.

TABLE 11-4

Return main and riser capacities for low-pressure systems (pounds per hour)

	1/32 Psi or 1/2 Oz Drop per 100 Ft			1/24 Psi or 2/3 Oz Drop per 100 Ft			1/16 Psi or 1 Oz Drop per 100 Ft			1/8 Psi or 2 Oz Drop per 100 Ft			1/4 Psi or 4 Oz Drop per 100 Ft			1/2 Psi or 8 Oz Drop per 100 Ft		
Pipe Size (in.)	Wet	Dry	Vac.	Wet	Dry	Vac.	Wet	Dry	Vac.	Wet	Dry	Vac.	Wet	Dry	Vac.	Wet	Dry	Vac.
(G)	(H)	(I)	(J)	(K)	(L)	(M)	(N)	(O)	(P)	(Q)	(R)	(S)	(T)	(U)	(V)	(W)	(X)	(Y)
Return Main																		
$\frac{3}{4}$						42			100			142			200			283
1	125	62		145	71	143	175	80	175	250	103	249	350	115	350			494
$1\frac{1}{4}$	213	130		248	149	244	300	168	300	425	217	426	600	241	600			848
$1\frac{1}{2}$	338	206		393	236	388	475	265	475	675	340	674	950	378	950			1,340
2	700	470		810	535	815	1,000	575	1,000	1,400	740	1,420	2,000	825	2,000			2,830
$2\frac{1}{2}$	1,180	760		1,580	868	1,360	1,680	950	1,680	2,350	1,230	2,380	3,350	1,360	3,350			4,730
3	1,880	1,460		2,130	1,560	2,180	2,680	1,750	2,680	3,750	2,250	3,800	5,350	2,500	5,350			7,560
$3\frac{1}{2}$	2,750	1,970		3,300	2,200	3,250	4,000	2,500	4,000	5,500	3,230	5,680	8,000	3,580	8,000			11,300
4	3,880	2,930		4,580	3,350	4,500	5,500	3,750	5,500	7,750	4,830	7,810	11,000	5,380	11,000			15,500
5						7,880			9,680			13,700			19,400			27,300
6						12,600			15,500			22,000			31,000			43,800
Riser																		
$\frac{3}{4}$		48			48	143		48	175		48	249		48	350			494
1		113			113	244		113	300		113	426		113	600			848
$1\frac{1}{4}$		248			248	388		248	475		248	674		248	950			1,340
$1\frac{1}{2}$		375			375	815		375	1,000		375	1,420		375	2,000			2,830
2		750			750	1,360		750	1,680		750	2,380		750	3,350			4,730
$2\frac{1}{2}$						2,180			2,680			3,800			5,350			7,560
3						3,250			4,000			5,680			8,000			11,300
$3\frac{1}{2}$						4,480			5,500			7,810			11,000			15,500
4						7,880			9,680			13,700			19,400			27,300
5						12,600			15,500			22,000			31,000			43,800

Source: *ASHRAE Handbook 1989 Fundamentals*, American Society of Heating, Refrigerating, and Air Conditioning Engineers, Atlanta, GA.
Note: Vac. values may be used for wet-return risers and main.

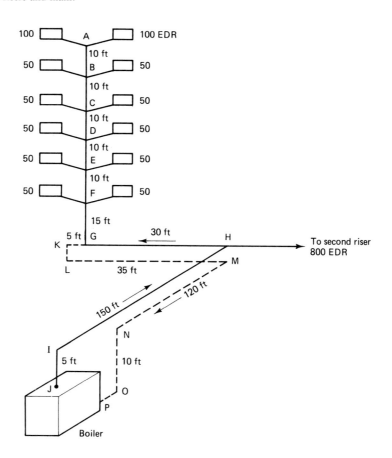

FIGURE 11-10 Sketch for Illustrative Problem 11-2.

TABLE 11-5

Results for Illustrative Problem 11-2

	Pipe Section	EDR	Pounds per Hour	Pipe Size Supply	Pipe Size Return
Branches to radiation		100	25	Table 11–3, 2 in.	
Branches to radiation		50	12	Table 11–3, $1\frac{1}{4}$ in.	
Riser	AB	200	50	Figure 11–8, $1\frac{1}{2}$ in.	
	BC	300	75	2 in.	
	CD	400	100	2 in.	
	DE	500	125	$2\frac{1}{2}$ in.	
	EF	600	150	$2\frac{1}{2}$ in.	
	FG	700	175	3 in.	
Runout to riser	GH	700	175	3 in.	
Supply to main	HI	1500	375	4 in.	
Riser to main	IJ	1500	375	4 in.	
Return (dry)	KLMN	700	175		Table 11–4, $1\frac{1}{2}$ in.
Return (wet)	NOP	700	175		Table 11–4, $1\frac{1}{2}$ in.

Notes: 1. Branches to radiation from column F of Table 11–3.

2. Table 11–2 does not list 1/24 psi per 100 ft of pipe. Use Figure 11–8.

3. Returns sized from Table 11–4, columns L and K.

6. Pitch the return main 1/4 in. per 10 ft toward the boiler.
7. Size the pipe from Tables 11-2 and 11-3.
8. Where horizontal piping must be reduced in size, use eccentric reducers.

Two-pipe low-pressure system: This type of system is used for commercial heating and ventilating systems.

1. Size the supply main and risers for a maximum pressure drop determined from Table 11-6, depending on the initial system steam pressure.
2. Size the supply main and risers for a maximum friction rate of 2 psi per 100 ft of equivalent pipe length.
3. Size the return main and risers for a maximum pressure drop determined from Table 11-6, depending on the initial steam pressure.

TABLE 11-6

Recommended total pressure drop for two-pipe low-pressure steam piping systems

Initial Steam Pressure (psig)	Total Pressure Drop in Supply Piping (psi)	Total Pressure Drop in Return Piping (psi)
2	$\frac{1}{2}$	$\frac{1}{2}$
5	$1\frac{1}{4}$	$1\frac{1}{4}$
10	$2\frac{1}{2}$	$2\frac{1}{2}$
15	$3\frac{3}{4}$	$3\frac{3}{4}$
20	5	5

4. Size the return main and risers for a maximum friction rate of 1/2 psi per 100 ft of equivalent pipe length.
5. Pitch the supply main 1/4 in. per 10 ft away from the boiler.
6. Pitch the return main 1/4 in. per 10 ft toward the boiler.
7. Use Tables 11-2 through 11-4 to size the pipe.
8. Use eccentric reducers as pipe size changers.

ILLUSTRATIVE PROBLEM 11-3

Figure 11-11 shows a sketch of a two-pipe low-pressure direct-return steam system. The sketch shows the supply main and risers, a riser diagram for the last THTU, and the return main. The initial steam pressure at the boiler is 2.0 psig. Heat loads are given in pounds of steam per hour. Determine the pipe sizes for the supply main, the risers to last heater, and the return main.

Solution:

Step 1. Determine the TEL and design friction rate. Table 11-6 indicates that the allowable pressure drop in the supply main should be 0.5 psi. The estimated TEL of the supply main is (115)(1.5) = 173 ft; therefore, the design friction rate is

$$f = \Delta P\left(\frac{100}{\text{TEL}}\right) = 0.5\left(\frac{100}{173}\right)$$

$$= 0.289 \text{ psi per 100 ft of pipe}$$

Since this friction rate is calculated using an assumed TEL, we will use a design friction rate of 0.25 psi per 100 ft of pipe as conservative.

The measured length of the return main is 122 ft. Using the same reasoning as for the supply main,

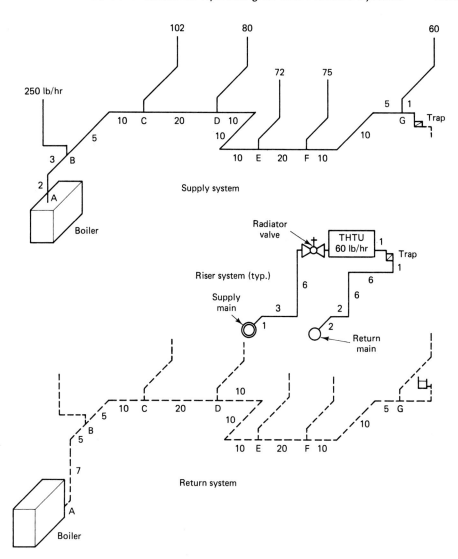

FIGURE 11-11 Sketch of piping system for Illustrative Problem 11-3.

we will use the same design friction rate for the return main.

Step 2. Determine the pipe sizes and estimate actual pressure loss in the supply main. See Table A for Illustrative Problem 11-3.

To estimate the actual pressure loss in the supply main, we need the actual TEL. See Table B for Illustrative Problem 11-3. Calculated TEL = 115 + 68.2 = 183.2 ft of pipe. Note that the *estimated* TEL was 173 ft. Therefore, the original estimate was very close. Actual pressure drop in the supply main is

$$\Delta P = f\left(\frac{\text{TEL}}{100}\right) = 0.25\left(\frac{183.2}{100}\right) = 0.46 \text{ psi}$$

Step 3. Determine pipe sizes and pressure loss for supply riser to THTU, 60 lb/hr of steam flowing. See Table C for Illustrative Problem 11-3. Supply riser TEL = 26.3 + 11 = 37.3 ft of pipe. Estimated pressure loss in riser = 0.25(37.3/100) = 0.09 psi. Estimated pressure loss from boiler to inlet of last heater is 0.46 + 0.09 = 0.55 psi.

Table A for Illustrative Problem 11-3—Pipe sizes for supply main

Pipe Section	Steam Flow (lb/hr)	Nominal Pipe Size (in.) (Table 11-2)
GF	60	$1\frac{1}{4}$
FE	135	2
ED	207	2
DC	287	$2\frac{1}{2}$
CB	389	3
BA	639	3

Note: It may be desirable to use a 5-in. riser from the supply outlet of the boiler (determined from column B of Table 11-3) to prevent moisture carry-over from the boiler.

Step 4. Determine pipe size and pressure loss for the return main. The return main will be sized with the same friction rate (0.25 psi/100 ft) as the supply main. The approximate TEL of the return is about the same as for the supply. See Tables D and E for Illustrative Prob-

Table B for Illustrative Problem 11-3—Equivalent length of pipe fittings (supply main)

	Fitting	EL (Table 10-2)
At *G*,	one $1\frac{1}{4}$-in. tee (used as an elbow)	= 8.0
FG,	two $1\frac{1}{4}$-in. 90° elbows (2 at 3.5)	= 7.0
At *F*,	one $1\frac{1}{4}$-in. reducing tee (D/d = 4/3)	= 3.1
At *E*,	one 2-in. flow-through tee	= 3.5
DE,	two 2-in. 90° elbows (2 at 5.3)	= 10.6
At *D*,	one 2-in. reducing tee (D/d = 4/3)	= 4.5
At *C*,	one $2\frac{1}{2}$-in. reducing tee (D/d = 4/3)	= 5.5
BC,	one 3-in. 90° elbow	= 8.0
At *B*,	one 3-in. flow-through tee	= 5.0
AB,	one 5-in. 90° elbow (estimated)	= 13.0
	EL of fittings	= 68.2 ft of pipe

Table C for Illustrative Problem 11-3—Supply riser size and equivalent length

Pipe Section	Pipe Size (in.) (Table 11-3)
Riser runout (horizontal)	2 (column C)
Riser (vertical)	2 (column B)

Fitting	EL (Table 10-3)
One 2-in. side-flow tee	= 12.0
One 2-in. 45° elbow	= 2.5
Two 2-in. 90° elbows (2 at 5.3)	= 10.6
One 2-in. gate valve (open)	= 1.2
EL of fittings	= 26.3 ft of pipe

Table D for Illustrative Problem 11-3—Pipe sizes for return main and riser

Pipe Section	Condensate Flow (lb/hr)	Nominal Pipe Size (in.) (Table 11-4)
Return riser to last THTU	60	1 (column U)
FG	60	1 (column U)
EF	135	$1\frac{1}{4}$ (column U)
DE	207	$1\frac{1}{4}$ (column U)
CD	287	$1\frac{1}{2}$ (column U)
BC	389	2 (column U)
AB	639	2 (column U)

lem 11-3. Calculated TEL of pipe from last THTU to boiler = 18 + 122 + 48.7 = 188.7 ft of pipe. Actual pressure drop in the return main is

$$\Delta P = f\left(\frac{\text{TEL}}{100}\right) = 0.25\left(\frac{188.7}{100}\right) = 0.47 \text{ psi}$$

The total pressure loss due to piping = 0.55 + 0.47 = 1.02 psi. This is close to one-half the supply pressure of 2 psig. Therefore, the design is satisfactory.

Two-pipe vacuum system: This type of system is used for commercial and industrial applications.

1. Size the supply main and risers for a maximum pressure drop of 1/8 to 1 psi.
2. Size the supply main and risers for a maximum friction rate of 1/8 to 1/2 psi per 100 ft of equivalent pipe length.
3. Size the return main and risers for a maximum pressure drop of 1/8 to 1 psi.
4. Size the return main and risers for a maximum friction rate of 1/8 to 1/2 psi per 100 ft of equivalent pipe length.
5. Pitch the supply main 1/4 in. per 10 ft away from the boiler.
6. Pitch the return main 1/4 in. per 10 ft toward the boiler.

Table E for Illustrative Problem 11-3—Equivalent length of fitting return main

	Fitting	EL (Table 10-2)
	four 1-in. 90° elbows (4 at 2.6)	= 10.4
	one 1-in. 45° elbow	= 1.3
At *G*,	one 1-in. side-flow tee	= 6.0
At *F*,	one increaser (d/D = 1/1$\frac{1}{4}$)	= 0.6
At *E*,	one flow-through tee	= 2.3
D to *E*,	two $1\frac{1}{4}$-in. 90° elbows (2 at 3.5)	= 7.0
At *D*,	one increaser (d/D = 1$\frac{1}{4}$/1$\frac{1}{2}$)	= 0.8
At *C*,	one increaser (d/D = 1$\frac{1}{2}$/2)	= 0.9
At *B*,	one 2-in. flow-through tee	= 3.5
A to *C*,	three 2-in. 90° elbows (3 at 5.3)	= 15.9
	EL of fittings	= 48.7 ft of pipe

7. Size the pipe from Tables 11-2 and 11-4, or Table 11-3 when it applies.

ILLUSTRATIVE PROBLEM 11-4

Using the same basic data of Illustrative Problem 11-3 and Figure 11-11, assume that the returns are to be resized for a vacuum-return system. Resize the return pipe.

Solution: Determine the design friction rate. The recommended design friction rate for the vacuum-return system is 1/8 to 1/2 psi per 100 ft. We will use 1/4 psi per 100 ft. See Table A for Illustrative Problem 11-4.

Two-pipe high-pressure and two-pipe medium-pressure systems: Due to limited space within this book, we will not discuss these two systems in any detail. The two-pipe medium-pressure and two-pipe high-pressure systems are used mostly in large industrial plants.

Medium-pressure systems are usually designed for a maximum supply main pressure drop of 5 to 10 psi with a design friction rate of 2 psi per 100 ft of equivalent pipe length. The return main pressure loss is set at a maximum of 5 psi with a design friction rate of 1 psi per 100 ft of equivalent pipe length.

Two-pipe high-pressure systems are usually designed for a maximum supply main pressure drop of 25 to 30 psi with a design friction rate of 2 to 10 psi per 100 ft of equivalent pipe length. The return main pressure drop may be limited to 20 psi and a maximum design friction rate of 2 psi per 100 ft of equivalent pipe length.

The best sources of information on the design of these two systems are *ASHRAE Handbook 1989 Fundamentals,* American Society of Heating, Refrigerating, and Air Conditioning Engineers, and *Handbook of Air Conditioning Systems Design,* Carrier Air Conditioning Company, 1965.

Table A for Illustrative Problem 11-4—Pipe sizes for vacuum return

Pipe Section	Condensate Flow (lb/hr)	Nominal Pipe Size (in.) (Table 11-4)
Return riser to last THTU	60	$\frac{3}{4}$ (column V)
FG	60	$\frac{3}{4}$ (column V)
EF	135	$\frac{3}{4}$ (column V)
DE	207	1 (column V)
CD	289	1 (column V)
BC	389	$1\frac{1}{4}$ (column V)
AB	639	$1\frac{1}{2}$ (column V)

11-12 FLASH STEAM

When hot condensate or boiler water, under pressure, is released to a lower pressure, part of it is reevaporated and becomes what is known as *flash steam*. This flash steam is important because it contains heat units that can be utilized for economy of operation. In order to take advantage of flash steam, however, it is necessary to know how it is formed and how much will be formed under given conditions.

Sometimes, leaking steam is confused with flash steam. Leaking steam "flows" from a leak producing a steady bluish stream, while flash steam (from steam traps) "floats" out intermittently (each time the trap discharges) in a whitish cloud.

We know from a review of thermodynamics that when water is heated under pressure, the boiling point is above 212°F, so the sensible heat is high, and for every pressure there is a corresponding saturation temperature. If the high-pressure saturated water is throttled to a lower pressure, the condition of the fluid at this lower pressure can be evaluated by determining its quality *x*, discussed briefly in Chapter 2. The quality may be defined as

$$x = \frac{m_v}{m_v + m_L} \tag{11-1}$$

where

m_v = mass of saturated vapor in the mixture
m_L = mass of saturated liquid in the mixture

Likewise, the volume fraction V_c of the vapor in the mixture is expressed as

$$V_c = \frac{V_v}{V_v + V_L} \tag{11-2}$$

where

V_v = volume of saturated vapor in the mixture
V_L = volume of saturated liquid in the mixture

If we consider the throttling process for the condensate passing through a steam trap, we may consider it a constant enthalpy process. The quality and the volume fraction of the mixture leaving the trap and entering the condensate return would be

$$h_1 = h_2 = h_{f2} + xh_{fg2}$$

or

$$x = \frac{h_1 - h_{f2}}{h_{fg2}} \tag{11-3}$$

and

$$V_c = \frac{xv_{g2}}{v_{f2}(1 - x) + xv_{g2}} \qquad (11\text{-}4)$$

where

h_1 = enthalpy of liquid condensate entering the trap evaluated at the supply pressure for saturated condensate or at the saturation pressure corresponding to the temperature of the subcooled liquid condensate

h_{f2} = enthalpy of saturated liquid at the return or downstream pressure of the trap

h_{fg2} = enthalpy difference $(h_{g_2} - h_{f_2})$ or latent heat of the fluid at the return or downstream pressure of the trap

v_{f2} = specific volume of saturated liquid at the return or downstream pressure of the trap

v_{g2} = specific volume of saturated vapor at the return or downstream pressure of the trap

ILLUSTRATIVE PROBLEM 11-5

Assume that saturated liquid water at 65 psia is being discharged from equipment to a pressure of 0 psig. Determine the quality and the volume fraction of the resulting mixture.

Solution: From Appendix Table A-2 (steam tables), determine the fluid properties. At 65 psia, $h_1 = h_{f1} = 267.67$ Btu/

lb. At 0 psig (14.696 psia), find $v_{f2} = 0.016715$ ft³/lb, $v_{g2} = 26.80$ ft³/lb, $h_{f2} = 180.15$ Btu/lb, and $h_{fg2} = 970.4$ Btu/lb. By Eq. (11-3),

$$x = \frac{h_1 - h_{f2}}{h_{fg2}} = \frac{267.67 - 180.15}{970.4}$$

$$= 0.090 \text{ or } 9.0\% \text{ vapor, mass basis}$$

By Eq. (11-4),

$$V_c = \frac{xv_{g2}}{v_{f2}(1 - x) + xv_{g2}}$$

$$= \frac{(0.09)(26.8)}{(0.016715)(1 - 0.09) + (0.09)(26.8)}$$

$$= 0.994 \text{ or } 99.4\% \text{ vapor, volume basis}$$

The results shown in Illustrative Problem 11-5 show that the percent of vapor on a mass basis, x, is small while the percent of vapor on a volume basis is very large. This would indicate that the return pipe cross section is predominantly occupied by vapor. This we will find true for most steam heating systems. Figure 11-12 is a working chart to determine the quality of the condensate entering the return line from the trap for various combinations of supply and return pressures. If the liquid is subcooled entering the trap, the saturation

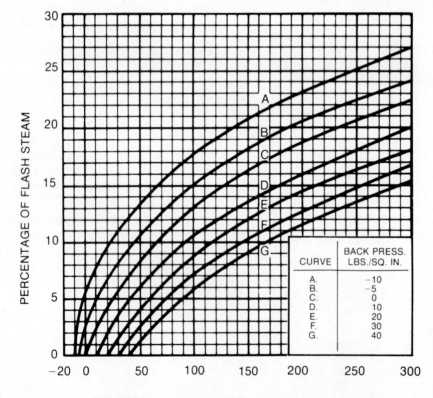

CURVE	BACK PRESS. LBS./SQ. IN.
A.	−10
B.	−5
C.	0
D.	10
E.	20
F.	30
G.	40

FIGURE 11-12 Percentage (x) of flash steam (mass basis) formed when discharging condensate to a reduced pressure (pressure in psig). (Courtesy of Armstrong Machine Works, Three Rivers, MI)

pressure corresponding to the liquid temperature should be used for supply pressure or upstream pressure.

11-13 METHODS OF USING FLASH STEAM

In Section 11-12, our discussion of flash steam was concerned with the fact that some of the condensate leaving a steam trap enters the return main as saturated steam at the lower pressure, which affects the required size of the return piping. We should also consider possible use of flash steam.

When steam traps drain medium- and high-pressure process equipment, substantial amounts of flash steam may be formed. The energy in this flash steam is identical to that of live steam at the same pressure. But, this energy is wasted when an excess amount of flash is allowed to escape through the vents in the receiver. *Note:* In smaller plants not employing deaerators, it is desirable to vent some flash from the receiver. This venting releases corrosion-producing oxygen and carbon dioxide commonly existent in steam systems.

The energy content of flash steam may be used for space heating, for heating or preheating water, oil or other liquids, and for low-pressure process heating.

The flash steam recovered may be used by itself to take care of the demands of one or more steam heating units or systems. Or, if exhaust steam is available, it may be combined with the flash. In other cases, the flash will have to be supplemented by live makeup steam at reduced pressure.

Flash Tank Hookup. Condensate return lines contain both flash steam and water. To recover the flash steam, the return main may be run to a flash tank where the liquid condensate is drained off. The flash steam is piped from the flash tank to points of use.

The demand for low-pressure steam should correspond with the availability of flash steam but should preferably be greater, the difference being made up with live steam through a pressure-reducing valve. The reducing valve is set to maintain the required low steam pressure and ensures a continuous supply.

Figure 11-13 illustrates how a reducing valve is used to combine flash and live steam. Remember that the amount of flash will never be as great as theoretical. Therefore, the reducing valve should be sized to pass enough makeup live steam under conditions of maximum demand and minimum supply of flash. Figure 11-13 also shows the proper way to hook up the flash tank to combine flash and live steam. Notice how the pipework is arranged so that the flash tank can be conveniently bypassed and condensate taken directly to the receiver. This is essential if flash steam cannot always be used when it is available. Notice the relief valve on the flash tank. It prevents pressure from building up and interfering with the operation of the high-pressure steam traps. A pressure gauge on the tank is also advisable.

Figure 11-14 illustrates flash steam recovery from a heater battery. Flash is taken from the flash tank and combined with live steam, the pressure of which is reduced to that of the flash by a reducing valve.

Sizing Flash Tanks. The flash tank can usually be conveniently constructed from a piece of large-diameter piping with the ends welded or bolted in position. The

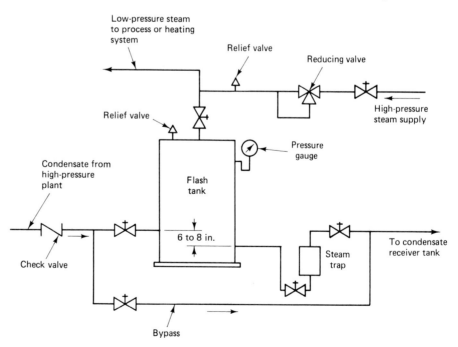

FIGURE 11-13 Flash steam tank hookup with live steam makeup, showing recommended fittings and connections.

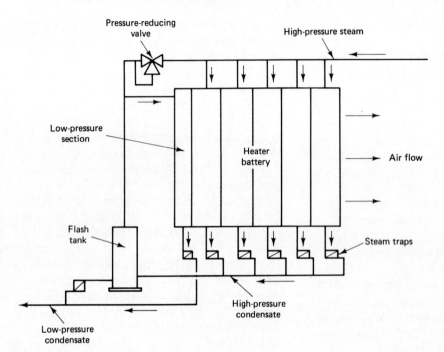

FIGURE 11-14 Flash steam recovery from air-heater battery.

tank should be mounted vertically. A steam outlet is required at the top and a water outlet a few inches from the bottom. The high-pressure condensate inlet connection should be 6 to 8 in. above the water outlet (see Figure 11-13). No internal fittings are needed.

The important dimension is the inside diameter of the tank. This should be such that the upward velocity of flash to the outlet is low enough to ensure that the amount of water carried over with the flash is kept low. If the upward velocity is kept low, the height of the tank is not important, but good practice is to use a height of 2 or 3 ft. It has been found that a steam velocity of 10 ft/s inside the flash tank will give good separation of steam and water. On this basis, the proper inside diame-

ters for various quantities of flash steam have been calculated; the results are plotted in Figure 11-15. This curve gives the smallest recommended internal diameters. If it is more convenient, a larger size tank may be used. Figure 11-15 does not take pressure into consideration, only weight. Although volume of steam and upward velocity are less at a higher pressure because the steam is more dense, there is an increasing tendency for priming (water carry-over). Thus, it is recommended that, regardless of the pressure, Figure 11-15 be used to find the internal tank diameter.

Practical Considerations in Installation of a Flash Tank. Bearing in mind that a flash tank causes back pressure on the steam traps feeding into the tank, the traps should be checked to see whether they still have sufficient capacity at the reduced differential pressure.

Condensate lines should be pitched down toward the flash tank. When more than one line feeds into the flash tank, each line should be fitted with a swing check valve. Then, if any line is not in use, it will be isolated from the others and will not condense and waste flash steam.

The steam trap draining the flash tank should be large enough to handle the maximum condensate load, which will usually occur when starting up the system. As this may be 3 times normal load, the trap should be sized to meet these conditions. Because the traps will work at low pressure, gravity drainage to the receiver should be provided.

Generally, the location chosen for the flash tank

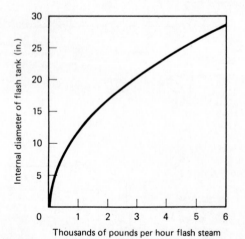

FIGURE 11-15 Flash tank diameter to handle given flash steam quantity.

should meet the requirement for maximum quantity of flash steam and minimum length of pipework.

Condensate lines, the flash tank, and the low-pressure steam lines should be insulated so as to prevent waste of flash steam through radiation heat transfer.

The fitting of a spray nozzle on the condensate inlet pipe inside the flash tank is *not* recommended. It may become choked, stop the flow of condensate, and set up a back pressure to the traps.

Condensate lines from low-pressure equipment us-

ing flash steam should not be connected to the lines feeding the flash tank. The reason is that the pressure will be about equal, and a slight drop in condensate line pressure will cause back-up flow.

In some cases, large volumes of air may need to be vented from the flash tank. A thermostatic air vent of adequate capacity will vent the air from the tank and keep it from passing through the low-pressure heating or process units.

REVIEW PROBLEMS

11.1. Dry saturated steam at 20 psia condenses and subcools to 180°F. Making use of steam tables, compute **(a)** the energy released during this process and **(b)** the change in specific volume during the condensation and subcooling.

11.2. The boiler pressure in a one-pipe steam system, and at entry to the supply main, is 16 psia. As the supply main reaches its end and drops into a wet return feeding the boiler, the pressure is 14.8 psia. Compute the height of the water in the drop leg with respect to the water level in the boiler. (Refer to Figure 11-1.)

11.3. A steam pipe line 120 ft long is installed at 50°F. When it is filled with steam at 5.3 psig pressure, what is the increase in length of the pipe **(a)** if the pipe is steel and **(b)** if the pipe is copper?

11.4. **(a)** What size steel pipe is required to carry steam at the rate of 6000 lb/hr through a main having a total equivalent length of 450 ft if the initial pressure is 100 psig and the final pressure is not less than 95 psig?
(b) What would be the actual pressure drop and flow velocity in the pipe selected in part (a)?

11.5. A central steam heating plant provides steam to a remote classroom building on a college campus through a 10-in. schedule 40 steel pipe. An addition to be the building is being considered, and the question has been raised as to whether the same steam main can serve the original building plus the addition. The linear run of pipe is 1400 ft, and it is estimated that the extension and equivalent length of fittings will add 600 ft. How many pounds of saturated steam can pass through the existing 10-in. steam main if a pressure drop of 15 psi is allowed for the 120 psig steam pressure at the plant?

11.6. Determine the pressure drop in 4-in. schedule 40 steel pipe having an equivalent length of 400 ft and carrying steam at the rate of 75 lb/min. if the inlet steam pressure is 60 psig.

11.7. A two-pipe vapor system serving a design load of 670,000 Btu/hr has an estimated total equivalent length of 200 ft and is to have a design pressure drop of 4 oz/in.².
(a) What size steam main and dry-return main should be used?

(b) To supply a terminal heat-transfer unit on the second floor delivering 18,000 Btu/hr, what size riser and riser runout should be used?

11.8. A vacuum steam heating system to deliver 1,950,000 Btu/hr has an estimated total equivalent length of 800 ft and is to be designed for a total pressure drop of 1.0 psi. What size steam main and return should be used?

11.9. A two-pipe low-pressure steam system has an initial steam pressure of 10 psig. The estimated total equivalent length (TEL) of the supply piping is 500 ft, and the TEL of the return piping is 550 ft. Determine a recommended design friction rate to size **(a)** the supply main and riser and **(b)** the return main and riser.

11.10. Determine the pressure loss in a 3-in. schedule 40 steam main that contains one globe valve, six 90° standard elbows, and 140 ft of straight pipe. The steam flow rate through the system is 15 lb/min. The initial steam pressure is 10 psig. What is the approximate steam velocity through the pipe?

11.11. What size steel pipe (SN 40) is needed to carry 300 lb/hr of saturated steam if the friction rate is not to exceed 1 oz per 100 ft of pipe? The initial steam pressure is 2.0 psig, saturated.

11.12. **(a)** Determine the pipe size required to carry a steam flow of 1000 lb/hr. The initial steam pressure is 10 psig, saturated. The maximum allowable pressure loss is 4 psi, and the total equivalent length of the pipe is 200 ft.
(b) What would be the actual pressure loss and flow velocity in the pipe selected in part (a)?

11.13. A steam supply main from a boiler must supply a large THTU having a capacity of 480,000 Btu/hr. The measured length of steel pipe from boiler to heater is 360 ft. It is estimated that an allowance of 25% will account for valves and fittings. Boiler supply pressure is 5 psig.
(a) Determine the pipe size to use for the supply main.
(b) Determine the actual pressure loss in the supply main selected in part (a).

11.14. A vacuum-return system is to serve THTU's that have a total rated capacity of 7600 sq. ft EDR. The equivalent length of the supply main is 800 ft, and the boiler supply pressure is to be 2 psig. Specify the nominal size of the steam supply main as it leaves the boiler header and the nominal size of the dry return as it approaches the vacuum pump. The pipe is SN 40. Equivalent length of return main is estimated to be 850 ft.

11.15. Determine a design friction rate (psi/100 ft) to use in the sizing of pipe for a two-pipe low-pressure steam system for the following conditions: (1) two-pipe reverse-return system; (2) initial steam pressure at boiler, 5 psig; (3) estimated TEL of supply piping, 500 ft, and return piping, 500 ft; and (4) total system load, 4800 EDR. Determine **(a)** design friction rate to use for sizing pipe, **(b)** pipe size for supply main, and **(c)** pipe size for dry-return main.

11.16. A steam main from a boiler supplies a large unit heater with a capacity of 500,000 Btu/hr. The measured length of piping from the boiler to the heater and back to the boiler is 360 ft. It is estimated that an allowance of 25% will account for valves and fittings. Boiler supply pressure is 5 psig.

   (a) What should be the size of the supply main if (1) Figure 11-8 is used and (2) Table 11-2 is used?

   (b) Using the pipe size selected in part (a1), what would be the estimated pressure drop in the supply main?

11.17. A one-pipe steam heating system has a total load of 600 sq. ft EDR and is laid out in such a way that the condensate flows toward the boiler in all parts of the main (counterflow to the steam). The boiler pressure under design conditions is 1 psig. Determine the nominal pipe size of the steam main.

11.18. A two-pipe vapor system serving 600,000 Btu/hr has a total equivalent length of 200 ft and a total pressure drop of 2 oz/in.^{2}

   (a) What size steam main and dry return should be used?

   (b) To supply a THTU on the second floor delivering 18,000 Btu/hr, what size riser and horizontal branch should be used?

11.19. Figure 11-16 shows a plan layout of the steam main for a one-pipe steam heating system. The THTU's have the following capacities:

THTU	Btu/hr	THTU	Btu/hr
A	5,000	E	4,000
B	12,000	F	8,000
C	6,000	G	9,000
D	8,000	H	8,000

Determine the pipe sizes for the main, risers, and run-outs for the system.

11.20. Figure 11-17 represents the plan layout of a two-pipe steam heating system. The steam pressure after the pressure-reducing valve (PRV) is 5 psig. All mains and riser runouts are dripped. All returns are dry returns. All steam loads are expressed in sq. ft EDR. Determine pipe sizes for all steam supply mains and all condensate return mains.

11.21. Estimate the pressure drop in 4-in. schedule 40 steel pipe having an equivalent length of 400 ft, carrying saturated steam at the rate of 100 lb/min., and having an initial pressure of 60 psig.

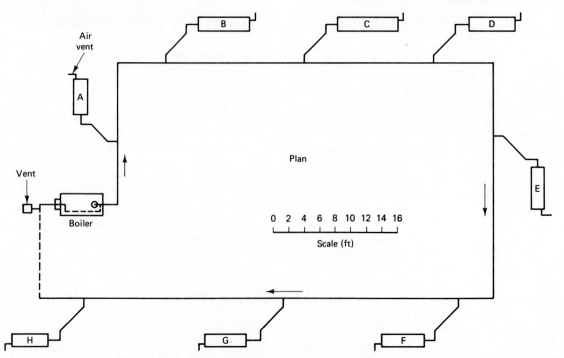

FIGURE 11-16 Piping system for Review Problem 11.19.

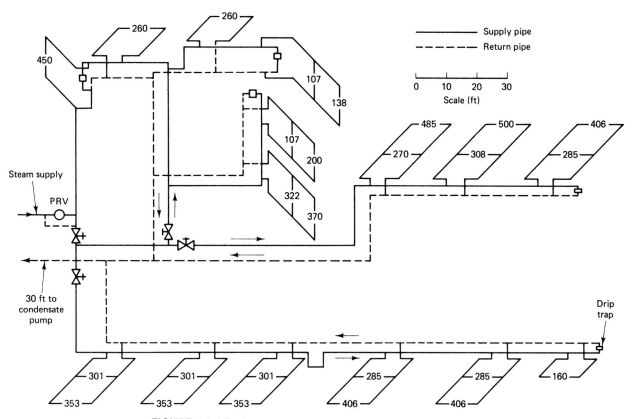

FIGURE 11-17 Piping system for Review Problem 11.20.

11.22. A 4-in. steam pipe is to carry steam with a pressure loss not to exceed 1.5 oz per 100 ft of equivalent length. The equivalent length of the line is 300 ft, and the initial steam pressure is 5.0 psig. If heat losses are neglected, estimate the weight of steam that it will handle in pounds per hour.

BIBLIOGRAPHY

11.1. *ASHRAE Handbook 1989 Fundamentals,* American Society of Heating, Refrigerating, and Air Conditioning Engineers, Atlanta, GA, 1989.

11.2. *ASHRAE Handbook 1987 HVAC Systems and Applications,* American Society of Heating, Refrigerating, and Air Conditioning Engineers, Atlanta, GA, 1987.

11.3. *Steam Heating Application Manual,* Bulletin No. TES-375, ITT Fluid Transfer Division, Morton Grove, IL, 1975.

12

Steam Specialties and Condensate-Handling Systems

12-1 INTRODUCTION

In this chapter, we discuss the different types of special fittings, valves, and steam traps. These devices are usually called *steam specialties* and are required for the proper operation of steam heating systems. We are concerned with what these devices are, how they function in the system, and how they are used to conserve energy in the proper handling of condensate.

When steam leaves the boiler and flows through the distribution piping system, the steam in the pipes is at a higher temperature than the air surrounding the pipes. Heat flows from the steam and through the pipe walls to the surrounding air. This heat loss causes a small portion of the steam to condense. Since this heat flow is normally wasted, steam distribution pipes are insulated to minimize the waste and undesirable heat transfer.

When steam reaches the terminal heat-transfer equipment, the transfer of heat from steam to air in an air heater or to water in a shell-and-tube heat exchanger is highly desirable. Nothing should interfere with this heat transfer.

Condensate is the by-product of a steam system. It forms in the distribution system because of unavoidable radiation and in heating and process equipment because of heat transfer from the steam to the substance heated. Once steam has condensed, giving up its latent heat, the hot condensate must be removed immediately as the available heat energy in a pound of condensate is negligible as compared to a pound of steam. However, this condensate is still valuable and should be returned to the boiler.

Steam moves rapidly through steam supply mains, often at 90 mph or more. A buildup of condensate in these lines can be pushed along by the fast-moving steam and turned into a battering ram. This can result in damaged pipe fittings and regulating valves and most likely cause annoying sound referred to as *water hammer*. It is essential that this condensate be removed as a "heavy dew" before it can grow into a dangerous slug moving at the same speed as the steam.

Obviously, condensate in the heat-transfer devices takes up space and, in effect, reduces the physical size and capacity of the equipment. It must be removed promptly so that the equipment can be kept full of steam.

Air is always present in piping and equipment during start-up and in the boiler feed water. In addition, the feed water may contain carbonates that dissolve and release carbon dioxide gas. The drainage problem involves more than just condensate removal. Air and carbon dioxide gas must be promptly cleared from the system. Air, with its excellent insulating properties, can "plate out" on heat-transfer surfaces as steam condenses and greatly reduces efficiency. Under certain conditions, as little as one-half of 1% by volume of air in steam can reduce heat-transfer efficiency by 50%. When carbon dioxide gas is released in the boiler as steam is produced, it can dissolve in condensate that has cooled below steam temperature and form carbonic acid. This acid is extremely corrosive and will eventually "eat through" piping and heat-transfer equipment. If oxygen is able to enter the system, it can cause pitting of iron and steel surfaces.

It is essential that condensate, air, and carbon dioxide gas be removed as quickly and as completely as possible. This is done with a *steam trap,* which is simply an automatic valve that opens for condensate, air, and carbon dioxide and closes for steam. For economic reasons, the trap should perform this function for long periods of time with minimum attention.

12-2   GENERAL REQUIREMENTS OF STEAM TRAPS

The job of the steam trap is to get condensate, air, and CO_2 out of the steam-heated unit as fast as they accumulate. In addition, for overall efficiency and economy, the trap must also provide the following:

1. *Minimum steam loss*—Table 12-1 shows how costly unattended steam leaks can be.

TABLE 12-1

Cost of various-sized steam leaks at 100 psig steam pressure (assuming steam costs $5.00 per 1000 lb)

Size of Orifice (in.)	Steam Wasted per Month (lb)	Total Cost per Month ($)	Total Cost per Year ($)
$\frac{1}{2}$	835,000	4,175.00	50,100.00
$\frac{7}{16}$	637,000	3,185.00	38,220.00
$\frac{3}{8}$	470,000	2,350.00	28,200.00
$\frac{5}{16}$	325,000	1,625.00	19,500.00
$\frac{1}{4}$	210,000	1,050.00	12,600.00
$\frac{3}{16}$	117,000	585.00	7,020.00
$\frac{1}{8}$	52,500	262.50	3,150.00

Source: Steam Conservation Guidelines for Condensate Drainage, Manual M101. Courtesy of Armstrong Machine Works, Three Rivers, MI.

2. *Long life and dependable service*—Rapid wear of parts quickly brings a trap to the point of undependability. By contrast, an efficient trap saves money by minimizing trap testing, repair, cleaning, and downtime.
3. *Corrosion resistance*—Working trap parts should be corrosion-resistant in order to combat the damaging properties of condensate.
4. *Air venting*—Air can be present in steam at any time, as well as on start-up, and must be vented for efficient heat transfer.
5. *CO$_2$ venting at steam temperature*—When CO_2 is vented and condensate is discharged at steam temperature, corrosive carbonic acid is unable to form in heat exchanger equipment.
6. *Operation against back pressure*—The possibility of back pressure in the condensate return line is always present. A steam trap should be able to operate against the actual back pressure in the return system.
7. *Freedom from dirt problems*—Condensate will probably pick up dirt and scale in the piping, and there may be a carry-over of solids from the boiler. The steam trap must be able to operate in the presence of the dirt, or *strainers* should be provided immediately before the trap.

A steam trap that delivers anything less than all the desirable operating and design features just listed will reduce the efficiency of the system and increase the costs. A trap that delivers all these features is a major factor in achieving (1) fast heat-up of heat-transfer equipment, (2) maximum equipment temperature, (3) maximum equipment capacity, (4) maximum fuel economy, (5) reduced labor per unit of output, and (6) minimum maintenance and long trouble-free trap life.

Sometimes, a specific application may demand a less-efficient trap. In the vast majority of applications, however, best results are obtained from a trap that meets all these basic requirements.

Steam apparatus, heat-transfer equipment, and piping vary widely in the character of the trapping needs. Some of the variables are (1) condensate flow, steady or variable; (2) operating pressure; (3) operating temperature; (4) air volume to be handled; (5) warm-up loads; (6) pressure difference between trap inlet and outlet; and (7) condensate temperature. No one trap can be used to meet all trapping situations because of these variables. Manufacturers have developed a number of trap types, each designed for specific trapping needs.

Steam traps may be classified as follows:

1. *Thermostatic traps* react to the difference in temperature of steam and condensate.
2. *Mechanical traps* operate by the difference in density between steam and condensate.
3. *Kinetic traps* rely on the difference in flow characteristics of steam and condensate.

The first step in steam trap selection is to determine the trap type best suited for the intended application. The overlapping operating characteristics of some trap types often make this a matter of personal preference rather than science. Once the trap is "typed," we may size it, taking into consideration the factors outlined earlier.

Steam traps, regardless of type, should be carefully sized for the application and condensate load to be handled since both undersizing and oversizing can cause serious problems. Undersizing can result in undesirable condensate backup and excessive cycling that can cause premature failure. Oversizing might appear to solve this problem and make selection much easier because fewer different sizes are required, but if the trap fails, excessive steam can be lost. The best approach is to become aware of the principles that make for good trap operation and then to use experience, judgment, and any required assistance from the manufacturer to select the trap type and size that will be most effective for a specific job.

12-3 THERMOSTATIC TRAPS

In thermostatic traps, a bellows or bimetallic element operates a valve that opens in the presence of condensate and closes in the presence of steam. Because condensate is initially at the same temperature as the steam from which it is condensed, the thermostatic element must be designed and calibrated to open at a temperature below the steam temperature; otherwise, the trap would blow live steam continuously. Therefore, the condensate must be subcooled by allowing it to back up in the trap and a portion of the upstream drip leg piping, both of which are left uninsulated. Some thermostatic traps operate with a continuous water leg behind the trap so that there is no steam loss; however, this inhibits the discharge of air and noncondensable gases and can cause excessive condensate to back up into the mains or terminal heat-transfer equipment, thus causing operating problems. Devices that operate without significant backup can lose steam before the trap closes.

Thermostatic Operation. Figure 12-1 shows a *bellows-type* thermal element mounted in a cage-type

FIGURE 12-1 Thermostatic steam trap by Hoffman. (Courtesy of ITT Fluid Transfer Division, Morton Grove, IL)

enclosure. The valve is fastened to the bellows, and the unit is inserted into the trap body.

The bellows action of the thermostatic trap is caused by steam pressure and temperature as condensate flows to the trap. When condensate temperature is close to steam temperature, vapor pressure inside the bellows causes the bellows to expand and tightly close the orifice. As the condensate backs up in the cooling leg, the temperature begins to drop and the bellows contracts and opens the orifice permitting condensate, air, and noncondensable gases to be discharged.

The bellows is partially filled with a liquid while under vacuum (usually distilled water to assure operation very near actual steam temperature and to follow the steam temperature curve). In its normal position, the valve is retracted from its seat and the trap is open.

Three forces tend to open the trap: (1) the pressure force due to inlet steam pressure, (2) the outlet or tail pipe pressure, and (3) the metal stress of the bellows. Opposing these forces and tending to close the valve is the bellows internal pressure due to evaporation of the volatile liquid. This force expands the thermal element and tends to close the trap. The pressure around the thermal element (steam pressure) tends to compress it and accordingly raises the pressure within it. The increased internal pressure raises the boiling temperature of the volatile fluid within the element.

The amount of fluid used to partially fill the element is such that enough will evaporate to cause the trap to close at a predetermined temperature. At this time, the internal pressure exceeds the total pressure forces external to the element. Since the internal and external pressure tend to offset each other, the bellows reacts only to temperature change. For this reason, thermostatic traps of this type are called *balanced-pressure* traps.

Thermostatic traps are designed to close at a temperature somewhat below the saturated steam temperature. They require a drop in temperature below this point to cause them to open. On a drop in temperature, the vapor pressure within the thermal element tends to drop. When the bellows internal pressure becomes less than the total external pressure forces, the bellows shrinks, pulling the valve off its seat. The difference between the closing and opening temperatures is called the *temperature drop* of the trap. This is usually between 5°F and 30°F, depending upon design and system pressure.

It is important that this characteristic of thermostatic traps be recognized when applying them to a trapping job. *The condensate reaching it must be sufficiently low in temperature to cause it to open.* If it is not, condensate will back up into the heating apparatus being served by the trap until it cools sufficiently to open the trap.

Bellows-type thermostatic traps are best suited for steady light loads on low-pressure service. They are most widely used on finned pipe and on convector radiation units in HVAC systems.

The *bimetallic* thermostatic trap has an element made from metals with different expansion coefficients. Heat causes the element to change shape, permitting the valve ports to open or close. Because a bimetallic element responds only to temperature change, most traps have the valve on the outlet so that steam pressure is trying to open the valve. Therefore, by properly designing the bimetallic element, the trap can operate on a pressure–temperature curve approaching the steam saturation curve (but not as closely as with the balanced-pressure bellows element).

Unlike the bellows thermostatic trap, bimetallic thermostatic traps are not adversely affected by superheated steam or subject to damage by water hammer; thus, they can be readily used for high-pressure applications. They are best suited for steam tracers, jacketed piping, and heat-transfer equipment, where some condensate backup is tolerable. If they are used on steam main drip legs, the element should not back up condensate.

12-4 FLOAT AND THERMOSTATIC TRAPS

The *float and thermostatic* steam trap, commonly known as the F&T trap, is one type of mechanical trap where operation is dependent on the difference in density between steam and condensate. The F&T trap is actually a combination of two types of traps in a single

trap body: (1) a bellows thermostatic element operating on temperature difference, which provides automatic venting, and (2) a float valve. See Figure 12-2.

The F&T trap has large venting capabilities, continuously discharges condensate without backup, handles intermittent loads very well, and can operate at extremely low pressure differentials. It is well suited for steam main and riser drip legs on low-pressure steam-heating systems. One of its most common applications is where the system being drained operates on a modulating steam pressure, such as temperature-regulated steam-heating coils. This means that the pressure in the heat exchanger unit being drained can vary anywhere from the maximum steam supply pressure down to vacuum under certain conditions. Thus, under conditions of zero gauge pressure, only the force of gravity is available to push condensate through the trap. To further complicate drainage, substantial amounts of air may be liberated under these conditions of low steam pressure. These specialized requirements are all filled through efficient operation of the F&T trap.

The actual operation of the F&T trap is quite simple, yet extremely efficient. A ball float is connected by a lever to the valve, and, once condensate reaches a certain level in the trap, the float begins to rise. This opens the valve and condensate is drained until the float is lowered and the valve rests tightly on the seat. The float is positioned above the orifice so that the condensate forms a water seal and eliminates the loss of any live steam. Since there are no pressure fluctuations due to intermittent operation, condensate is discharged at a constant temperature. At the top of the trap is a thermostatic air vent, which discharges all air and noncondensable gases as soon as they reach the trap and allows for maximum condensate drainage. This thermostatic air vent is responsive to temperature and will open a few degrees below saturation. Therefore, this type of trap handles a large quantity of air due to its separate orifice, but at slightly reduced temperature.

SHEMA Ratings. Float and thermostatic traps for pressures up to 15 psig may be selected by pipe size on the basis of ratings established by the Steam Heating Equipment Manufacturers Association (SHEMA). SHEMA ratings are the same for all makes of F&T traps since they are an established measure of the capacity of a pipe flowing half full of condensate under specific conditions of pressure, length of pipe, pitch of pipe, and so forth.

Various specification-writing authorities indicate different procedures for sizing traps by SHEMA ratings, and their procedures must be followed to assure compliance with their specifications. However, as SHEMA ratings provide for continuous air elimination

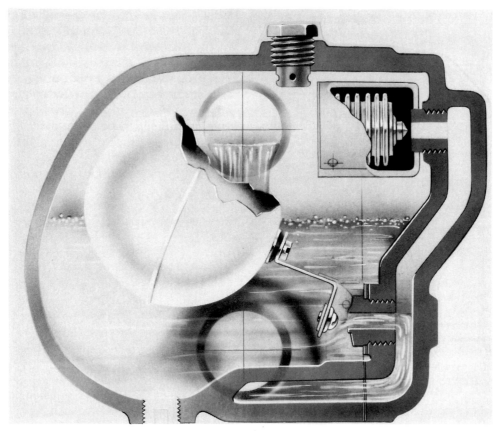

FIGURE 12-2 Float and thermostatic steam trap (section). (Courtesy of Armstrong Machine Works, Three Rivers, MI)

when the trap operates at maximum condensate capacity rating and provision is made for overload conditions, no trap capacity selection factor is necessary.

12-5 INVERTED BUCKET TRAPS

Bucket traps operate on buoyancy, but they use a bucket that is either open at the top or inverted instead of a closed float. Figure 12-3 shows the inverted bucket trap. Steam and/or air enters the submerged inverted bucket and causes it to float and close the discharge valve. As more condensate enters the trap, it forces air and steam out of the vent on top of the inverted bucket into the trap body where the steam condenses by cooling. When the weight of the bucket exceeds the buoyancy effect, the bucket sinks and opens the discharge valve; steam pressure forces the condensate out, and the cycle repeats.

Unlike most cycling-type traps, the inverted bucket trap continuously vents air and noncondensable gases. Although it discharges condensate intermittently, there is no condensate backup in a properly sized trap.

High discharge line pressure affects the force required to open the trap valve just the same as lowered steam pressure. As back pressure approaches the inlet pressure, discharge becomes continuous, just as it does on very low pressure differentials. Back pressure has no adverse effect on inverted bucket trap operation other than capacity reduction due to low differential. The trap will not fail to close and will not blow steam due to high back pressure.

Inverted bucket traps are made for all pressure ranges and are well suited for steam main drip legs and most HVAC applications. Although they can be used for temperature-regulated steam coils, the F&T trap is usually better because it has the high venting capability desirable for such applications.

12-6 KINETIC TRAPS

Numerous devices operate on the difference between flow characteristics of steam and condensate and on the fact that condensate discharging to a lower pressure contains more heat than necessary to maintain its liquid

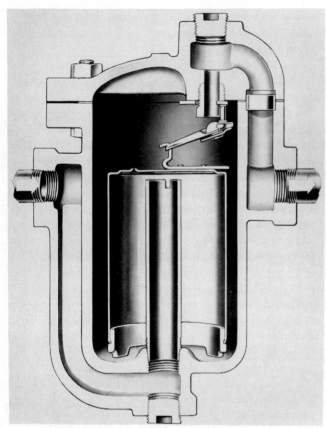

Figure 12-3 Armstrong 800 Series inverted bucket steam trap (section). (Courtesy of Armstrong Machine Works, Three Rivers, MI)

orifice. This flow lifts the disc off the inlet orifice, and the condensate flows through to the outlet passages. When steam reaches the disc, increased flow velocity across the face of the disc reduces pressure at this point and increases pressure in the control chamber, and the disc closes the orifice. Controlled bleeding of steam from the control chamber causes the trap to open. If condensate is present, it will be discharged. The trap recloses in the presence of steam and then continues to cycle at a controlled rate.

The minimum operating pressure of 10 psig is usually required to assure positive operation of the controlled disc. In order to meet the ratio discharge capacities, the back pressure should not exceed 50% of the inlet pressure. The safety factor range for these traps is between 1 and 1.2. Thermodynamic traps need not and, in most cases, should not be oversized.

One of the advantages of a disc trap is its ability to operate at high working pressures. The controlled disc trap by Armstrong Machine Works has a maximum working pressure of 600 psig. Also, trap operation is independent of gravity, and it will operate in any position. When installed vertically, the trap is self-draining when shut down. This feature makes it suitable for outdoor installations, where freezing could be a problem.

In general, disc traps are best used where tolerance to light loads and the ability to operate with high pressure are needed. One of the principal areas for use of this type of trap is the draining of steam tracer lines. Tracer lines are used to heat piping carrying viscous fluids through outdoor areas. They consist of tubing fastened to the process piping so as to provide heat by conduction. Tubing is supplied with steam as the heating medium, at pressures ranging from 15 to 200 psig.

The tracer lines must be drained at regular intervals every 150 ft to keep the lines free of condensate. Since these lines can be run outdoors, the freeze-proof feature of the trap makes them a good choice for this service. The refining industry uses tracer lines in much of their process piping.

The traps are also used for dripping high-pressure steam supply lines, draining high-pressure submerged heating coils, and in sterilizers, dryers, and autoclaves.

Disc traps apply to problem installations, where limited space, water hammer, corrosion, or superheat (not recommended) make other traps difficult to use.

12-7 TRAP SELECTION

In order to obtain the full benefits from the steam traps described in the preceding sections, it is necessary that

phase. This excess heat causes some of the condensate to flash to steam at the lower pressure. Several traps fall into this category of kinetic traps, such as piston (or impulse) traps, orifice traps, and thermodynamic (or disc) traps. We shall discuss the disc trap because it is more common. See Figure 12-4.

The *controlled disc* steam trap is very light and compact and is consequently used in many applications where space is quite limited. It contains only one moving part, the disc itself. Beside the disc trap's simplicity and small size, it also offers advantages such as resistance to hydraulic shock, the complete discharge of all condensate when open, and intermittent operation for a steady purging action.

The design and operation of controlled disc traps as manufactured by Armstrong Machine Works may be explained as follows (refer to Figure 12-4). Condensate and air entering the trap pass through the heating chamber, around the control chamber, and through the inlet

DESIGN AND OPERATION OF MODEL CD-42 DISC TRAPS

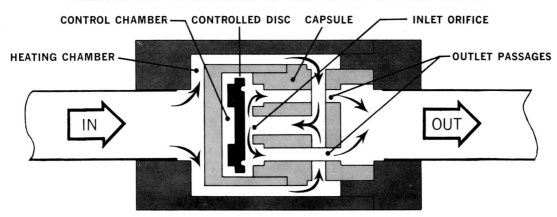

CONTROL CHAMBER — CONTROLLED DISC   CAPSULE   — INLET ORIFICE

HEATING CHAMBER — OUTLET PASSAGES

IN   OUT

Condensate and air entering the trap pass through the heating chamber around the control chamber and through the inlet orifice. This flow lifts the disc off the inlet orifice and the condensate flows through to the outlet passages. When steam reaches the disc, increased flow velocity across the face of the disc reduces pressure at this point and increases pressure in the control chamber and the disc closes the orifice. Controlled bleeding of steam from the control chamber allows the trap to open. If condensate is present it will be discharged. The trap recloses in the presence of steam.

FIGURE 12-4 Design and operation of model CD-42 disc traps. (Courtesy of Armstrong Machine Works, Three Rivers, MI)

the correct size and pressure of the trap be selected for each application and that it be properly installed and maintained. Selection or installation should always be accompanied by competent technical assistance or advice. Engineers with little experience are encouraged to seek assistance from product manufacturers.

Basic Considerations. Unit trapping is the use of a separate steam trap on each steam condensing unit. If a single device has multiple condensate outlets, each outlet should be trapped separately because possible *short circuiting* may occur. If more than one drain point is connected by a single trap, condensate and air from one or more of the units may fail to reach the trap. Any difference in condensing rates will result in a difference in the steam pressure drop. A pressure drop difference too small to register on a pressure gauge is enough to let steam from the higher-pressure unit block the flow of air or condensate from the lower-pressure unit. The net result is reduced heating output and fuel waste. See Figure 12-5.

Most traps are selected on the basis of experience. This may be (1) your personal experience, (2) the experience of the manufacturer's representative, or (3) the ex-

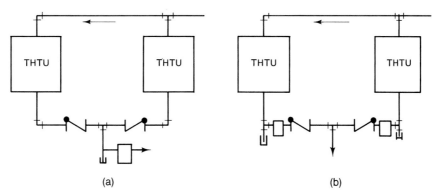

(a)   (b)

FIGURE 12-5 Draining two THTU's: (a) Two THTU's drained by a single trap may result in short circuiting; (b) short circuiting is impossible when each unit is drained by its own trap (higher efficiency is assured).

perience of thousands of others in trapping identical equipment. Fortunately, trap sizing is simple when you know or can determine the following:

- Condensate load in lb/hr.
- Safety factor to use.
- Pressure differential.
- Operating differential.
- Maximum allowable pressure.

Condensate load: We shall see in the following sections formulas and useful information on steam condensing rates and proper sizing procedures.

Safety factor or experience factor: Assume a steam heating device condenses 500 lb/hr. Also assume users have found that a 1500-lb/hr-capacity trap will enable the device to turn out more and better work than when a 1000-lb/hr-capacity trap is used. For this particular device, the trap selection safety factor is 3:1 for best machine performances.

Safety factors will vary from a low of 2:1 to a high of 10:1. The safety factors here are user-experience factors. A 500-lb/hr-capacity trap would hardly be enough for a 500-lb/hr-capacity heating coil operating at 100-psi pressure differential. The condensate formed might be more than 500 lb/hr, or the differential pressure might drop. Extra steam trap capacity is needed and costs very little.

Pressure differential: Maximum differential is the difference between boiler or steam main pressure, or the downstream pressure of a pressure-reducing valve (PRV) and return line pressure. The steam trap must be able to open against the differential pressure. *Note:* Because of flashing condensate in the return lines, do not assume a decrease in pressure differential due to static head when elevating.

Operating differential: When the plant is operating at capacity, the steam pressure at the trap inlet may be lower than steam main pressure, and the pressure in the condensate return header may go above atmospheric pressure.

If the operating differential is at least 80% of the maximum differential, it is safe to use maximum differential in selecting steam traps.

Modulated control of the steam supply causes wide changes in pressure differential. The pressure in the unit drained may fall to atmospheric pressure or even lower. This does not prevent condensate drainage if proper installation practices are followed.

In summary, we may say that inlet pressure can be (1) boiler or steam main pressure or (2) reduced pressure controlled by a pressure-reducing valve station. Dis-

charge pressure from the trap can be (1) atmospheric; (2) below atmospheric (under vacuum); add vacuum to inlet pressure to get the pressure differential; or (3) above atmospheric due to pipe friction, controlled back pressure with receiver relief valve or differential valve, or elevating of the condensate.

Figure 12-6 illustrates a trap discharging into an elevated return main. Every 2 ft of lift reduces trap pressure differential by approximately 1 psi when the trap discharge is only condensate. However, with flash steam present, the static head could be reduced to zero.

Maximum allowable pressure: The trap must be able to withstand the maximum allowable pressure of the system, or design pressure. It may not have to operate at this pressure but must be able to contain it. As an example, assume the maximum inlet pressure is 350 psig and the return line pressure is 150 psig. This results in a pressure differential of 200 psi; however, the trap must be able to withstand the 350-psig maximum allowable pressure.

12-8 HOW TO TRAP STEAM DISTRIBUTION SYSTEMS

Steam distribution systems make up the vital link between the boilers and the many steam-using terminal devices. They represent the method by which steam is actually transmitted to all parts of the heating system when steam is demanded by the terminal devices.

The three primary components of the steam distribution system are (1) boiler headers, (2) steam mains, and (3) branch lines. They each fill certain requirements of the system and together with steam separators and steam traps contribute to efficient steam utilization. Common to all steam distribution systems is the need

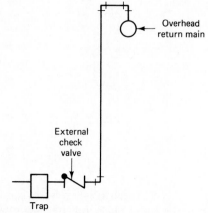

FIGURE 12-6 Steam trap discharging to overhead return main: When trap valve opens, steam pressure will elevate the condensate.

for *drip legs* at various intervals, which let condensate escape by gravity from the fast-moving steam and store the condensate until the pressure differential can discharge it through the steam trap.

Boiler Headers. A boiler header is a specialized type of steam main that can receive steam from one or more boilers. It is most often a horizontal line that is fed from the top and that in turn feeds the steam mains. It is important that the boiler header be properly trapped to assure that any carry-over (boiler water and solids) be removed from the steam before distribution to the system.

Small steam boilers usually have one steam outlet connection sized to reduce steam velocity to minimize carry-over of water into supply lines. Large boilers have several outlets that minimize boiler water entrainment. Figure 12-7 shows piping connections to the steam header. Although some engineers prefer to use an enlarged steam header for additional storage space, since there usually is not a sudden demand for steam except during the warm-up period, an oversized header is a disadvantage. The boiler header can be the same size as the boiler connection or the pipe used on the main. The horizontal runouts from the boiler (or boilers) to the header should be sized by calculating the heaviest load

that will be placed on the boiler. The runouts should be sized on the same basis as the building steam mains. Any change in size after the vertical uptakes should be made by reducing elbows.

Steam traps that serve the steam header must be capable of discharging large carry-over slugs of water as soon as they are present. Resistance to hydraulic shock is also a consideration in the selection of traps.

A 1.5:1 safety factor is recommended by Armstrong Machine Works for virtually all boiler steam header applications. The required trap capacity can be obtained by using the following relationship:

required trap capacity = safety factor × load connected to boiler(s) × anticipated carry-over (typically 10%) (12-1)

ILLUSTRATIVE PROBLEM 12-1

What size steam trap will be required on a connected load of 50,000 lb/hr with an anticipated carry-over of 10%?

Solution: By Eq. (12-1),

required capacity = (1.5)(50,000)(0.10)
= 7500 lb/hr

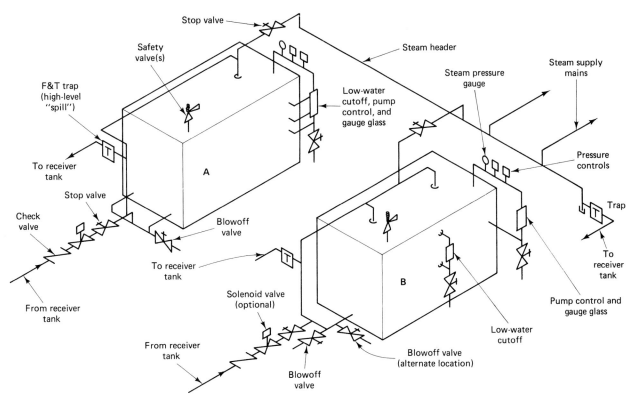

FIGURE 12-7 Boiler connections for steam header and return piping for mechanical-return systems: For boiler *A*, return piping is typical for cast-iron boiler; for Boiler *B*, return piping is typical for steel boilers.

The ability to respond immediately to slugs of condensate, excellent resistance to hydraulic shock, dirt-handling ability, and efficient operation on very light loads are features that make the inverted bucket trap most suitable for this application.

If the steam flow through the header is in one direction only, a single steam trap is sufficient at the downstream end (see Figure 12-7). With a midpoint feed to the steam header or a similar two-directional steam flow arrangement, each end of the steam header should be trapped.

Steam Mains. One of the most common uses of steam traps is the trapping of steam mains. These lines need to be kept free of air and condensate in order to keep the supplied equipment operating properly. Inadequately trapped steam mains results in water hammer and slugs of condensate that can damage control valves and other equipment.

Two methods are commonly used for the warm-up of steam mains, namely, *supervised* and *automatic.* Supervised warm-up is widely used for initial heating of large-diameter and/or long steam mains. A suggested method is for drip valves to be opened wide for free blow to the atmosphere before steam is admitted to the main. These drip valves are not closed until all or most of the warm-up condensate has been discharged. Then, the steam traps take over the job of removing condensate that may form under operating conditions. Warm-up of principal piping in a steam power plant will follow much of this same procedure.

Automatic warm-up occurs when the boiler is fired and allows the mains and some or all equipment to come up to pressure and temperature without manual help or supervision.

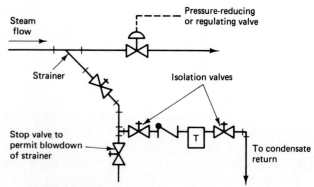

FIGURE 12-8 Trap draining strainer ahead of pressure-reducing valve.

Caution: Regardless of the warm-up method, sufficient time should be taken during the warm-up cycle to minimize thermal stress and prevent any damage to the system.

Both methods of warm-up utilize drip legs and traps at all low spots or natural damage points such as (1) ahead of valves and regulators (see Figure 12-8); (2) at end of steam main or rise in steam main (see Figure 12-9); and (3) for long horizontal steam mains with no natural drainage points, in which case drip legs should be spaced at intervals not exceeding 300 ft (90 m) when the pipe is pitched so that condensate flow is opposite of steam flow (see Figure 12-10). The distance should be reduced by about half in systems with automatic warm-up rather than supervised warm-up.

On supervised warm-ups, drip leg length should be at least $1\frac{1}{2}$ times the diameter of the main but never less than 10 in. Drip legs on automatic warm-ups should be a minimum of 28 in. long. For both methods, it is good practice to use a drip leg the same diameter as the

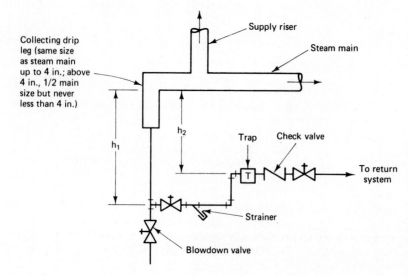

Figure 12-9 Rise in steam main: For automatic warm-up, dimension h_1 should be 28 in. minimum; for supervised warm-up, h_1 should be $1\frac{1}{2}$ times pipe diameter, with minimum 10 in. Distance h_2 in inches divided by 28 gives static pressure head for forcing water through steam trap.

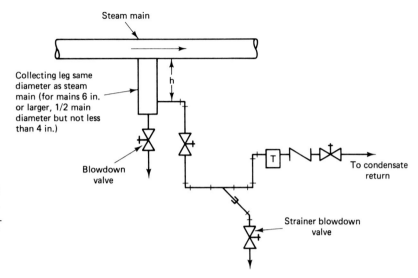

Figure 12-10 Dripping horizontal steam main: For automatic warm-up, height *h* should be 28 in. minimum; for supervised warm-up, *h* should be $1\frac{1}{2}$ times pipe diameter, with minimum 10 in.

main up to 4 in. pipe size and at least 1/2 the diameter of the main above that but never less than 4 in.

Trap selection and safety factor for steam mains (saturated steam only): The trap should be selected to discharge condensate produced by radiation losses and running loads. Condensate loads of insulated pipe can

be found in Table 12-2. All figures in Table 12-2 assume the insulation is 75% effective. For pressure or pipe sizes not included in Table 12-2, use the following relationship:

$$C = \frac{A \times U \times (t_1 - t_2) \times E}{h_{fg}} \qquad (12\text{-}2)$$

TABLE 12-2

Condensation in insulated pipes carrying saturated steam in quiet air at 70°F (insulation assumed to be 75% efficient)

Pipe Size (in.)	Sq. Ft per Linear Ft	15	30	60	125	180	250	450	600	900
		Lb Condensate per Hr per Linear Ft								
1	.344	.05	.06	.07	.10	.12	.14	.186	.221	.289
$1\frac{1}{4}$	.434	.06	.07	.09	.12	.14	.17	.231	.273	.359
$1\frac{1}{2}$	.497	.07	.08	.10	.14	.16	.19	.261	.310	.406
2	.622	.08	.10	.13	.17	.20	.23	.320	.379	.498
$2\frac{1}{2}$	.753	.10	.12	.15	.20	.24	.28	.384	.454	.596
3	.916	.12	.14	.18	.24	.28	.33	.460	.546	.714
$3\frac{1}{2}$	1.047	.13	.16	.20	.27	.32	.38	.520	.617	.807
4	1.178	.15	.18	.22	.30	.36	.43	.578	.686	.897
5	1.456	.18	.22	.27	.37	.44	.51	.698	.826	1.078
6	1.735	.20	.25	.32	.44	.51	.59	.809	.959	1.253
8	2.260	.27	.32	.41	.55	.66	.76	1.051	1.244	1.628
10	2.810	.32	.39	.51	.68	.80	.94	1.301	1.542	2.019
12	3.340	.38	.46	.58	.80	.92	1.11	1.539	1.821	2.393
14	3.670	.42	.51	.65	.87	1.03	1.21	1.688	1.999	2.624
16	4.200	.47	.57	.74	.99	1.19	1.38	1.927	2.281	2.997
18	4.710	.53	.64	.85	1.11	1.31	1.53	2.151	2.550	3.351
20	5.250	.58	.71	.91	1.23	1.45	1.70	2.387	2.830	3.725
24	6.280	.68	.84	1.09	1.45	1.71	2.03	2.833	3.364	4.434

Source: Steam Conservation Guidelines for Condensate Drainage, Manual M101. Courtesy of Armstrong Machine Works, Three Rivers, MI.

where

C = condensate (lb/hr per ft of pipe)

A = external area of pipe (sq. ft) (see Table 12-2)

U = Btu per sq. ft per hour per degree temperature difference (see Figure 12-11)

t_1 = steam temperature (°F)

t_2 = ambient air temperature (°F)

E = 1 minus efficiency of insulation (example, 75% efficient insulation: $1 - 0.75 = 0.25$, or $E = 0.25$)

h_{fg} = latent heat of saturated steam at steam pressure (Btu/lb)

For traps installed between the boiler and the end of the steam main, use a safety factor of 2:1. Traps installed at the end of the main or ahead of reducing and shutoff valves that are closed part of the time should have a safety factor of 3:1.

For very short warm-up times at minimum pressure differentials, Table 12-3 may be used for determining the amount of condensate produced in warming schedule 40 pipe to 219°F or 2 psig. For steam pressures and pipe schedules not covered by Table 12-3, the warm-up load can be calculated using the following relationship:

$$C = \frac{0.114W(t_1 - t_2)}{h_{fg}} \qquad (12\text{-}3)$$

where

C = weight of condensate (lb)

W = total weight of pipe (lb) (see Table 12-4 for unit pipe weights)

t_1 = final pipe temperature (°F)

t_2 = initial pipe temperature (°F)

0.114 = specific heat of steel pipe

h_{fg} = latent heat of steam at final temperature (Btu/lb)

Divide the warm-up load by the number of minutes required to reach 219°F or 2 psig to get lb/min., and multiply by 60 to obtain lb/hr. Pressure differential for trap sizing will be 1 psig for every 28 in. of head between bottom of main and top of trap (see Figure 12-9).

The ability to handle dirt and slugs of condensate and to resist hydraulic shock makes an inverted bucket trap the recommended type to use. In addition, should an inverted bucket trap fail, it will most often do so in an open position.

TABLE 12-3

The warm-up load, schedule 40 pipe

Pipe Size (in.)	Wt. of Pipe per Ft (lb)	Steam Pressure (psig)						
		2	15	30	60	125	180	250
		Lb Water per Linear Ft						
1	1.69	.030	.037	.043	.051	.063	.071	.079
$1\frac{1}{4}$	2.27	.040	.050	.057	.068	.085	.095	.106
$1\frac{1}{2}$	2.72	.048	.059	.069	.082	.101	.114	.127
2	3.65	.065	.080	.092	.110	.136	.153	.171
$2\frac{1}{2}$	5.79	.104	.126	.146	.174	.215	.262	.271
3	7.57	.133	.165	.190	.227	.282	.316	.354
$3\frac{1}{2}$	9.11	.162	.198	.229	.273	.339	.381	.426
4	10.79	.190	.234	.271	.323	.400	.451	.505
5	14.62	.258	.352	.406	.439	.544	.612	.684
6	18.97	.335	.413	.476	.569	.705	.795	.882
8	28.55	.504	.620	.720	.860	1.060	1.190	1.340
10	40.48	.714	.880	1.020	1.210	1.500	1.690	1.890
12	53.60	.945	1.170	1.350	1.610	2.000	2.240	2.510
14	63.00	1.110	1.370	1.580	1.890	2.340	2.640	2.940
16	83.00	1.460	1.810	2.080	2.490	3.080	3.470	3.880
18	105.00	1.850	2.280	2.630	3.150	3.900	4.400	4.900
20	123.00	2.170	2.680	3.080	3.690	4.570	5.150	5.750
24	171.00	3.020	3.720	4.290	5.130	6.350	7.150	8.000

Source: Steam Conservation Guidelines for Condensate Drainage, Manual M101. Courtesy of Armstrong Machine Works, Three Rivers, MI.

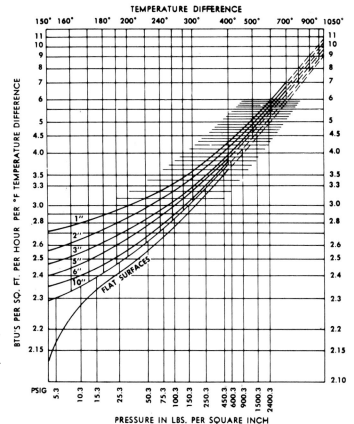

TEMPERATURE DIFFERENCE

Figure 12-11 BTU heat loss curves: Unit head loss per sq. ft of surface of insulated pipe of various diameters (also flat surfaces) in quiet air at 75°F for various saturated steam pressures or temperature differences. (Courtesy of Armstrong Machine Works, Three Rivers, MI)

ILLUSTRATIVE PROBLEM 12-2

A $2\frac{1}{2}$-in. schedule 40 steel pipe steam main is 120 ft long. A total of 60 ft of 1-in. supply risers feed the heat-transfer units from this section of main. Operating pressure is 1 psig. Initial pipe temperature is 70°F. Warm-up time is 15 min. Calculate the warm-up load and pressure differential for a trap if the effective length of the drip leg is 8 in.

Solution: From Table 12-4, the weight of the pipe would be

$$(120)(5.79) = 695 \text{ lb}$$
$$(60)(1.69) = \underline{101 \text{ lb}}$$
$$= \overline{796 \text{ lb}}$$

The piping will warm up to approximately 216°F at 1 psig from an initial temperature of 70°F. By Eq. (12-3),

$$C = \frac{(0.114)(796)(216 - 70)}{968} = 13.9 \text{ lb}$$

Thus,

$$\text{hourly condensate load} = \frac{(13.9)(60)}{15} = 55 \text{ lb/hr}$$

TABLE 12-4

Pipe weight per foot (lb)

Pipe Size (in.)	Schedule 40	Schedule 80	Schedule 160	XX Strong
1	1.69	2.17	2.85	3.66
$1\frac{1}{4}$	2.27	3.00	3.76	5.21
$1\frac{1}{2}$	2.72	3.63	4.86	6.41
2	3.65	5.02	7.45	9.03
$2\frac{1}{2}$	5.79	7.66	10.01	13.69
3	7.57	10.25	14.32	18.58
$3\frac{1}{2}$	9.11	12.51	—	22.85
4	10.79	14.98	22.60	27.54
5	14.62	20.78	32.96	38.55
6	18.97	28.57	45.30	53.16
8	28.55	43.39	74.70	72.42
10	40.48	54.74	116.00	—
12	53.60	88.60	161.00	—
14	63.00	107.00	190.00	—
16	83.00	137.00	245.00	—
18	105.00	171.00	309.00	—
20	123.00	209.00	379.00	—
24	171.00	297.00	542.00	—

If the distance between the bottom of the main and the top of the trap is 8 in., the differential pressure for the trap is 8/28 = 0.29 psi if back pressure was zero gauge. The trap would be selected for a pressure differential of 0.25 psi.

Branch Lines. Branch lines are takeoffs from steam mains that supply specific pieces of steam-utilizing equipment. The entire system must be designed and hooked up to prevent accumulation of condensate at any point.

Equations 12-2 and 12-3 may be used for computing condensate loads in branch lines. Branch lines also have a recommended safety factor of 3:1.

Recommended piping from the main to the control valve is shown in Figure 12-12 for horizontal runouts under 10 ft in length. Figure 12-13 shows piping for horizontal runouts longer than 10 ft. Figure 12-14 shows branch connections where the control valve must be below the main. A full pipe size strainer should be installed ahead of each control valve as well as ahead of a pressure-reducing valve, if used. Blowdown valves, or preferably inverted bucket traps, should also be provided.

Steam Separators. Steam separators are designed to remove any condensate that forms within steam distribution systems. They are most often used ahead of equipment where especially dry steam is essential and on secondary steam lines, which by their very nature have a large percentage of entrained condensate. Of major importance in trap selection is the ability to handle slugs of condensate, operate at light load, and provide good resistance to hydraulic shock.

Although different types of traps are recommended depending on condensate and pressure levels, a 3:1 safety factor should be applied in all cases. The required trap capacity can be obtained by using the following relationship:

required trap capacity (lb/hr) = safety factor
× steam flow rate (lb/hr) × anticipated percent
of condensate (typically 1% to 20%) (12-4)

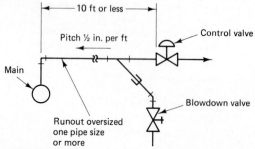

FIGURE 12-12 Piping for horizontal runout less than 10 ft long: No trap is required unless pitch back to supply main is less than $\frac{1}{2}$ in. per ft. (See also Figures 11-3 and 11-4)

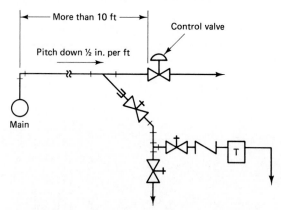

FIGURE 12-13 Piping for horizontal runout greater than 10 ft long: Drip pocket and trap are required ahead of control valve. Strainer ahead of control valve can serve as dirt pocket if blowdown connection runs to an inverted bucket trap. This will also minimize strainer cleaning problem. Trap should be equipped with an internal check valve or a swing check valve installed ahead of trap.

ILLUSTRATIVE PROBLEM 12-3

What size steam trap will be required on a steam separator handling 10,000 lb/hr and assuming 10% moisture?

Solution: By Eq. (12-4)

$$\text{required capacity (lb/hr)} = (3)(10,000)(0.10)$$
$$= 3000 \text{ lb/hr}$$

The inverted bucket trap with a large vent is recommended for use on steam separators. When dirt and

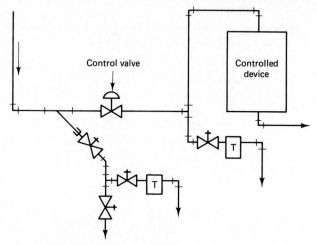

FIGURE 12-14 Regardless of length of runout, drip pocket and trap are required ahead of a control valve located below steam supply. If controlled device is above control valve, a trap should also be installed at downstream side of control valve.

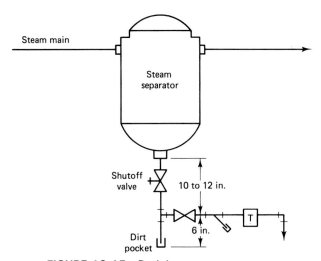

FIGURE 12-15 Draining steam separator.

hydraulic shock are not significant problems, the F&T-type trap is an acceptable alternative. The trap should be connected to the separator drain line 10 to 12 in. below the separator with the drain pipe running full size of the drain connection down to the trap takeoff. See Figure 12-15.

12-9 HOW TO TRAP SPACE HEATING EQUIPMENT

Space heating equipment, such as unit heaters, air-handling units, finned radiation, and pipe coils are found in virtually all industries. This type of equipment is quite basic and requires very little routine maintenance; consequently, the steam traps are usually neglected for long periods of time. One of the problems encountered is residual condensate in the heating device, which can result in damage from freezing, corrosion, and water hammer.

Different application requirements involving constant or variable steam pressure determine which type of trap should be used. To help determine the proper type and size, two standard methods of sizing traps for coils have been developed:

1. *For constant steam pressure*—Use inverted bucket traps and/or float and thermostatic traps and a safety factor of 3:1 at operating pressure differentials.
2. *For modulating steam pressure*—Use float and thermostatic traps and/or inverted bucket traps with thermic buckets.
    (a) For 0 to 15 psig steam pressure, use a 2:1 safety factor at $\frac{1}{2}$-psi pressure differ-

ential (on F&T traps, SHEMA ratings can also be used).
    (b) For 16 to 30 psig steam pressure, use a 2:1 safety factor at 2-psi pressure differential.
    (c) For above 30 psig steam pressure, use a 3:1 safety factor at 1/2 the maximum pressure differential across the trap.
Or use inverted bucket traps without thermic buckets. Above 30 psig steam pressure only, use a 3:1 safety factor at 1/2 the maximum pressure differential across the trap.

Trap Selection for Unit Heaters and Air-Handling Units Three methods are commonly used for computing the amount of condensate to be handled. The particular method to be utilized depends on the known operating conditions.

Btu method: The standard rating for unit heaters and other air-heating coils is Btu output with steam pressure of 2 psig in the heater and entering air temperature at 60°F. (See Table 8-14.) To convert from standard to actual rating, use the conversion factors in Table 12-5. Once the actual operating conditions are known, multiply the condensate load by the proper safety factor and consult a steam trap capacity chart for the specific model required.

To illustrate the use of Table 12-5, assume that you wish to determine the capacity (Btu/hr) and condensate load (lb/hr) of an HS-165 unit heater when the steam pressure is 15 psig and the entering air temperature is 50°F. From Table 8-14, the HS-165 has a capacity of 165,000 Btu/hr at standard conditions. From Table 12-5, for 15 psig steam pressure and 50°F entering air, find correction factor = 1.28. The corrected capacity is (165,000)(1.28) = 211,200 Btu/hr. From Appendix Table A-2, at 15 psig (30 psia approximately), find h_{fg} = 945.4 Btu/lb. Then, condensate load (lb/hr) = 211,200/945.4 = 223.4 lb/hr.

Cfm and air temperature rise method: When only cfm capacity of the fan and the air temperature rise are known, the actual Btu output can be found by using Eq. (4-10). To illustrate, assume that you wish to determine the trap capacity to drain a 3500-cfm heater that produces an 80°F air temperature rise. Steam pressure is 60 psig. Using Eq. (4-10), Btu/hr = (1.10)(3500)(80) = 308,000 Btu/hr. From Appendix Table A-2, at 60 psig (75 psia approximately), find h_{fg} = 904.8 Btu/lb. Then, condensate flow (lb/hr) = 308,000/904.8 = 340 lb/hr. Applying a 3:1 safety factor, the trap capacity should be 1020 lb/hr.

Condensate method: As we have seen in the previous sections, once the Btu/hr output has been deter-

TABLE 12-5

Steam heating capacity conversion factors (conditions other than standard, 60°F EAT, and 2 psig steam pressure)

Unit Heater Type	Steam Press (psig)	Temperature of Entering Air (°F)											
		−10°	0°	10°	20°	30°	40°	50°	60°	70°	80°	90°	100°
	0	1.54	1.45	1.37	1.27	1.19	1.11	1.03	0.96	0.88	0.81	0.74	0.67
	2	1.59	1.50	1.41	1.32	1.24	1.16	1.08	1.00	0.93	0.85	0.78	0.71
	5	1.64	1.55	1.46	1.37	1.29	1.21	1.13	1.05	0.97	0.90	0.83	0.76
	10	1.73	1.64	1.55	1.46	1.38	1.29	1.21	1.13	1.06	0.98	0.91	0.84
	15	1.80	1.71	1.61	1.53	1.44	1.34	1.28	1.19	1.12	1.04	0.97	0.90
	20	1.86	1.77	1.68	1.58	1.50	1.42	1.33	1.25	1.17	1.10	1.02	0.95
	30	1.97	1.87	1.78	1.68	1.60	1.51	1.43	1.35	1.27	1.19	1.12	1.04
Horizontal	40	2.06	1.96	1.86	1.77	1.68	1.60	1.51	1.43	1.35	1.27	1.19	1.12
Delivery	50	2.13	2.04	1.94	1.85	1.76	1.67	1.58	1.50	1.42	1.34	1.26	1.19
	60	2.20	2.09	2.00	1.90	1.81	1.73	1.64	1.56	1.47	1.39	1.31	1.24
	70	2.26	2.16	2.06	1.96	1.87	1.78	1.70	1.61	1.53	1.45	1.37	1.29
	75	2.28	2.18	2.09	1.99	1.90	1.81	1.72	1.64	1.55	1.47	1.40	1.32
	80	2.31	2.21	2.11	2.02	1.93	1.84	1.75	1.66	1.58	1.50	1.42	1.34
	90	2.36	2.26	2.16	2.06	1.97	1.88	1.79	1.71	1.62	1.54	1.46	1.38
	100	2.41	2.31	2.20	2.11	2.02	1.93	1.84	1.75	1.66	1.58	1.50	1.42
	125	2.51	2.41	2.31	2.21	2.11	2.02	1.93	1.84	1.76	1.68	1.59	1.51
	150	2.60	2.50	2.40	2.30	2.20	2.11	2.02	1.93	1.84	1.76	1.67	1.59
	0	1.49	1.41	1.33	1.25	1.18	1.11	1.03	0.96	0.90	0.83	0.76	0.69
	2	1.52	1.45	1.37	1.29	1.22	1.15	1.07	1.00	0.93	0.86	0.80	0.73
	5	1.58	1.50	1.42	1.34	1.27	1.20	1.12	1.05	0.98	0.91	0.85	0.78
	10	1.64	1.57	1.49	1.41	1.34	1.27	1.19	1.12	1.05	0.98	0.91	0.85
	15	1.70	1.62	1.55	1.47	1.40	1.32	1.25	1.18	1.11	1.04	0.97	0.90
	20	1.75	1.67	1.60	1.52	1.45	1.37	1.30	1.23	1.16	1.09	1.02	0.96
Vertical	30	1.83	1.75	1.68	1.61	1.53	1.46	1.39	1.32	1.25	1.18	1.11	1.04
Delivery	40	1.90	1.82	1.75	1.68	1.61	1.53	1.46	1.39	1.32	1.25	1.18	1.11
and	50	1.96	1.87	1.81	1.74	1.67	1.59	1.52	1.45	1.38	1.31	1.24	1.17
"Power-	60	2.02	1.94	1.87	1.79	1.72	1.64	1.57	1.50	1.43	1.36	1.29	1.22
Throw"	70	2.07	1.99	1.92	1.84	1.76	1.69	1.62	1.55	1.47	1.40	1.33	1.27
	75	2.10	2.02	1.94	1.86	1.79	1.71	1.64	1.57	1.49	1.42	1.36	1.29
	80	2.11	2.04	1.96	1.88	1.80	1.73	1.66	1.59	1.51	1.44	1.38	1.31
	90	2.15	2.08	2.00	1.92	1.84	1.77	1.69	1.62	1.55	1.48	1.41	1.34
	100	2.19	2.11	2.03	1.95	1.88	1.80	1.73	1.66	1.59	1.52	1.45	1.38
	125	2.27	2.19	2.11	1.99	1.91	1.88	1.81	1.74	1.67	1.60	1.53	1.46
	150	2.34	2.26	2.18	2.10	2.03	1.95	1.88	1.81	1.74	1.67	1.60	1.53

Source: Catalog 1–150.5. Courtesy of Modine Manufacturing Company, Racine, WI.
*To apply, multiply the standard Btu capacity rating of heater by indicated factor.

mined, the output is divided by the latent heat of steam at the steam pressure to obtain required trap capacity in lb/hr. To obtain the actual continuous discharge capacity for the trap, multiply by the safety factor.

Figure 12-16 is a sketch showing the trapping and venting used for an air-heating coil. The "safety drain" trap discharges into an open drain and keeps the heat exchanger free of condensate when insufficient supply pressure, due to modulating steam valve, prevents discharging the condensate into a pressurized return or elevating it to an overhead return. (Refer also to Figure 8-25.)

Trap Selection for Pipe Coils and Finned Radiation

Pipe coils: When possible, trap each pipe individually to avoid short circuiting.

Single pipe coils: To size steam traps for single pipe coils or individually trapped pipes, find the condensing rate per linear foot in Table 12-6. Multiply the condensing rate per linear foot by the length in feet to get the normal condensate load.

For quick heating, apply a trap safety factor of 3:1 and use a trap with a thermic vent bucket. Where

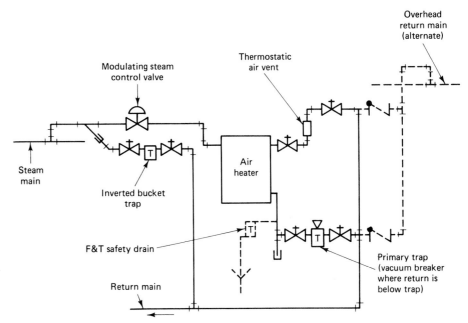

Figure 12-16 Trapping and venting air-heating coil. *Note:* Dashed line applies to overhead return. (See also Figure 8-25)

quick heating is not required, use a trap safety factor of 2:1 and select a standard trap.

Multiple pipe coils: To size traps to drain coils consisting of multiple pipes, proceed as follows: (1) Multiply the linear feet of pipe in the coil by the condensing rate given in Table 12-6 to obtain normal condensate load; (2) from Table 12-5, find the multiplier for the specific service conditions; and (3) multiply normal condensate load by the Table 12-5 factor to obtain the required continuous discharge capacity required by

the trap. The safety factor is not included in the multiplier.

Finned radiation: The condensing rate for finned radiation may be determined as follows. When the Btu/hr output is not known and cannot be determined from engineering or purchasing records, condensing rates can be estimated from Tables 12-7 and 12-8 with sufficient accuracy for trap selection purposes. To enter Table 12-7, observe the size of pipe, size of fins, number of fins, and material. Determine con-

TABLE 12-6

Condensing rates in bare steel pipe carrying saturated steam

Steam Pressure (psig) Temp. Rise from 70°		15 180°	30 204°	60 237°	125 283°	180 310°	250 336°
Pipe Size (in.)	Sq. Ft per Linear Ft	Lb Condensate per Hr per Linear Ft					
$\frac{1}{2}$	.220	.13	.15	.19	.26	.30	.35
$\frac{3}{4}$	.275	.15	.19	.24	.33	.38	.45
1	.344	.19	.23	.28	.39	.46	.54
$1\frac{1}{4}$	.434	.23	.28	.36	.49	.57	.67
$1\frac{1}{2}$	.497	.26	.32	.41	.55	.65	.76
2	.622	.33	.40	.50	.68	.80	.93
$2\frac{1}{2}$	.753	.39	.47	.59	.81	.95	1.11
3	.916	.46	.56	.70	.96	1.13	1.31
$3\frac{1}{2}$	1.047	.52	.63	.80	1.08	1.27	1.50
4	1.178	.58	.70	.89	1.21	1.43	1.72

Source: Steam Conservation Guidelines for Condensate Drainage, Manual M101. Courtesy of Armstrong Machine Works, Three Rivers, MI.

TABLE 12-7

Finned radiation condensing rates (65°F entering air, 215°F steam) for trap selection purposes only

	Pipe Size (in.)	Fin Size (in.)	Fins per Foot	No. of Pipes High on 6-In. Centers	Lb Condensate per Hr per Ft Pipe
Steel pipe,	$1\frac{1}{4}$	$3\frac{1}{4}$	32 to 40	1	1.1
steel fins,				2	2.0
painted				3	2.6
black	$1\frac{1}{4}$	$4\frac{1}{4}$	32 to 42	1	1.6
				2	2.4
				3	3.1
	2	$4\frac{1}{4}$	24 to 34	1	1.5
				2	2.4
				3	3.1
Copper pipe,	$1\frac{1}{4}$	$3\frac{1}{4}$	40 to 48	1	1.6
aluminum				2	2.2
fins,				3	2.8
unpainted	$1\frac{1}{4}$	$4\frac{1}{4}$	40 to 50	1	2.2
				2	3.0
				3	3.6

Source: Steam Conservation Guidelines for Condensate Drainage, Manual M101. Courtesy of Armstrong Machine Works, Three Rivers, MI.

densing rate per foot and standard conditions from Table 12-7. Convert to actual conditions by using Table 12-8.

Safety factor recommendations in nearly all applications are to (1) overcome the short-circuiting hazard created by the multiple tubes of the heater; (2) ensure adequate trap capacity under severe operating conditions (in extremely cold weather, the entering air temperature is likely to be lower than calculated and the increased demand for steam in all parts of the system may result in lower steam pressures and higher return line pressures, all of which cut trap capacity); and (3) ensure the removal of air and other noncondensibles.

Caution: For low-pressure heating, use a safety factor at the actual pressure differential, not necessarily

TABLE 12-8

Finned radiation conversion factors for steam pressures and air temperatures other than 65°F air and 215°F steam

Pressure (psig)	Steam Temp. (°F)	Entering Air Temperature (°F)						
		45	55	65	70	75	80	90
0.9	215.0	1.22	1.11	1.00	0.95	0.90	0.84	0.75
5.	227.1	1.34	1.22	1.11	1.05	1.00	0.95	0.81
10.	239.4	1.45	1.33	1.22	1.17	1.11	1.05	0.91
15.	249.8	1.55	1.43	1.31	1.26	1.20	1.14	1.00
30.	274.0	1.78	1.66	1.54	1.48	1.42	1.37	1.21
60.	307.3	2.10	2.00	1.87	1.81	1.75	1.69	1.51
100.	337.9	2.43	2.31	2.18	2.11	2.05	2.00	1.81
125.	352.9	2.59	2.47	2.33	2.27	2.21	2.16	1.96
175.	377.4	2.86	2.74	2.60	2.54	2.47	2.41	2.21

Source: Steam Conservation Guidelines for Condensate Drainage, Manual M101. Courtesy of Armstrong Machine Works, Three Rivers, MI.

the steam supply pressure, remembering that the trap must also be able to function at the maximum pressure differential it will experience.

12-10 HOW TO TRAP SHELL-AND-TUBE HEAT EXCHANGERS AND SUBMERGED COILS

Submerged coils are heat-transfer elements that are immersed in the liquid to be heated, evaporated, or concentrated. This type of coil is found in virtually every plant or institution that uses steam. Common examples are water heaters, reboilers, suction heaters, evaporators, and vaporizers. These are used in heating water for process or domestic use; vaporizing industrial gases such as propane and oxygen; concentrating in-process fluids such as sugar, black liquor, and petroleum; and heating fuel oil for easy transfer and atomization.

Different applications requirements involving constant or variable steam pressure determine which type of trap should be used. Factors most commonly used in trap selection include the ability to handle air at low differential pressures, the conservation of energy, and removal of dirt and slugs of condensate. To help determine the proper type and size, three standard methods of sizing traps for coils have been developed:

1. *For constant steam pressure*—Use inverted bucket traps or float and thermostatic traps and a safety factor of 2:1 at operating pressure differentials.
2. *For modulating steam pressure*—Use float and thermostatic traps or inverted bucket traps.
    (a) For 0 to 15 psig steam pressure, use a 2:1 safety factor at $\frac{1}{2}$-psi pressure differential (on F&T traps, SHEMA ratings can also be used).
    (b) For 16 to 30 psig steam pressure, use a 2:1 safety factor at 2-psi pressure differential.
    (c) Above 30 psig steam pressure, use a 3:1 safety factor at 1/2 the maximum pressure differential across the trap.
3. *For constant or modulating steam pressure with syphon drainage*—Use an inverted bucket trap with a large vent and a 5:1 safety factor. Apply the safety factor at full differential on constant steam pressure. Apply the safety factor at 1/2 the maximum differential for modulating steam pressure.

Shell-and-Tube Heat Exchangers. One type of submerged coil is the *shell-and-tube heat exchanger*. In these heat exchangers, numerous tubes are installed in a housing or shell with a confined free area. This assures positive contact with the tubes by any fluid flowing in the shell. Although the term "submerged coil" implies that steam is in the tubes and the tubes are submerged in the liquid being heated, the reverse can also be true where steam is in the shell and a liquid is in the tubes (see Figure 8-26).

Figure 12-17 illustrates a typical piping diagram for a shell-and-tube heat exchanger where water is being heated in the tubes and steam is condensing in the shell. The shell-and-tube heat exchanger, sometimes called a *converter*, is used with a temperature regulator for controlling the temperature of the heated water. When the modulating control valve throttles hard or closes off, a vacuum forms in the heater shell. Air is drawn in through a vacuum breaker at this time. This air must be vented by the steam trap on the next heating cycle.

The modulating discharge of float and thermostatic traps makes them an excellent choice for this type of service. They will act to continuously discharge condensate at the exact rate at which it forms, within the limits of their capacity. This helps to prevent overloading of the return lines with sudden large slugs of condensate. Oversizing of these traps should be avoided because of this characteristic.

To determine the condensate load on shell-and-tube heaters, use the following formula when the actual rating is known:

$$\text{lb/hr} = \frac{500 \times \text{gpm} \times c_p \times \text{SG} \times \Delta t}{h_{fg}} \quad (12\text{-}5)$$

where

500 = 60 min./hr × 8.33 lb/gal of water
gpm = liquid flow
c_p = specific heat of liquid
SG = specific gravity of liquid
Δt = temperature rise of liquid (°F)
h_{fg} = latent heat of steam at steam pressure in the shell (Btu/lb)

ILLUSTRATIVE PROBLEM 12-4

Determine the condensate load for a shell-and-tube heat exchanger that will heat 50 gpm of water from 40°F to 140°F. Steam pressure in the shell is 15 psig.

Solution: From Appendix Table A-2, at 30 psia (15 psig), find h_{fg} = 945.4 Btu/lb. By Eq. (12-5),

$$\text{lb/hr} = \frac{(500)(50)(1)(1)(140 - 40)}{945.4}$$

$$= 2644 \text{ lb/hr}$$

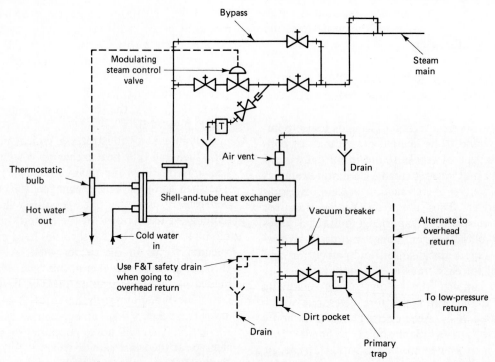

FIGURE 12-17 Shell-and-tube heat exchanger (typical piping diagram).

Embossed Coils and Pipe Coils. Although these devices are not used in steam heating systems, they are frequently found in industry. We shall discuss them briefly to determine their trapping needs.

Very often, open tanks of water or chemicals are heated by means of *embossed coils*. See Figure 12-18. Passages for steam are produced by upsetting grooves in the sheet metal of two halves. The two halves are mirror images and, when welded together, produce the passages for steam entry, heat transfer, and condensate evacuation.

The condensate load on embossed coils can be obtained with the following relationship and by dividing the result by the latent heat of vaporization for the steam at the steam pressure:

$$\dot{q} = UA(\text{LMTD}) \qquad (12\text{-}6)$$

where

$\dot{q}$ = total heat transfer (Btu/hr)
A = area of outside surface of coil (sq. ft)
U = overall heat-transfer coefficient (Btu/hr-ft²-°F)
$LMTD$ = logarithmic mean temperature difference between steam and liquid, as between inlet and outlet of a heat exchanger (°F)

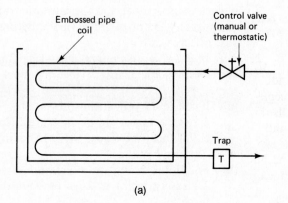

(a)

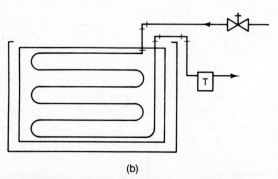

(b)

FIGURE 12-18 Embossed pipe coils: (a) continuous coil, gravity drained; and (b) continuous coil, siphon drained.

TABLE 12-9

U-values for embossed coils (Btu/hr-ft²-°F)

Type of Service	Circulation	
	Natural Convection	Forced Convection
Steam to watery solutions	100–200	150–275
Steam to light oil	40–45	60–110
Steam to medium oil	20–40	50–100
Steam to bunker C oil	15–30	40–80
Steam to tar asphalt	15–25	18–60
Steam to molten sulfur	25–35	35–45
Steam to molten paraffin	25–35	40–50
Steam to molasses or can syrup	20–40	70–90
Dowtherm to tar asphalt	15–30	50–60

Source: Steam Conservation Guidelines for Condensate Drainage, Manual M101. Courtesy of Armstrong Machine Works, Three Rivers, MI.

The LMTD is defined as follows:

$$\text{LMTD} = \frac{\Delta t_{\max} - \Delta t_{\min}}{\ln\left(\dfrac{\Delta t_{\max}}{\Delta t_{\min}}\right)} \qquad (12\text{-}7)$$

where

$\Delta t_{\max}$ = maximum temperature difference (°F)
$\Delta t_{\min}$ = minimum temperature difference (°F)

U-values are determined by tests under controlled conditions. Tables 12-9 and 12-10 show the commonly accepted ranges for embossed coils and submerged pipe coils. For trap selection purposes, use a *U*-value that is slightly greater than the conservative *U*-value selected for estimating actual heat transfer. To determine trap capacity required, multiply condensate rate by the recommended safety factor.

TABLE 12-10

U-values for pipe coils (Btu/hr-ft²-°F)

Type of Service	Circulation	
	Natural Convection	Forced Convection
Steam to water	50–200	150–1200
1½-in. tube heaters	180	450
¾-in. tube heaters	200	500
Steam to oil	10–30	50–150
Steam to boiling liquid	300–800	—
Steam to boiling oil	50–150	—

Source: Steam Conservation Guidelines for Condensate Drainage, Manual M101. Courtesy of Armstrong Machine Works, Three Rivers, MI.

ILLUSTRATIVE PROBLEM 12-5

A steam-to-water embossed plate heater has a surface of 20 sq. ft and a *U*-value assumed at 175 Btu/hr-ft²-°F. Water in is 40°F, water out is 150°F, and steam pressure is 125 psig saturated. Determine the condensate rate.

Solution: From Appendix Table A-2, with steam pressure at 140 psia, find saturation temperature of 353.08°F (say, 353°F) and h_{fg} = 868.7 Btu/lb. By Eq. (12-7),

$$\text{LMTD} = \frac{(353 - 40) - (353 - 150)}{\ln\left(\dfrac{313}{203}\right)}$$

$$= 254°F$$

By Eq. (12-6),

$$\dot{q} = UA(\text{LMTD}) = (175)(20)(254)$$
$$= 889{,}000 \text{ Btu/hr}$$

Thus,

$$\text{condensate (lb/hr)} = \frac{889{,}000}{868.7}$$

$$= 1023 \text{ lb/hr}$$

Submerged pipe coils are heat-transfer tubes (pipes) immersed in tanks that are large in volume compared to the coils themselves. This is their primary difference when compared with shell-and-tube heat exchangers. Submerged pipe coils are sometimes gravity drained, with the trap installed below the coil. Figure 12-19 shows a sketch of such an installation. A safety factor of 2:1 is recommended for such an installation. Some pipe coils are immersed in open tanks as illustrated in Figure 12-20, which shows a siphon-drained continuous coil.

The condensate load for pipe coils is determined by applying Eq. (12-5) if the capacity is known. When the physical dimensions of coils are known, Eq. (12-6) may be used with an appropriate *U*-value selected from Table 12-10.

When gravity drainage is utilized on shell-and-tube heat exchangers, embossed coils, and pipe coils, the steam trap should be located below the heating coil. Under modulating pressure control, a vacuum breaker should be used. This can be integral in F&T traps or mounted off the inlet piping on an inverted bucket trap. An ample drip leg should be placed ahead of the trap to act as a reservoir. This assures coil drainage when there is a maximum condensate load and a minimum steam pressure differential.

Avoid lifting condensate from a shell-and-tube heat exchanger, embossed coils, or pipe coils under

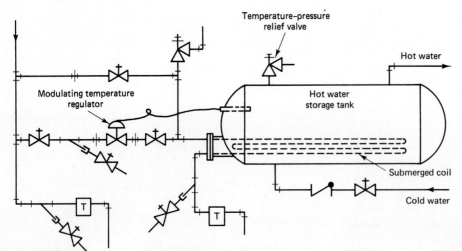

Figure 12-19 Submerged pipe coil for heating water.

modulated control. However, if it must be done, the following suggestions are given by Armstrong Machine Works:

1. Do not attempt to elevate condensate more than 1 ft for every 1 psi of normal pressure differential, either before or after the trap.
2. If the condensate lift takes place after the steam trap, install a low-pressure safety drain (see Figure 12-17).
3. If the condensate lift takes place ahead of the steam trap (siphon lift), an automatic differential condensate controller (see data from Armstrong Machine Works) should be installed to efficiently vent all flash steam.
4. If an I.B. trap is used, an internal check valve is required.

Siphon Drainage. Figure 12-21 illustrates an external siphon that lifts condensate from a gravity drain to an overhead return. Flash steam tends to form in siphons, and the trap must be able to operate properly

with a certain amount of it present in the condensate. The trap shown in Figure 12-21 is draining a steam main. Condensate first drains into the water seal. Steam in the siphon above the water seal condenses, dropping the pressure. Condensate rises in the siphon as this takes place. The siphon may form and break several times before it is established and condensate enters the trap.

A check valve must be used to hold the siphon while it is forming. This should be installed after the strainer as shown. Once the siphon is established, the

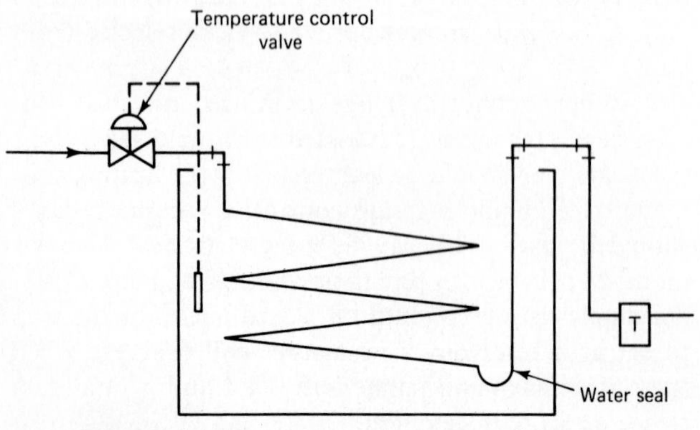

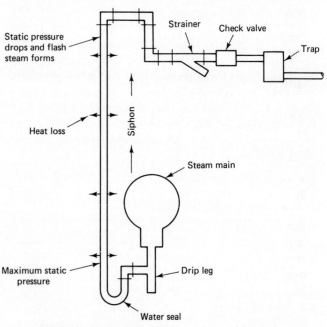

FIGURE 12-21 Principle of condensate line siphon: Condensate from gravity drain point is lifted to trap by siphon. Every 2 ft of lift reduces pressure differential on trap by approximately 1 psi. (Note the water seal at low point and the check valve)

FIGURE 12-20 Continuous pipe coil heater, siphon drained.

drop in static pressure as the elevation increases (every 2 ft of lift in siphon reduces pressure differential on trap by approximately 1 psi) causes some of the hot condensate to "flash off." The presence of steam in the condensate decreases its density and actually assists the flow. An external siphon loses some heat by radiation to the ambient air, and the condensate within it tends to cool. As a result, the amount of flash steam is less than that in an internal siphon, which absorbs heat from the steam surrounding it.

In either case, the trap must be of a type capable of handling flash steam without closing off, or condensate drainage will be adversely affected. For example, flash steam can float the bucket in an inverted bucket trap and close it off while condensate remains to be drained. The bucket vent hole is sometimes drilled oversize to allow for venting flash steam. The trap manufacturer should be consulted for instructions before this is done.

12-11 STEAM SUPPLY CONTROLS

The selection of the proper steam control for a given application requires the consideration of a number of variables. The control type involved also deserves separate consideration. These factors will be explained in the following sections.

Steam Supply Valves. The choice of steam supply valves, usually called *radiator supply valves,* for terminal heat-transfer units is governed by four factors: (1) system type (one-pipe or two-pipe), (2) operating pressure, (3) pipe size, and (4) body type.

In *one-pipe systems,* steam must enter and condensate must leave the THTU through a common port. The valve must therefore be installed at the bottom of the THTU and have a large enough port to accept both flows. The valve cannot be throttled because this would restrict condensate flow. It must be either fully open or fully closed.

A typical valve for a THTU in a one-pipe system is shown in Figure 12-22. This is an angle-type valve. Straightway-patterned valves cannot be used for one-pipe systems because they would not permit adequate two-way flow of steam and condensate. The stem of the valve illustrated is of the *packing* type.

The valve should be full riser size. If the system is fitted with vacuum-type radiator vents, the valve should be of the spring-loaded packing or packless type to prevent loss of vacuum past the valve stem. Conventional packing-type valves may be used where nonvacuum systems are involved.

In *two-pipe systems,* the supply valves must be full

FIGURE 12-22 Radiator supply valve, packed-stem type. (Courtesy of ITT Fluid Transfer Division, Morton Grove, IL)

riser size and may be of the angle or the straightway type. Vacuum systems should be fitted with valves of the spring-loaded packing or packless type. The standard packed-type construction may be used on nonvacuum systems. Modulating-type valves may be used where control of the THTU output is desired. Figure 12-23 illustrates a *packless* type of valve, applicable to either one-pipe or two-pipe systems. These valves open fully with one turn or less and are equipped with a dial and pointer for visual regulation. They use a diaphragm stem seal.

12-12 VENT VALVES AND STRAINERS

Vent valves, used for elimination of air from one-pipe steam systems, are classified into two general types: (1) radiator vent valves and (2) end-of-main vent valves.

The choice of vent valves requires that the following five factors be considered: (1) type of equipment to be vented, (2) operating pressure, (3) maximum working pressure, (4) venting rate required, and (5) vacuum or nonvacuum system.

FIGURE 12-23 Radiator supply valve, packless type. (Courtesy of ITT Fluid Transfer Division, Morton Grove, IL)

Radiator Vent Valves. These valves are available in a wide range of air-venting capacities. They are furnished in special constructions depending on whether they will be used in nonvacuum systems. One-pipe systems that operate at atmospheric pressure to pressures of about 2 to 3 psig use an "open" type of vent. As the steam in the system condenses on the off cycle, these vents allow air to be drawn into the system. This air is vented on the next firing cycle. A typical open vent is shown in Figure 12-24. These vents are furnished with either nonadjustable or adjustable venting rates. The vent shown in Figure 12-24 is an adjustable type.

Large one-size steam heating systems often heat at a very uneven rate during start-up. The first THTU's off the main will vent their air first and heat quickly, with those at the far end being the last to vent and heat. If adjustable vents are used, the venting rate of the first THTU's can be set lower than that of the far end THTU's, resulting in better heat distribution. Small systems, where uneven heating on start-up is not a problem, may use vents with nonadjustable air ports.

One-pipe steam heating systems that operate in the vacuum range during a portion of their heating cycle have been in use for many years. These systems operate well when coal-fired. These types of systems require vents that will not admit air if the THTU is under vacuum. They are essentially the same as the nonvacuum type with one exception; they have a check valve in the vent port. Figure 12-25 shows a vent of this type. The vent valve illustrated is of the adjustable type. Any one of six venting rates is made available by rotating the disc containing the venting ports until the desired port is

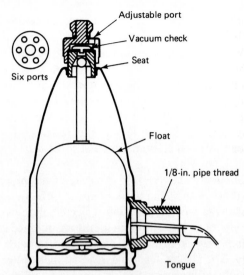

FIGURE 12-25 Adjustable-type radiator vent for vacuum systems. (Courtesy of ITT Fluid Transfer Division, Morton Grove, IL)

over the vent opening. A check valve in the vent port closes if a vacuum occurs in the THTU. This prevents entry of air. These vents are also available without the adjustable venting feature.

For coal-fired systems, when enough heat is present in the fuel bed, the system operates at above atmospheric pressure. As the fire diminishes, the rate of steam generation decreases. The check valves in the vents of the THTU's prevent air from being aspirated as the steam condenses, and a general vacuum forms in the system. The decreased pressure allows the continued production of steam at a lower temperature. The THTU's remain warm over a longer period of time, resulting in good temperature control.

This system, commonly called a "vapor vacuum system," does not lend itself well to gas- or oil-fired boiler operation. During mild weather, the firing cycles may be too short to allow the steam to completely purge the system of air. On the off cycle, this air will expand as the system drops to a vacuum, often to the extent that it will creep into the mains. Repeated short cycling in this fashion will result in an air-bound system with poor heat distribution. For this reason, the use of vacuum-type vents with one-pipe oil- or gas-fired systems is not usually recommended.

Vents used with conventional free-standing cast-iron radiators are of the angle type, being screwed into a side tapping of the radiator. The vent must be rated at an operating pressure equal to or greater than that of the system. Radiator vents have two pressure ratings: (1) operating pressure, the maximum pressure at which the vent will perform its function; and (2) maximum

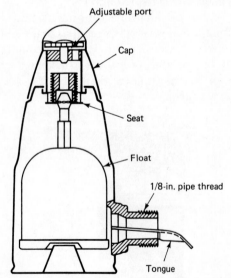

FIGURE 12-24 Adjustable-type radiator vent. (Courtesy of ITT Fluid Transfer Division, Morton Grove, IL)

pressure, the maximum pressure that can be applied to the valve.

Vents for convector-type THTU's are of the straight-shank type. The nature of this THTU type makes it necessary to install the vent at its top. In order to provide quick venting, an air chamber should be installed. The air collecting in the chamber will permit cooling of the thermostatic element in the vent, necessary for proper venting action.

End-of-Main Vents. The major difference between these vents and radiator vents is the venting rate. End-of-main vents, or, simply, *main vents,* have much larger venting rates. Main vents are always of the non-adjustable type since their function is to remove air as quickly as possible. They are available in various venting capacities to fit the requirements of small-, medium-, and large-size systems. As a rule, their venting capacity is so described in manufacturers' catalog information.

Main vents are furnished for either open or vacuum operation, the difference being in the check valve furnished with the vacuum vent. The construction of the vacuum vent is shown in Figure 12-26. This vent employs a bellows to close off the port on steam entry. The vent is normally open. If water enters the valve, the float will rise from its support and seal off the valve port.

These vent valves are of the straight-shank type. The shanks are both tapped and threaded for either male or female installation. The usual size is $\frac{1}{2}$-in. female by $\frac{3}{4}$-in. male on the shank of the vent. The connecting

nipple for the installation should be at least 6 to 10 in. long to provide a cooling leg for the vent.

Pipe Line Strainers. Pipe line strainers are effective, inexpensive devices that protect steam system components against costly shutdowns and heavy repair bills caused by dirt, scale, and metal chips in pipe lines. They are installed ahead of the equipment, pressure and temperature regulators, steam traps, and boiler water feed controls.

Strainers are not provided ahead of thermostatic traps used on radiator draining service. Dirt and scale tend to settle out in the radiators, and the velocities through the trap seat opening are low. The abrasive action of the dirt particles is therefore at a minimum. Traps used for high-pressure dripping, industrial service, and similar applications should have strainers provided. The velocities through such traps can be high. As a result, severe abrasive action by dirt particles can cause rapid wear of the seat and pin.

Boiler water contains a great amount of sediment. It is important that larger particles be kept out of the working parts of the boiler controls. Strainers, either integral or external, are extensively used for this purpose.

External strainers are available in many shapes and sizes to fit various applications. One thing they must all have in common is accessibility of the strainer for cleaning and provision for blowdown of accumulated sediment. The most commonly used steam strainer is the Y-type strainer illustrated in Figure 12-27. These strainers are furnished with either mesh-type or perforated metal screens in a wide variety of openings. The tapping in the strainer cover permits installation of a blowdown valve. The strainer screen may be removed for inspection or replacement by taking off the screen

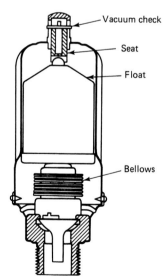

FIGURE 12-26 End-of-main vent, vacuum type. (Courtesy of ITT Fluid Transfer Division, Morton Grove, IL)

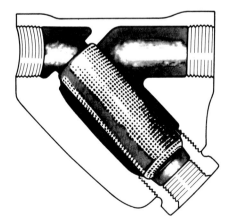

FIGURE 12-27 Typical Y-type pipeline strainer. (Courtesy of ITT Fluid Transfer Division, Morton Grove, IL)

cover. Strainers are identified by line size and the mesh or perforation size required.

12-13 BOILER ACCESSORIES

Steam heating systems lose some water during operation due to venting, blowdown, and possible leakage. As a result, provision must be made for the addition of makeup water. The boiler must also be protected from damage should a low-water condition occur while the boiler is being fired. The accessories used to *automatically* perform these functions are boiler water feeders, low-water cutoffs, or a combination of both in one control.

Some systems add makeup water to the condensate receiver and use a water-level sensing switch *on the* boiler to actuate the boiler feed pump to bring that water to the boiler. This switch is known as a *pump control*. Both low- and high-pressure steam systems incorporate this type of makeup arrangement. A steam boiler with water feeder and pump control is shown in Figure 12-28.

Boiler Water Feeder. A combination water feeder and low-water cutoff is shown in Figure 12-29. This is installed in an equalizer line on the boiler so that the float can sense the boiler water level.

A makeup water feeder does not act to maintain the normal boiler waterline, which should be at the center of the gauge glass. On initial firing, the water level tends to drop somewhat below normal until the conden-

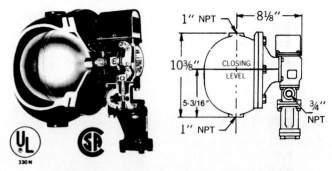

FIGURE 12-29 Combination water feeder and low-water cutoff by McDonnell & Miller. (Courtesy of ITT Fluid Transfer Division, Morton Grove, IL)

sate begins to return, and provisions must be made for this in establishing the feeder operating level. A marker on the float bowl casting indicates the feeder closing level. Feeders are usually installed with a closing level 2 in. to $2\frac{1}{2}$ in. below the normal waterline but not lower than 1 in. of water in the gauge glass.

Should the boiler water level drop below this point, the float of the feeder will drop and open the valve. Water enters through the strainer, stopping when the feeder closing level is reached. The valve as just described would comprise a makeup water feeder.

A low-water cutoff switch can be added to such a water feeder to make it a combination control. A linkage from the float lever operates this switch so as to open its contacts on a drop in boiler water level. A drop in boiler water level to $\frac{3}{4}$ in. below the feeder operating

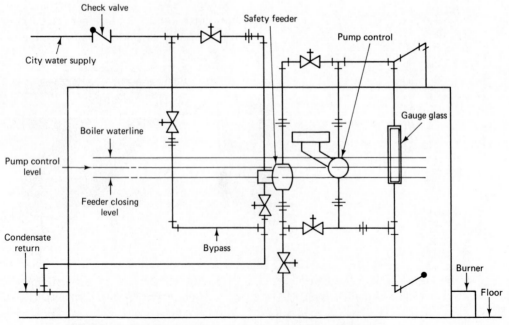

FIGURE 12-28 Steam boiler controls (typical installation).

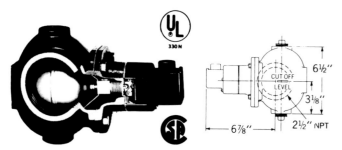

FIGURE 12-30 Low-water cutoff by McDonnell & Miller. (Courtesy of ITT Fluid Transfer Division, Morton Grove, IL)

level causes the switch to open and shut down the burner until the water feeder reestablishes a safe operating level. A rise in water level of $\frac{1}{2}$ in. over the cutoff level causes this switch to make its contacts and restart the burner.

Low-Water Cutoffs. Some boilers require separate low-water cutoffs. One of these is shown in Figure 12-30. The control should be installed so as to have the cutoff level marker on the float bowl about $\frac{1}{2}$ in. higher than the lowest visible point in the sight glass. The control will shut down the burner when the boiler water

reaches this level and restart it on a $\frac{1}{2}$ in. increase in boiler water level over this point.

Pump Control. Pump controls are usually furnished with auxiliary switches that also enable them to function as low-water cutoffs. An example of such a control is shown in Figure 12-31. The control has two mercury switches, each of which operates independently. One starts and stops the boiler feed pump; the other serves as a low-water cutoff and alarm switch.

A marker on the float bowl casting indicates the low-water cutoff level. When used as a pump control, the marker should be about $1\frac{1}{2}$ in. to 2 in. below the normal boiler water level but never lower than $\frac{3}{4}$ in. of water in the gauge glass. The boiler feed "cutoff" level will be $1\frac{1}{2}$ in. above the marker, and the "pump-on" level will be $\frac{3}{4}$ in. lower than this. Where the boiler feed pump is operated by a pump switch, a makeup water feeder is installed in the receiver.

With this arrangement, the pump switch and low-water cutoff act to control condensate return to the boiler and also to shut down the burner should a low-water condition occur. As the need for makeup water appears, this will show up as a drop in water level in the receiver. The makeup water feeder located at this point

Figure 12-31 Pump control and low-water cutoff by McDonnell & Miller. (Courtesy of ITT Fluid Transfer Division, Morton Grove, IL)

will then add water as needed to ensure that there will always be a reservoir of water for the pump.

12-14 CONDENSATE-HANDLING SYSTEMS

The purpose of this section is to discuss the equipment required to collect the condensate from a steam system and to return it to the steam boiler. Many older, small steam heating systems used a gravity condensate return to the boiler (see Chapter 11). These systems are seldom employed in modern installation and will not be covered in this section.

The selection of condensate-handling equipment should be given careful consideration to maintain a properly balanced, efficient steam system. Several factors should be considered in the system to determine the proper selection. These factors include the following:

- System size.
- Temperature of the returned condensate.
- Operating pressure of the boiler.
- Amount of makeup water required to replace steam lost through leaks.
- Vent valves.
- Flash steam and steam consumed in industrial processes.
- Change in load rate during various time periods.

In many applications, condensate return temperatures may exceed the saturation temperature at the existing pressure. The selection of return equipment should give consideration to operating efficiencies, and flash steam should be avoided.

The condensate return rate can be easily calculated by using the conversion factors in Table 12-11. Many

TABLE 12-11

Conversion factors

	Multiply	By	To Get
1.	Boiler horsepower	34.5	lb of steam (water) per hour
2.	Boiler horsepower	0.069	gpm water
3.	Boiler horsepower	33,479	Btu
4.	Boiler horsepower	139	Sq. ft of equivalent direct radiation (EDR)
5.	Sq. ft of equivalent direct radiation (EDR)	0.000496	gpm water
6.	Pounds of steam per hour	0.002	gpm water

steam heating systems are rated in square feet of equivalent direct radiation (sq. ft EDR). We have previously defined the unit as "each sq. ft EDR will give off 240 Btu when filled with a heating medium at 215°F and surrounded by 70°F ambient air temperature." For purposes of estimating the condensate return rate, each 1000 sq. ft EDR will condense and return condensate at the rate of 0.5 gpm.

In some applications, steam consumption is expressed in lb/hr. The condensate return rate would be estimated by using 2.0 gpm for each 1000 lb/hr of steam.

Condensate Transfer Units. There are two steps in the handling of condensate from a steam heating system. The first step is the returning of condensate to the boiler room. The second step is the feeding of condensate into the boiler.

1. *Returning condensate to the boiler room*—As the steam is distributed through the supply system and into the THTU's, it gives up its latent heat and condensate is formed. The condensate is drained through steam traps and into the condensate return pipes. When the elevation of the return pipes permits gravity flow of the condensate to the boiler room, or boiler feed unit, condensate pumping equipment is not required. However, in many applications, gravity flow of condensate cannot be provided and condensate transfer units must be used. These transfer units collect condensate at remote locations and pump it to the boiler feed unit located in the boiler room.

2. *Feeding condensate into the boiler*—The second step in condensate handling is the feed or condensate return into the boiler. Condensate feed into the boiler deserves careful consideration to maintain a proper water level in the boiler during various load rates and operating cycles.

In any steam system, there is a *time lag* between when the steam leaves the boiler until it is returned in the form of condensate. The greatest time lag exists during cold start-up. In a cold system, the steam mains, THTU's, and return piping are completely drained. When the system is put into operation, the steam mains and THTU's require a volume of steam to fill the system. This volume of steam must come from the boiler during this period and causes a drop in the water level in the boiler. Additional time is required for the condensate to flow through the return lines back to the boiler room, either by gravity or by being pumped back from condensate transfer units. The opposite condition occurs when the system is shut down; all the steam in the mains and THTU's is returned in the form of condensate and must be stored for the next system start-up. During normal operation, fluctuating load rates will cause surges of steam output and condensate return in the system.

The steam boiler system, including the condensate return unit, should be designed to have a storage capacity to store an ample supply of water for the variation in flow rate.

12-15 CONDENSATE-HANDLING EQUIPMENT

There are many types of condensate pumping systems. In general, they may be classified as follows:

1. Condensate pump units.
2. Boiler feed pump units.
3. Vacuum pump units.

All types have the function of accumulating condensate from the system and delivering it to the boiler.

Condensate Pump Units. Condensate pump units are used for several reasons such as (1) when boiler pressure is too high to permit gravity return of the condensate; (2) when the return mains are below the water level in the boiler; (3) in large systems where the condensate must be returned at a controlled rate; and (4) when condensate must be pumped over a high point.

Condensate pump units are available as single or duplex units. *Single units* have a single receiver tank and one pump. The pump operates from a float switch in the receiver tank.

Duplex units have a single receiver tank and two pumps. Each pump is capable of handling the full rated system capacity so as to provide standby capability. The pumps are operated by float switches in the receiver tank. Should the lead pump allow the receiver level to rise to the float setting of the lag pump, the lag pump will cut in. Manual or automatic switching of the pumping sequence can be provided.

Figure 12-32 shows a duplex-type condensate return unit with a cast-iron receiver tank (also available with a steel tank). Figure 12-33 illustrates a typical hookup for a single condensate return unit for a two-pipe steam heating system. The return mains must pitch at least $\frac{1}{4}$ in. per 10 ft of length to the receiver inlet for gravity return. If adequate pitch for gravity return is not possible, a condensate pump transfer unit must be used to pump the return. The condensate pump unit of Figure 12-33 discharges into the boiler return header—never to a Hartford loop. Such an arrangement results in very noisy operation. The receiver vent pipe is brought up to an elevation above the boiler waterline to

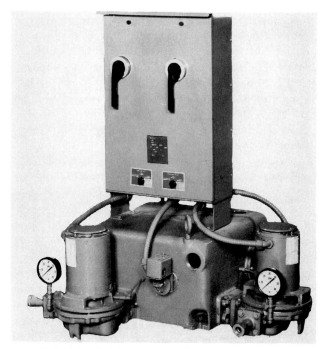

FIGURE 12-32 Domestic® Series CC condensate return unit with cast-iron receiver. (Courtesy of ITT Fluid Transfer Division, Morton Grove, IL)

prevent discharge of the condensate at this point during shutdown in the event of a leaking check valve at the pump discharge.

When the condensate pump and receiver tank must be below floor level to ensure condensate drainage, underground-type condensate pump units are available.

As explained before, the time lag is the time interval required for the condensate to begin returning to the boiler once steaming has started. If the boiler has sufficient water storage capacity to furnish steam until the time lag has been satisfied, a condensate pump unit can be used to return condensate to it. The storage capacity of the boiler is the available volume between the high and low water levels.

At design conditions, the condensing rate corresponding to 1000 sq. ft EDR is 0.496 gpm (see Table 12-11). At start-up, the condensing rate will be greater than that at design but will always be somewhat less than 150% of design. Old cast-iron boilers may be capable of handling 150% of rated load. Therefore, the old standard was to size pumps for 3 times the design condensing rate. This also allowed for loss of pump capacity due to cavitation. The standard sizing of condensate pump units is to select the pump for a rate equal to 2 or 3 times the system design condensing rate.

The condensate receiver tank should be sized as

small as practical to provide prompt return of water to the boiler, or boiler feed unit in the case of a condensate transfer unit. Standard sizing of receiver tanks for condensate return units provide 1 min. *net* storage capacity based on the system return rate. The net storage capacity is defined as the capacity between where the float switch starts the pump on high water level and where it stops the pump on low water level. The total receiver volume will be larger than the net.

The condensate pump unit feed system has been successful in small space heating systems. On larger installations, the boiler does not have an adequate storage volume to handle the system surges. When the condensate pump unit is applied to a system where the boiler does not have adequate storage volume, the makeup water feed valve on the boiler will add additional makeup during start-up or heavy system steam demand. When the condensate is later returned, the boiler will flood and shut off on high water level.

A practical system size limit for a condensate pump feed system may be about 8000 sq. ft EDR (60 boiler horsepower). This maximum system size will vary with different types of boilers having various storage capacities between high- and low-level operating levels. The boiler storage or working capacity can be obtained from the boiler manufacturer. Quite frequently, when older boilers having a large storage volume are replaced with modern boilers, generally having smaller storage capacity, the existing condensate feed system causes problems as previously stated. Also, this type of feed system is not practical in feeding multiple boilers or where a large percentage of makeup water is required.

Larger steam systems and multiple boiler installations require a boiler feed unit to balance the flow. In these larger systems, condensate pump units are frequently used as transfer units to collect the condensate at remote locations and pump it to the boiler feed pump system.

Boiler Feed Pump Units. The boiler feed pump unit is preferable over a condensate pump unit for all systems. The boiler feed unit consists of a condensate receiver tank sized to store an adequate volume of water to handle the system surges or system time lag and a pump or pumps wired to operate in response to a level control switch on the boiler. The normal boiler water level control maintains a boiler water level within 1 in. differential. When a boiler feed unit is used, the makeup water for the system is added into the condensate receiver tank on low water level and pumped into the boiler water level and the system surges occur in the condensate receiver tank.

The condensate receiver tank should be sized large enough to prevent overflow of returned condensate. The normal receiver sizing is to provide 5 min. storage volume for systems up to 30,000 sq. ft EDR or about 200 boiler horsepower. Larger systems should provide a minimum of 10 min. storage volume. Oversizing the boiler feed unit receiver does not affect the system operation; it only adds slightly to the initial system cost. Undersizing the receiver can cause overflow of returned condensate, which must be replaced with raw makeup water. This wastes heat, makeup water, and chemical treatment.

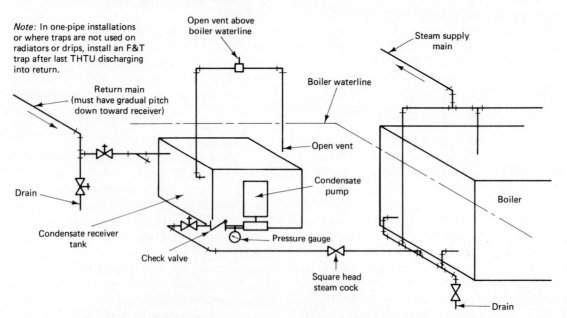

FIGURE 12-33 Typical condensate pump installation where return mains are above condensate receiver inlet.

The boiler feed pump should be sized the same as those for condensate pump units providing for 2 to 3 times the system condensing rate. The temperature of the returning condensate has an important bearing on the type of pump selected for a boiler feed pump or a condensate pump. Where the condensate temperature approaches saturation, pumps designed to operate at a low NPSH should be used. Such pumps are available that have a required NPSH of only 2 ft and that can handle 210°F condensate in a heightless receiver or 212°F condensate with a receiver elevated only 2 ft above the floor.

When the elevation of all return lines permits gravity return of condensate to the boiler feed unit receiver, only the boiler feed pump is required to feed the boiler. In many installations, some of the return lines may not permit gravity, and condensate transfer units may be required to pump the return back to the boiler feed unit.

Figure 12-34 shows a sketch of a typical hookup for a boiler feed unit with a combination pump control and low-water cutoff operating the pump. The low-water cutoff in the receiver shuts down the feed water pump if the water level in the receiver drops to an unsafe level. When this installation is used, the makeup water control may be a float-operated mechanical type in smaller systems. Large systems often require more makeup water than this type can provide. In such cases, a diaphragm valve operated by water pressure from a water feeder may be used, or a liquid level controller may be used to operate an electric valve as an alternative.

Vacuum Pump Units. Vacuum pumps may be used for the sole purpose of producing a vacuum on

FIGURE 12-35 Domestic® Series VLR® vacuum heating unit. (Courtesy of ITT Fluid Transfer Division, Morton Grove, IL)

the system, or they may be of the dual purpose type, producing a vacuum and returning condensate to the boiler. These dual-role units are called *vacuum return units.* Figure 12-35 shows the Domestic® VLR vacuum pump unit with a single pump. Duplex models are also available. Figure 12-36 shows a cutaway view of the single unit, which is typical of all single and duplex models.

The vacuum pump units shown employ the dependable multijet vacuum producer. This proven device assures better system performance, quicker system warm-up, fuel and power savings, and long life at peak efficiency. Refer to Figure 12-36 as we discuss the principle of operation.

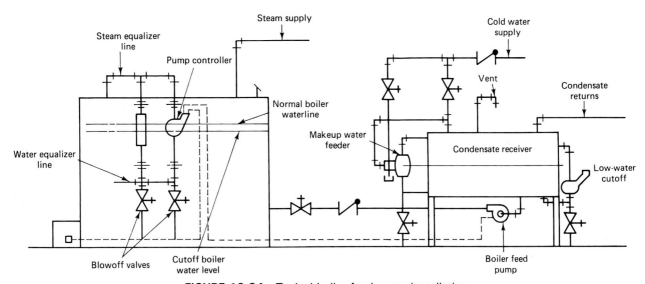

FIGURE 12-34 Typical boiler feed pump installation.

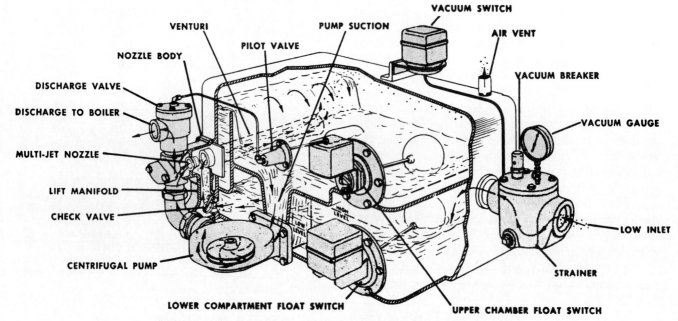

FIGURE 12-36 Cutaway view of Domestic™ VLR single unit, typical of all single and duplex models. *Note:* Current VLR unit uses a simpler solenoid valve to sense upper chamber level and opens discharge valve (replaces pilot valve). (Courtesy of ITT Fluid Transfer Division, Morton Grove, IL)

Units of this type have a two-compartment receiver, which is divided horizontally. The lower compartment acts as an accumulator and is under the same vacuum as that maintained in the return piping. The upper compartment is used to separate the air from the condensate and must always be open to the atmosphere.

For initial starting, the upper compartment must be half-filled with water as indicated on a gauge glass. With the line disconnect switch closed, the pump motor then operates in response to the float and vacuum switches, while the discharge of condensate is controlled by float-operated pilot valves and hydraulically actuated discharge valves.

Condensate and air entering the lower compartment through the strainer actuate either the lower float switch or vacuum switch, or both, to start the pump motor. When the pump is operating, "hurling water" enters the pump suction from the upper compartment, is discharged through an elbow into the forward part of the nozzle body through the multijet nozzle and venturi, and is returned to the upper compartment. (The upper compartment water circulated to produce a vacuum is called "hurling water.") The water driving at high velocity across the gap between nozzle and venturi creates a vacuum that draws condensate, air, and gases from the lower compartment and heating system through the left manifold and rear section of the nozzle body. This mixture is entrained in the jet streams and discharged through the venturi into the upper compartment of the

receiver, where the air and gases separate from the condensate and are vented. See Figure 12-37.

As the condensate level rises in the upper compartment, it causes the float-actuated pilot valve to snap open. This releases the pressure inside the bellows of the discharge valve, allowing the pump pressure to override the closing spring, and causes the valve to open. The centrifugal pump then pumps condensate through the discharge valve as well as through the multijet nozzles. Thus, the unit continues to pump air at approximately the normal rate while discharging condensate. When the water level drops to a predetermined level, the pilot valve closes the discharge valve, retaining a volume of hurling water in the upper compartment even though the pump may continue operation as an air pump.

When the vacuum and float switches have been satisfied and the pump stops, the check valve in the lift manifold closes, preventing the return of air and water to the lower compartment. The purpose of the float switch in the upper compartment is to continue the operation of the pump or pumps to return the maximum amount of condensate to the boiler each cycle.

Since the vacuum pump unit has the dual role of pumping condensate and of maintaining a vacuum in the system, the sizing of the unit depends on the required water pumping rate and on the air removal rate.

The condensate pumping capacity would be selected as discussed in the section on condensate pump units.

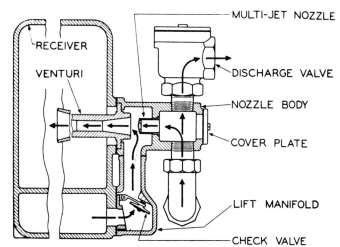

MULTI-JET NOZZLE

DISCHARGE VALVE

NOZZLE BODY

COVER PLATE

LIFT MANIFOLD

CHECK VALVE

RECEIVER

VENTURI

Figure 12-37 Cross section of Domestic™ vacuum producer, VLR unit. (Courtesy of ITT Fluid Transfer Division, Morton Grove, IL)

Pumps for tight vacuum systems should have a capacity of 0.3 cfm to 0.5 cfm of air removal for each 1000 sq. ft EDR served. The larger air capacity is suggested for systems up to 10,000 EDR. The pump selection basis should be $5\frac{1}{2}$ in. Hg at 160°F, which is representative of actual system conditions. These air removal rates are based on the vacuum range of 0 in. to 10 in. Hg, with the air pumps usually operating between 3 in. and 8 in. Hg.

Some conditions require larger air removal rates. Systems with excessive air-in leakage are sometimes difficult to correct. It has been found that air removal rates of about 1.0 cfm per 1000 EDR have been adequate for most systems of this type operating in the standard 3 in. to 8 in. Hg vacuum range.

Table 12-12 offers suggested vacuum pump unit sizing data.

Vacuum pump units are available in a number of types and capacities. Some commonly used types are as follows:

1. Single units with a common pump for pumping water and inducing a vacuum (see Figure 12-35).
2. Duplex units with two common water and vacuum pumps.
3. Single units with separate water and vacuum pumps.
4. Duplex units with two water pumps and two vacuum pumps.
5. Semi-duplex units with two water pumps and a single vacuum pump.

The purpose of the semi-duplex and duplex pump units is to provide standby protection. Each water or vacuum pump is selected to take care of the full system requirements. Integral control systems provide for the operation of the lag pump should the lead pump fail or be unable to meet the system requirements.

In steam systems where the return lines are below the level of the vacuum pump unit, provision must be made for lifting the condensate to the vacuum pump unit receiver. A condensate pump unit may be used to collect condensate and deliver it to the vacuum pump. An installation of this type is shown in Figure 12-38. The vacuum return line drains the condensate from the radiation above the boiler water line. The low return for the radiation below the boiler water lines drains to the condensate pump unit receiver. The condensate pump unit receiver is sized for this radiation load. The condensate pump unit is kept under vacuum by a connection to the vacuum return as shown. The vacuum pump unit is sized for the total condensate load.

The required pumping head for each of the pump units is dictated by the job conditions. In the case of the pump units of Figure 12-38, two conditions prevail. The

TABLE 12-12

Summary of water requirements (single pump, 1 gpm/1000 EDR; duplex and semi-duplex pump, 0.75 gpm/1000 EDR)

Suggested Guide to Air Removal Requirements		
System	Vac. Range (in. Hg)	Cfm/1000 EDR
Tight systems through 10,000 EDR	0–10*	0.5
Tight systems in excess of 10,000 EDR	0–10*	0.3
All systems, some air-in leakage	0–10*	1.0
All systems	10–15	1.5
All systems	15–20	2.0
Replacing steam pumps, all systems	0–8*	1.0
Replacing steam pumps, all systems	above 8	2.0

Source: Steam Heating Application Manual, Bulletin No. TES-375. ITT Fluid Transfer Division.

* Air pumps are usually adjusted to operate between 3 in. and 8 in. Hg.

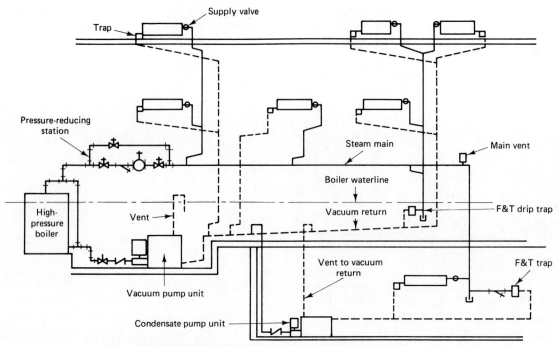

FIGURE 12-38 Condensate pump unit collecting low returns for vacuum pump unit.

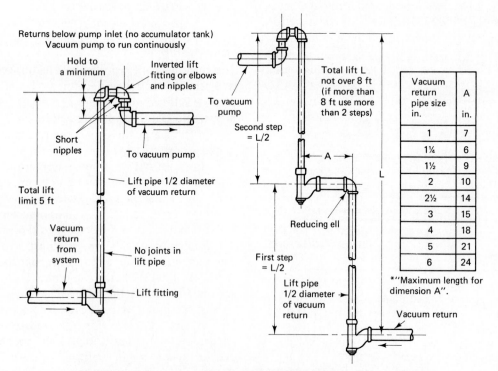

FIGURE 12-39 One-step and two-step lift fitting constructions. (From *Heating, Ventilating,* and *Air Conditioning,* 1959. Used with permission of American Society of Heating, Refrigerating, and Air Conditioning Engineers, Atlanta, GA)

condensate pump unit head would be the static elevation between the pump discharge opening and the vacuum return line plus the pressure drop in the pipe and fittings of the discharge line. The vacuum pump unit head would be the pressure drop of the discharge line plus the boiler pressure against which the pump must operate.

Lift fittings are sometimes used to elevate condensate from the low return vacuum systems to the vacuum pump unit. Commercial lift fittings can be used to make a single-step lift or a series of "lifting steps," as shown in Figure 12-39. The total lift is divided by an equal number of steps, each of which is trapped by a lift fitting. Water and air are drawn up by the vacuum pump from the low returns in a series of "slugs."

Additional vacuum above the system requirements is needed to produce this lift. The additional required vacuum is 1 in. Hg per 1 ft of lift. The condensate temperature must be well below its saturation temperature at the system vacuum, or "flashing" will take place and destroy the syphon action of the lift fitting.

It is necessary to operate the vacuum pump continuously for the successful operation of the lift fitting. Therefore, the vacuum pump unit must have separate water and air pumps. The air pump will operate continuously, and the condensate pump will be cycled by its float switch as required.

BIBLIOGRAPHY

12.1. *ASHRAE Handbook 1989 HVAC Systems and Applications,* American Society of Heating, Refrigerating, and Air Conditioning Engineers, Atlanta, GA, 1989.

12.2. *Handbook of Air Conditioning System Design,* Carrier Air Conditioning Company, Syracuse, NY, 1965.

12.3. *Steam Heating Application Manual,* Bulletin No. TES-375, ITT Fluid Transfer Division, Morton Grove, IL, 1975.

12.4. *Domestic Pump Product Application and Sizing Man-* *ual,* Bulletin No. TES-676, ITT Fluid Transfer Division, Morton Grove, IL, 1976.

12.5. *Hoffman Specialty Steam Trap Application and Sizing Manual,* Bulletin No. TES-776, ITT Fluid Transfer Division, Morton Grove, IL, 1980.

12.6. *Basic Controls for Low Pressure Steam Boilers,* Bulletin SL-BCS, McDonnell & Miller, ITT Fluid Transfer Division, Morton Grove, IL.

13

Principles of Room
Air Distribution

13-1 INTRODUCTION

Air distribution is one of the most important branches of the engineering science of HVAC. The purpose of this chapter is to provide a source of accurate engineering information in the field of air distribution within conditioned spaces. An attempt is made to reduce somewhat intricate formulas to simple equations and to introduce a concept of the subject based on authentic facts discussed in simple language.

Air mechanically supplied to interior spaces of a building to provide for ventilation, heating, and/or cooling requires careful selection of terminal air distribution devices, a well-designed duct system, and a fan or fans. This chapter deals with some fundamentals of terminal air distribution devices, a very important link in the chain of load estimating, duct design, fan equipment selection, and control. Chapter 14 will deal with duct system design and fan performance and selection.

The success of a good heating, ventilating, or air conditioning system depends upon its air distribution. Most of the complaints arising out of faulty room air distribution can be classified as either *drafts* or *stuffiness*. It is appropriate to analyze these two sensations because they illustrate the interrelationship of the ac-

cepted standards of room temperatures, humidity, air motion, and airflow direction.

Perhaps a third source of complaints, *noise,* should be added. We will certainly be concerned with taking the necessary precautions to prevent noise problems when we select outlets. However, for purposes of this discussion, we will consider that outlet velocities, which are the biggest factor in noisy air distribution, are within the recommended limits.

A draft is a current of air that, due to its temperature, humidity, rate of movement, or any combination of these factors, removes more heat from the surface of a body than that surface normally dissipates. The complaint of drafts may arise whenever there is a local or general sensation of feeling too cool.

Conversely, stuffiness is usually thought of as a stagnant condition of the air. Actually, the complaint of stuffiness generally arises when less heat is removed from the surface of the body than it normally generates.

This indicates that drafts and stuffiness are not functions of the single property of air velocity but are due to the properties of motion, temperature, and humidity in combination—or in other words, the effective temperature. To these properties may be added a fourth factor, air direction, whenever the sensation is localized, such as a draft on the back of the neck.

This concept explains difficulties frequently encountered in practice. Comparatively high air motion may be tolerated with temperature and humidity adjusted upward to result in the optimum effective temperature. This same higher temperature and humidity with lower air motion may be considered stuffy even though the residual velocity may be what is normally considered optimum. The effect of inequities and fluctuations in either temperature or velocity serves to call the occupant's attention to differences in sensations. Good room air distribution must avoid temperature stratification, fluctuating gusts of air, warm or cold spots, and local high velocities.

13-2 TERMINOLOGY

Air distribution is one of the least understood branches of HVAC systems. This is true because of many factors, not the least of which is the lack of standard terminology and methods of rating the performance of similar air distribution devices in the industry. The following definitions are largely from *ASHRAE Handbook 1989 Fundamentals*.

Air distribution outlet: Any mechanical device for distributing air within an enclosure, such as a grille, register, ceiling diffuser, or slot.

Aspect ratio: The ratio of length to width of an opening or the core of a grille. Also, with regard to rectangular sheet metal ducts, it is the ratio of the long dimensions to the short dimensions of the cross section.

Aspiration: See *induction*.

Attenuation: The process of absorbing or dissipating sound waves between point of origin and the human ear.

Axial-flow jet: A stream of air whose motion is approximately symmetrical along a line, although some spreading and drop or rise may occur due to diffusion and buoyancy effects.

Ceiling diffuser: A circular, square, or rectangular outlet located in the ceiling of an enclosure through which supply air is discharged on a plane approximately parallel to the ceiling.

Coefficient of discharge: The ratio of the area of the vena contractor to the area of the opening.

Core area: The total plane area of that portion

of a grille included within lines tangent to the outer edges of the opening through which air can press.

Damper: A device used to vary the volume of air passing through a confined cross section by varying the cross-sectional area.

Diffuser: A supply outlet discharging supply air in various directions and planes.

Diffusion: Distribution of air within a space by an outlet discharging supply air in various directions and planes.

Drop: The vertical distance that the lower edge of the horizontally projected airstream drops between the outlet and the end of its throw where terminal velocity is reached. It is affected by the length of throw and by the temperature differential between the primary and secondary air. Drop is generally associated with cooling.

Duct area: The area of the cross section of a duct based on its inside measurements at the point of installation of a device.

Effective area: The net area of an outlet or inlet device through which air can pass, equal to the free area times a coefficient of discharge. In field practice, it is the area that, when multiplied by the average jet velocity, gives the volume flow rate (cfm) passing through the device.

Entrainment: The mixing of room air and the primary air after the primary air leaves the air supply outlet (called *secondary air motion,* also *induction*).

Entrainment ratio: The total air divided by the discharge air. See *induction ratio*.

Envelope: The outer boundary of an airstream moving at a perceptible velocity.

Exhaust opening or inlet: Any opening through which air is removed from a space.

Free area: The total area of the openings in an outlet or inlet through which air can pass. In the era of gravity warm-air systems, free area was of prime importance. Today, with forced-air systems, except in sizing return-air grilles, free area is secondary to "total pressure loss."

Grille: A covering for an opening through which air passes.

High-pressure system: A supply- or return-air system where the maximum static pressure at any outlet on the system exceeds 3.5 in. WG, but does not exceed 6.0 in. WG.

Induction: The process of drawing room air

(secondary air) into the primary airstream (commonly called *aspiration*).

Induction ratio: A value that may be established for any air distribution device by measuring the relative quantities of primary and secondary air in the mixed airstream at a predetermined distance from the device.

Jet velocity: The velocity of air in fpm measured at a point in the vena contractor between the bars of a grille or register or at the lips of a ceiling outlet. See *outlet velocity*.

Low-pressure system: A supply- or return-air system where the maximum static pressure at any outlet on the system does not exceed 0.5 in. WG.

Medium-pressure system: A supply- or return-air system where the maximum static pressure on the system does not exceed 3.5 in. WG.

Multilouver damper: A damper having a number of blades used for the purpose of throttling the flow of supply air or exhaust air passing through a duct.

Opposed-blade damper: A multilouver-type damper with center-pivoted blades, where the action moves adjacent blades in opposite directions in closing and opening.

Outlet velocity: The average velocity of air emerging from a supply-air outlet or entering a return-air outlet, measured normal to the plane of the opening (normally called *face velocity*).

Perimeter system: A heating or cooling forced-air system where the diffusers are installed to blanket the outside walls of a space. Returns are usually located in one or more centrally located places. High sidewalls or ceiling returns are preferred especially for cooling, although low returns are acceptable for heating. This design is highly recommended for combination heating and cooling installations.

Plaque: A flat ceiling outlet in which the supply air impinges against a flat plate or series of parallel plates and is deflected horizontally.

Primary air: The air delivered to the supply outlet by the supply-air duct.

Radial-flow jet: A stream of air distributed outwardly from a center so that the area of jet increases approximately in proportion to the distance from the center. Ceiling diffusers usually produce this type of jet.

Radius of diffusion: The horizontal axial distance an airstream travels after leaving an air outlet before the maximum stream velocity is reduced to a specified terminal level (i.e., 200, 150, 100 fpm).

Register: A grille equipped with a damper or control valve.

Residual velocity: The residual air velocity in the occupied zone of the conditioned space (i.e., 65, 50, or 35 fpm).

Secondary air: The air that the primary airstream entrains by the passage of the primary air through the enclosure.

Spread: The divergence of an airstream in a horizontal or vertical plane (measured in degrees) after it leaves the supply outlet. Sometimes, it is defined as the measurement in feet of the maximum width of the air pattern at the point of terminal velocity.

Static pressure: The outward force of air within a duct. This pressure is measured in inches water gauge (in. WG).

Supply opening or outlet: Any opening through which air is delivered into a space that is being heated, cooled, humidified, dehumidified, or ventilated.

Temperature differential: The temperature difference between the primary air and room air.

Temperature variation: The temperature difference between points within a space.

Terminal velocity: The maximum airstream velocity at the end of the throw. It is normally taken as 50 fpm for grilles and registers and as 100, 150, or 200 for ceiling outlets, depending on the application.

Throw: The distance measured in feet that the airstream travels from the supply outlet to the point of terminal velocity. The throw is measured vertically from floor and low sidewall diffusers and horizontally from high sidewall and ceiling diffusers.

Total air: The mixture of primary air and entrained air.

Total pressure: The sum of the static pressure and velocity pressure (sometimes called *impact pressure*). This pressure is expressed in inches water gauge (in. WG). Total pressure is directly associated with the sound level of an outlet. Therefore, anything that increases the total pressure, such as undersizing of outlets or increasing the speed of the fan, will also increase the sound level.

Vane: A thin plate in the opening of a grille. It may be flat, curved, or airfoil shaped.

Vane ratio: The ratio of the depth of vane to the minimum width between two adjacent vanes.

Velocity pressure: The forward-moving force of

air within a duct. This pressure is measured in inches water gauge (in. WG).

Ventilating ceilings: Multiple ceiling supply-air openings with vertical discharge, in close proximity with one another, covering a significant part of the ceiling area and acting as a whole not as individual units.

13-3 STATIC, VELOCITY, AND TOTAL PRESSURES

It is desirable at this point to briefly discuss some of the characteristics of airflow. This should help us to better understand the design, performance, and selection of the various air distribution devices for different applications.

Flow of any fluid is produced by a pressure difference. It is not important whether the regions in which the pressure difference occurs are above or below atmospheric pressure; it is the magnitude of the difference in pressure that determines the characteristics of flow. This phenomenon of fluid flow from one point to another under pressure difference is common throughout heating, ventilating, and air conditioning system design. In air conditioning and industrial ventilation applications, the pressure differences to be dealt with are usually quite low, and, as a result, air is considered incompressible. Therefore, Bernoulli's equation for incompressible fluid flow applies. Bernoulli's equation was developed in Chapter 9 from the steady-flow energy equation Eq. (9-8) which is repeated here:

$$\left(\frac{g}{g_c}\right)Z_1 + \frac{p_1}{\rho} + \frac{V_1^2}{2g_c} = \left(\frac{g}{g_c}\right)Z_2 + \frac{p_2}{\rho} + \frac{V_2^2}{2g_c} \quad (9\text{-}8)$$

Recall that each term of Eq. (9-8) has units of ft·lbf/lbm or J/kg, which is *energy per unit mass*.

If each term of Eq. (9-8) is multiplied by the fluid density ρ, we would have

$$\left(\frac{g}{g_c}\right)(\rho Z_1) + p_1 + \frac{\rho V_1^2}{2g_c}$$

$$= \left(\frac{g}{g_c}\right)(\rho Z_2) + p_2 + \frac{\rho V_2^2}{2g_c} \quad (13\text{-}1)$$

Each term in Eq. (13-1) has units of ft·lbf/ft³, or pressure in lbf/ft², or pascals, and is *energy per unit volume*.

Recall that the symbols and dimensions used in the steady-flow equations, Eqs. (9-8), (9-9), and (13-1), are as follows:

g_c = dimensional constant (32.3 lbm ft/lbf s²)
g = acceleration due to gravity (ft/s²)
Z = elevation (ft)
V = mean flow velocity (ft/s)
ρ = mass density (lbm/ft³)
p = pressure (lbf/ft²)
$\gamma = \rho(g/g_c)$ = specific weight (lbf/ft³)

The first equation, Eq. (9-8), is the general form of Bernoulli's equation and states that in ideal flow the *total energy* remains constant. The second form, Eq. (9-9), states that the *total head* remains constant. And, the third form, Eq. (13-1), states that the *total pressure* remains constant.

In gas flow and air conditioning duct system design, the pressure form, Eq. (13-1), is usually used; however, Eq. (9-8) should be used when significant density variations ($\rho_1 \neq \rho_2$) occur. For liquid flow, the head form, Eq. (9-9), is commonly used. Identical results are obtained no matter which of the three forms are used if proper care is taken with units and if the fluid is homogeneous.

Head and *pressure* are often used interchangeably, but these terms have specific meanings. *Head* is the height of a column of fluid supported by fluid pressure, while *pressure* is the normal force per unit area. With a liquid, it is frequently convenient to measure the head of a fluid in terms of "feet of fluid flowing." With a gas or with air, however, "feet of gas" would be a large number, and it is practical and customary to measure pressure by a column of liquid. In the case of airflow, the liquid is water at a constant specified condition. Water at 62°F has a mass per unit volume of 62.36 lbm/ft³; thus, a 1-in. column of water creates a pressure of 5.19 lbf/ft².

In Eq. (13-1), the combined term $(g/g_c)(\rho)(Z_2\text{-}Z_2)$ represents elevation pressure change, which in most cases is negligibly small. Therefore, it is frequently omitted from consideration in airflow problems. The terms p and $\rho V^2/2g_c$ are the *static pressure* (p_s) and *velocity pressure* (p_v), respectively. The sum of static pressure and velocity pressure is called *total pressure* (p_t).

As stated previously, it is practical and customary in airflow work to express pressures in terms of the height of a column of water, or in *inches water gauge* (in. WG). Conversion of units is accomplished simply by stating that

$$p_a = p_w$$

or

$$h_a(\rho_a)\left(\frac{g}{g_c}\right) = \frac{h_w}{12}(\rho_w)\left(\frac{g}{g_c}\right)$$

where

p_a, p_w = pressures of air and water, respectively (lbf/ft²)

h_a = pressure head (ft of air)

h_w = pressure head (in. of water)

ρ_a = air density (lbm/ft³)

ρ_w = water density (lbm/ft³, usually assumed to be 62.36 lbm/ft³)

Rearrangement gives

$$h_a = \left(\frac{h_w}{12}\right)\left(\frac{62.36}{\rho_a}\right) = 5.19\left(\frac{h_w}{\rho_a}\right) \qquad (13\text{-}2)$$

For velocity head, we know that $h_a = V^2/2g$; therefore,

$$\frac{V^2}{2g} = 5.19\left(\frac{h_w}{\rho_a}\right)$$

or

$$h_w = \frac{\rho_a V^2}{2g(5.19)} = \rho_a\left(\frac{V}{18.28}\right)^2 \qquad (13\text{-}3)$$

If V is to be expressed in feet per minute (fpm), then

$$h_v = h_w = \rho_a\left(\frac{V/60}{18.28}\right)^2 = \rho_a\left(\frac{V}{1096.5}\right)^2 \qquad (13\text{-}4)$$

where

h_v = velocity pressure head (in. WG)

ρ_a = air density at given conditions of pressure and temperature (lbm/ft³)

V = mean flow velocity ($V = Q/A$) (fpm)

By rearrangement of Eq. (13-4), we would have

$$V = 1096.5\sqrt{\frac{h_v}{\rho_a}} \qquad (13\text{-}5)$$

In much design work, "standard air" is assumed. Standard air is assumed to be at 70°F and 29.92 in. Hg, with a density of 0.07494 (say, 0.075) lbm/ft³. Therefore, Eq. (13-4) would become

$$h_v = \left(\frac{V}{4005}\right)^2 \qquad (13\text{-}6)$$

and Eq. (13-5) becomes

$$V = 4005\sqrt{h_w} \qquad (13\text{-}7)$$

It must be noted that, in Eqs. (13-4) through (13-7), flow velocity is in fpm and we have now called h_v the velocity pressure head in inches water gauge (in. WG). Therefore, we should refer to the static pressure head as h_s in inches water gauge (in. WG). Again, the sum of static and velocity pressure heads will be the total pressure head, h_t, or

$$h_t = h_s + h_v = h_s + \rho_a\left(\frac{V}{1096.5}\right)^2 \qquad (13\text{-}8)$$

with all terms expressed in inches water gauge. The total pressure head in a duct system remains constant except for friction and shock losses. These losses will be discussed in detail in the following sections and in Chapter 14.

If fluid flow through a sharp-edged, round orifice is analyzed, much can be learned about the principles of airflow through air distribution devices. Let Figure 13-1 represent such an orifice with a total pressure in the airstream at section 1 of, say, 1.0 in. WG higher than that at section 2, which is in free space. The flow will produce a convergence toward the orifice opening, and the air will reach a theoretical maximum velocity of about 4000 fpm at section 3 [Eq. (13-7)], the smallest area of the stream, technically called the *vena contractor*. The area of the stream at the vena contractor will be approximately 0.6 times the area of the orifice opening. Its location will be a distance downstream equal to about 0.5 times the diameter of the pipe, depending on the air velocity and size of the orifice.

The analogy of airflow through a sharp-edged orifice located at the end of a duct can be made to demonstrate the performance of air distribution devices. Any diffuser or grille performs in exactly the same way as in Figure 13-2, where airflow is shown through the members of a ceiling diffuser and the bars of a grille. Note that the only difference between Figures 13-1 and 13-2 is that the latter shows the action of multiple orifices. In other words, there is a vena contractor existing in every ring of a circular ceiling diffuser and between pairs of bars of a grille instead of a single vena contractor as in the sharp-edged orifice.

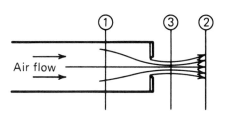

FIGURE 13-1 Vena contractor in sharp-edged orifice.

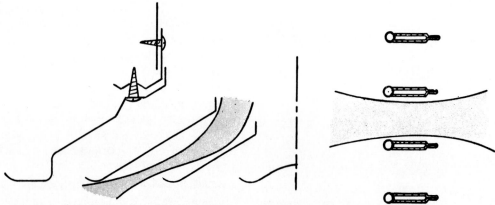

FIGURE 13-2 Vena contractor in (a) ceiling diffuser and (b) grille.

The sum of the areas of the vena contractors existing in a circular ceiling diffuser is sometimes referred to as the *effective area, A_e.* In the case of a grille, the effective area would be the sum of the vena contractors existing between the bars. In considering a sharp-edged orifice, the effective area will be the area of a single stream at the vena contractor. The shape, size, and number of bars or rings of a distribution device will affect the effective area and location of the vena contractors, but the basic concept applies regardless of the individual design.

Bernoulli's equation is applicable to any air diffuser in that the total pressure in the airstream behind the outlet is converted to velocity pressure at the points of the vena contractors since the friction loss by air in contact with the rings and bars is negligible.

13-4 MEASUREMENT OF STATIC, VELOCITY, AND TOTAL PRESSURES

Any fluid, including air, will exert pressure on the walls of the duct in which it is confined. Thus, a gauge connected to a tank of compressed air will indicate what is called *static pressure*—that is, the outward push of the air against the walls. In a duct in which the fluid is at rest because a damper at its outlet has been shut tightly, the static pressure of the air in the duct can be measured by means of a gauge connected to a tube that is inserted through the wall of the duct as shown in Figure 13-3.

In this same figure, a hinged vane is shown in the airstream. If the air were at rest, the vane would hang vertically. However, if the damper at the duct outlet is opened and the air starts to move, the vane will be tipped back due to the impact of the moving air against its face. The pressure on the face of the vane, which holds it up at an angle, is due altogether to the velocity of the moving stream of air.

If the vane in Figure 13-3 were hung parallel to the flow of air, it would not be affected by the movement of the air. The pressure on the two faces of the vane would be equal because only the static pressure of the air would be exerted against these two faces. Similarly, even though the air is in motion, a tube with its opening perpendicular to the air movement as in Figure 13-3 will not be affected by the movement of air. A gauge connected to it will indicate only static pressure or a sideward push of the air exactly as it would when the air was at rest.

On the other hand, if a tube is inserted into the duct with its opening pointing upstream as in Figure 13-4, the pressure indicated by the gauge will be higher than when the static tube was used. The reason for this is apparent when the cause for the tilt of the vane is considered. The vane swings up at an angle because of the pressure exerted upon its face by the moving stream of air. Similarly, a gauge connected to a tube with its open end facing the stream of air would indicate not only the static pressure but, in addition, a pressure due to the moving stream of air. The greater the velocity, the greater the pressure exerted against the tube opening and the face of the vane.

The action of the moving stream of air offers a

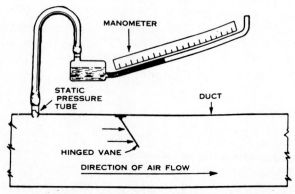

FIGURE 13-3 Measuring static pressure.

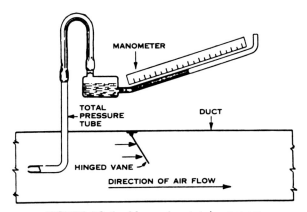

FIGURE 13-4 Measuring total pressure.

means of measuring the velocity of the air. The pressure measured in this way is known as the *total pressure*. It is greater than the static pressure because of the added pressure due to the moving stream of air.

However, if measurements are made of the total and static pressures by means of two separate tubes, the difference in the measurements of the two gauges would give the excess pressure that is due only to the velocity. This excess pressure above the static pressure is known as *velocity pressure*.

As indicated by Eq. (13-8), we are interested in three pressures—static, velocity, and total. Actually, only the total and static pressures can be measured directly, and the velocity pressure is found from their difference. After the velocity pressure has been found in this way, the air velocity can be computed by using Eqs. (13-5) or (13-7).

In actual test work, the manometer gauges used in Figures 13-3 and 13-4 may be used to indicate the velocity pressure directly by connecting the total and static tubes to the opposite ends of a manometer, as shown in Figure 13-5. In this way, the displacement of the fluid in the manometer is due only to the *difference* between the total and static pressures; it indicates the velocity pressure only ($p_v = p_t - p_s$).

The arrangement in Figure 13-5 is unsatisfactory for actual work. Both for convenience and accuracy, the static and total pressure tubes are combined in one instrument called a *Pitot tube*. Such an instrument consists of two tubes, one inside the other, as shown in Figure 13-6. The inner tube measures the total pressure. Eight holes 0.04 in. in diameter are drilled into the side of the outer tube for measuring the static pressure.

The Pitot tube is satisfactory for measuring velocities in excess of 700 fpm. Below this velocity, the deflection indicated on the inclined manometer scale is too small to read accurately. For laboratory work, ultrasensitive manometers have been developed that permit accurate velocity measurements below 700 fpm. These manometers are called *micromanometers*.

In summary,

1. Static pressure is the amount of compressive or expansive energy contained in a fluid or gas and is a measure of its potential energy. Static pressure may exist in a fluid or gas at rest or in motion and is the means of producing and maintaining flow against resistance.
2. Velocity pressure is the pressure corresponding to the velocity of flow and is a measure of the kinetic energy in the fluid or gas.
3. Total pressure is the sum of the static and velocity pressures and is a measure of the total energy within the fluid or gas. Static pressure may be transformed into velocity pressure, and vice versa, but only at the expense of total energy.

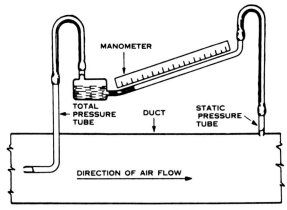

FIGURE 13-5 Measuring velocity pressure.

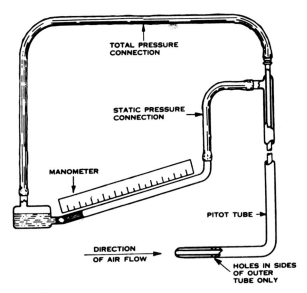

FIGURE 13-6 Pitot tube used to measure velocity pressure.

13-5 PRESSURE CHANGES IN A DUCT SYSTEM

An air duct of varying cross-sectional flow area is illustrated in Figure 13-7. If the airflow through the duct were ideal, there would be no friction, turbulence, or dynamic losses, and the total pressure (energy) is constant everywhere along the duct regardless of the changes in flow area. On the other hand, the static pressure and the velocity pressure will change in magnitude with each change in flow area. Figure 13-7 shows graphically the changes in static and velocity pressures (heads) as the air flows through the ideal system.

Figure 13-8 illustrates an actual airflow system that includes a fan, an intake duct, and an exit duct, again illustrating the changes in static and velocity pressures. Illustrated also is the decrease in total pressure in the direction of fluid flow caused by friction and dynamic (shock) losses. The following points should be noted:

1. The total pressure, h_t, at any point in the system is a measure of the total mechanical energy at that point.

2. In any actual flow system, the total pressure decreases in the direction of flow.

3. Static and velocity pressures are mutually convertible and can either increase or decrease in the direction of flow.

4. In section *AB,* the static pressure and total pressure decrease because of fan entrance losses and inlet duct friction loss. (Duct friction loss is dependent on Reynolds number; entrance losses do *not* depend directly on Reynolds number but are functions of velocity pressure or head.)

5. At *B,* the fan adds mechanical energy to the fluid, increasing both its velocity and static pressure heads.

6. In section *BC,* the expansion of the duct causes a deceleration of the flow. The reduction of velocity causes a *conversion* of velocity pressure head to

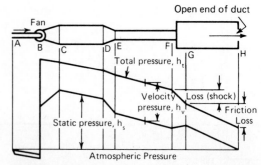

FIGURE 13-8 Actual changes in pressure with changes in duct flow area and energy supplied by fan.

static pressure head. The loss in total head is caused by shock losses associated with the change in flow area.

7. In section *CD,* the decrease in total head is caused by duct friction. The decrease in static head follows at the same rate because the velocity head is constant with the uniform cross section of the duct.

8. In section *DE,* the flow is accelerated by the reduction in cross section of the duct. In this section, there is a *conversion* of static to velocity head. The change in total head is due to shock losses associated with the change in flow area.

9. In section *EF,* there is a decrease in total and static pressure heads due to duct friction. The velocity head is constant.

10. In section *FG,* there is a sudden expansion of the duct. There is a *conversion* of velocity head to static head with a drop in total head due to shock losses.

11. In section *GH,* there is a uniform decrease in total and static heads due to duct friction.

12. The total pressure drops continually until at the outlet end of the duct, the static pressure is zero gauge (atmosphere) and the total and velocity pressures are equal. The reason for this is that at the outlet of the duct the pressure of the air is the same as the surrounding atmospheric pressure—that is, the static pressure is zero. However, the velocity pressure is still the same because the air issues from the mouth of the duct with the same velocity that it had while flowing in the duct. The velocity pressure is finally dissipated after the moving air issues from the duct and diffuser through the surrounding atmosphere. If the area at the outlet end of the duct is the same as the area at its inlet end, the velocity pressure will have the same value at both ends of the duct.

It is for this reason that the term *static pressure loss* is often used in practical work. While the total pressure loss is the actual loss in the system, the change in static pressures from inlet to outlet is numerically equal

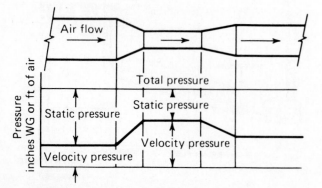

FIGURE 13-7 Theoretical changes in pressure with changes in duct flow area.

to the friction loss when the duct inlet and outlet areas are identical.

In many installations, the area at the outlet end of the duct is larger than it is at the inlet. Hence, the velocity at the outlet end of the duct is smaller than it is at the inlet. Because of this, a small part of the initial velocity pressure is converted into static pressure, which may then be lost in friction. As a result, not only the initial static pressure but also a portion of the initial velocity pressure will be lost in friction. However, the initial velocity pressure is usually small compared to the static pressure. For this reason, the small part of the velocity pressure converted into static, which is lost in friction, can usually be disregarded. It is evident that for ordinary work the use of static pressure is of greater utility than total pressure because its magnitude is a very close measure of the pressure available to deliver the air against the friction of a given duct system.

Although Figure 13-8 shows a simple duct with one outlet, the preceding discussion is true regardless of the number and size of the outlets connected to the duct system.

13-6   AIR DIFFUSION

Conditioned air is normally supplied to air outlets at velocities much greater than those acceptable in the occupied spaces. Conditioned air-supply temperature may be above, below, or equal to the air temperature in these spaces. Proper air diffusion, therefore, calls for (1) entrainment of room air by the primary airstream outside the zone of occupancy to reduce air motion and temperature differences to acceptable limits before the air enters the occupied space and (2) counteraction of the natural convection and radiation effects within the room.

Practical room air diffusion problems are complex because of the countless variations in building construction and in system design and operating requirements. Therefore, it is desirable to know the basic character of air diffusion.

The exchange of thermal and kinetic energy, in the form of heating, cooling, and air motion, from the primary to the secondary air may be done by induction only. The problem of properly distributing air to a conditioned space lies largely in completing the induction process above the occupied zone in the space.

Air emerging from a supply outlet at once commences to entrain room air. The induction effect increases the volume and decreases the velocity of the mixture stream according to Newton's second law, or the conservation of momentum theorem, which may be stated as follows:

$$m_{a_1} V_1 + m_{a_2} V_2 = (m_{a_1} + m_{a_2}) V_3 \qquad (13-9)$$

where

$$
\begin{aligned}
m_{a_1} &= \text{mass of primary air} \\
m_{a_2} &= \text{mass of secondary air} \\
V_1 &= \text{velocity of primary air} \\
V_2 &= \text{velocity of secondary air} \\
V_3 &= \text{velocity of mixture}
\end{aligned}
$$

The term V_2 may, for practical purposes, be considered zero, and, since the density is essentially constant, the volume rate may be substituted for mass. Eq. (13-9) then becomes

$$Q_1 V_1 = (Q_1 + Q_2) V_3 \qquad (13-10)$$

where

$$
\begin{aligned}
Q_1 &= \text{cfm of primary air} \\
Q_2 &= \text{cfm of secondary air}
\end{aligned}
$$

Induction decreases gradually at lower velocities giving way to turbulence and eddies.

The induction ratio (R) is defined as the ratio of the total air to primary air, or

$$R = \frac{\text{total air}}{\text{primary air}} = \frac{\text{primary + secondary air}}{\text{primary air}} \qquad (13-11)$$

The practical operating induction ratio of a diffuser or grille may be determined by velocity or temperature measurement. Readings of either temperature or velocity should be taken as follows:

t_1 or V_1 = temperature or velocity of primary airstream at last confines of diffuser or grille

t_2 or V_2 = average temperature or velocity of room air

t_3 or V_3 = average mixed airstream temperature or velocity at a reference point

The readings must be taken within the confines of the envelope of the mixed airstream. The accuracy of determining the induction rate of a diffuser or grille depends upon taking temperature or velocity measurements at a multiplicity of points in a plane at right angles to the mixed airstream at a reference point. Figure 13-9 illustrates a schematic of a setup for determining the induction rate for a wall-mounted rectangular diffuser. The plane grid network is placed at a reference distance from the face of the diffuser (say, 3 ft). The grid network is composed of a series of squares, the centers of which serve as a velocity-measuring station or the point location of a thermocouple for temperature measurement. Readings are taken in the center of each square,

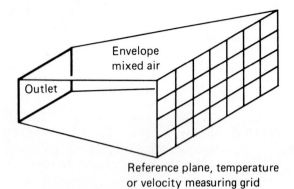

Reference plane, temperature
or velocity measuring grid

FIGURE 13-9 Setup for determining induction ratio of a rectangular diffuser.

either temperature or velocity, and averaged. Then the induction ratio is determined from

$$R = \frac{t_1 - t_3}{t_3 - t_2} = \frac{V_1 - V_3}{V_3 - V_2} \qquad (13\text{-}12)$$

Where a potentiometer and thermocouples are available, the temperature method is preferred since it gives a more stable reading than a velocity meter due to inertia and other characteristics generally inherent with velocity meters.

It must be borne in mind that an induction ratio must be referred to a given point in the stream travel. For example, assume a diffuser discharging air under conditions illustrated in Figure 13-9. Assume the reference point is established at 3.0 ft from the diffuser. The following average temperatures were recorded: $t_1 = 65°F$, primary air; $t_2 = 74.5°F$, secondary air; $t_3 = 73.5°F$, mixed-air temperature (average) at reference point. By Eq. (13-12),

$$R = \frac{t_1 - t_3}{t_3 - t_2} = \frac{65 - 73.5}{73.5 - 74.5} = 8.5 \text{ to } 1$$

If the reference point is taken at 6.0 ft from the diffuser, the temperatures are: $t_1 = 65°F$; $t_2 = 74.5°F$; $t_3 = 74.2°F$. Then,

$$R = \frac{65 - 74.2}{74.2 - 74.5} = 30.7 \text{ to } 1$$

The difference in density between the primary air and the secondary air for extreme differences of temperature between the two will cause some error due to gravity effects. Therefore, such tests should be run with normal difference of temperature between the primary and secondary air. The preceding example illustrates the necessity of establishing a common reference point when

comparing the induction rates of competitive equipment or of different types of equipment.

It will be seen that a given volume of air discharged at a given velocity from a low induction ratio device will carry a greater distance before reaching a specified mixture velocity as compared with the same volume discharged at the same velocity from a device with a high induction ratio characteristic.

Various types of air distribution devices have different induction characteristics and vary quite widely in the rapidity with which induction takes place. Table 13-1 shows the relative induction characteristics of present-day equipment.

Each of these types has its application, and the success of an air distribution system depends on the selection of equipment with induction characteristics best suited to the installation. It is obvious that an air outlet with a low induction characteristic must not be used where the volume to be distributed is high and the throws are short. Conversely, a high induction device should not be used where the volume to be distributed is low and the throws are long. Ceiling height and type of occupancy will affect the selection, of course, but the general statement may be made that the higher the cfm per square foot of area to be conditioned, the higher the induction characteristic of the air distribution device should be.

Effective area *(A_e):* One of the most important things to know about any air distribution outlet is its effective area. The entire air distribution performance and any pressure calculations involving the device are predicted upon its effective area. The effective area may be defined by the following expression:

$$A_e = A_g \times C \qquad (13\text{-}13)$$

where

A_g = geometrically measured area
C = combined coefficient of entry and discharge

The value of C can only be determined from tests for various outlets, many of which are quite complicated and difficult to measure geometrically. A typical laboratory test arrangement for such a test is shown in Figure 13-10.

The term *effective area* applies to any air distribution outlet. An analogy between air distribution equipment and a sharp-edged orifice located at the end of a duct may be drawn. As air passes down the duct, it contains both static pressure and velocity pressure, the sum of which is, of course, total pressure. As the air approaches the sharp-edged orifice, it assumes a stream shape that, at some point in its travel, is smaller than

TABLE 13-1

Relative induction characteristics of grilles and diffusers

Type	Mounting	Induction
Rectangular or linear diffuser	On ceiling	Low
Straight-flow grille	In sidewall at ceiling	Low
Straight-flow grille	In sidewall below ceiling	Low
Side deflection grille	In sidewall at ceiling	Medium
Half-round ceiling diffuser	In ceiling	Medium
Full-round ceiling diffuser	In ceiling	High
Full-round ceiling diffuser	Located down from ceiling	High
Perforated ceiling	—	High

Source: Used by permission. Tuttle & Bailey, Holland, MI.

the aperture through which the air passes (see Figure 13-1). This area is the vena contractor, as previously discussed. It may be said, then, that the vena contractor area is the *effective area* of the sharp-edged orifice.

Exactly the same comparison may be made with a ceiling outlet. Air does not completely fill the passages of a ceiling outlet, and, at some point within the passages, the air assumes a stream shape smaller than the actual opening of the passage (see Figure 13-2). The sum of the vena contractors existing between each ring is called the *effective area* of the complete diffuser.

The same fact applies to a grille. Air does not completely fill the space between the bars of a grille, and, at some point within the passages, the air assumes a stream shape smaller than the actual aperture opening, no matter how streamlined the bars may be. Again, the sum of the vena contractors is called the *effective area* of the grille.

Referring to Figure 13-10, the effective area may be found as follows:

1. Determine the average flow velocity in duct section F, from $Q = AV$, with Q through orifice.
2. Calculate velocity pressure head by Eq. (13-6).

3. Given the static pressure head at F, find total pressure head from Eq. (13-8).
4. Assume total pressure head is converted 100% to velocity pressure head at vena contractor.
5. Determine jet velocity in vena contractor by Eq. (13-7).
6. Determine effective area, A_e, by $A_e = Q/V_{jet}$.

ILLUSTRATIVE PROBLEM 13-1

Assume that a test setup as shown in Figure 13-10 is used to test a ceiling diffuser connected to a 12-in. round duct. The orifice Δp indicates a flow rate of 300 cfm. The static pressure head in section F is 0.096 in. WG. Determine the effective area of the ceiling diffuser.

Solution:

$$\begin{aligned}
\text{velocity at section } F &= 300/0.7854 \\
&= 382 \text{ fpm} \\
\text{velocity pressure head at } F &= (V/4005)^2 \\
&= (382/4005)^2 \\
&= 0.009 \text{ in. WG} \\
\text{static pressure head at } F &= 0.096 \text{ in. WG (given)} \\
\text{total pressure head at } F &= 0.096 + 0.009 \\
&= 0.105 \text{ in. WG} \\
\text{velocity in jets} &= 4005(0.105)^{1/2} \\
&= 1298 \text{ fpm} \\
A_e &= 300/1298 = 0.231 \text{ ft}^2
\end{aligned}$$

Since many outlets are of complicated shapes, this method is often the only way possible to determine the effective area of the diffuser. The procedure used here is primarily restricted to laboratory use due to the equipment required but is applicable to all types of air distribution equipment.

Average room velocity: Eq. (13-11) defines induction ratio R as the total air divided by the primary air. By definition, the entire airstream is composed of

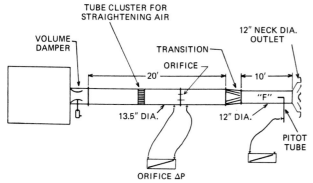

FIGURE 13-10 Equipment hookup for testing effective area of a ceiling diffuser.

primary and secondary air, which is called the *total air*. Primary air is supplied from the outlet, and, by induction, the secondary air is entrained in the airstream.

Average room velocity is defined as the total air divided by the area of the wall opposite to the outlet. To take into consideration room velocity at approximately the level of the head and shoulders, the area of the wall is multiplied by 0.7. Essentially, this means 0.7 of the wall height times the length, or

average room velocity

$$= \frac{\text{total air}}{0.7 \times \text{area of wall opposite outlet}}$$

$$= \frac{\text{total air} \times 1.4}{\text{area of wall opposite outlet}} \quad (13\text{-}14)$$

Some manufacturers of supply outlets utilize a room circulation factor K in the rating of such devices. The K factor is expressed as primary air cfm per square foot of wall area opposite the outlet. Note how this factor takes into account the average room velocity, primary and secondary airflows, and the vertical area opposite the outlet. From Eq. (13-11), we see that total air equals primary air times R. Substituting into Eq. (13-14), we have

average room velocity

$$= \frac{1.4 \times R \times \text{primary air}}{\text{area of wall opposite outlet}}$$

or

$$K = \frac{\text{average room velocity}}{1.4 \times R}$$

or

$$K = \frac{\dfrac{1.4 \times \text{total air}}{\text{area of wall opposite outlet}}}{1.4 \times \dfrac{\text{total air}}{\text{primary air}}}$$

$$= \frac{\text{primary air}}{\text{area of wall opposite outlet}} \quad (13\text{-}15)$$

Once the required primary air is determined, it may be divided by the wall area opposite the outlet to determine the K factor. Knowing the application, we will be able to determine the outlet velocity that will be acceptable for a given average room velocity.

13-7 SUPPLY-AIR OUTLETS

Supply-air outlets are important components of heating, ventilating, and air conditioning systems. The correct types of outlets, properly sized and located, control the air pattern within a given space to obtain air motion and temperature equalization.

Four basic types of supply outlets are commonly available: (1) ceiling diffuser outlets, (2) grille outlets, (3) slot diffuser outlets, and (4) perforated ceiling panels. These types differ in their construction features, physical configurations, and performance characteristics (air patterns).

Accessory devices used in connection with these outlets regulate the volume of supply air and control its flow pattern. The outlet cannot discharge air properly and uniformly unless the air is conveyed to it in a uniform flow. Accessory devices may also be necessary for proper air distribution in a space, and they must be selected and used in accordance with manufacturers' recommendations. Outlets should be sized to project air in a way that permits its velocity and temperature to reach acceptable levels before entering the occupied zone (0 to 6 ft above the floor).

Where an outlet without vanes is located directly against the side of the duct, the direction of blow of the air from the outlet is the vector sum of the duct velocity and the outlet velocity (Figure 13-11a). This may be modified by the peculiarity of the duct opening.

Where an outlet is applied to the face of the duct, the resultant velocity V_c can be modified by adjustable vanes behind the outlet. Whether they should be applied or not depends on the amount of divergence from straight blow that is acceptable.

Often, outlets are mounted on short extension collars away from the face of the duct (Figure 13-11b). Whenever the duct velocity exceeds the outlet discharge

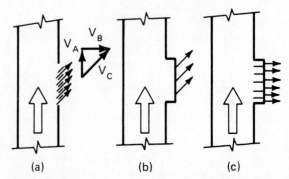

(a)   (b)   (c)

FIGURE 13-11 Supply outlet grilles mounted on face of duct.

velocity V_b (velocity due to pressure difference across the outlet), vanes should be used where the collar joins the duct (Figure 13-11c).

Importance of correct blow: Normally, it is not necessary to blow the entire length or width of a room. A good rule of thumb to follow is to blow three-fourths of the distance to the opposite wall. Exceptions occur, however, when there are local sources of heat at the end of the room opposite the outlet. These sources can be equipment heat and open doors. Under these circumstances, overblow may be required, and caution must be exercised to prevent draft conditions.

Supply-air temperature differential: The allowable supply temperature difference that can be tolerated depends to a great extent on (1) outlet induction ratio, (2) obstructions in the path of the primary air, and (3) the ceiling effect (see Figure 13-12).

The four basic types of supply outlets vary widely in the manner in which they diffuse or disperse supply air into a conditioned space and induce or entrain room air into a primary airstream.

Outlet types with high induction rates result in short throws and rapid temperature equalization. Ceiling diffusers with radial patterns have shorter throws and obtain more rapid temperature equalization than do slot diffusers. Grilles, with long throws have the

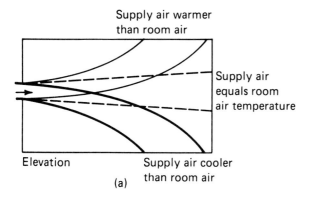

Supply air warmer than room air

Supply air equals room air temperature

Elevation

Supply air cooler than room air

(a)

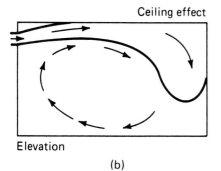

Ceiling effect

Elevation

(b)

FIGURE 13-12 (a) Effect of supply-air temperature differences and (b) ceiling effects.

lowest diffusion and induction rates. Therefore, round or square ceiling diffusers will deliver more air to a given space than grilles and slot diffuser outlets in cases that require recommended residual velocities of 25 to 35 fpm (0.13 to 0.18 m/s). In some spaces, higher residual velocities can be tolerated or the ceilings may be high enough to permit a longer throw that will still result in the recommended residual velocities.

Outlets with high induction characteristics can also be used to advantage in air conditioning systems with low supply-air temperatures and consequently high temperature differentials between room air and supply air. Therefore, ceiling diffusers may be used in systems with cooling temperature differentials from 30°F to 35°F (17°C to 19°C) and still provide satisfactory temperature equalizations within the spaces. Slot diffusers may be used in systems with cooling temperature differentials as high as 25°F (14°C). Grilles may generally be used in well-designed systems with cooling differentials up to 20°F (11°C). (See Figure 13-12.)

The induction or entrainment characteristics of a moving airstream leads to a phenomenon referred to as *ceiling effect* (sometimes called *surface effect* because it can apply to any flat surface, walls as well as ceilings): An airstream moving adjacent to, or in contact with, a wall or ceiling surface that creates a low-pressure area immediately adjacent to that surface causes the air to remain in contact with the surface substantially throughout the length of throw. The surface effect counteracts the drop of horizontally projected cool airstreams (see Figure 13-12).

Ceiling diffusers exhibit ceiling effect to a very high degree because a circular air pattern blankets the entire ceiling area surrounding each outlet. Slot diffusers, which discharge the airstream across the ceiling, exhibit ceiling effect only if they are sufficiently long to blanket a ceiling area large enough to produce the effect. Grilles exhibit varying degrees of surface effect, depending on the spread of the particular air pattern.

In many installations, the outlets must be mounted on an exposed duct and must discharge the airstream into free space. This type of installation affects all factors previously discussed. Since the airstream entrains air on both its upper and lower surfaces, a higher rate of entrainment is obtained; this results in a throw shortened by about 33%. Air loadings (cfm/square foot of floor space, or air changes per hour) for these outlets can therefore be increased. Air drop from ceiling diffusers is increased by installation on exposed ducts because there is no surface effect, and temperature differentials in air conditioning systems must be restricted to a range of 15°F to 20°F (8°C to 11°C). Airstreams from slot diffusers and grilles show a marked tendency to drop because of the lack of surface effect.

13-8 SELECTION PROCEDURE FOR SUPPLY-AIR OUTLETS

The following procedure may be used in the selection of supply-air outlet locations, types, and proper size:

1. Determine the amount of air to be supplied to each space to be conditioned.
2. Select the type and number of outlets for each space, considering such factors as air quantity required, distance available for throw or radius of diffusion, space structural characteristics, and architectural concepts. Table 13-2 from *ASHRAE Handbook 1988 Equipment* is based on experience and typical ratings of various outlets. It may be used as a guide to the outlets applicable for use with various room air loadings. Special conditions, such as ceiling heights greater than the normal 8 to 12 ft (2.4 to 3.6 m) and exposed duct mounting, as well as product modifications and unusual conditions of room occupancy, can modify this table. Manufacturers' rating data should be consulted for final determination of the suitability of the outlet used.
3. Locate outlets in the room to distribute the air as uniformly as possible. Outlets may be sized and located to distribute air in proportion to the heat gain or loss in various portions of the room.
4. Select the proper outlet size from manufacturers' ratings according to air quantities, discharge velocities, distribution pattern, and sound levels. Note manufacturers' recommendations with regard to use. In an open space configuration, the interaction of

TABLE 13-2

Guide to use of various outlets

Type of Outlet	Air Loading of Floor Space		Approx. Max. Air Changes per Hour for 10-Ft (3-m) Ceiling
	cfm/ft²	(L/s per m²)	
Grille	0.6–1.2	(3–6)	7
Slot	0.8–2.0	(4–10)	10
Perforated panel	0.9–3.0	(5–15)	18
Ceiling diffuser	0.9–5.0	(5–25)	30
Perforated ceiling	1.0–10.0	(5–50)	60

Source: ASHRAE Handbook 1988 Equipment, American Society of Heating, Refrigerating, and Air Conditioning Engineers, Atlanta, GA.

airstreams from multiple diffuser sources may alter single diffuser throw data or single diffuser air temperature/air velocity data, and it may not be sufficient to predict particular levels of air motion in a space. Also, obstructions to the primary air distribution pattern require special consideration.

13-9 CEILING DIFFUSERS

By definition, a *ceiling diffuser* is a round, square, or rectangular air distribution device mounted in or near the ceiling, discharging air in a plane approximately parallel with the ceiling. The first ceiling diffusers were single-plane plaques mounted under ducts or conduits projecting through the ceiling. Some single-passage plaques are commercially available today, although most ceiling diffusers are of the multipassage type to provide greater efficiency and better appearance (see Figure 13-13).

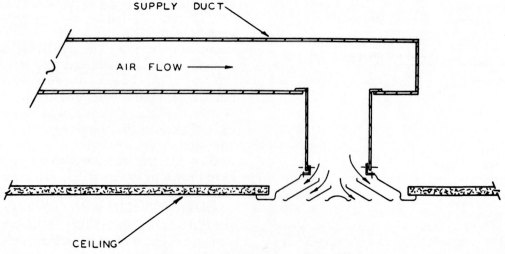

FIGURE 13-13 Multipassage ceiling diffuser.

The variable-pattern diffuser, the linear and rectangular diffusers, and the perforated ceiling tile diffuser must be placed in the category of ceiling diffusers, although the discharge air patterns of each does not necessarily conform to the definition just given.

In a homogeneous group of diffusers of any manufacturer, the effective areas of the diffusers in the group usually have a definite ratio to the neck areas. However, in comparing the diffusers of competitive manufacturers, the same ratios do not apply, and comparison must be made on the basis of effective areas. Diffusers of the same effective area will produce equivalent performance regardless of the neck area.

Full-round diffuser with fixed effective area and fixed airflow pattern: Figure 13-14 shows the Series 65 round ceiling diffuser manufactured by Lima Register Company. This diffuser handles full pipe capacity. The step-down tiers with advanced ring spacing provide maximum free area, enough to handle all the capacity of ducts of similar size. This design delivers air outward and slightly downward in all directions, blanketing the room without drafts or whistling. The larger outside ring provides greater smudge protection and covers irregularly cut holes. The recessed screw holes in this diffuser present a smooth flush design. A thick sponge-rubber gasket forms a tight seal that prevents air leakage and streaking. The Series 68 damper can be used for volume control. If a damper is not used, the Series 67 installation ring will facilitate installing.

Round diffuser with fully adjustable pattern: This type of round diffuser with mechanical adjustments enables the angle of discharge pattern to be changed and thus has become somewhat popular. Users feel that such adjustability enables the air distribution pattern of a given unit to be varied to suit conditions. In general, this is true *if* the diffuser is mounted on exposed ductwork and is not closer than $1\frac{1}{2}$-neck diameters to the ceiling. If the diffuser is mounted on a ceiling, it has only two patterns: (1) a horizontal blow along the ceiling typical of a fixed-pattern diffuser due to the ceiling effect or (2) a straight columnar discharge downward into the occupied zone. A columnar downblow pattern rarely accomplishes good air distribution, and the use of this style is primarily restricted to spot heating, ventilating, or cooling where high ceiling heights are involved. A downblow diffuser changes its pattern by mechanically changing the components of force of the airstream. When the air pattern is changed from horizontal to vertical, the total effective area is decreased, and the velocity for a given flow rate (cfm) is increased by perhaps 30%. When the diffuser is mounted on exposed ductwork, however, intermediate patterns may be obtained.

(a)

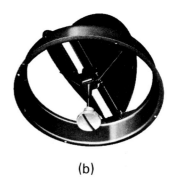

(b)

(c)

FIGURE 13-14 (a) Series 65 round, steel ceiling diffuser; (b) Series 68 knob-operated damper; and (c) Series 67 installation ring. (Courtesy of Lima Register Company, Lima, OH)

Square and rectangular diffusers: Architectural demand has produced many variations in the design of ceiling diffusers. Due to structural conditions, it is sometimes impossible to mount a ceiling diffuser in the center of spaces to be conditioned. Also, the need to harmonize with acoustical tile ceilings had to be satisfied.

Figure 13-15 shows the Series 60 square ceiling diffuser manufactured by Lima Register Company. Square step-down diffusers like this one handle full pipe flow with greater free area and are excellent systems. They

(a)

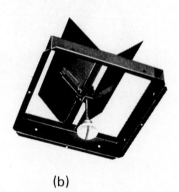

(b)

(c)

FIGURE 13-15 (a) Series 60 square, steel ceiling diffuser; (b) Series 64 knob-operated damper; and (c) Series 62 installation frame. (Courtesy of Lima Register Company, Lima, OH)

are also recommended for use as return-air inlets in perimeter systems. A thick sponge-rubber gasket forms a tight seal that prevents air leakage and streaking. The heavy-duty construction of the unit eliminates whistling and vibration at high air velocity. The Series 64 damper can be installed for volume control. If a damper is not used, the Series 62 installation frame will facilitate installing.

FIGURE 13-16 Series 90 CB, aluminum, curved-blade ceiling grille. (Courtesy of Lima Register Company, Lima, OH)

Figure 13-16 shows the Series 90 CB, aluminum, curved-blade ceiling grille manufactured by Lima Register Company. Lima's aluminum, curved-blade registers and grilles are alike in design. The only difference is the positioning of the streamlined blades for one-, two-, three-, or four-way throw. When one of these units is selected, the unit may be specified as a grille without a damper, a grille with an attached multi-louver damper, or a grille with an attached opposed-blade damper. The face and blades of the unit are constructed of aluminum extrusions, and each has an attractive anodized finish.

Figure 13-17 shows the Series 95, aluminum, directional diffuser with removable core manufactured by Lima Register Company. The aluminum frames can be interchanged with any of the 23 removable core patterns shown in Figure 13-18. This huge selection of patterns plus an almost unlimited range of sizes provides the specific airflow pattern that may be required. The core is

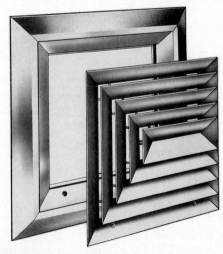

FIGURE 13-17 Series 95, aluminum, directional diffuser with removable core. (Courtesy of Lima Register Company, Lima, OH)

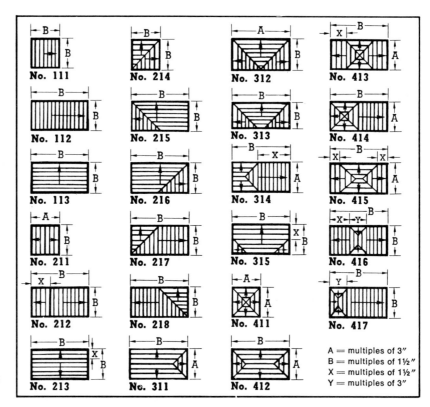

FIGURE 13-18 Core patterns for Series 95 directional diffusers (plan view neck size as viewed from neck side). (Courtesy of Lima Register Company, Lima, OH)

FIGURE 13-19 Series 80, steel, deluxe T-bar lay-in ceiling diffuser. (Courtesy of Lima Register Company, Lima, OH)

attached to the frame by means of stud hangers and special quarter-turn fasteners. After the frame is installed, cores can be inserted or removed quickly by turning the quarter-turn fasteners only 90 degrees.

Figure 13-19 shows the Series 80, steel, deluxe T-bar lay-in ceiling diffuser manufactured by Lima Register Company. Units like this one are designed to fit into T-bar ceilings. The cutaway view in Figure 13-19 shows the deflector plates that are used to deflect the airflow across the grille face in a one-, two-, three-, or four-way pattern.

13-10 ENGINEERING PERFORMANCE DATA (CEILING DIFFUSERS)

In predicting the performance of a ceiling diffuser, it is first necessary to determine its effective area, A_e (see Section 13-6). Certain empirical formulas have been derived (see *ASHRAE Handbook 1989 Fundamentals*) by which the performance of a ceiling diffuser may be predicted after its effective area has been determined. Exhaustive tests have been run by various laboratories to establish and substantiate these formulas. Some investigators write the formulas in slightly different forms, but practically they are the same. For ceiling diffusers, the *throw,* sometimes called the *radius of diffusion,* can be predicted from the following formula:

$$\text{throw } (T) = \frac{K \times \text{cfm}}{\sqrt{A_e}} \qquad (13\text{-}16)$$

where

T = distance in feet the airstream travels to a point at which its motion falls to a predetermined value (terminal velocity V_T)

cfm = flow volume through diffuser

K = constant determined by test and based on induction characteristics of diffuser (see Table 13-3)

A_e = effective area of diffuser in square feet

ILLUSTRATIVE PROBLEM 13-2

Find the throw (T) of a ceiling diffuser discharging air on a 360-degree periphery with an effective area of 0.229 ft^2 when installed in a ceiling and handling 300 cfm, with 100-fpm terminal velocity required.

Solution: By Eq. (13-16) and from Table 13-3, $K = 0.0120$. So, we have

$$T = \frac{K \times \text{cfm}}{\sqrt{A_e}} = \frac{(0.0120)(300)}{\sqrt{0.229}} = 7.5 \text{ ft}$$

For a diffuser mounted on an exposed duct, Table 13-3 gives $K = 0.0080$. Then,

$$T = \frac{(0.0080)(300)}{\sqrt{0.229}} = 5.0 \text{ ft}$$

By Eq. (13-16), the required effective area could be found if cfm, K, throw, and terminal velocity are given.

Illustrative Problem 13-2 shows that a diffuser of the same size handling the same cfm when installed on a ceiling will throw farther than one mounted on exposed ductwork. The airstream after discharge is slowed and

TABLE 13-3

K values for diffusers discharging 360° circular air pattern

Terminal Velocity V_T (fpm)	Average Residual Velocity V_R (fmr)	Flush Ceiling Mounting	Exposed Mounting
100	25–40	0.0120	0.0080
150	40–60	0.0100	0.0066
200	60–85	0.0075	0.0050

Source: Used by permission. Tuttle & Bailey, Holland, MI.

its energy absorbed by inducing room air into the primary airstream. Since, with a ceiling, no air is induced on the ceiling side of the stream, the kinetic energy is not dissipated as rapidly and the airstream throws farther. This is true, not only of a ceiling diffuser, but of any air distribution device located at or near the ceiling.

It will also be readily seen that, if a ceiling diffuser with a given effective area has that effective area reduced by some means while still discharging the same volume of air, the throw will be longer. In Illustrative Problem 13-2, if the effective area of 0.229 ft^2 were reduced to 0.150 ft^2 by some mechanical adjustment while the air flow remained at 300 cfm and K was 0.0120, the throw would be

$$T = \frac{(0.0120)(300)}{\sqrt{0.150}} = 9.3 \text{ ft}$$

Therefore, it is possible to secure a variation in throw while using the same diffuser and the same cfm—if the effective area is changed. The total pressure required at discharge will vary, of course.

In the preceding examples, a constant volume was discharged through a varying effective area. Now consider a constant effective area and a varying air volume. If the volume is reduced from 300 to 150 cfm by dampering, the throw would be

$$T = \frac{(0.0120)(150)}{\sqrt{0.229}} = 3.75 \text{ ft}$$

Next, it must be understood that restricting the volume flow by dampering in the collar will not change the effective area of the diffuser. Dampering does nothing but reduce the cfm. The number of passages through which the air is discharged and their size is unaffected.

It is impossible to get the same throw, and hence the same performance, when the cfm is reduced by dampering unless the effective area is also changed. It is also apparent that on many systems where the fan cfm is reduced the performance of the diffusers on the job is indiscriminately altered.

ILLUSTRATIVE PROBLEM 13-3

Assume an example of an actual installation where the design conditions for a diffuser are 300 cfm, 7.5-ft throw, 100-fpm terminal velocity, and effective area of 0.229 ft^2. The jet velocity would be 300/0.229 = 1310 fpm.

After installation, it is found that the jet velocity is 1960 fpm, giving a cfm of (0.229)(1960) = 450 cfm, which does not correspond to 300 cfm required. The throw is

$$T = \frac{(0.0120)(450)}{\sqrt{0.229}}$$

$$= 11.9 \text{ ft at 100-fpm terminal}$$

instead of 7.5 ft required. How can all the design requirements be met?

Solution: If the diffuser effective area is reduced to 0.153 ft^2 by mechanical means, then the cfm becomes (0.153)(1960) = 300 cfm, which is the volume required; but the throw becomes $T = (0.012)(300)/\sqrt{0.153} = 9.2$ ft, which does not meet the throw requirement. However, the cfm can be reduced to 300 cfm by dampering in the neck or the diffuser so that the total pressure at this point corresponds to 1310 fpm instead of 1960 fpm. Therefore, the effective area of 0.229 ft^2 can be made to produce the required flow, or

$$T = \frac{(0.012)(300)}{\sqrt{0.229}} = 7.5 \text{ ft}$$

Thus, all requirements of the design are satisfied. A damper in the collar allows the total pressure available at the diffuser to be adjusted to the proper degree; changing the effective area only does not.

It can also be seen that volume-control means must be available *in addition* to effective area control in order for a diffuser to perform all the requirements for which it may be called upon. It is also rarely possible to calculate the friction loss in any system closely enough to eliminate the need for volume-control dampers to equalize the friction losses between the diffusers on various branch ducts (Chapter 14).

Presentation of engineering performance data for air diffusion devices is not standard in the industry. Manufacturers have their own particular method of presenting data, which leads to different procedures for the selection of the type and size of diffusion equipment. In general, such procedures are relatively easy to apply.

Table 13-4 may be used for converting cfm into Btu/hr, or vice versa, at various diffuser supply-air temperatures. The table values shown are the solution of Eq. (4-10). Note that the table values are corrected to standard air and based on 70°F return air for heating and on 80°F return air for cooling.

Sound caused by an air supply outlet varies in direct proportion to the velocity of the air flowing through it. The air velocity can be controlled by selecting outlets of proper sizes. Table 13-5 suggests recommended outlet velocities that are within safe sound limits for the applications shown.

Tables 13-6 and 13-7 show performance data for Series 65 round, steel ceiling diffusers and Series 60 square, steel ceiling diffusers, respectively (see Figures 13-14 and 13-15). Lima Register Company presents the performance data for their steel line of diffusers in the manner shown in Tables 13-6 and 13-7. The double vertical lines divide the performance data in two parts. Keeping to the left of these double lines will ensure lower total pressure loss and quieter operation. We must also note that the data in these tables are limited to the following:

1. Heating Btu/hr is computed at 140°F air temperature behind the louvers.
2. Cooling Btu/hr is computed at 60°F air temperature behind the louvers.
3. Air volume is expressed in cfm.
4. Total pressure loss (static plus velocity) is for stackheads and diffusers.
5. Throw is based on terminal velocity of 50 fpm.
6. Velocity is the average calculated face velocity.

13-11 GENERAL APPLICATION DATA (CEILING DIFFUSERS)

The application of ceiling diffusers is a matter of engineering judgment and careful analysis of the job conditions. Because round or square ceiling diffusers project air equally in all directions, they best serve a square area. Therefore, they should be located as nearly as possible in the center of equal squares within the conditioned space. Divide the ceiling area as nearly as possible into squares, the sides of which should not exceed 3 times the ceiling height.

In many cases, because of structural conditions or for other reasons, it is impossible to locate the diffuser in the exact center of equal squares. When the ceiling area cannot be divided into squares, divide it into rectangles; the long side of such a rectangle should not exceed $1\frac{1}{2}$ times the short side.

The ceiling diffusers are then located in the center of those squares or rectangular areas. When this is not possible, the distance from the diffuser center to one side of the square or rectangle should not exceed $1\frac{1}{2}$ times the distance to the other side.

Multipattern diffusers (see Figures 13-17 and 13-18) can be used in the center of the spaces or adjacent to partitions, depending on the discharge pattern. By using different core assemblies, their air pattern can be changed to suit particular requirements.

Half-round diffusers and multipattern diffusers, when installed high in sidewalls, should generally discharge air toward the ceiling.

Some ceiling diffuser types, particularly *stepped-down* units, perform satisfactorily on exposed ducts. Manufacturers' catalogs should be consulted for specific types.

TABLE 13-4

Heating and cooling Btu/hr capacity table

Cfm	Heating (approximate Btu/hr with following air temperature inside of diffuser or register taken directly behind louvers)[a]				Cooling (approximate sensible Btu/hr with following air temperature inside of diffuser or register taken directly behind louvers)[b]		
	120°	140°	160°	180°	65°	60°	55°
40	2,175	3,045	3,915	4,785	640	855	1,070
60	3,260	4,565	5,870	7,170	960	1,280	1,600
80	4,350	6,090	7,825	9,565	1,280	1,710	2,135
100	5,435	7,610	9,785	11,955	1,600	2,135	2,670
125	6,795	9,515	12,230	14,945	2,000	2,670	3,335
150	8,155	11,415	14,675	17,930	2,400	3,200	4,000
175	9,515	13,320	17,120	20,920	2,800	3,735	4,670
200	10,870	15,220	19,570	23,920	3,200	4,270	5,340
225	12,230	17,125	22,015	26,910	3,600	4,805	6,005
250	13,590	19,025	24,460	29,900	4,000	5,340	6,670
275	14,950	20,930	26,905	32,885	4,405	5,870	7,340
300	16,310	22,830	29,350	35,875	4,805	6,405	8,005
325	17,665	24,735	31,800	38,865	5,205	6,940	8,670
350	19,025	26,635	34,245	41,855	5,605	7,470	9,340
375	20,385	28,540	36,690	44,845	6,005	8,005	10,005
400	21,745	30,440	39,135	47,835	6,405	8,540	10,675
450	24,460	34,245	44,030	53,815	7,205	9,605	12,010
500	27,180	38,050	48,920	59,795	8,005	10,675	13,345
600	32,615	45,660	58,705	71,755	9,605	12,810	16,010
700	38,050	53,270	68,490	83,715	11,210	14,945	18,680
800	43,490	60,880	78,270	95,670	12,810	17,080	21,350
900	48,925	68,490	88,055	107,630	14,410	19,215	24,015
1000	54,360	76,100	97,840	119,590	16,010	21,350	26,685
1200	65,230	91,320	117,410	143,510	19,215	25,620	32,025
1400	76,105	106,540	136,975	167,425	22,415	29,890	37,360
1600	86,975	121,760	156,545	191,345	25,620	34,160	42,700
1800	97,850	136,980	176,110	215,260	28,820	38,430	48,035
2000	108,720	152,200	195,680	239,180	32,025	42,700	53,370
2200	119,590	167,420	215,260	263,095	35,225	46,970	58,710
2400	130,460	182,640	234,830	287,010	38,430	51,235	64,045

Source: Courtesy of Lima Register Company, Lima, OH.

[a] Figures are corrected to standard air and based on 70° return air for heating.

[b] Figures are corrected to standard air and based on 80° return air for cooling.

Selecting the Throw. Throw is the linear distance from the center of the diffuser to the point at which the desired terminal velocity (V_t) is reached. Terminal velocities of 50 to 100 fpm may be selected as conservative standards for ceiling diffuser installations. However, it should be obvious that the nature of the occupancy in the space to be conditioned should determine the terminal velocity and may allow the use of considerably higher terminal velocities, such as 150 to 200 fpm, with associated changes in room velocity (V_R). Satisfactory results will be obtained because of the rapid rate of temperature and energy equalization provided by the circular air pattern issuing from the diffuser. This performance differs from that of a sidewall grille or lin-

ear diffuser, where the primary airstream, because of its mass and slower induction rate, may still be several degrees below room temperature (cooling) at the end of its throw.

Table 13-8 shows some suggested maximum terminal velocities varying with length and time of occupancy for installations using diffusers providing a circular air pattern.

It is apparent that all other factors being equal, the use of different terminal velocities will affect the throw of a given size diffuser and also the room velocity (V_R) in the occupied space. For example, the manufacturer's selection data may indicate the use of a diffuser that provides a 9-ft throw at 100-fpm terminal velocity

TABLE 13-5
Recommended delivery velocities for various applications

Application	Recommended Face Velocity (fpm)
Broadcasting studios	500
Residences	500–750
Apartments	500–750
Churches	500–750
Hotel bedrooms	500–750
Legitimate theaters	500–1000
Private offices, acoustically treated	500–1000
Motion picture theaters	1000–1250
Private offices, not treated	1000–1250
General offices	1250–1500
Stores, upper floors	1500
Stores, main floors	1500
Industrial buildings	1500–2000

when the measured throw is 12 ft. In this case, the remaining 3 ft will still be properly conditioned, although the terminal velocity at the 12-ft point will be lower than that originally chosen for the installation. The room velocity will also be somewhat reduced but not to a critical degree. In cases where the measured throw is somewhat less than that provided by the selected diffuser, satisfactory results will still be obtained, even though the terminal and room velocities at the end of the measured throw will be somewhat higher than that originally selected.

Most ceiling diffusers are installed in rooms with ceiling heights of 8 to 12 ft. No correction must be made for this variation in selecting the diffuser. However, when ceiling heights are above 12 ft, a corrected procedure should be used, as follows:

1. Measure the distance from the diffuser to the nearest wall or opposing airstream.
2. Measure the distance from the ceiling to a point 12 ft from the floor. Add to the plan throw (step 1) 75% of the difference between the actual ceiling and the 12-ft height.
3. Use this value as the actual throw and select the diffuser for the desired terminal velocity.

For installations with excessive ceiling heights, it is better to increase the cfm per diffuser and use fewer diffusers. For ceiling heights below 9 ft, it is better to reduce the cfm per diffuser to a minimum and use more diffusers.

Accessories. Dampering and air turning and straightening accessories of various types are available as standard equipment. Collars, round or square corre-

sponding to a particular ceiling diffuser, should always be used to connect a branch duct to the diffuser. A duct ring should be used to provide a means for leveling the diffuser against the ceiling and allow for variation in collar length due to construction (see Figures 13-14c and 13-15c). Dampers such as the Series 68 and Series 64 (see Figures 13-14b and 13-15b) can be used for volume control of the Series 65 and Series 60 ceiling diffusers.

Functionally, a diffuser is designed and rated to distribute air with the supposition that the air supplied to it has a total pressure and a direction within the limits that will enable it to completely perform its function. No diffuser can be expected to correct unreasonable conditions of flow in the air supplied to it.

Figure 13-20 shows a ceiling diffuser mounting assembly at the end of a branch duct. Note the cushion head, branch takeoff, connecting collar, straightening grid, damper, and diffuser. Omission of any of these component parts will result in less than ideal air distribution.

Accessories for the Series 95, aluminum, directional ceiling diffusers consist of the Series 95 OB opposed-blade damper, Series 95 DC directional control, and Series 95 EX extractor shown in Figures 13-21, 13-22, and 13-23, respectively.

The Series 95 OB damper is available in a range of sizes to fit any Lima directional diffuser. It controls the volume of air from fully open to fully closed, ensuring uniform air distribution over the diffuser face. It can be regulated by a lever well recessed just behind the diffuser louvers to discourage tampering. It is equipped

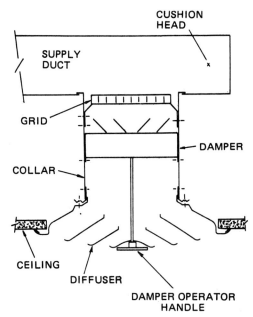

FIGURE 13-20 Ideal ceiling diffuser assembly including grid and damper.

TABLE 13-6

Performance data for Series 65 round, steel ceiling diffuser (white finish)

Product Number and Size	Free Area (sq. in.)	Performance Data																				
65—6"	23	Heating Btu/hr	3045	3805	4565	5710	6090	7610	9515	11415	13320	15220	17125	19025	20930	22830	24735	26635				
		Cooling Btu/hr	855	1070	1280	1600	1710	2135	2670	3200	3735	4270	4805	5340	5870	6405	6940	7470				
		Cfm	40	50	60	75	80	100	125	150	175	200	225	250	275	300	325	350				
		TP loss	.005	.008	.012	.018	.021	.032	.050	.072	.098	.130	.164	.201	.242	.291	.341	.396				
		Throw (ft)	2.9	3.3	3.9	4.6	4.8	5.8	7.0	8.2	9.3	10.7	11.8	13.2	14.2	15.5	16.7	17.9				
		Outlet velocity	250	313	375	470	501	626	783	939	1096	1252	1409	1565	1722	1878	2034	2191				
65—8"	40	Heating Btu/hr	5710	6090	7610	9515	11415	13320	15220	17125	19025	20930	22830	24735	26635	28540	30440	34245	38050	45660	53270	
		Cooling Btu/hr	1600	1710	2135	2670	3200	3735	4270	4805	5340	5870	6405	6940	7470	8005	8540	9605	10675	12810	14945	
		Cfm	75	80	100	125	150	175	200	225	250	275	300	325	350	375	400	450	500	600	700	
		TP loss	.005	.006	.009	.014	.020	.027	.036	.045	.055	.067	.080	.094	.109	.125	.142	.181	.223	.319	.436	
		Throw (ft)	3.6	3.9	4.8	6.7	7.0	8.1	9.1	10.2	11.3	12.4	13.5	14.6	15.7	16.8	17.9	20.1	22.4	26.7	31.2	
		Outlet velocity	270	288	360	450	540	630	720	810	900	990	1080	1170	1260	1350	1440	1620	1800	2160	2520	
65—10"	61	Heating Btu/hr	9515	11415	13320	15220	17125	19025	20930	22830	24735	26635	28540	30440	34245	38050	45660	53270	60880	68490	76100	
		Cooling Btu/hr	2670	3200	3735	4270	4805	5340	5870	6405	6940	7470	8005	8540	9605	10675	12810	14945	17080	19250	21350	
		Cfm	125	150	175	200	225	250	275	300	325	350	375	400	450	500	600	700	800	900	1000	
		TP loss	.005	.008	.011	.014	.018	.022	.026	.031	.036	.042	.049	.055	.070	.086	.124	.158	.220	.278	.343	
		Throw (ft)	4.8	6.0	7.1	8.1	9.3	10.3	11.4	12.4	13.5	14.6	15.6	16.7	18.8	20.9	25.3	29.5	33.8	38.2	42.4	
		Outlet velocity	295	354	413	472	531	590	649	708	767	826	885	944	1062	1181	1417	1653	1889	2125	2361	
65—12"	88	Heating Btu/hr	13320	15220	17125	19025	20930	22830	24735	26635	28540	30440	34245	38050	45660	53270	60880	68490	76100	91320	106540	121760
		Cooling Btu/hr	3735	4270	4805	5340	5870	6405	6940	7470	8005	8540	9605	10657	12810	14945	17080	19215	21350	25620	29890	34160
		Cfm	175	200	225	250	275	300	325	350	375	400	450	500	600	700	800	900	1000	1200	1400	1600
		TP loss	.005	.007	.009	.010	.013	.015	.018	.020	.023	.027	.034	.042	.060	.082	.106	.134	.166	.240	.323	.424
		Throw (ft)	4.6	5.3	6.1	7.3	7.7	8.5	9.2	10.0	10.8	11.6	13.2	14.6	17.7	20.8	23.9	27.0	30.2	36.4	42.7	48.9
		Outlet velocity	286	327	368	409	450	491	532	573	614	654	736	818	982	1145	1309	1472	1636	1963	2290	2618

Source: Courtesy of Lima Register Company, Lima, OH.

Note: For overall outside dimensions, add 4 in. to listed size.

with four mounting feet (see Figure 13-21) that allow the damper to be secured in any of the frames.

The Series 95 DC directional control (Figure 13-22) has two rows of individually adjustable vanes for controlling the airflow pattern. The upper row directs the air from the main duct to branch ducts. The lower row directs the flow of air and distributes it evenly over the neck of the diffuser. The vanes are made of aluminum extrusions and are securely held in position by press-fit nylon bushings for constant tension.

The Series 95 EX extractor (Figure 13-23) diverts air from a main duct into branch ducts. Gang-operated turning vanes remain parallel in both airstreams throughout the full range of adjustment that provides both directional and volume control. Branch duct volume control is controlled by the angle of the extractor.

TABLE 13-7

Performance data for Series 60 square, steel ceiling diffuser (white finish)

Product Number and Size	Free Area (sq. in.)	Performance Data												
60—6" × 6"	35	Heating Btu/hr	3805	5710	7610	9515	11415	13320	15220	17125	19025			
		Cooling Btu/hr	1070	1600	2135	2670	3200	3735	4270	4805	5340			
		Cfm	50	75	100	125	150	175	200	225	250			
		TP loss	.006	.014	.026	.040	.058	.080	.102	.126	.152			
		Horiz. throw (ft)	4	5	7	9	11	12	14	16	18			
		Outlet velocity	206	309	412	515	618	721	824	927	1030			
60—8" × 8"	63	Heating Btu/hr	7610	9515	11415	13320	15220	17125	19025	22830	26635	30440	34245	
		Cooling Btu/hr	2135	2670	3200	3735	4270	4805	5340	6405	7470	8540	9605	
		Cfm	100	125	150	175	200	225	250	300	350	400	450	
		TP loss	.007	.012	.017	.023	.030	.037	.045	.064	.088	.110	.140	
		Horiz. throw (ft)	5	6	8	9	10	11	13	15	18	20	23	
		Outlet velocity	228	285	342	399	456	513	570	684	798	912	1026	
60—10" × 10"	89	Heating Btu/hr	11415	13320	15220	17125	19025	22830	26635	30440	34245	38050	45660	53270
		Cooling Btu/hr	3200	3735	4270	4805	5340	6405	7470	8540	9605	10675	12810	14945
		Cfm	150	175	200	225	250	300	350	400	450	500	600	700
		TP loss	.006	.008	.011	.013	.015	.024	.032	.042	.054	.065	.095	.125
		Horiz. throw (ft)	6	7	8	9	10	12	14	16	18	20	24	28
		Outlet Velocity	243	284	324	365	405	486	567	648	729	810	972	1134
60—12" × 12"	133	Heating Btu/hr	15220	17125	19025	22830	26635	30440	34245	38050	45660	53270	60880	68490 76100
		Cooling Btu/hr	4270	4805	5340	6405	7470	8540	9605	10675	12810	14945	17080	19215 21350
		Cfm	200	225	250	300	350	400	450	500	600	700	800	900 1000
		TP loss	.004	.005	.007	.009	.013	.017	.021	.026	.038	.051	.065	.084 .100
		Horiz. throw (ft)	7	8	9	11	12	14	16	18	21	25	29	32 36
		Outlet velocity	216	243	270	324	378	432	486	540	648	756	864	972 1080

Source: Courtesy of Lima Register Company, Lima, OH.

Note: For overall outside dimensions, add 5 1/2 in. to listed size.

Since diffusers mounted on a ceiling produce a relatively thin, flat airstream moving parallel with the ceiling, they will produce direct impingement on suspended light fixtures, exposed beams, and so forth. The airstream will be deflected down into the occupied zone, creating objectionable drafts. Manufacturers' catalogs should be studied and manufacturers should be consulted when these problems exist.

The location of lighting fixtures and other ceiling obstructions in relation to high sidewall grilles should also be checked. This is often a more difficult problem than it is with ceiling diffusers since a greater mass of air is likely to strike a given obstruction. With the lower induction characteristics of grilles, the airstream may lack several degrees of being up to room temperature (cooling) at the point of impingement and may fall rap-

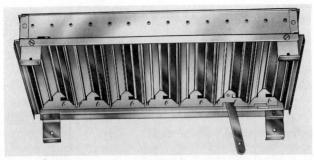

FIGURE 13-21 Series 95 OB opposed-blade damper. (Courtesy of Lima Register Company, Lima, OH)

idly into the occupied zone after striking the obstruction.

13-12 GRILLES AND REGISTERS

The oldest form of commercially manufactured air distribution device is a grille. Many types and styles have been developed over the years to meet the many requirements. Much research work has been done concerning the laws governing the performance of these supply-air grilles and diffusers. There are more variables to be considered in selecting grilles than in selecting any other air distribution device, particularly since the adjustable-bar grille has practically supplanted the fixed-bar type.

Adjustable-Bar Grille. The adjustable-bar grille is the most common type of grille used as a supply outlet. The *single-deflection* grille consists of a frame enclosing a set of individually adjustable vertical or horizontal vanes. Vertical vanes deflect the airstream in a horizontal plane; horizontal vanes deflect the airstream in a vertical plane. The grille may be combined with a multilouver damper to form a register. Figure 13-24 shows the Series 90V ML single-deflection register manufactured by Lima Register Company. The individually adjustable vertical airfoil-shaped vanes are made of aluminum as is the frame. Press-fit nylon bushings provide proper and consistent tension on the vanes to prevent loosening or spinning. Opposed-blade dampers may

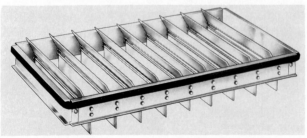

FIGURE 13-22 Series 95 DC directional control. (Courtesy of Lima Register Company, Lima, OH)

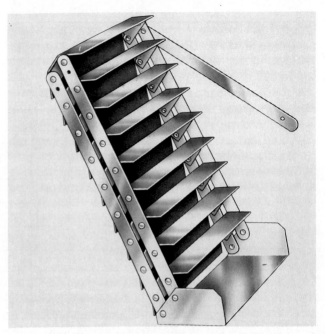

FIGURE 13-23 Series 95 EX extractor. (Courtesy of Lima Register Company, Lima, OH)

also be used with single-deflection grilles. Ideally, damper blades should run horizontally with vertical-front bar grilles and vertically with horizontal-front bar grilles so that all the blade passages fill with air.

Single-deflection grilles and registers are used most frequently in high sidewall installations.

The *double-deflection* bar grille has a second set of adjustable vanes installed behind and at a right angle to the face vanes. This grille controls the airstream in both the horizontal and vertical planes. This type of diffuser is a most versatile and functionally satisfactory combination, particularly if cooling is used. The vertical front bars allow throw control by obtaining variable

TABLE 13-8

Suggested maximum terminal velocities

$V_T = 100$ fpm $V_R = 35$ fpm	Recommended where people are located adjacent to walls of structures for extended periods of time at sedentary occupations; private offices, apartments, residences, hotel bedrooms, hospital (private rooms, and wards).
$V_T = 150$ fpm $V_R = 50$ fpm	Recommended where people are not located adjacent to walls of structures for extended periods of time; general offices, restaurants, theaters, department stores, clothing stores, churches, operating rooms.
$V_T = 200$ fpm $V_R = 65$ fpm	Recommended where people are not located adjacent to walls of structures at any time; industrial plants, corridors, process areas.

FIGURE 13-24 Series 90V ML aluminum air-foil register with multilouver damper. (Courtesy of Lima Register Company, Lima, OH)

side spread of the airstream. The rear blades, called *deflectors,* permit the airstream to be arched to control the drop of the airstream caused by temperature differential in a cooling application (see Figure 13-25b). For high sidewall installations, this also increases the ceiling effect. The assembly may be combined with a vertical opposed-blade damper to form a register.

Double-deflection grilles with horizontal front bars (see Figure 13-25a) may be more pleasing architecturally; however, they are less effective than vertical front bars because maximum side spread cannot be secured from the rear (deflector) bars. At maximum deflection of the rear bars, the airstream will strike the sides of the grille frame, thus reducing the control of the

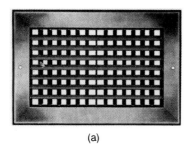

(a)

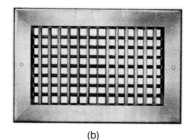

(b)

FIGURE 13-25 Double-deflection airfoil grilles: (a) Series 90 HV and (b) Series 90 VH. (Courtesy of Lima Register Company, Lima, OH)

throw. The application of this style of grille is therefore limited to medium- and straight-deflection settings. These grilles may be equipped with a horizontal opposed-blade damper to form a register.

Fixed-Bar Grille. Fixed-bar grilles are similar to single-deflection grilles, except the vanes are not adjustable. The vanes are set at the factory and may be straight or fixed at an angle. The angle at which the air is discharged from the grille face depends on the type of deflection vanes.

One type of fixed-bar grille is shown in Figure 13-26. Linear diffusers like this one, manufactured by Lima Register Company, are designed specifically for floor and sill installations. First, the F 93 unit is pencil- and heel-proof. Second, the mullions, which are spaced at 9-in. intervals, are anchored to the border and backed by reinforcing crossbars that provide additional support for the core (core not removable). Third, the Series F diffusers have duct spring clips that seat the diffuser snugly into position in the floor or sill. Maximum sectional length of this series is 96 in. Lengths are available in 1-in. increments. For continuous lengths, sections are customized to provide for butting. Alignment of continuous lengths is accomplished by means of an alignment key. Blank-off strips are available for inactive sections.

The Series C linear diffuser is similar to the F series, but it is designed for installation in a ceiling or sidewall.

Another type of fixed-bar grille recommended for use as a return-air grille is the Series 100 HF fixed-deflection grille manufactured by Lima Register Company and shown in Figure 13-27. The fixed louvers on these units are factory set for 0° or 35° deflection. The streamlined airfoil louvers increase free area and reduce airflow resistance. Sometimes, a damper may be required.

Another type of fixed-vane diffuser is the Series 40 floor unit manufactured by Lima Register Company. This type of diffuser is popular for both heating and cooling applications. The unit is similar to the Series 41 (stamped-face grille), except all parts of the Series 40 unit are assembled and then every joint is welded; all vanes are resistance welded to withstand traffic and abuse. The unit has a stop-screw adjustment that permits the damper to be closed and reopened to a balanced position. Provision is made for a locking screw concealed beneath the floor for tamper-proof valve settings, which is particularly important in public buildings.

Stamped Grilles. Stamped grilles are stamped from a single sheet of metal to form a fretwork through which air can pass. Various designs are used, varying

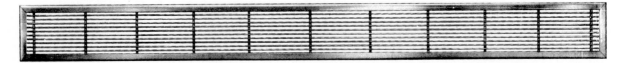

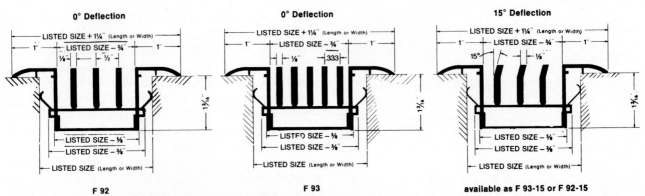

FIGURE 13-26   Series F linear diffuser for floor and sill installations. (Courtesy of Lima Register Company, Lima, OH)

from square or rectangular holes to intricate ornamental designs.

Figure 13-28 shows the Series 41 shallow-valve floor diffuser manufactured by Lima Register Company. This unit combines a stamped-face grille and a damper to form a very economical floor diffuser. The 4-in.-wide unit features an opposed-blade damper.

Variable-Area Grilles.   Variable-area grilles are similar to the adjustable, double-deflection grilles but can vary the discharge area to achieve an air volume change (variable volume outlet) at constant pressure so that the variation in throw is minimized for a given change in supply-air volume.

13-13   PARAMETERS AFFECTING GRILLE PERFORMANCE

The application and selection of a grille or register involves job condition requirements, selection judgment, and performance data analysis. Grilles and registers

FIGURE 13-27   Series 100 HF fixed-deflection return-air grille. (Courtesy of Lima Register Company, Lima, OH)

FIGURE 13-28   Series 41 floor diffuser. (Courtesy of Lima Register Company, Lima, OH)

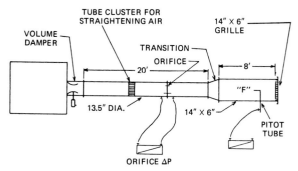

FIGURE 13-29 Rig for testing effective area of grilles.

should be selected and sized for (1) type and style; (2) function; (3) air volume requirement; (4) throw, spread, and drop requirement; (5) sound requirement; and (6) pressure requirement. We have previously discussed some of the various types and styles of grilles and regis-

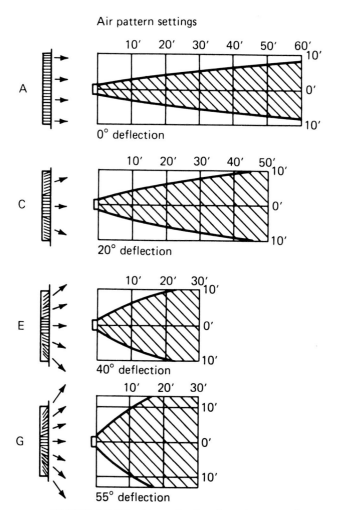

FIGURE 13-30 Spread related to throw and deflection setting: Required settings are made after installation using a key that gauges the bars as they are deflected.

ters and how they function. In the following sections, we shall discuss the other items.

To predict the performance of grilles and registers, the effective area must be determined. In principle, the procedure for determining effective area for a grille is the same as that for ceiling diffusers except that the test installation is made with square or rectangular duct-work instead of round (see Figure 13-29). A difference exists between ceiling diffusers and grilles in that ceiling diffusers usually have a fixed effective area (except those with adjustable vanes), whereas adjustable-bar grilles theoretically have an infinite number of effective areas. Anytime a grille bar (vane) is deflected one degree, the effective area is minutely changed. Practically, such a small change does not affect performance. In order to rate a grille properly, however, it is necessary to test at several deflection settings. Four to six such tests will establish the effective area limits of any grille line.

Each manufacturer of adjustable-bar grilles has its own established angular settings. For example, Lima Register Company gives ratings at 0°, 22°, and 45° deflection, whereas Tuttle & Bailey normally establish their ratings on 0°, 20°, 40°, and 55° deflection, identified by letters *A, C, E,* and *G,* respectively, (in Figure 13-30).

ILLUSTRATIVE PROBLEM 13-4

As an example of a laboratory procedure for determining the effective area of a grille, assume that it is desired to find the effective area of a 14-by-6-in. adjustable-bar grille with the bars set at straight flow (0° deflection). Use a test rig such as shown in Figure 13-29. Given: cfm through orifice = 500 cfm, static pressure at F = 0.0359 in. WG, duct size at F = 14 in. by 6 in.

Solution:

$$\text{duct area at } F = (14)(6)/144 = 0.583 \text{ ft}^2$$

$$\text{velocity at } F = 500/0.583 = 858 \text{ fpm}$$

$$\text{velocity pressure at } F = (858/4005)^2$$

$$= 0.0459 \text{ in. WG}$$

$$\text{static pressure at } F = 0.0359 \text{ in. WG}$$

$$\text{total pressure at } F = 0.0359 + 0.0459$$
$$= 0.0818 \text{ in. WG}$$

velocity at vena contractor between grille bars
$$= 4005 \sqrt{0.0818} = 1145 \text{ fpm}$$

$$\text{effective area } (A_e) = 500/1145 = 0.437 \text{ ft}^2$$

The core area (A_c) may be found by consulting manufacturers' catalogs for the particular grille. The core area is usually the listed length minus a unit of measurement (0.5 to 1.0 in.) multiplied by the listed width minus the same unit of measurement. For example, the 14-by-6-in. grille just discussed may have a core area of

$$A_c = \frac{(14 - 0.5)(6 - 0.5)}{144} = 0.516 \text{ ft}^2$$

The effective area by test was 0.437 ft². Therefore,

$$\frac{A_e}{A_c} = \frac{0.437}{0.516} = 0.85$$

This means that the effective area of the grille, when the bars are set for straight flow, is 85% of the core area. The relationship between core area and effective area can be found for any deflection setting by a similar computation. In the grille lines of most manufacturers, the relative proportion of effective area to core area is usually maintained for all sizes.

So far, the discussion of effective area of grilles and registers has centered around straight-flow or "no-deflection" vane setting. It can readily be seen that, as the vanes are deflected, the effective area of any grille or register is reduced, depending on the extent of the deflection. For every deflection setting, it follows that there is a definite effective area factor (A_{ef}).

For the deflection settings shown in Figure 13-30, Table 13-9 gives the effective area factor. Therefore, the discharge velocity from the grille for a given cfm passing through it may be obtained from

$$V_K = \frac{\text{cfm}}{A_c \times A_{ef}} \qquad (13\text{-}17)$$

It should be noted that the effective area factors shown in Table 13-9 are based on single-deflection grilles (also, Tuttle & Bailey). The use of a double-deflection core reduces the given effective area factors

by about 5%. The use of an opposed-blade damper in the fully open position with a single-deflection grille reduces the effective area by about 4%. Usually, these reductions are not significant unless the installation is a very critical one. Most manufacturers' rating tables do not differentiate between the various combinations of component register parts, but it may be calculated if necessary.

Air Volume Requirement. The air volume requirement is usually accepted as the prime requisite following the grille type and style selection. The air volume per grille is that which is necessary for design heating, cooling, or ventilation requirements of the specific area served by the grille. The air volume required, when related to throw, sound, or pressure design limitations, determines the proper grille size. Single-deflection grilles and registers perform efficiently with temperature differentials of up to 18–20°F cooling, 20–50°F heating, and for ventilating while handling 0.75 to 1.75 cfm per square foot of room area. Double-deflection supply grilles and registers perform efficiently with temperature differentials of 20–22°F cooling, 20–50°F heating, and for ventilating while handling 1.0 to 2.0 cfm per square foot with draftless distribution.

Throw Requirement. As with ceiling diffusers, an empirical formula substantiated by tests [see Eq. (13-18)] may be used to determine the distance a grille or register will throw. Values of the K factor are shown in Table 13-10.

$$T = \frac{K \times \text{cfm}}{\sqrt{A_c}} \qquad (13\text{-}18)$$

where

T = throw in feet
K = constant developed from experiment
cfm = volume air flow from device
A_c = core area in square inches

TABLE 13-9

Effective area factors for single-deflection grilles (refer to Figure 13-30)

Angle of Deflection		Effective Area Factor, A_{ef}
A	Straight 0°	0.850
C	20°–0°–20°	0.750
E	40°–20°–0°–20°–40°	0.675
G	55°–40°–20°–0°–20°–40°–55°	0.630

Source: Used by permission. Tuttle & Bailey, Holland, MI.

TABLE 13-10

K values for adjustable-bar grilles and registers

Angle of Deflection	K Values
Straight	0.80
20°–0°–20°	0.68
40°–20°–0°–20°–40°	0.55
55°–40°–20°–0°–20°–40°–55°	0.43

Source: Used by permission. Tuttle & Bailey, Holland, MI.

ILLUSTRATIVE PROBLEM 13-5

Determine the throw for a 24-by-12 in. adjustable-bar grille (core area = 23.5 in. by 11.5 in.) discharging 400 cfm with maximum deflection.

Solution: By Eq. (13-18),

$$T = \frac{K \times \text{cfm}}{\sqrt{A_c}} = \frac{(0.43)(400)}{\sqrt{(23.5)(11.5)}}$$

$$= 10.5 \text{ ft}$$

If the same grille is used to discharge the same cfm with straight-bar deflection,

$$T = \frac{(0.80)(400)}{\sqrt{(23.5)(11.5)}} = 19.5 \text{ ft}$$

This illustrates the fact that a grille of a given size handling a given cfm can be made to satisfactorily have a throw from 10.5 to 19.5 ft by simply adjusting the face bars. This explains why the adjustable-bar grille has become so popular. By manipulation of Eq. (13-18), the required core area can be found when the cfm, deflection required, and throw is given. Also, the deflection setting may be found when cfm, throw, and grille size are known by solving for the value of *K*.

Throw and air motion in the occupied area are closely related, and both should be considered in the analysis of specific area requirements. The rated throw (*T*) for a given condition of operation is based on a stated terminal velocity (V_T) at that distance (*T*) from the grille. The residual room velocity (V_R) is a function of throw, terminal velocity, and the deflection setting for the outlet discharge pattern. The rated V_R is obtained with an unobstructed air path along the throw dimension. Ceiling protrusions, such as soffits, beams, surface-mounted luminairs, and so forth, may prematurely divert the stream into the occupied area.

Room air motion (V_R) is rated in the occupied zone of the room and is based on measured residual velocities at a given condition of supply air volume, terminal velocity (V_T), and temperature difference (ΔT). A definite relationship exists between terminal velocity and occupied zone room velocity. An increase in V_T at the rated throw (*T*) dimension will result in an increased V_R. Figure 13-31 illustrates throw and room air motion.

Terminal velocity (V_T) is a value in fpm that exists at the end of the throw (*T*) from the grille. When the V_T occurs at the juncture of a ceiling and wall or at another specified location or throw dimension, its value determines the resultant V_R. Room air motion values of sidewall grilles (A50–60, see Figure 13-32) in performance

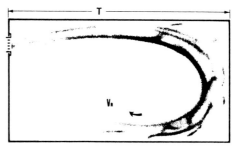

FIGURE 13-31   Relationship of throw (*T*) and room air velocity (V_R).

data ratings range from 35 to 40 fpm at corresponding terminal velocities of 75 to 100 fpm.

The specific values of V_T and V_R of each grille type are stated on the respective engineering performance data page (see Table 13-11). Performance data are based on an 18°F and 25°F cooled air-supply temperature differential, which, for the rated condition, results in a maximum ± 1°F variation in the occupied zone.

Throw values in feet for terminal velocities (V_T) other than the base of $V_T = 75$ fpm performance table rating can be obtained from Table 13-12.

Sidewall supply grilles have horizontal throw (*T*) ratings based on a grille mounting height of 8 to 10 ft. For a given listed throw, the room air motion (V_R) will increase or decrease inversely with the mounting height; or for given conditions of size, capacity, deflection, and V_R, the listed throw can be decreased 1 ft for each 1-ft increase in mounting height above 10 ft.

The rated throw value is based on the full distance the airstream is to travel to a wall or opposing airstream. When grilles are located on opposite walls discharging toward each other, they should be sized for equal capacities and to throw one-half the distance between them. Selecting for throws up to one-third greater than these dimensions will provide slightly increased V_R values up to 50 fpm for noncritical occupied areas.

Sill supply grilles find use in perimeter application involving moderate-to-large glass areas. The standard single- or double-deflection grille will perform satisfactorily in these applications. Grilles should generally be selected for outlet velocities (V_K) in the range of 700 to 1200 fpm and for throws (*T*) equal to the vertical distance to the ceiling plus the required horizontal projection (normally 10 to 15 ft). The rated throw value can be decreased 1 ft for each 1-ft increase in ceiling height above 10 ft.

Ceiling supply grilles have horizontal throw (*T*) ratings based on a ceiling height of 8 to 10 ft. For a given listed throw, the room air motion (V_R) will increase or decrease inversely with the ceiling height; or for given conditions of size, capacity, pattern, and V_R,

SERIES A60

**VERTICAL
FACE BARS**

Single
Deflection

Double
Deflection

SERIES A50

**HORIZONTAL
FACE BARS**

Single
Deflection

Double
Deflection
(Shown with
45 Frame)

FIGURE 13-32 Series A60 and A50 single-deflection grilles and Series A64 and A54 double-deflection grilles (aluminum). (Used by permission. Tuttle & Bailey, Holland, MI)

the listed throw can be decreased 1 ft for each 1-ft increase in mounting height above 10 ft.

The minimum throw results in room air motion higher than that for the maximum throw, generally 65 fpm versus 35 fpm. The minimum throw is one-half the recommended minimum distance between ceiling grilles discharging toward each other, and the maximum throw is one-half the maximum recommended distance between grilles. The minimum–maximum throw is also the recommended distance to a wall or other air path obstruction.

Spread Requirement. The spread of an unrestricted airstream is determined by the grille bar deflection and is expressed in feet at a given throw dimension (T) from the face of the grille. Air discharged from a grille with vertical face bars at zero-degree deflection will spread about 7 degrees per side in both horizontal and vertical planes or about 1.0 ft in every 8.0 ft of throw due only to the natural expansion of the free airstream.

As the deflection setting of the vertical bars is increased, the airstream covers a wider area and the throw (T) decreases. The shorter throw results from the greater stream surface that entrains more room air, dissipating stream energy and equalizing temperatures more rapidly. For the same size, capacity, V_T, and V_R, a 55-degree deflection setting will throw about one-half the distance of a zero-degree deflection setting. The actual spread for several standard deflections is shown in Figure 13-30.

The distance between the centers of individual grilles mounted in the same wall (see Figure 13-33) should not be less than the indicated spread of a single grille at a design throw (T) dimension. The distance from the center of an individual grille to an adjoining sidewall should not be less than one-half the indicated spread at the required throw.

The vertical-face bar grille is generally recommended for applications where spread deflections are required. When vertical rear bars are used, a 40-degree deflection setting is the maximum available because of the confining action of the side margins. When very narrow grilles are used, this may be reduced to 20 degrees maximum.

Airstream adjustment in the vertical plane (trajectory) can be obtained with horizontal face or rear bars to avoid direct impingement with beams, pendant lighting fixtures, or other obstructions.

Drop Requirement. If the air supplied from a sidewall grille is cooled below room temperature, the airstream will tend to drop into the occupied zone due to the density differential between primary and second-

TABLE 13-11

Engineering performance data for supply-air grilles and registers, single- and double-deflection, adjustable-bar type (Series T60 and T50—steel, Series A60 and A50—aluminum)

LISTED HEIGHT (4)	(5)	(6)	(8)	(10)	(12)	(14)	(16)	(18)	(20)	(24)	Outlet Velocity 450-600 FPM CFM	0° (Vk 450, Pт .01)	20° (Vk 500, Pт .01)	40° (Vk 550, Pт .02)	55° (Vk 600, Pт .02)	Outlet Velocity 550-700 FPM CFM	0° (Vk 550, Pт .02)	20° (Vk 600, Pт .02)	40° (Vk 650, Pт .03)	55° (Vk 700, Pт .03)	Outlet Velocity 600-800 FPM CFM	0° (Vk 600, Pт .02)	20° (Vk 700, Pт .03)	40° (Vk 750, Pт .04)	55° (Vk 800, Pт .04)	Outlet Velocity 700-1000 FPM CFM	0° (Vk 700, Pт .03)	20° (Vk 800, Pт .04)	40° (Vk 900, Pт .05)	55° (Vk 1000, Pт .06)
6	5										40	6	5	4	3	45	8	7	6	4	60	10	9	7	5	65	11	10	8	6
8	6										50	7	6	5	4	60	9	11	6	5	70	11	10	8	6	80	12	11	9	7
10	8	6									75	9	7	6	5	90	10	9	7	6	100	12	11	9	7	125	13	12	10	8
12	10	8									100	10	8	6	5	125	11	9	8	7	150	13	12	9	8	175	14	13	11	8
14	12	10	8								125	11	9	7	6	150	12	11	9	7	175	14	13	10	8	200	16	14	11	9
18	14	12									150	11	10	8	7	175	13	11	10	8	200	15	14	11	9	250	17	15	12	10
20	16	14	10								175	12	11	9	7	200	15	13	11	9	250	17	15	12	9	275	19	17	14	11
24	20	16	12	10							225	13	12	10	8	275	16	14	12	10	300	19	16	13	10	350	21	18	15	12
30	24	20	14	12							300	14	13	11	9	350	18	15	13	10	425	20	17	15	11	475	24	20	17	13
38	30	24	18	14	12						350	17	14	12	9	425	20	17	13	11	500	22	20	17	12	550	26	22	18	14
40	32	26	20	16	14						400	18	16	13	10	475	21	18	15	12	550	24	21	18	13	650	28	24	20	15
44	36	30	24	18	16	14					450	19	17	13	11	550	22	19	17	13	625	26	22	19	14	725	30	25	22	17
	44	36	26	22	18	16					500	20	17	14	12	600	24	20	18	13	700	27	24	20	15	800	32	27	23	18
	48	40	30	24	20	18	16				600	21	18	15	12	700	25	22	18	14	850	29	25	22	16	950	34	29	25	19
		48	38	30	24	22	18				700	23	20	17	13	850	28	24	20	15	975	33	28	24	18	1125	37	31	27	21
			40	32	28	24	20	18			800	25	21	18	14	950	31	26	22	17	1125	35	30	26	20	1275	39	34	29	22
			42	36	30	26	24	22	20		900	27	22	19	15	1100	32	27	23	18	1250	36	31	27	21	1450	42	36	31	23
			46	42	36	30	26	24	22		1000	29	25	21	16	1200	34	29	25	19	1400	40	34	29	22	1600	45	39	34	25
					44	38	34	30	28	24	1250	32	27	23	18	1500	36	30	27	23	1750	43	37	31	24	2000	48	43	36	27
						48	42	36	34	28	1500	34	29	25	19	1800	39	32	29	25	2100	45	40	34	26	2400	52	46	39	30
							44	40	36	30	1750	37	32	27	21	2100	42	35	32	27	2450	48	44	37	28	2800	55	49	42	32
								48	42	36	2000	39	34	29	22	2400	45	38	34	29	2800	52	46	40	30	3200	58	52	44	34
									48	40	2250	41	36	30	24	2700	48	40	36	31	3150	54	48	42	32	3600	61	54	46	37
										42	2500	43	38	32	25	3000	51	42	38	33	3500	57	51	44	33	4000	64	57	49	39
										48	3000	46	40	34	27	3600	54	45	40	35	4200	60	53	47	35	4800	68	60	52	42

Throw (T) VT 75

≡NC30 □   NC35 ▨   NC40 ▦   ≡NC45 ■

SYMBOLS
VT Terminal Velocity in FPM
VR Room Velocity in FPM
Vk Outlet Velocity in FPM

Ak Outlet Area in Sq. Ft.
Pт Total Pressure in. H₂O
Ps Static Pressure in. H₂O

NC re 18db room attenuation
T Throw in Feet

Source: Used by permission. Tuttle & Bailey, Holland, MI.

ary air. When heated air is supplied at a temperature above that of the room air, the drop due to temperature is a negative one because the primary warm airstream will rise, again due to density differences. Therefore, both of these conditions must be evaluated to correctly design a grille installation (see Figure 13-12).

If there is no significant temperature differential between the primary and secondary air (ventilating), the airstream still exhibits a minor drop due to the vertical expansion of the stream as it entrains room air. This drop is approximately 1.0 ft for every 8.0 ft of throw.

The drop of an airstream is normally computed

TABLE 13-11

Data (continued)

Header key for each Outlet Velocity group — Deflection columns 0°, 20°, 40°, 55°:

- Outlet Velocity 900–1200 FPM: V_K = 900 / 1000 / 1100 / 1200; P_T = .05 / .06 / .08 / .09; Throw (T), V_T 75
- Outlet Velocity 1100–1450 FPM: V_K = 1100 / 1200 / 1300 / 1450; P_T = .08 / .09 / .10 / .13; Throw (T), V_T 75
- Outlet Velocity 1250–1700 FPM: V_K = 1250 / 1400 / 1550 / 1700; P_T = .09 / .12 / .15 / .18; Throw (T), V_T 75
- Outlet Velocity 1450–1900 FPM: V_K = 1450 / 1600 / 1750 / 1900; P_T = .13 / .16 / .19 / .23; Throw (T), V_T 75

LISTED HEIGHT / LISTED WIDTH

4	5	6	8	10	12	14	16	18	20	24	CFM	0°	20°	40°	55°	CFM	0°	20°	40°	55°	CFM	0°	20°	40°	55°	CFM	0°	20°	40°	55°
6	5										80	13	11	9	7	100	15	13	11	8	125	19	16	13	10	150	22	19	15	12
8	6										100	15	13	11	8	125	18	15	13	10	150	21	18	14	11	175	25	21	17	13
10	8	6									150	17	14	12	9	175	20	17	14	11	200	24	20	17	13	225	27	23	19	15
12	10	8									200	18	15	13	10	250	22	19	15	12	275	25	21	18	14	325	29	25	21	17
14	12	10	8								250	21	18	15	11	300	25	21	18	14	350	28	25	20	15	400	33	28	24	18
18	14	12									300	24	20	16	13	350	27	24	20	15	425	31	27	22	17	475	36	32	26	20
20	16	14	10								350	25	21	18	14	425	30	25	21	17	500	34	30	25	19	550	39	32	27	21
24	20	16	12	10							450	27	23	19	15	550	32	28	24	18	625	37	32	27	22	725	42	36	31	24
30	24	20	14	12							600	29	25	21	16	725	35	30	26	20	850	41	35	30	23	950	45	40	34	26
38	30	24	18	14	12						700	32	28	24	18	850	39	33	28	21	975	45	39	33	25	1125	49	43	38	28
40	32	26	20	16	14						800	34	30	25	19	950	41	35	30	22	1125	48	41	35	27	1275	52	46	40	31
44	36	30	24	18	16	14					900	36	31	26	20	1075	43	37	32	24	1250	51	43	37	28	1450	55	49	42	32
	44	36	26	22	18	16					1000	39	33	28	22	1200	45	40	34	26	1400	53	45	39	30	1600	58	51	44	34
	48	40	30	24	20	18	16				1200	42	36	31	24	1450	48	42	36	28	1675	57	48	42	32	1925	61	54	46	36
		48	38	30	24	22	18				1400	45	39	34	26	1675	52	45	40	31	1950	60	51	45	35	2250		58	49	40
			40	32	28	24	20	18			1600	47	43	37	28	1925	56	48	43	34	2250		54	47	39	2500		62	53	43
			42	36	30	26	24	22	20		1800	49	44	40	31	2150	59	51	46	35	2500		57	50	41	2750			56	46
			46	42	36	30	26	24	22		2000	52	47	42	32	2400		54	49	38	2800		61	53	43	3200			59	48
					44	38	34	30	28	24	2500	56	50	45	34	3000		59	53	41	3500			58	47	4000				54
			48	42	36	34	28				3000	60	54	48	36	3500			58	44	4200				52	4800				59
				44	40	36	30				3500		59	53	40	4200				49	4900				56					
					48	42	36				4000			57	43	4800				54	5600				60					
						48	40				4500			60	45	5400				58										
							42				5000				49															
							48				6000				54															

Deflecting Key — 20°, 40°, 55°

The indicated air paths may be easily obtained by using the appropriate notch on the bar deflecting key. The bars should be adjusted as shown for the required pattern.

only for low induction devices such as grilles, registers, rectangular diffusers, and slots. The circular air pattern with which most ceiling outlets discharge air provides a high induction rate between primary and secondary air. This high rate of induction eliminates the temperature differential before the airstream can fall into the occupied zone.

The drop varies directly with the temperature differential (ΔT) and the throw (T) and inversely with outlet velocity (V_K). It can be calculated by an empirical formula substantiated by the tests of many investigators.

For drop due to temperature differentials, we have

$$D_t = \frac{(N)(t_r - t_a)(T)^{1.2}}{V_K} \quad (13\text{-}19)$$

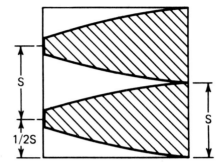

FIGURE 13-33 Location of sidewall outlets considering spread (s) and throw (T). (Plan view.)

ILLUSTRATIVE PROBLEM 13-6

Assume that we wish to estimate the total drop for a 36-by-8 in. grille with straight-vane deflection. The volume flow rate is 1000 cfm, cooling differential $t_r - t_a = 15$ degrees, and throw is 49 ft.

Solution:

$$V_K \text{ (jet velocity)} = \frac{\text{cfm}}{A_e}$$

Thus,

$$A_e = (\text{core area})(\text{effective area factor})$$

$$= \frac{(35.5)(7.5)(0.85)}{144} = 1.57 \text{ ft}^2$$

$$V_K = \frac{1000}{1.57} = 637 \text{ fpm}$$

By Eq. (13-19),

$$D_t = \frac{5(15)(49)^{1.2}}{637} = 12.6 \text{ ft}$$

By Eq. (13-20),

$$D_s = \tan 7° \times 49 = 6.0 \text{ ft}$$

By Eq. (13-21),

$$D = 12.6 + 6.0 = 18.6 \text{ ft} \quad \text{(for cooling)}$$

where

> D_t = drop in feet due to temperature differential
> N = constant numerically equal to 5
> t_r = temperature of room air (°F)
> t_a = temperature of supply air (°F)
> T = throw in feet
> V_K = grille jet velocity measured in vena contractor

For drop due to spread angle, we have

$$D_s = \text{tangent of spread angle times throw}$$

or

$$D_s = \tan 7° \times T \qquad (13\text{-}20)$$

where

> D_s = drop in feet due to spread angle
> T = throw in feet

The total drop is additive; therefore,

$$\text{total drop } D = D_t + D_s \quad \text{(for cooling)} \qquad (13\text{-}21)$$

$$\text{total drop } D = D_s - D_t \quad \text{(for heating)} \qquad (13\text{-}22)$$

The fact that air supplied at the top of the room will fall by density differential toward the occupied zone of the room is advantageous in aiding general room air distribution in one respect, but extremely disadvantageous if the airstream falls into the occupied zone.

Table 13-13 shows tabulated values of total drop for a cooled airstream (including the vertical expansion) for two typical conditions of cooling temperature differentials and a range of throws and outlet velocities. The table values correspond reasonably well with calculation from Eqs. (13-19) through (13-21).

Figure 13-34a illustrates the calculated drop at certain conditions. It is obvious that unsatisfactory conditions will result in the occupied zone (below the 6-ft level). The minimum separation between mounting height and ceiling height must equal 60% of the total air drop (minimum separation of 3 ft). The minimum mounting height should be 7 ft. The recommended mounting height is equal to the ceiling height minus the minimum separation distance.

There are two ways of locating the sidewall grille to reduce an excessive drop, which will create an objec-

TABLE 13-12

Throw factors* ($V_T = 75$ fpm base)—multiply throw (T) by factor

	$V_T 50$	$V_T 75$	$V_T 100$	$V_T 150$
Factor	1.5	1.0	0.75	0.50

Source: Used by permission. Tuttle & Bailey, Holland, MI.
*Applies to data of Table 13-11.

TABLE 13-13

Total air drop from sidewall outlet (ft)*

V_K in fpm	Sidewall Throw in Feet													
	10		15		20		25		30		40		50	
	−18F	−25F	−18F	−25F	−18F	−25F	−18F	−25F	−18F	−25F	−18F	−25F	−18F	−25F
500	3.5	4.0	5.5	6.0	7.5	8.5	9.0	10.0	10.5	13.5	15.5	18.0	18.5	23.0
750	2.5	3.5	4.0	5.5	6.0	6.5	7.0	8.0	8.5	10.5	11.5	14.5	15.0	18.5
1000	2.0	3.0	3.5	4.0	5.0	5.5	6.0	6.5	7.0	8.5	10.0	12.0	12.5	16.0
1250	2.0	2.5	3.0	3.5	4.5	5.0	5.5	6.0	6.5	7.5	9.0	11.0	11.5	13.5
1500	1.5	2.0	3.0	3.5	4.0	4.5	5.0	5.5	6.0	7.0	8.5	9.5	10.5	12.5
1750	1.0	2.0	2.5	3.0	3.5	4.0	4.5	5.0	5.5	6.5	8.0	9.0	10.0	11.5
2000	1.0	1.5	2.5	3.0	3.5	4.0	4.0	4.5	5.0	6.0	7.5	8.5	9.5	10.5

*Total drop = drop due to temperature differential plus airstream expansion.

Source: Cat. RG-1. See catalog for complete data. Courtesy of Tuttle & Bailey, Holland, MI.

tionable draft in the occupied zone. One way is to locate a single-deflection grille high enough on the sidewall so that most of the velocity is lost, and the airstream, utilizing the ceiling effect, reaches the opposite wall before it falls into the occupied zone (see Figure 13-34b). The more flexible way is to locate a double-deflection grille just above the 6-ft level and deflect the air upward 15 to 20 degrees toward the ceiling (see Figure 13-34c). This is a somewhat longer induction travel and causes the air to be brought up to room temperature on the way up to the ceiling and also on the way down again. In effect, there is twice as much height above the occupied zone as actually exists. This allows an amount of drop equal to twice the distance between the ceiling and the grille to be absorbed plus the distance from the grille to the 6-ft level of the occupied zone.

When the air is arched in this manner, there is no appreciable effect on the throw of the airstream. Generally, an upward setting of 15 to 20 degrees of the deflector bars will suffice. Table 13-14 shows values of drop correction to be applied to the values of drop from Table 13-13.

Sound Requirement. Acoustical control is as important as temperature and humidity control. To

cope with noise problems, it is necessary to have a basic understanding of sound, for noise is simply unwanted sound.

Sound may be defined in two ways, physically and physiologically. Physical sound is a particular form of

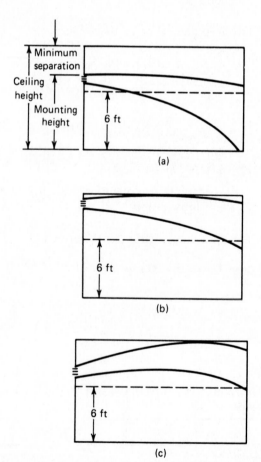

FIGURE 13-34 Air patterns for sidewall grilles showing effect of airstream drop.

TABLE 13-14

Drop correction with deflected* grille bars

	Throw in Feet						
	10	15	20	25	30	40	50
Drop reduction in feet, apply to Table 13-13	2.5	3.5	4.5	6.0	7.0	9.0	11.5

Source: Cat. RG-1. See catalog for complete data. Courtesy of Tuttle & Bailey, Holland, MI.

*15 to 20 degrees upward deflection.

wave motion taking place in matter due to an original vibration or disturbance set up in the surrounding body. Physiologically, or psychologically, it is a sensory interpretation in the brain, conveyed from the ear, and differing to a great extent with each individual. It is in the attempt to rationalize physical sound with the psychological aspect that the complicated phases of sound study arise.

Sound is invariably produced by a compressive and expansive motion set up in a gaseous, liquid, or solid body. If the motion is sharp and discontinuous, such as a pistol shot, or if it is continuous, as when the fingernail is rubbed across a rough surface, the sound may be harsh and unmusical. On the other hand, if the vibrations are more or less defined and regularly spaced and have one or more definite frequencies, the sound presented may be musical in nature, such as the tones of a violin or a piano. In the language of the layman, the former represents noise to the human ear, whereas the latter represents a musical tone, although in the technical sense both represent sound.

Waveforms where the motion of particles in a gaseous medium are parallel to the direction in which the wave travels are considered to be longitudinal waves. Transverse waves, which occur perpendicular to longitudinal waves, cannot exist in a gaseous atmosphere, as the molecule structure is widely spaced and individual particles cannot transmit the force magnitude required to generate such motion. As the molecule structure vibrates with simple harmonic motion, the particles vibrate about their equilibrium point, creating a configuration or arrangement of particles moving in a given direction. A series of compression and rarefaction areas occur, which propagate this type of longitudinal or compressional wave motion. Sound is created by a vibrating body and is transferred through a medium having mass and elasticity where compressional waves can be generated. Sound cannot be distributed in a vacuum.

In discussing sound, one must consider magnitude and frequency, measurement of sound magnitude and frequency, and human reaction to sound of varying frequency and magnitude.

Sound, for practical purposes, is any vibration that is audible to the human ear. As noted, sound waves may be transmitted through solids, liquids, or gases. Any pure sound is completely described by two terms, frequency and magnitude. The frequency of a sound corresponds to the pitch of the sound. The frequency of a sound is the same as the frequency of the vibrating source. The magnitude of a sound relates to the energy contained in the sound wave per unit area perpendicular to the direction of travel.

The human ear responds to sound due to the variation of pressure from the normal atmospheric pressure. This regular pattern of increasing and decreasing air pressure is the result of energy expended to the air. The amount of energy contained in a sound is the factor used in measuring the magnitude of the sound. The magnitude of a sound is expressed as the ratio of a measured sound level above an energy reference level.

The unit of sound measurement is the dimensionless *decibel* (db). The decibel expresses the ratio of magnitude of measured sound to magnitude of reference-level sound. The sound level in decibels is a logarithmic function of that ratio:

$$\text{level in db} = 10 \log_{10} \left(\frac{\text{measured quantity}}{\text{reference value}} \right) \quad (13\text{-}23)$$

In sound measurement, the decibel is used to narrow the range of values of measured sound for ease of mathematical handling. The range of audible sound in directly measured units varies from 0.000 000 001 W for a whisper to 100,000 W for a jet plane. The corresponding units of decibels varies from 40 db for a whisper to 180 db for a jet plane.

In sound-level measurement, there are two types of sound levels used. These types are (1) sound pressure level, L_p, and (2) sound power level, L_w. These two sound levels use the same unit of measurement, the decibel, measuring sound magnitude, but they are two different quantities.

To understand the definitions of sound pressure and sound power, first assume a given source of sound. Assume this source is a fan located in a room. If the magnitude of the sound produced by this fan is measured throughout the room, this magnitude will obviously vary with distance from the sound source and the acoustical properties of the room. This sound magnitude, which is affected by acoustical environment and distance from the source, is designated as sound pressure level, L_p. Sound pressure level is the sound level a person hears from a source of sound at the given point of reference. This sound level is the value of the ratio of the measured sound level to the reference sound level. This value is expressed in decibels by the following relationship:

$$L_p = 10 \log_{10} \left(\frac{p}{0.0002} \right)^2$$

$$= 20 \log_{10} \left(\frac{p}{0.0002} \right) \quad (13\text{-}24)$$

where p is sound pressure in microbars. Energy reference for sound pressure level is 0.0002 microbar, which is the same as 0.0002 dyne per square centimeter. The sound pressure level is not used to rate the sound level produced by a given source because sound pressure varies with the distance from the source and the acoustical environment.

Sound power level, L_w, is used to rate the sound output of a sound source. The sound output or sound power level of a given sound source is a constant. Its sound pressure level at a given observation point may be varied by introducing varying acoustical arrangements. For example, for a given sound source, a bare hard-walled room will give a higher sound pressure level than a carpeted room with walls covered by drapes. However, the sound power level of this sound source is unchanged regardless of this change in room environment. A sound power level may be determined for any sound source such as a fan. Sound power levels must be measured under ideal acoustical conditions, and in this instance the measured sound pressure level equals the sound power level. This sound power rating may then be used to determine the resulting sound pressure level in a given acoustical environment and distance from the source.

When a manufacturer of air conditioning equipment rates equipment noise, this rating must be in sound power rather than sound pressure to have specific meaning. Sound power level, L_w, in decibels is defined as

$$L_w = 10 \log_{10} \left(\frac{W}{10^{-12}} \right) \qquad (13\text{-}25)$$

where W is the measured sound power level in watts and the reference energy level is 10^{-12} W.

In order to provide a comfortable acoustical atmosphere, the frequency as well as the magnitude of the sound must be given careful study. Sound frequency is directly involved in noise control for such reasons as (1) the human ear is more sensitive to high frequencies than low frequencies and (2) the design of material used to attenuate sound depends on the frequency of the sound.

The audible frequency range includes frequencies from 20 to 10,000 cycles per second. For ease of mathematical handling, this vast frequency range has been subdivided into eight separate frequency ranges called *octaves*. The frequency of the upper limit of each octave band is twice that of the lowest frequency in each octave band. Frequency measurement and application of frequency criteria to noise-level problems is done in octave bands. Since sounds of the same magnitude but of different frequencies will have different degrees of loudness to the human ear, a graph of sound level versus frequency is often used. This type of graph is called a *frequency spectrum*.

Sound measurement: The reception of sound by the human ear varies with each individual. This fact in-

creases the difficulty of establishing absolute values for sounds generated by air conditioning equipment in order to predict what the environmental noise level of a given system design will be.

Modern sound measurement is obtained by means of a sound-level meter used with a microphone-amplifier circuit. A sound-level meter reading is the actual sound pressure level in decibels when the frequency network of the meter is set for "flat response," which is the "C" network of the meter. A flat-response rating means that the meter reads the actual sound pressure level experienced by the meter and not a "loudness rating" simulating the reaction of the human ear, which emphasizes high-frequency sound.

When it is necessary to know the magnitude of sound at certain octaves, an octave band analyzer is used in conjunction with a sound-level meter. By using the octave band analyzer to electronically filter out all frequencies except those in a certain octave band, a sound pressure level of this certain octave band can be determined in decibels. This reading will be an octave band sound pressure level in decibels when the meter is set for flat response.

Sound-level meters may electronically deemphasize the low-frequency end of the frequency spectrum to approximate the "loudness-level" reaction of the human ear in decibels. This type of measurement is done on the A-scale of a standard meter and is therefore called the *A-scale noise level*.

The response of the human ear to a sound level can be approximated by direct measurement or by a combination of octave band sound pressure measurements and mathematical calculation of the loudness level. The A-scale noise level, just explained, is a direct-measurement estimate of the human ear's response. Two other methods of loudness-level determination of human ear response use the units of *sone* and *phon* for total loudness and loudness level, respectively. These methods use a combination of sound pressure level measurement and an equation to calculate the loudness response. In addition to these methods, a widely used method of representing loudness criteria is the *noise criteria curve* (NC curve), which defines the maximum allowable sound pressure level for each frequency band.

However, for sound-control problems where the noise level in certain octave bands is especially important, design based on octave band sound levels is necessary. Noise criteria (NC) curves provide this information. NC curves plot octave band frequencies versus sound level in decibels. NC curves are numbered and have increasing NC numbers as the allowable magnitude of the sound spectrum increases. Areas such as a concert hall have a low NC number; factories, a relatively high NC number.

Absorptive characteristics of rooms: When a sound wave strikes a solid barrier, a part of its energy is reflected, part is absorbed, and part is transmitted through the barrier. If the sound wave originates within the room, the portions absorbed and transmitted by the walls are not returned to the room so that the two may be taken together under the single heading of absorption. The fraction that is returned to the room is the absorption coefficient.

It is obvious that when a given sound pressure is released within a room, the construction of the room, the depth of its walls, the acoustic properties of the walls, ceiling, and floor, and the furnishings and occupants within the space all affect the ratio of the reflective fraction to the absorption fraction. Absorption of sound concerns the dissipation in the form of heat of the vibrational energy of sound waves. Materials that are absorbent to sound waves to any degree are either porous, or inelastically flexible or inelastically compressible, or may possess two or more of these properties in varying degrees. Therefore, in any given room, the sound waves returned to the ear of an occupant from an air distribution device operating at a given outlet velocity will be different than if that same device operating at the same velocity were placed in a room whose absorption coefficient was different.

An accurate determination of the sound level in any given room would involve the calculation of the sound-absorbing rate of that room. For a complete air conditioning installation, this would obviously be a long and tedious procedure. Inasmuch as the majority of air-conditioned spaces for human occupancy concern themselves with relatively similar building materials and con-

TABLE 13-15

Recommended maximum jet velocities for diffusers and grilles

Application	Recommended Maximum Jet Velocities (fpm)
Broadcast studios	500
Residences	750
Apartments	750
Churches	750
Hotel bedrooms	750
Legitimate theaters	1000
Private offices, acoustically treated	1000
Motion picture theaters	1250
Private offices, not treated	1250
General offices	1500
Stores, upper floors	1500
Stores, main floors	1500
Industrial buildings	2000

struction procedures, it is much simpler to avoid the complexity of noise problems by selecting the air distribution devices to operate at velocities that are known to be quiet or at NC levels that are acceptable. Practical experience indicates that noise from fan apparatus, ducts, and accessories such as dampers is more likely to cause sound problems than from outlets conservatively selected. Table 13-15 gives recommended outlet velocities for various applications and applies to both ceiling diffusers and grilles.

In summary, the intensity of sound created by a grille or diffuser for a given condition of operation is not generally the sound that is heard by an observer in a typical building environment. The sound created by a grille is the total sound generated, independent of environment and is termed *sound power level* (L_w). The sound actually heard in a given environment is the power-level value minus room attenuation and is termed *sound pressure level* (L_p). The amount of reduction from L_w to L_p is determined by the acoustic absorption of the environment. An 18-db room reduction is considered typical, and performance data ratings are values of sound pressure based on $L_w - 18$ db with the power level referenced to 10^{-13} W. A change in the 18-db room factor will inversely change the apparent NC rating by the same amount. Room factors commonly range from 15 to 22 db.

A given power level contains various intensity levels in prescribed frequencies ranging from 75 to 10,000 cps; this range is commonly divided into seven octave bands for convenience. The power-level intensity within frequency-limited octave bands is numerically reduced by 18 db and plotted on NC charts. The point on the plot tangent to the highest numerical NC curve designation is the NC rating for a specific condition of operation (see Figure 13-35).

Performance ratings of grilles in NC values are levels that will not be exceeded for a given condition of operation (not including system noise) in an environment having an 18-db absorption. Noise created by or transmitted through the duct system to the grille can add to the total noise being introduced or heard within the room, falsely indicating an increased grille sound level. Removal of the grille quickly establishes whether the offending noise is duct system noise, which will persist, or grille-generated sound, which will cease. The noise in the duct system at the grille collar should be at least 5 db lower than the grille NC rating.

Effect of combining two sound levels: The decibel scale is a logarithmic function and decibel values, of course, cannot be added directly. The values to be added must be reverted back to the actual intensities, added, and then converted back to the decibel values.

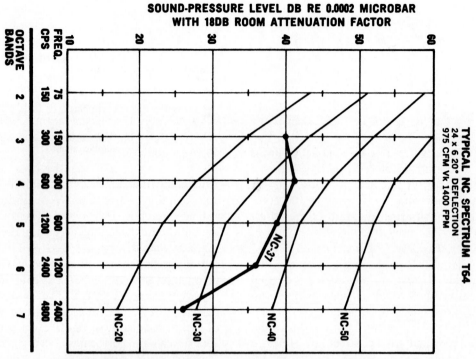

FIGURE 13-35 Typical NC Spectrum T64: 24 in. by 6 in.; 20-degree deflection; 975 cfm; $V_K = 1400$ fpm. (Used by permission. Tuttle & Bailey, Holland, MI)

For example, in a space with a background sound level of 42 db (perhaps a private office), the use of a diffuser (grille) selected at 40 db will produce a combined noise level of 44 db (see Table 13-16).

Grille sound ratings will be increased by integral or localized air volume dampering, by nonuniform airflow in the grille collar, and by addition of inherent system noise. Table 13-17 indicates register throttling sound correction.

The grille NC rating will also be increased if two or more outlets are located in an area of 200 ft² or less (see Table 13-18).

Grilles should be selected for the recommended NC rating for a specific application (see Table 13-19) or for recommended velocities (see Table 13-15) dependent on a particular manufacturer's method of rating. Grilles may also be selected for sound levels higher than low

ambient background levels to mask extraneous and occupant-disturbing noises. A grille with a low sound rating is not necessarily a properly selected grille for a good "sound-conditioned" application.

Pressure Requirement. The *grille minimum pressure* for a given air volume reflects itself in ultimate system fan horsepower requirements. A grille with a lower pressure rating requires less total energy than a grille with a higher pressure rating for a given air volume and effective area. Grilles of a given size having lower pressure ratings usually have a low sound-level rating at a specified air volume.

Velocity pressure (p_v), discussed in Section 13-4, exerts a force only in the direction of air motion. The only energy available for diverting the air at right angles to the direction of motion into the connecting duct and

TABLE 13-16

Effect of combining two sound levels

Difference between two sound levels, dB	0	2	5	8
dB addition to higher level to obtain combined level	3	2	1	0

TABLE 13-17

Register throttling sound correction

Damper Throttling Effect	Pressure Drop in WG		
	0.05	0.15	0.25
Approximate damper opening	¾	⅔	½
NC addition to single outlet sound rating	5	10	15

TABLE 13-18

Multiple-outlet sound correction

Number of outlets in 200 ft² area	2	3	4 to 6
NC addition to single-outlet sound to obtain total sound	3	5	6

through the grille is static pressure (p_s) in the main duct, which exerts a force equally in all directions. The static pressure in the main duct must be generally equal to or greater than the total pressure (p_t) required in the connecting duct ($p_t = p_v + p_s$).

The engineering performance data (typical of Table 13-11) provide both total pressure and static pressure required in the collar behind the grille to effect rated performance. The additional static pressure loss (turning loss) encountered in transferring flow from the duct into the grille collar is approximately 20% of the rated total pressure. This "turning loss" must be *added* to the *grille-only total pressure* (p_t) in order to establish the total static pressure required in the branch duct connected to the collar.

The *static pressure loss* encountered in the duct system (see Chapter 14) between the outlet closest to the fan and the outlet on the longest equivalent duct run is important. The last outlet on the duct run must have the minimum pressures as determined above. Of necessity then, the outlet closest to the fan will have available a pressure equal to the sum of (1) minimum total static pressure at last outlet and (2) duct friction loss between last grille and first grille.

13-14 GENERAL APPLICATION DATA (GRILLES AND REGISTERS)

As with ceiling diffusers, the selection and application of grilles and registers is a matter of judgment and analysis. The quantity and velocity of air movement within the space and the proper mixing of supply air with space air affect comfort levels. Supply air should be directed to the sources of greatest heat loss (or heat gain) to offset their effects. Registers and grilles for the supply air should accommodate all aspects of the supply-air distribution pattern such as throw, spread, and drop; also, the outlet grille velocities must be held within reasonable limits. Any noise generated at the grille is of equal or greater importance than duct noise.

A few rules should be borne in mind when applying grilles and registers:

1. The throw (T) from a straight-flow grille varies with the square root of the core area of the grille and with the face velocity.

TABLE 13-19

Recommended NC criteria

NC Curve	Communication Environment	Typical Occupancy
Below NC 25	Extremely quiet environment, suppressed speech is quite audible, suitable for acute pickup of all sounds.	Broadcasting studios, concert halls, music rooms.
NC 30	Very quiet office, suitable for large conferences; telephone use satisfactory.	Residences, theaters, libraries, executive offices, directors' rooms.
NC 35	Quiet office; satisfactory for conference at a 15 ft. table; normal voice 10–30 ft.; telephone use satisfactory.	Private offices, schools, hotel rooms, courtrooms, churches, hospital rooms.
NC 40	Satisfactory for conferences at a 6–8 ft. table; normal voice 6–12 ft.; telephone use satisfactory.	General offices, labs, dining rooms.
NC 45	Satisfactory for conferences at a 4–5 ft. table; normal voice 3–6 ft.; raised voice 6–12 ft.; telephone use occasionally difficult.	Retail stores, cafeterias, lobby areas, large drafting & engineering offices, reception areas.
Above NC 50	Unsatisfactory for conferences of more than two or three persons; normal voice 1–2 ft.; raised voice 3–6 ft.; telephone use slightly difficult.	IBM rooms, stenographic pools, print machine rooms, process areas.

Source: Used by permission. Tuttle & Bailey, Holland, MI.

2. The aspect ratio (the ratio of the height to the width) has no appreciable effect on the distance of the throw from grilles whose aspect ratio is less than 25:1.

3. The convergence of airstreams from a grille

results only in cutting down the effective area of the grille.

4. Breaking the airstream into jets has no effect on either the rate of air mixing or the throw.
5. Deflecting the airstream by turning the vanes outward to increase the spread shortens the throw, depending on the amount of deflection (see Figure 13-30).
6. The drop for a given throw varies inversely as the face velocity for an airstream below room temperature and varies directly as the temperature differential.

Properly selected grilles operate satisfactorily from high sidewall and perimeter locations in the sill, curb, or floor. Ceiling-mounted grilles, which discharge the airstream down, are generally not acceptable in comfort air conditioning installations in interior zones and may cause draft in perimeter applications.

Unlike ceiling diffusers, grilles have a comparatively low induction rate. Moreover, the airstream moving in just one direction induces room air into the primary airstream in such a manner that this room air must be replaced by noninduced room air. This causes a counterflow room air movement that has a velocity of about one-half the terminal velocity. Since ideal room air velocity is limited to 50 fpm or less, the terminal velocity should not exceed 100 fpm. Higher velocities may be used in industrial applications.

For *high sidewall* installations, the use of a double-deflection grille usually provides the most satisfactory solution. The vertical-face louvers of a well-designed grille deflect the air up to 50 degrees to either side (see Figure 13-30) and amply cover the conditioned space. The rear horizontal-deflector louvers deflect the air at least 15 degrees in the vertical plane, which is ample to control the elevation of the discharge pattern. Proper spacing of high sidewall grilles (see Figure 13-33) is important. If the airstreams from adjacent grilles strike too soon, undesirable turbulence in the occupied zone may occur.

Collars should be used to connect branch ducts and the high sidewall grilles and registers. A collar should not be less than 12 in. in length. A cushion head should be used at the end of a branch run to air distribution in the last grille.

Dampers should be used to provide the correct total pressure behind the grille so that the required volume is discharged. Air-turning devices with adjustable blades should be used to control the volume and correct imperfections in the condition of approach of air to the grille. Figure 13-36 shows an ideal register assembly for sidewall installation.

For *perimeter* installation, the grille selected must

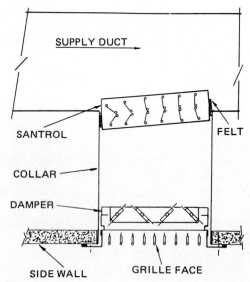

FIGURE 13-36   Ideal register assembly showing collar, air extractor (Santrol by Tuttle & Bailey), damper, and grille for sidewall installation. (Used by permission. Tuttle & Bailey, Holland, MI)

fit the specific job. When small grilles are used, adjustable-vane grilles improve the coverage of perimeter surfaces. When the perimeter surface can be covered with long grilles (see Figure 13-26), the fixed-vane grille is satisfactory. When grilles are located more than 8 in. from the perimeter surface, it is usually desirable to deflect the airstream toward the perimeter wall. This can be done with adjustable or fixed deflecting-vane grilles.

A very popular type of perimeter system uses floor diffusers such as the Series 41 diffuser shown in Figure 13-28. Supply outlets, if possible, should always be located to blanket every window and every outside wall (see Figures 13-37 and 13-38 as opposed to Figure 13-39).

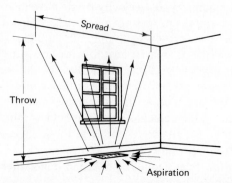

FIGURE 13-37   Application of floor diffuser showing aspiration (induction), throw, and spread. (Courtesy of Lima Register Company, Lima, OH)

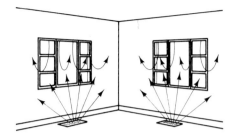

FIGURE 13-38 Perimeter system showing location of floor diffusers to blanket exposed walls and windows. (Courtesy of Lima Register Company, Lima, OH)

Table 13-20 gives performance data for the Series 40 and 41 floor diffusers (see Figure 13-28). For residential applications, the diffuser size selected should be large enough so that the Btu/hr capacity in the table falls to the left of the double vertical lines and to still have a minimum vertical throw of 6 ft where cooling is involved. The data found in Table 13-20 are restricted to the following limitations:

1. Heating Btu/hr is computed at 140°F air temperature behind the louvers.
2. Cooling Btu/hr is computed at 60°F air temperature behind the louvers.
3. Air volume delivered through the diffuser is expressed in cfm.
4. Total pressure loss (static plus velocity) is for stackheads and diffusers.
5. Throw is based on terminal velocity of 50 fpm.
6. Velocity is the average calculated face velocity in fpm.

Outlets in Variable-Air-Volume Systems. The performance of a particular outlet or diffuser is generally independent of the terminal box that is upstream. For a given supply-air volume and temperature differential (to meet a particular load), a standard outlet does

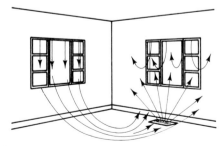

FIGURE 13-39 Cold downdrafts from exposed windows. (Courtesy of Lima Register Company, Lima, OH)

not recognize whether the terminal box is of constant-volume, variable-volume, or indirection type. However, any diffuser, or system of diffusers, gives optimum air diffusion at some particular load condition and air volume. For a VAV system, as the load changes, so does the level of air movement in the conditioned space. At minimum load conditions, when the volume of air is lowest, the level of air movement in the conditioned spaces may be low.

Grilles versus Ceiling Outlets. The question often arises in the designer's mind, when *cooling* applications are considered, as to when to use ceiling outlets and when to use sidewall grilles. Frequently, for architectural reasons, one is preferred over the other. However, there is no clear line of demarcation, but it may be said that, for distributing large quantities of air (over $1\frac{1}{2}$ cfm per square foot of floor space) into rooms with low ceilings or short throws, ceiling diffusers work somewhat better than sidewall grilles (refer to Table 13-2). Other than these points, properly designed installation using either method will produce satisfactory results.

For *heating* applications, floor registers located at the base of cold walls and below windows are normally best. The air should be blown upward with sufficient velocity to obtain distribution across the ceiling toward the opposite wall. This is particularly true if the floor registers are used for cooling as well as heating to prevent stratification of air within the conditioned space. Double-deflection, high sidewall grilles may also be satisfactory in relatively mild climates. However, for more severe winter conditions, auxiliary radiation may be required, located under windows and along cold walls. Ceiling diffusers are seldom used in heating applications.

In summary, we may say that supply outlets fall into four different groups, defined by their air discharge patterns: (1) horizontal high, (2) vertical nonspreading, (3) vertical spreading, and (4) horizontal low. Table 13-21 lists the general characteristics of supply outlets. The table includes the performance of various types for cooling, as well as for heating, since one of the advantages of forced-air systems is that they may be used for both heating and cooling. However, as shown in the table, no single outlet type is best for both heating and cooling.

The best outlets for heating are located near the floor at outside walls and provide a vertical-spreading air jet, preferably under windows, to blanket cold areas and counteract cold downdrafts. Called *perimeter heating,* this arrangement mixes the warm supply air with both the cool air from the area of high heat loss and the cold air from infiltration, thus preventing drafts.

TABLE 13-20

Performance data for Series 40 and 41 steel floor diffusers

Nominal Size	Free Area (sq. in.)		Heating Btu/hr	3045	4565	6090	7610	9515	11415	13320	15220	17125	19025	20930	22830	24735	26635	30440	34245	38050	45660
			Cooling Btu/hr	855	1280	1710	2135	2670	3200	3735	4270	4805	5340	5870	6405	6940	7470	8540	9605	10675	12810
			Cfm	40	60	80	100	125	150	175	200	225	250	275	300	325	350	400	450	500	600
2 1/4" × 10"	19	TP loss		.013	.022	.035	.045	.060	.092	.120	.150										
		Vert. throw (ft)		3.0	4.5	5.5	6.5	8.5	10.5	13.0	15.5										
		Vert. spread (ft.)		6	8	10	12	15	18	23	26										
		Face velocity		309	463	617	771	964	1159	1352	1545										
2 1/4" × 12"	21	TP loss		.009	.015	.027	.037	.050	.080	.105	.134										
		Vert. throw (ft)		3	4	5	6	8	10	12	14										
		Vert. spread (ft)		6	8	10	11	14	17	22	25										
		Face velocity		280	420	565	705	880	1050	1230	1400										
2 1/4" × 14"	24	TP loss		.006	.010	.021	.031	.042	.070	.093	.121	.150									
		Vert. throw (ft)		3	4	4.5	5.5	8	9.5	11	12.5	14									
		Vert. spread (ft)		6	8	9	11	14	16	19	22	25									
		Face velocity		245	365	490	610	760	915	1065	1220	1370									
4" × 10"	32	TP loss			.008	.021	.026	.032	.045	.062	.084	.110	.134	.163							
		Vert. throw (ft)			3	4	5	7	8.5	10	11	12	13	14							
		Vert. spread (ft)			6	8	9	12	14	17	19	22	24	26							
		Face velocity			265	355	445	555	665	775	890	1000	1120	1220							

Size	Parameter																
4" × 12" (39)	TP loss	.004	.010	.016	.023	.033	.042	.058	.075	.089	.107	.128	.159				
	Vert. throw (ft)	2	3	4	6.5	8	9	10	11	12	13	14	15				
	Vert. spread (ft)	4	6	8	12	14	16	18	20	23	24	26	28				
	Face velocity	220	295	370	460	555	645	735	830	925	1020	1110	1200				
4" × 14" (46)	TP loss	.006	.010	.016	.021	.028	.039	.051	.060	.080	.101	.124	.137	.167			
	Vert. throw (ft)	3	4	6	7	8	9	10	11	12	13	14	15	16			
	Vert. spread (ft)	6	8	11	12	14	16	18	20	22	24	26	28	30			
	Face velocity	255	320	395	475	555	635	715	790	870	950	1025	1110	1270			
6" × 10" (52)	TP loss	Series 40 only	.009	.014	.019	.027	.035	.044	.054	.064	.078	.090	.104	.135	.171	.210	.314
	Vert. throw (ft)		4	5.5	6.5	7.5	8.5	10	11	12	13	14	15	17.5	19	21.5	26
	Vert. spread (ft)		8	9.5	11.5	13.5	15.5	17	19	21	23	25	27	31	34.5	38.5	46
	Face velocity		278	348	417	487	556	626	695	765	834	904	973	1112	1251	1390	1668
6" × 12" (59)	TP loss	Series 40 only	.007	.011	.015	.021	.027	.035	.042	.050	.060	.071	.083	.107	.134	.165	.242
	Vert. throw (ft)		4	5	6	7	8	9	10.5	11.5	12.5	13.5	14	16.5	18.5	20.5	29
	Vert. spread (ft)		7	9	11	12.5	14.5	16	18	20	21.5	23.5	25	29	32.5	36	43
	Face velocity		245	307	368	429	491	552	613	675	736	797	859	981	1103	1227	1472
6" × 14" (66)	TP loss	Series 40 only	.005	.009	.012	.017	.022	.028	.034	.041	.050	.058	.067	.088	.110	.132	.194
	Vert. throw (ft)		4	5	6	7	8	8.5	9.5	10.5	11.5	12	13	15	17	19	23
	Vert. spread (ft)		7	8.5	10	12	13.5	15	17	18.5	20.5	22	24	27	30.5	34	41
	Face velocity		219	274	329	384	439	494	549	604	659	713	768	878	988	1098	1317

Source: Courtesy of Lima Register Company, Lima, OH.

Note: For overall dimensions, add $1\frac{5}{16}$ in. to width and height listed except for 6-in. models. For overall dimensions of 6-in. models, add $1\frac{7}{8}$ in. to width and height listed.

TABLE 13-21
General characteristics of supply outlets

Group	Outlet Type	Outlet Flow Pattern	Most Effective Application	Preferred Location	Size Determined By
1	Ceiling and high sidewall	Horizontal	Cooling	Not critical	Major application, heating or cooling
2	Floor registers, baseboard, and low sidewall	Vertical non-spreading	Cooling and heating	Not critical	Maximum acceptable heating temperature differential
3	Floor registers, baseboard, and low sidewall	Vertical spreading	Heating and cooling	Along exposed perimeter	Minimum supply velocity (velocity differs with type and differential acceptable temperature)
4	Baseboard and low sidewall	Horizontal	Heating only	Long outlet perimeter; short outlet; not critical	Maximum supply velocity (velocity should be less than 300 fpm or 1.5 m/s)

Source: ASHRAE Handbook 1987 HVAC Systems and Applications, American Society of Heating, Refrigerating, and Air Conditioning Engineers, Atlanta, GA.

The best outlets for cooling are located in the ceiling and high sidewall and have a horizontal discharge pattern. Standard 90B of the National Fire Code (NFPA No. 90B-84) provides the following summary regarding outlet type:

For year-round operation the correct choice of a system depends on the principal application. If heating is of major or equal importance, perimeter diffusers (the floor diffuser is but one type of perimeter diffuser) should be selected. The system should be designed for the optimum supply velocity during cooling. If cooling is to be the primary application and heating is of secondary importance because of relatively mild outdoor conditions, ceiling diffusers will perform most satisfactorily.

13-15   RETURN- AND EXHAUST-AIR INLETS

Return-air inlets may either be connected to a duct or be simple vents that transfer air from one area to another. Exhaust-air inlets remove air directly from the building and, therefore, are always connected to a duct. Whatever the arrangement, inlet size and configuration determine velocity and pressure requirements for the required airflow.

In general, the same type of equipment, grilles, slot diffusers, and ceiling diffusers used for supplying air may also be used for air return and exhaust. Return inlets, however, do not require the deflection, flow equalizing, and turning devices necessary for supply outlets. Dampers, however, are necessary when it is desirable to balance the airflow in the return duct system.

Types of Inlets

Adjustable-bar grilles: These are the same grilles that are used for air supply to match the deflection setting of the bars with that of the supply outlets.

Fixed-bar grilles: One type of fixed-bar grille was shown in Figure 13-27. Some of the most common return-air grilles are those that have the same general outside appearance as the supply-air outlets. Figure 13-40 shows the Series A80 and A70 return-air grilles manufactured to match the Series A60 and A50 supply-air grilles shown in Figure 13-32. The Series A80 return grille has fixed vertical bars spaced on 2/3-in. centers with 0° straight deflection. The Series A70 return grille has fixed horizontal bars spaced on 2/3-in. centers with 0° straight or 42° face deflection. The streamlined airfoil-shaped bars and open bar spacing maintain an effective area capacity of greater than 75%, which minimizes intake velocity, reduces inlet pressure, and provides quiet operation. The smooth bar shapes do not accumulate lint and plug up as do sharp-edged, core-type returns. Deflected-bar grilles installed in a low or high sidewall location are sight-proof with the grille deflection facing away from the line of sight.

Another common type of return is the cubed-core return. Figure 13-41 shows the Series 90 CC, 2-TB cubed-core return grille manufactured by Lima Register Company. The cubed core is made of fabricated mill-

SERIES A80

**VERTICAL
FACE BARS**

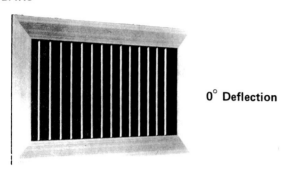

0° **Deflection**

SERIES A70

**HORIZONTAL
FACE BARS**

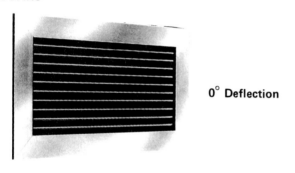

0° **Deflection**

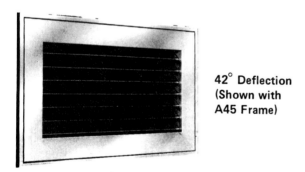

42° **Deflection
(Shown with
A45 Frame)**

FIGURE 13-40 Series A80 and A70 return-air grilles. (Used by permission. Tuttle & Bailey, Holland, MI)

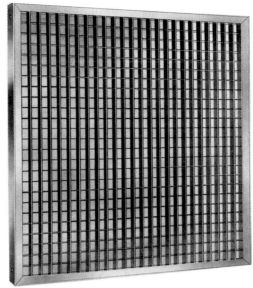

FIGURE 13-41 Series 90 CC, 2-TB, cubed-core return grille. (Courtesy of Lima Register Company, Lima, OH)

finish aluminum strips spaced to form 1/2-in. squares. It is designed for the ultimate free area (approximately 90%) with a minimum of see-through. The Series 90 opposed-blade damper can be factory attached if required. The unit shown is for T-bar ceiling installation. This has a special channel frame (instead of the trim border) to fit T-bar spacing.

V-bar grilles: These units have bars in the shape of inverted V's stacked within the grille frame. They have the advantage of being sight-proof. Door grilles are usually V-bar grilles. The capacity of the grille decreases with increased sight tightness.

Stamped grilles: These grilles are frequently used as return and exhaust inlets, particularly in restrooms and utility areas. They are usually the least expensive type of grille.

13-16 GENERAL APPLICATION DATA (RETURN-AIR GRILLES)

The selection of return and exhaust inlets depends on (1) air velocity in the occupied zone near the inlet, (2) permissible pressure drop through the inlet, (3) noise, and (4) location.

Velocity. Control of room air motion to maintain comfort conditions depends on proper *supply* outlet selection and location. The effect of airflow through return inlets is slight. Air handled by the inlet approaches the opening from all directions, and its velocity decreases rapidly as the distance from the opening increases. Figure 13-42 illustrates a typical velocity fall-off chart. If the general room air velocity 2 ft from the face of the grille is 10 fpm, one-half that distance would indicate a velocity of only 40 fpm, even though the velocity through the inlet is 500 fpm. Also, as long as the general drift of the room air toward the return-air grille is not more than 50 fpm, then no objectionable drafts would arise. Table 13-22 contains recommended return-inlet face velocities.

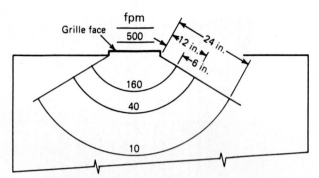

FIGURE 13-42 Velocity falloff with distance from return grille.

Pressure Drop Through Inlet. The design of the return-air system and the selection of return grilles require the same consideration as the supply system. The pressure drop across the return grilles will be the least at the end of a duct system and greatest near the riser and fan due to friction and turning losses as the air flows toward the fan. It does no good to select a return grille size to operate at a low pressure drop and low sound level and install it in a system where a large negative static pressure exists. This will result in a large pressure drop across the grille causing excess return airflow and high sound levels. If registers are installed, the damper may be throttled to control the airflow, but the large pressure drop and accompanying high sound level still exist.

It is extremely important that the negative static pressure be controlled by designing for minimum friction and turning losses. This is done by use of appropriate friction factors, radius elbows or turning vanes in square elbows, static pressure regain, and the use of branch duct balancing dampers.

Noise. The problem of return-inlet noise is the same as that for supply outlets. In computing resultant room noise levels from operation of an air conditioning system, the return inlet must be included as a part of the total grille area. The major difference between supply outlets and return inlets is the frequent installation of the latter at ear level. When they are so located, the return-inlet velocity should not be in excess of 75% of maximum permissible outlet velocity.

Return-air grilles should be selected for static pressures that will provide the required NC rating and conform to the return system performance characteristics. Since fan sound power is transmitted through the return-air system as well as the supply system, fan silencing may be necessary or desirable in the return side, particularly if silencing is being considered on the supply side.

When excessive pressure must be throttled through the grille damper, the resulting NC rating increases sharply. Large pressure drops should be handled with remote duct dampers with intervening sound attenuation treatment in the duct. Return grilles up to 3 ft² nominal size can be close-coupled to pressure-reducing valves (normally used in a high-pressure supply duct system) to produce an intake unit capable of providing low noise levels at high duct pressures without supplementary duct attenuation.

Transfer grilles venting into a ceiling plenum should be located remote from plenum noise sources. The use of a lined sheet metal elbow can reduce transmitted sound as much as 12 to 15 db in octave bands 5 to 7 and 3 to 5 db in bands 3 to 4. Lined elbows on vent grilles and lined common ducts on ducted return grilles can minimize "crosstalk" between private offices.

Location of Return Inlets. It remains the function of supply-air outlets to establish room air distribution, air motion, and thermal equilibrium. It is erroneous to feel that the supply air can be "pulled" from the supply opening to the return grille location. As previously noted, a return grille creates a low-pressure region next to its face and atmospheric pressure "pushes" air into the grille. Because of this, the location of the return-air grilles is not critical, and their placement can

TABLE 13-22

Recommended return-inlet face velocities

	Velocity Over Gross Area	
Inlet Location	fpm	(m/s)
Above occupied zone	800 up	(4.0 up)
Within occupied zone, not near seats	600–800	(3.0–4.0)
Within occupied zone, near seats	400–600	(2.0–3.0)
Door or wall louvers	200–300	(1.0–1.5)
Undercutting of doors (through) undercut area	200–300	(1.0–1.5)

Source: ASHRAE Handbook 1989 Fundamentals, American Society of Heating, Refrigerating, and Air Conditioning Engineers, Atlanta, GA.

be largely a matter of convenience in many cases. Specific locations in the ceiling may be desirable for local heat loads or smoke exhaust, or a location in the perimeter sill may be desirable for an exterior zone intake under a window wall section. Despite the localized effects of a return grille (see Figure 13-42), it is not advisable to locate large centralized return grilles in an occupied area because the large mass of air moving toward the grille can cause objectionable air motion for nearby occupants.

A wall return near the floor may be a good compromise location for both heating and cooling applications. Stratification due to poor mixing in winter is counteracted by a low return since the cool air near the floor is withdrawn first and is replaced by the warmer upper air strata. Also, a return near the floor makes it possible for the supply air to release its full capacity without any possibility of short circuiting to the return inlet.

On heating installations, return and supply ceiling grilles should never be used unless auxiliary heating is used to raise the mean radiant temperature of the outside wall. In this way, cold air falling down the exposed wall will not pass across the feet and legs of the occu-

TABLE 13-23

Engineering performance data for return-air grilles and registers, fixed-bar type, 0° straight bars (Series T80 and T70—steel; Series A80 and A70—aluminum)

		Return Air Capacity					
		NC 20–25 Application Nonducted		NC 25–30 Application Ducted		NC 30–40 Application Ducted	
		p_s		p_s		p_s	
Listed Size (W × H)	A_k	−.02″ cfm	−.03″ cfm	−.08″ cfm	−.10″ cfm	−.15″ cfm	−.20″ cfm
8 × 4	.26	85	105	170	190	230	265
8 × 6	.34	110	135	220	245	300	345
10 × 6	.42	140	170	275	310	380	435
12 × 6	.50	170	205	335	375	455	525
10 × 8	.53	195	235	385	430	525	605
12 × 8	.63	240	290	470	525	640	735
10 × 10	.64	245	300	490	550	670	770
18 × 6	.75	265	325	530	595	725	835
12 × 12	.89	375	455	740	830	1010	1160
18 × 12	1.3	570	695	1130	1265	1545	1775
22 × 10	1.4	615	750	1220	1365	1665	1915
24 × 12	1.7	770	940	1530	1715	2090	2405
18 × 18	1.9	880	1075	1750	1960	2390	2750
34 × 10	2.1	940	1145	1865	2090	2550	2930
30 × 12	2.2	975	1190	1940	2170	2645	3040
24 × 18	2.5	1190	1450	2365	2650	3235	3720
22 × 22	2.8	1300	1585	2585	2895	3530	4060
30 × 18	3.2	1495	1825	2975	3330	4060	4670
24 × 24	3.3	1610	1965	3200	3585	4375	5030
36 × 18	3.8	1810	2210	3600	4030	4915	5650
30 × 24	4.1	2020	2465	4015	4495	5485	6305
34 × 22	4.3	2150	2625	4280	4795	5850	6725
36 × 24	4.9	2440	2975	4850	5430	6625	7620
46 × 22	5.9	2980	3635	5925	6635	8095	9310
36 × 30	6.1	3090	3770	6145	6880	8395	9655
48 × 24	6.6	3250	3965	6460	7235	8825	10150
48 × 30	8.1	4170	5085	8290	9285	11325	13025
48 × 36	9.7	4950	6040	9845	11025	13450	15465

Source: Cat. RG-1. Used by permission. Tuttle & Bailey, Holland, MI.
Note: p_s = static pressure (in. WG)
A_k = outlet area, 1 in. out from face (ft²)
NC = Re 18-dB room attenuation

pants when it is returning to the inlet. In some cases, for economic reasons, such radiation must be omitted. Several return inlets should be located at the base of the exposed wall to trap as much cold air as possible. This installation does not compare in quality of results, however, with an installation where radiation is placed along the outside wall.

Floor returns should be avoided whenever possible because they tend to be a catchall for dirt. This imposes a severe handicap on central system filters and coils. When there are common returns in corridors and so

forth, return-air grilles are frequently located in the bottom of doors, or many times the doors are undercut. The pressure drop through the door returns should not be excessive; otherwise, the air diffusion in the room may be seriously unbalanced by the opening or closing of the doors. Outward leakage through doors and windows cannot be counted on for dependable results.

Some ceiling outlets combine supply and return openings in a single unit. This method is used for heating and cooling applications. However, application of these combination supply and return fittings for com-

TABLE 13-24

Engineering performance data for return-air grilles and registers, fixed-bar type, 42° deflected bars (Series T70D; Series A70D)

Listed Size W × H	A_k	Return Air Capacity					
		NC 20–25 Application Nonducted		NC 25–30 Application Ducted		NC 30–40 Application Ducted	
		p_s		p_s		p_s	
		−.02″	−.03″	−.08″	−.10″	−.15″	−.20″
		cfm	cfm	cfm	cfm	cfm	cfm
8 × 4	.26	55	65	105	115	140	160
8 × 6	.34	75	90	145	160	195	225
10 × 6	.42	90	110	180	200	245	280
12 × 6	.50	110	135	220	245	300	345
10 × 8	.53	125	150	245	275	335	385
12 × 8	.63	155	190	310	345	420	485
10 × 10	.64	160	195	315	350	425	490
18 × 6	.75	170	205	335	375	455	525
12 × 12	.89	240	290	470	525	640	735
18 × 12	1.3	370	450	735	825	1005	1155
22 × 10	1.4	380	465	755	845	1030	1185
24 × 12	1.7	505	615	1000	1120	1365	1570
18 × 18	1.9	575	700	1140	1275	1555	1790
34 × 10	2.1	600	732	1195	1340	1635	1880
30 × 12	2.2	630	770	1255	1405	1715	1970
24 × 18	2.5	770	940	1530	1715	2090	2405
22 × 22	2.8	880	1075	1750	1960	2390	2750
30 × 18	3.2	970	1185	1930	2160	2635	3030
24 × 24	3.3	1040	1270	2070	2320	2830	3255
36 × 18	3.8	1170	1425	2320	2600	3170	3645
30 × 24	4.1	1320	1610	2625	2940	3585	4120
34 × 22	4.3	1355	1655	2695	3020	3685	4235
36 × 24	4.9	1580	1925	3135	3510	4280	4920
46 × 22	5.9	1940	2365	3855	4315	5265	6055
36 × 30	6.1	2010	2450	3995	4475	5460	6280
48 × 24	6.6	2130	2600	4240	4750	5795	6665
48 × 30	8.1	2700	3295	5370	6015	7340	8440
48 × 36	9.7	3220	3930	6405	7175	8755	10065

Source: Cat. RG-1. Used by permission. Tuttle & Bailey, Holland, MI.

Note: p_s = static pressure (in. WG)
  A_k = outlet area, 1 in. out from face (ft²)
  NC = RE 18-dB room attenuation

fort heating in temperate and cold climates will probably produce large ceiling-to-floor temperature differentials.

Exhaust inlets in bars, kitchens, lavatories, dining rooms, and club rooms should be located near ceiling level to collect warm air, odors, and fumes.

In summary, it should be remembered that, even though many locations of return grilles are possible, wall returns located near the floor are usually the best.

Also, objectionable drafts may result if return-air velocities through the outlet become excessive.

Tables 13-23 and 13-24 show performance data for return-air grilles and registers of the type shown in Figure 13-40. Performance of any size grille not shown in the tables will be the same as for the size shown having the same listed size area. For example, a 36 × 16 (not shown) has an area of 576 in.^{2}. Its performance would be the same as a 24 × 24 grille, which is listed.

BIBLIOGRAPHY

13.1. *ASHRAE Handbook 1989 Fundamentals,* American Society of Heating, Refrigerating, and Air Conditioning Engineers, Atlanta, GA, 1989.

13.2. *ASHRAE Handbook 1988 Equipment,* American Society of Heating, Refrigerating, and Air Conditioning Engineers, Atlanta, GA, 1988.

13.3. *ASHRAE Handbook 1987 HVAC Systems and Applications,* American Society of Heating, Refrigerating, and Air Conditioning Engineers, Atlanta, GA, 1987.

13.4. Nivens, Ralph G., *Air Diffusion Dynamics, Theory, Design and Applications,* Business News Publishing Company, Birmingham, MI, 1976.

13.5. *Lima Engineering Data,* Catalog, Lima Register Company, Lima, OH.

13.6. *Ceiling Diffusers,* Catalog CD-1, Tuttle & Bailey, Hart & Cooley, Inc., Holland, MI.

13.7. *Grilles/Registers,* Catalog RG-1, Tuttle & Bailey, Hart & Cooley, Inc., Holland, MI.

14

Airflow in Duct Systems and Fans

14-1 INTRODUCTION

Air mechanically supplied to a building for ventilation, heating, and/or cooling requires a well-designed air duct system extending from the fan or fans to the ultimate points of discharge at distribution diffusers and grilles, as well as returning all or part of the air to the conditioning apparatus from the return grilles. This chapter describes the methods used in the design of sheet metal duct systems and the performance characteristics and selection of fans used to force air through such systems.

The purpose of an air duct system is to convey conditioned air from one location to another within prescribed limits of space, noise, and cost. Good duct design takes into consideration not only the fundamentals of fluid flow in closed conduits but also the aesthetics and economics of the finished job.

All the necessary steps in the design of a duct system are summarized below. Different listed steps may overlap and would be combined in practice. With experience, each HVAC engineer will usually develop his or her own technique, but the following breakdown will serve as a convenient checklist.

1. Calculate the volume of air required for each space within the building envelope to meet the heat losses, heat gains, and/or ventilation loads.

2. Determine the sizes and location of all supply-air outlets and return inlets in each space or area to be conditioned using the techniques presented in Chapter 13.

3. Study the building plans and draw a tentative system of supply-air ducts connecting all supply-air outlets with the central-fan unit. (This would normally be a single-line drawing, the line representing the centerline of the proposed supply duct system.) In this layout, remember to avoid obstructions and any direct contact with steel work and concrete. The best system is usually the simplest and most direct.

4. Lay out a return-air duct system using a single-line drawing. Do not return air from spaces where contaminated air may exist, such as bathrooms, kitchens, laundries, and so forth. Provide for suitable direct exhaust or vents for such areas.

5. Size the supply- and return-air duct systems using an accepted method as directed in this chapter.

6. Determine the friction loss of the supply-air duct offering the greatest resistance to airflow. To that

friction loss, add an amount equal to the *total* pressure required at the last supply outlet, plus an amount required for the turning loss of the air into the diffuser. The sum of all these pressures represents the *total* pressure required at the entrance to the duct system, after the supply-air fan, to flow the supply-air system.

7. Make a similar analysis of the return-air system. This will be a suction pressure on the fan.

8. With the required discharge and suction pressure known, select a fan to flow the total system.

14-2 BASIC PRINCIPLES OF GOOD DUCT DESIGN AND CONSTRUCTION

Keep the following principles in mind when designing any sheet metal, air duct system.

1. Get the air where it is wanted as directly as possible, using the least material, power, space, and cost.

2. Stay within the maximum recommended velocities of Table 14-1 unless there is good reason for a deviation. The recommended velocities provide the quietest, most trouble-free operations.

3. Avoid sharp elbows and bends in the duct. Splitters and turning vanes should be used to reduce elbow and outlet pressure losses.

4. Avoid both sudden enlargements and abrupt contractions. The angle of divergence of enlargements should not exceed 20 degrees. The angle of convergence in contractions should not be greater than 60 degrees. Special care should be taken to avoid restriction of flow.

5. For the *greatest* air-carrying capacity per square foot of sheet metal, use *round* ducts. In installations where rectangular ducts must be used, make the ducts as square as possible, and select the dimensions on the basis of equal friction with the round ducts from the table of circular equivalents (Table 14-2). Rectangular ducts with aspect ratio (ratio of long side to short side) greater than 6 to 1 should be avoided.

6. Make ducts as tight as possible. All laps should be in the direction of airflow. Avoid raw edges on splitters and vanes.

7. Provide expansion joints in long duct runs.

8. Provide adequate bracing and secure supports for all ductwork.

9. Use plated sheet metal screws, rivets, or bolts. Hardware for aluminum air ducts should be made of aluminum or should be zinc or cadmium-plated.

10. Use duct insulation where necessary to prevent excessive heat loss, heat gain, or condensation.

11. Provide dampers in all duct branches for final balancing.

12. Provide adequate access openings where apparatus and components may be serviced.

14-3 LOSS IN PRESSURE DUE TO FRICTION

As noted in Section 13-5, there is a decrease in the total and static pressures as air flows through a system of ducts. This loss is due to fluid friction and to fluid turbulence. In this section, we shall confine our discussion to fluid friction losses. This loss is due to internal fluid friction and movement of the air over the interior surfaces of the duct.

In Chapter 9, we developed an equation for head loss due to friction in a round, closed conduit for any fluid as

TABLE 14-1
Recommended and maximum duct velocities

Designation	Recommended Velocity (feet per minute)			Maximum Velocity (feet per minute)		
	Residences	Schools Theaters Public Buildings	Industrial Buildings	Residences	Schools Theaters Public Buildings	Industrial Buildings
Outside air intakes[a]	500	500	500	800	900	1200
Filters[a]	250	300	350	300	350	350
Heating coils[a]	450	500	600	500	600	700
Air washers	500	500	500	500	500	500
Suction connections	700	800	1000	1000	1400	1400
Fan outlets	1000–1600	1300–2000	1600–2400	1500–2000	1700–2800	1700–2800
Main ducts	700–900	1000–1300	1200–1800	800–1200	1100–1600	1300–2200
Branch ducts	600	600–900	800–1000	700–1000	800–1300	1000–1800
Branch risers	500	600–700	800	650–800	800–1200	1000–1600

[a]The velocities are for total face area; not the net free area. Other velocities are for net free area.
Source: Used by permission, Reynolds Metal Company.

$$h_L = f\left(\frac{L}{D}\right)\left(\frac{V^2}{2g}\right) \text{ ft of fluid} \qquad (9\text{-}11)$$

Since $V^2/2g$ has been defined as a velocity head, we may substitute Eq. (13-4) into Eq. (9-11) and obtain

$$h_L = f\left(\frac{L}{D}\right)\rho_a\left(\frac{V}{1096.5}\right)^2 \text{ in. WG} \qquad (14\text{-}1)$$

or, if standard air is considered, we may use Eq. (13-6) and obtain

$$h_L = f\left(\frac{L}{D}\right)\left(\frac{V}{4005}\right)^2 \text{ in. WG} \qquad (14\text{-}1a)$$

The symbols and units of Eqs. (14-1) and (14-1a) are as follows:

where

h_L = friction loss (in. WG)
f = friction factor (dimensionless)
L = duct length (ft)
D = duct diameter (ft)
V = mean airflow velocity (fpm)
ρ_a = air density (lbm/ft^3)

As was the case with water, the greater the quantity of air flowing through a given sized duct, the greater will be the loss in pressure due to friction, which means greater operating costs. If the size of a duct is increased to reduce the flow velocity and decrease the friction loss, the first cost increases. Therefore, the principles of good duct design must be known in order to secure satisfactory operation of the system with the proper balance existing between first cost and final operating cost. It will be readily seen that a poorly designed system with inherently large friction losses will cost considerably more to operate than a well-designed system where the pressure losses are kept to a minimum. From the fan laws (discussed later), it can be seen that a fan of a given size selected for a given pressure must be speeded up if it is to overcome an additional resistance. Since the power consumption varies as the cube of the speed change, it is apparent that pressure losses must be kept to a minimum if economical operation is to be obtained.

In practice, this balancing of costs is a lengthy process and sometimes a useless one because of the lack of reliable cost data. Duct designers invariably depend upon past experience with air conditioning and ventilating systems in selecting the pressure losses to be allowed in the duct system. In addition to this, in conventional low-velocity duct design, the problem is further complicated in commercial buildings by the fact that air veloc-

ities in ducts must be maintained below some definite point if quiet operation is to be obtained. Recommended practice is shown in Table 14-1. If the maximum velocity is thus set, this immediately fixes the maximum friction loss for a given air quantity without any other considerations of cost. In industrial systems where noise is not as much a factor, the problem is only one of balancing the operating cost against the first cost.

The friction loss in straight, round ducts may be determined from charts such as those in Figures 14-1a and 14-1b, which are based on Eq. (14-1a). Therefore, these charts are based on standard air (ρ = 0.0750 lb/ft^3) flowing through average, clean, round, aluminum ducts. Ducts are also constructed of galvanized sheet metal. Friction loss in galvanized metal ducts is slightly more than that in aluminum ducts. However, considering the degree of accuracy with which we can read the charts and the accuracy in fabrication of the ducts, we may use the design charts (Figures 14-1a and 14-1b) for either aluminum or galvanized sheet metal.

Referring to Figures 14-1a and 14-1b, we may note that the friction loss shown on the horizontal scale is expressed in inches water gauge (in. WG) per 100 ft of straight duct. This will be referred to as the "friction rate" (f_{100}). The vertical scale is the airflow quantity in cubic feet per minute (cfm). Inclined lines from lower left to upper right are lines of constant duct diameter in inches. Inclined lines from upper left to lower right are lines of constant velocity in feet per minute (fpm). Although these charts are designed for standard air flowing, they may be used with little error for air temperatures between 50°F and 90°F. Also, a correction is not necessary for humidity or for small deviations in barometric pressure (not exceeding ± 0.5 in. Hg).

ILLUSTRATIVE PROBLEM 14-1

A sheet metal duct 150 ft long is 14 in. in diameter. The quantity of standard air to be delivered through the duct is 2000 cfm. Find (1) the friction loss in the duct and (2) the velocity of air flowing through the duct.

Solution:

1. Referring to Figure 14-1b, at the intersection of the horizontal line for 2000 cfm and the diagonal line for 14-in. diameter, drop vertically down to the friction rate scale and read 0.34 in. WG per 100 ft of length. For a total length of 150 ft,

$$\text{friction loss} = 0.34(150/100)$$
$$= 0.51 \text{ in. WG}$$

2. The air velocity in the round duct may be read from Figure 14-1b at the same point of intersection as in part (1) and is 1900 fpm.

TABLE 14-2

Circular equivalents of rectangular ducts for equal friction and capacity*
(Dimension in inches, feet, or meters)

Side Rectangular Duct	4.0	4.5	5.0	5.5	6.0	6.5	7.0	7.5	8.0	8.5	9.0	9.5	10.0	10.5	11.0	11.5	12.0	12.5	13.0	13.5
3.0	3.8	4.0	4.2	4.4	4.6	4.8	4.9	5.1	5.2	5.4	5.5	5.6	5.7	5.9	6.0	6.1	6.2	6.3	6.4	6.5
3.5	4.1	4.3	4.6	4.8	5.0	5.2	5.3	5.5	5.7	5.8	6.0	6.1	6.3	6.4	6.5	6.7	6.8	6.9	7.0	7.1
4.0	4.4	4.6	4.9	5.1	5.3	5.5	5.7	5.9	6.1	6.3	6.4	6.6	6.8	6.9	7.1	7.2	7.3	7.5	7.6	7.7
4.5	4.6	4.9	5.2	5.4	5.6	5.9	6.1	6.3	6.5	6.7	6.9	7.0	7.2	7.4	7.5	7.7	7.8	8.0	8.1	8.2
5.0	4.9	5.2	5.5	5.7	6.0	6.2	6.4	6.7	6.9	7.1	7.3	7.4	7.6	7.8	8.0	8.1	8.3	8.4	8.6	8.7
5.5	5.1	5.4	5.7	6.0	6.3	6.5	6.8	7.0	7.2	7.4	7.6	7.8	8.0	8.2	8.4	8.6	8.7	8.8	9.0	9.2

Side Rectangular Duct	6	7	8	9	10	11	12	13	14	15	16	17	18	19	20	22	24	26	28	30
6	6.6																			
7	7.1	7.7																		
8	7.5	8.2	8.8																	
9	8.0	8.6	9.3	9.9																
10	8.4	9.1	9.8	10.4	10.9															
11	8.8	9.5	10.2	10.8	11.4	12.0														
12	9.1	9.9	10.7	11.3	11.9	12.5	13.1													
13	9.5	10.3	11.1	11.8	12.4	13.0	13.6	14.2												
14	9.8	10.7	11.5	12.2	12.9	13.5	14.2	14.7	15.3											
15	10.1	11.0	11.8	12.6	13.3	14.0	14.6	15.3	15.8	16.4										
16	10.4	11.4	12.2	13.0	13.7	14.4	15.1	15.7	16.3	16.9	17.5									
17	10.7	11.7	12.5	13.4	14.1	14.9	15.5	16.1	16.8	17.4	18.0	18.6								
18	11.0	11.9	12.9	13.7	14.5	15.3	16.0	16.6	17.3	17.9	18.5	19.1	19.7							
19	11.2	12.2	13.2	14.1	14.9	15.6	16.4	17.1	17.8	18.4	19.0	19.6	20.2	20.8						
20	11.5	12.5	13.5	14.4	15.2	15.9	16.8	17.5	18.2	18.8	19.5	20.1	20.7	21.3	21.9					
22	12.0	13.1	14.1	15.0	15.9	16.7	17.6	18.3	19.1	19.7	20.4	21.0	21.7	22.3	22.9	24.1				
24	12.4	13.6	14.6	15.6	16.6	17.5	18.3	19.1	19.8	20.6	21.3	21.9	22.6	23.2	23.9	25.1	26.2			
26	12.8	14.1	15.2	16.2	17.2	18.1	19.0	19.8	20.6	21.4	22.1	22.8	23.5	24.1	24.8	26.1	27.2	28.4		
28	13.2	14.5	15.6	16.7	17.7	18.7	19.6	20.5	21.3	22.1	22.9	23.6	24.4	25.0	25.7	27.1	28.2	29.5	30.6	
30	13.6	14.9	16.1	17.2	18.3	19.3	20.2	21.1	22.0	22.9	23.7	24.4	25.2	25.9	26.7	28.0	29.3	30.5	31.6	32.8
32	14.0	15.3	16.5	17.7	18.8	19.8	20.8	21.8	22.7	23.6	24.4	25.2	26.0	26.7	27.5	28.9	30.1	31.4	32.6	33.8
34	14.4	15.7	17.0	18.2	19.3	20.4	21.4	22.4	23.3	24.2	25.1	25.9	26.7	27.5	28.3	29.7	31.0	32.3	33.6	34.8
36	14.7	16.1	17.4	18.6	19.8	20.9	21.9	23.0	23.9	24.8	25.8	26.6	27.4	28.3	29.0	30.5	32.0	33.0	34.6	35.8
38	15.0	16.4	17.8	19.0	20.3	21.4	22.5	23.5	24.5	25.4	26.4	27.3	28.1	29.0	29.8	31.4	32.8	34.2	35.5	36.7
40	15.3	16.8	18.2	19.4	20.7	21.9	23.0	24.0	25.1	26.0	27.0	27.9	28.8	29.7	30.5	32.1	33.6	35.1	36.4	37.6
42	15.6	17.1	18.5	19.8	21.1	22.3	23.4	24.5	25.6	26.6	27.6	28.5	29.4	30.4	31.2	32.8	34.4	35.9	37.3	38.6
44	15.9	17.5	18.9	20.2	21.5	22.7	23.9	25.0	26.1	27.2	28.2	29.1	30.0	31.0	31.9	33.5	35.2	36.7	38.1	39.5
46	16.2	17.8	19.2	20.6	21.9	23.2	24.3	25.5	26.7	27.7	28.7	29.7	30.6	31.6	32.5	34.2	35.9	37.4	38.9	40.3
48	16.5	18.1	19.6	20.9	22.3	23.6	24.8	26.0	27.2	28.2	29.2	30.2	31.2	32.2	33.1	34.9	36.6	38.2	39.7	41.2
50	16.8	18.4	19.9	21.3	22.7	24.0	25.2	26.4	27.6	28.7	29.8	30.8	31.8	32.8	33.7	35.5	37.3	38.9	40.4	42.0
52	17.0	18.7	20.2	21.6	23.1	24.4	25.6	26.8	28.1	29.2	30.3	31.4	32.4	33.4	34.3	36.2	38.0	39.6	41.2	42.8
54	17.3	19.0	20.5	22.0	23.4	24.8	26.1	27.3	28.5	29.7	30.8	31.9	32.9	33.9	34.9	36.8	38.7	40.3	42.0	43.6
56	17.6	19.3	20.9	22.4	23.8	25.2	26.5	27.7	28.9	30.1	31.2	32.4	33.4	34.5	35.5	37.4	39.3	41.0	42.7	44.3
58	17.8	19.5	21.1	22.7	24.2	25.5	26.9	28.2	29.3	30.5	31.7	32.9	33.9	35.0	36.0	38.0	39.8	41.7	43.4	45.0
60	18.1	19.8	21.4	23.0	24.5	25.8	27.3	28.7	29.8	31.0	32.2	33.4	34.5	35.5	36.5	38.6	40.4	42.3	44.0	45.8
62	18.3	20.1	21.7	23.3	24.8	26.2	27.6	29.0	30.2	31.4	32.6	33.8	35.0	36.0	37.1	39.2	41.0	42.9	44.7	46.5
64	18.6	20.3	22.0	23.6	25.2	26.5	27.9	29.3	30.6	31.8	33.1	34.2	35.5	36.5	37.6	39.7	41.6	43.5	45.4	47.2
66	18.8	20.6	22.3	23.9	25.5	26.9	28.3	29.7	31.0	32.2	33.5	34.7	35.9	37.0	38.1	40.2	42.2	44.1	46.0	47.8
68	19.0	20.8	22.5	24.2	25.8	27.3	28.7	30.1	31.4	32.6	33.9	35.1	36.3	37.5	38.6	40.7	42.8	44.7	46.6	48.4
70	19.2	21.1	22.8	24.5	26.1	27.6	29.1	30.4	31.8	33.1	34.3	35.6	36.8	37.9	39.1	41.3	43.3	45.3	47.2	49.0
72															39.6	41.8	43.8	45.9	47.8	49.7
74															40.0	42.3	44.4	46.4	48.4	50.3
76															40.5	42.8	44.9	47.0	49.0	50.8
78															40.9	43.3	45.5	47.5	49.5	51.5
80															41.3	43.8	46.0	48.0	50.1	52.0
82															41.8	44.2	46.4	48.6	50.6	52.6
84															42.2	44.6	46.9	49.2	51.1	53.2
86															42.6	45.0	47.4	49.6	51.6	53.7
88															43.0	45.4	47.9	50.1	52.2	54.3
90															43.4	45.9	48.3	50.6	52.8	54.8
92															43.8	46.3	48.7	51.1	53.4	55.4
94															44.2	46.7	49.1	51.6	53.9	55.9
96															44.6	47.2	49.5	52.0	54.4	56.3

Source: Reprinted from *ASHRAE Handbook 1989 Fundamentals*, with permission from American Society of Heating, Refrigerating, and Air Conditioning Engineers, Atlanta, GA.

14.0	14.5	15.0	15.5	16
6.6	6.7	6.8	6.9	7.0
7.2	7.3	7.4	7.5	7.6
7.8	7.9	8.1	8.2	8.3
8.4	8.5	8.6	8.7	8.9
8.9	9.0	9.1	9.3	9.4
9.4	9.5	9.6	9.8	9.8

32	34	36	38	40	42	44	46	48	50	52	56	60	64	68	72	76	80	84	88	Side Rectangular Duct
																				6
																				7
																				8
																				9
																				10
																				11
																				12
																				13
																				14
																				15
																				16
																				17
																				18
																				19
																				20
																				22
																				24
																				26
																				28
																				30
35.0																				32
36.0	37.2																			34
37.0	38.2	39.4																		36
38.0	39.2	40.4	41.6																	38
39.0	40.2	41.4	42.6	43.8																40
39.9	41.1	42.4	43.6	44.8	45.9															42
40.8	42.0	43.4	44.6	45.8	46.9	48.1														44
41.7	43.0	44.3	45.6	46.8	47.9	49.1	50.3													46
42.6	43.9	45.2	46.5	47.8	48.9	50.2	51.3	52.6												48
43.5	44.8	46.1	47.4	48.8	49.8	51.2	52.3	53.6	54.7											50
44.3	45.7	47.1	48.3	49.7	50.8	52.2	53.3	54.6	55.8	56.9										52
45.0	46.5	48.0	49.2	50.6	51.8	53.2	54.3	55.6	56.8	57.9										54
45.8	47.3	48.8	50.1	51.5	52.7	54.1	55.3	56.5	57.8	58.9	61.3									56
46.6	48.1	49.6	51.0	52.4	53.7	55.0	56.2	57.5	58.8	60.0	62.3									58
47.3	48.9	50.4	51.8	53.3	54.6	55.9	57.1	58.5	59.8	61.0	63.3	65.7								60
48.0	49.7	51.2	52.6	54.2	55.5	56.8	58.0	59.4	60.7	62.0	64.3	66.7								62
48.7	50.4	52.0	53.4	55.0	56.4	57.7	59.0	60.3	61.6	62.9	65.3	67.7	70.0							64
49.5	51.1	52.8	54.2	55.8	57.2	58.6	59.9	61.2	62.5	63.9	66.3	68.7	71.1							66
50.2	51.8	53.5	55.0	56.6	58.0	59.5	60.8	62.1	63.4	64.8	67.3	69.7	72.1	74.4						68
50.9	52.5	54.2	55.8	57.3	58.8	60.3	61.7	63.0	64.3	65.7	68.3	70.7	73.1	75.4						70
51.5	53.2	54.9	56.5	58.0	59.6	61.1	62.6	63.9	65.2	66.6	69.2	71.7	74.1	76.4	78.8					72
52.1	53.9	55.6	57.2	58.8	60.4	61.9	63.3	64.8	66.1	67.5	70.1	72.7	75.1	77.4	79.9					74
52.7	54.6	56.3	57.9	59.5	61.2	62.7	64.1	65.6	67.0	68.4	71.0	73.6	76.1	78.4	80.9	83.2				76
53.3	55.2	57.0	58.6	60.3	62.0	63.4	64.9	66.4	67.9	69.3	71.8	74.5	77.1	79.4	81.8	84.2				78
53.9	55.8	57.6	59.3	61.0	62.7	64.1	65.7	67.2	68.7	70.1	72.7	75.4	78.1	80.4	82.8	85.2	87.5			80
54.5	56.4	58.2	60.0	61.7	63.4	64.9	66.5	68.0	69.5	71.0	73.6	76.3	79.0	81.4	83.8	86.2	88.6			82
55.1	57.0	58.9	60.7	62.4	64.1	65.7	67.3	68.8	70.3	71.8	74.5	77.2	79.9	82.4	84.8	87.2	89.6	91.9		84
55.7	57.6	59.5	61.3	63.0	64.8	66.4	68.0	69.5	71.1	72.6	75.4	78.1	80.8	83.3	85.8	88.2	90.6	92.9		86
56.3	58.2	60.1	62.0	63.7	65.4	67.0	68.7	70.3	71.8	73.4	76.3	79.0	81.6	84.2	86.8	89.2	91.6	93.9	96.3	88
56.9	58.8	60.7	62.6	64.4	66.0	67.8	69.4	71.1	72.6	74.2	77.1	79.9	82.5	85.1	87.8	90.2	92.6	94.9	97.3	90
57.4	59.4	61.3	63.2	65.0	66.8	68.5	70.1	71.8	73.3	74.9	77.8	80.8	83.4	86.0	88.7	91.2	93.6	95.9	98.3	92
57.9	60.0	61.9	63.8	65.6	67.5	69.2	70.8	72.5	74.1	75.6	78.6	81.7	84.3	86.9	89.6	92.1	94.6	96.9	99.3	94
58.4	60.5	62.4	64.4	66.2	68.2	69.8	71.5	73.2	74.8	76.3	79.4	82.6	85.2	87.8	90.5	93.0	95.6	97.9	100.3	96

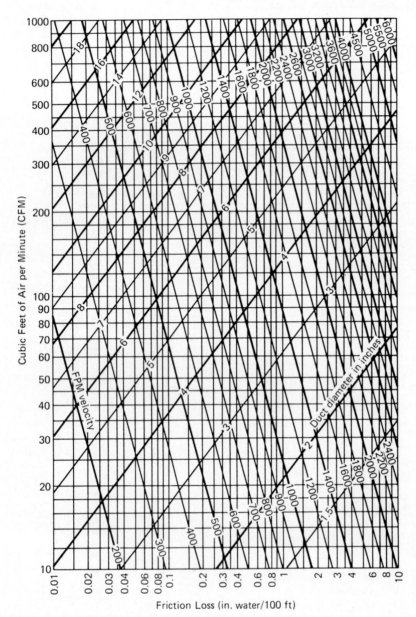

FIGURE 14-1a Friction loss in straight, aluminum sheet metal ducts for volumes 10 to 1000 cfm. (based on standard air density of 0.0750 lb/ft³ flowing through average, clean, round, aluminum ducts having approximately 40 joints per 100 ft). (Courtesy of Reynolds Metal Company)

ILLUSTRATIVE PROBLEM 14-2

The static pressure available to overcome friction loss in a duct is 0.25 in. WG. The duct is 22 in. in diameter and is 315 ft long. Find (1) the maximum quantity of standard air that can flow through the duct and (2) the airflow velocity.

Solution:

1. Friction rate = 100(0.25/315) = 0.079 (say, 0.08) in. WG/100 ft. Referring to Figure 14-1b, locate on the bottom friction rate scale the value of 0.08 and proceed vertically to the sloping line for 22-in. diameter. At this point of intersection, read on the vertical scale 3000 cfm.
2. The airflow velocity at this same point of intersection is 1150 fpm.

ILLUSTRATIVE PROBLEM 14-3

The maximum velocity of air in a duct is limited to 1600 fpm. The total length is 155 ft. If the total quantity of standard air to be delivered through the duct is 10,000 cfm, find (1) the required diameter of the duct and (2) the friction loss in the duct.

Solution:

1. Referring to Figure 14-1b, at the intersection of the horizontal line of 10,000 cfm and the diagonal velocity line of 1600 fpm, the size of the required round duct is 34 in. in diameter.
2. The friction loss per 100 ft of duct found in Figure 14-1b is 0.086 in. WG per 100 ft of duct length.

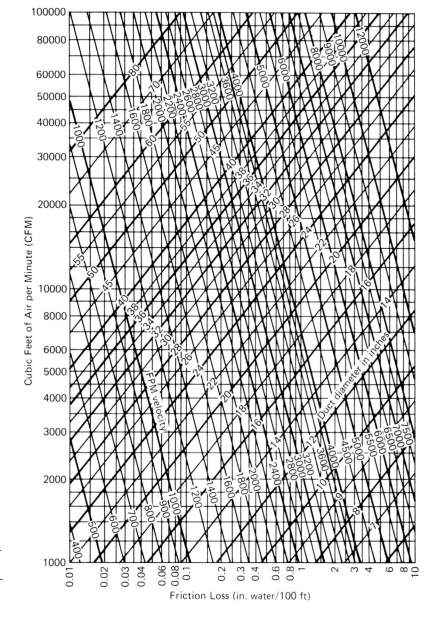

FIGURE 14-1b Friction loss in straight, aluminum sheet metal ducts for volumes 1000 to 100,000 cfm (based on standard air density of 0.0750 lb/ft³ flowing through average, clean, round, aluminum ducts having approximately 40 joints per 100 ft). (Courtesy of Reynolds Metal Company)

$$
\begin{aligned}
\text{total friction loss} &= 0.086(155/100) \\
&= 0.133 \text{ in. WG}
\end{aligned}
$$

$$
\frac{(h_L)_2}{(h_L)_1} = \frac{\rho_{a2}}{\rho_{a1}} \qquad (14\text{-}2)
$$

14-4 CHANGES IN DUCT FRICTION LOSS WITH CHANGE IN AIR DENSITY

When the air flowing in ducts is at a density significantly different from that for which the design charts (Figures 14-1a and 14-1b) were constructed, a correction for density must be made.

With a *constant volume of air flowing* (not a constant mass), the friction loss for a duct system varies directly as the density. Therefore, we may say that

where $(h_L)_1$ and ρ_{a_1} are the static head loss and density, respectively, of the air at condition 1—that is, standard air; and $(h_L)_2$ and ρ_{a_2} are the static head loss and density, respectively, at a condition 2.

From this, then, the mass flow changes with any variation in density since Eq. (14-2) is based on volume flow through the duct. It is usually desirable to maintain a definite mass flow of air to absorb or supply sufficient heat in the conditioned space.

When this is the case, confusion can be avoided if

the friction loss is first determined for a volume of standard air equal to the volume of actual nonstandard air. The friction loss is then modified by means of Eq. (14-2) to the loss for the nonstandard air.

The actual volume and actual pressure loss obtained by means of these computations are those for which a fan must be selected. This selection procedure will be discussed in a later section.

ILLUSTRATIVE PROBLEM 14-4

Find the fan selection volume and actual friction loss to maintain a *constant-mass flow* corresponding to a standard air volume of 11,050 cfm in a 32-in. diameter round duct 500 ft long, when the air density is 0.0637 lbm/ft^3.

Solution: The actual volume of air flowing at a density of 0.0637 lbm/ft^3, maintaining constant mass, is

$$11,050(0.0750/0.0637) = 13,000 \text{ cfm}$$

From Figure 14-1b, the friction rate with 13,000 cfm of standard air in a 32-in. round duct is 0.190 in. WG per 100 ft. The friction loss in 500 ft is

$$\begin{aligned} \text{friction loss} &= 0.190(500/100) \\ &= 0.95 \text{ in. WG} \end{aligned}$$

By Eq. (14-2),

$$\begin{aligned} (h_L)_2 &= (h_L)_1(\rho_{a2}/\rho_{a1}) \\ &= 0.95(0.0637/0.0750) \\ &= 0.81 \text{ in. WG (which is the actual loss)} \end{aligned}$$

ILLUSTRATIVE PROBLEM 14-5

A sheet metal duct 75 ft long is 18 in. in diameter. The quantity of standard air to be delivered through the duct is 3000 cfm. Find (1) the friction loss in the duct and (2) if the air temperature were increased to 120°F, what the friction loss would be.

Solution:

1. Referring to Figure 14-1b, with 3000 cfm and an 18-in. round duct, find the friction rate of 0.22 in. WG per 100 ft. Therefore, the friction loss for a 75-ft length is

$$\begin{aligned} \text{friction loss} &= 0.22(75/100) \\ &= 0.165 \text{ in. WG} \end{aligned}$$

2. Since there is no indication in the problem statement concerning constant-mass flow, constant-volume flow is assumed. Only the temperature changes, and we know that the density varies inversely with absolute temperature. Therefore, Eq. (14-2) may be modified as follows:

$$\frac{(h_L)_2}{(h_L)_1} = \left(\frac{460 + 70}{460 + t}\right)$$

Therefore,

$$\begin{aligned} (h_L)_2 &= 0.165\left(\frac{460 + 70}{460 + 120}\right) \\ &= 0.151 \text{ in. WG} \end{aligned}$$

Density of air through a fan and duct differ: Occasionally installations are made in which a heating coil is installed between the fan and the ductwork. As a result of this, the air flowing through the duct is at a different temperature and has a different density than the air flowing through the fan. Evidently, the mass of air flowing through the duct is exactly the same as the mass delivered by the fan. However, the volume of air flowing through the duct is greater than the volume of air flowing through the fan.

If the duct had been designed on the basis of the volume of standard air needed, the static pressure against which the fan must actually work will be different from the static pressure computed on the basis of standard air flowing in the duct. For a *constant mass of air flowing through the duct,* the static pressure against which the fan must work can be computed by means of the following relationship:

$$\frac{(h_L)_2}{(h_L)_1} = \frac{\rho_{a1}}{\rho_{a2}} \qquad (14\text{-}2a)$$

where $(h_L)_1$ and ρ_{a1} are the static pressure and density, respectively, of standard air; and $(h_L)_2$ and ρ_{a2} are the actual static pressure and density of the air flowing through the duct.

ILLUSTRATIVE PROBLEM 14-6

The static pressure loss in a duct with standard air flowing through the duct is 0.77 in. WG. Find the static pressure loss in the duct if the air flowing through the duct has a density of 0.0675 lb/ft^3, maintaining constant-mass flow.

Solution: By Eq. (14-2a),

$$\begin{aligned} (h_L)_2 &= (h_L)_1(\rho_{a1}/\rho_{a2}) \\ &= 0.77(0.075/0.0675) \\ &= 0.856 \text{ in. WG} \end{aligned}$$

For the condition of this problem, a fan handling standard air should be selected for the volume of standard air needed, but against a static pressure of 0.856 in. WG, not 0.77 in. WG. Thus, if 15,000 cfm of standard air were needed for the conditions of the problem, the fan would be selected for 15,000 cfm, operating against a static pressure of 0.856 in. WG.

14-5 EQUIVALENT RECTANGULAR DUCTS

Rectangular ducts are used more frequently than round ducts. Although round ducts require the least metal in order to carry a given quantity of air, rectangular ducts are used in most installations because of space consideration. As a rule, rectangular ducts fit into available spaces in buildings and can be installed less conspicuously than can round ducts.

A greater friction loss occurs in a rectangular duct than in a circular duct of equal cross-sectional flow area. If the friction loss per foot of length in a circular duct is to be made equal to that in a rectangular duct, fundamental expressions for the friction loss for the two types of ducts are equated; and, with suitable friction factors being used, the following expression is obtained:

$$D_e = 1.3 \left[\frac{(HW)^5}{(H+W)^2} \right]^{1/8} = 1.3 \frac{(HW)^{0.625}}{(H+W)^{0.25}} \quad (14\text{-}3)$$

where H and W represent height and width of a rectangular duct (ft or in.) and D_e is the diameter (ft or in.) of a circular duct having the same friction loss per foot as a rectangular duct delivering the same quantity of air.

Table 14-2 gives the circular equivalents of rectangular ducts based on Eq. (14-3). It should be noted that the mean velocity in rectangular ducts is less than that in an equivalent round duct.

Rectangular duct dimensions are seldom quoted in fractions of inches because of the additional time and labor cost for fabrication and also because in using the design charts it is difficult to work with such accuracy. It has become common practice to specify one dimension of the duct (usually the vertical dimension) as 4, 8, 12, 14, 16, 18, or 20 in. and to let the duct width vary as required to obtain the flow area.

ILLUSTRATIVE PROBLEM 14-7

If 4000 cfm of standard air flows through a 20-in. diameter duct, determine the friction loss for a 75-ft section of straight duct, the flow velocity in the round duct, the equivalent rectangular duct size, and the flow velocity in the rectangular duct.

Solution: Referring to Figure 14-1b, for 4000 cfm flowing through a 20-in. diameter duct, find $V = 1880$ fpm and the friction rate of 0.22 in. WG per 100 ft. For 75 ft of length, the friction loss is $0.22(75/100) = 0.165$ in. WG.

By referring to Table 14-2, an equivalent rectangular duct may be selected. We find that many different duct dimensions could be selected, three of which are 50 by 8 in. ($D_e = 19.9$ in.), 30 by 12 in. ($D_e = 20.2$ in.), or 22 by 16 in. ($D_e = 20.4$ in.). If we select the 30-by-12-in. duct, the mean flow velocity in the rectangular duct would be

$$V = \frac{Q}{A} = \frac{4000(144)}{(30)(12)} = 1600 \text{ fpm}$$

ILLUSTRATIVE PROBLEM 14-8

A duct 24 by 12 in. is to deliver 3000 cfm of standard air. Find the friction loss per 100 ft of duct length.

Solution: Referring to Table 14-2, for a 24-by-12-in. duct, find the equivalent round duct to be 18.3 in. Referring to Figure 14-1b, for 3000 cfm and 18.3-in. diameter, find the friction loss of 0.20 in. WG per 100 ft of duct length.

Aspect Ratio. When rectangular ducts are used, particular care should be exercised to maintain a low duct aspect ratio (AR). Aspect ratio of a rectangular duct is the ratio of the long side to the short side of the duct. This ratio should be kept below 6 to 1 for the following reasons, all of which are economic:

1. Heat gain or loss to the air in the duct increases as the aspect ratio increases.
2. Material quantities increase with an increase in aspect ratio, as well as labor costs.
3. Operating costs increase because friction losses increase with increasing aspect ratios.

These facts can be illustrated in the following manner. Refer to Figure 14-2, where two duct cross sections are represented, each having the same cross-sectional area of 144 in.^{2}. For section *a*, the aspect ratio is 1:1 and the perimeter is 48 in. For section *b*, the aspect

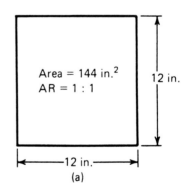

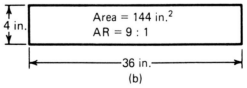

FIGURE 14-2 Rectangular duct cross sections illustrating duct aspect ratio.

ratio is 9:1 and the perimeter is 80 in. Therefore, the higher aspect ratio section requires more material per foot of length and has greater exposed surface area for heat transfer. Furthermore, the long side of the duct may need cross bracing to support the sheet metal.

Assume that a rectangular duct is to be sized to carry 5000 cfm of standard air at a velocity of 2000 fpm. This would require a flow area of $5000/2000 = 2.5 \text{ ft}^2 = 360 \text{ in.}^2$. Table 14-3 shows the effect on friction rate as the aspect ratio changes.

14-6 DYNAMIC (SHOCK) LOSSES IN DUCTWORK

Dynamic losses result from disturbances in the flow caused by fittings that cause a change in direction and/or area of the airflow path such as inlets, outlets, diffusers, nozzles, orifices, tees, elbows, and other obstructions. Much data have been presented in past years in manufacturers' data books, handbooks on fluid flow, and scientific papers. In some cases, the data are contradictory and limited to particular flow rates and shapes of fittings. Also, the data are reported in the format of the author's choice, resulting in the presentation of losses in terms of total equivalent length, additional equivalent length, equivalent length expressed as dimensional multipliers, static regain, or velocity pressure head multipliers called loss coefficients.

In this book, we will use the loss coefficients presented in Tables 14-4 through 14-7. The dynamic loss caused by the fitting shown may be calculated from

$$\text{dynamic loss, } (h_L)_d = C_o \, h_{vo} \qquad (14\text{-}4)$$

where, $(h_L)_d$ is the dynamic pressure loss in in. WG, C_o is the corrected loss coefficient and equal to either $K_\theta C_o'$ or $K_\theta K_{RE} C_o'$ dependent on data given, and h_{vo} is the velocity head in in. WG at the section o. K_θ is a correction factor for angle of fitting, K_{RE} is a correction factor for Reynolds number, and C_o' is the uncorrected loss coefficient.

TABLE 14-3

Effect of aspect ratio on friction rate

Rectangular Dimensions H × W	AR	Equivalent Diam., D_e (Table 14-2)	Fraction Rate Figure 14-1b
18 × 20	1 : 1	20.7	0.28
12 × 30	2.5 : 1	20.2	0.32
10 × 36	3.6 : 1	19.8	0.36
8 × 45	5.6 : 1	19.1	0.43
6 × 60	10 : 1	18.1	0.54

TABLE 14-4

(A) Loss coefficient for *round* duct elbows, smooth radius

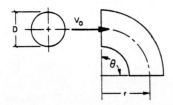

$C_o = K_\theta C_o'$

Coefficients for 90° Elbows

r/D	0.5	0.75	1.0	1.5	2.0	2.5
C_o'	0.71	0.33	0.22	0.15	0.13	0.12

Angle Correction Factors K_θ

θ	0	20	30	45	60	75	90	110	130	150	180
K_θ	0	0.31	0.45	0.60	0.78	0.90	1.00	1.13	1.20	1.28	1.40

(B) Loss coefficients for *round* duct elbows of 3,4,5 pieces

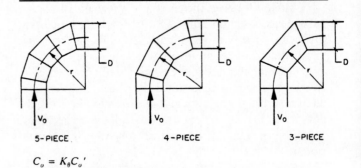

5-PIECE   4-PIECE   3-PIECE

$C_o = K_\theta C_o'$

Coefficients for 90° Elbows (C_o')

No. of Pieces	r/D			
	0.75	1.0	1.5	2.0
5	0.46	0.33	0.24	0.19
4	0.50	0.37	0.27	0.24
3	0.54	0.42	0.34	0.33

Angle Correction Factors K_θ

θ	0	20	30	45	60	75	90	110	130	150	180
K_θ	0	0.31	0.45	0.60	0.78	0.90	1.00	1.13	1.20	1.28	1.40

Source: Reprinted from *ASHRAE Handbook 1989 Fundamentals,* with permission from American Society of Heating, Refrigerating, and Air Conditioning Engineers, Atlanta, GA.

TABLE 14-5

Loss coefficients for *rectangular* duct elbows without vanes

Smooth Radius

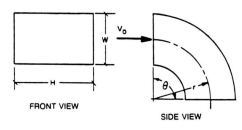

FRONT VIEW

SIDE VIEW

$$C_o = K_\theta K_{Re} C_o'$$

90°, Sharp Throat Radius Heel ($r/W = 0.5$)

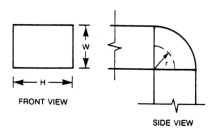

FRONT VIEW

SIDE VIEW

$$C_o = K_{Re} C_o'$$

Coefficients for 90° Elbows (C_θ')

	H/W										
r/W	0.25	0.5	0.75	1.0	1.5	2.0	3.0	4.0	5.0	6.0	8.0
0.5	1.3	1.3	1.2	1.2	1.1	1.0	1.0	1.1	1.1	1.2	1.2
0.75	0.57	0.52	0.48	0.44	0.40	0.39	0.39	0.40	0.42	0.43	0.44
1.0	0.27	0.25	0.23	0.21	0.19	0.18	0.18	0.19	0.20	0.21	0.21
1.5	0.22	0.20	0.19	0.17	0.15	0.14	0.14	0.15	0.16	0.17	0.17
2.0	0.20	0.18	0.16	0.15	0.14	0.13	0.13	0.14	0.14	0.15	0.15

Angle Correction Factor

θ	0	20	30	45	60	75	90	110	130	150	180
K_θ	0	0.31	0.45	0.60	0.78	0.90	1.00	1.13	1.20	1.28	1.40

Reynolds Number Correction Factor (K_{Re})

	Re × 10⁻⁴								
r/W	1	2	3	4	6	8	10	14	≥ 20
0.5	1.40	1.26	1.19	1.14	1.09	1.06	1.04	1.0	1.0
≥ 0.75	2.0	1.77	1.64	1.56	1.46	1.38	1.30	1.15	1.0

Source: Reprinted from *ASHRAE Handbook 1989 Fundamentals,* with permission from American Society of Heating, Refrigerating, and Air Conditioning Engineers, Atlanta, GA.

TABLE 14-6

(A) Loss coefficients for *round* duct elbows, mitered

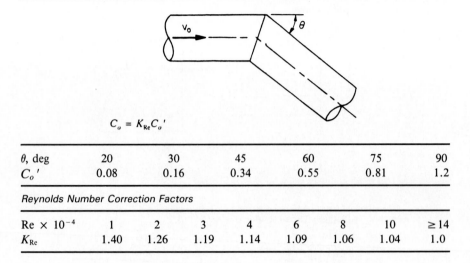

$$C_o = K_{Re}C_o'$$

θ, deg	20	30	45	60	75	90
C_o'	0.08	0.16	0.34	0.55	0.81	1.2

Reynolds Number Correction Factors

Re × 10⁻⁴	1	2	3	4	6	8	10	≥ 14
K_{Re}	1.40	1.26	1.19	1.14	1.09	1.06	1.04	1.0

(B) Loss coefficients for *rectangular* duct elbows, mitered

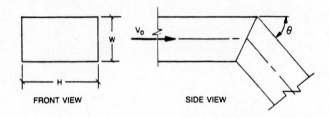

FRONT VIEW   SIDE VIEW

$$C_o = K_{Re}C_o'$$

C_o'

θ, deg					H/W						
	0.25	0.5	0.75	1.0	1.5	2.0	3.0	4.0	5.0	6.0	8.0
20	0.08	0.08	0.08	0.07	0.07	0.07	0.06	0.06	0.05	0.05	0.05
30	0.18	0.17	0.17	0.16	0.15	0.15	0.23	0.13	0.12	0.12	0.11
45	0.38	0.37	0.36	0.34	0.33	0.31	0.28	0.27	0.26	0.25	0.24
60	0.60	0.59	0.57	0.55	0.52	0.49	0.46	0.43	0.41	0.39	0.38
75	0.89	0.87	0.84	0.81	0.77	0.73	0.67	0.63	0.61	0.58	0.57
90	1.3	1.3	1.2	1.2	1.1	1.1	0.98	0.92	0.89	0.85	0.83

Reynolds Number Correction Factors

Re × 10⁻⁴	1	2	3	4	6	8	10	≥ 14
K_{Re}	1.40	1.26	1.19	1.14	1.09	1.06	1.04	1.0

Source: Reprinted from *ASHRAE Handbook 1989 Fundamentals,* with permission from American Society of Heating, Refrigerating, and Air Conditioning Engineers, Atlanta, GA.

TABLE 14-7

(A) Loss coefficients for *round* duct transitions

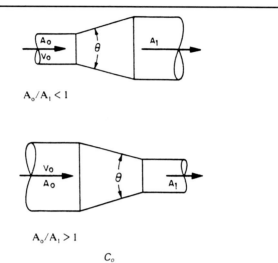

$A_o/A_1 < 1$

$A_o/A_1 > 1$

C_o

A_o/A_1	\u03b8, degrees									
	10	15	20	30	45	60	90	120	150	180
0.06	0.21	0.29	0.38	0.60	0.84	0.88	0.88	0.88	0.88	0.88
0.1	0.21	0.28	0.38	0.59	0.76	0.80	0.83	0.84	0.83	0.83
0.25	0.16	0.22	0.30	0.46	0.61	0.68	0.64	0.63	0.62	0.62
0.5	0.11	0.13	0.19	0.32	0.33	0.33	0.32	0.31	0.30	0.30
1	0	0	0	0	0	0	0	0	0	0
2	0.20	0.20	0.20	0.20	0.22	0.24	0.48	0.72	0.96	1.0
4	0.80	0.64	0.64	0.64	0.88	1.1	2.7	4.3	5.6	6.6
6	1.8	1.4	1.4	1.4	2.0	2.5	6.5	10	13	15
10	5.0	5.0	5.0	5.0	6.5	8.0	19	29	37	43

(B) Loss coefficients for *rectangular* duct transitions, pyramidal

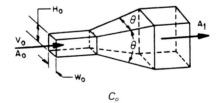

C_o

A_o/A_1	\u03b8, degrees									
	10	15	20	30	45	60	90	120	150	180
0.06	0.26	0.30	0.44	0.54	0.53	0.65	0.77	0.88	0.95	0.98
0.1	0.24	0.30	0.43	0.50	0.53	0.64	0.75	0.84	0.89	0.91
0.25	0.20	0.25	0.34	0.36	0.45	0.52	0.58	0.62	0.64	0.64
0.5	0.14	0.15	0.20	0.21	0.25	0.30	0.33	0.33	0.33	0.32
1	0	0	0	0	0	0	0	0	0	0
2	0.23	0.22	0.21	0.20	0.22	0.2	0.49	0.74	0.99	1.1
4	0.84	0.68	0.68	0.64	0.88	1.1	2.7	4.3	5.6	6.6
6	1.8	1.5	1.5	1.4	2.0	2.5	6.5	10	13	15
10	5.0	5.0	5.1	5.0	6.5	8.0	19	29	37	43

Source: Reprinted from *ASHRAE Handbook 1989 Fundamentals,* with permission from American Society of Heating, Refrigerating, and Air Conditioning Engineers, Atlanta, GA.
[a] $A_o/A_1 > 1$ is tentative (adapted from TABLE 14-7a).

Chapter 32 of *ASHRAE Handbook 1989 Fundamentals* contains loss coefficients for many types of fittings that may be found in ductwork design.

14-7 DYNAMIC LOSSES IN ELBOWS, TRANSITIONS, AND BRANCH TAKEOFFS

The elbows in duct systems must be correctly designed if the friction losses are to be held to a minimum. In addition to keeping the friction losses to a minimum, it is very important that the airflow be uniformly distributed over the cross section of the elbow if a register or diffuser supply grille is placed at the outlet of an elbow. Excessive friction loss in elbows is due to the fact that air is not uniformly distributed over the duct flow area. The portion of the airstream that travels along the outside edge of the elbow is deflected around the turn. However, the portion of the air traveling along the inside tends to continue in a straight line until it impinges against the airstream on the outer edge. As a result of the turbulence created, the dynamic loss in elbows is likely to be high unless this turbulence is prevented or reduced to a minimum.

Before proceeding with the methods of designing elbows, the meaning of a few terms that will be used must be understood. In straight lengths of horizontal ductwork, the width (W) of the duct is the horizontal dimension of the ductwork and the depth (H) is the vertical dimension. However, when we are discussing elbows, the terms *width* and *depth* of the elbow have a different meaning. The width (W) of an elbow is always the dimension of the elbow that lies in the same plane as the radius of the elbow. Thus, with reference to Figure 14-3, the width of elbow A is 20 in. and its depth is 10 in. For elbow A of this illustration, the conventional definitions of width and depth of straight duct apply.

However, for elbow B, the width of the duct is 10 in. and its depth is 20 in. The width of an elbow is *always* the dimension in the same plane as the elbow radius.

A term that is widely used in elbow design is *radius ratio*. For a round duct, the radius ratio is r/D. For a rectangular duct, the radius ratio is found by dividing the centerline radius of an elbow by its width (r/W). Thus, for Figure 14-3, the radius ratio of elbow A would be $(12 + 10)/20 = 1.1$ and, for elbow B, it would be $(12 + 5)/10 = 1.7$.

Another factor that is important in the design of rectangular elbows is the aspect ratio of the elbow. The aspect ratio for elbows and for straight ducts is defined differently. The aspect ratio of a straight duct is always equal to the long side divided by the short side of the rectangular duct, as previously noted. However, in the case of elbows, the aspect ratio is always equal to the depth of the duct divided by its width (H/W). Thus, the aspect ratio of elbow A of Figure 14-3 is $10/20 = 0.5$ and, for elbow B, the aspect ratio is $20/10 = 2.0$.

For elbow turns other than 90 degrees, the loss is not proportional to the angle. Correction factors, K_θ, for loss at angles other than 90 degrees are shown in Tables 14-4 and 14-5.

The dynamic loss coefficient, C_o, is not affected by the roughness of the duct walls except in elbows. Here, it affects the dynamic loss coefficient, thus introducing another friction factor besides dynamic loss and normal friction loss.

Static pressure losses in 90-degree and angular takeoffs for round ducts (Figure 14-4) and for 90-degree takeoffs for rectangular ducts (Table 14-8) depend upon the relative flows in the main and branch. These takeoff losses are given as static pressure losses from the main to the branch. There is also a dynamic loss in the main (upstream to downstream) at the branch takeoff. This loss is normally very small in well-designed takeoffs and can be neglected for small and medium-sized systems. However, for large, more complex systems, this loss should be investigated.

ILLUSTRATIVE PROBLEM 14-9

Determine the dynamic pressure loss in a horizontal 90-degree elbow having a width of 12 in., a depth of 24 in., and a radius ratio of 0.75. Standard air flows through the duct at a flow rate of 4000 cfm.

Solution: The aspect ratio of the elbow (H/W) = 24/12 = 2.0. The radius ratio is given as 0.75. From Table 14-5a, we find $C_o = 0.39$. The velocity of flow in the elbow is

$$V = \frac{Q}{A} = \frac{4000(144)}{(24)(12)} = 2000 \text{ fpm}$$

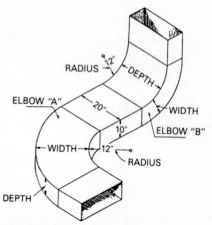

FIGURE 14-3 Duct elbow nomenclature.

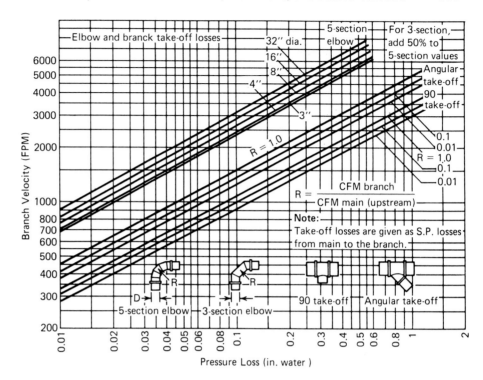

FIGURE 14-4 Elbow and branch takeoff losses for round ducts. (Reprinted from *Trane Air Conditioning Manual* with permission from The Trane Company, La-Crosse, WI)

By Eq. (13-6),

$$h_v = \left(\frac{V}{4005}\right)^2 = \left(\frac{2000}{4005}\right)^2 = 0.249 \text{ in. WG}$$

By Eq. (14-4),

$$(h_L)_d = C_o \times h_v = (0.39)(0.249)$$
$$= 0.097 \quad (\text{say, } 0.10) \text{ in. WG}$$

ILLUSTRATIVE PROBLEM 14-10

A 16-in.-diameter duct contains a five-piece, 90-degree elbow with a radius ratio of 2.0. Flow rate of standard air through the elbow is 3000 cfm. Determine the dynamic loss through the elbow by using (1) Table 14-4 and (2) Figure 14-4.

Solution:

1. The velocity of airflow is

$$V = \frac{Q}{A} = \frac{3000}{(\pi/4)(16/12)^2} = 2150 \text{ fpm}$$

or, from Figure 14-1b, approximately 2150 fpm. By Eq. (13-6),

$$h_v = \left(\frac{V}{4005}\right)^2 = \left(\frac{2150}{4005}\right)^2 = 0.288 \text{ in. WG}$$

Using Table 14-4b, for a five-piece elbow with $r/D = 2.0$, find $C_o = 0.19$. By Eq. (14-4),

$$(h_L)_d = C_o \times h_v = (0.19)(0.288)$$
$$= 0.0547 \text{ (say, } 0.055) \text{ in. WG}$$

2. For a flow velocity of 2150 fpm and a 16-in. round section, the static pressure loss for a five-piece, 90-degree elbow is read as 0.055 approximately.

The radius ratio used for the data of Figure 14-4 is not indicated. If the data for Figure 14-4 is to be compared with that of Table 14-4b, one would have to assume that the radius ratio used for data in Figure 14-4 was equal to 2.0. Such comparison is risky. As a suggestion, for future work regarding losses in round elbows, we shall refer to the data of Table 14-4.

ILLUSTRATIVE PROBLEM 14-11

A 90-degree, round branch takeoff occurs in a round duct main. The flow in the 20-in. main is 4000 cfm and in the 8-in. branch is 400 cfm. What is the static pressure loss from the main to the branch?

Solution: By reference to Figure 14-1a, the branch velocity is 1150 fpm. From Figure 14-4, we need the ratio of the flow in the branch to the flow in the main (upstream), or

$$R = \frac{\text{cfm branch}}{\text{cfm main (upstream)}} = \frac{400}{4000} = 0.10$$

In Figure 14-4, with $R = 0.10$ and velocity in the branch of 1150 fpm, find pressure loss equal to 0.125 in. WG.

TABLE 14-8

Branch takeoff losses for rectangular ducts

	Aspect Ratio: H/W									
	.25—.50		.75—3.0			4—6				
	Velocity Ratio: V_B/V_U									
	1.2	1.3	1.1	1.2	1.3	.9	1.0	1.1	1.2	1.3
Branch Velocity, V_B, fpm	Area Ratio: A_B/A_U									
	.7—1.2		.4—.6			.15—.3				
	Static Pressure Loss. Inches wg									
600	.011	.013	.007	.011	.018	.008	.011	.015	.018	.02
800	.02	.024	.013	.02	.032	.015	.02	.027	.032	.036
1000	.032	.038	.020	.032	.051	.023	.031	.042	.05	.057
1200	.046	.055	.028	.046	.073	.033	.045	.061	.073	.082
1400	.062	.074	.038	.062	.099	.045	.061	.082	.098	.111
1600	.081	.097	.05	.081	.129	.059	.08	.108	.129	.145
1800	.102	.123	.063	.102	.163	.074	.101	.136	.163	.183
2000	.126	.151	.078	.126	.20	.091	.125	.168	.20	.23
2100	.139	.167	.086	.139	.22	.101	.138	.185	.22	.25
2200	.153	.184	.095	.153	.24	.111	.151	.20	.24	.27
2300	.166	.20	.104	.166	.27	.121	.165	.22	.27	.30
2400	.182	.22	.113	.182	.29	.131	.18	.24	.29	.33
2500	.197	.24	.122	.197	.31	.142	.195	.26	.31	.35
2600	.21	.26	.133	.21	.34	.154	.21	.28	.34	.38
2700	.23	.28	.143	.23	.37	.166	.23	.31	.37	.41
2800	.25	.30	.153	.25	.39	.179	.24	.33	.39	.44
2900	.27	.32	.165	.27	.42	.192	.26	.35	.42	.48
3000	.28	.34	.176	.28	.45	.21	.28	.38	.45	.51
3100	.30	.36	.188	.30	.48	.22	.30	.40	.48	.54
3200	.32	.39	.20	.32	.52	.23	.32	.43	.52	.58
3300	.34	.41	.21	.34	.55	.25	.34	.46	.55	.62
3400	.36	.44	.23	.36	.58	.26	.36	.49	.58	.65
3500	.39	.46	.24	.39	.62	.28	.38	.51	.62	.69
3600	.41	.49	.25	.41	.65	.30	.40	.54	.65	.73
3700	.43	.52	.27	.43	.69	.31	.43	.58	.69	.78
3800	.46	.55	.28	.46	.73	.33	.45	.61	.73	.82
3900	.48	.58	.30	.48	.77	.35	.47	.64	.77	.86
4000	.51	.61	.31	.51	.81	.36	.50	.67	.81	.91
4100	.53	.64	.33	.53	.85	.38	.52	.71	.85	.95
4200	.56	.67	.35	.56	.89	.40	.55	.74	.89	1.00
4300	.58	.70	.36	.58	.93	.42	.58	.78	.93	1.05
4400	.61	.73	.38	.61	.97	.44	.60	.81	.97	1.10
4600	.67	.80	.41	.67	1.07	.48	.66	.89	1.06	1.20
4800	.73	.87	.45	.73	1.16	.53	.72	.97	1.16	1.30
5000	.79	.95	.49	.79	1.26	.57	.78	1.05	1.26	1.41
5200	.85	1.03	.53	.85	1.36	.62	.84	1.14	1.36	1.53
5400	.92	1.11	.57	.92	1.47	.67	.91	1.22	1.46	1.65
5600	.99	1.19	.61	.99	1.58	.72	.98	1.32	1.57	1.77
5800	1.06	1.27	.66	1.06	1.69	.77	1.05	1.41	1.69	1.90
6000	1.14	1.36	.70	1.14	1.81	.82	1.12	1.51	1.81	2.04

Example: Upstream duct, 24 by 12 inches, handling 5000 cfm; branch duct, 12 by 12, handling 3000 cfm; A_B/A_U = 0.5; H/W = 1.0; V_B/V_U = 3000/2500 = 1.2. Static pressure drop for 3000-fpm branch velocity = 0.28 inches wg.

Note: Radius ratio (R/W) should equal or exceed unity. For branch take-offs with lower velocity ratios (V_B/V_U) uses values for elbows of comparable aspect ratio (H/W) and radius ratio of 1.0. See Table 14-5.

Source: Reprinted from Handbook of Air Conditioning, Heating, and Ventilating, 3rd ed., with permission from Industrial Press, New York, 1979.

ILLUSTRATIVE PROBLEM 14-12

A rectangular duct has a 90-degree branch takeoff located where the upstream flow in the main is 5000 cfm. The branch flow is 2000 cfm. The duct dimensions upstream of the branch takeoff are 22 in. in width and 12 in. in depth. The branch duct is 8 in. wide by 12 in. deep. The radius ratio is 1.0. Determine the static pressure loss for the 2000-cfm branch takeoff.

Solution: Use Table 14-8.

$$A_B/A_U = (8 \times 12)/(22 \times 12) = 0.36$$
$$\text{branch } H/W = 12/8 = 1.5$$
$$V_B = (2000 \times 144)/(8 \times 12)$$
$$= 3000 \text{ fpm}$$
$$V_U = (5000 \times 144)/(22 \times 12)$$
$$= 2727 \text{ fpm}$$
$$V_B/V_U = 3000/2727 = 1.1$$

From Table 14-8, find loss in branch takeoff of 0.176 in. WG.

14-8   ELBOWS WITH SPLITTERS

When an elbow is followed by a duct whose length is at least 4 times the equivalent diameter of the elbow, no splitters are required if the radius ratio of the elbow has a value of 1.5. The friction loss in an ordinary elbow with a radius ratio of 1.5 is about the same as the friction loss in an elbow of smaller radius ratio with splitters. Hence, where space permits, an ordinary elbow of this radius ratio should be used.

When space limitations make it necessary to install elbows having a radius ratio less than 1.5, splitters should be installed in the elbow. For elbows of a small radius ratio, the installation of such splitters materially reduces the pressure loss.

When elbows discharge directly into the atmosphere or are followed directly by a discharge grille or diffuser, splitters should always be used. The splitters help to distribute the airflow evenly across the discharge so that the airflow leaving the elbow is more uniform.

The dimensions of an elbow cannot be changed because they are determined by the duct in which the elbow is installed. However, if the aspect ratio of an elbow is low, the aspect ratio may be increased by installing splitters as shown in Tables 14-9a and 14-9b. The effect of installing splitters in an elbow is to make a number of separate, smaller elbows out of the original elbow.

Proper location of the splitters installed in an elbow is important. The data in Tables 14-9a and 14-9b can be used to determine the location of either one or two splitters. The use of one splitter is quite common, and two splitters are used occasionally. The splitters are usually located so that the radius ratio of each of the

separate elbows formed by the splitters is about the same. A word of caution: When splitters are used in an elbow to improve the aspect ratio, there is a possibility that the radius ratio of each of the individual elbows formed may be considerably greater than 2. In such a case, the friction loss of the elbow with splitters may be greater than the friction loss of a plain elbow due to the increased length of the elbow.

14-9   SQUARE (MITER) ELBOWS WITH VANES

Referring to Table 14-6, we may note that loss coefficients for square, or miter, elbows are high compared with loss coefficients for elbows of a round-heel radius. Frequently, however, because of requirements of appearance or space, a square elbow must be used. If this is the case, vanes should be installed as in Figure 14-5 to keep turbulence losses as low as possible. In addition, when such elbows are used to discharge air directly to the atmosphere or through a grille, vanes are very valuable for uniformly distributing the airflow.

The space between each pair of vanes in Figure 14-5 forms an individual elbow. Obviously, the aspect ratio of each of the individual elbows thus depends on the space between the vanes, and this space, in turn, depends upon the number of vanes used. As a rule, vanes are spaced so that the aspect ratio of each of the individual elbows formed by the vanes will be about 5. Thus, elbow *A* of Figure 14-3 is 20 in. wide and 10 in. deep. If this elbow were made square and the vanes were spaced 2 in. apart, the aspect ratio of each elbow formed would be $10/2 = 5$. Consequently, if the elbow were designed like the one in Figure 14-5, since the distance between each vane is 2 in. and since the elbow is 20 in. wide, the number of vanes required would be $(20/2) - 1 = 9$.

An extended lip is usually installed on each end of the vanes. This extension is usually made equal to about half the radius on the entering side and equal to the radius on the leaving side, as shown in Figure 14-5.

Because the blades illustrated in Figure 14-5 are each drawn with a different center, the space between corresponding points of adjacent blades is not the same. Consequently, the velocity of the air will change as it flows through each of the individual elbows formed by adjacent blades. Because of this change in velocity, eddy losses occur that increase the loss of such elbows. It is possible, by using airfoil-shaped blades, to maintain a uniform spacing and hence a uniform velocity. These blades are more expensive, and frequently the gain by use of them is not sufficient to warrant the additional cost.

Ducturns are nonadjustable, 90-degree air turns

TABLE 14-9a

Loss coefficients for smooth, rectangular elbows with *one* splitter vane

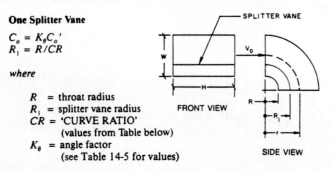

One Splitter Vane

$C_o = K_\theta C_o'$
$R_1 = R/CR$

where

R = throat radius
R_1 = splitter vane radius
CR = 'CURVE RATIO'
       (values from Table below)
K_θ = angle factor
       (see Table 14-5 for values)

Coefficients for elbows with 1 splitter vane (C_o')

								H/W					
R/W	r/W	CR	0.25	0.5	1.0	1.5	2.0	3.0	4.0	5.0	6.0	7.0	8.0
0.05	0.55	0.362	0.26	0.20	0.22	0.25	0.28	0.33	0.37	0.41	0.45	0.48	0.51
0.10	0.60	0.450	0.17	0.13	0.11	0.12	0.13	0.15	0.16	0.17	0.19	0.20	0.21
0.15	0.65	0.507	0.12	0.09	0.08	0.08	0.08	0.09	0.10	0.10	0.11	0.11	0.11
0.20	0.70	0.550	0.09	0.07	0.06	0.05	0.06	0.06	0.06	0.06	0.07	0.07	0.07
0.25	0.75	0.585	0.08	0.05	0.04	0.04	0.04	0.04	0.05	0.05	0.05	0.05	0.05
0.30	0.80	0.613	0.06	0.04	0.03	0.03	0.03	0.03	0.03	0.03	0.04	0.04	0.04

Source: Reprinted from *ASHRAE Handbook 1989 Fundamentals,* with permission from American Society of Heating, Refrigerating, and Air Conditioning Engineers, Atlanta, GA.

TABLE 14-9b

Loss coefficients for smooth, rectangular elbows with *two* splitter vanes

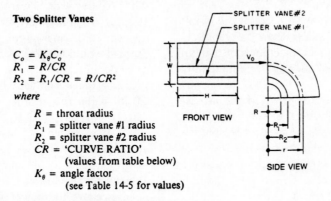

Two Splitter Vanes

$C_o = K_\theta C_o'$
$R_1 = R/CR$
$R_2 = R_1/CR = R/CR^2$

where

R = throat radius
R_1 = splitter vane #1 radius
R_2 = splitter vane #2 radius
CR = 'CURVE RATIO'
       (values from table below)
K_θ = angle factor
       (see Table 14-5 for values)

Coefficients for elbows with 2 splitter vanes (C_o')

								H/W					
R/W	r/W	CR	0.25	0.5	1.0	1.5	2.0	3.0	4.0	5.0	6.0	7.0	8.0
0.05	0.55	0.362	0.26	0.20	0.22	0.25	0.28	0.33	0.37	0.41	0.45	0.48	0.51
0.10	0.60	0.450	0.17	0.13	0.11	0.12	0.13	0.15	0.16	0.17	0.19	0.20	0.21
0.15	0.65	0.507	0.12	0.09	0.08	0.08	0.08	0.09	0.10	0.10	0.11	0.11	0.11
0.20	0.70	0.550	0.09	0.07	0.06	0.05	0.06	0.06	0.06	0.06	0.07	0.07	0.07
0.25	0.75	0.585	0.08	0.05	0.04	0.04	0.04	0.04	0.05	0.05	0.05	0.05	0.05
0.30	0.80	0.613	0.06	0.04	0.03	0.03	0.03	0.03	0.03	0.03	0.04	0.04	0.04

Source: Reprinted from *ASHRAE Handbook 1989 Fundamentals,* with permission from American Society of Heating, Refrigerating, and Air Conditioning Engineers, Atlanta, GA.

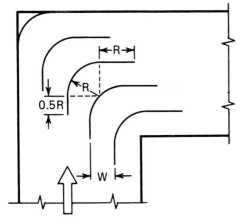

FIGURE 14-5 Turning vanes in rectangular (miter) elbows. (Reprinted from *Trane Air Conditioning Manual* with permission from The Trane Company, LaCrosse, WI)

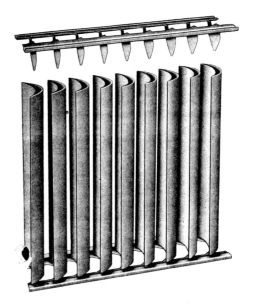

FIGURE 14-6 Ducturns. (Used by permission. Tuttle & Bailey, Holland, MI)

designed to reduce the pressure loss in square duct elbows. Such devices, as manufactured by Tuttle & Bailey, have galvanized steel blades, are double walled, and are formed to assure that any point on one blade is equidistant from the same point on an adjacent blade (see Figure 14-6). This precise blade shape maintains constant duct area and assures constant air velocity around the elbow, eliminates loss due to velocity changes, and results in low friction loss, as seen in Table 14-10. The blade is rigid and eliminates the need for cross bracing or reinforcing rods. Roll-formed from a single sheet of metal, surfaces and edges are smooth and free from edge friction and blade turbulence. Blades assemble over precision-formed tenons on the sidepieces, which ensures precise blade alignment, proper spacing, and predictable engineering performance. Blades and sidepieces are finished in 6-ft lengths. They are cut to size and assembled in the field. Assembly is simple and economical since no special tools and no fasteners are needed. The complete unit is screwed or riveted into the duct elbow.

14-10   DIVERGING AND CONVERGING DUCT SECTIONS

Special care must be exercised in making transition connections to and from equipment. Usually the cross-sectional area of the conditioning equipment (duct-mounted coils, etc.) will be greater than the cross-sectional area leading to or from the equipment, as shown in Figure 14-7, necessitating a diverging section of ductwork ahead of the equipment and a converging section after it. Angle A should be made as small as possible in order to spread the airflow evenly across the

face of the equipment and also to reduce the loss. When the angle is unduly large, the losses will be high because the air is unable to expand and fill the diverging section as rapidly as the metal sides diverge. Experiments have shown that the minimum loss is obtained when angle A is about 3 degrees. However, such small angles can rarely be used in actual installations because the length of the diverging section would be too great. An effort should be made to keep angle A less than 10 degrees. If it must be made larger because of space limitations, then splitters should be installed. These splitters will

TABLE 14-10

Pressure loss in ducturns, in. WG

Type of Elbow	Duct Velocity (fpm)					
	500	1000	1500	2000	2500	3000
Ducturns	<0.01	0.02	0.04	0.07	0.11	0.16
No vanes*	0.02	0.08	0.19	0.35	0.52	0.75

Source: Courtesy of Tuttle & Bailey, Holland, MI.
*See Table 14-5 or 14-6.

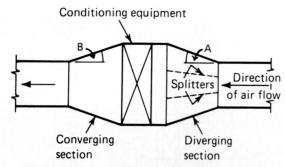

FIGURE 14-7   Transition pieces in ducts. (Reprinted from *Trane Air Conditioning Manual* with permission from The Trane Company, LaCrosse, WI)

serve the double purpose of reducing the loss and, at the same time, will distribute the air evenly over the face of the conditioning equipment. See Table 14-7 for losses in diverging sections.

The losses in converging sections are very much less than those of diverging sections. An airstream can be forced into a smaller area with almost no eddy loss if angle *B* of Figure 14-7 can be made less than 15 degrees. If the angle is greater than 15 degrees, splitters could be used to reduce losses, but because of cost it is seldom done, unless angle *B* exceeds 30 degrees. See Table 14-7 for losses in converging sections.

14-11 DUCT DESIGN METHODS

The most common methods of air duct system design are (1) velocity reduction, (2) equal friction rate, (3) static regain, and (4) constant velocity. Air conditioning and ventilation systems and exhaust systems conveying vapors, gases, and smoke generally are designed by the equal-friction-rate and static-regain methods. Exhaust systems conveying particulates are designed with a constant air velocity.

The choice of method generally depends on the designer's choice, which is usually influenced by his or her past experience and the size of the system. Small duct systems for residences and small commercial applications where airflow velocities seldom exceed 1000 fpm are frequently designed using a modification of the velocity-reduction and equal-friction-rate methods (see Section 14-15). Large systems using high-velocity airflow are most frequently designed by the static-regain method with the assistance of specially designed computer programs. Duct arrangements between the small and large systems are nearly always designed by the equal-friction-rate method. Sometimes, a duct system will be designed by a combination of two methods. For instance, the trunk (main) will be laid out by the static-

regain method and the branch ducts will be designed by the equal-friction-rate method.

There are no set rules concerning the airflow velocities to be used in duct system design. However, velocities recommended in Table 14-1 for small-capacity (5000 cfm or less) systems or velocity ranges shown in Figure 14-8 for low-velocity and high-velocity systems having greater volume flow may be used as guides in selecting velocities for various systems.

14-12 THE VELOCITY-REDUCTION METHOD

The velocity-reduction method of duct design requires an arbitrary assignment of velocities to the various sections of the duct system. The highest velocity is chosen at the entrance to the duct system, immediately following the fan outlet, in accordance with velocities shown in Table 14-1 or Figure 14-8. The velocities in succeeding sections are reduced as various branches are taken off the main duct. The velocity will be lowest at the end of the duct. Since the air quantity for each section of the duct is already known, assuming a velocity permits the flow area to be calculated, the diameter of the duct may be easily calculated. A more direct method would be to use the flow volume (cfm) and assumed velocity (fpm) and enter Figure 14-1a or 14-1b and determine the duct diameter and friction rate per 100 ft of duct.

The velocity-reduction method is empirical and requires good judgment by the designer to select appropriate velocities. In some cases, such as sizing exhaust duct systems, a modification of the velocity-reduction method is fastest and best suited for design purposes. This modified method, sometimes called *balanced-pressure-loss method,* which considers total pressure losses, makes it possible to design each branch run to have the same pressure loss from the fan in order that minimum dependence on dampering would be required.

ILLUSTRATIVE PROBLEM 14-13

The supply ductwork for a small office complex is shown in Figure 14-9. Ceiling diffusers will be used with the duct system above the ceiling. Supply-air volumes are represented at each outlet. Determine the following:

1. Size the ductwork by the velocity-reduction method, using a uniform duct depth of 14 in. in the main and branches.
2. Calculate the friction loss for the system, assuming that all branch takeoffs are 90-degree, rectangular elbows and that all elbows have a radius ratio of 1.0.

Solution: Tabulation of various quantities is helpful in the solution of problems in duct design. Table 14-11 will be used for this problem.

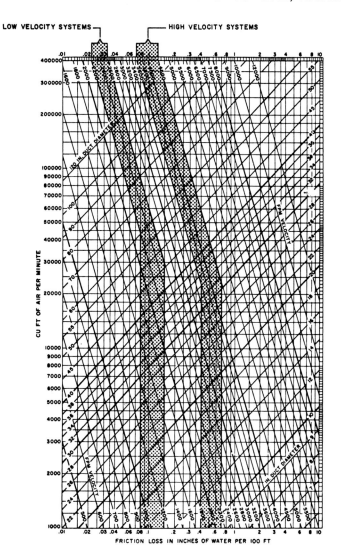

FIGURE 14-8 Recommended velocity ranges for airflow systems. (From *ASHRAE Handbook 1989 Fundamentals.* Used with permission of American Society of Heating, Refrigerating, and Air Conditioning Engineers, Atlanta, GA)

TABLE 14-11

Tabulation of results for Illustrative Problem 14-13

Duct Section	Capacity (cfm)	Round Duct Velocity (fpm)	Round Duct Size (in.)	Rectangular Duct W × H	Friction Rate (in. WG per 100 ft)
ABC	6000	1400	28.0	52 × 14	0.085
CD	4300	1300	25.0	40 × 14	0.085
DE	2500	1200	20.0	24 × 14	0.090
E–7	1000	900	14.3	12 × 14	0.085
E–5	1500	900	17.5	18 × 14	0.065
5–6	800	800	13.7	11 × 14	0.070
D–3	1800	900	19.3	24 × 14	0.059
3–4	1000	800	15.2	14 × 14	0.064
C–1	1700	900	19.0	22 × 14	0.060
1–2	900	800	14.3	12 × 14	0.066

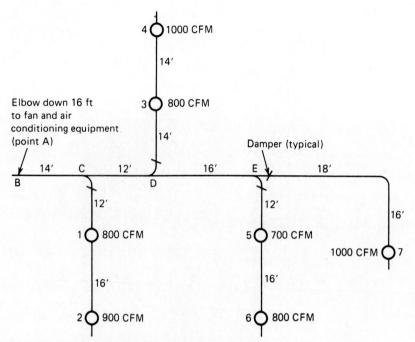

FIGURE 14-9 Diagram for Illustrative Problem 14-13 (does not represent an actual system).

Step 1. Tabulate the flow quantities in Table 14-11 for the various duct sections.

Step 2. Using Table 14-1 or Figure 14-8, select velocities of various duct sections and enter in Table 14-11.

For main ducts, Table 14-1 indicates a range of velocities between 1000 and 1300 fpm (maximum 1100 to 1600 fpm). Figure 14-8, for a flow of 6000 cfm, indicates a range of 1400 to 1750 fpm for the main *ABC*. We shall select 1400 fpm for section *ABC* and reduce it to 1300 fpm for section *CD*, 1200 fpm for section *DE*, and 900 fpm for section *E*–7.

For the branch ducts, Table 14-1 recommends 600 to 900 fpm. We shall select 900 fpm for sections *E*–5, *D*–3, and *C*–1; 800 fpm, for sections 5–6, 3–4, and 1–2. These velocities are on the high side, but remember that they are for the round ducts. That airflow velocities in the rectangular ducts selected for final design will be lower.

Step 3. Using Figures 14-1a and 14-1b for the cfm and selected velocities, determine the round sizes for the various sections and the friction rate and enter values in Table 14-11.

Step 4. Using Table 14-2, convert the round duct diameters to rectangular equivalents, maintaining a uniform duct depth of 14.0 in. Tabulate rectangular sizes in Table 14-11.

Step 5. Determine the maximum pressure loss for the system. The longest run (usually the one found to have the greatest loss) is *ABCDE* to outlet 7. In that section, we have an elbow at *B* and one in section *E*–7. In addition, we have the friction loss in the various duct sections.

For the elbow at B (vertical):

$$\text{radius ratio } (r/W) = 1.0 \text{ (given)}$$
$$\text{aspect ratio } (H/W) = 52/14 = 3.7$$

From Table 14-5,

$$C_o = 0.19 \text{ (approximate)}$$

The velocity in the rectangular duct is

$$V = Q/A = (6000)(144)/(52)(14)$$
$$= 1187 \text{ fpm}$$

Assuming standard air, the velocity head is

$$h_v = (V/4005)^2 = (1187/4005)^2$$
$$= 0.0878 \text{ in. WG}$$

Therefore, the loss for the elbow is

$$(h_L)_d = C_o \times h_v = (0.19)(0.0878)$$
$$= 0.017 \text{ in. WG}$$

For the elbow between E and 7 (horizontal):

$$\text{radius ratio } (r/W) = 1.0 \text{ (given)}$$
$$\text{aspect ratio } (H/W) = 14/12 = 1.16$$

From Table 14-5,

$$C_o = 0.21 \text{ (approximate)}$$

The velocity in the rectangular duct is

$$V = Q/A = (1000)(144)/(12)(14)$$
$$= 857 \text{ fpm}$$

Assuming standard air, the velocity head is

$$h_v = (V/4005)^2 = (857/4005)^2$$
$$= 0.0458 \text{ in. WG}$$

Therefore, the loss for the elbow is

$$(h_L)_d = C_0 \times h_v = (0.21)(0.0458)$$
$$= 0.0096 \text{ (say, 0.010) in. WG}$$

The various duct sections (*ABCDE* to 7) have slightly different friction rates (in. WG per 100 ft); therefore, we shall determine the friction loss in each section and sum the results:

$$
\begin{aligned}
ABC &= 0.085(30/100) = 0.0255\\
CD &= 0.085(12/100) = 0.0102\\
DE &= 0.090(16/100) = 0.0144\\
E\text{-}7 &= 0.085(34/100) = \underline{0.0289}\\
\text{straight duct friction} &= 0.0790 \text{ in. WG}
\end{aligned}
$$

The total loss for the duct run *ABCDE* to 7 is

$$
\begin{aligned}
\text{straight duct friction} &= 0.0790\\
\text{elbow loss at } B &= 0.0170\\
\text{elbow loss at } E\text{-}7 &= \underline{0.0100}\\
\text{total loss } ABCDE\text{-}7 &= 0.1060\\
&\quad (\text{say, } 0.110) \text{ in. WG}
\end{aligned}
$$

14-13 THE EQUAL-FRICTION-RATE METHOD

In the equal-friction-rate method of duct design, the design friction rate (in. WG per 100 ft) is maintained constant throughout the system. The design friction rate depends upon the allowable velocity in the duct system. In commercial installations, this is determined wholly by considerations of noise. For ordinary commercial and industrial systems, the maximum velocities shown in Table 14-1 or Figure 14-8 should not be exceeded.

ILLUSTRATIVE PROBLEM 14-14

The supply ductwork for a portion of an industrial system is shown in Figure 14-10. Determine the rectangular size for each section of the system using the equal-friction-rate method. The duct depth will be maintained at 14 in. throughout the system. Standard air is flowing in the system at the flow rates indicated. Determine the maximum static pressure required at point A to flow the system. Assume that the static pressure loss at each of the supply outlets is 0.05 in. WG; r/W for elbows will be taken as 1.0.

Solution:

Step 1. Determine the uniform design friction rate. Table 14-1 allows a velocity range of 1300 to 2200 fpm for main ducts and 1000 to 1800 fpm for branch lines in industrial applications. Figure 14-8 allows comparable velocity ranges.

Reference to Figure 14-1b, with our total flow rate of 9000 cfm and a velocity of 2200 fpm, gives the corresponding friction rate as 0.21 in. WG per 100 ft. Also, for 9000 cfm and a velocity of 1800 fpm, the corresponding friction rate is 0.125 in. WG per 100 ft. Since a vertical friction rate line of 0.150 in. WG per 100 ft appears on Figure 14-1b, and because this rate is between the above two values, we shall select a design friction rate of 0.150 in. WG, which will be held constant for the system.

Step 2. Construct a results table for the problem data. (See Table 14-12.)

Step 3. Tabulate flow quantities (cfm) and the design friction rate in Table 14-12.

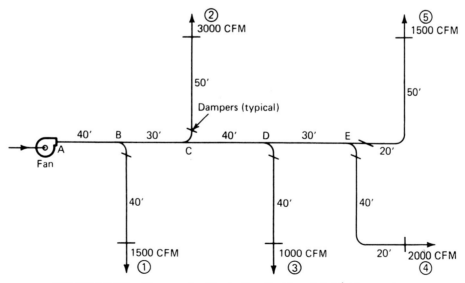

FIGURE 14-10 Diagram for Illustrative Problem 14-14 (does not represent an actual system).

TABLE 14-12
Tabulation of results for Illustrative Problem 14-14

Duct Section	Capacity (cfm)	Design Friction Rate (in. WG per 100 ft)	Round Duct Size (in.)	Rectangular Duct ($W \times H$)	Rectangular Duct Velocity (fpm)	Velocity Head (in. WG)
AB	9000	0.150	29.2	58 × 14	1596	0.159
BC	7500		27.0	48 × 14	1607	0.161
CD	4500		22.5	32 × 14	1446	0.130
DE	3500		20.5	26 × 14	1385	0.120
E–5	1500		15.0	14 × 14	1102	0.076
E–4	2000		16.6	17 × 14	1210	0.091
D–3	1000		12.8	10 × 14	1028	0.066
C–2	3000		19.4	22 × 14	1402	0.123
B–1	1500		15.0	14 × 14	1102	0.076

Step 4. Using Figures 14-1a and 14-1b with cfm and constant design friction rate, find round duct diameters and insert in Table 14-12.

Step 5. Using Table 14-2, convert the round duct diameters to rectangular equivalents and insert results in Table 14-12.

Step 6. Calculate airflow velocity in each section of the duct system from $Q = AV$ and insert results in Table 14-12. Two reasons for this are to determine whether the velocities are within limits and to determine the velocity heads in the sections for use as required in finding losses.

Step 7. Determine the greatest pressure loss for the system. It is difficult by inspection to establish whether the greatest loss occurs from point A to outlet 5 or to outlet 4. Therefore, we shall check each one.

(a) *Loss A to outlet 4:*

Branch takeoff at E—Use of Table 14-8 requires that we know the branch velocity V_B, aspect ratio H/W, velocity ratio V_B/V_U, and the area ratio A_B/A_U.

$$V_B = 1210 \text{ fpm} \quad \text{and} \quad V_U = 1385 \text{ fpm}$$
$$\text{(results table)}$$

$$H/W = 14/17 = 0.82$$
$$V_B/V_U = 1210/1385 = 0.87$$
$$A_B/A_U = (17 \times 14)(26 \times 14) = 0.65$$

From Table 14-8, find $(h_L)_d = 0.028$ in. WG (estimated).

Elbow between E and outlet 4—To find C_o from Table 14-5, we need r/W and H/W.

$$r/W = 1.0 \text{ (given)}$$
$$H/W = 14/17 = 0.82$$

From Table 14-5a, find $C_o = 0.22$ (approximate). Therefore, the elbow loss $(h_L)_d = C_o \times h_v = (0.22)(0.091) = 0.02$ in. WG.

duct loss $(h_L)_{A-4} = 0.150(200/100)$
$$= 0.300 \text{ in. WG}$$

total loss A to $4 = 0.300 + 0.028 + 0.02$
$$= 0.348 \text{ in. WG}$$

(b) *Loss A to outlet 5:*

Elbow between E and outlet 5—$r/W = 1.0$ (given) and $H/W = 14/14 = 1.0$. From Table 14-5, find $C_o = 0.21$. Therefore, the elbow loss $(h_L)_d = (0.21)(0.076) = 0.016$ in. WG.

duct loss $(h_L)_{A-5} = 0.150(210/100)$
$$= 0.315 \text{ in. WG}$$

total loss A to $5 = 0.315 + 0.016$
$$= 0.331 \text{ in. WG}$$

The loss A to 4 is slightly higher than the loss A to 5. Therefore, the static pressure required at point A depends on the loss A to outlet 4.

Step 8. Determine the static pressure required at point A. The total pressure at A is equal to

$$h_{tA} = \text{duct loss } A \text{ to } 4 + \text{diffuser loss} + \text{velocity head in section } E\text{-}4$$

Therefore,

$$h_{tA} = 0.348 + 0.05 + 0.091 = 0.489 \text{ in. WG}$$

Since

$$h_{tA} = h_{sA} + h_{vA}, \text{ then } h_{sA} = h_{tA} - h_{vA}$$

or

$$h_{sA} = 0.489 - 0.159 = 0.330 \text{ in. WG}$$

Use of the equal-friction-rate method requires that dampers be used in each of the branch lines because the available static pressure in the main at the branch take-offs is higher than that required to flow in the branches. For example, the total pressure at point D will be equal to the total pressure at A minus the friction loss A to D, or

$$h_{tD} = h_{tA} - \text{losses}$$
$$= 0.489 - (0.150)(110/100)$$
$$= 0.324 \text{ in. WG}$$

The static pressure at D will be

$$h_{sD} = h_{tD} - (h_v)_{CD}$$
$$= 0.324 - 0.130$$
$$= 0.194 \text{ in. WG}$$

If the pressure losses in branch D–3 total less than 0.194 in. WG, then a damper must be installed in the branch. An alternative to the use of a damper would be to resize branch D to 3 to use up the available pressure at D of 0.194 in. WG. This technique is called the *balanced-pressure method* and is sometimes employed in system design to reduce the necessity of severe dampering, which otherwise might be required. The balanced-pressure method requires trial-and-error techniques that may be time consuming, and dampers should still be used to provide a means for final balancing of the system after it is installed. The following example illustrates this principle of balancing pressure losses in the system.

ILLUSTRATIVE PROBLEM 14-15

Figure 14-11 shows a ventilation system with three supply outlets. Determine the required rectangular sizes of the duct sections such that the static pressure loss from the fan (point A)

to each outlet will be approximately the same. The duct section AB is to have a depth of 18 in., and all other duct sections will have a depth of 14 in. Assume that all elbows and branch takeoffs have an $r/W = 0.5$. Start the problem by sizing the duct section $ABCD$ by the equal-friction-rate method, using a design friction rate of 0.250 in. WG per 100 ft.

Solution:
Step 1. Determine duct sizes by equal-friction-rate method for duct run $ABCD$ (see Table 14-13).
Step 2. Determine pressure loss in section BC.

Because of the change in duct depth from section AB to section BC, an equal area transition (EAT) would normally be used. *For any equal area transition,* the loss coefficient $C_o = 0.15$. Because of the 3000-cfm flow into branch BF, 7000 cfm flows into section BC. If the duct depth immediately following point B is 18 in., then, from Table 14-2, the width would be 27 in. Therefore, the EAT would be from 27 by 18 to 36 by 14 (approximately).

$$V_{BC} = (7000)(144)/(36)(14) = 2000 \text{ fpm}$$
$$(h_v)_{BC} = (V/4005)^2 = (2000/4005)^2$$
$$= 0.249 \text{ in. WG}$$
$$\text{loss } EAT = C_o(h_v)_{BC} = (0.15)(0.249)$$
$$= 0.037 \text{ in. WG}$$
$$\text{duct loss } BC = 0.250(100/100) = 0.250 \text{ in. WG}$$
$$\text{total loss } BC = 0.250 + 0.037 = 0.287 \text{ in. WG}$$

Step 3. Determine pressure loss for section CD. For the single elbow between C and D,

$$r/W = 0.5 \text{ (given)}$$
$$H/W = 14/28 = 0.5$$

From Table 14-5, $C_o = 1.3$. Velocity in section CD is

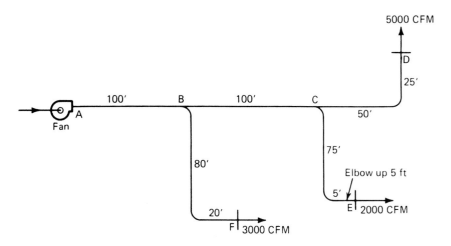

FIGURE 14-11 Diagram for Illustrative Problem 14-15 (does not represent an actual system).

TABLE 14-13

Data for Illustrative Problem 14-15

Duct Section	Capacity (cfm)	Design Friction Rate (in. WG/100 ft)	Round Duct Size (in.)	Rectangular Duct Size ($W \times H$)	Rectangular Duct Velocity (fpm)
AB	10,000	0.250	27.8	38 × 18	2100
BC	7,000	0.250	24.0	36 × 14	2000
CD	5,000	0.250	21.2	28 × 14	1837

$$V_{CD} = (5000)(144)/(28)(14) = 1837 \text{ fpm}$$

$$(h_v)_{CD} = (1837/4005)^2 = 0.210 \text{ in. WG}$$

$$\begin{aligned}\text{elbow loss} &= C_o(h_v)_{CD} = (1.3)(0.210) \\ &= 0.27 \text{ in. WG}\end{aligned}$$

$$\text{duct loss } CD = 0.250(75/100) = 0.188 \text{ in. WG}$$

$$\text{total loss } CD = 0.188 + 0.27 = 0.458 \text{ in. WG}$$

Step 4. Make pressure loss in section *CE* equal to pressure loss in section *CD* (0.458 in. WG). The procedure here is to assume a friction rate for section *CE* such that the loss will be equal to 0.458 in. WG. Noting that the length of section *CE* is about the same as section *CD*, but there are three elbows and a branch takeoff at *C*.

As a *first trial* in our trial-and-error solution, assume the friction rate at, say, 0.20 in. WG. From Figure 14-1b, with 2000 cfm, find a trial diameter of 15.6 in. From Table 14-2, an equivalent rectangular size would be 15 by 14.

$$V_{CE} = (2000)(144)/(15)(14) = 1371 \text{ fpm}$$

$$(h_v)_{CE} = (1371/4005)^2 = 0.117 \text{ in. WG}$$

For the *branch takeoff,* using Table 14-8,

$$H/W = 14/15 = 1.0 \text{ (approximate)}$$

$$V_B/V_U = 1371/2000 = 0.69$$

$$A_B/A_U = (14 \times 15)/(36 \times 14) = 0.42$$

The footnote of Table 14-8 recommends treating the takeoff as an elbow of $r/W = 1.0$. Referring to Table 14-5, with $H/W = 1.0$ and $r/W = 1.0$, find $C_o = 0.21$.

$$\begin{aligned}\text{branch takeoff loss} &= C_o(h_v)_{CE} = (0.21)(0.117) \\ &= 0.025 \text{ in. WG}\end{aligned}$$

For each of the 3 elbows, use of Table 14-5 with r/W = 0.5 (given), and H/W = 1.0 (approximate) gives C_o = 1.2.

$$\text{elbow loss} = (3)(1.2)(0.117) = 0.420 \text{ in. WG}$$

$$\text{duct loss} = (0.20)(85/100) = 0.170 \text{ in. WG}$$

$$\text{total loss } CE = 0.170 + 0.025 + 0.420 = 0.615 \text{ in. WG}$$

This is significantly larger than the desired value of 0.458 in. WG.

For a *second trial,* assume a friction rate of 0.150 in. WG per 100 ft. From Figure 14-1b, with a flow of 2000 cfm and the new friction rate, find $D = 16.5$ in. From Table 14-2, the equivalent rectangular size would be 16 by 14 in. Referring to Tables 14-8 and 14-5, the values of C_o for the branch takeoff and the 3 elbows remain the same as in the first trial, or 0.21 and 1.2, respectively.

$$V_{CE} = (2000)(144)/(16)(14) = 1285 \text{ fpm}$$

$$(h_v)_{CE} = (1285/4005)^2 = 0.103 \text{ in. WG}$$

$$\text{takeoff loss} = (0.21)(0.103) = 0.022 \text{ in. WG}$$

$$\text{elbow loss} = (3)(1.2)(0.103) = 0.370 \text{ in. WG}$$

$$\text{duct loss } CE = (0.15)(85/100) = 0.13 \text{ in. WG}$$

$$\text{total loss } CE = 0.13 + 0.022 + 0.37 = 0.522 \text{ in. WG}$$

This loss is only $0.522 - 0.458 = 0.064$ in. WG higher than the desired value. For the degree of accuracy with which we can work, this is satisfactory.

Step 5. Make the pressure loss in section BF equal to the loss in section BCD, or $0.287 + 0.458 = 0.745$ in. WG. Note: The procedure here would be essentially the same as followed in step 4. To use up the loss of 0.745 in. WG in this short section BF may require rather high velocities. If this becomes a problem, it may be desirable to redesign the duct system using a design friction rate less than 0.250 in. WG per 100 ft for the section ABCD.

14-14   THE STATIC-REGAIN METHOD

The intent of this method is to maintain the static pressure at practically a constant value throughout the system. The advantage of doing this is that static pressure determines the rate of discharge through outlets; hence, if the static pressure remains constant, the size of outlet for a given volume of discharge would also be constant. For installations such as hotels and hospitals, there would be an obvious advantage in having the same size outlet in a series of like rooms.

The static-regain method utilizes a velocity reduction at the end of each section of duct; the magnitude of the reduction is sufficient to provide a loss of velocity pressure equal to the loss of total pressure that occurred in the preceding section of duct. Thus, for a given length of duct, the method of application would be as follows:

1. Determine the friction loss per 100 ft from friction charts (Figures 14-1a or 14-1b).
2. Calculate the static pressure loss in the given length of duct by multiplying friction rate per 100 ft (step 1) by the duct length divided by 100.
3. For known velocity in the duct section, calculate the velocity pressure $(h_v)_U$.
4. Calculate the required velocity pressure in the downstream section $(h_v)_D$ from

$$SPR = 0.5[(h_v)_U - (h_v)_D] \qquad (14\text{-}5)$$

or

$$(h_v)_D = (h_v)_U - 2(SPR) \qquad (14\text{-}5a)$$

where SPR is the static pressure regain equal to the friction loss as determined in step 2; $(h_v)_U$ is the velocity pressure from step 3. The coefficient reflects the assumption that regain occurs at 50% efficiency. The velocity pressure $(h_v)_D$ is the necessary velocity pressure in the downstream section of the duct to provide complete regain of the static pressure lost in the upstream section.

5. By knowing $(h_v)_D$, one may easily calculate the necessary velocity in the downstream section and thus the required duct diameter.

The static-regain method is likely to be more time consuming than other duct design methods, but in many cases it will be found to justify the added effort through more effective air distribution. If this method is to be employed much of the time, it is recommended that one refer to *ASHRAE Handbook 1989 Fundamentals*.

A simple problem follows to illustrate the principle of static-pressure regain.

ILLUSTRATIVE PROBLEM 14-16

Figure 14-12 represents a section of the main of a round duct system where three branch ducts are shown equally spaced along the main. The entering flow rate in this section of the main is 10,000 cfm at 2000 fpm, and each of the branches will carry 2000 cfm. The duct main (*ABCD*) will be maintained the same size throughout its length. Determine the values of the total, static, and velocity pressures at points *A, B, C,* and *D*. Assume the total pressure at station *A* is 1.00 in. WG.

Solution:
Step 1. Construct a table for data and results (see the summary table for Illustrative Problem 14-16).
Step 2. Determine results to complete the table.

Referring to Figure 14-1b, for an assumed velocity in section *AB* of 2000 fpm and a flow rate of 10,000 cfm, find the round duct diameter of 30.2 in. and a friction rate of 0.150. Insert values in summary table. The friction loss then is 0.150(40/100) = 0.06 in. WG.

If the trunk main is maintained at a diameter of 30.2 in. throughout its length, follow this line of constant diameter on Figure 14-1a; at the intersection of 8000 and 6000 cfm, find the velocity and friction rate for sections *BC* and *CD*. Enter these values in summary table. Again, the friction loss (line 5) may be calculated as friction rate times length divided by 100. Enter values in summary table.

Note that the volume flow rate decreases along the duct (line 1). Also, note that the velocity decreases along the duct (see line 3). This occurs because the duct size has been kept constant (line 2) even though the volume flow rate decreases. The friction rate for each section decreases (line 4), as well as the friction loss (line 5).

Summary table for Illustrative Problem 14-16

Line	Item	Section AB	Section BC	Section CD
1	Flow quantity (cfm)	10,000	8,000	6,000
2	Duct diameter (in.)	30.2	30.2	30.2
3	Flow velocity (fpm)	2,000	1,600	1,200
4	Friction rate (in. WG/100 ft)	0.150	0.10	.059
5	Friction loss (in. WG)	0.06	0.04	.024
		Station B	Station C	Station D
6	Velocity pressure (in. WG)	0.249	0.160	0.090
7	Total pressure (in. WG)	0.940	0.900	0.876
8	Static pressure (in. WG)	0.691	0.740	0.786

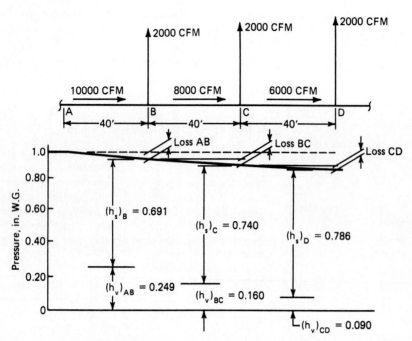

FIGURE 14-12 Diagram for Illustrative Problem 14-16 (does not represent an actual system).

Since the velocity in each section (line 3) is known, the velocity pressures may be calculated:

$$(h_v)_{AB} = (V_{AB}/4005)^2 = (2000/4005)^2$$
$$= 0.249 \text{ in. WG}$$

$$(h_v)_{BC} = (V_{BC}/4005)^2 = (1600/4005)^2$$
$$= 0.160 \text{ in. WG}$$

$$(h_v)_{CD} = (V_{CD}/4005)^2 = (1200/4005)^2$$
$$= 0.090 \text{ in. WG}$$

Enter these values in summary table (line 6).

Now the total pressure (h_t) and friction should be calculated. The total pressure at station *A* is 1.00 in. WG (given). The friction loss in section *AB* was 0.06 in. WG; therefore, the total pressure at station *B* would be

$$(h_t)_B = (h_t)_A - (\text{loss})_{AB} = 1.00 - 0.06$$
$$= 0.94 \text{ in. WG}$$

Also,

$$(h_t)_C = (h_t)_B - (\text{loss})_{BC} = 0.94 - 0.04$$
$$= 0.90 \text{ in. WG}$$

and

$$(h_t)_D = (h_t)_C - (\text{loss})_{CD} = 0.90 - 0.024$$
$$= 0.876 \text{ in. WG}$$

Since by Eq. (13-8), $h_t = h_s + h_v$, the static pressures at *B, C,* and *D* may be calculated as follows:

$$(h_s)_B = (h_t)_B - (h_v)_{AB} = 0.940 - 0.249$$
$$= 0.691 \text{ in. WG}$$

$$(h_s)_C = (h_t)_C - (h_v)_{BC} = 0.90 - 0.160$$
$$= 0.740 \text{ in. WG}$$

$$(h_s)_D = (h_t)_D - (h_v)_{CD} = 0.876 - 0.090$$
$$= 0.786 \text{ in. WG}$$

Notice, in Illustrative Problem 14-16, that along the main duct the velocity pressure decreased and the static pressure increased. In other words, there was a *conversion* of velocity pressure to static pressure. This is called *static-pressure regain* (SPR). And here is the key to the static-regain method of duct design: Select the velocity in section *BC* such that the static pressure at station *C* is 0.691 in. WG; select the velocity in section *CD* such that the static pressure at station *D* is 0.691 in. WG.

Now it is evident what this means. The static pressure is 0.691 in. WG at each branch takeoff. Balancing the duct system should be simple; dampering will rarely be required. Some duct runs have outlets stubbed into the side or bottom of the duct (in place of the branch takeoffs). In such designs, the static pressure behind each outlet is the same and the system is well balanced.

Theoretically, static pressure regain is independent of the manner in which the velocity is reduced, but the actual regain depends greatly on the physical configuration of the duct system. Illustrative Problem 14-16 obtained static regain by using a constant-diameter duct with decreasing flow after each branch. Another technique would be to alter the duct diameter after each branch takeoff to obtain the required velocity reduction.

ILLUSTRATIVE PROBLEM 14-17

Using Figure 14-12 with the associated data and required calculated results from Illustrative Problem 14-16, determine the required duct diameters in sections *BC* and *CD* by use of Eq. (14-5a) if the diameter of section *AB* is 30.2 in.

Solution:

Diameter of section BC—From Illustrative Problem 14-16, the pressure loss in section $AB = 0.06$ in. WG, and the velocity head $(h_v)_{AB} = 0.249$ in. WG. The static-pressure regain in section *BC* must overcome losses in section *AB*. Therefore, by Eq. (14-5a),

$$(h_v)_{BC} = (h_v)_{AB} - 2(SPR) = 0.249 - 2(0.06)$$
$$= 0.129 \text{ in. WG}$$

or

$$(V_{BC}/4005)^2 = 0.129$$

So,

$$V_{BC} = 1398 \text{ fpm (required velocity in}$$
$$\text{section } BC \text{ to regain static pressure)}$$

Then, the flow area section *BC* is

$$Q/V_{BC} = 8000/1398 = 5.72 \text{ ft}^2$$

$$A = \pi D^2/4 = 5.72$$

$$D_{BC} = 2.699 \text{ ft} = 32.4 \text{ in. (required diameter}$$
$$\text{of section } BC)$$

Diameter of section CD—Referring to Figure 14-1b with $D_{BC} = 32.4$ in. and $Q = 8000$ cfm, find friction rate of 0.07 in. WG per 100 ft. Therefore, the friction loss in section *BC* is $(0.07)(40/100) = 0.028$ in. WG. By Eq. (14-5a),

$$(h_v)_{CD} = (h_v)_{BC} - 2(SPR) = 0.129 - 2(0.028)$$

$$= 0.073 \text{ in. WG}$$

$$(V_{CD}/4005)^2 = 0.073$$

$$V_{CD} = 1082 \text{ fpm}$$

$$A_{CD} = 6000/1082 = 5.55 \text{ ft}^2$$

$$\pi D^2/4 = 5.55$$

$$D_{CD} = 2.66 \text{ ft} = 31.9 \text{ in. (required diameter}$$
$$\text{of section } CD)$$

14-15 DUCT DESIGN FOR SMALL HEATING SYSTEMS

We have discussed the methods generally used for the design of large commercial and industrial duct systems; however, any duct system, large or small, may be designed by these methods. As we may note, these methods in general use specified limits of airflow velocity and friction rate. *After* the duct sizes are determined, the friction losses are evaluated to determine the fan size and capacity.

Large numbers of residential and small commercial air conditioning systems (heating and/or cooling) are designed and installed. Since the airflow volume is seldom greater than 3000 cfm, the friction and shock losses are relatively low. Furthermore, most of these smaller systems have a package-type furnace, air conditioner, or heat pump that has, as one of its components, a fan (blower) having a set capacity and maximum stated external static pressure capabilities, up to 0.25 in. WG maximum. Because of these conditions, design methods are somewhat modified and simplified.

Shock losses for fittings are evaluated by determining the *equivalent lengths* of such fittings, which are added to the scaled length of straight duct to obtain a *total equivalent length* (TEL) of straight duct. Static pressure losses of devices such as supply- and return-air grilles, filters, and coils are evaluated and summed. This sum is subtracted from the blower static pressure rating. The remaining blower static pressure is proportioned between the supply and return duct systems (approximately 75% for supply and 25% for return). The design friction rate for sizing the ducts is then found by using available static pressure = $f_{100} \times$ TEL/100. The friction chart (Figures 14-1a and 14-1b) may then be used to determine duct sizes. If the velocity exceeds recommended values, increase the duct size and specify a damper.

A residence's characteristics determine the location and the type of forced-air system than can be installed. The presence or absence of particular areas within a residence has a direct influence on equipment and ductwork location. The structure's size, room or area use, and air distribution system determine how many central systems will be needed to maintain comfort temperatures in all areas.

A full basement is an ideal location to install equipment and ductwork. If a residence has a crawl space, the ductwork and equipment can be placed in a closet or utility room. When the equipment is within the conditioned space, a nonducted return-air system may be adequate. The equipment's enclosure must meet all fire and safety code requirements; adequate service clearance must also be provided. In a house built on a concrete slab, the equipment should be located in the conditioned space (for systems that do not require combustion air), an unconditioned closet, an attached garage, the attic space, or outdoors. The ductwork normally is located in a furred-out space, in the slab, or in the attic.

Construction materials for a slab duct system must meet specific corrosion, fire resistance, crushing and bending strength, deterioration, odor, delamination, and moisture absorption requirements specified by Federal Specifications SS-A-701 of the Federal Housing Administration and Standard 90B of the National Fire Code (NFPA No. 90B-84).

Weather conditions should be considered when locating heating and air conditioning equipment, as well as ductwork. Packaged outdoor units for houses in severely cold climates must be carefully installed according to the manufacturer's recommendations to operate satisfactorily. Most houses in cold climates have basements, making them well suited for indoor furnaces and split-system air conditioners or heat pumps. In mild and moderate climates, the ductwork is frequently placed in the attic or crawl space. Ductwork located outdoors, in attics, crawl spaces, and basements must be insulated.

Many design manuals are available that may be used to assist the HVAC engineer in the design and layout of heating and/or cooling duct systems. Of particular interest would be the manuals published by National Environmental Systems Contractors Association and the manuals published by Air Conditioning Contractors of America.

14-16 FANS

A fan, as it is considered here, is a gas flow-producing machine that operates on the same basic principles as a centrifugal pump or compressor. A fan is similar to these machines in that they all convert mechanical rotative energy, applied at their shafts, to gas (or fluid) energy. The conversion of mechanical rotative energy to gas energy is accomplished by means of a wheel (impeller), which imparts a spin to the gas. As a result of this spin, and, provided there is no entering spin, the tendency of the gas is always to leave the wheel with a forward spiral motion.

The difference between fans and pumps is readily apparent; however, the distinction between fans and compressors is not so obvious. Generally speaking, this distinction is one of output pressures with the lower-pressure machines called *fans*, while the higher-pressure machines are called *compressors*. Even this dividing line is not distinct.

Fans for heating, ventilating, and air conditioning applications, even on high-velocity, high-pressure systems, rarely encounter more than 10 to 12 in. WG pressure. Industry standards on classes of fans have been established by the Air Moving and Conditioning Association (AMCA) as follows:

Class I: Max. total pressure $= 3\frac{3}{4}$ in. WG
Class II: Max. total pressure $= 6\frac{3}{4}$ in. WG
Class III: Max. total pressure $= 12\frac{1}{4}$ in. WG

14-17 TYPES OF FANS

Most commercial fans used in HVAC work may be placed in one of two general types based upon construction and airflow patterns. These two types are (1) *centrifugal* or *radial flow* and (2) *axial flow* and are illustrated in Figure 14-13. Centrifugal fans have flow within the rotating wheel (rotor) that is substantially radial to the shaft, with the rotor operating in a scroll-type housing. Axial fans have flow within the wheel that is substantially parallel to the shaft and operates within a cylindrical or ring-type housing.

Centrifugal fans are designed to move air or gas over a wide range of volumes. Usual static pressures go up to 12 in. WG but, with special design, may go as high as 60 in. WG. The rotor may have straight, forward-curve, backward-curve, radial-tip, or airfoil-type blades. The housing is sheet or case metal, and the rotor may have belt or direct drive.

Propeller fans are designed to move air from one enclosed space to another or from outdoors to indoors, or vice versa, in a wide range of volumes at low pressure (0 to 1 in. WG, static). The automatic shutter on discharge, shown in Figure 14-13, is not part of the fan. It protects the fan from wind, rain, snow, and cold during shutdown. These fans have either belt or direct drive.

Tubeaxial fans have the axial-flow rotor in a cylinder and are designed to move a wide range of air volumes at medium pressures ($\frac{1}{4}$ to $2\frac{1}{2}$ in. WG, static). The fans may be mounted in the air duct and have either belt or direct drive. Rotor blades may be disc or airfoil type.

Vaneaxial fans have a set of air guide vanes mounted in a cylinder before or behind the airfoil-type rotor. It moves air over a wide range of volumes and pressures. Usual pressure range is $\frac{1}{2}$ to 6 in. WG, static, but special designs can go as high as 60 in. WG, static. The rotor is often made of sheet metal but cast metal is also used. The fans have either belt or direct drive.

Fans are often designated as *boosters, blowers,* or *exhausters.* As they will be considered here, a booster is a fan with ducts connected to both inlet and outlet; a blower has a discharge duct only; an exhauster has an inlet duct only.

available differ from one another in structural details and in the details of their design and assembly. But the principal feature that distinguishes one type of centrifugal fan from another is the inclination of the impeller blades. Figure 14-14 shows the most common types of centrifugal fan elements. The fan impellers produce pressure from two related sources: (1) the centrifugal force created by the rotating air column enclosed between the blades and (2) the kinetic energy imparted to the air by virtue of its absolute velocity (V_2) leaving the impeller. This velocity in turn is a combination of rotative velocity (V_t) of the impeller and air velocity relative to the impeller (V_r). When the blades are inclined forward (Figure 14-14c), these two velocities are cumulative; when the blades are inclined backward (Figure 14-14a), oppositional. Backward-inclined blade fans are generally somewhat more efficient than forward-inclined blade fans, with radial blade fans between them.

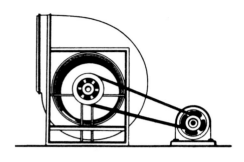

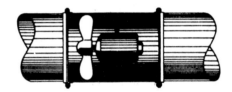

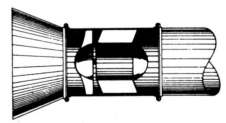

FIGURE 14-13 Types of fans. (Reprinted with permission from *POWER* magazine, © McGraw-Hill, Inc., 1981)

14-18 PRINCIPLES OF FAN OPERATION

All fans produce pressure by altering the velocity vector of the fluid flow. The fans produce pressure and/or flow due to the rotating blades of the impeller, which imparts kinetic energy to the fluid (air or gas) in the form of velocity changes. These velocity changes are resultants of tangential and radial velocity components in the case of centrifugal fans and of axial and tangential velocity components in the case of axial-flow fans.

The various types of centrifugal fans that are

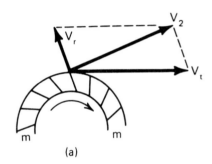

(a)

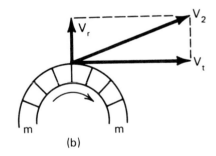

(b)

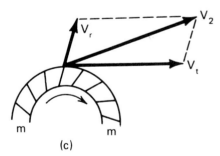

(c)

FIGURE 14-14 Centrifugal blade impeller types: (a) backward-inclined, (b) radial, and (c) forward-inclined.

Forward-curved blade fans operate at relatively low wheel peripheral velocities, V_t. The air absolute velocity, V_2, leaving the wheel is greater than the wheel peripheral speed. Forward-curved blade fans are very dependent upon the discharge casing design for conversion of velocity energy into static. This type of fan obtains about 20% of its static pressure from the wheel and about 80% from the casing. It needs a long duct run to slow down the air. The principal advantages of a forward-curved blade fan are that it occupies much less space, weighs less for a given duty, and is generally the least expensive.

Radial-blade fans operate with air speeds, V_2, about equal to the wheel peripheral velocities V_t. The radial-blade fan operates at medium speeds, with medium noise, and requires a fairly large space. In general, the peak efficiency is higher than for forward-curved blade fans. This type of fan is not as dependent upon the discharge casing for conversion of velocity pressure to static pressure. About 45% of its energy is converted to static pressure; about 55% of it, to velocity pressure.

Backward-curved blade fans operate with air speed, V_2, lower than the peripheral wheel velocity and are driven at high speeds, which makes this type of fan most suitable for electric-motor drive. This type of fan produces the highest efficiency (70% to 80%). It is much less dependent on the casing design for conversion of velocity pressure to static pressure because about 70% of the energy added develops static pressure in the wheel. Its chief disadvantages are higher cost and larger size for a given duty.

Axial-flow fans produce pressure from the change in velocity of the air passing through the blades, with none being produced by centrifugal force.

14-19 FAN PERFORMANCE

Commercial fan performance data are calculated from laboratory tests conducted in accordance with the strict requirements of ASHRAE 51-75 plus AMCA 210-74.

The ASHRAE standard specifies in detail the procedures and test setups to be used in testing the various types of fans and other air-moving devices. The codes establish uniform testing methods to determine a fan's flow rate, pressure, power, speed of rotation, and efficiency. The codes are designed for testing centrifugal and axial fans with various duct arrangements. The fans are tested in either the blower or exhauster configuration. Figure 14-15 shows a typical AMCA code test setup. When a fan is tested, the Pitot tube makes a prescribed traverse of the cross-sectional area of the duct measuring velocity pressure, static pressure, and total pressure. From the results of these tests and from the computations outlined by the AMCA code, fan performance can be completely established.

The following terms are generally used when discussing fan performance.

Fan volume: The fan volume is defined as the volume, in cfm, passing through the fan outlet. In normal applications, the volume leaving the fan is substantially equal to that entering since the change in specific volume of the air between fan inlet and outlet is negligible.

Fan outlet velocity: The outlet velocity of a fan is obtained by dividing the air volume by the fan outlet area. It is the average velocity that would occur at a point removed from the fan in a discharge duct having the same cross-sectional area as the fan outlet. It is important to note that the fan outlet velocity does not describe the velocity conditions that exist at the fan outlet since all fans have nonuniform outlet velocity. The fan outlet velocity is in reality a theoretical value; it is the velocity that would exist in the fan outlet if the velocity were uniform across the outlet area.

Fan velocity pressure: The fan velocity pressure is the pressure corresponding to the fan outlet velocity. The velocity pressure may be calculated by use of Eq. (13-4) or for standard air by Eq. (13-6).

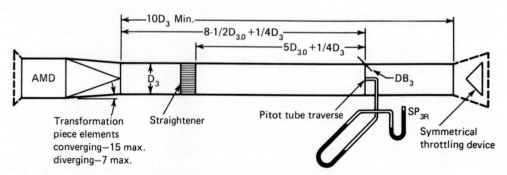

FIGURE 14-15 Fan test arrangement.

Fan total pressure: The fan total pressure is the difference between the total pressure at the fan outlet and the total pressure at the fan inlet. The fan total pressure is a measure of the total mechanical energy added to the air or gas by the fan. Fan total pressure may be measured as illustrated in Figure 14-16.

Fan static pressure: The fan static pressure is the fan total pressure minus the fan velocity pressure. It can be calculated by subtracting the total pressure at the fan inlet from the static pressure at the fan outlet. It can be measured as illustrated in Figure 14-17. By definition, the fan static pressure is

$$\text{fan } h_s = (h_t)_o - (h_t)_i - (h_v)_o \qquad (14\text{-}6)$$

but, from Eq. (13-8), rearranged, we have $(h_t)_o - (h_s)_o = (h_v)_o$. Therefore, with substitution into Eq. (14-6), we have

$$\text{fan } h_s = (h_s)_o - (h_t)_i \qquad (14\text{-}6a)$$

which is the method of measurement where

h_s = static pressure
$(h_s)_o$ = static pressure at fan outlet
$(h_s)_i$ = static pressure at fan inlet
$(h_t)_o$ = total pressure at fan outlet
$(h_t)_i$ = total pressure at fan inlet
$(h_v)_o$ = velocity pressure at fan outlet
$(h_v)_i$ = velocity pressure at fan inlet

The definitions of fan total pressure and fan static pressure hold regardless of the manner of duct connec-

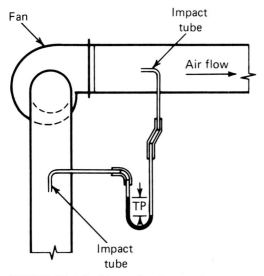

FIGURE 14-16 Measuring fan total pressure.

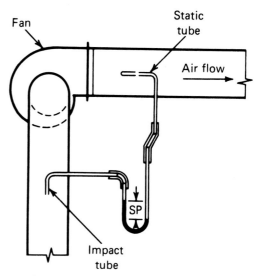

FIGURE 14-17 Measuring fan static pressure.

tion—that is, whether the fan is connected as a blower, booster, or exhauster.

ILLUSTRATIVE PROBLEM 14-18

The following pressure measurements are reported on a fan connected as a booster—that is, with inlet and outlet ducts as follows:

At fan inlet: $(h_s)_i$ = −1.5 in. WG (below atmospheric)
$(h_v)_i$ = +0.25 in. WG

At fan outlet: $(h_s)_o$ = +0.75 in. WG
$(h_v)_o$ = +0.50 in. WG

What is the fan static pressure?

Solution: Fan static pressure = fan total pressure minus fan velocity pressure. By Eq. (14-6),

$$\text{fan } h_s = (h_t)_o - (h_t)_i - (h_v)_o$$

but

$$(h_t)_i = -1.5 + 0.25 = -1.25 \text{ in. WG}$$
$$(h_t)_o = +0.75 + 0.50 = +1.25 \text{ in. WG}$$

Therefore,

$$\text{fan } h_s = 1.25 - (-1.25) - 0.50$$
$$= 2.0 \text{ in. WG}$$

or, by Eq. (14-6a),

$$\text{fan } h_s = (h_s)_o - (h_t)_i = +0.75 - (-1.25)$$
$$= 2.0 \text{ in. WG}$$

Fan horsepower: The theoretical horsepower required to drive a fan is the power that would be required if there were no losses in the fan—that is, if its efficiency were 100%. The theoretical horsepower to drive a fan can be calculated from

$$\text{air horsepower (AHP)} = \frac{\text{cfm}[(h_t/12)(62.3)]}{33,000}$$

$$= \frac{(\text{cfm})(h_t)}{6356} \quad (14\text{-}7)$$

where cfm is the fan volume and h_t is the fan total pressure (in. WG).

The brake horsepower (BHP) is the power actually required to drive a fan. The brake horsepower is always larger than the theoretical horsepower because of energy losses in the fan. The brake horsepower required can be determined only by an actual test of the fan.

Fan efficiency: After testing a fan and determining the brake horsepower, both the *total efficiency* and *static efficiency* can be determined.

The total efficiency, or, as it is frequently called, the *mechanical efficiency,* is calculated by using the relationship

$$\text{fan total efficiency } (\eta_f)_t = \frac{(\text{cfm})(h_t)}{(6356)(\text{BHP})} \quad (14\text{-}8)$$

The fan static efficiency is calculated in a similar manner using fan static pressure instead of fan total pressure:

$$\text{fan static efficiency } (\eta_f)_s = \frac{(\text{cfm})(h_s)}{(6356)(\text{BHP})} \quad (14\text{-}9)$$

Fan total efficiency is the total output of useful energy divided by the power input. Fan static efficiency is the static energy output divided by the power input.

The total efficiency of a fan always provides a true indication of fan performance whenever the total pressure is known or can be accurately determined, while the static efficiency is not always satisfactory for measuring the actual work accomplished by the fan. For example, when the outlet velocity is higher than the duct velocity, a velocity pressure is available that may be converted to static pressure provided the duct system is properly designed. The static efficiency, as normally determined, would not take into account the increase in static pressure thus provided and therefore would not be an accurate portrayal of the fan performance. Again, when a fan operates at "free delivery," the static efficiency becomes zero and is therefore valueless.

In most circumstances, however, static efficiency

is an entirely accurate indicator of fan performance. It has an added advantage in that the fan static pressure is known in practically every case whereas the fan total pressure frequently is not known. For this reason, fan static efficiency is probably used more frequently than fan total efficiency.

It is evident that, with the fan efficiency known, the required brake horsepower may be obtained by rearranging Eqs. (14-8) and (14-9). It is important to note, however, that the proper efficiency should be used with the corresponding pressure. As indicated above, fan total efficiency must be used with fan total pressure and fan static efficiency with fan static pressure.

ILLUSTRATIVE PROBLEM 14-19

A fan is to deliver 8000 cfm of standard air against a fan static pressure of 0.75 in. WG. The static efficiency of the fan is 73%. Find the brake horsepower required to drive the fan.

Solution: By Eq. (14-9),

$$(\eta_f)_s = \frac{(\text{cfm})(h_s)}{(6356)(\text{BHP})}$$

or

$$\text{BHP} = \frac{(\text{cfm})(h_s)}{6356(\eta_f)_s} = \frac{(8000)(0.75)}{(6356)(0.73)} = 1.29 \text{ hp}$$

14-20 FAN PERFORMANCE CURVES AND MULTIRATING TABLES

Fan performance curves are obtained from a series of laboratory tests on a fan with various levels of restrictions at the end of test duct (see Figure 14-15). The curves connecting the test points form the performance curves. Such curves, sometimes called *characteristic curves,* show the relationship between the quantity of standard air that a fan will deliver and the fan pressure, horsepower, efficiency, and, occasionally, noise level. Constant-speed (rpm) performance curves are presented for a specific fan (type and size) operating at a stated speed and handling air of stated density, which is usually standard air having a density of 0.0750 lb/ft³ (70°F, barometric pressure 29.92 in. Hg).

Typical performance curves for a backward-inclined airfoil blade and forward-curved blade fan are shown in Figures 14-18 and 14-19, respectively. Notice that in the figures the pressure–volume curves for both backward-inclined and forward-curved fans have a peak point at which the pressure is a maximum. To the right of this peak point, both types of fans have a steadily falling, steep pressure–volume performance curve.

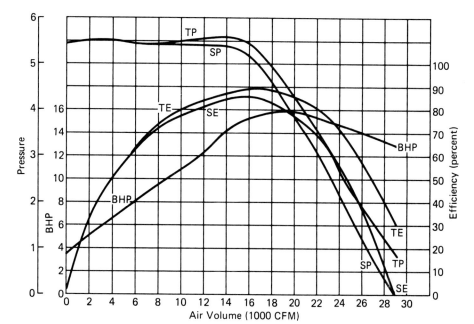

FIGURE 14-18 Sample performance curve for backward-inclined airfoil blade fan (36 AF SW, 1042 rpm, 0.0750 lb/ft³ air density). *Note:* TP = total pressure; SP = static pressure; TE = total efficiency; SE = static efficiency; BHP = brake horsepower. (Courtesy of The Trane Company, La-Crosse, WI)

This steeply falling performance curve is desirable in fans. If a fan has a steep performance curve, a slight change in pressure from the one at which the fan is operating will not greatly affect the air delivery. For example, referring to the backward-inclined airfoil fan represented in Figure 14-18, if the static pressure is 4 in. WG, the fan will deliver 19,600 cfm. However, if the static pressure required of the fan should increase to 4.5 in. WG at 19,600 cfm, the air delivery will fall only to 18,200 cfm, at which point the fan will develop 4.5 in. WG static pressure. The procedure for determining this second point of operation is explained later in this chapter.

The forward-curved blade fan represented in Figure 14-19 has a peak static pressure that corresponds to the region of maximum efficiency whereas, with the backward-inclined airfoil fan, this maximum pressure

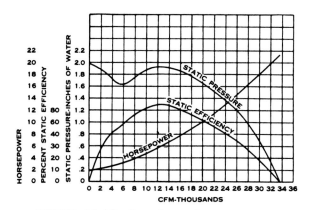

FIGURE 14-19 Sample performance curve for forward-curved blade fan. (Courtesy of The Trane Company, LaCrosse, WI)

occurs somewhat to the left of the region of maximum efficiency.

As shown in Figures 14-18 and 14-19, the horsepower for both backward-inclined and forward-curved fans is a minimum at no delivery. The horsepower for the forward-curved fan increases continuously with increasing volume flow with maximum horsepower occurring at free delivery.

The horsepower for the backward-inclined airfoil fan increases with increasing airflow only up to a point to the right of maximum efficiency and then gradually decreases. Fans with a horsepower curve of this shape are often referred to as having a nonoverloading horsepower characteristic because the maximum power that can be absorbed is usually not more than 10% above the power required at a normal selection point.

Performance curves of a typical propeller fan are illustrated in Figure 14-20a and for a typical vaneaxial fan in Figure 14-20b. As mentioned previously, a propeller fan has minimum horsepower requirements at free delivery. As can be seen from inspection of Figure 14-18, this is distinctly different from a centrifugal fan. Tubeaxial fans and vaneaxial fans have similar horsepower characteristics, as shown in Figure 14-20b. This horsepower characteristic makes an axial fan a logical choice when large volumes at low pressures are necessary.

Fan performance curves are useful in that they show the general characteristics of a fan operating at a single speed and help in visualizing results of changes in operating conditions when the fan is installed in a system of ducts, as we shall see in following sections.

In order to make the selection as easy as possible,

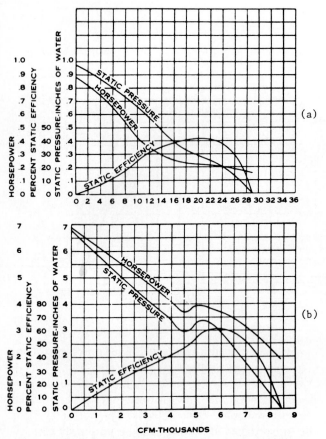

FIGURE 14-20 Axial-flow fan performance curves: (a) propeller fan and (b) vaneaxial fan.

practically all fan manufacturers publish what are known as *multirating tables*. Such tables show the operating speed (rpm) and brake (BHP) required for various combinations of air quantity (cfm) and static pressure (SP or h_s) for each fan size and type. Such tables are always based on standard air ($\rho = 0.0750 \, lb/ft^3$). Tables 14-14 and 14-15 are typical of such multirating tables. Table 14-14 is for a $24\frac{1}{2}$-in.-diameter, single-width, belt-driven, backward-inclined utility fan as manufactured by The Trane Company. Table 14-15 is for a geometrically similar fan, but one having a wheel diameter of $36\frac{1}{2}$ in.

Since the multirating tables give fan performance for standard air, we may use the tables directly to obtain the operating rpm and BHP if we know the required standard cfm and static pressure.

ILLUSTRATIVE PROBLEM 14-20

Select a utility fan to deliver 12,000 cfm of standard air against a static pressure of 1.0 in. WG.

Solution: Refer to Table 14-15. The capacity of 12,000 cfm is between 11,490 and 12,256 cfm. Interpolation is required to determine the operating rpm and BHP.

$$rpm = 539 + \left(\frac{12,000 - 11,490}{12,256 - 11,490}\right)(557 - 539)$$

$$= 539 + \frac{510}{776}(18)$$

TABLE 14-14

Multirating table (24BI utility fan)

		(Wheel Dia. 24½″		Tip Speed FPM = 6.41 x RPM				Maximum RPM 1430				Outlet Area 3.46 Sq. Ft.)									
VOL. CFM	OUT. VEL. FPM	⅛″ S.P.		¼″ S.P.		⅜″ S.P.		½″ S.P.		¾″ S.P.		1″ S.P.		1½″ S.P.		2″ S.P.		2½″ S.P.		3″ S.P.	
		RPM	BHP	RPM	BHP	RPM	BHP	RPM	BHP	RPM	BHP	RPM	BHP	RPM	BHP	RPM	BHP	RPM	BHP	RPM	BHP
2768	800	364	0.12	415	0.18	462	0.24	504	0.30	586	0.44										
3114	900	397	0.15	443	0.22	487	0.28	528	0.35	599	0.49	675	0.66								
3460	1000	431	0.19	473	0.26	514	0.34	552	0.41	622	0.56	684	0.73								
3806	1100	466	0.23	505	0.31	542	0.40	579	0.48	646	0.64	705	0.81	827	1.22						
4152	1200	502	0.29	537	0.37	571	0.47	606	0.56	670	0.73	728	0.91	834	1.31	946	1.77				
4498	1300	538	0.35	570	0.44	603	0.54	634	0.64	696	0.83	753	1.02	853	1.42	957	1.91				
4844	1400	575	0.42	605	0.52	635	0.62	664	0.73	723	0.94	778	1.14	876	1.57	965	2.03	1065	2.58		
5190	1500	612	0.51	640	0.60	667	0.71	696	0.83	751	1.07	803	1.28	899	1.72	985	2.19	1073	2.74	1163	3.32
5536	1600	649	0.60	675	0.70	701	0.82	728	0.94	779	1.19	831	1.44	924	1.89	1008	2.38	1086	2.91	1172	3.52
5882	1700	686	0.71	711	0.81	735	0.93	760	1.06	809	1.32	858	1.59	949	2.07	1031	2.58	1106	3.11	1180	3.70
6228	1800	723	0.82	747	0.93	770	1.05	793	1.19	840	1.47	886	1.75	974	2.27	1055	2.80	1129	3.35	1199	3.93
6574	1900	761	0.96	784	1.08	806	1.19	827	1.33	872	1.63	915	1.92	1001	2.49	1080	3.03	1152	3.60	1219	4.19
6920	2000	798	1.10	820	1.23	841	1.35	862	1.49	904	1.80	945	2.11	1028	2.73	1105	3.29	1177	3.88	1243	4.49
7266	2100	836	1.26	857	1.39	877	1.52	897	1.66	937	1.98	977	2.31	1056	2.97	1131	3.56	1201	4.16	1266	4.79
7612	2200	874	1.44	894	1.58	913	1.71	933	1.85	970	2.17	1009	2.52	1084	3.21	1157	3.85	1226	4.47	1291	5.13
7958	2300	912	1.63	931	1.77	950	1.92	968	2.06	1004	2.39	1041	2.74	1113	3.46	1185	4.18	1252	4.80	1316	5.47
8304	2400	950	1.84	968	1.99	987	2.14	1004	2.28	1038	2.62	1074	2.98	1143	3.73	1212	4.49	1278	5.16	1340	5.84
8650	2500	988	2.06	1006	2.22	1023	2.38	1040	2.52	1074	2.86	1107	3.24	1174	4.02	1240	4.81	1305	5.56	1366	6.24
8996	2600	1026	2.31	1043	2.47	1060	2.64	1076	2.79	1109	3.13	1140	3.51	1206	4.32	1269	5.14	1332	5.95	1392	6.65
9342	2700	1064	2.57	1081	2.74	1097	2.91	1113	3.08	1144	3.41	1175	3.81	1238	4.64	1298	5.48	1360	6.34	1419	7.10
9688	2800	1103	2.85	1118	3.03	1134	3.21	1150	3.38	1180	3.72	1209	4.12	1270	4.98	1328	-5.85	1388	6.74		
10034	2900	1141	3.15	1156	3.34	1171	3.52	1186	3.71	1216	4.05	1244	4.45	1302	5.33	1359	6.23	1416	7.15		
10380	3000	1179	3.48	1194	3.67	1209	3.86	1223	4.05	1251	4.41	1280	4.81	1335	5.71	1392	6.65				

V-BELT DRIVE #1, RPM RANGE 364-466; #2, 461-542; #3, 543-720; #4, 681-778; #5, 744-898; #6, 876-1007; #7, 931-110?;
#8, 1043-1151; #9, 1079-1281; #10, 1250-1450.

PERFORMANCE IS FOR UTILITY FAN WITH OUTLET DUCT. BHP DOES NOT INCLUDE LOSSES.

Source: Reprinted with permission, courtesy of The Trane Company, La Crosse, Wis. See manufacturer's catalog for complete data.

TABLE 14-15

Multirating table (36BI utility fan) (wheel diam. = 36 1/2 in.; top speed
fpm = 9.55 × rpm; max. rpm = 950; outlet area = 7.66 sq. ft)

(Wheel Dia. 36½″   Tip Speed FPM = 9.55 x RPM   Maximum RPM 950   Outlet Area 7.66

VOL. CFM	OUT. VEL. FPM	1/8″ S.P.		1/4″ S.P.		3/8″ S.P.		1/2″ S.P.		3/4″ S.P.		1″ S.P.		1 1/2″ S.P.		2″ S.P.		2 1/2″ S.P.		3″ S.P.	
		RPM	BHP	RPM	BHP	RPM	BHP	RPM	BHP	RPM	BHP	RPM	BHP	RPM	BHP	RPM	BHP	RPM	BHP	RPM	BHP
6128	800	244	0.26	278	0.40	310	0.52	338	0.66	393	0.97										
6894	900	266	0.33	297	0.48	327	0.63	354	0.77	402	1.09	453	1.47								
7660	1000	289	0.41	317	0.58	345	0.75	370	0.91	417	1.24	459	1.61								
8426	1100	312	0.51	338	0.70	363	0.89	388	1.06	433	1.42	473	1.79	555	2.69						
9192	1200	336	0.63	360	0.82	383	1.03	406	1.24	449	1.61	488	2.01	560	2.89	635	3.92				
9958	1300	360	0.77	382	0.97	404	1.19	425	1.42	467	1.84	505	2.26	572	3.15	642	4.23				
10724	1400	385	0.93	405	1.14	425	1.37	445	1.62	485	2.10	521	2.53	587	3.47	647	4.50	715	5.72		
11490	1500	410	1.12	429	1.33	447	1.57	466	1.84	503	2.36	539	2.83	603	3.80	660	4.85	720	6.06	781	7.35
12256	1600	434	1.32	452	1.54	470	1.80	488	2.07	522	2.63	557	3.17	619	4.18	676	5.27	729	6.44	787	7.80
13022	1700	459	1.56	476	1.79	493	2.05	509	2.34	542	2.92	575	3.52	636	4.58	691	5.70	742	6.89	791	8.20
13788	1800	484	1.82	500	2.06	516	2.32	532	2.62	563	3.25	594	3.87	653	5.02	708	6.19	757	7.42	804	8.71
14554	1900	509	2.11	525	2.37	540	2.63	554	2.94	584	3.59	613	4.25	671	5.50	724	6.70	773	7.96	818	9.28
15320	2000	535	2.43	549	2.71	564	2.98	577	3.29	606	3.97	634	4.65	689	6.04	741	7.26	789	8.57	834	9.93
16086	2100	560	2.78	574	3.08	588	3.35	601	3.66	628	4.36	655	5.10	708	6.56	758	7.86	805	9.21	849	10.60
16852	2200	585	3.17	599	3.48	612	3.76	625	4.08	650	4.80	676	5.56	726	7.09	776	8.50	822	9.88	866	11.34
17618	2300	611	3.59	624	3.91	636	4.23	648	4.54	673	5.27	698	6.06	746	7.65	794	9.23	839	10.62	882	12.10
18384	2400	636	4.05	649	4.39	661	4.72	672	5.03	696	5.78	719	6.58	766	8.24	813	9.92	857	11.40	899	12.91
19150	2500	662	4.54	674	4.90	685	5.25	697	5.57	719	6.31	741	7.15	786	8.87	831	10.63	875	12.29	916	13.79
19916	2600	687	5.08	699	5.45	710	5.81	721	6.14	743	6.90	764	7.76	808	9.55	850	11.35	893	13.15	933	14.70
20682	2700	713	5.66	724	6.04	735	6.42	745	6.79	766	7.53	787	8.41	829	10.26	870	12.12	912	14.01		
21448	2800	738	6.28	749	6.68	760	7.07	770	7.46	790	8.21	810	9.10	851	11.00	890	12.92	930	14.90		
22214	2900	764	6.95	774	7.36	784	7.77	795	8.17	814	8.94	834	9.82	873	11.78	911	13.77				
22980	3000	790	7.66	800	8.09	809	8.51	819	8.93	838	9.72	857	10.61	894	12.60	932	14.69				

V-BELT DRIVE #1. RPM RANGE 233-304; #2. 303-362; #3. 350-422; #4. 370-418; #5. 419-502; #6. 485-580; #7. 552-648; #8. 649-694; #9. 643-746; #10. 746-850; #11. 826-955.

PERFORMANCE IS FOR UTILITY FAN WITH OUTLET DUCT. BHP DOES NOT INCLUDE LOSSES.

Source: Reprinted with permission, courtesy of The Trane Company, La Crosse, Wis. See manufacturer's catalog for complete data.

$$BHP = 2.83 + \left(\frac{12{,}000 - 11{,}490}{12{,}256 - 11{,}490}\right)(3.17 - 2.83)$$

$$= 2.83 + \frac{510}{776}(0.34) = 3.05$$

If the air being handled by the fan is at other than standard, an air density correction is required. As we know, air density is affected by both temperature and barometric pressure. Barometric pressure varies with altitude. Table 14-16 gives barometric pressures at various altitudes.

14-21 FAN LAWS

Fortunately, the performance of a fan at varying speeds, varying geometric size, and varying air density may be predicted by certain basic *fan laws* when performance test data are available for a given fan. The test data for a given fan are normally obtained at a certain constant fan speed and at standard air conditions. The data are then plotted to produce performance curves similar to those of Figures 14-18, 14-19, and 14-20.

The fan laws show what happens to fan operating characteristics (1) when speed varies, (2) when fan size is changed in geometric proportion, and (3) when air density changes. These laws apply to all types of fans and are derived from the principles of dimensional anal-

ysis and dynamic similarity. The fan laws hold true as long as it can be assumed that efficiency remains the same for different operating conditions.

Summarized next are the fan laws. Symbols used are as follows: Q = air volume, cfm; h = static pressure, in. WG; hp = horsepower input; d = rotor diameter.

Fan Law No. 1. Variation in rpm (constant air density and constant system).

TABLE 14-16

Barometric pressures at various altitudes

Elevation	Barometric Pressure
Sea level	29.92 in. Hg
1,000 ft	29.86
2,000	27.82
3,000	26.81
4,000	25.84
5,000	24.89
6,000	23.98
7,000	23.09
8,000	22.22
9,000	21.38
10,000	20.58

$$Q_2 = Q_1\left(\frac{rpm_2}{rpm_1}\right)$$

$$h_2 = h_1\left(\frac{rpm_2}{rpm_1}\right)^2$$

$$hp_2 = hp_1\left(\frac{rpm_2}{rpm_1}\right)^3$$

Fan Law No. 2. Change in rotor diameter (tip speed, air density, fan proportions constant; fixed point of rating).

$$Q_2 = Q_1\left(\frac{d_2}{d_1}\right)^2$$

$$rpm_2 = rpm_1\left(\frac{d_1}{d_2}\right)$$

$$hp_2 = hp_1\left(\frac{d_2}{d_1}\right)^2$$

$$h = constant$$

Fan Law No. 3. Change in rotor diameter (rpm, air density, and fan proportions constant; fixed point of rating).

$$Q_2 = Q_1\left(\frac{d_2}{d_1}\right)^3$$

$$h_2 = h_1\left(\frac{d_2}{d_1}\right)^2$$

$$hp_2 = hp_1\left(\frac{d_2}{d_1}\right)^5$$

Fan Law No. 4. Variation in air density (constant volume, rpm, and system with fixed fan size).

$$h_2 = h_1\left(\frac{\rho_2}{\rho_1}\right)$$

$$hp = hp\left(\frac{\rho_2}{\rho_1}\right)$$

Fan Law No. 5. Variation in air density (constant pressure, system with fixed fan size and variable rpm).

$$Q_2 = Q_1\left(\frac{\rho_1}{\rho_2}\right)^{1/2}$$

$$rpm_2 = rpm_1\left(\frac{\rho_1}{\rho_2}\right)^{1/2}$$

$$hp_2 = hp_1\left(\frac{\rho_1}{\rho_2}\right)^{1/2}$$

Fan Law No. 6. Variation in air density (constant *weight* flow, system with fixed fan size and variable rpm).

$$Q_2 = Q_1\left(\frac{\rho_1}{\rho_2}\right)$$

$$rpm_2 = rpm_1\left(\frac{\rho_1}{\rho_2}\right)$$

$$h_2 = h_1\left(\frac{\rho_1}{\rho_2}\right)$$

$$hp_2 = hp_1\left(\frac{\rho_1}{\rho_2}\right)^2$$

The best way to see how fan laws apply is to solve a few illustrative problems similar to those usually encountered.

ILLUSTRATIVE PROBLEM 14-21

A fan delivers 10,000 cfm of air against a static pressure of 2.0 in. WG when the speed is 500 rpm and the horsepower input is 6.0 BHP. What speed, static pressure, and horsepower will be obtained for a delivery of 14,000 cfm?

Solution: Fan law 1 applies. For speed,

$$rpm_2 = rpm_1\left(\frac{cfm_2}{cfm_1}\right) = 500\left(\frac{14,000}{10,000}\right) = 700 \text{ rpm}$$

For static pressure,

$$(h_s)_2 = (h_s)_1\left(\frac{rpm_2}{rpm_1}\right)^2 = 2.0\left(\frac{700}{500}\right)^2 = 3.92 \text{ in. WG}$$

For horsepower,

$$hp_2 = hp_1\left(\frac{rpm_2}{rpm_1}\right)^3 = 6.0\left(\frac{700}{500}\right)^3 = 16.5 \text{ hp}$$

ILLUSTRATIVE PROBLEM 14-22

A fan delivers 8000 cfm of air at 70°F and 29.92 in. Hg (density = 0.0750 lb/ft³) against a static pressure of 2.0 in. WG when the speed is 600 rpm and the power input is 5.0. If the inlet air temperature is raised to 200°F (density = 0.602 lb/ft³) but fan speed stays the same, what is the new static pressure and horsepower?

Solution: Fan law 4 applies. For static pressure,

$$(h_s)_2 = (h_s)_1\left(\frac{\rho_{a2}}{\rho_{a1}}\right) = 2.0\left(\frac{0.0602}{0.0750}\right) = 1.6 \text{ in. WG}$$

For horsepower,

$$\text{hp}_2 = \text{hp}_1\left(\frac{\rho_{a2}}{\rho_{a1}}\right) = 5.0\left(\frac{0.0602}{0.0750}\right) = 4.0 \text{ hp}$$

ILLUSTRATIVE PROBLEM 14-23

For the fan of Illustrative Problem 14-22, what speed would be required to give a static pressure of 2.0 in. WG at 200°F? What will be the cfm and power?

Solution: Fan law 5 applies. For rpm,

$$\text{rpm}_2 = \text{rpm}_1\sqrt{\frac{\rho_{a1}}{\rho_{a2}}} = 600\sqrt{0.0750/0.0602} = 670 \text{ rpm}$$

For cfm,

$$\text{cfm}_2 = \text{cfm}_1\sqrt{\frac{\rho_{a1}}{\rho_{a2}}} = 8000\sqrt{0.0750/0.0602} = 8930 \text{ cfm}$$

For hp,

$$\text{hp}_2 = \text{hp}_1\sqrt{\frac{\rho_{a1}}{\rho_{a2}}} = 5.0\sqrt{0.0750/0.0602} = 5.58 \text{ hp}$$

ILLUSTRATIVE PROBLEM 14-24

Ratings in fan performance tables (Tables 14-14 and 14-15) are based on standard air. If air being introduced into the fan is at other than standard conditions, an air density correction is required.

Before air density corrections are made, the cfm and static pressure must be stated in terms of actual density, not in terms of standard air. It should also be noted that most air friction charts for duct design, filters, coils, and so forth, are based on standard air and must be corrected to actual conditions to determine proper system losses.

Problem statement: A 36BI utility fan is to deliver 15,300 cfm of air at 2.0 in. WG static pressure. The air is at 200°F, and the installation is at a 5000-ft elevation. At start-up, the fan will be handling zero-degree air. Determine the operating rpm and BHP for the fan. What horsepower motor would be required for this installation?

Solution: Since two air densities are involved, we will calculate them first. Table 14-16 indicates the barometric pressure is 24.89 in. Hg. Therefore, at 0°F,

$$\rho_o = \frac{P_o}{R(T_o)} = \frac{144(0.491)(24.89)}{(53.34)(460)}$$

$$= 0.0717 \text{ lb/ft}^3$$

At 200°F,

$$\rho_{200} = \frac{P_{200}}{R(T_{200})} = \frac{144(0.491)(24.89)}{(53.34)(660)}$$

$$= 0.045 \text{ lb/ft}^3$$

Using fan law 4, we may correct for nonstandard air and then refer to Table 14-14:

$$h_2 = h_1\left(\frac{\rho_2}{\rho_1}\right)$$

$$2.0 = h_1(0.045/0.075)$$

$$h_1 = 3.0 \text{ in. WG (standard air)}$$

Refer to Table 14-14 with 15,300 cfm and 3.0 in. WG and find $\text{rpm}_1 = 834$ and $\text{BHP}_1 = 9.93$. Therefore, at 200°F,

$$\text{BHP}_2 = \text{BHP}_1\left(\frac{\rho_2}{\rho_1}\right) = 9.93(0.045/0.075)$$

$$= 5.96(\text{operating horsepower})$$

At 0°F,

$$\text{BHP}_2 = 9.93(0.0717/0.075)$$

$$= 9.49 \text{ (start-up horsepower)}$$

The fan will require a 10-hp motor for this installation.

14-22 SYSTEM CHARACTERISTICS

So far, the flow of only a fixed quantity of air through a duct system has been discussed. This volume of air is that which will satisfy the design heating, cooling, ventilating, or process requirements. The pressure head losses in the system have been calculated and corrected for air density when required.

The head loss through a duct system also varies with the volume flow rate. By computing the head loss for a particular duct system with various air quantities and plotting the points on a graph, a curve similar to the one in Figure 14-21a is obtained. Such a curve is called a *system characteristic curve*. It can be drawn not only for a duct of uniform diameter but also for an entire ductwork system. Once the head loss has been determined for a given flow rate, the head loss for any other flow rate can be calculated *approximately* by means of Eq. (14-10):

$$\frac{(h_s)_2}{(h_s)_1} = \left(\frac{\text{cfm}_2}{\text{cfm}_1}\right)^2 \qquad (14\text{-}10)$$

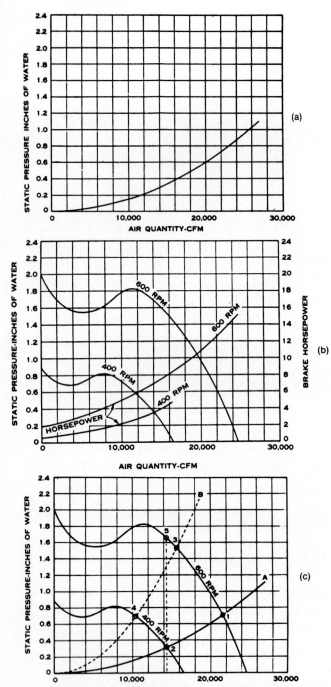

FIGURE 14-21 System and fan characteristic curves: (a) system characteristic, (b) fan characteristic, and (c) system and fan characteristics. (Courtesy of The Trane Company, La-Crosse, WI)

A fan operating at a given speed in a given system is a *constant-volume machine:* It will handle a constant volume of air. When the density of air differs from standard, it is necessary to determine the actual cfm requirement in order to select the proper fan.

Static pressure across the fan is determined from

the resistance of the air distribution system. Duct layout, duct size, heating and/or cooling coils, filters, and dampers all affect the system resistance. Careful selection and sizing of these components can reduce the static pressure and result in lower fan horsepower and noise.

As will be shown next, system characteristic curves are of value in helping to analyze the performance of a duct system and fan operating together.

14-23 FAN SELECTION

After the system pressure loss curve of the air distribution system has been defined as described in Section 14-22, a fan can be selected to meet the system requirements.

Fan manufacturers present fan performance data in either graphic (curve) form or in multirating tables as noted in Section 14-20. Multirating tables usually provide performance data only within the recommended (safe) operating range. The optimum selection area or peak efficiency point is normally identified in various ways by the different manufacturers.

In selecting a fan to meet system requirements, the following factors should be considered: (1) fan type, (2) fan size, (3) available space, (4) fan arrangement, (5) effect of installation on fan performance, (6) first cost versus operating cost, (7) allowable noise level, (8) capacity control, (9) effect of air density, (10) drive selection, and (11) motor starting torque.

One may observe, when studying the multirating tables, that several fan sizes would provide the requirements for flow and static pressure. When possible, select the fan size such that it will be operating in the area of maximum efficiency. Selections above and below this area take more power and are more noisy. The size selected depends upon operating factors that are considered most important. If space is limited, it may be necessary to use a smaller fan and accept the power penalty.

In addition to fan performance curves and multirating tables, some manufacturers make available a computerized fan selection program to assist the system designer in evaluating the many fan selection alternatives, at the specific operating conditions, to arrive at the best fan selection.

Increased use of systems requiring fan capacity modulation, such as variable-air-volume (VAV) systems, make fan and accessory selection more important. These systems require efficient, stable operations over the full capacity range. It is important that the fan not be allowed to operate in the surge region of the fan performance curves. The surge portion of these curves is the region in the vicinity of the peak point of the head-capacity curve. Fan operation should always be to the right of this area. With complete knowledge of the cfm and static pressure variations at all modulation conditions, proper fan and accessory selection can be made.

Several methods and accessories are available for capacity modulation. The choice of the proper method for any particular case is influenced by two basic considerations: (1) the frequency with which changes must be made and (2) the balancing of reduced power consumption against increases in first cost.

To achieve control of the flow, the characteristic of either the system or the fan must be changed. The system characteristic curve may be altered by installing dampers or orifice plates in the supply or return ducts. This technique reduces flow only by increasing the system resistance and therefore can lead only to increased power consumption.

Changing the fan characteristic curve results in a mode of control that results in savings in power consumption. From the standpoint of power consumption, the most desirable method of control is to vary the fan speed (rpm) to produce the desired performance. Variable-speed motors or variable-speed drives may be used when frequent variations are desired.

In order to visualize the results of changes in the operating conditions, a combination graph on which both fan and system characteristic curves are plotted is invaluable. As was the case with pipe and pump characteristic curves, the use of such combination graphs permits a great many otherwise complex problems to be analyzed in a readily understandable manner.

Fan performance curves of Figure 14-21b have been plotted in Figure 14-21c. These curves are for a fan running at 600 and 400 rpm. In addition, the system characteristic curve of Figure 14-21a has been plotted on this same graph. The point at which the fan will operate on its characteristic when the fan is running at 600 rpm is found at the intersection of its performance curve for this speed and the system characteristic curve—in this case, point 1 of Figure 14-21c. The fan will deliver 21,500 cfm and will operate against a static pressure of 0.70 in. WG as read at point 1. However, if the fan speed is reduced to 400 rpm, the quantity of air delivered by the fan will be reduced to 14,400 cfm and the static pressure will be 0.30 in. WG as read at point 2.

If the resistance of a duct system is increased because of the closing of a damper or the clogging of air filters with dirt, the characteristic curve for the system will move up to some such position as indicated by the dotted line cf Figure 14-21c. In such a case the fan, when operating at 600 rpm, will deliver 15,600 cfm against a static pressure of 1.53 in. WG as read at point 3. If, however, the speed should be reduced to 400 rpm,

the fan will deliver 10,500 cfm against 0.70 in. WG of static pressure as read at point 4.

Frequently, the volume of air delivered by a fan must be reduced. The fan volume can be reduced by (1) reducing the fan speed and (2) throttling the airflow by a damper.

The effect on the power consumption of a fan by reducing the air volume by either of these two methods can be studied advantageously by using the combination diagram. For example, assume that the volume of air is to be reduced from 21,500 cfm at point 1 of Figure 14-21c to 14,400 cfm by reducing the fan speed from 600 to 400 rpm. The fan will, therefore, operate at point 2 of Figure 14-21c. The brake horsepower required to operate the fan at point 1 of Figure 14-21c is 12.3 hp as read in Figure 14-21b. On the other hand, the brake horsepower required to operate the fan at point 2 of Figure 14-21c is 3.7 hp as read in Figure 14-21b.

If, however, the fan is allowed to operate at constant speed and the air volume is reduced by partially closing a damper, it is evident that the system characteristic curve of Figure 14-21c will move up on the fan curve. If the air volume is finally reduced to 14,400 cfm by closing a damper while the fan is still running at 600 rpm, the fan will finally operate at, say, point 5 of Figure 14-21c. The power required to operate the fan at point 5 of Figure 14-21c is 7.2 hp as read in Figure 14-21b, compared to 3.7 hp obtained by speed reduction.

From the foregoing discussion, it is evident that if the fan speed can be reduced without sacrificing the efficiency of the motor or without excessive losses in electric speed-control equipment, more economical operation can be obtained at reduced air volumes by reducing the fan speed than by a throttling damper.

When a fan is connected to a given duct system, all points of intersection between the system characteristic curve and the fan performance curves at different speeds, such as points 1 and 2 of Figure 14-21c, are called *corresponding points.* This statement would be strictly true if the friction in the system varied as the square of the air quantity, Eq. (14-10). In Figure 14-21c, this is not strictly true. The friction varies as some exponent of the air quantity that is slightly less than the square of the air quantity. However, the "square" relationship is close enough for most practical purposes.

Inlet vanes are often used for capacity modulation. They give accurate modulation and power savings at reduced flow. As the inlet vanes are closed, they impart a spin to the air in the direction of wheel rotation. This reduces cfm, static pressure, and brake horsepower.

As shown in Figure 14-22, separate cfm–static

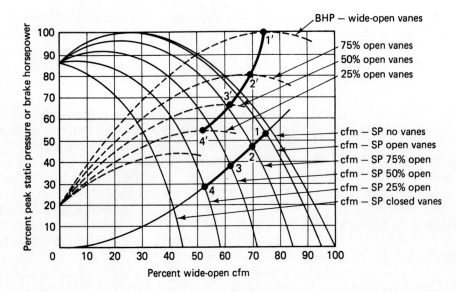

FIGURE 14-22 Inlet vane performance: Solid lines represent cfm–static pressure curves; dashed lines, cfm–brake horsepower curves.

pressure and cfm–brake horsepower curves are generated for each vane position. As the vanes are closed, the brake horsepower curves become lower than with the vanes open. The heavy black lines on Figure 14-22 indicate inlet vane modulation along a constant system curve and the resulting brake horsepower reduction. Points 1 through 4 along this curve show the cfm, static pressure, and horsepower at vane positions between open and 25% open.

Variable-air-volume (VAV) systems modulate along a curve that is not a constant system. As the terminal boxes close, static pressure in the ductwork increases, creating a new system resistance curve. The inlet vanes modulate along this system curve until the constant pressure sensor in the ductwork is satisfied. The resulting modulation curve, shown in Figure 14-23, has a square relationship between cfm and static pres-

sure. The modulation curve passes through design static pressure and cfm and through the static pressure sensor setting at zero cfm.

Due to the additional pressure drop across the inlet vanes, fan speed and horsepower must be increased to achieve catalog fan performance with vanes full open. Fan manufacturers should be consulted to determine the magnitude of these corrections.

14-24 CENTRIFUGAL FAN DESIGNATIONS

In order to designate the direction in which a fan wheel will rotate, the designations *clockwise* (CW) and *counterclockwise* (CCW) are used. To determine the direction in which the wheel of a double-inlet fan wheel will rotate, the fan should be viewed from the pulley end of

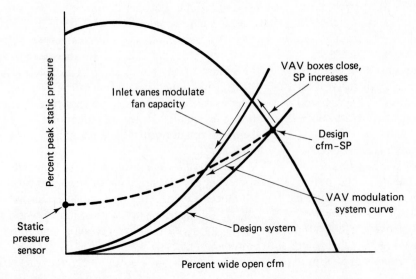

FIGURE 14-23 Variable-air-volume system modulation curve.

the shaft. To determine the direction in which the wheel of a single-inlet fan will rotate, the fan should be viewed from the side opposite the inlet.

There are also a series of terms used in designating the air discharge of a fan. Fan discharge is designated as *horizontal* if the air leaves in a horizontal direction, as *up blast* if the air is discharged vertically upward from the fan, and as *down blast* if the air discharges vertically downward from the fan.

The location of the fan outlet is said to be either *top* or *bottom*, depending upon whether the air outlet is located above or below the shaft of the fan. In a bottom horizontal discharge fan, the air outlet is located below the shaft of the fan and the air leaves in a horizontal direction. The centrifugal fan shown in Figure 14-13 is said to have a top horizontal discharge because the air leaves in a horizontal direction from the top of the fan.

In addition to horizontal and vertical discharges, there are *angular* discharges. In a fan with an angular discharge, the air leaves at an angle of 45 degrees with the vertical and horizontal axes of the fan. (See Figure 14-24.)

Centrifugal fans also have standardized nomenclature according to the arrangement of drive. These arrangements are illustrated in Figure 14-25. Usually, for ventilation and general air conditioning work, the belt-drive arrangements are preferred. This eliminates the necessity of selecting a direct-connected fan for a definite motor speed. The modern V-belt drive is inexpensive and reliable. Also, it permits fan speed adjustment on the installation if an adjustable-pitch motor sheave is used. Even with a fixed-speed drive, adjustments of speed can be made to meet changing conditions by substituting sheaves of different size.

Single-inlet (SI) or double-inlet (DI) fan: Double-inlet fans, while usually less expensive for a given duty than single-inlet fans, should be located in a plenum

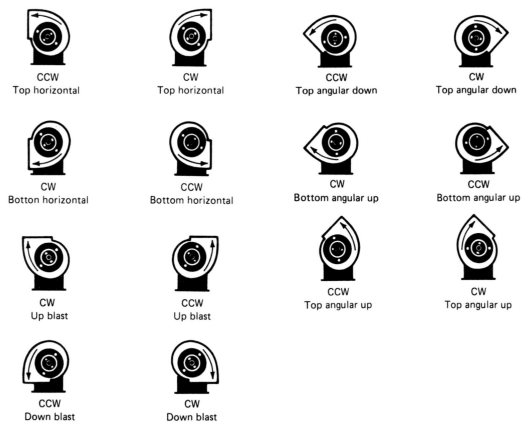

FIGURE 14-24 Fan rotation and discharge: Direction of rotation is found from drive side for either single- or double-inlet or single- or double-width fans. Drive side of single-inlet fans is side opposite inlet regardless of actual location of drive. For a fan inverted for ceiling suspension, direction of rotation and discharge is determined when fan is resting on floor. Designations here apply to all centrifugal fans. (Reprinted with permission from *POWER* magazine, © McGraw-Hill, Inc., 1981)

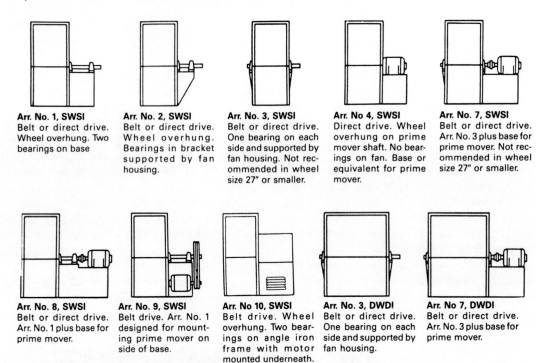

Arr. No. 1, SWSI
Belt or direct drive. Wheel overhung. Two bearings on base

Arr. No. 2, SWSI
Belt or direct drive. Wheel overhung. Bearings in bracket supported by fan housing.

Arr. No. 3, SWSI
Belt or direct drive. One bearing on each side and supported by fan housing. Not recommended in wheel size 27" or smaller.

Arr. No 4, SWSI
Direct drive. Wheel overhung on prime mover shaft. No bearings on fan. Base or equivalent for prime mover.

Arr. No. 7, SWSI
Belt or direct drive. Arr. No. 3 plus base for prime mover. Not recommended in wheel size 27" or smaller.

Arr. No. 8, SWSI
Belt or direct drive. Arr. No. 1 plus base for prime mover.

Arr. No. 9, SWSI
Belt drive. Arr. No. 1 designed for mounting prime mover on side of base.

Arr. No 10, SWSI
Belt drive. Wheel overhung. Two bearings on angle iron frame with motor mounted underneath.

Arr. No. 3, DWDI
Belt or direct drive. One bearing on each side and supported by fan housing.

Arr. No 7, DWDI
Belt or direct drive. Arr. No. 3 plus base for prime mover.

FIGURE 14-25 Drive arrangements: With standard arrangements for fan drive, when ordering or referring to a fan, you can easily identify by number its drive arrangement, width, and number of inlets. Thus, Arr. No. 1 SWSI means belt- or direct-driven unit with overhung wheel, two bearings on base, single width (SW), single inlet (SI). DW and DI mean double width and double inlet, respectively. (Courtesy of The Trane Company, LaCrosse, WI)

since duct connections to both inlets cannot be made unless the fan is furnished with inlet boxes. Thus, the majority of ventilating and air conditioning centrifugal fans used are of the single-inlet type.

When a duct is connected to the inlet of a single-inlet fan, it is desirable to use drive arrangements other than No. 3 or 7 because, with the other single-inlet drive arrangements, there is no bearing in the fan inlet. A bearing located inside the duct is more difficult to inspect and service than a bearing mounted on the drive side of the fan. Also, on smaller fans, the bearing mounted in the inlet of single-inlet fans is detrimental to fan performance. Most manufacturers test and catalog single-inlet fans without an inlet bearing. The addition of an inlet bearing and bearing support presents additional air friction, which reduces fan output.

The effect of an inlet bearing on fan performance depends upon the size of the fan, the size of the bearing, and the design of the bearing support. This effect is not serious on large fans, and ratings on large single-inlet fans are usually made from tests with an inlet bearing.

Single-width (SW) or double-width (DW) fan:
The available space, duct connections, air temperature,

and degree of air contamination must be considered in choosing an SW or a DW fan. The cost of a DW fan will generally be less than an equivalent SW fan for the same duty. However, the DW fan is not used normally when inlet duct connections must be made or when bearings must be out of the airstream. Figure 14-26 illustrates a double-width backward-inclined fan wheel.

Effect of fan installation on performance: The effect of inlet and discharge configurations on fan performance should always be considered when a fan selection is made. With any fan, restricted or unstable flow into the fan wheel or turbulent flow in the discharge duct before flow stabilization may reduce fan performance and increase sound wheel.

Static pressure losses due to inlet or discharge conditions can be expressed as a percent of velocity pressure as follows:

$$\text{static pressure loss} = K_t \left(\frac{\text{fan outlet velocity}}{4005} \right)^2$$

where K_t defines the loss factor for the inlet or discharge conditions being considered. Since loss factors, K_t, vary

FIGURE 14-26 Double-width fan wheel. (Courtesy of The Trane Company, LaCrosse, WI)

DOUBLE WIDTH FAN WHEEL

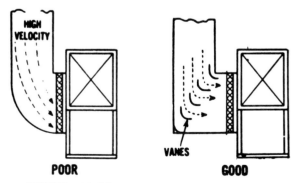

FIGURE 14-27 Compound curves on fan inlet duct (left) cause spin; vanes in elbow and at fan (right) eliminate spin.

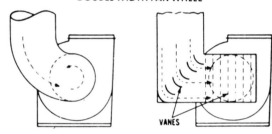

FIGURE 14-28 Examples of spin at fan inlet.

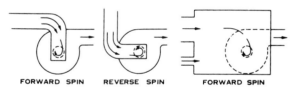

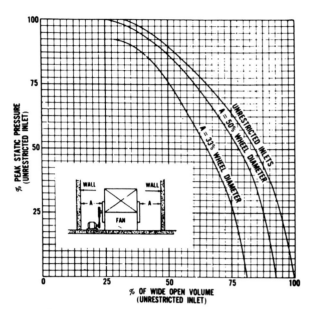

FIGURE 14-30 Space restriction reduces fan capacity as indicated by these curves from tests on double-inlet fan. (Courtesy of The Trane Company, LaCrosse, WI)

some with type of fan and different designs, one should refer to manufacturers' catalogs for appropriate values.

The effective static pressure for selection of the fan becomes the system static pressure plus the static pressure loss as determined above. When more than one inlet and discharge condition exists that affects fan performance, the losses should be summed to arrive at the total loss.

At the fan inlet, the following conditions will reduce the fan performance: (1) spinning airstreams, (2) nonuniform air distribution, and (3) insufficient space between fans or from the fan inlet to the wall. Figures 14-27 through 14-32 show good and poor methods of designing inlet and outlet conditions for fans.

The spin of the air at the inlet, as shown in Figure 14-27, may be due to the design of the duct connection at the inlet. This spin of the air into a fan is either for-

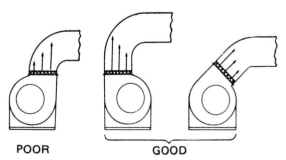

FIGURE 14-31 Proper discharge connections can minimize capacity reduction.

FIGURE 14-29 Uniform inlet distribution, through proper use of vanes, improves fan operation and reduces noise.

ward or reverse spin with respect to the direction of rotation of the fan wheel, as shown in Figure 14-28. If the spin is in the direction of wheel rotation, capacity will be reduced and the power consumption will drop. If the spin is counter to wheel rotation, the power consumption will increase, the noise level will increase, but there will be practically no increase in pressure.

When there is nonuniform airflow into the fan inlet, as shown in Figure 14-29, this may also be corrected by using guide vanes as shown. When there are no inlet ducts, fan capacity is affected when the space between

fans in a multiple-fan installation or the space between fan inlet and the wall is too small. General practice is to locate multiple fans at least one fan diameter apart and not to place a fan inlet closer than a distance of one-half fan diameter from the nearest obstruction. Test results, shown in Figure 14-30, show that even these distances result in a slight decrease in capacity. Greater distances should be used if possible.

The discharge duct should be a straight section for at least several duct diameters to allow the conversion of fan energy from velocity pressure to static pressure.

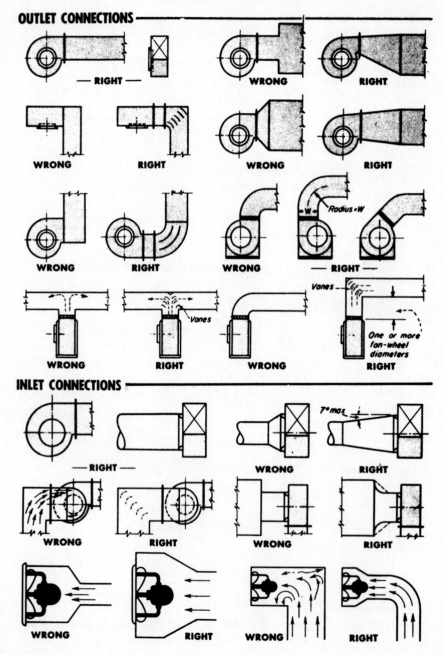

FIGURE 14-32 Right and wrong ways to install fans. (Reprinted with permission from *POWER* magazine, © McGraw-Hill, Inc., 1981)

When job conditions dictate that elbows be installed near the fan outlet, the loss of capacity and static pressure may be minimized by the use of vanes and proper direction of fan rotation relative to the direction of the bend in the elbow. The high-velocity side of the fan outlet should be directed at the outer radius of the

elbow rather than the inside. Examples of correct elbow arrangements are shown in Figure 14-31.

To summarize, Figure 14-32 illustrates some hints on fan installation. Refer to them when installing a new fan or relocating an old one.

REVIEW PROBLEMS

14.1. A round, aluminum duct must be selected to handle 2000 cfm of standard air at a maximum velocity of 1200 fpm. The duct is 40 ft long. Determine the required duct diameter, friction rate, and friction loss for the section.

14.2. A section of duct is 65 ft long and 16 in. in diameter. If 2000 cfm of standard air is delivered through the duct, determine **(a)** the velocity of the air flowing and **(b)** the friction loss in the duct.

14.3. The static pressure available to overcome friction in a section of round duct 125 ft long is 0.20 in. WG. The air quantity to be delivered is 3000 cfm (standard air). Determine **(a)** the required duct diameter and **(b)** the flow velocity.

14.4. A section of round duct 85 ft long is 16 in. in diameter and carries 2500 cfm of air at 120°F. Determine the friction in the duct section, assuming constant-mass flow.

14.5. Determine the actual friction loss for a constant-mass flow corresponding to a standard airflow volume of 4000 cfm in a 20-in. round duct that is 100 ft long when the actual air density is 0.0650 lb/ft³.

14.6. Figure 14-33 shows a sketch of a small ventilation system supplying standard air to four outlets in the quantities shown. Scale lengths are given in feet. Using a design friction rate of 0.10 in. WG per 100 ft and the equal-friction-rate method, determine the following:
(a) The round duct size for each section of duct shown.
(b) The equivalent rectangular sizes for each section for equal friction and maintaining a duct depth of 8 in.

(c) Which rectangular section has the highest velocity, its magnitude, and whether it is a reasonable velocity.

14.7. Figure 14-34 shows a sketch of a small, warm-air heating system for a house. The lengths of each section of duct are indicated in feet, and diffusers have capacities and pressure drops as indicated. Using a design friction rate of 0.060 in. WG per 100 ft and the equal-friction-rate method, determine the following:
(a) The rectangular duct sizes for each section for a uniform depth of 8 in.
(b) The static pressure required at point *A*. (Assume all branch takeoffs are 90-degree elbows and all elbows have an *r/W* = 0.50.)

14.8. **(a)** Figure 14-35 shows a portion of a ventilating system for a building. Duct section *ABCD* is to be sized by the equal-friction-rate method using a design friction rate of 0.20. in. WG per 100 ft and a duct depth of 16 in.
(b) Determine the rectangular duct size for section *CE* such that the pressure loss will be approximately equal to that in section *CD*, maintaining a 16-in. duct depth.
(c) What will be the static pressure required at point *A*? (Assume each outlet grille requires a static pressure of 0.05 in. WG.)

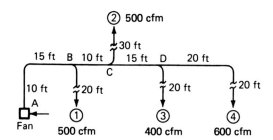

FIGURE 14-33 Schematic for Review Problem 14.6.

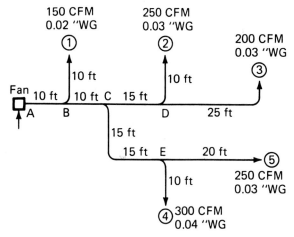

FIGURE 14-34 Schematic for Review Problem 14.7.

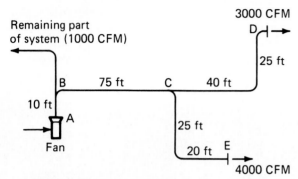

FIGURE 14-35 Schematic for Review Problem 14.8.

(d) Refer to Table 14-14 and determine the required fan operating rpm and horsepower. Assume that the pressure loss through the fan inlet system and filter is 0.6 in. WG and the gradual duct expansion (fan outlet) has a 40-degree included angle. *Note:* Assume all duct angles have an $r/W = 0.5$.

14.9. A duct system has been designed for a total airflow of 4875 cfm of 130°F air and a design friction rate of 0.06 in. WG per 100 ft. The velocity of air in the longest branch of the system was found to be 800 fpm and, at the start of the duct system, was 1225 fpm. The calculated equivalent length of the longest duct run was 200 ft. The static pressure loss in the return-air system was 0.624 in. WG.
  (a) Determine the required static pressure rise through the fan.
  (b) Select a fan for the system and determine the actual fan operating conditions.

14.10. A fan delivers 12,000 cfm of standard air against a static pressure of 1.0 in. WG when operating at a speed of 400 rpm and requires an input of 4.0 brake horsepower. If, in the same system, 15,000 cfm are desired, what would be the required rpm, static pressure, and horsepower at this new flow rate?

14.11. The fan of Review Problem 14.10 handles 12,000 cfm of standard air (density 0.0750 lb/ft³) against a static pressure of 1.0 in. WG rotating at 400 rpm. If the air temperature is increased to 200°F and the speed remains the same, what will be the static pressure and horsepower?

14.12. If the fan of Review Problem 14.11 has its speed increased to produce a static pressure of 1.0 in. WG with the 200°F air (same as with the standard air), what will be its speed, capacity, and power?

14.13. If the speed of the fan of Review Problem 14.12 is increased to deliver the same *weight* of air at 200°F as at standard conditions (70°F), what will be the speed, capacity, static pressure, and power?

14.14. A fan develops 1.0 in. WG static pressure and 0.4 in. WG velocity pressure when delivering 8000 cfm of 100°F air. The measured brake horsepower input was 2.0, and the speed was 1750 rpm. Calculate the fan static and mechanical efficiencies.

14.15. A fan is to be selected from a manufacturer's catalog to handle 70,000 lb/hr of air at 90°F against a static pressure of 1.75 in. WG when the barometric pressure is 29.00 in. Hg.
  (a) What volume of air (cfm) and static pressure (in. WG) should be used to select the fan from Table 14-15?
  (b) What would be the actual brake horsepower and rpm required when handling the air at 90°F?
  (c) What would be the fan static efficiency?

14.16. A fan delivers 9192 cfm of standard air against a static pressure of 1.0 in. WG. The speed and brake horsepower are 488 rpm and 2.01, respectively. Calculate the fan performance when the speed is reduced to 450 rpm.

14.17. A BI utility fan (Table 14-15) is to deliver 17,800 cfm of air at a static pressure of 1.0 in. WG when the air temperature is 200°F and the barometric pressure is 24.89 in. Hg (5000 ft above sea level). At start-up, the fan will be handling 0°F air.
  (a) Determine the operating rpm and brake horsepower for the fan.
  (b) What will be the start-up brake horsepower when 0°F air is flowing through the fan?

14.18. (a) A fan discharges 130°F air through a system of round ducts shown in Figure 14-36. The barometric pressure is 29.92 in. Hg. The quantity of air delivered at each supply outlet is shown. Branch duct velocity is not to exceed 900 fpm. Dampers are to be used in each branch to control the flow. Determine the duct diameter for each section of the system using the equal-friction-rate method. The design friction rate will be that determined by the size of duct branch section D–4.
  (b) What is the pressure head loss between point A and outlet 4?
  (c) What is the pressure head loss between point A and outlet 1?
  (d) If the total pressure required at point 4 is 0.10 in. WG, what should be the static pressure at point A?

14.19. (a) If the round duct sections of Review Problem 14.18 were converted to rectangular cross sections maintaining a 12-in. depth, what would be the width of each section?
  (b) Assume that the elbow between points D and 4 has an $r/W = 1.0$ and that the total pressure at point 4 is 0.10 in. WG. What should be the static pressure at point A?

14.20. Using data and results of Review Problems 14.18 and 14.19 where the system is handling 4800 cfm of 130°F air, select a fan from Table 14-14 to use in the system. Assume that the return-air duct system, heater, and

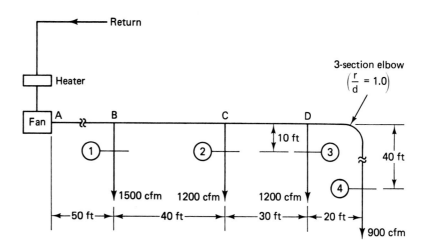

FIGURE 14-36 Schematic for Review Problems 14.18 and 14.19.

filters produce a static pressure at the fan inlet of −0.59 in. WG. Determine the actual fan speed and operating brake horsepower.

14.21. Figure 14-37 illustrates a simple ventilation system using standard air. Determine the rectangular duct size for each section of ductwork such that the pressure head loss from point *A* to each outlet will be approximately equal. The duct section *AB* will have a depth of 18 in.; all other sections will have a depth of 14 in. Assume that the branch takeoffs at points *B* and *C* have an *r*/*W* = 1.0 and all elbows have an *r*/*W* = 0.5. Start this problem by sizing the section *ABCD* by the equal-friction-rate method using a design friction rate of 0.25 in. WG per 100 ft of duct.

14.22. **(a)** Figure 14-38 represents a ventilation system handling standard air. Determine the rectangular duct size for each duct section using the equal-friction-rate method and a design friction rate of 0.20 in. WG per 100 ft of duct. Branch take-

offs have an *r*/*W* = 1.0, and all elbows have an *r*/*W* = 0.5.

(b) If each of the supply outlets requires a static pressure of 0.05 in. WG, what would be the required static pressure at the start of the duct system (point *A*)?

(c) If the static pressure at the entrance to the fan is −0.75 in. WG and the fan outlet velocity is 3400 fpm, what should be the static pressure rise through the fan?

14.23. A duct system has been designed for an airflow leaving the fan of 10,100 cfm at a temperature of 140°F. The barometric pressure is 27.30 in. Hg. The airflow velocity in the last section of the ductwork (greatest pressure loss section) was found to be 800 fpm and, at the start of the duct system, was 1500 fpm. Both velocities were for the rectangular section of the respective duct sections. The estimated total equivalent length of the duct system (greatest pressure loss sys-

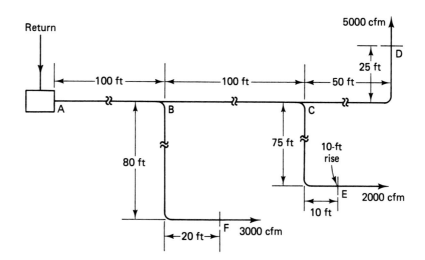

FIGURE 14-37 Schematic for Review Problem 14.21.

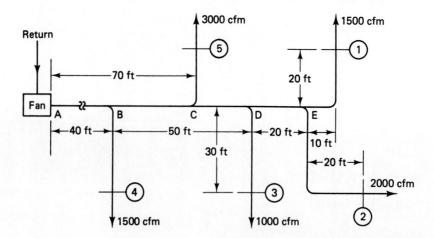

FIGURE 14-38 Schematic for Review Problem 14.22.

tem) was 265 ft. The design friction rate for sizing the ducts carrying standard air was 0.20 in. WG per 100 ft. The pressure head loss in the return-air duct system was calculated to be 0.59 in. WG. The static pressure required at the last outlet is 0.06 in. WG.

(a) Determine the required static pressure at the start of the duct system.

(b) Select a fan from Table 14-14 to flow the system

and determine the actual operating rpm and brake horsepower with the 140°F air. *Note:* If a transition piece is required between the fan and supply duct inlet, use a 30-degree included angle.

14.24. Using Table 14-15, select a fan to deliver 15,320 cfm of air at 150°F with a barometric pressure at 29.92 in. Hg. What will be the actual rpm and brake horsepower required when the fan is installed?

BIBLIOGRAPHY

14.1. *ASHRAE Handbook 1989 Fundamentals,* American Society of Heating, Refrigerating, and Air Conditioning Engineers, Atlanta, GA, 1989.

14.2. *ASHRAE Handbook 1988 Equipment,* American Society of Heating, Refrigerating, and Air Conditioning Engineers, Atlanta, GA, 1988.

14.3. *ASHRAE Handbook 1987 HVAC Systems and Applications,* American Society of Heating, Refrigerating, and Air Conditioning Engineers, Atlanta, GA, 1987.

14.4. *The Engineer's Reference Library, POWER,* McGraw-Hill, Inc., New York, 1968.

14.5. *Trane Air Conditioning Manual,* The Trane Company, LaCrosse, WI, 1965.

14.6. *Aluminum Air Ducts,* Reynolds Metal Company, 1956.

14.7. Aberback, Robert J., *Fans, POWER* Special Report, McGraw-Hill, Inc., New York, 1968.

14.8. *Centrifugal Fans,* Catalog DS-FAN6-1175, The Trane Company, LaCrosse, WI, 1975.

14.9. *Centrifugal Fans,* Catalog PL-AH-FAN-000-DS-6-1083, The Trane Company, LaCrosse, WI, 1983.

15

Fundamentals of Automatic Control and Control Components

15-1 INTRODUCTION

In previous chapters, we have considered accepted methods for the design and layout of HVAC systems that will meet the requirements to maintain comfort conditions in buildings if such systems are properly controlled. The HVAC system, its control system, and the building in which they are installed are inseparable parts of a whole. They must work together. The neglect of any one element may cause a partial or complete failure to provide for human comfort. Any well-designed HVAC system is useless without adequate control. Also, it should be stated that the best control system cannot overcome poor design of the HVAC system.

This, and the following chapter, will discuss automatic controls. In this limited space, we will consider the fundamentals of automatic control and present some of the typical control systems in common use.

In our modern-day living, the term *automatic control* has come to be applied to almost everything we come in contact with—from simple household items such as an electric coffee pot to the more complicated automatic industrial machines capable of continuous operation without attention. As time has passed and technical developments have increased, so also has the need for automatic control of our complex machines

and devices increased. The field of automatic control is a vast one and, like all of our other advancements, has become a series of highly specialized industries. One of these specialized industries is the field of automatic control of heating, ventilating, and air conditioning equipment. It is a field that deals not only with human comfort and productivity but is concerned also with industry—by developing controls that increase safety, economy, and production.

Automatic controls can be used wherever a variable condition must be controlled. Almost any variable condition may be controlled automatically. That variable condition may be pressure, temperature, humidity, or rate or volume of flow, and it may exist in a gas, liquid, or solid. Of these four general conditions, some are controlled directly; others, indirectly as a result of controlling another condition.

The purpose of an automatic control system is to provide stable operation of a process by maintaining the desired values of variable conditions. An automatic system is a collection of components, each with a definite function and designed to interact with the others, organized so that the system regulates itself. The control system senses disturbances and takes action to restore conditions to normal.

A control system is successful if the process consistently and economically produces an end product of good quality. Its success can be evaluated by how well it achieves three basic goals: (1) minimum deviation from normal conditions following a disturbance, (2) minimum time to restore conditions to normal, and (3) minimum offset due to changes in operating conditions.

Automatic control systems for heating, ventilating, and cooling fall into several basic categories according to application:

1. *Establishment of final conditions*—This is control of the end result in temperature, humidity, pressure, and so forth, which the control system is designed to provide.
2. *Provision for safe operation*—Many applications require automatic controls to ensure that temperatures, humidities, pressures, liquid levels, and so forth, are kept within certain limits for safety reasons. Also, in many applications, it is required that safe conditions exist before mechanical equipment, such as a cold or heat generator, is permitted to operate. The control equipment is chosen to protect mechanical parts from damage due to freezing or overheating if the control system fails. This is referred to as "failsafe" operation.
3. *Assurance of economical operation*—The correct application and proper correlation of a complete automatic control system provides proper operation of a control system at optimum operating cost.

15-2 COMPONENTS OF A CONTROL SYSTEM[1]

To simplify the explanation of a control system, we can represent it in block diagram form (see Figure 15-1). Each block represents an essential component of the *closed-loop system.*

The major requirement of a control system is the *process.* If there is nothing to control, there is no control system and no need for controlling elements. An elaborate steam pressure control system is no longer a system if the steam is removed from the pipes. The process can take numerous forms, such as the flight of an airplane, the maintenance of a required temperature in a hospital operating room, or the operation of a power boiler.

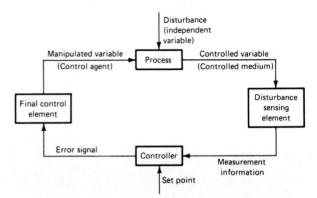

FIGURE 15-1 Simple block diagram of a closed-loop control system: Each step responds to action of step before and affects step following.

Disturbance-Sensing Element. The process is subject to a *disturbance* or external influence of some sort—a change in set point, supply, demand, or environment—or else there would be no need for a control system. The disturbance itself is an *independent variable* and cannot be changed by the system.

The disturbance causes a change in the *controlled variable,* which is the quantity or condition being controlled. The controlled variable is a characteristic of the *controlled medium,* which is the substance being controlled. For example, if water temperature is being controlled, the controlled variable is the temperature and the controlled medium is the water. In order to counteract the disturbance, it must first be detected and measured. The system must know how big a change in the controlled variable the disturbance caused. This is the job of the *disturbance-sensing element.* It is impossible for a human operator to tell the pressure within a pipe by looking at the outside of it or to tell the temperature of a furnace accurately by observing the color of the flame. Fortunately, there are transducers of many kinds to get this information for the system. Pressure-, temperature-, and humidity-sensing transducers are only a few of the many disturbance-sensing elements available to provide *measurement information.*

Controller. The *controller* receives the *measurement information* from the disturbance-sensing element and interprets it to see how well the process is going. (In many cases, the disturbance-sensing element may be a part of the controller, such as the bimetallic element in a thermostat.) A *set point* (or reference standard) is established in the controller. If conditions match this set point, the process is doing fine; but if a comparison of the measurement information with the set point shows a difference, the controller produces an *error signal* to initiate corrective action.

[1]Figures 15-1 through 15-6 and the associated descriptive matter (Sections 15-2 and 15-3) are from *Automatic Control Principles,* Honeywell, Inc., Milwaukee, WI, 1972.

The set point may be manually set (for example, a temperature setting), in which case it will stay constant for a long period of time. Or it may be adjusted automatically (for example, a boiler system in which steam demand is the reference input and airflow is adjusted accordingly), in which case the reference standard is varying continuously.

Final Control Element. The *final control element* is a mechanism that changes the value of the manipulated variable in response to the error signal from the controller. The *manipulated variable* is the quantity or condition (a characteristic of the control agent) that causes the desired change in the controlled variable. For example, if steam is flowing through a heating coil to heat a room, the steam is the control agent and its flow through the coil is the manipulated variable. The flow of steam varies in response to changes in the temperature (controlled variable) of the air (controlled medium) in the room.

The final control element causes the manipulated variable and control agent to exert a correcting influence on the process operation. Thus, the automatic control system is an error-sensitive, self-correcting system, often called a *closed-loop feedback control system.*

In summary, three basic parts must be considered when putting together a closed-loop control system:

1. *Control agent*—This source of energy supplied to the system can be either *hot* or *cold,* such as steam, hot water, heated air, chilled water, or chilled air.
2. *Controlled device*—A valve or damper can be either *normally open* (NO) or *normally closed* (NC) to regulate the flow of the control agent. It is chosen primarily for fail-safe operation, as mentioned previously.
3. *Controller action*—A controller is furnished with either *direct action* or *reverse action.* This will allow the balance mentioned above.

The proper combination of these three parts must be applied, or the closed-loop system will not operate.

The closed-loop control system is by far the most commonly used system. Nevertheless, the *open-loop system* finds application in certain instances. For example, more heat is required to maintain a suitable room temperature as the outdoor air temperature drops. It is possible to arrange a thermostat to measure outdoor air temperature and cause the heat input of a building to increase as the outdoor air temperature decreases. The outdoor thermostat cannot measure the result of heat input to the room; hence there is no feedback. Figure 15-2 illustrates such a control system.

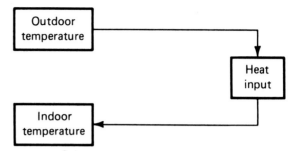

FIGURE 15-2 Simple block diagram of an open-loop control system: No direct relation exists between steps and hence no feedback.

It should be clear that all control systems are comprised of one or more loops. Most of them will be closed loops, but some may be open loops or a combination of both. It is necessary to understand the difference in order to appreciate what results can be expected. When controllers are studied, it will be seen that they comprise one or more control loops within themselves.

It should be pointed out that there are no perfect closed loops—that is, loops that must account only for the effect of the parts as described. In actual control conditions, outside forces constantly work on the various parts to change the balance and set the loop cycle in operation to reestablish balance. If this were not the case, there would be no need for automatic control.

In conclusion, it can be said that a closed-loop system is one in which all parts have an effect on the next step in the loop and are affected by the action of the previous step. An open-loop system is one in which one or more of the steps has no direct effect or action imposed on the following step or is not affected by other steps in the loop.

15-3 MODES OF CONTROL (CONTROL ACTION)

A mode of control is the manner by which a control system makes corrections in response to a disturbance. It relates the operation of the final control element to the measurement information provided by the disturbance-sensing element. Thus, it is generally a function of the controller. The proper matching of the mode to the process determines the overall performance of the control system. The basic modes are (1) on–off (two-position), (2) multiposition (multistage), (3) floating, and (4) proportioning (modulating). More complex variations of the proportioning mode include the addition of reset action, rate action, or both.

On–Off (Two-Position) Control. On-off control, as the name suggests, provides only two positions—either full ON or full OFF. There are no intermediate positions. When the controlled variable deviates a predetermined amount from the set point, the final control element moves to either of its extreme positions. The time travel to ON or OFF is varied according to the load demand.

On–off control is the simplest mode of control, but it has definite disadvantages. It allows the controlled variable to vary over a range (sometimes a wide range) instead of letting it settle down to a near-steady condition. If this range becomes too narrow, the controller will wear itself out by continually switching on and off. Figure 15-3 illustrates schematically the system response and controller action.

Multiposition (Multistage) Control. Multiposition control is an extension of the on–off control to two or more stages. Where the range between full ON to full OFF is too wide to achieve the operation desired, multiple stages with much smaller ranges can be used. Each stage has only two positions (ON or OFF), but there are as many positions as there are stages, resulting in a steplike operation. The greater the number of stages or steps, the smoother will be the operation. As the demand or load increases, more stages are turned on. Normally, multiposition control provides for two to ten operating stages. Two-stage firing of oil burners is quite common. Figure 15-4 illustrates schematically the system response and controller action.

Floating Control. Floating control differs from on–off or multiposition control in that the final control element may assume any position between its extremes. The controller itself has two positions with a neutral zone in between. With the controller in the neutral zone, the final control element will not move. When a large enough disturbance occurs, the controller moves to one of its two positions. This causes the final control element to move in one direction or the other, depending on the position of the controller. The final element moves at a *constant speed*. Therefore, this mode is more properly called *single-speed floating control*. When the controller returns to the neutral zone, the final control element stops. It will stay at this position until the con-

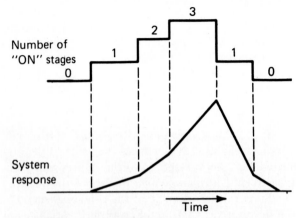

FIGURE 15-4 Multiposition (multistage) mode of control.

troller again moves to one of its two positions. Thus, this mode is called *floating* because the final control element comes to rest when the controller is floating between its two positions.

Floating control is used in applications with gradual load changes or where there is little lag between a disturbance and its detection. The speed of the final control element should be fast enough to keep pace with the most rapid load changes. If not, hunting (instability) and excessive cycling will occur. Figure 15-5 illustrates schematically the system response, controller action, and position of the final control element.

Proportioning (Modulating) Control. In proportioning control, as in floating control, the final control element may assume any position between its extremes. Unlike floating control, the controller may also assume any position. It has no neutral zone; so any small disturbance will cause it to move. For each movement of the controller, there is a proportional amount

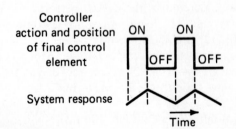

FIGURE 15-3 On–off (two-position) mode of control.

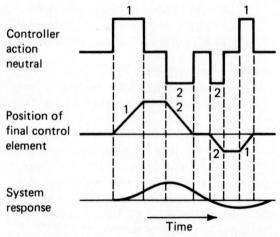

(1) and (2) indicate positions of controller

FIGURE 15-5 Floating mode of control.

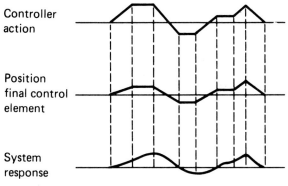

FIGURE 15-6 Proportioning (modulating) mode of control.

of movement of the final control element. These movements can be made as frequently as disturbances occur. Proportional control produces a continuous linear relation between the amount of disturbance, controller action, and position of the final control element. Figure 15-6 illustrates schematically the system response, controller action, and position of the final control element.

More complex variations of proportioning control: On a sustained load change, an *offset* can occur between the value of the controlled variable and the set point. Continual adjustment of the set point is necessary to keep the controlled variable at the same point throughout the load range of the equipment. In sophisticated controllers, two types of automatic action are frequently added to assure the best possible control under the most severe process conditions.

With *reset action,* the error signal from the controller is varied in proportion to the offset and to the time that it persists. The final control element continues to move in a direction to correct the error. It will stop only when the error signal becomes zero, at which time the controlled variable is at the set point.

With *rate action,* the error signal from the controller is varied in proportion to the rate at which the disturbance takes place. It is used (1) to accelerate (speed up) the return of the controlled variable to the set point in a part of the process that is slow in responding and (2) to anticipate a disturbance that will occur and start corrective action before any change is detected.

15-4 TYPES OF CONTROL SYSTEMS

A control system can be classified by its source of power (energy). Control systems may be pneumatic, hydraulic, electromechanical, electronic, fluidic, or combinations of these.

A *pneumatic system* is operated by air pressure supplied by a compressor, and its components are connected by air lines. A *hydraulic system* is operated by movement and force of a liquid under pressure, and its components are connected by pipes and/or flexible tubes. An *electromechanical system* is operated by electricity, and its components are connected by electric circuits. An *electronic system* is also operated by electricity, but it uses electronic elements, such as transistors and vacuum tubes. In most cases, it also uses an electronic amplifier to increase the minute voltage variations produced by the disturbance-sensing element to values required for operation of standard electromechanically controlled devices. *Fluidic systems* use air or gas and are all similar in operating principles to electronic as well as pneumatic systems. In reality, many control systems use combinations of these several power sources. An example is a system using an electronic sensing element, an electromechanical controller, and a pneumatic valve.

In the following sections, we will discuss the electromechanical, electronic, and pneumatic system components. These are the most commonly used for HVAC systems.

15-5 BASIC ELECTRICITY[2]

Before studying electromechanical and electronic control systems and the required component devices, a brief review of basic electricity will be helpful to better understand the function of these systems.

Ohm's Law. Three factors are present in every electrical circuit: current, voltage, and resistance.

A common electrical power source is a battery. The battery in an electrical circuit is the same as the pump in a water circuit. Just as the pump produces a source of pressure measured in psi, the battery produces a source of electromotive force (EMF) measured in volts.

The amount of flow (gpm) in a water circuit depends on the pump pressure and the amount of restriction in the circuit. Similarly, the amount of current (amperes) in an electrical circuit depends on the battery voltage and the amount of resistance (ohms) in the circuit. The voltage is the force, the resistance is the opposition to the force, and the current is the flow or movement of electrons that results from the combination. Current is the flow rate of electrons per second past a point in the wire. One volt (V) is required to force 1 ampere (A) of current through 1 ohm (Ω) of resistance. Expressed mathematically,

[2]Extracted from *Basic Electricity,* Training Manual, Johnson Controls, Inc., Milwaukee, WI, 1980.

E (voltage) = I (current in amperes)
$\times$ R (resistance in ohms)

This is the equation on which all electric and electronic circuit theory is based and is called *Ohm's law*. It can be arranged as required to find one of the three factors, such as

$$E = IR \quad I = \frac{E}{R} \quad R = \frac{E}{I}$$

Figure 15-7 shows a light bulb connected across a 10-V battery. The light bulb has 1 Ω of electrical resistance. To determine the current through the bulb, Ohm's law is stated and used as follows:

$$I = \frac{E}{R} = \frac{10\ V}{1\ \Omega}$$
$$= 10\ A$$

Thus, a current of 10 A is flowing through the bulb.

Series Connections. Figure 15-8 shows two light bulbs connected in series across a battery. Two resistances are said to be connected in series when the current flowing through one resistance must also flow through the other. A series connection used to be used for Christmas tree lights; when one light bulb burned out, the circuit was broken and the whole string went out. The total resistance (R_T) in a series circuit is the sum of the individual resistances:

$$R_T = R_1 + R_2 + R_3 + \cdots$$

For Figure 15-8, $R_T = 4\ \Omega + 1\ \Omega = 5\ \Omega$. To determine the current through the light bulbs, we have

$$I = \frac{E}{R} = \frac{10\ V}{5\ \Omega} = 2\ A$$

Therefore, a current of 2 A is flowing through the bulbs.

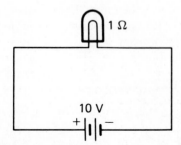

FIGURE 15-7   Light bulb in a battery circuit.

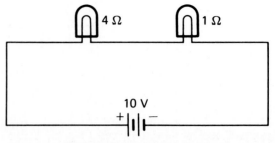

FIGURE 15-8   Series circuit.

The current is the same through all the resistances, but the voltage drop across each is dependent on the resistance value. When the current through a bulb and the resistance are known, Ohm's law can be used to determine the voltage drop across the bulb. For the 1-Ω bulb,

$$E = IR = 2 \times 1 = 2\ V$$

For the 4-Ω bulb,

$$E = 2 \times 4 = 8\ V$$

Parallel Connections. Figure 15-9 shows two light bulbs connected in parallel across a battery. In parallel connection, the voltage drops across the resistances will always be equal; the current flow through each resistance is dependent on the resistance values, or

$$I_1 = \frac{E}{R_1} = \frac{10}{4} = 2.5\ A$$

$$I_2 = \frac{E}{R_2} = \frac{10}{1}\ 10.0\ A$$

Both bulbs draw current from the battery; so the total current supplied is the current flow through both bulbs, or $I_T = I_1 = I_2 = 2.5 + 10.0 = 12.5\ A$.

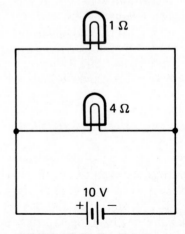

FIGURE 15-9   Parallel circuit.

If Christmas tree lights were connected in parallel, one light burning out would not break the circuit because each bulb would have a direct connection to the power source.

The total resistance of the circuit can be determined by using the voltage across the circuit and the total current flow into the circuit, or $R_T = E/I_T = 10.0/12.5 = 0.8 \, \Omega$. The total resistance of a parallel circuit is always less than the lowest resistance in the circuit and can be calculated as follows:

$$\frac{1}{R_T} = \frac{1}{R_1} + \frac{1}{R_2} + \frac{1}{R_3} + \cdots$$

The total current is the sum of the individual currents through each lamp:

$$I_T = I_1 + I_2 + I_3 + \cdots$$

Using Ohm's law, $I = E/R$,

$$\frac{E_T}{R_T} = \frac{E_1}{R_1} + \frac{E_2}{R_2} + \frac{E_3}{R_3} + \cdots$$

But the voltage across each lamp is the same:

$$E_T = E_1 = E_2 = E_3 = \cdots$$

Substituting this into the current equation gives

$$\frac{E_T}{R_T} = \frac{E_T}{R_1} + \frac{E_T}{R_2} + \frac{E_T}{R_3} + \cdots$$

In summary, for DC series circuits,

1. The same current flows through all resistances.
2. Total voltage is the sum of all voltage drops across all resistances.
3. Total resistance is the sum of all resistances.

In summary, for DC parallel circuits,

1. The total current is the sum of all currents.
2. Voltage is the same across all resistances.
3. The reciprocal of the total resistance is the sum of the reciprocals of the individual resistances.
4. Total resistance is always less than the lowest resistance in the circuit.

AC and DC Power. Electrical power can be supplied in two different forms, the difference being the characteristics of the current flow. A power source that

causes current to flow in one direction only is referred to as a *direct current* (DC) source. A power source that causes current to flow alternately in one direction and then in the other is referred to as an *alternating current* (AC) source.

Batteries and automobile DC generators are common examples of DC electrical power sources. Normally, a DC current flow is thought of as a continuous unidirectional flow that is constant in magnitude; however, a pulsating current flow that changes in magnitude but not direction is also considered direct current.

The power supplied by power companies throughout the country is the most common example of an AC power source. If the magnitude of the current is recorded as it varies with time, the shape of the resultant curve is called the *waveform*. The waveform produced by the power companies' generators is a sine wave, as shown in Figure 15-10. When the waveform of an AC voltage or current passes through a complete set of positive and negative values, it completes a cycle. The frequency of an AC voltage or current is the number of cycles that occur in 1 s; the frequency of the voltage supplied by U.S. power companies is 60 cycles per second (60 hertz).

All calculations using AC voltage are based on sine waves. Four values of sine waves are of particular importance.

1. *Instantaneous value*—The voltage or current in an AC circuit is continually changing. The value varies from zero to maximum and back to zero. If we measure the value at any given instant, we will obtain the instantaneous voltage or current value.
2. *Maximum value*—For two brief instants in each cycle, the sine wave reaches a maximum value; one is a positive maximum, the other negative. This value is often referred to as the *peak* value. The two terms have identical meanings and are interchangeable.
3. *Average value*—Since the positive and negative halves of a sine wave are identical, the average value can be found by determining the area below the wave and calculating what

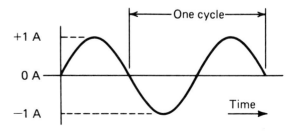

FIGURE 15-10 AC-cycle waveform.

DC value would enclose the same area over the same amount of time. For either the positive or negative half of the sine wave, the average value is 0.636 times the maximum value.

4. *Effective value* (rms)—The effective value of an AC voltage or current required to provide the same average power or heating effect. Heating effect is independent of direction of electron flow and therefore is the same for a full cycle as it is for a half cycle of a sine wave. The effective value of an AC sine wave is equal to 0.707 times the maximum value. Thus, an alternating current of a sine wave having a maximum value of 14 A produces the same amount of heat in a circuit that a direct current of 10 A produces.

The heating effect varies as the square of the voltage or current. If we square the instantaneous values of a voltage or current sine wave, we obtain a wave that is proportional to the instantaneous power or heating effect of the original sine wave. The average of this new waveform represents the average power that will be supplied. The square root of this average value is the voltage or current value that represents the heating effect of the original sine wave of voltage or current. This is the effective value of the wave, or the rms value (root-mean-square)—the square root of the average of the squared waveform.

The effective values of voltage and current are more important than the instantaneous, maximum, or average values. Most AC voltmeters and ammeters are calibrated in rms values.

The *phase* of an AC voltage refers to the relationship of its instantaneous polarity to that of another AC voltage. Figure 15-11 shows two AC sine waves in phase but unequal in amplitude. Figure 15-12 shows the two

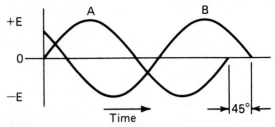

FIGURE 15-12 AC sine waves 45 degrees out of phase.

sine waves 45 degrees out of phase. Figure 15-13 shows the two sine waves 180 degrees out of phase. The length of a sine wave can be measured in angular degrees because each cycle is a repetition of the previous one, much like going in a circle. One complete cycle of a sine wave is 360 angular degrees.

Power in DC Circuits. Whenever a force causes motion, work is performed. Electrical force is expressed as voltage, and when voltage causes a movement of electrons (current) from one point to another, energy is expended. The rate of work, or the rate of producing, transforming, or expending energy, is generally expressed in watts or kilowatts (1000 W). In a DC circuit, 1 V forcing a current of 1 A through a 1-Ω resistance results in 1 W of power being expended. The equation for this is

$$P \text{ (watts)} = E \text{ (volts)} \times I \text{ (amperes)}$$

To illustrate, what is the DC power used by the light bulbs in a series circuit such as the one in Figure 15-8? The current is $I = E/R_T = 10/5 = 2$ A. Then, from $P = E \times I$, we have $P = 10 \times 2 = 20$ W of power.

The same calculations can be made for the light bulbs in a parallel circuit such as the one in Figure 15-9. The current is $I = E/R_T = 10/0.8 = 12.5$ A. Then, the power is $P = E \times I = 10 \times 12.5 = 125$ W.

Power can be computed if any two of the three

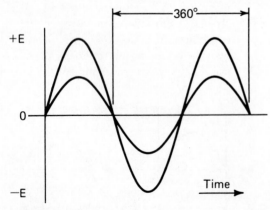

FIGURE 15-11 AC sine waves in phase but unequal in amplitude.

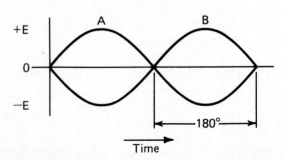

FIGURE 15-13 AC sine waves 180 degrees out of phase.

values of current, voltage, and resistance are known, as follows:

1. When resistance is unknown, $P = E \times I$.
2. When voltage is unknown, $P = (IR)I = I^2R$.
3. When current is unknown, $P = E(E/R) = E^2/R$.

Power Rating of Equipment. Most electrical equipment is rated for both voltage and power (volts and watts). Electric lamps rated at 120 V are also rated in watts; they are usually identified by wattage rather than by voltage because the voltage is always assumed to be 120. The wattage rating of a light bulb or other electrical device indicates the rate at which electrical energy is changed into other forms of energy, such as light and heat. The greater the amount of electrical power a lamp changes to light, the brighter the lamp will be; therefore, a 100-W bulb furnishes more light than a 75-W bulb.

Similarly, the power rating of motors, resistors, and other electrical devices indicate the rate at which the devices are designed to change electrical energy into some other form of energy. If the rated wattage is exceeded, the excess energy is usually converted to heat, and the equipment will overheat and perhaps be damaged. Some devices will have maximum DC voltage and current ratings instead of wattage; multiplied, these values give the effective wattage.

Resistors are rated in watts dissipation in addition to ohms of resistance; resistors of the same resistance value are available with different wattage ratings. Usually, carbon composition or ceramic resistors are rated from about 1/10 W to 2 W. Wire-wound resistors are used when a higher wattage rating is required. Generally, the larger the physical size of the resistor, the higher its wattage rating since a larger amount of material will absorb and give up heat more easily.

Capacitance. Capacitance, like resistance, is a physical property of an electrical circuit. Resistance is opposition to current flow in an electrical circuit the same as friction is the opposition to motion in a mechanical system. Capacitance is the property of an electrical circuit (or component) that opposes any change in voltage across it. Capacitance, in an electrical circuit, allows electrons to be stored the same as a liquid or gas would be stored in a tank in a mechanical system. The capacity of the tank is rated in gallons or cubic feet; the capacity of a capacitor is rated in farads—a farad (F) being the unit of capacitance. A capacitor has a capacitance of 1 F when a voltage change of 1 V per second across its terminals produces a displacement current flow of 1 A. Figure 15-14 illustrates an electrical circuit

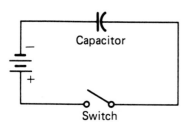

FIGURE 15-14   Circuit with capacitor.

that includes a capacitor. Because the usual size of a capacitor is only a small portion of 1 F, the capacity is usually expressed in microfarads (millionths of a farad) or picofarads (millionths of a microfarad).

A capacitor has two conducting surfaces (material with low electrical resistance) separated by a dielectric (an insulating material with almost infinite electrical resistance).

Capacitors are used in electronic circuits for one of three basic purposes:

1. To couple an AC signal from one section of a circuit to another.
2. To block out and/or stabilize any DC potential from some component.
3. To bypass or filter out the AC component of a complex wave.

Inductance. Another factor to consider in electrical circuits is inductance. Inductance is the property of a circuit (or component) to oppose any change in current through it. When current attempts to change in an inductor, a voltage is self-induced in the coil. In discussing an inductor, the word *coil* is often used to describe it. A coil is a series of rings or a spiral of wire. Automobile and relay coils are common examples. The polarity of the induced voltage is such as to oppose the change in current. In this way, inductance can be compared to kinetic inertia in a mechanical system, which tends to keep a body moving at a constant velocity. Figure 15-15 illustrates an electrical circuit that includes an inductor (coil).

Power in AC Circuits. Ohm's law for an AC circuit states that the current flowing through an AC cir-

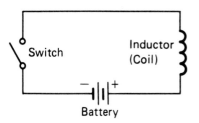

FIGURE 15-15   Circuit with inductor.

cuit is equal to the voltage impressed across that circuit divided by the impedance of the circuit. This is the same as Ohm's law for DC circuits, except that the word *impedance* is substituted for *resistance*. Impedance (*Z*) is the total opposition to the flow of current in an AC circuit offered by resistance, capacitance, and inductance. The ohm is the unit of impedance in an AC circuit. The three Ohm's law formulas for AC circuits are as follows:

$$ I = \frac{E}{Z} \qquad Z = \frac{E}{I} \qquad E = IZ $$

Power in a purely resistive AC circuit is calculated the same as in a DC circuit; that is, $P = E \times I$.

When we add capacitance or inductance to the circuit, voltage × current does not indicate the actual wattage being taken from the power source. In a circuit, when capacitance predominates, the current will lead the voltage; where impedance predominates, the current will lag the voltage. The measurement of the phase difference between current and voltage in a circuit is called *power factor* and is simply the cosine of the shared angle. When the current is greatly out of phase with the voltage, the power factor is said to be low. When the current is nearly in phase with the voltage, the power factor is said to be high. When the angle between current and voltage is 90 degrees, the power factor is zero. When the voltage is exactly in phase with the current, the power factor is unity. A high power factor is desirable in AC power systems because circuit losses are reduced. Even though the inductive or capacitive component in an AC circuit does not use any electrical power from the power source, it does draw current from the source. This extra current must flow through the generators, transformers, and wires that make up a power system and will cause extra losses in these devices. Also, the increased current may force the use of larger devices than would be necessary with a higher power factor, and this costs money. For these reasons, power companies will penalize a customer with a lower power factor load by increasing the cost per kilowatt hour.

If two AC motors have the same horsepower rating, the electrical power used by both will be equal when they do the same amount of mechanical work. However, if the power factor of one motor is larger than the other, it will take less current to do that work. If the power factor of one motor is 0.9 and the other 0.7, but both are 1-hp (746-W), 120-V AC motors,

$$ P = EI \times \text{power factor} = 746 $$
$$ 746 = 120 \times I \times 0.9 $$
$$ I = 6.92 \text{ A} $$

and

$$ P = EI \times \text{power factor} = 746 $$
$$ 746 = 120 \times I \times 0.7 $$
$$ I = 8.9 \text{ A} $$

The preceding example shows why a higher power factor is desirable. A low power factor is a disadvantage in power circuits and should be corrected if possible. Since the properties of inductance and capacitance have the opposite affect in AC circuits, in highly inductive circuits (such as those containing many electric motors) placing large capacitors across the power line will raise the power factor.

Usually, electrical machinery is rated in volt-amperes (VA) instead of watts to aid in determining the power factor. A wattmeter reading across the power lines will give the actual power taken from the source. The wattmeter reading divided by the full load VA for a motor is the power factor.

Semiconductors. An insulator has a very high resistance to current flow, and a conductor has a very low resistance. Therefore, as the name implies, a semiconductor has a "medium" resistance.

Semiconductors used in making diodes and transistors are basically germanium or silicon crystals with controlled amounts of impurities added. When arsenic or antimony is the added impurity, *N*-type semiconductor material is formed (an excess number of electrons). When gallium or indium is the added impurity, *P*-type semiconductor material is made (a lack of electrons).

Rectifier (Diode). When *N*- and *P*-type materials are joined together, they form a rectifier, also referred to as a diode. The *PN* junction (diode) acts as a one-way valve to the flow of current. There is a forward, or low-resistance, direction through the junction. Current flowing in the low-resistance direction is called *forward bias* (see Figure 15-16); current flow in the opposite or high-resistance direction is called *reverse bias* (see Figure 15-17).

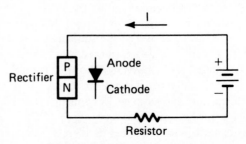

FIGURE 15-16 Forward bias.

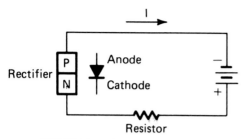

FIGURE 15-17   Reverse bias.

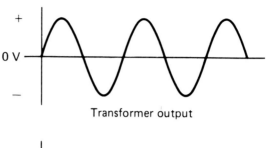

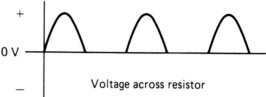

FIGURE 15-19   Rectifier action in circuit of Figure 15-18.

Diodes are used as rectifiers, converting alternating current to direct current and isolating one circuit from another circuit. A simple rectifier circuit is shown in Figure 15-18. The output from the transformer is an AC voltage (see Figure 15-19), but because the rectifier blocks current flow when the transformer output is negative, only pulsating DC voltage appears across the resistor.

Zener Diode.   Compared with pneumatic equipment, the zener diode is the equivalent of a nonadjustable pressure-reducing valve. It is used when a constant DC voltage is required.

When one polarity of voltage is applied to a rectifier, it blocks the flow of current. However, if the voltage is raised high enough, the rectifier breaks down, allowing current to flow. Normal rectifiers would be destroyed by this breakdown, but a zener diode is specially designed to operate in the breakdown region.

Figure 15-20 shows a typical zener diode circuit. The breakdown voltage on the diode is 10 V. As long as the battery voltage is 10 V or higher, the output across the diode will be 10 V. Any battery voltage above 10 V is dropped across the resistor; so, if the voltage changes from 12 V to 14 V, the voltage drop across the resistor changes from 2 V to 4 V while the output voltage remains at 10 V.

Transistor.   A transistor is a device used to amplify electronic signals. It consists of three layers of *P*-

and *N*-type semiconductor material arranged in either of two ways, as shown in Figure 15-21.

The theory of transistors has been known almost as long as vacuum tubes have been practical. However, it was not until 1948 that research laboratories developed a practical, high-quality transistor to meet production standards of cost and utility. Now, the transistor is rapidly replacing the vacuum tube in many applications of electronics, plus opening up many more.

The transistor has several advantages over the vacuum tube:

1. Small size reduces the bulk and weight of electronic equipment.
2. Low power consumption means economical operation.
3. No warm-up time results in faster equipment response.
4. Ruggedness enables it to better withstand mechanical shock and abuse.
5. Long life eliminates frequent maintenance and replacement.

Figure 15-22 shows a simple transistor amplifier. Battery 1 and adjustable resistor R_1 determine the input

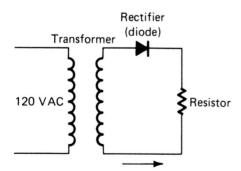

FIGURE 15-18   Simple rectifier circuit.

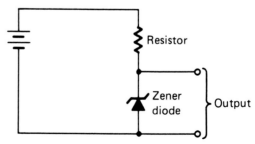

FIGURE 15-20   Zener diode circuit.

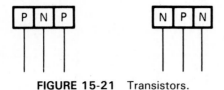

FIGURE 15-21 Transistors.

current to the transistor. When R_1 is high in resistance, the current flowing from the base to the emitter is very small. When the base-to-emitter current is small, the collector-to-emitter resistance appears as a very high resistance, limiting the current flow from battery 2 and limiting the voltage drop across R_2. As the resistance of R_1 is lowered, the current flowing through the base-to-emitter junction increases. As the base-to-emitter current is increased, the resistance of the transistor to collector to emitter is decreased. More current is flowing from battery 2 through R_2, and the voltage drop across R_2 is increased. A very small change in the current from battery 1 causes a large change in the current from battery 2. The ratio of the large change to the small change is defined as the *gain* of the transistor.

Silicon-Controlled Rectifier (SCR).   The silicon-controlled rectifier (SCR) is a four-layered *PNPN* device. The SCR can be defined as a high-speed semiconductor switch. It requires only a short voltage pulse to turn it on, and it remains on as long as current is flowing through it.

Referring to Figure 15-23, assume that the SCR is off (has a very high resistance); therefore, there is no current flowing through the resistor. When switch S_1 is closed for a time, just long enough to turn the SCR on (changes to a very low resistance), a current will flow through the resistor and SCR. The SCR will remain on until switch S_2 is opened. Opening S_2 stops the current flowing through the resistor and SCR, and the SCR will turn off. When S_2 is again closed, the resistance of the SCR remains high, and no current will flow through the resistor until S_1 is reclosed.

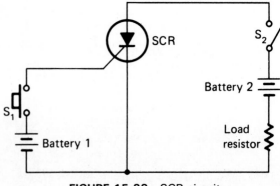

FIGURE 15-23 SCR circuit.

Figure 15-24 shows an SCR represented (a) schematically and (b) pictorially. The anode is the positive terminal, and the gate is the terminal used to turn the SCR on.

Bridge Theory.   A bridge circuit is a network of resistances and capacitive or inductive impedances usually used to make precise measurements. The most common is the Wheatstone bridge, which consists of variable and fixed resistances. This is simply a series-parallel circuit, redrawn as shown in Figure 15-25. The branches of the circuit forming the diamond shape are called *legs*.

If 10 V direct current were applied to the bridge

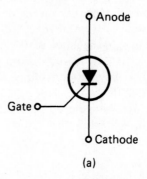

(a)

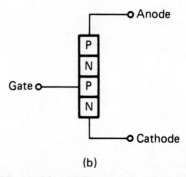

(b)

FIGURE 15-24 SCR symbol: (a) schematic and (b) pictorial.

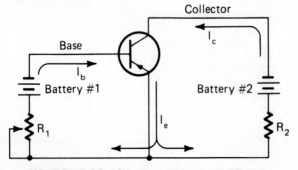

FIGURE 15-22 Simple transistor amplifier circuit.

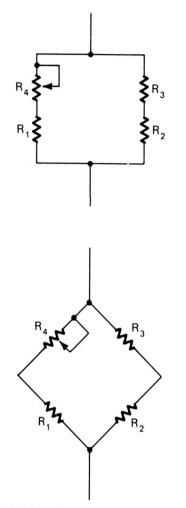

FIGURE 15-25   Series–parallel bridge circuit.

in Figure 15-26, one current would flow through R_1 and R_2 and another through R_3 and R_4. Since R_1 and R_2 are both fixed 1000-Ω resistors, the current through them is constant and each resistor will drop one-half of the battery voltage, 5 V.

However, as R_4 varies, the division of the battery voltage between R_3 and R_4 changes. In Figure 15-26, 5 V is dropped across each resistor. The voltmeter senses the sum of the voltage drops across R_2 and R_3. Both are 5 V—the R_2 drop is a ± 5-V drop and the R_3 drop is ± 5 V—but they are opposite in polarity and cancel each other. This is called a *balanced* bridge. The relationship is usually expressed as a ratio of $R_1/R_2 = R_3/R_4$. The actual resistance values are not important; what is important is that this ratio is maintained and the bridge is balanced.

In Figure 15-27, the variable resistor R_4 has changed to 950 Ω; the rest of the resistors are at the same value. Using Ohm's law, the voltage drop across R_4 is found to be 4.9 V; the rest of the voltage, 5.1 V, is dropped across R_3. In this illustration, the voltmeter senses the algebraic sum of the voltage drops across R_2 and R_3, ± 5 V and ± 5.1 V, registering a total of -0.1 V.

Conversely, in Figure 15-28, the value of R_4 has changed to 1050 Ω; the voltage drop across R_3 is then 4.9 V. The voltmeter senses the sum of ± 5 V and ± 4.9 V, or $+0.1$ V.

When R_4 changes the same amount above or below the balanced bridge resistance, the magnitude of the DC output, measured by the voltmeter, is the same but the polarity is reversed.

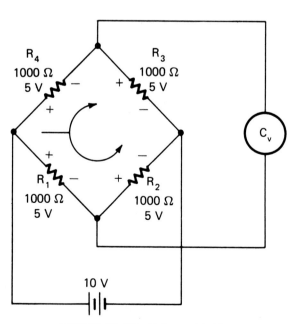

FIGURE 15-26   Balanced bridge.

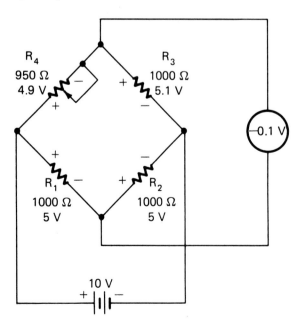

FIGURE 15-27   Bridge circuit with variable resistance reduced.

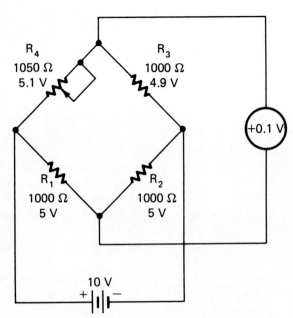

FIGURE 15-28 Bridge circuit with variable resistance increased.

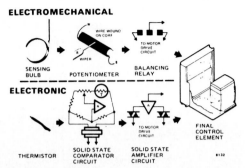

FIGURE 15-29 Comparison of electromechanical and electronic devices.

Electronic components have been developed for use in applications where the use of standard electromechanical components is difficult or inconvenient. They are intended to complement existing lines of control.

15-7 DISTURBANCE-SENSING ELEMENTS

A disturbance-sensing element is a transducer that detects and measures changes in the controlled variable. Electromechanical elements, using moving parts, measure the motion produced by these changes. Electronic elements have no moving parts; they measure other characteristics. Table 15-1 lists the common disturbance-sensing elements used in electric control systems.

Liquid-Level-Sensing Elements. A change in liquid level, which itself is produced by motion, is perhaps the simplest change to measure. An electromechanical float mechanism is usually used. A common application is the low-water cutoff (LWCO) used on a steam heating boiler to prevent operation when the boiler is nearly dry. A sensitive pressure-sensing element can also be used.

Pressure-Sensing Elements. Pressure is defined as force per unit area. In electromechanical pressure-sensing elements, this force is easily applied to cause motion. Figure 15-30 illustrates some of the common pressure-sensing elements. Electronic pressure-sensing elements, such as piezoelectric crystals, are not commonly used in industrial or commercial applications. (Piezoelectric crystals, such as quartz or barium titanite, produce a voltage when subjected to mechanical stress such as compression, expansion, or twisting.) Flow rate (quantity per unit time), velocity, static pressure, and liquid level can also be measured using variations of pressure-sensing elements.

Flexible diaphragm: A flexible diaphragm is distorted by increased pressure (Figure 15-30). A linkage

15-6 ELECTRIC CONTROL SYSTEMS[3]

An electric control system consists of the basic control system components—disturbance-sensing element, controller, and final control element—interconnected by electrical wiring, relays, and amplifiers (where required). It uses electrical energy provided by an electric power supply.

Electric systems are used in all types of producing and processing industries and in HVAC applications everywhere. They are particularly applicable when the final control element is electric or where relatively long distances separate the controller and the final control element.

Electromechanical versus electronic control systems: Electric control systems include both electromechanical and electronic systems. The main difference between them is that electronic systems use solid-state devices (vacuum tubes in older systems). Figure 15-29 shows comparable electromechanical and electronic devices. In an electronic system, a thermistor could replace a temperature-sensing bulb, a solid-state comparator circuit could replace a potentiometer, and a solid-state amplifier could replace a balancing relay. In actual practice, many systems and components are *hybrid*—they use both electromechanical and electronic devices.

[3]Figures 15-29 through 15-56 and the associated descriptive matter (Sections 15-6 through 15-12) are from *Automatic Control Principles,* Honeywell, Inc., Milwaukee, WI, 1972.

TABLE 15-1

Disturbance-sensing elements

Controlled Variable	Electromechanical Elements	Electronic Elements
Liquid level	1. Float mechanism 2. Static pressure elements	—
Pressure	1. Flexible diaphragm 2. Bellows 3. Pressure bell	1. Piezoelectric crystal (not commonly used)
Temperature	1. Bimetal element 2. Rod-and-tube element 3. Sealed bellows 4. Remote bulb 5. Fast-response element 6. Averaging element	1. Thermistor 2. Resistance bulb 3. Thermocouple
Humidity	1. Nylon ribbon 2. Human hair 3. Wood 4. Leather, horn, silk, etc.	1. Hygroscopic (gold-foil grid)
Enthalpy	1. Remote bulb plus nylon ribbon	—
Flame Detection	1. Bimetal element	1. Flame rod (ionization) 2. Photoelectric cell 3. Phototube 4. Thermocouple
Smoke Opacity	—	1. Photoconductive cell

Source: Automatic Control Principles, Honeywell, Inc., Milwaukee, WI, 1972.

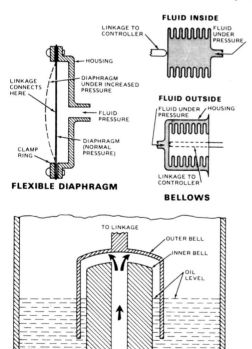

FIGURE 15-30 Electromechanical pressure-sensing elements.

connected to the diaphragm translates this distortion into motion. One of the most common applications is in a gas pressure regulator used to maintain a constant gas pressure in a fuel-gas pipeline.

Bellows: A bellows is nothing more than a stack of washer-shaped diaphragms joined alternately at their outer and inner circumferences. The bellows allow more movement over a given range of pressure without increasing the circumference of the diaphragms.

The fluid can be admitted to the *inside* of the bellows; in which case, the other end of the bellows is closed by a solid diaphragm and the linkage makes contact outside. Or, the fluid can exert pressure on the *outside* of the bellows, in which case the bellows is enclosed in a sealed, bell-shaped housing; the bellows is filled with air, and the linkage makes contact inside.

One end of the bellows is firmly anchored. The other end (restrained by a spring) moves in or out as an increase in pressure causes the bellows to expand or contract. The mechanical linkage transmits this motion to a switch or potentiometer in the controller.

Pressure bell: A pressure bell is used for maximum sensitivity in the measurement of small changes in pressure (or vacuum) of air or some other gas. An outer bell (inverted cup) is suspended from the linkage over an inner bell, which is fixed to the bottom of the case or tank. The buoyant force of the gas in the ring-shaped space between the bells causes the outer bell to "float" above the inner bell. The oil provides a nearly frictionless seal of the space between the bells. Changes in gas pressure cause variations in the buoyant force so that the outer bell floats higher or lower. The linkage transmits these changes to the controller or measuring instrument.

In normal practice, two identical bells are suspended from a beam like a pharmacist's scale. The weight of each bell is balanced by that of the other. One bell is subjected to a constant *reference* pressure, and the other bell to the pressure being measured. The difference of these pressures determines the position of the beam. This balanced system minimizes friction and the mechanical loads that must be moved, thus resulting in an extremely sensitive device.

Electromechanical Temperature-Sensing Elements. Electromechanical temperature-sensing elements operate on the familiar principle that all substances tend to expand or contract with increases or decreases in temperature. The rate of expansion and contraction (the coefficient of thermal expansion) is different for different substances. This *differential expansion* causes motion, which is measured. Electromechanical elements are large and bulky and respond relatively slowly to temperature changes. Figure 15-31 illustrates

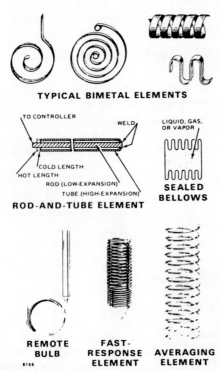

TYPICAL BIMETAL ELEMENTS

TO CONTROLLER   WELD   LIQUID, GAS, OR VAPOR

COLD LENGTH
HOT LENGTH
ROD (LOW-EXPANSION)
TUBE (HIGH-EXPANSION)

ROD-AND-TUBE ELEMENT

SEALED BELLOWS

REMOTE BULB   **FAST-RESPONSE ELEMENT**   **AVERAGING ELEMENT**

FIGURE 15-31 Electromechanical temperature-sensing elements.

the more common electromechanical temperature-sensing elements.

Bimetal element: A bimetal element is composed of two thin layers or strips of dissimilar metals, welded or brazed together. Because the two metals have different coefficients of thermal expansion, the element bends as the temperature varies. The most common application of a bimetal element is in room thermostats.

A change in length of either strip, for a given temperature change, is always proportional to the original length of the strip (although the ratio varies for different metals). Therefore, the corresponding movement of the free end of the element is likewise proportional to the length of the element. Thus, a certain minimum length is required to provide a measurable movement for a small temperature change. To increase the sensitivity of bimetals, designers have resorted to curves and spirals, enabling much longer elements to be packaged in smaller spaces.

Rod-and-tube element: A rod-and-tube element consists of a high-expansion metal tube inside of which is a low-expansion rod with one end attached to the rear of the tube. As the temperature of the liquid or gas surrounding the tube changes, the length of the tube changes, causing the free end of the rod to move. This element is commonly used in certain types of insertion and immersion temperature controllers, particularly those mounted in boilers or storage tanks.

Sealed bellows: A bellows, similar to that used in pressure-sensing elements, is evacuated of air; filled with liquid, gas, or vapor; and sealed at both ends. The volume or pressure of the liquid, gas, or vapor changes as the temperature changes, causing the bellows to expand or contract lengthwise. A linkage connected to the free end of the bellows transmits this movement to the controller. This type of element is often used in room thermostats.

Remote bulb: A remote bulb is a capsule connected to a sealed bellows or diaphragm by a capillary tube. The entire system—bulb, capillary, and bellows (or diaphragm)—is filled with a liquid, vapor, or gas. Temperature changes at the bulb result in changes of volume and pressure in the fluid, which are transmitted to the bellows or diaphragm through the capillary tubing. The remote bulb is useful where the temperature measuring point is at a distance from the desired location of the controller. It is usually provided with fittings to allow it to be installed in a pipe, duct, or tank.

Fast-response element: A fast-response element is a tightly coiled capillary that can be used in place of the bulb in a remote-bulb element. The surface area to volume ratio is about 7 times greater than that of a standard bulb; so its response time is 7 times faster.

Averaging element: An averaging element is used in place of the bulb in a remote-bulb element to obtain the *average* temperature in a duct. It is similar to the capillary except that it has a larger bore in order to hold the same amount of liquid as a standard bulb. The averaging element is distributed evenly over the cross section of the duct by winding it back and forth several times across the duct.

Electronic Temperature-Sensing Elements. Electronic temperature-sensing elements possess certain properties, such as resistance, that vary as the temperature changes. Generally, these elements are very small and can respond quickly to temperature changes. The speed of response to temperature changes is defined by the *time constant* of the sensing element—the shorter the time constant, the faster the response. Figure 15-32 illustrates some of the commonly used electronic temperature-sensing elements.

Thermistor: A thermistor is a solid-state semiconductor, the electrical resistance of which varies with temperature. Its temperature coefficient of resistance is high, nonlinear, and negative—its resistance decreases as temperature increases, and vice versa.

The primary advantage of a thermistor is the large change in its resistance with a small change in temperature. It provides a large input signal, which makes the design of the controller easier and less expensive. Its small size is another advantage.

Disadvantages include its nonlinear temperature coefficient of resistance, which limits its operating temperature range to a linear portion of the curve—generally less than 800°F (427°C). Also, because of its negative temperature coefficient of resistance, it is not failsafe; an open thermistor would cause the controller to keep calling for more heat. Another drawback is that the controller has to be calibrated to the individual sensing element, and sometimes it has to be recalibrated if aging changes the resistance of the thermistor.

Resistance bulb: A resistance bulb is a coil of fine wire wound around a bobbin. The resistance of the wire *increases* as the temperature increases. Thus, it would be fail-safe—a break in the wire would indicate a maximum temperature and the heating system would shut down. Other advantages include a wide operating range, up to 1400°F (760°C), interchangeability with other resistance bulbs, high output, excellent linearity, and excellent stability. Its primary disadvantage is its comparatively higher cost than the thermistor, although it is less expensive than the thermocouple.

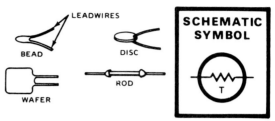

TYPICAL THERMISTORS

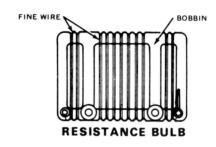

RESISTANCE BULB

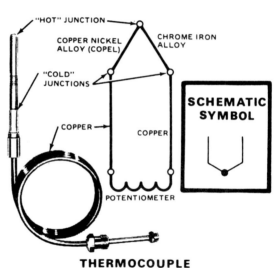

THERMOCOUPLE

FIGURE 15-32 Electronic temperature-sensing elements.

Thermocouple: A thermocouple is a junction of two dissimilar metals joined at the point of heat application—the "hot" junction. A resulting millivoltage difference, directly proportional to the temperature at the hot junction, is developed across the free ends—the "cold" junctions. A potentiometer is usually used to measure the voltage. Different types of thermocouples are used, depending on the operating temperature range required and the operating environment.

The primary advantage of a thermocouple is the extreme temperatures it is capable of sensing, up to 3000°F (1649°C). One disadvantage is its low output, requiring a more expensive amplifier. Also, the controller must have a cold junction compensator to offset changes in ambient temperature.

Electromechanical Humidity-Sensing Elements. Humidity is the moisture content in air. Electromechanical sensing elements used are made of *organic* materials that absorb moisture (stretch or swell) when the humidity increases and release moisture (shrink) when the humidity decreases. Figure 15-33 illustrates some of the commonly used humidity-sensing elements.

Nylon ribbon or human hair element: Human hair has the property of changing length with changes in its moisture content. Because of its small diameter, it rapidly absorbs and dissipates moisture as humidity changes. The hairs used in humidity-sensing elements are carefully selected and matched to ensure the greatest possible uniformity. The strands are bunched, and several bunches are combined in a band or ribbonlike element.

Nylon has the same property. Because the manufacturing of nylon can be closely controlled, the selection and matching process required for human hair is eliminated, thus reducing the cost. Thus, most elements are now made of the less expensive nylon ribbon.

Although many mechanical arrangements are used, the one shown in Figure 15-33 provides relatively large movement with a small change in humidity. The rocker arm at the top doubles the length of the nylon ribbon or hair for a given length of the element by, in effect, connecting the two ribbons in series. (Elements with only one ribbon and without the rocker are also used extensively.) As the humidity decreases, the air circulating over the nylon or hair absorbs some moisture,

and the ribbons shorten. As the ribbon on the left shortens, it rotates the rocker arm counterclockwise. This shifts the top end of the right ribbon upward a distance equal to the amount that the left ribbon has shortened. Since the right ribbon has also shortened the same amount, the movement of its bottom end is *twice* the amount of shortening. This motion is transmitted through a lever mechanism to the controller.

Wooden element: A sticky desk drawer in humid weather demonstrates the principle of the wooden element. It consists of a strip of wood cut *across* the grain. One end of the strip is anchored, and the other is connected through a linkage to the controller. As the humidity increases, the wood absorbs moisture and swells; as the humidity decreases, the wood releases moisture and shrinks. These changes in the length of the strip move the linkage to the controller. A plain wooden element exhibits considerable lag in moisture absorption and dissipation, and it is also subject to cracking in dry atmospheres. For these reasons, wooden elements are seldom used in modern controls. However, they make a good teaching tool to demonstrate the principle of electromechanical humidity-sensing elements.

Other elements: Other materials that may be used in electromechanical humidity-sensing elements include leather, horn, and silk. All are more sensitive than wood.

Electronic Humidity-Sensing Element. The electronic humidity-sensing element commonly used is a resistance element consisting of a thin slab of plastic or glass to which is fused strips of gold foil to form two interleaved, comblike grids (see Figure 15-34). The grids are coated with a hygroscopic salt (one that readily absorbs moisture from the atmosphere). The salt forms a conductive path between the adjacent strips of foil in the two grids. The electrical resistance of the salt coating changes as it absorbs and releases moisture, depend-

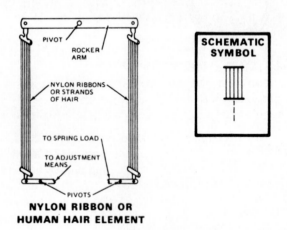

NYLON RIBBON OR HUMAN HAIR ELEMENT

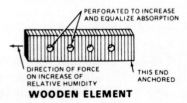

WOODEN ELEMENT

FIGURE 15-33 Electromechanical humidity-sensing elements.

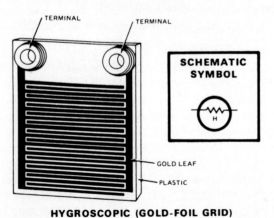

HYGROSCOPIC (GOLD-FOIL GRID)

FIGURE 15-34 Electronic humidity-sensing elements.

ing on the humidity. Thus, the resistance between the two terminals changes. This change in electrical resistance is used to actuate the controller.

The large surface area and small volume of this element allow it to respond rapidly to humidity changes. The high resistance of the circuit (thousands of ohms) results in a relatively large change in resistance for a small change in humidity, thus providing high accuracy.

Enthalpy-Sensing Elements. Enthalpy is dependent on both temperature and humidity; so, from a comfort standpoint, it is desirable to be able to control it. An enthalpy-sensing element simply combines temperature- and humidity-sensing elements in a single device. Usually, a remote bulb senses temperature and a nylon ribbon senses humidity. The controller combines the results and interprets them as enthalpy.

Flame Detection. A flame detector may be looked upon as a form of temperature-sensing element since it is the heat of the flame that produces the ions and light detected. All flame safeguard systems require that a flame be detected to keep a burner in operation. If no flame were present, fuel could accumulate to form an explosive mixture.

Types of flame detectors commonly used are electronic. (A bimetal element, previously described, may be used on small, domestic burners.) Electronic flame detectors include flame rods, photoelectric cells, phototubes, and thermocouples (previously described).

Flame rod (ionization): A flame rod is simply a metal or ceramic rod inserted into the flame envelope to function as the positive electrode in a flame detection circuit. The ground electrode is usually the burner itself. When a flame is present, the heat causes molecules between the electrodes to collide with one another so forcibly as to knock some electrons out of the atoms, producing ions. This is called *flame ionization.* When the ions are present and a voltage is applied across the electrodes, a current will flow between them. Thus, a current indicates that a flame is present.

Photoelectric cell: A photoelectric cell is simply a device that is sensitive to light. When used in a flame detector, it is located so that it is sensitive to the light emitted by the flame. The cell can be photoconductive, photoemissive, or photovoltaic. It can be sensitive to ultraviolet, infrared, or visible light rays.

A *photoconductive cell* is made of a material, such as cadmium sulfide or lead sulfide, the electrical resistance of which varies inversely with the intensity of the light that strikes it. Thus, if a flame were present, the resistance of the cell would be low and a measurable current would flow.

A *photoemissive cell* has a cathode coated with a material, such as caesium oxide, that emits electrons when light strikes it. When voltage is applied to the cell and a flame is present, electrons flow from the cathode to the anode, and a current flows in the detection circuit. Rectifying photocells, which respond to visible light, are examples of photoemissive cells.

A *photovoltaic cell* generates a voltage when light strikes it. It is not used in flame detectors because the voltage generated is too small to be useful without more expensive amplification than that required by other types of flame detectors.

Phototube: A phototube is an electron tube containing a photocathode that releases electrons when exposed to light. When used in a flame detector, the tube is pointed at the flame so that light from the flame falls upon the photocathode. When sufficient voltage is applied across the cathode and anode and a flame is present, current flows through the tube and through the external detection circuit.

Smoke Opacity. Smoke opacity is smoke density or blackness measured in Ringelmanns. Ringelmann numbers 0 and 5 represent 0 and 100% blackness, respectively. The sensing element used in a smoke detector does not detect smoke opacity directly. It detects the amount of light from a source that shines through the smoke. A photoconductive cell, made of cadmium sulfide, is used as a sensing element. The electrical resistance of the cell varies inversely with the intensity of the light that strikes it. Thus, the greater the smoke opacity, the less the light, the greater the resistance, and the less the current flowing in the detection circuit.

15-8 CONTROLLERS

A controller receives the measurement information from the disturbance-sensing element and produces an error signal to initiate corrective action. The controller mechanism translates the measurement information into a form of energy that can be used by the control system. Electronic controllers are generally more complex than electromechanical controllers, but they provide *all* modes of control. Table 15-2 lists the common types of controllers used in electric control systems.

15-9 ELECTROMECHANICAL CONTROLLERS

An electromechanical controller directly utilizes the motion produced by a disturbance-sensing element to operate a switch or potentiometer. In many controllers, the switch, in turn, energizes a relay to increase the current-handling capacity or to perform additional functions.

TABLE 15-2

Controllers

Modes of Control	Electromechanical Mechanisms	Electronic Mechanisms
On-off (two-position)	Mercury switches	Galvanometric (on-off only)
Multiposition	Snap-acting switches	Potentiometric
Floating	Mercury plunger relays	Resistance bridge
	Electromagnetic relays	
Proportioning (modulating)	Potentiometers	Potentiometric Resistance bridge
Reset		
Rate	—	

An electromagnetic controller is relatively inexpensive and is generally used for simple on–off or proportioning control.

Electromechanical Mechanisms. An electromechanical mechanism uses the motion produced by the sensing element to open or close an electric circuit or to set up a varying resistance in an established circuit. Switches, relays, and potentiometers are the basic mechanisms used.

Mercury switches: A mercury switch is a glass tube in which fixed contacts and a pool of loose mercury are hermetically sealed (see Figure 15-35). Insulated lead wires are attached to the contacts at the end(s) of the tube. The tube is usually mounted on a plate that is rotated by linkage from the sensing element. As the mounting plate rotates, the tub tilts and the mercury pool flows to the lower end to close or open the gap between the contacts. This action *makes* or *breaks* the contacts and any electrical circuits connected to the lead wires. The mechanical force needed to tilt the switch is low; so the make-or-break action is rapid and repeats accurately.

The contacts may be rated for currents from 0.1 A up to 10 A. They are arranged for SPST (single-pole, single-throw) or SPDT (single-pole, double-throw) switching action. SPDT switches make *R* to *B* when the value of the controlled variable falls and make *R* to *W* when it rises.

Advantages of mercury switches include (1) sealed construction so that the contacts are protected from dirt, dust, and other contamination; (2) no moving parts (other than the mercury pool) so that they provide maintenance-free, dependable operation; and (3) visible contacts so that a person can see whether they are made or broken.

Limitations include the following: (1) They must be carefully leveled to obtain operation at the proper point(s); (2) in locations with much vibration, the mercury slops around in the glass tube, causing contact chatter; and (3) since mercury freezes at $-35°F$ ($-37°C$), mercury switches cannot be used below this temperature.

Snap-acting switches: In a snap-acting switch, metal-to-metal contacts are made or broken by the movement of the sensing element against a reciprocat-

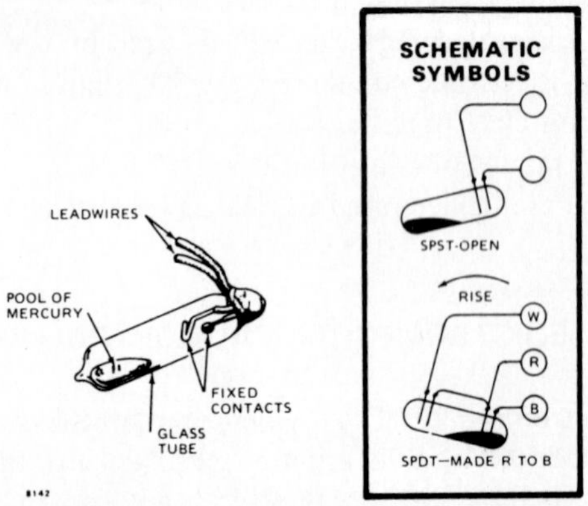

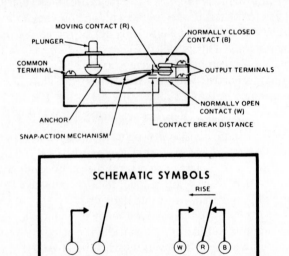

FIGURE 15-35 Mercury switches.

FIGURE 15-36 Snap-acting switches.

ing plunger (see Figure 15-36). The plunger actuates a snap-action spring mechanism built into a plastic case. The moving contact snaps from one fixed contact to the other with repeatable accuracy. (An SPST switch has only one fixed contact.) The contacts have about the same ratings as mercury switches (0.1 A up to 10 A) and are also arranged for SPST or SPDT switching action.

Advantages of snap-acting switches include the following: (1) They are fast and positive acting; (2) since only a slight mechanical motion is needed for actuation, they provide precise and accurate control where small tolerances are required; (3) no leveling is required (unlike mercury switches); and (4) they are unaffected by moderate vibration.

Limitations include the following: (1) They are not sealed (because of the plunger), and are thus not as dust-tight as mercury switches; (2) they have lower DC ratings; and (3) since the contacts are not visible, an ohmmeter must be used to determine whether they are open or closed.

Mercury plunger relays: In a mercury plunger relay (see Figure 15-37), the contacts are pools of mercury into which electrodes have been inserted. A ferromagnetic plunger, actuated by a solenoid coil, accomplishes the switching. The use of mercury pools as contacts allows the switching of large currents—as high as 60 A. A mercury switch or snap-acting switch with a lower rating is usually used to energize the solenoid coil.

The plunger, mercury, and conducting electrodes are sealed in a glass tube surrounded by a solenoid coil. When the coil is deenergized, the plunger floats (partially submerged) on the main mercury pool. Inside the cylindrical plunger, just above the main pool, is a small

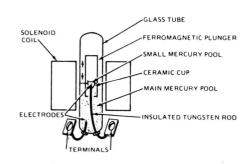

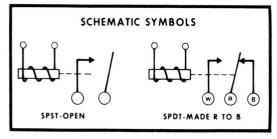

FIGURE 15-37 Mercury plunger relays.

mercury pool in a ceramic cup, supported by an insulated tungsten rod. When the solenoid coil is energized, the plunger moves downward, displacing the main mercury over the top of the ceramic cup. The two pools of mercury are now in contact and the switch is closed. When the coil is deenergized, the plunger moves upward and the surface of the main mercury pool falls below the top of the ceramic cup, breaking the contact and opening the switch.

Advantages of mercury plunger relays include (1) heavy current switching capacity; (2) sealed construction permitting operation in dirty and explosive environments; and (3) mercury-to-mercury contact, eliminating contact maintenance.

Electromagnetic relays: Electromagnetic relays (see Figure 15-38) are merely electrically operated switches. Simple relays and relay combinations are extensively used to perform one or more of the following functions: (1) switching to line voltage or heavy current load when activated by a low-voltage controller of limited current-carrying capacity, (2) switching two or more separate load circuits from a single-pole controller switch, and (3) coordinating two or more control circuits to provide a desired sequence of load switching.

The coil-and-armature mechanism is similar to that used in the familiar doorbell or buzzer. It consists of (1) a magnetic circuit with fixed core, movable armature and two air gaps; (2) one or more actuating coils that establish the magnetic flux; (3) one or more sets of contacts, with one contact in each set movable and coupled to the armature, and the others fixed; and (4) restoring springs and limit stops for armature positioning.

In the deenergized position shown for the SPDT relay, the force of the restoring spring overcomes the magnetic force, and movable contact *R* is making contact with fixed contact *B*. When the actuating coil is energized, flux in the magnetic circuit increases and the magnetic force overcomes the spring force (plus friction). The armature moves to close the active air gap, and the movable contact *R* breaks with fixed contact *B* and makes with fixed contact *W*.

An SPST relay can be either normally open (no *B* contact) or normally closed (no *W* contact). The DPST (double-pole, single-throw) relay shown can be represented schematically by two sets of normally open, SPST relay contacts.

Potentiometers: A potentiometer (see Figure 15-39) typically consists of fine wire wound around a core. A wiper rests on the windings and can move along them. The wire ends are connected to terminals labeled *W* and *B*. The wiper is connected to terminal *R*. As motion of the sensing element moves the wiper back and

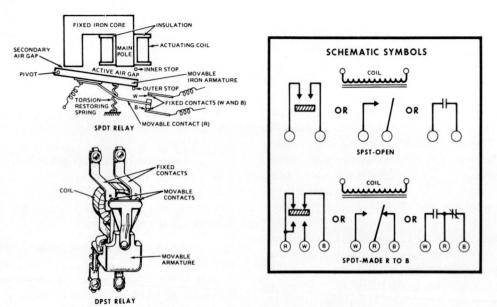

FIGURE 15-38 Electromagnetic relays.

forth, the amount of resistance between *R* and *W* and between *R* and *B* changes. If the wiper moves toward *W,* resistance increases between *R* and *B* and decreases between *R* and *W,* and vice versa. The *R, W,* and *B* designations stand for red, white, and blue lead wire colors.

Potentiometers are used in proportioning (modulating) controllers. They determine how far the final control element must move to counteract a change in the controlled variable.

15-10   ELECTROMECHANICAL MODES OF CONTROL

Section 15-3 presented a brief discussion of the various modes of control. Here, we will discuss some of the circuitry for electromechanical systems. Switches and relays are used for on–off (two-position), multiposition, and floating modes of control. Potentiometers are used for the proportioning (modulating) mode of control. The figures in the following discussions show line voltage circuits. Low-voltage devices would require a transformer if used in line voltage circuits.

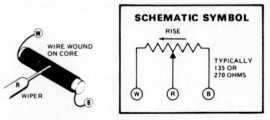

FIGURE 15-39 Potentiometer.

On–Off (Two-Position) Mode of Control.   In the on–off (two-position) mode of control, the controller is in one or the other of two possible positions, except for a brief period when it is passing in between. Each of these positions, in turn, determines which of two possible positions is occupied by the final control element. There are two values of the controlled variable that determine the position of the controller. Between these values is a zone, called *differential,* in which the controller cannot switch.

In the simplest mode of control, an SPST controller is connected in a two-wire circuit (see Figure 15-40). When the controlled variable rises to the higher of the two values (upper end of the differential), the controller contacts will either open or close, depending on the design. Let us assume they will close. This applies power to the final control element, which will be *electrically* driven to one of its two positions. The controller contacts will stay closed, and the final control element will remain in this position until the controlled variable falls to the lower of the two values (lower end of the differential). Then, the controller contacts will open and the final control element will be *mechanically* driven (usually by a spring) to its other position. The controller contacts will stay open and the final control element will remain in this position until the controlled variable again rises to the high end of the differential. Had the controller been designed to *open* at the high end and *close* at the low end, the positions of the final control element would be reversed.

An SPDT controller is used in a three-wire circuit (see Figure 15-40). The final control element has three terminals corresponding to those of the controller. Usu-

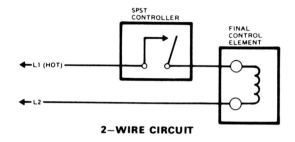

2—WIRE CIRCUIT

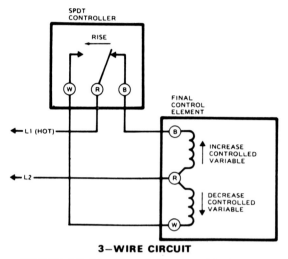

3—WIRE CIRCUIT

FIGURE 15-40 On–off (two-position) mode of control (electromechanical).

ally, the movable contact of the controller is labeled *R,* and the fixed contacts are labeled *W* and *B.* The *R, W,* and *B* designations stand for red, white, and blue lead wire colors.

When the value of the controlled variable *rises* to the high end of the differential, the controller contacts will *break R to B* and *make R to W.* This applies power to the section of the final control element between terminals *R* and *W.* When this section is energized, the final control element will be *electrically* driven to a position that will *decrease* the value of the controlled variable. The controller contacts and final control element will remain in this position until the controlled variable *falls* to the low end of the differential. Then, the controller contacts will *break R to W* and *make R*

to *B.* This removes power from the *R–W* section and applies power to the *R–B* section of the final control element, which will now be *electrically* driven to a position that will *increase* the value of the controlled variable. Thus, the controlled variable is kept within the values determined by the differential of the controller.

If, as is often the case, a *heating* system of some type is being considered, the following rhyme can be memorized to help remember the operation just described: ''Red to Blue for BTU.'' This rhyme indicates that when the temperature falls and there is a call for heat (Btu's), the controller will make *R* to *B,* and vice versa. It also applies to other modes of control.

Multiposition Mode of Control. The multiposition mode of control (see Figure 15-41) simply consists of two or more on–off stages that operate at different values of the controlled variable. In the example shown (Figure 15-41), the mercury switches are all rotated together by the same sensing element. They are rotated *clockwise* as the value of the controlled variable falls and *counterclockwise* as it rises. Because they are mounted at different angles, the switches *close in sequence* (stage 1 first and stage 5 last) as the value of the controlled variable falls. The final control elements are energized in the same order. Similarly, the switches *open in the opposite sequence* (stage 5 first and stage 1 last), and the final control elements are deenergized as the value of the controlled variable rises.

This mode of control allows as many stages to be energized as required to maintain the desired value of the controlled variable. Each stage may include an *entire* final control element that takes one of two positions as described for the on–off mode of control. Or, each stage may include a *section* of the mechanism of one final control element. In the latter case, as each stage is energized, the final control element will be driven to a different position.

Floating Mode of Control. The floating mode of control (see Figure 15-42) provides bidirectional action of the final control element anywhere over its full range. The controller uses an SPDT switch that has a

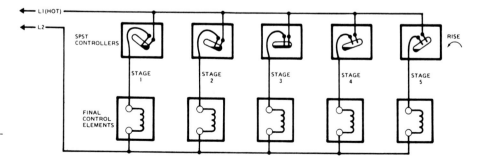

FIGURE 15-41 Multiposition mode of control (electromechanical).

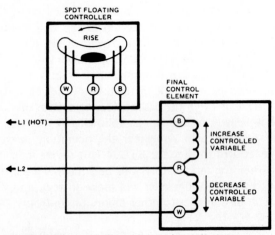

FIGURE 15-42 Floating mode of control (electromechanical).

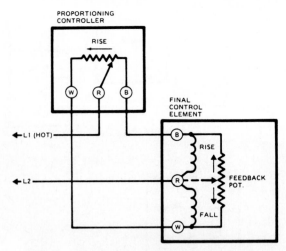

FIGURE 15-43 Proportioning (modulating) mode of control (electromechanical).

neutral zone in which no contact is made. As long as the controlled variable stays within this neutral zone, no contacts are closed; so no power is applied to the final control element and it stays where it is.

If the controlled variable *rises above* the neutral zone, the controller contacts will make *R* to *W*, applying power to the *R-W* section of the final control element. The final control element will be *electrically* driven in a direction that will *decrease* the controlled variable. When the controlled variable falls back into the neutral zone, the *R-W* contacts will break and the final control element will stop.

Similarly, if the controlled variable *falls below* the neutral zone, the controller contacts will make *R* to *B*, applying power to the *R-B* section of the final control element and *electrically* driving it in a direction that will *increase* the controlled variable. When the controlled variable rises back into the neutral zone, the *R-B* contacts will break and the final control element will stop.

In summary, the final control element can occupy any position between its two extremes, and the controller "floats" to keep the controlled variable in the neutral zone. Floating control has largely been replaced by proportioning control, but it may be found in older systems.

Proportioning (Modulating) Mode of Control. The proportioning mode of control (see Figure 15-43) is a refined version of floating control. The controller has a potentiometer instead of a switch; so it can *continuously vary* the position of the final control element, anywhere between its extremes, to keep the controlled variable at the desired value. This process is called *modulation*.

To simplify the explanation of proportioning control, much of the circuitry in the final control element has been omitted in Figure 15-43. With the potentiom-

eter wiper as shown, there is less resistance between terminals *R* and *B* of the controller than between *R* and *W;* so more current will flow in the *R-B* section of the final control element than in the *R-W* section. This will *electrically* drive the final control element in a direction to increase the controlled variable. Simultaneously, the wiper on a feedback potentiometer will drive toward the *W* end. When the *R-W* resistance of the feedback potentiometer equals the *R-B* resistance of the controller, the currents will be balanced and the final control element will stop.

If the controlled variable rises, the wiper will move toward the *W* end of the potentiometer, decreasing the resistance between terminals *R* and *W* of the controller. More current will flow in the *R-W* section of the final control element, *electrically* driving it in a direction to decrease the controlled variable. The wiper on the feedback potentiometer will drive to the *B* end until the *R-B* resistance of the feedback potentiometer equals the *R-W* resistance of the controller. The current will then be balanced and the final control element will stop.

Thus, for each position of the wiper on the controller, there is a corresponding position of the wiper on the feedback potentiometer and for the final control element. The final control element will be driven a distance proportional to the change in the controlled variable. This results in the most precise type of controller, used where the controlled variable must be kept within small tolerances.

15-11 ELECTRONIC CONTROLLERS

An electronic controller compares the signal from a disturbance-sensing element to a standard reference to obtain an error signal. It then amplifies the error signal to provide enough energy to drive the final control element. Electronic controllers provide all modes of control, including reset and rate, plus the functions of alarm and limiting.

Electronic Mechanisms. A basic electronic controller (see Figure 15-44) consists of a null detector and amplifier, with input adjustments and an output switching element to drive the final control element. It utilizes solid-state components in place of electromechanical switches, although electromagnetic relays are frequently used as output switching elements.

The *null detector* compares the difference between the signal from the disturbance-sensing element and the standard reference (set point). When they are equal, there is a balanced condition that usually results in zero output, called a *null*. The different mechanisms used to detect the null result in three types of electronic controllers: (1) galvanometric, (2) potentiometric, and (3) resistance bridge. The resistance bridge controller will be emphasized because it is the most common.

Input adjustments include set point, differential, and proportioning range (throttling range) adjustments. Calibration adjustments may also be included. In most cases, these adjustments take the form of a wire-wound potentiometer with a knob to adjust the position of the wiper.

The *solid-state amplifier* multiplies the output from the null detector to produce an amplified error signal. The amount of magnification is referred to as the *gain* of the amplifier. Most amplifiers also *discriminate* between signals of different polarity or phase. Discrimination determines whether the signal to the amplifier is caused by an increase or a decrease in the controlled variable; so the error signal will drive the final control element in the proper direction.

The *output switching element* may be an electromagnetic relay, a saturable-core reactor, an SCR (silicon-controlled rectifier), or another device that drives the final control element. In the proportioning mode of control, the amplifier may drive the final control element directly, eliminating the output switching element.

Galvanometric controller: A galvanometer (see Figure 15-45) is an instrument for measuring an electric current by measuring the mechanical motion produced by electromagnetic forces set up by the current. A galvanometric controller is basically a galvanometer that has been converted to a controller. It was introduced before it was technologically feasible to mass-produce electronic amplifiers. It uses the *motion* produced by the sensing element to produce a signal that activates an SPST or SPDT relay. A galvanometric controller is limited to the on–off mode of control.

The basic mechanism is a vane-oscillator null detector. The lightweight metal vane is connected by a linkage to the sensing element. As it moves between (without touching) two oscillator coils, mutual inductance decreases and the modified Hartley oscillator is detuned. When the set point is reached (at the null), oscillation ceases. When oscillation stops, current is fed to the relay coil, causing precise snap action of the relay contacts. Vane motions as small as 0.002 in. (0.05 mm) initiate controller action. The set point adjustment knob moves the oscillator coil assembly upscale or downscale.

Potentiometric controller: A potentiometric controller (see Figure 15-46) operates on the null balance principle. The signal from the sensing element is

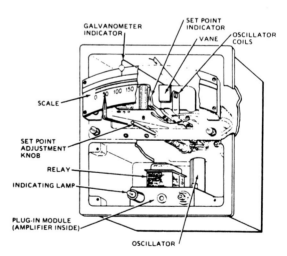

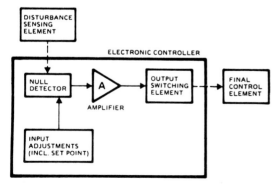

FIGURE 15-44 Basic electronic controller.

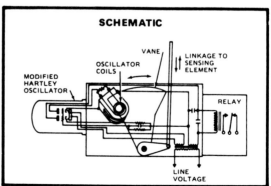

FIGURE 15-45 Galvanometric controller.

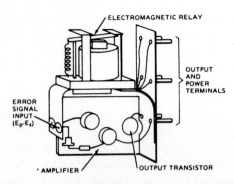

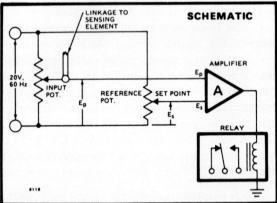

FIGURE 15-46 Potentiometric controller.

tion, the voltage E_p is developed across the input potentiometer. The wiper on the reference potentiometer is manually adjusted to the desired set point, developing the voltage E_s. When the controlled variable differs from the set point, an error signal equal to the difference between E_p and E_s is produced and fed to the amplifier. The amplified error signal energizes the relay, which drives the final control element. When the controlled variable reaches the set point, E_p equals E_s and the relay is deenergized.

Resistance bridge controller: The resistance bridge controller is by far the most used electronic controller. It also operates on the null balance principle. The input circuitry generates an error signal based on the difference between the resistance of an electronic sensing element and the set point resistance. The bridge circuit was discussed in Section 15-5, and most electronic controllers are adaptations of this circuit.

For control purposes, the basic Wheatstone bridge is somewhat modified in order to provide signal to the controlled equipment. Figure 15-47 illustrates a typical modified Wheatstone bridge circuit. The key changes are as follows:

1. The galvanometer (voltmeter of Figures 15-26, 15-27, and 15-28) is replaced by a solid-state difference amplifier. The amplifier compares the voltage at point A with the reference voltage (set point) at point B.
2. The battery is replaced by a regulated DC power supply.
3. The R_4 resistor in the left half of the bridge is replaced by an electronic sensing element (varying-resistance type, such as a thermistor).

The difference amplifier multiplies the voltage difference between points A and B of the bridge. The

balanced against a set point voltage in the input circuit to produce an error signal. The error signal is simplified and used to position the final control element to keep the controlled variable at the desired value. For precise control, this system is intrinsically more accurate than the galvanometer controller.

The potentiometric controller shown in Figure 15-46 consists of a three-stage transistorized amplifier and an SPDT electromagnetic relay mounted in a plug-in module. Linkage from the sensing element moves the wiper on the input potentiometer to a position representing the value of the controlled variable. In this posi-

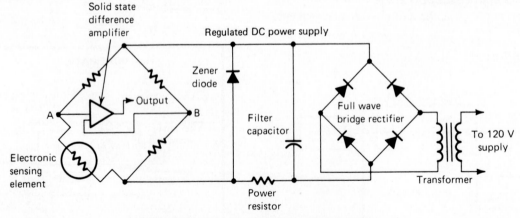

FIGURE 15-47 Modified Wheatstone bridge.

amount of magnification is referred to as the *gain* of the amplifier. For instance, a gain of 10 means that if the difference between *A* and *B* is 1.0 V, the output of the amplifier will be 10.0 V. The amplifier has two functions in the system. The first is to isolate the bridge from voltage changes in the rest of the control circuit. The second is to make the very small signal from the bridge strong enough to energize the output switching element (such as a relay coil or solid-state motor).

15-12 ELECTRONIC MODES OF CONTROL

The potentiometric and resistance bridge controllers can be used for *all* modes of control, including reset and rate. Galvanometric controllers are limited to the on–off mode of control. The figures in this section show line voltage circuits. Low-voltage devices would require a transformer if used in line voltage circuits.

On–Off (Two-Position) Mode of Control. In the on–off (two-position) mode of control (see Figure 15-48), the output switching element is an SPST or SPDT electromagnetic relay. The set point is adjusted so that the relay switches at the null or at some point on either side of the null. Operation is the same as for electromechanical on–off control discussed previously. Remember the rhyme—''Red to Blue for BTU.''

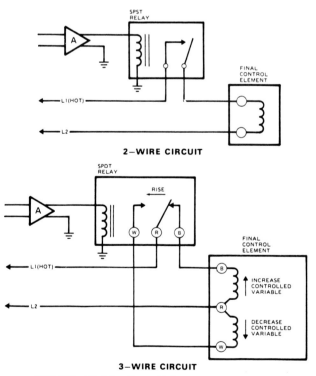

FIGURE 15-48 On–off (two-position) mode of control (electronic).

Multiposition Mode of Control. The multiposition mode of control (see Figure 15-49) for electronic controllers is similar in concept to that for electromechanical controllers—two or more on–off stages operate at different values of the controlled variable. The error signal (amplifier output voltage) increases as the controlled variable departs from the set point. The SPST relays, used as output switching elements, are set to pull in at different values of the amplifier output voltage. The relay switches close in sequence (stage 1 first and stage 5 last) as the controlled variable departs from the set point. The final control elements are energized in the same order. Thus, the farther the controlled variable departs from the set point, the higher the amplifier output voltage, and the more stages will be energized. For departure from the set point in the opposite direction (e.g., decreasing instead of increasing), a similar set of stages could be used. Polarity or phase discrimination in the amplifier would determine which set of stages to energize. In Figure 15-49, departure from the set point and the resulting amplifier output are indicated by shading. For the departure shown, three stages are energized.

Floating Mode of Control. Although seldom used anymore, floating control (see Figure 15-50) can be accomplished in several ways. One method is the use of two stages in a manner similar to multiposition control. The amplifier incorporates polarity or phase discrimination. As long as the controlled variable is ''floating'' in the neutral zone (shaded area) around the set point, neither relay is energized. If the controlled variable *falls below* the neutral zone, relay 1 will pull in, the switch will close, section *B–R* of the final control element will be energized, and it will drive in a direction to *increase* the controlled variable. When the controlled variable rises back into the neutral zone, relay 1 will drop out and the final control element will stop. Similarly, if the controlled variable *rises above* the neutral zone, relay 2 will pull in, section *W–R* will be energized, and the final control element will drive in a direction to *decrease* the controlled variable.

Proportioning (Modulating) Mode of Control. When properly tuned, an electronic controller operating in the proportioning mode automatically adjusts its output to a value that will cause the controlled variable to stabilize at the set point. There are three types of proportioning control associated with electronic controllers: (1) position proportioning, (2) time proportioning, and (3) current proportioning.

Position-proportioning control (see Figure 15-51), the most common, is the same type used with electromechanical controllers (see Figure 15-43). The discrimina-

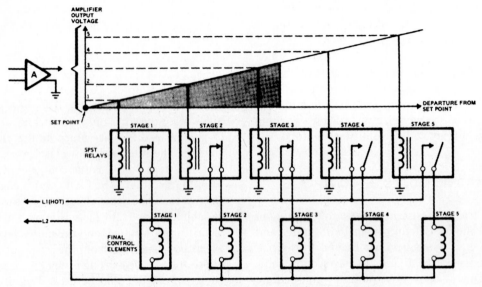

FIGURE 15-49 Multiposition mode of control (electronic).

tor is actually part of the amplifier circuitry but is shown separately to explain the operation. The discriminator determines whether the controlled variable is rising or falling. If it is *rising above the set point,* relay 1 pulls in, and the *R–W* section of the final control element is energized. This electrically drives the final control element in a direction to *decrease* the controlled variable. Simultaneously, the wiper on the feedback potentiometer will drive toward the *B* end. The feedback potentiometer is electrically connected into the null detector circuit of the controller. When the wiper has driven far enough to balance the null detector, the amplifier output is zero, relay 1 drops out, and the final control element stops.

Similarly, if the controlled variable is *falling below the set point,* relay 2 pulls in, and the *R–B* section of the final control element is energized. The final control element will be electrically driven in a direction to *increase* the controlled variable, and the feedback potentiometer wiper will drive toward the *W* end until the null

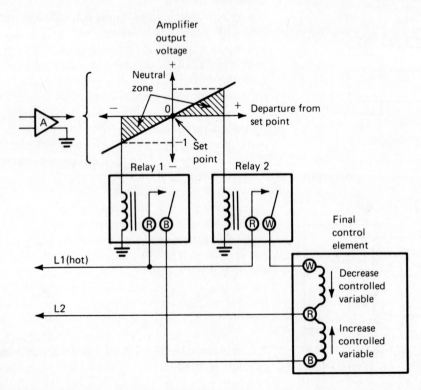

FIGURE 15-50 Floating mode of control (electronic).

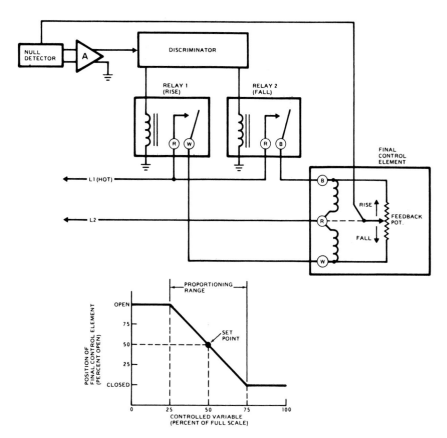

FIGURE 15-51 Position-proportioning control (electronic).

detector is balanced. Then, relay 2 will drop out, and the final control element will stop.

Thus, the final control element will be driven open or closed as necessary to keep the controlled variable at the set point. The graph in Figure 15-51 for a typical controller shows the positions taken by the final control element as the controlled variable departs from the set point.

Time-proportioning control (see Figure 15-52) is available only on electronic controllers. Special circuitry, too complex to describe here, automatically switches the controller on and off, driving the final control element to either of two possible positions—fully open or fully closed. The ratio of on-time to off-time is proportional to the departure of the controlled variable from the set point. The length of a complete cycle (on-time plus off-time) remains constant, but the percentage of on-time varies to keep the controlled variable at the set point.

In the example of Figure 15-52, a complete cycle is 10 min. At *set point,* the final control element is on as much as it is off—5 min. each per cycle. *Below* the set point, it is on 7.5 min. and off only 2.5 min. *Above* the set point, the periods are reversed—off 7.5 min. and on 2.5 min.

The graph of Figure 15-52 for a typical controller shows how the on-time varies as the controlled variable

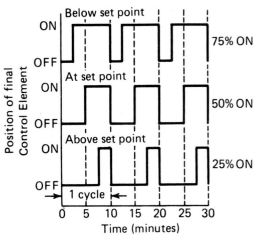

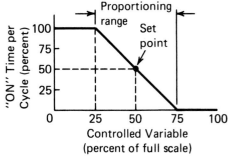

FIGURE 15-52 Time-proportioning control (electronic).

departs from the set point. Notice that outside of the proportioning range of the controller, the final control element will stay either on or off. As the controlled variable nears the set point, the on-time approaches 50%.

In *current-proportioning control,* the controller output is a DC milliampere signal that varies as the controlled variable changes. This signal drives a saturable-core reactor, an SCR (silicon-controlled rectifier), a current relay, or a similar current-operated device.

Reset action: A proportioning controller will not necessarily stabilize at the set point. The deviation from the set point—called *offset, droop,* or *drift*—may be small or large. If the offset were the same at all times, it could be remedied by a simple recalibration of the controller. However, it varies with the load conditions. Continual adjustment of the set point is necessary to keep the controlled variable at the same point throughout the load range of the equipment. There are two ways to eliminate offset: (1) manual reset and (2) automatic reset. Automatic reset is required for an automatic control system.

Manual reset is simply a potentiometer that can be adjusted to change the amplifier output in either a negative or positive direction. In effect, it recalibrates the controller to eliminate offset. Manual reset is most effective when there is no large change in the process, ambient condition, line voltage, or load.

Automatic reset results if the potentiometer is replaced by a secondary electronic sensing element that automatically adjusts the set point as the load changes. This method is shown in Figure 15-53 for a resistance bridge controller. Let us assume that the sensing elements are thermistors; their resistance increases as the temperature decreases. Let us also suppose that R_2 is sensing indoor temperature and R_3 is sensing outdoor temperature. If the outdoor temperature falls, the resistance of R_3 increases. More voltage is dropped across R_3, and the reference voltage at point B (which is actu-

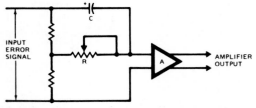

FIGURE 15-54 Automatic reset by integral action.

ally the set point) decreases. To rebalance the bridge, the voltage at point A must also decrease. This requires a decrease in the resistance of R_2, or an increase in the indoor temperature. Thus, as the outdoor temperature falls, the set point is automatically adjusted for the higher indoor temperature, and vice versa. This method is limited to systems in which the secondary sensing element can be located so that it detects changes in ambient conditions (as in the example), line voltage, the process, or the load—whatever is causing the offset.

Automatic reset can also be accomplished by adding a capacitor C (see Figure 15-54), which continuously charges to a value proportional to the *offset*. This charge is added to the input error signal to automatically produce an amplifier output that has been corrected for offset. The manipulation of the signal in this manner is a form of advanced positive feedback called *integral action.* Integral action is proportional to the time constant RC and is initially calibrated by adjusting the resistor R.

Rate action: If a system has a long time delay between a change in the controlled variable and a corrective response by the controller, poor control may result. This time delay is called *lag.* A common cause of lag is having a sensing element located too far away from the controlled medium. If the lag is too great, the system becomes unstable.

The lag problem can be overcome by rate action (see Figure 15-55), also called *derivative action.* Rate action is delayed negative feedback, accomplished by adding a capacitor C, which continuously changes to a value proportional to the *rate of change* of the input error signal. (Note that the *location* of the capacitor in

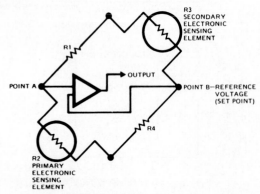

FIGURE 15-53 Automatic reset by adding a secondary electronic sensing element.

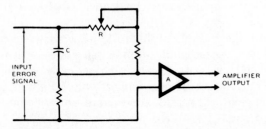

FIGURE 15-55 Rate action (derivative action).

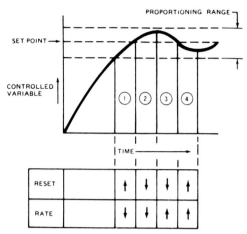

FIGURE 15-56 Three-mode control.

an *RC* network determines how it charges and hence the type of action—integral or derivative.) This charge is added to the input error signal to produce an amplifier output that has been corrected for lag. Rate action is proportional to the time constant *RC* and is initially calibrated by adjusting the resistor *R*.

Three-mode control: A proportioning controller with *both* reset and rate action is also called a *three-mode* (see Figure 15-56) controller (proportioning-plus-reset-plus-rate). This is the most sophisticated control mode, resulting in the best possible control under the most severe process conditions. In the example (Figure 15-56), the controlled variable, on start-up, enters the proportioning range, overshoots the set point, and then undershoots below the set point. The curve is divided into four zones. The table shows the effect of adding reset and rate action, with the arrows indicating the direction in which each mode tends to change the controlled variable. Whenever the controlled variable approaches the set point, whether from below (zone 1) or from above (zone 3), rate action opposes reset in an effort to prevent overshoot and undershoot. After the set point has been passed, and with the controlled variable still moving farther away from the set point (zones 2 and 4), all three control modes combine to minimize deviation from the set point.

15-13 PNEUMATIC CONTROLLERS[4]

In our discussion of electromechanical and electronic controllers, we discussed the disturbance-sensing element and the controller as two separate devices (which

[4]Figures 15-57 through 15-62 and the associated descriptive matter (Sections 15-13 and 15-14) are from *Fundamentals of Pneumatic Control,* Johnson Controls, Inc., Milwaukee, WI, 1979.

they are). However, we have seen that these two devices are frequently "packaged" in a single container, which may be called simply a *controller.* Therefore, a controller has two main parts: (1) the measuring or sensing element and (2) the relay that produces the output signal. A pneumatic controller contains the same two parts.

The measuring or disturbance-sensing elements of a pneumatic controller are similar in most respects to the electromechanical devices discussed previously, such as the bimetal elements and remote bulbs for temperature, the membrane or biwood element for humidity, and the diaphragm, bellows, and Bourdon spring for pressure.

15-14 PNEUMATIC RELAY

In all pneumatic controllers, supply air is piped to the controller at a constant pressure of either 15 or 20 psig. This supply flow provides (1) *volume,* to fill large areas within the controlled devices and connecting piping and (2) *pressure,* which provides the force to do the required work. The controller is designed to use the air internally through two separate circuits—pilot and volume amplifier.

Pilot circuit: The pilot circuit (see Figure 15-57) has a small volume and a reduced airflow that is restricted by a fixed orifice set to a value of 5 to 7 in. WG pressure and a flow of approximately 20 scim (standard cubic inches per minute) with the element or lid away from the control port. (Older models of controllers are fitted with an adjusted orifice set to the same value.) Pilot pressure is then regulated by the position of the element relative to the control port, increasing to the same value as supply pressure with no flow when the port is fully closed.

Volume amplifier circuit: The amplifier circuit is regulated by the action of the pilot circuit but has a much larger capacity since it admits supply air directly to the output line and the controlled device. It is therefore referred to as a *volume* amplifier.

There are four basic sections in a complete pneumatic relay (see Figure 15-58): (1) The pilot chamber admits supply air through an orifice to the control port and pilot diaphragm; (2) the exhaust chamber, between the pilot and output chambers, includes the exhaust seat assembly and exhaust port, which releases the excess output air to the atmosphere; (3) the output chamber incorporates a spring to oppose the pilot diaphragm, the output diaphragm, and an outlet connection to provide output air to the controlled devices; and (4) the supply chamber includes the supply valve and supply valve seat

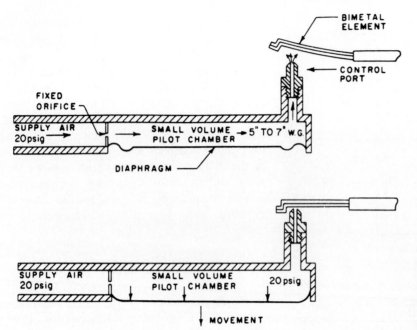

FIGURE 15-57 Pilot chamber action: The pilot chamber uses a restricted flow in a small area to respond quickly to changes dictated by the sensing element.

assemblies, which admit supply air to the output chamber.

The relay of a pneumatic controller translates the action of the sensing element into a useful output signal. The relay can be designed to produce proportional or two-position output, proportional being the more popular of the two. Any of the disturbance-sensing elements, involving motion, may be used for either mode.

Sequence of operation: In all controllers, the sensing element senses and reacts to changes in the con-

trolled variable. The sensing element is mounted or coupled to the relay so that it acts on a pilot circuit, modulating the pilot signal in proportion to changes in the controlled variable. The pilot signal applies its proportional signal to the amplifier, which produces a useful signal. This signal is the output of the controller and is used to actuate the controlled device.

In a final step, a feedback gain loop is included, which allows the output signal to act on the pilot signal (see Figure 15-59). Thus, the sensing element and relay together perform a total of five interrelated functions in

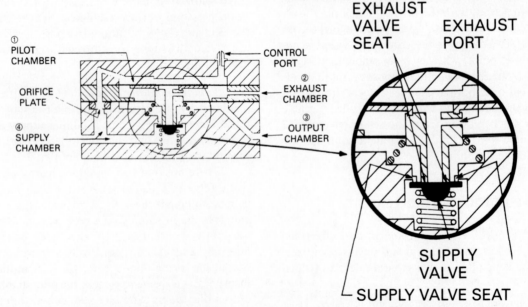

FIGURE 15-58 Four basic sections of a pneumatic relay (4000 Series proportional relay).

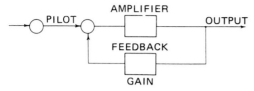

FIGURE 15-59 Relay components: pilot, amplifier, output, and feedback gain loop shown schematically.

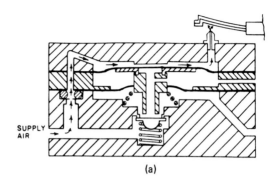

(a)

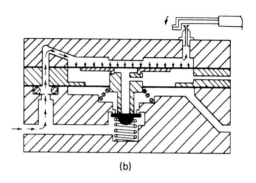

(b)

sequence. The way the relay components are assembled determines whether the controller is proportional or two-position.

In Figure 15-60, the various operating conditions are shown for a Johnson Controls 4000 Series pneumatic controller. Supply air is fed through the orifice into the pilot chamber (Figure 15-60a). As the lid or sensing element closes against the control port, the pilot pressure increases gradually to maximum or supply pressure. The low-volume pilot pressure has several functions in the proportional relay. As pilot pressure increases, it moves a diaphragm to close the exhaust valve seat, which prevents output air from exhausting to the atmosphere (Figure 15-60b). A further increase in pilot pressure will then open the supply valve to admit supply air pressure to the output air chamber and the output air line (Figure 15-60c). The relay is designed with a high ratio to produce a large initial change in output pressure for a small change in pilot pressure caused by a minute movement of the lid against the control port. In the balanced condition (Figure 15-60d), the exhaust seat is closed against the supply valve so that no output air can escape to the atmosphere and the supply valve is seated against the supply valve seat so that no supply air can enter the output chamber. The air pressure in the output chamber and air line will remain at the established pressure until the relay acts in response to a new change in lid position. The spring in the output chamber pushes against the pilot chamber diaphragm so that pilot pressure must build up to about 3 psig before any action will occur.

Proportional pneumatic feedback action: Negative feedback produces the proportional action in a proportional controller. Negative feedback is arranged so that, when a given force or pilot signal works in one direction, the resultant output force counteracts the original force and produces a balanced condition proportional to the value of the pilot signal. This can be referred to as a *reverse-acting feedback loop.* Figure 15-61 is a complete graph showing the output pressure changes that occur due to the 1.0 psig change in pilot signal. A formula is also provided to determine the input-to-output ratio. The relay shown in Figure 15-61 is referred to as a *7:1 relay.*

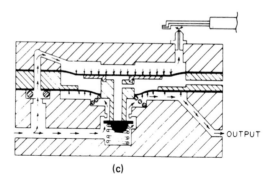

(c)

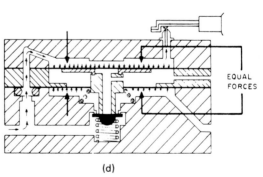

(d)

FIGURE 15-60 Operation of the 4000 Series proportional relay.

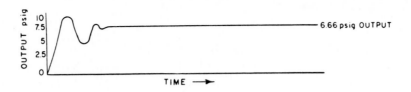

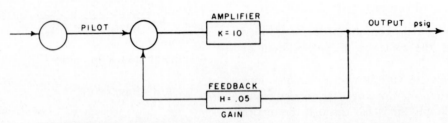

FIGURE 15-61 Proportional pneumatic feedback action: The cycle of response to any pilot signal change is similar to the graph. It has a high initial response to provide a high energy level with a relatively gradual settling or balancing at an output value with a precise relation to input signal. The resulting point of balance can be calculated by the formula:

$$p_o = \frac{K}{1 + KH} = \frac{10}{1 + (10)(0.05)} = 6.67 \text{ psig}$$

where p_o is the output pressure, psig; K is the forward loop gain (amplifier ratio); and H is the feedback loop gain ratio.

To observe how pneumatic feedback works in a proportional controller, assume that the sensitivity setting of a room thermostat is one point per degree Fahrenheit. When the bimetal element moves toward the control port in response to a 1-degree rise in temperature, the output pressure of the relay will increase 10 psig. This high volume is to provide a large flow to fill the system. However, as this flow fills the feedback chamber, pressure increases against the feedback diaphragm in the controller (see Figure 15-62). The diaphragm raises the feedback bar and sensitivity slider and moves the bimetal element away from the control port, cutting down the initial high output to a balance point of 1.0 psig output per 1.0 degree change in temperature.

This action is referred to as *negative feedback* in

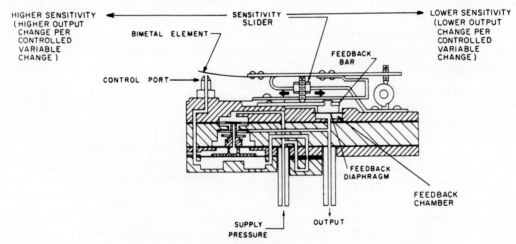

FIGURE 15-62 Cutaway drawing of a complete 4000 Series pneumatic proportional controller showing controller feedback diaphragm and sensitivity adjustment.

that, when the sensing element moves toward the control port, due to a change in temperature the resulting increase in output pressure tends to move the sensing element away from the control port. This counteraction provides a balanced condition for the output signal with respect to the temperature change measured at the sensing element.

Most proportional controllers used today incorporate negative pneumatic feedback with a *sensitivity adjustment,* which transmits more or less feedback into the controller. If the sensitivity slider shown in Figure 15-62 is moved toward the feedback diaphragm, more of the diaphragm movement would be transferred to the element, canceling out a greater portion of the change in output pressure initially created by the temperature change at the element. If the slider is moved away from the feedback diaphragm, the diaphragm movement and higher output pressure would have less canceling effect on the movement of the element. The result then is a greater change in output pressure in response to a given temperature change. It can be seen then that, as more controller feedback is introduced, the sensitivity of the instrument becomes lower, resulting in a smaller overall change in output pressure per unit change in the controlled variable. Conversely, as the amount of controller feedback is reduced, sensitivity increases, which allows a greater change in output pressure per unit change in the controlled variable.

The *action* of a proportional controller is determined by the way the sensing element is mounted. It can be mounted to bend or move toward the control port and thereby increase output pressure on an increase in measured variable (temperature, humidity, etc.). This combination is called *direct acting.* Or, the sensing element may be mounted to move away from the control port on an increase in the variable, allowing output pressure to drop. This combination is called *reverse acting.*

15-15   FINAL CONTROL ELEMENTS[5]

A final control element changes the value of the manipulated variable in response to the error signal from the controller. The manipulated variable and control agent then exert a correcting influence on the process operation to return the controlled variable to the set point.

The mechanism consists of two parts: (1) an actuator (or operator) and linkage, which translate the controller output signal into sufficient power and motion

[5]Figures 15-63 through 15-72 and the associated descriptive matter (Sections 15-15 and 15-16) are from *Automatic Control Principles,* Honeywell, Inc., Milwaukee, WI, 1972.

to operate the final control element and (2) the final control element, which is a device to adjust the value of the manipulated variable by controlling the flow of the agent. Mechanisms commonly used include motors, valves, and dampers.

Depending on the process, these mechanisms at times must work in extreme temperatures and pressures, be resistant to chemical action, be nearly maintenance free, and respond rapidly to error signals from the controller. The mechanisms vary, but in all cases they depend upon controller signals that are pneumatic, hydraulic, or electric in nature.

15-16   ACTUATORS AND LINKAGES

An *actuator* (sometimes called an *operator*) is a device that starts and stops or varies the operation of the final control element in response to the controller. A *linkage* is a series of connecting rods with movable joints that transfer the motion of the actuator to the final control element. For instance, the linkage used to connect a motor to a damper usually consists of a push rod, two crank arms, and two ball joints. The linkage can usually be adjusted to vary the response of the final control element to the actuator.

The actuator in an automatic control system may be electric, pneumatic, or hydraulic. Solenoids and motors are electric, diaphragm types are pneumatic, and oil-operated cylinders are hydraulic. We will briefly discuss electric and pneumatic actuators in this section.

Characterizing Methods.   In most final control elements (valves and dampers), flow versus opening follows a square root relationship. This is disadvantageous for most control systems because it makes it difficult to control the flow accurately, especially at low flow rates. For example, in one type of valve, a change from 0 to 10% valve opening results in a flow rate change from 0 to 31.6% of maximum. At the high end, a change from 90 to 100% valve opening results in a change of flow from 95 to 100%. It is also characteristic of many final control elements that they offer the greatest resistance to motion at the nearly closed position. Obviously, it is desirable to have a linear relationship between the error signal and the flow rate. This can be obtained by altering, or *characterizing,* the linkage or actuator.

Linkage adjustment.   One method of characterizing is by *adjusting the angularity of the linkage* between the actuator and the final control element. For example, let us look at a typical damper linkage as shown in Figure 15-63. For a standard parallel linkage, the crank arms are parallel so that the angle of travel of

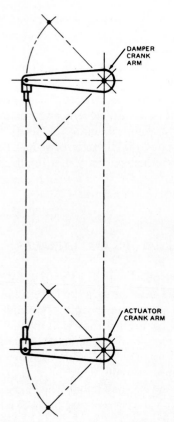

FIGURE 15-63   Parallel linkage.

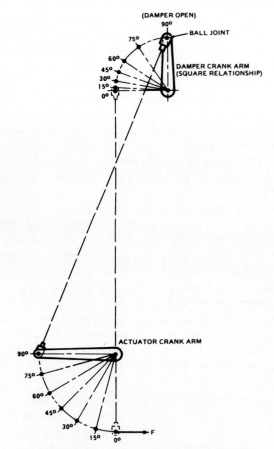

NOTE: DEGREE MARKINGS ALL REFER TO CRANK ARM POSITION.

FIGURE 15-64   Characterized linkage.

the damper crank arm is always the same as the angle of travel of the actuator crank arm. Without characterization, the square root relationship will exist between flow rate and error signal; that is, flow rate will change more when the error signal is small and the damper is just beginning to open.

If the linkage is adjusted as shown in Figure 15-64, the angular rotation of the damper crank arm will approximate the square of the rotation of the actuator crank arm. This will result in a small amount of damper travel when it is just beginning to open, as shown by the dashed curve in Figure 15-65. The *square* relationship of the characterized linkage will compensate for the *square root* relationship of the flow rate versus damper opening ($X^2 \cdot \sqrt{X} = X$), resulting in a *linear* relationship for flow rate versus error signal. If required, the linkage can be adjusted to obtain other curves, such as a square or even a fourth-power relationship between the flow rate and the error signal.

Cam shaping: For actuators with a cam-operated feedback mechanism, the actuator can be characterized by *shaping the cam*.

The *template method* of cam shaping is, as the name implies, the development of a template under actual operating conditions. At each of several values of

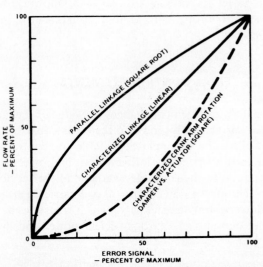

FIGURE 15-65   Result of characterizing the linkage.

error signal, the feedback mechanism is adjusted to obtain the desired position of the final control element. For each of these positions, the position of the cam roller is marked on a template. A cam of the required shape can then be made from the template.

The *drafting method* allows the cam shape to be determined by a simple procedure that saves installing a template. The straight cam is left on the actuator and various error signals are introduced. The manipulated-variable condition is read from a meter and plotted against the error signal. Differences between this curve and the desired straight line can then be transferred to a drawing of the straight cam to plot the shape of the required cam face.

Solenoids. A solenoid, as shown in Figure 15-66, is simply a coil of wire that, when current flows through it, will act as a magnet to pull a movable iron core into the coil. The iron core is attached to the final control element. The most common use of a solenoid as an actuator is with a valve. The iron core is attached to the stem of the valve. A disc on the stem is positioned in an orifice. When the solenoid is energized, the disc either opens or closes the orifice, depending on the valve model (normally closed or normally open).

The solenoid is limited to the on–off (two-position) mode of operation or to one stage in a multiposition mode. It provides fast response and is easily field replaceable.

Control Motors. For heavier loads or for greater movement than a solenoid actuator can provide, specially designed electric motors are used. Single-phase induction motors are the most common. The magnetic action of the field coils *induces* currents in the rotor, causing it to turn. The rotor needs no direct electrical connections and so requires neither a commutator nor brushes.

Automatic control motors differ from ordinary small motors. Ordinary motors are designed to drive *continuously rotating* machinery, usually at *high* speed. Most of them consist only of field coils, an armature or rotor, and a shaft.

Control motors, on the other hand, must be capa-

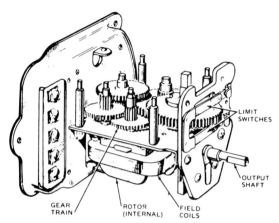

FIGURE 15-67 Typical control motor.

ble of producing *slow* movement *from one definite position to another,* and they must hold this definite position under load. Therefore, besides the field coils and rotor, a control motor (see Figure 15-67) also normally includes (1) a *gear train* between the rotor and the output shaft, for reducing speed and increasing torque, and (2) mechanically operated switches, called *light switches,* for stopping at definite positions of the output shaft. A proportioning motor also includes a *balancing relay* and *feedback potentiometer* to proportion the travel of the motor to the change at the controller.

Unidirectional motors: A *unidirectional (nonreversible) motor* (see Figure 15-68) is used for the on–off (two-position) mode of operation. Current through both field coils is always in phase; so the motor always runs in the same direction. A cam with only one node operates a maintaining switch. The cam is mounted on the motor shaft and turns with it. Both contacts, A and B, are closed when the motor is running. When the cam has rotated halfway, the node is at the bottom and switch B is open (shown by dashed line). Controller circuitry starts the motor again. When it has rotated

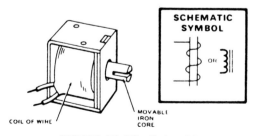

FIGURE 15-66 Solenoid.

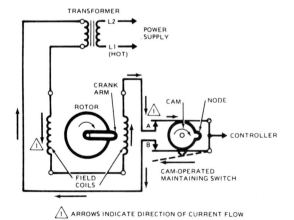

FIGURE 15-68 Unidirectional motor.

another 180 degrees, the node is at the top of the cam and contact *A* is open. Thus, the motor is stopped after each half-turn of the motor shaft and crank arm.

Reversible motors: For floating and proportioning modes of operation, the motor must be able to reverse its direction. Capacitor-type and shaded-pole-type motors are the two types most often used in control applications.

In the *capacitor-type motor* (see Figure 15-69), the two field coils are connected together at one end, and the other ends are connected through a capacitor. The capacitor causes the magnetic fields produced by the two coils to be out-of-phase, resulting in a rotating field that causes the rotor to turn.

The power source may be connected directly across either of the field coils. If the power source is connected between terminals *A* and *C,* power is supplied to coil 1 directly and to coil 2 through the capacitor. If the power source is connected between terminals *B* and *C,* power is supplied to coil 2 directly and to coil 1 through the capacitor. Consequently, the phase relationship between the coils is reversed, and the direction of rotation of the rotor reverses. This change in power source connections may be directly accomplished by the controller or indirectly by a balancing relay in the motor.

In the *shaded-pole-type motor* (see Figure 15-70), the stationary field coil is embedded in a laminated block of iron, called the *stator.* When the power is turned on, the field coil generates a magnetic field in the iron. The rotor moves freely in the hollowed-out stator. Heavy strips of copper, called *shades,* are attached to the poles of the stator on either side of the rotor. The copper magnetizes less rapidly than the iron in the stator. Therefore, the magnetic field of the shades always lags behind the field of the stator. This condition, combined with the swift reversal of the magnetic poles due to the alternating current of the power source, establishes a rotating magnetic field that causes the rotor to turn.

Reversing action is accomplished by using two

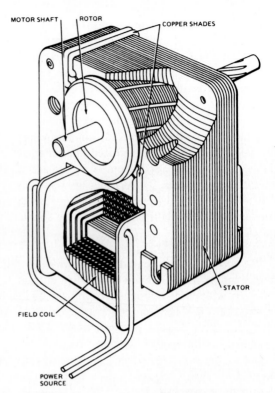

FIGURE 15-70 Shaded-pole-type motor.

pairs of coils for shades (see Figure 15-71). Shortening one pair of coils will cause the magnetic field induced in the shades to *lag* behind the field of the stator, while shortening the other pair will cause it to *lead,* thus reversing the direction of the rotating magnetic field.

Spring-return motors: Many motors have a helical spring attached to the motor shaft to mechanically return the motor to the starting (closed) position when power is removed. The spring mechanism is contained in a housing (see Figure 15-72), which projects from one end of the motor, identifying it as a spring-return type. When used in the on–off (two-position) mode, a spring-

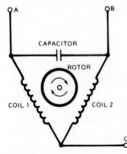

FIGURE 15-69 Capacitor-type reversible motor.

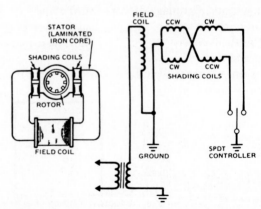

FIGURE 15-71 Reverse action in a shaded-pole motor.

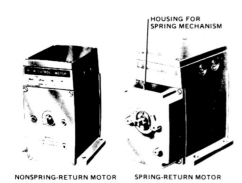

FIGURE 15-72 Identifying a spring-return motor.

return motor only requires electrical power to drive it in one direction; thus, one less terminal or wire is needed. Spring-return motors are often used in the floating or proportioning mode, even when electrically driven in both directions, to provide fail-safe action in case of a power interruption.

Electronic motors: Electronic motors are similar to other reversible motors except that they have solid-state control circuits and solid-state switching circuits to control the direction of motor shaft rotation. Mechanical contacts are eliminated; therefore, these motors will not experience contact chatter, and vibration is not an important consideration during installation.

Electronic motors provide proportioning (modulating) or sequencing control of valves, dampers, step controllers, and economizers. They are used with central processors to control heating, cooling, and economizer operation in multizone control systems. They are also used with electronic sequencers to provide sequencing control of heating–cooling equipment in commercial or industrial applications.

Many electronic motors are designed for specific applications or with special functions, such as automatic reset with an adjustable reset ratio. For economy and convenience of installation, some of these motors combine the functions of an electronic controller and actuator. They include a null detector, input adjustments, an amplifier, and output switching elements. Most of these special motors are controlled by an electronic (thermistor-type) sensing element that is one leg of a bridge circuit.

Pneumatic Actuators.[6] Pneumatic *damper actuators* (operators) of the *piston type* have a long, powerful straight stroke, which requires no lever arrangements. Air from the controller is applied to the molded

[6]Figures 15-73 through 15-75 and the associated descriptive matter are from *Fundamentals of Pneumatic Control,* Johnson Controls, Inc., Milwaukee, WI, 1979.

diaphragm, which has a positive seal to prevent air leakage (see Figure 15-73). This air pressure expands the diaphragm, forcing the piston and stem outward against the force of the spring. The movement of the piston varies proportionally with the air pressure applied to the diaphragm. This air pressure, from the controller, varies over the full pressure range.

The spring returns the actuator to its normal position when the air pressure is removed from the diaphragm. Full movement of the actuator can be restricted to set limits by using various spring ranges. Assume a spring range of 5 to 10 psig. With this spring, the actuator is in its normal position when the air pressure applied to the diaphragm is 5 psig or less. Between 5 and 10 psig, the stroke will be proportional to the air pressure on the diaphragm. Above 10 psig, the actuator will be at its maximum stroke.

Damper actuators can be mounted on the damper frame and coupled directly to the damper blades. In some cases, the actuator is mounted on the ductwork and coupled to the damper blade axis through a crank arm and linkage arrangement (Section 16).

Pneumatic *valve actuators* generally furnished with valves are the *piston top* (see Figure 15-74) and the *rubber diaphragm* (see Figure 15-75). The former is used mainly on small valves controlling terminal heating units and terminal air conditioning units, while the latter is used to control large valves regulating flow through heating or cooling coils on central or zoned systems.

Valve actuators function in a manner similar to damper actuators. Air pressure, or hydraulic fluid, is applied to the diaphragm, which expands and opposes

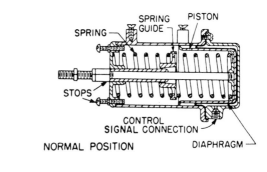

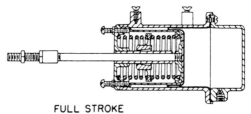

FIGURE 15-73 Components of a piston-type damper operator.

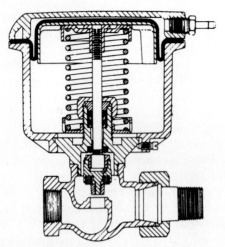

FIGURE 15-74 Piston-top valve operator on a normally open valve.

the force of the spring. This causes the inner valve (stem and disc or plug) to move toward the seat of a normally open valve (Figures 15-74 and 15-75), stopping the flow, or away from the seat of a normally closed valve, allowing flow. When air pressure is removed from the dia-

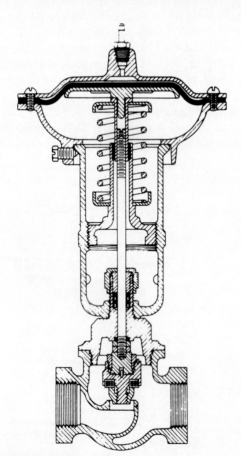

FIGURE 15-75 Rubber-diaphragm valve operator on a normally open valve.

phragm, the spring will return the inner valve to its normal position.

15-17 CONTROL VALVES[7]

A control valve is any device by which the flow of liquid, air, other gas, loose bulk material, and so forth, may be started, stopped, or regulated by a movable part that opens or obstructs passage. The simplest definition is one used by valve designers—"an engineered obstruction in a pipe." A valve acts as a variable orifice in a fluid flow line based on some planned pattern of change in the rate of flow. In an automatic control system, it is a final control element that directly changes the value of the manipulated variable by changing the rate of flow of the control agent.

Control valves are the most common and basic final control element in industry. They are widely used in regulating pressure, temperature (blending), and rate of flow of water, steam, chemicals, petroleum products, and just about anything that flows. They are used in processes, in power generation, in transmission pipelines—wherever flow of liquids and gases is controlled.

At this point in our work, we should review Sections 9-7 through 9-9 for some particular information on flow resistance in valves.

Control Valve Components. Figure 15-76 is a sketch showing the various valve components defined in the following sections.

Body: The portion of the valve through which the medium flows.

Stem: The cylindrical shaft that is moved by the actuator and to which the disc is attached.

Packing: Material that seals the fluid flowing through the valve body, while permitting the valve stem to pass through as freely as possible. Standard packing is usually in the form of a nest of V-shaped rings of lubricated asbestos fiber or Teflon.

Bonnet: The removable cover or top of the valve body (attached by screws or bolts). It provides a means for joining the actuator to the body, and it houses the packing. The bonnet assembly may include the valve trim.

Disc: The movable portion of a valve that comes into contact with the seat as the valve closes and that controls the flow of the medium through the valve.

---

[7]Figures 15-76 through 15-89 and the associated descriptive matter (Sections 15-17 through 15-22) are from *Automatic Control Principles,* Honeywell, Inc., Milwaukee, WI, 1972.

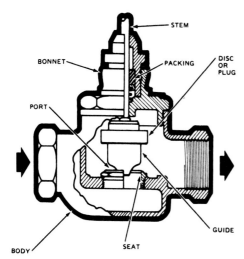

STEM

BONNET

PACKING

DISC OR PLUG

PORT

GUIDE

BODY

SEAT

FIGURE 15-76 Components of a control valve.

Three common types of control discs are the contoured, the V-port, and the quick-opening.

Contoured disc: Controls the flow by a shaped end, sometimes called a valve *plug,* and is usually end-guided at the top or at the bottom (or both) of the valve body.

V-port disc: Has a cylinder (a skirt or guide) that rides up and down in the seat ring. The skirt guides the disc, and the shaped openings in the skirt vary the flow area.

Quick-opening disc: Is machined so that maximum flow is quickly achieved when the disc lifts from the seat. It is either end-guided or guided by wings riding in the seat ring.

Some discs are built so that the part of the disc that contacts the seat is replaceable. This type is known as the *renewable* disc. A common example is a kitchen sink hot water faucet. Renewable discs are usually made of a composition material. Many valves having all metal, or nonrenewable, discs have to be "ground-in" to restore the seating surface, but valves with flat-type metal discs do not.

Guide: The part of the valve disc that keeps the disc aligned with the valve seat. Top or bottom guides on a valve disc usually do not influence flow; they merely perform a centering function. Valve guides often determine the valve flow characteristic. These guides are known as *characterized guides* or *skirted guides* and usually have notches or V's cut into them to characterize the flow through the valve.

Plug: A term used to describe a disc that has a solid, shaped end (instead of a skirted guide) to characterize flow. In some valves, the plug serves as a guide.

Seat: The stationary portion of a valve with which the movable portion (disc) comes in contact to stop the flow completely.

Port: The flow-controlling opening between the seat and disc of a valve. It does not refer to the body size or to the end connection size of the valve. Regardless of the pipe size, the valve port size determines the flow. For example, a 1-in. valve with a $\frac{3}{4}$-in. reduced port size has the same flow control as a $\frac{3}{4}$-in. valve with a standard port size.

Trim: All parts of a valve that are in contact with the flowing medium but are not part of the valve shell or casting. Seats, discs, packing rings, and so forth, are all trim components.

15-18 IMPORTANT VALVE CHARACTERISTICS

To select the proper valve for a given application, several flow-related characteristics of the valve must be known.

Valve Flow Characteristic: The factor that is most useful in selecting a valve type for a given application is the flow characteristic. This characteristic is the relationship that exists between the flow rate through the valve and the valve stem travel as the latter is varied from 0 to 100%. Different valves have different flow characteristics, depending primarily on internal construction. This flow relationship is usually shown in the form of a graph. The characteristic that is usually graphed is the "inherent flow characteristic," which is found under laboratory conditions and is the characteristic observed with constant pressure drop across the valve. However, the pressure drop across a valve in a system is not constant; it varies with flow and other changes in the system. We will discuss this in Chapter 16.

The three most common valve flow relationships are quick-opening, linear, and equal percentage. The ideal flow relationships are shown graphically in Figure 15-77. The *quick-opening valve plug* gives the steepest initial flow characteristic possible. A small initial lifting of the plug produces a large increase in flow, and close to maximum flow is reached at a relatively low percentage of maximum stroke. This type of plug is generally used for on–off types of applications but may be used successfully in some linear valve applications. This is possible because of its initial linear characteristic at low percentage stem travel. The slope of this linear region is very steep, which produces a much higher gain than an ordinary linear plug and consequently is much more likely to be unstable.

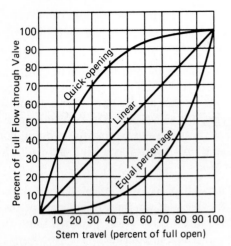

FIGURE 15-77   Valve flow characteristics.

The *linear plug* produces a characteristic with essentially a constant slope so that with a constant pressure drop across the valve the valve gain will be the same at any flow. This type of plug would be the easiest and most logical plug to use if everything else in the control loop were linear. However, this is not true in most systems, as will be shown in Chapter 16.

The third type of plug is the *equal-percentage plug*. The characteristic for this plug shows that for each unit step increase in valve stroke the flow increases an "equal percentage" over the previous flow. For example, if 30% stem lift produces 5 gpm flow and an increase of 10% lift, to 40%, produces 8 gpm, or a 60% increase over the previous 5 gpm, then a further stroke of 10%, to 50% stroke, now produces a 60% increase over the previous 8 gpm for a total flow of 12.8 gpm. Because of this constant relative increase in flow per unit stroke, the valve gain at a specified flow is independent of both the valve flow coefficient and the pressure drop. Because of this, system stability (at a specified flow) is independent of the valve size selected.

A valve may be *characterized* for the desired relationship between stem travel and flow by shaping the sides of the port and the openings in the skirted guide. A linkage between actuator and stem can also be characterized, as discussed previously in the section on actuators and linkages.

Valve gain: In the preceding discussion of valve flow characteristics, the term *gain* was used. Valve gain is the incremental change in flow rate produced by an incremental change in plug position. This gain is a function of valve size and type, plug configuration, and system operating conditions. The gain at any point in the stroke of a valve is equal to the slope of the characteristic curve at that point.

Travel Coefficient.   Travel coefficient is the ratio between the flow at a given valve stem position and the flow through the valve at its wide-open position. Travel coefficient is expressed as a decimal fraction. For example, if a valve with a lift of 1 in. passes 100 gpm fully open and passes 66 gpm at a valve lift of 0.5 in., this valve is said to have a 0.66 travel coefficient at 0.5-in. lift. Valve travel coefficients can be read directly from any plot showing valve flow characteristics.

Rangeability.   The ratio of the maximum *controllable* flow to the minimum *controllable* flow is called *rangeability*. For example, a valve with a rangeability of 50 to 1 having a total flow capacity of 100 gpm fully open can accurately control a flow as low as 2 gpm. Generally speaking, rangeabilities in the area of 50 to 1 or 40 to 1 are excellent for precision control. Valves with high rangeability are very expensive to manufacture since very close tolerances are required between the disc and the seat.

Turndown.   The ratio of the maximum *usable* flow to the minimum *controllable* flow is called *turndown*. Turndown is usually somewhat less than the rangeability since the maximum usable or required flow is usually less than the total flow capacity of a valve. For instance, in the rangeability example, after the 100-gpm valve has been applied to a job, it might turn out that the most flow you would ever need through the valve is 68 gpm. Since the minimum controllable flow is 2 gpm, the turndown of the valve is 34 to 1. In comparing rangeability and turndown, rangeability is the measure of the *predicted* stability of a control valve and turndown is the measure of the *actual* stability of the valve.

Tight Shutoff.   A valve having tight shutoff has virtually no flow or leakage in its closed position. Generally speaking, only single-seated valves have tight shutoff. Double-seated valves can be expected to have 2 to 5% leakage in the closed position.

Cavitation.   Cavitation is a two-stage phenomenon that can greatly shorten the life of the valve trim in a control valve. Whenever a given quantity of liquid passes through a restricted area such as an orifice or a valve port, the velocity of the fluid increases. As the velocity increases, the static pressure decreases. If this velocity continues to increase, the pressure at the orifice will decrease below the vapor pressure of the liquid, and vapor bubbles will form in the liquid. This is the first stage of cavitation. Figure 15-78 shows schematically the variation of velocity and pressure as liquid flows through a restriction such as an orifice or valve port.

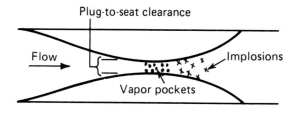

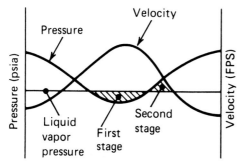

FIGURE 15-78 Variations of pressure and velocity for points along a flow stream through a restriction.

As the liquid moves downstream, the velocity decreases with a resultant increase in pressure. If the downstream pressure is maintained above the vapor pressure of the liquid, the voids or cavities of vapor will collapse or implode. This is the second stage of cavitation (see Figure 15-78).

The second stage of cavitation is detrimental to valves. Because of the tremendous pressures created by these implosions (sometimes as high as 100,000 psi), tiny shock waves are generated in the liquid. If these shock waves strike the solid portions of the valve, they act as hammer blows on these surfaces. Repeated implosions on a minute surface will eventually cause fatigue of the metal surface and chip a portion of this surface off. Tests show that only those implosions close to the solid surfaces of a valve act on the valve in this manner.

Low degrees of cavitation are tolerable in a control valve. Minimum damage to the valve trim and little variation in flow occur at these levels. However, there is a point where the increasing cavitation becomes detrimental to the valve trim and possibly to the valve body. Also at this point, the cavitation is beginning to "choke" the flow through the valve. At some point, the flow rate will stay the same regardless of increases in pressure drop. The point at which cavitation becomes damaging can be expressed by

$$K_m = \frac{\Delta p_{max}}{p_1 - p_v} \qquad (15\text{-}1)$$

where

K_m = valve recovery coefficient
p_1 = absolute inlet pressure (psia)
p_v = absolute vapor pressure of liquid (psia)
Δp_{max} = maximum allowable pressure drop (psia)

The valve recovery coefficient differs among the various types of valves. A conservative value from K_m is 0.5. Equation (15-1) can be used to estimate the allowable maximum pressure drop through a valve when warm liquids are flowing.

15-19 VALVE RATINGS

To specify a valve for a given job, several ratings must be considered.

Capacity Index (C_v). Capacity index (also called *flow coefficient* in Chapter 11) is the quantity of water, in gallons per minute at 60°F (16°C), that flows through a given valve with a pressure drop of 1 psi. Once the C_v of the valve has been determined, the flow of any fluid through the same valve can be calculated, provided the fluid properties and the pressure drop through the valve are known. This will be illustrated in Chapter 16.

Closeoff Rating. The closeoff rating of a valve is the maximum allowable pressure drop to which the valve may be subjected while fully closed. This rating is usually a function of the power available from the valve actuator for holding the valve closed and is independent of the actual valve body rating. Structural parts, such as the stem, are sometimes the limiting factor. To illustrate, a valve having a closeoff rating of 10 psi could operate with an upstream pressure of 40 psi and a downstream pressure of 30 psi. However, in applications where failure of the valve to close is hazardous, the maximum upstream pressure should not exceed the closeoff rating regardless of the downstream pressure.

The closeoff rating of a three-way valve is the maximum pressure difference between the outlet and either of the two inlets. For a three-way diverting valve, it is the pressure difference between the inlet and either of the two outlets.

Maximum Fluid Pressure and Temperature. These ratings are limitations placed upon the valve by the maximum pressure and temperature to which the valve may be subjected. Determining factors may be packing, body material, disc material, or actuator limitations. For example, a valve having an actual body rating of 125 psig at 355°F (179°C) may have a packing with a 100 psi rating; this limits the valve to a fluid pressure of 100 psi maximum. The temperature of 115-psia steam is 338°F (170°C). This fact should limit the above valve to a maximum temperature rating of 338°F (170°C) rather than 355°F (179°C). However, if the valve has a composition disc with a maximum fluid temperature rating of 240°F (116°C), consider this temperature as the maximum fluid temperature.

Pressure Drop.   Pressure drop of a valve is the difference between the upstream pressure and the downstream pressure of the fluid passing through a valve. Pressure drop across control valves may be assigned by the control system designer using values recommended by control valve manufacturers. We shall look at this in more detail in Chapter 16.

Valve Body Rating.   This rating is the maximum pressure the valve body can withstand.

The *nominal body rating* is the nominal pressure rating of the valve body (exclusive of packing, disc, etc.) expressed in psig. This is a specified rating (often cast on the valve body) based on the specified body material and universally accepted construction characteristics (such as wall thickness, flange dimensions, etc.) determined by specifications issued by approval bodies and other authorities having jurisdiction. It provides a convenient method of classifying the valve by pressure for identification purposes (see Chapter 11). Note that the *nominal* valve body rating is *not* the same as the *actual* valve body rating.

The *actual body rating* is the maximum safe permissible fluid pressure at a given temperature. Each nominal valve body rating has definite corresponding permissible pressures at various temperatures. Thus, the permissible pressure is usually dependent on the temperature. For instance, a 125-psi (nominal body rating) cast-iron screwed body may have actual body ratings of 125 psi at 353°F (178°C) and 175 psi at 150°F (66°C). However, the maximum rating of the valve may be limited to a valve below the body rating by the packing and disc material.

15-20   VALVE CLASSIFICATION BY TYPE OF CONSTRUCTION

Single-Seated Valve.   A valve with only one seat and one disc is cheaper to construct than a double-seated valve. It is generally suitable for service requiring tight shutoff. Figure 15-79 shows schematically two types of single-seated valves. Since there is nothing in the single-seated valve to balance the force exerted by the fluid pressure against the disc, the single-seated valve requires more force to close than a double-seated valve of the same size.

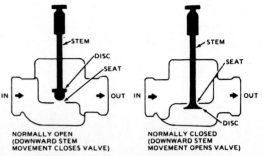

FIGURE 15-79   Single-seated valves.

Double-Seated Valve.   A valve with two seats and two discs (see Figure 15-80) is arranged so that in the closed position there is very little fluid pressure forcing the stem toward either the open or closed position. For a valve of given size and port area, a double-seated valve will always require less power to operate than a single-seated valve. An additional advantage of a double-seated valve is that it often has a larger port area for a given pipe size. A disadvantage is that it does *not* have tight shutoff since both discs are rigidly connected. Changes in temperature of the fluid being controlled cause either the disc or the valve casting to expand, allowing one disc to seat before the other.

Three-Way Mixing Valve.   This type of valve always has two inlets and one outlet (see Figure 15-81). The proportion of fluid entering from each inlet may be controlled by moving the valve stem. A valve designed for mixing service is not suitable for diverting applications because it has only one disc and two seats. If it is piped for diverting applications, the inlet pressure will slam the disc against the seat when it nears closing. This

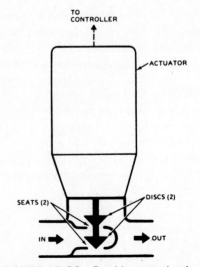

FIGURE 15-80   Double-seated valve.

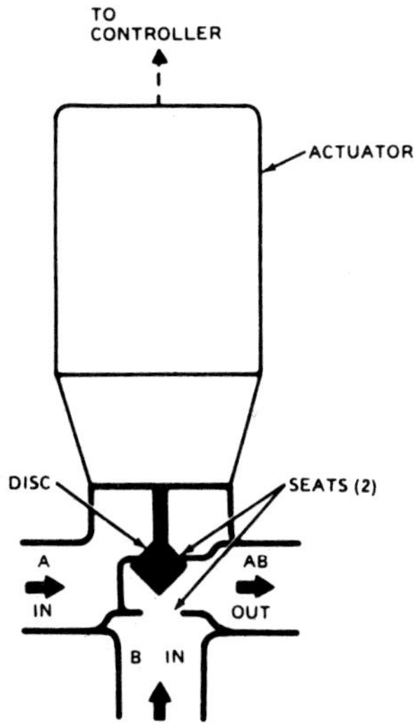

FIGURE 15-81 Three-way mixing valve.

causes loss of control, oscillations, vibrations, excessive wear, and noise.

Three-Way Diverting Valve. This type always has one inlet and two outlets (see Figure 15-82). A fluid entering the inlet port is diverted to either of the two outlet ports in any proportion desired by moving the stem. A valve designed for diverting service can generally be used in mixing applications also because it is double seated with a disc for each seat. When the diverting valve is piped for a mixing application, the two inlet pressures oppose and cancel each other. Under these conditions, the pressures on the discs are balanced and the diverting valve responds much the same as a mixing valve.

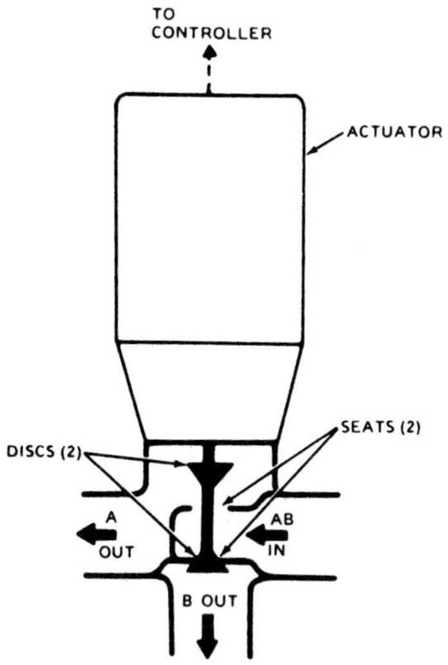

FIGURE 15-82 Three-way diverting valve.

15-21 VALVE CLASSIFICATION BY METHOD OF CONTROLLING FLOW

Sliding Plug Valve. In a sliding plug valve (see Figure 15-83), the disc or plug is attached to the end of the stem, and the orifice size is adjusted by moving the stem up or down (or in and out) along its axis. The valve flow characteristic is determined by the shape of the plug and port. A disc-shaped plug, as in an ordinary globe valve, gives a linear characteristic, but it has the disadvantage that its entire turndown range is expired within a very small range of axial movement. A V-port or a parabolic plug permits a more gradual change of valve opening and gives an equal-percentage characteristic.

Rotary Plug Valve. A rotary plug valve (see Figure 15-84) consists of a ported sleeve or plug that is rotated past an opening in the body. The valve flow characteristic is determined by the shape of the port and the opening in the body. The figure shows a rotary plug valve with a round body opening and a conical plug with a V-shaped, tapered port, which gives an equal-percentage flow characteristic similar to that of a V-port sliding plug valve. A cylindrical plug with a straight-through rectangular port gives a characteristic approaching a straight line.

Butterfly Valve. A butterfly valve (see Figure 15-85) has a rotating hinged disc that somewhat resembles the opened wings of a butterfly. The pressure drop across a wide-open butterfly valve is very small. The amount of leakage depends on the degree of machining of the disc and valve casing (which is the seat). The valve gives an equal-percentage flow characteristic similar to that of a V-port sliding plug valve. A butterfly valve has excellent high-volume flow characteristics, making it useful for close modulation of the supply of air or fuel-gas to large furnaces. It is often used as a

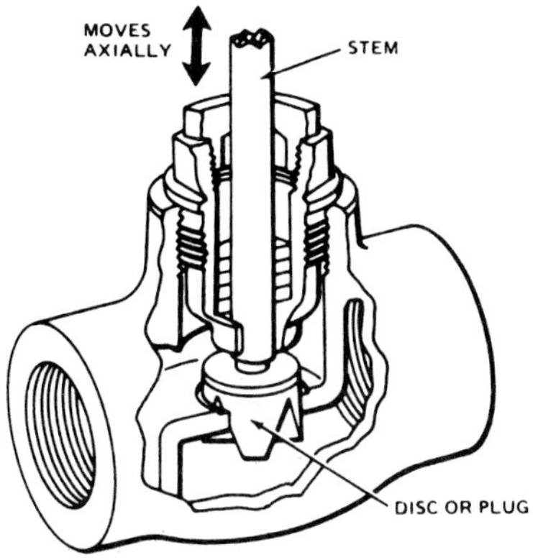

FIGURE 15-83 Typical sliding plug valve.

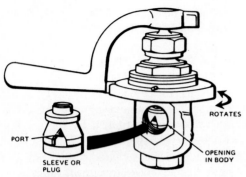

FIGURE 15-84 Typical rotary plug valve.

FIGURE 15-85 Typical butterfly valve.

firing rate valve. This type of valve does not close tightly; so a separate valve must be used where tight shutoff is required.

15-22 DAMPERS

Regulation of airflow is a vital factor in every HVAC control system as well as in many industrial processes—from small domestic systems to large industrial or commercial systems. These systems use dampers installed in air-carrying ducts to control airflow.

Applications. Dampers are made in many shapes and sizes for installation in *round* or *rectangular* ducts. In individual ducts, they are used for brute-force control of air volume. For *diverting* or *mixing action,* similar to three-way valves, two dampers are installed with synchronous operation in two-branch ducts. An example of this application is the simultaneous control of return air and outdoor air in a ventilation control system. The two dampers are connected so that one opens as the other closes. Another important application is the control of combustion air in a burner system.

Many types of electric, pneumatic, and hydraulic actuators are used to provide on–off (two-position), floating, or proportioning (modulating) operation. Two or more actuators are often used to operate two or more dampers in unison or sequence, or one actuator may be used to operate several damper sections simultaneously.

Blade Designs. A *flap-type* damper consists of one or more metal plates, each hinged at one edge, and all linked together for simultaneous operation (see Figure 15-86). The *splitter* damper is a single-blade, flap-type damper, usually located at a branch connection of a rectangular duct or outlet to divert flow into the branch. It is easy to operate but often causes irregular airflow (turbulence) in the duct. The *pinch* damper, or double-door type, is also a flap type used in horizontal ducts.

A *single-blade* damper is *center pivoted* for installation in rectangular ducts. The *butterfly* damper is also a single-blade, center-pivoted damper, but the blade is disc shaped for use in round ducts. A *louver* damper is a series of center-pivoted blades with edges mating in the closed position. The louver damper is the most used for controlling inlet and exhaust flow because it provides more precise control.

Louver dampers: There are two types of louver dampers—parallel blade and opposed blade. Each type is designed to do a specific job.

Parallel-blade dampers (see Figure 15-87) are designed so that all blades rotate in the *same* direction.

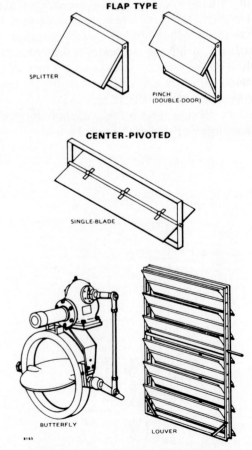

FIGURE 15-86 Damper blade designs.

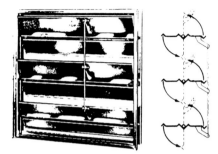

FIGURE 15-87 Parallel-blade damper.

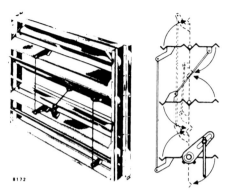

FIGURE 15-89 Opposed-blade damper.

They are most commonly seen in ventilation applications because they are less expensive and easier to install. Also, they cause more turbulence and hence better mixing. Therefore, parallel-blade dampers perform better for most ventilation applications.

The airflow characteristic for parallel-blade dampers is nonlinear (see Figure 15-88). Changes in damper blade rotation do *not* cause proportional changes in airflow. As the damper starts to open, a small rotational change causes a large increase in airflow. Therefore, parallel-blade dampers should be adjusted so that the *actual* airflow meets system and code requirements. Improper adjustment of dampers can result in substantial waste of energy. Also, for better control from one end of the proportioning range of the controller to the other, the damper opening should be limited to a maximum of 60 degrees. Further opening of the damper will provide only a small additional amount of air.

Opposed-blade dampers (see Figure 15-89) are designed so that alternate blades rotate in *opposite* directions. They have a more linear flow characteristic and are intended for variable-air-volume (VAV) applications such as zone or individual room control. In many applications, they tend to stratify the air, which results in poor mixing and less control.

Opposed-blade dampers have an equal-percentage airflow characteristic (see Figure 15-88). Successive equal increments of rotation produce equal-percentage increases in flow. This allows close control at low airflow. Because these dampers are more expensive than parallel-blade dampers, they are usually used only where accurate control of low airflow is necessary.

Damper Leakage. For the most economical operation, dampers must allow only the amount of airflow required by the system and by local codes. If dampers do not close tightly, the extra air that leaks into the system must be conditioned. Newer dampers have special seals and low-friction bearings to reduce leakage. Damper leakage is expressed in two ways: (1) percentage of the airflow through an open damper or (2) cubic feet of air per minute per square foot of damper area; both at a specified pressure differential.

Damper Actuators. Damper actuators are rated by the square feet of damper they can reliably operate. If the area of a damper (or combined dampers) is larger than the rating of the actuator, the installer has two options: (1) use an actuator with a larger rating or (2) use two or more smaller dampers with an actuator for each.

If the second option is chosen, the dampers must usually be driven in unison. A motor with auxiliary potentiometers can be used to drive others (called "slaving"). Also, some actuators can be driven in parallel from the appropriate controller.

Damper actuators may be mounted either outside or inside the air duct. When the actuator is mounted outside the duct, the axle driving the blades of the damper is extended out through the duct wall. When mounted inside, the actuator is mounted on the bottom of the duct. If the duct (or air passage) is stiff enough, the actuator may be bolted directly to it; otherwise, a mounting bracket is used. A damper linkage is used to connect the actuator to the damper, as was discussed in Section 16.

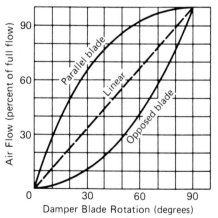

FIGURE 15-88 Airflow characteristics for parallel-blade and opposed-blade dampers.

REVIEW PROBLEMS

15.1. Describe the function of each of the following components of an electric control system: **(a)** power supply, **(b)** controller, **(c)** primary control, **(d)** high limit, **(e)** low limit, and **(f)** manual controls.

15.2. What is the primary element in a temperature controller?

15.3. Name a type of primary control that **(a)** controls system action directly and **(b)** controls system action indirectly.

15.4. A differential gap, or differential, occurs in (two-position/modulating) systems. How is it defined?

15.5. What is the purpose of reset in a proportional control system?

15.6. What is a delay in response from one part of a system to change occurring in another part of the system called? It is most obvious in (temperature/pressure) systems.

BIBLIOGRAPHY

15.1. *Automatic Control Principles,* Honeywell, Inc., Milwaukee, WI, 1972.

15.2. *Basic Electricity,* Training Manual, Johnson Controls, Inc., Milwaukee, WI, 1980.

15.3. *Fundamentals of Pneumatic Control,* Johnson Field Training Handbook, Johnson Controls, Inc., Milwaukee, WI, 1979.

15.4. *Fundamentals of: Heating and Cooling Systems, Pneumatic Control, Electronic Control,* Training Manual, Johnson Controls, Inc., Milwaukee, WI, 1975.

16

Automatic Control Systems

16-1 INTRODUCTION

The preceding chapter discussed the various control components, how they function with different types of energy, and the various modes of control. In this chapter, we will consider the application of these components to heating and ventilating systems. We will focus on the elementary control systems, with the understanding that the more complex systems are made up of combinations of these elementary systems.

Although our primary concern in this chapter will be the control of winter heating and ventilation systems, we will find that in some cases we must consider the control of summer cooling systems as well. This is necessary because so many systems are designed for year-round operation.

We deal primarily with control system function, with the assumption that this function can be accomplished by means of any of the various energy sources—electric (electromechanical or electronic), pneumatic, or fluids.

Again, let us repeat the definition of *air condition* as *the process of treating air so as to control simultaneously its temperature, humidity, cleanliness, and distribution.* Air conditioning systems are designed to meet the comfort and health needs of a building's occupants and also to provide the correct environment for the

equipment and processes in use. Occupant health and comfort are directly dependent on temperature, humidity, and cleanliness. In turn, these qualities usually depend on air distribution.

The demands placed on an air conditioning system vary with the design, construction, and use of the building in which they are installed. The equipment used to meet these demands may be classified into three general groups: (1) central-fan systems, (2) unitary systems, and (3) packaged systems.

Unitary and packaged systems are, by and large, complete with controls when they leave the factory. At the most, some type of space thermostat may have to be added. Their application is mostly a matter of selecting the proper functions and capacity. Central-fan systems, on the other hand, are normally built up on the job site. Each system is specifically designed for the building in which it is to be used.

16-2 ELEMENTARY CENTRAL-FAN SYSTEMS[1]

Use of a central air conditioning system requires some means of moving air throughout the building.

[1]Figures 16-1 through 16-13 and Figures 16-17 through 16-19 and the associated descriptive matter (Section 16-2) are from *Basic Control Systems,* Honeywell, Inc., Milwaukee, WI, 1972.

493

Ventilation, technically defined, is *the process of supplying or removing air by natural or mechanical means, to or from any space.* Such air may or may not have been conditioned. This air may be either outdoor air, recirculated air, or a combination of the two, depending on the needs of the building and its occupants.

Many of the same principles that are used in central-fan systems are the same as, or similar to, those used in unitary and packaged systems. The special features associated with unitary and packaged systems will be covered later in this chapter.

Ventilation. The most elementary ventilation system is the one that recirculates building air. No special provision is made for the introduction of outdoor air except infiltration. This is the ventilation system typically used in a residential warm-air heating system. Sufficient fresh outdoor air normally leaks into a house to meet the needs of the occupants. The purpose of bringing in outdoor air is to replenish the oxygen supply and dilute airborne contamination. When these needs are fairly small, as in a house, they usually can be met by natural infiltration of outdoor air. Figure 16-1 illustrates schematically a duct system and fan handling recirculated air only.

In larger systems, however, definite arrangements normally must be made to mechanically bring outdoor air into the building. Building codes usually specify a definite amount of outdoor air to be used. One simple way of doing this would be to use a duct sized to admit the correct ratio of outdoor air (see Figure 16-2). Theoretically, this system would do the job of getting outdoor air into the building. It does, however, have some practical limitations in actual practice.

A practical system should have a means of closing the outdoor air entrance when the system is not in use. The system illustrated in Figure 16-3 provides this feature. In this system, a damper has been installed in the outdoor air inlet. The actuator that controls the damper is powered through the fan starter. When the fan is turned on, the damper opens; when the fan is turned off, power to the damper actuator is interrupted allow-

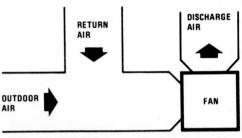

FIGURE 16-2 Outdoor air.

ing the spring-loaded actuator to close the damper. Here, the amount of outdoor air brought in is still constant, but a means is provided to close the outdoor air entrance.

A further refinement to the ventilation system is the addition of a means to vary the amount of outdoor air brought in or a means to control the ratio of outdoor air to return air (see Figure 16-4). In this system, dampers are installed in both the return and outdoor air ducts. The dampers are connected together by a linkage so that as one opens the other closes. As in the previous system, the power to the damper actuator is controlled by the fan starter so that the outdoor air damper will close when the system shuts down (also the return air damper opens). A manual potentiometer (electric) is used to control the position of the damper actuator and thus the two dampers. The potentiometer can be located in a convenient location where the amount of outdoor air can be regulated to meet the needs of the occupants.

Exhaust: Some of the previous systems examined have brought outdoor air into the building, but so far we have not considered removing or exhausting the air. Typically, an exhaust fan and exhaust air damper are connected into the ventilation system, as indicated in Figure 16-5. The ventilation system shown here is similar to the one in Figure 16-4 in that it provides a manually adjustable quantity of outdoor air. To this has been added an exhaust fan and exhaust damper. The provision of these added components allows the amount of air exhausted to be adjusted according to the amount

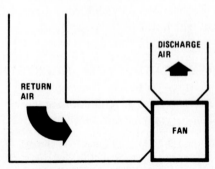

FIGURE 16-1 Recirculation.

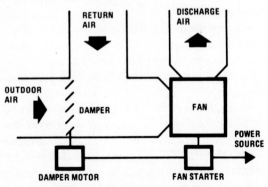

FIGURE 16-3 Outdoor air damper.

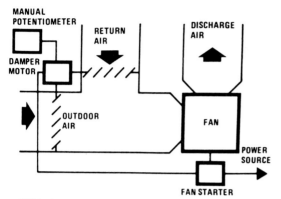

FIGURE 16-4 Outdoor air (manual control).

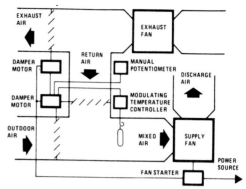

FIGURE 16-6 Outdoor air (automatic control).

of outdoor air used. Normally, it is desirable to maintain a slight positive pressure within the building. The relationship between the position of the outdoor air damper and the exhaust air damper controls the amount of positive pressure. The two damper actuators, for outdoor air and exhaust air, are connected so that they operate together. They are also both spring loaded to close the dampers when the system is shut down.

The ventilation systems we have considered so far have had either fixed or manually adjustable arrangements for bringing in outdoor air. In many applications, it is necessary or desirable to provide automatic control of the ventilation system. One common method is to put an automatic controller in the mixed airstream (see Figure 16-6). It is used to adjust the outdoor and return air dampers to maintain a given mixed-air temperature. This system, a winter ventilation control, admits more outdoor air to lower the mixed-air temperature. The manual potentiometer permits a certain amount of outdoor air to enter regardless of the demands of the mixed-air controller.

The next step in the evolution of a ventilation system is to limit the amount of outdoor air used during the summertime. The system shown in Figure 16-6 uses more and more outdoor air as the mixed-air temper-

ature exceeds the controller set point. This may require additional cooling system operation to cool the hot outdoor air while already cool return air is being exhausted.

The answer, then, is to limit the amount of outdoor air brought in when it is very warm outdoors. The system that does this is called an *economizer* ventilation control system and is shown schematically in Figure 16-7. The only difference between Figure 16-7 and Figure 16-6 is the addition of a control sensor in the outdoor air. Now, when the outdoor air temperature rises too high, the outdoor air controller acts to close the outdoor air damper to the minimum position. The minimum position is set by the manual potentiometer.

Additional ventilation control functions and features may be provided in more complex systems. However, the systems covered here demonstrate the basic control functions that are necessary to ventilation systems. Additional control functions that tie the ventilation system into the heating and/or cooling system will be discussed later.

Heating Systems. There are a number of heating systems used in central-fan-type air conditioning systems. The ones to be considered in this discussion include the following:

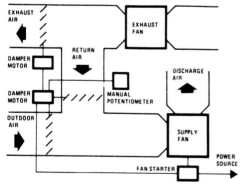

FIGURE 16-5 Outdoor air/exhaust air (manual control).

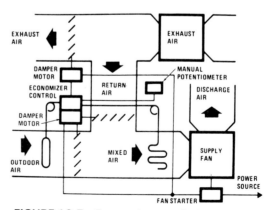

FIGURE 16-7 Economizer ventilation system.

- Steam coil
- Hot water coil
- Gas or oil-fired duct heater
- Gas direct-fired makeup air heater
- Electric duct heater

Methods of control: A control system may be designated by the controlled medium and the location of the sensors. Normally, the controlled medium will be hot water, steam, or air. The system sensors will be placed to control one of these media from one of the following locations:

1. *Space control*—The primary sensor (thermostat) is located somewhere in the controlled space. The controlled medium is adjusted to meet the requirements of the space.
2. *Control from discharge air* (*medium control*)—The primary sensor is located in the discharge air duct. The controlled medium is adjusted to maintain constant discharge air temperature.
3. *Control from outdoor air*—The primary sensor is located either outdoors or in the duct bringing in outdoor air. The control works as a changeover switch or else controls a preheater coil.

In typical systems, two or more of these types of control may be combined—for example, space control of a steam valve with a discharge air low-limit sensor or control from discharge air with reset by outdoor air temperature. Figure 16-8 shows the relative locations of the control sensors.

Steam coils: Steam is generated in a boiler at a central location and distributed to one or several central systems within the building. Distribution is by means of supply and return mains at supply steam pressures of 2 to 15 psig, with 5 psig most common.

Normally, steam coils are designed for a maximum air temperature rise of 35°F through the coil. Two or more coils in series (in airstream) may be used where a greater temperature rise is required.

Steam coils may be used with either an on–off (two-position) control valve or a modulating control valve. A steam coil with a modulating control valve is used only in applications where the air entering the coil is always above freezing. This eliminates the possibility of coil freeze-up under light-load operating conditions.

One relatively simple heating system for a central-fan system consists of a steam coil controlled by a sensor in the discharge air and a space thermostat, as illustrated in Figure 16-9. In this illustration, part of the ventilation system discussed previously is omitted for simplicity. This omitted part would have a means for bringing in outdoor air and exhausting air, using the associated controls already discussed.

In Figure 16-9, the flow of steam to the coil is controlled by a modulating valve in the steam supply line. This valve is connected to a thermostat in the conditioned space and is run open or closed to maintain the desired space temperature. As a precaution against discharging uncomfortably cold air into the building, a low-limit sensor has been installed in the discharge airstream. The low-limit controller will partially open the steam valve, even though the space thermostat is not calling for heat, to prevent the discharge air temperature from dropping below 65°F to 70°F.

As stated before, since this system employs a modulating valve to control the steam coil, the temperature of the mixed air entering the coil must not be allowed to drop below freezing. In applications where an unusually large temperature rise through the coil is required, two steam coils may be used.

Water coils: As with the steam coils, the water coil is supplied by a central hot water boiler. Water temperatures commonly used range from 180°F to 220°F. Water coils may be controlled by either an on–off (two-position) or a modulating control valve. A system using

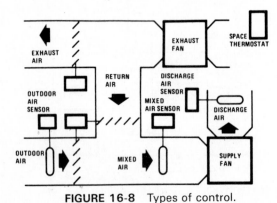

FIGURE 16-8 Types of control.

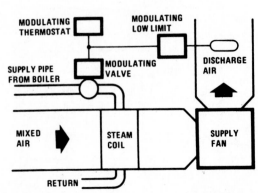

FIGURE 16-9 Steam coil (modulating control).

a hot water coil controlled by space and discharge air temperature looks substantially the same as the steam coil system and is shown schematically in Figure 16-10. In this case, hot water is pumped from the boiler to the central system. The control valve is a modulating, three-way mixing valve. In other words, it can adjust the proportion of the water allowed to go through the coil. As the space thermostat calls for more heat, the valve opens to let more supply water through the coil and less through the bypass. The low limit in the discharge air functions, as in the steam system, to prevent cold air from being blown into the building.

Face-and-bypass dampers: Another heating system using a hot water or steam coil is the face-and-bypass system, illustrated in Figure 16-11. The new feature demonstrated by this system is on–off (two-position) control of the coil and modulating control of the airflow through the coil.

The mixed-air flow is divided into two duct sections, one of which contains the steam or hot water coil. Dampers at the entrance to each section are connected by linkage to work in opposition to each other. In other words, when one opens, the other closes. The actuator that drives the face-and-bypass dampers is controlled by a space thermostat. As the space temperature drops, the actuator drives the face damper open and the bypass damper closed. More air goes through the heating side of the duct, increasing the discharge air temperature.

The two-position control valve on the steam or hot water coil is controlled by an auxiliary switch on the modulating damper actuator. When the face damper (in front of the coil) starts to open, the valve opens. It remains fully open as long as the face damper is open any amount. As with the preceding heating systems controlled by space temperature, the modulating low limit prevents the temperature of the discharge air from getting too low.

Gas direct-fired makeup air heaters: A direct-fired makeup air heater is similar to a duct heater except

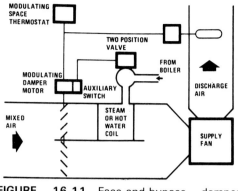

FIGURE 16-11 Face-and-bypass damper system.

that combustion takes place in the air being circulated. These heaters are fueled by gas and use 100% outdoor air. Products of combustion are mixed with the outdoor air and vented into the controlled space. Makeup air heaters are used to replace the air exhaust from a building to remove contamination such as that from some industrial process. Makeup air heaters are supplied by the manufacturer complete with controls and in a wide range of sizes. A typical schematic control setup is illustrated in Figure 16-12.

Gas supply to the burner is controlled by a modulating valve that allows the output of the burner to be adjusted automatically. The gas valve, and in turn the burner output, are controlled to maintain a constant discharge air temperature. To compensate for increasing offset, as the system approaches full capacity, an outdoor reset control is sometimes used. The reset control raises the discharge air temperature control point as the outdoor air temperature drops. In this way, the discharge air temperature remains the same whether the system is under light or heavy load.

Duct heaters: A duct heater, sometimes called a *duct furnace,* is installed directly in the distribution duct of central air conditioning systems. They are supplied by the manufacturer complete with controls already in-

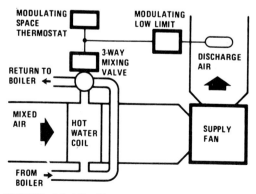

FIGURE 16-10 Hot water coil (modulating control).

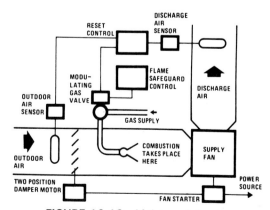

FIGURE 16-12 Makeup air heater.

stalled. Duct heaters may be either gas or oil fired. Gas-fired units are easily vented through the roof of a single-story building and are most common in applications such as supermarkets, garages, and warehouses. Basically, a gas-fired duct heater is the same as a residential gas-fired furnace.

Electric heaters: Electric heaters consist of one or more stages of resistive heating elements in a duct or central furnace. One method of using electric heating in a central system is to put a heating element in the discharge duct leading to each room or zone. This heater is controlled by a space thermostat. Electric duct heaters of this type provide the advantage of individual room control and at the same time allow the use of the central air conditioning, air cleaning, outdoor ventilation air, and humidification.

Figure 16-13 illustrates a system using electric resistance heating. To prevent the discharge of very cold air to conditioned spaces, a heating element is sometimes installed at a central location. This heater brings the mixed-air temperature up to at least 65°F to 75°F when necessary. Even with the zone heaters turned off, the discharge air will not get uncomfortably cold.

Humidifiers[2]:   Humidifiers are usually located downstream from a heating unit of some kind in a central system. This means that the air in contact with the humidifier will be warmer and will be able to absorb the amount of moisture necessary to meet the needs to the conditioned space. Four basic types of humidifiers are commonly used with central-fan systems: (1) steam jet, (2) water pan, (3) water spray or atomizer, and (4) water or air washer.

The proper location, installation, and control of humidifiers are essential to obtain totally satisfactory, trouble-free performance. The primary objective in this regard is to provide the required relative humidity without dripping, spitting, carry-over, or condensation. Liquid moisture, even in the form of damp spots, cannot be tolerated in the system. Aside from the hazards to the structure caused by water in the ducts, there is an even more critical health hazard if breeding grounds are provided for bacteria.

In addition to the need for properly designed, properly performing humidifiers, it is essential that the humidifiers (1) have the proper capacity for the system, (2) be properly located in relationship to other components of the system, and (3) be properly installed and piped in a manner that will not nullify all the other precautions taken.

In the sizing of humidifiers, you should be sure that they will deliver the amount of moisture per hour called for in the design calculations. Steam pressure to steam humidifiers must be kept relatively constant to assure sufficient capacity. Double-check to be sure that you are not trying to put more moisture into the airstream than it can hold at its existing temperature.

Proper location of humidifiers in the system is most important, although sometimes in the design of the system optimum location cannot be achieved.

The *steam-jet humidifiers* of systems 1 through 3, to be discussed next, demonstrate proper steam humidifier location. Features of these systems include (1) accurate control, which is possible because of immediate response of the steam humidifier; (2) control that is modulating electric or pneumatic (illustrated), or two-position; (3) no need for drain pans or eliminator plates, which makes the location of the humidifier more flexible; (4) addition of moisture accomplished with no appreciable change in air dry-bulb temperature; and (5) accurate sizing of the humidifier's integral steam-jacketed control valve with parabolic plug to meet capacity requirements.

System 1 (see Figure 16-14) is a simple ventilation system with primary humidification. We assume that the final duct air temperature is slightly above the desired room temperature. The desirable location of the steam-jacketed distribution manifold of the primary humidifier is downstream of the supply fan. This humidifier would be sized for maximum design load. If the humidifier were located between the heating coil and the fan, it might interfere with the temperature-sensing bulb. The indicated use of a high-limit duct humidity controller as shown is optional. The use of such a controller is advisable if the capacity of the humidifier at design loads could possibly overload the air when outside air moisture content is higher than the design. The high-limit controller should be 10–12 ft downstream from the humidifier. Place the high-limit controller so that it will ''see'' the same temperature as the humidifier. A cooler temperature at the humidifier would allow

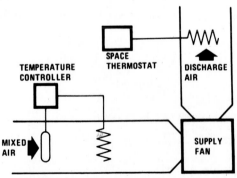

FIGURE 16-13   Electric duct heater.

[2]Figures 16-14 through 16-16 and the associated descriptive matter are from *The Armstrong Humidification Handbook,* Armstrong Machine Works, Three Rivers, MI.

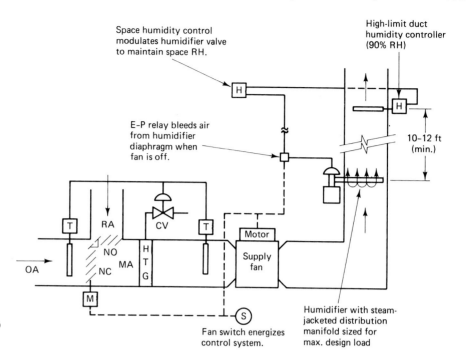

FIGURE 16-14 Simple ventilation system with primary humidification:

EA = exhaust air
E–P = electric–pneumatic relay
Ⓗ = humidity controller
Ⓜ = damper motor
MA = mixed air
NC = normally closed
NO = normally open
OA = outside air
RA = return air
Ⓢ = switch
Ⓣ = temperature controller

Note: These symbols also apply to Figures 16-15 and 16-16.

saturation if the high-limit controller were in warmer air.

The system shown in Figure 16-14 uses a pneumatic control system. The fan switch activates the control system, and the electric–pneumatic relay bleeds air from the humidifier actuator diaphragm when the fan is off. If the control system were electric, control component locations would remain the same.

System 2 (see Figure 16-15) is a typical 100% outside air system with preheat and reheat coils. The pre-

heat coil heats the air from outside design to a duct temperature controlled at, say, 50–60°F. The reheat coil adds more sensible heat depending on the space heat requirement. Here, the desirable location for the primary humidifier is downstream from the reheat coil so as to introduce the moisture into the air where the highest dry-bulb temperature exists. Note the humidity controller location in the exhaust air duct. When a good pilot location for a humidity controller is not available in the space to be humidified, one placed in the exhaust air

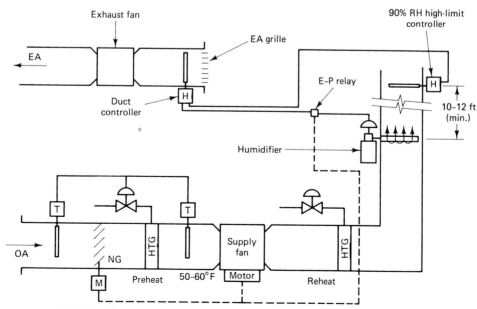

FIGURE 16-15 Typical 100% outside air heating and ventilating system with primary humidification.

duct as close to the outlet grille as possible serves the purpose very well. Again, the high-limit controller is optional but generally recommended.

System 3 (see Figure 16-16) is another 100% outside air system. In this case, the air leaving the preheat coil is held at a constant dry-bulb temperature in the 50–60°F range. This system indicates the use of two humidifiers—one as a primary humidifier and the second as a booster or secondary humidifier.

This system allows a primary humidifier to be controlled directly from a duct humidity controller at a level high enough to maintain a space condition of about 35% RH at a space temperature of 70°F. The booster unit, located downstream from a reheat coil and fan, can then be sized and controlled to produce the necessary moisture to raise the space RH from 35% to some higher condition, say, 55%, where and when desired. This allows individual humidity control for each zone at a higher level than otherwise possible.

This is an important combination because the use of the primary unit allows the capacity of the booster unit to be small enough so that supersaturation and visible moisture will not occur, even when the units are located as close as 3 ft from the discharge grille.

In this typical air-handling system, it would not be psychrometrically possible to introduce enough moisture into the 50–60°F air downstream of the preheat coil to give the maximum required condition in excess of

35% RH in the space. The use of both primary and booster humidifiers is the only method for controlling the relative humidity in a space at any level above approximately 35%.

A *pan-type humidifier* consists of an open pan of water located in the duct of a central air-handling system where the water surface is open to the airstream. The pan also contains a means of heating the water, such as a steam coil, hot water coil, or an electric immersion coil, and a means for controlling the water temperature by varying the heat input. The pan must also have an automatic means for maintaining water level in the device. Rate of evaporation, humidifier capacity, is determined by the water temperature in the open pan. In the pan-type humidifier illustrated in Figure 16-17, the steam coil in the water pan is controlled by a modulating valve. A modulating space humidistat is used to sense room humidity requirements and adjust the valve accordingly. The valve is interconnected to the system supply fan so that it will shut off the steam supply when the central system is off. Pan-type humidifiers have a limited capacity and are used only when humidification requirements are fairly small.

A *water spray,* or *atomizer, humidifier* (see Figure 16-18) consists of a series of nozzles in the duct, through which water is sprayed into the air. The fine mist produced is completely absorbed by the air. Uniform water pressure is necessary to provide proper atomization of

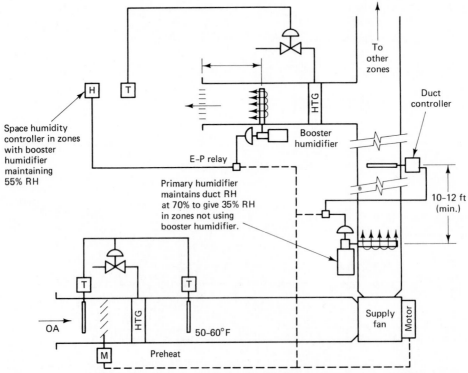

FIGURE 16-16 Typical 100% outside air heating and ventilating system with primary and booster humidification.

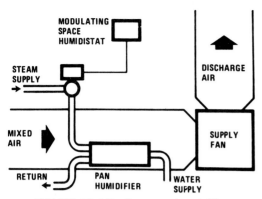

FIGURE 16-17 Pan-type humidifier.

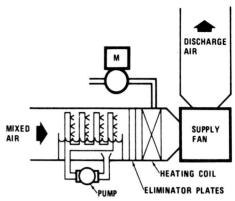

FIGURE 16-19 Air washer humidifier.

the water. For this reason this type of humidifier is controlled only on an on–off mode of control.

The evaporation of water removes a significant amount of sensible heat from the airstream. As the water spray is turned on and off, heating requirements vary widely. This makes close temperature control difficult.

The *air washer humidifier,* illustrated in Figure 16-19, also sprays water into the airstream but is *not* designed to evaporate all the water. Air leaving the washer is nearly saturated so that it is necessary to provide a means of *reheat* to get the correct temperature and humidity. The essential parts of an air washer include the following:

- A mechanism (pump) for circulating and spraying water through the airstream
- An eliminator to remove water droplets from the airstream before reaching the heating coil or fan
- A heating coil to adjust the discharge air temperature

Other arrangements of equipment in air washer types of humidifiers provide a preheat coil and/or heat exchanger to heat the spray water.

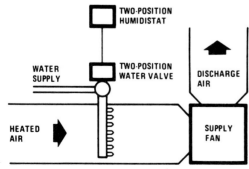

FIGURE 16-18 Water spray humidifier.

16-3 ELEMENTARY UNITARY SYSTEMS[3]

A unitary system is a room unit used in conjunction with a central system. Unitary systems are usually classified according to the heating medium delivered to them by the central system. Three groups are as follows:

- All water and no air
- Part water and part air
- All air and no water

There may or may not be a central-fan system associated with the unitary equipment. A unitary system (see Figure 16-20), in effect, moves part of the equipment and some of the air conditioning functions from the central location to the conditioned area. An advantage of this arrangement is that the air need not be returned to the central system for conditioning. The only air that must be distributed through the building is ventilation (outdoor) air. Sometimes, the function is also accomplished by the unitary system, called a *unit ventilator,* eliminating the need for any central air distribution system.

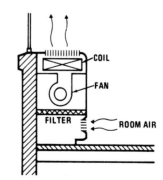

FIGURE 16-20 Unitary system (all water).

[3]Figures 16-20 through 16-23 and the associated descriptive matter (Section 16-3) are from *Basic Control Systems,* Honeywell, Inc., Milwaukee, WI, 1972.

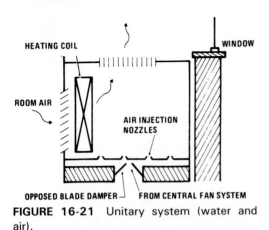

FIGURE 16-21 Unitary system (water and air).

All-Water and No-Air System. Typically called a *fan coil,* this type of unit is normally installed under a window, although many other arrangements are available. A common way of controlling a fan coil unit is to use a room thermostat to control a valve leading to the coil and operate the fan manually by a start-and-stop switch (frequently with speed selection). Quite a number of other control systems are also available. When the unit is used for both heating and cooling, a multiple-speed fan can be used to provide the required air circulation rate in each mode of operation.

Part-Water and Part-Air System. This type of unitary system (see Figure 16–21) receives both air and hot water or steam from a central system. They are also called *induction systems,* either high pressure or low pressure. Air from the central-fan system induces the flow of room air through the unit. No fan is used in the unit.

All-Air and No-Water System. These systems condition the air at a central location and deliver it through two ducts, one cold and one warm, to a mixing unit in the room. The system (see Figure 16-22) is usually high pressure and high velocity and frequently is called "high-velocity, dual-duct." The mixing box is located in the room and mixes the hot and cold air as re-

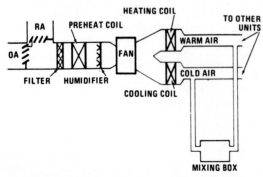

FIGURE 16-22 Unitary system (all air).

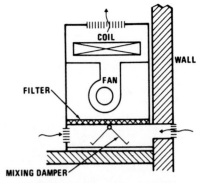

FIGURE 16-23 Unit ventilator.

quired to maintain the desired room temperature. A room thermostat controls a pair of coupled dampers in the two ducts to control the amount of hot and cold air admitted to the room.

Because distribution of conditioned air is at high pressure and high velocity, small-size ducts can be used in the system, saving space and expense. However, silencers must be used to reduce the noise level of the high-velocity air.

Unit Ventilators. Unit ventilators are similar to fan coil units but have the added feature of ventilation (outdoor air) capability (see Figure 16-23). They generally contain a set of outdoor and return air dampers, coil, fan, and filters. The dampers are arranged to provide varying amounts of outdoor air for ventilation, depending on the load. The controls, usually supplied with the unit, are designed to provide a definite sequence of operation. The units are almost always located on outside walls under windows, but other arrangements are available. Some are designed to provide cooling, using chilled water.

16-4 ELEMENTARY PACKAGED SYSTEMS

The term *packaged systems* applies to a complete set of components for air conditioning, or one or more air conditioning functions, manufactured and shipped as a unit for easy installation. Included are warm-air furnaces, boilers, and unit air conditioners. A *packaged multizone* unit is typically installed on a rooftop and provides heating, cooling, and ventilation for a number of different zones within the building.

16-5 TYPICAL AUTOMATIC CONTROL SYSTEMS

In the preceding sections, we have considered some of the possible elementary automatic control systems. In this section, we will discuss some of the more commonly

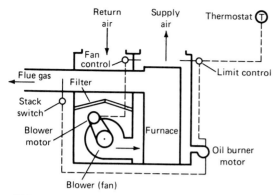

FIGURE 16-24 Control components for forced-circulation warm-air heating system.

found complete systems using representative diagrams with descriptions of such things as special features and control sequences.

Forced-Circulation Warm-Air Furnace. The automatic controls for a forced-circulation warm-air furnace may fall in the category of packaged equipment because most of the controls are installed and wired at the factory, with the exception of the space thermostat. An almost endless number of control arrangements is possible, but the simplest and most practical is that shown in Figure 16-24. The furnace may be gas fired or oil fired. The control components include the following:

1. *Room thermostat* should be within the house in a central location where a representative temperature exists. The thermostat should be provided with a differential adjustment setting not to exceed about $1.0°F$.

2. *Limit control* shuts down the burner whenever the bonnet air temperature becomes excessively high. The recommended setting is between $175°F$ and $200°F$. In the normal control sequence, the limit control is not called upon for action but is available in case of possible overheating of the furnace (safety control). Overheating may occur if airflow through the unit is restricted, perhaps by a partially clogged filter.

3. *Burner motor* is directly connected to the room thermostat circuit. Obviously, the frequency of operation of the burner is dependent upon the setting and sensitivity of the room thermostat. In the warm-air furnace system, unlike a hot water or steam system, there is little lag and at the same time no heat storage. In other words, the heat must be generated only as needed and must be distributed as soon as it is generated. For this reason, the sensitivity of the room thermostat must be acute, and the differential settings must be of the order of $0.5°F$ to $1.0°F$. The purpose of the small differential setting is to cause frequent and short operations of the burner. The length of each burner

cycle should not, however, be less than about 2 min. for a gas burner or less than about 4 min. for an oil burner because shorter operating periods might contribute to large combustion losses due to incomplete combustion at the beginning of each burner operation.

4. *Fan control* located in the supply-air bonnet is a device that turns on the blower motor when the air temperature in the bonnet is high and turns off the blower motor when the bonnet air temperature is low. The recommended setting of the fan control is about $110°F$ for the cut-in point and about $85°F$ for the cut-out point. These settings are adjustable and are of great importance in the successful operation of the entire system.

5. *Stack switch* is another safety device installed in the unit to delay the operation of the burner after a general power failure has occurred. If the power was immediately restored, and if the burner had been in operation, oil sprayed into the combustion chamber could explode.

The normal *control sequence* for the forced warm-air system would be as follows: When room thermostat calls for heat, the oil burner starts; when the bonnet air temperature reaches the cut-in point for the fan control, the fan starts; when room temperature reaches thermostat set point, the oil burner stops. The fan continues to operate until the bonnet air temperature drops to the cut-out point of the fan control.

A statement was made in the section concerning the fan control that settings were important in the successful operation of the system. With a forced-air system, the flow of air is not dependent upon the weather or the heating demands but is fixed only by the blower speed, the blower size, and the frictional restrictions in the duct system. It is possible with any given duct system to vary the airflow rate over a considerable range. Furthermore, it is possible to control over a wide range the number of hours that the blower will operate during the day. For example, on a day in which the outdoor air temperature is $30°F$, the fan control can be adjusted so that the blower runs continuously during the 24-hr period, or for 16 hrs, or for as little as 8 hrs. This wide range of blower operating periods is dependent mainly upon the settings of the fan control and to some extent upon the blower speed. Longer operating periods of the blower are desirable since, when the blower is operating, the control of distribution of heated air is possible; whereas when the blower is not operating, and the system is under the influence of gravity action, no control exists as far as air distribution to the various rooms is concerned.

The argument might be raised that, if continuous air circulation is desirable, the system should be operated to give continuous blower operation by replacing the fan control with a manual switch, which would be used to turn the blower on at the start of the heating season and have it run continuously during the season. Or, the fan control could be left in place and the cut-in point lowered to about 70°F. The two possible objections to this arrangement are

1. That the blower does not need to be operated in extremely mild weather when little heat demand occurs and the electrical cost of blower operation could be saved.
2. That the temperature of the air issuing from the registers can be as low as room air temperature and might create objectionable drafts in front of the registers.

With high sidewall registers, with floor registers, or with baseboard registers that discharge the air sharply upward toward the ceiling, this draft possibility does not exist. With such register installations, the fan control cut-in settings can be lowered below 110°F, and such a lower setting would prove most advantageous from the standpoint of temperature control and uniform distribution of heated air.

In conclusion, the fan control should be set to give nearly continuous fan operation during the coldest weather, with intermittent oil burner operation controlled by the room thermostat.

Forced-Circulation Hot-Water System.   Figure 16-25 represents a schematic layout of the controls normally used for the simple hot-water heating system for small residential applications. Again, as with the forced-circulation hot-air system, the primary controls come with the packaged boiler installed at the factory.

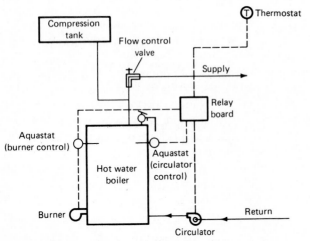

FIGURE 16-25 Control components for forced-circulation hot-water heating system.

The normal *control sequence* for this system would be as follows: Boiler water temperature is maintained at the set point established by the aquastat (burner control); when space thermostat calls for heat, the circulator starts, which forces water through the heating system components. The circulator is also under the control of the aquastat (circulator control), which prevents circulator operation if the boiler water is not up to the aquastat setting. When space temperature is at the space thermostat set point, the circulator stops.

The flow-control valve is a mechanical device that prevents gravity circulation of the water in the system when the circulator is not operating. It is particularly necessary in a summer–winter system where the low-limit control aquastat (aquastat circulator control) maintains a relatively high boiler water temperature, whether heating is required or not.

Zone Control Systems.   As the size of buildings increases, it becomes more difficult to provide good temperature regulation from a single thermostat. Therefore, it is advisable to split the building into two or more *zones,* each controlled by its own thermostat. The zones would be selected according to the conditioning (heating and/or cooling) load imposed by the type of occupancy or use and by the effects of different exposures to sun and prevailing winds. This might be and sometimes is done by providing a separate conditioning plant and control system for each zone. However, *zone control* implies the independent control of the conditioning medium (hot air, chilled air, hot water, chilled water, steam) to each zone from the same central conditioning plant.

The most accurate and satisfactory type of control system for any building, whether residence, school, apartment house, office, or industrial building, is individual control of each conditioned space. A control system of this type includes, in each conditioned space, a thermostat that controls a heating and/or cooling device located in that space. Heating and cooling medium is supplied to these devices from a central plant. Thus, without regard to conditions in other spaces, the controller in each space regulates the amount of heating and/or cooling required by that space only. This form of control, primarily because of the number of devices required throughout the building, is usually most expensive. However, where maximum flexibility and the most accurate control are desired, individual space control provides the most satisfactory results.

Figure 16-26 illustrates the individual room control for a hot-water heating unit. This type of control could be applied to each room throughout a building. It is designed for two-position (on–off) control from a room thermostat. It has the special feature, provided by

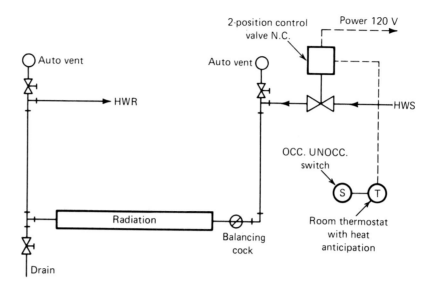

FIGURE 16-26 Control components for an individually controlled hot-water radiation unit.

the *OCC–UNOCC switch,* of permitting the radiation unit to be operated to maintain a lower room temperature when the space is unoccupied (nighttime, holidays, etc.). The sequence of operation would be as follows:

1. *Occupied*—A heat-anticipating room thermostat will open or close the two-position motorized valve to control the room temperature at the set point.
2. *Unoccupied*—The heat-anticipating room thermostat will control the two-position motorized valve to maintain a lower set point temperature (say, 10°F below the occupied set point).

Zoned hot-water heating system using multiple circulators or multiple-zone valves: Two popular methods for zone control of forced-circulation hot-

water systems for large residential or small commercial buildings are illustrated in Figures 16-27 and 16-28. Both represent only two zones but could easily be expanded to include other zones.

Figure 16-27 represents a portion of a two-zone system using a separate thermostat and circulator for each zone. Basically, it represents two hot-water systems connected to the boiler that operate independently of each other.

Figure 16-28 represents a portion of a two-zone system using a separate thermostat and two-position motorized valve for each zone and one circulator for the combined system. Each motorized control valve is provided with an auxiliary switch. The auxiliary switches are connected in parallel to the single circulating pump. A call for heat by either zone thermostat causes the respective zone valve to open and start the circulator. If neither zone is calling for heat, the two zone valves would be closed and the circulator stops.

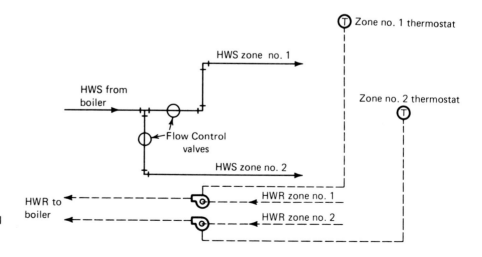

FIGURE 16-27 Zone control components for two-zone hot-water heating system (individual circulators).

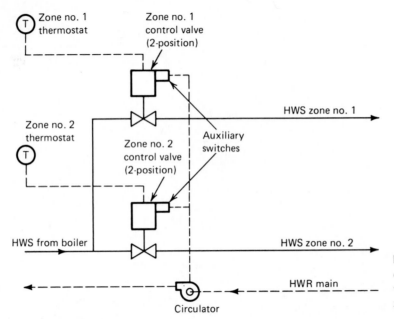

FIGURE 16-28   Zone control components for two-zone hot-water heating system (individual zone valves).

Typical Reset Hot-Water and Pump Control (Electronic). Figure 16-29 illustrates a control system used to adjust the temperature of the supply water in a hot-water heating system (upward as outdoor air temperature drops, downward as outdoor air temperature rises). This type of system is frequently used as part of the control system for hot-water heating in commercial and industrial buildings. It is normally not used in residential applications because of the additional cost of the system. The sequence of operation of the system shown in Figure 16-29 is as follows:

1. An outdoor sensor will reset an immersion sensor in the supply water to provide 80°F

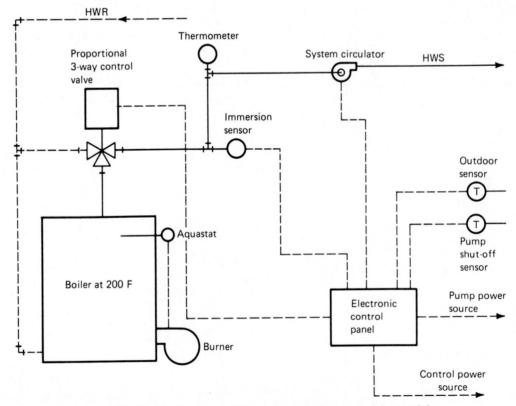

FIGURE 16-29   Reset hot-water control system (electronic).

supply water at 70°F outdoor air temperature, increasing to 200°F supply water as outdoor temperature drops to −20°F. Bridge, amplifier, and solid-state drive for proportioning three-way control valve and pump relay will be mounted in prewired panel.

2. Boiler water temperature is maintained at 200°F by aquastat recycling of oil burner.

3. Outdoor air temperature sensor will start hot-water circulating pumps when outdoor air temperature falls below 70°F and stop pump when outdoor air temperature rises above 70°F.

Control of Fan Coil Units (Typical). Many types of fan coil units are available; however, as the name implies, they all have a heat-transfer coil and a fan to circulate air through the coil and to distribute the conditioned air. Some units are designed for heating only, and some may be arranged for either heating or cooling.

Controls for fan coil units can either be (1) *on–off operation* of the fan or (2) *continuous fan operation* with modulation of the heating or cooling medium.

For on–off operation, a room thermostat is used to start and stop the fan motor, as illustrated in Figure 16-30. A limit thermostat, often strapped to the supply or return pipe, prevents fan operation in the event that heat (heating unit) is not being supplied to the unit. An auxiliary switch that energizes the fan only when power is applied to open the motorized supply valve may also be used to prevent undesirable cool air being discharged by the unit.

Continuous fan operation eliminates the intermittent blasts of hot air resulting from on–off operation, as well as the stratification of temperature from floor to ceiling that often occurs during the off period. In this arrangement, a proportioning room thermostat and valve modulate the heat supply to the coil (see Figure 16-31) or a bypass around the heating element. A limit

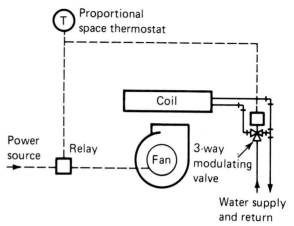

FIGURE 16-31 Control system for fan coil unit (constant fan operation, modulated heating-medium temperature).

thermostat, or auxiliary switch on the control valve, stops the fan when heat is not available in the heating coil.

The unit controls shown in Figures 16-30 and 16-31 would also apply if chilled water was supplied to the coils, assuming that the thermostat provided for heating–cooling changeover.

One type of control, illustrated in Figure 16-32, used with blow-down unit heaters is designed to automatically return the warm air, which would normally stratify at the higher level, down to the occupied zone. Two thermostats, a two-position control valve, and an auxiliary switch are required. The lower thermostat is placed in the occupied zone and used to control the two-position supply valve to the heating element. The auxiliary switch is used to stop the fan when the supply valve

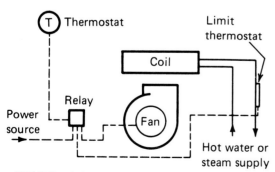

FIGURE 16-30 Control system for fan coil unit (cycle fan, constant heating-medium flow).

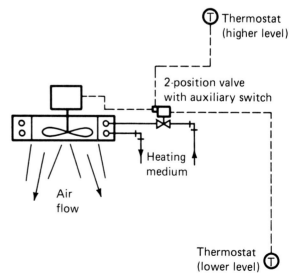

FIGURE 16-32 Control system for down-flow unit heater.

is closed. The higher thermostat is placed near the unit heater at the ceiling or roof level where the warm air tends to stratify. The lower thermostat will automatically close the steam (water) two-position valve when its set point is reached, but the higher thermostat will override the auxiliary switch to let the fan continue to run until the temperature at the higher level falls below a point sufficiently high to produce a heating effect.

Control of Unit Ventilators (Typical).   For many years, unit ventilators have been used for the winter heating–ventilating cycle for school classrooms. This has led to the general name *classroom unit ventilator.* They have also found application in many commercial and industrial buildings. With the addition of cooling coils and appropriate control, the unit is equipped for year-round control of the environment within building spaces.

The basic construction and essential operating elements of a typical unit ventilator are illustrated schematically in Figure 16-33. The other types in common use are functionally the same, although they may differ slightly in the arrangement of dampers, coils, blower units, and so forth. For instance, a bypass damper may be used to bypass air around the coil when no heat is required.

Referring to Figure 16-33, note that the unit ventilator has all the necessary elements to deliver air to the room at any temperature desired. At the bottom of the unit is found a dual damper arranged so as to vary the relative volume of outdoor air and recirculated air admitted to the unit. As the outdoor air damper opens, the return air damper closes proportionally. By positioning the dampers, it is possible to admit any proportion from 0 to 100% of outdoor air to the unit. A blower unit, consisting of two or more centrifugal blowers, draws a constant volume of air through the unit.

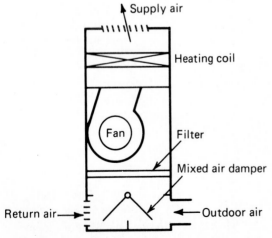

FIGURE 16-33   Unit ventilator components.

A heating coil, supplied with steam or hot water and equipped with a control valve, heats the air leaving the unit. Electric heating coils may also be used. It is obvious that changing the position of the heating coil and placing it below the fan (draw-through) does not alter its effect or change the control requirements for the unit.

Many different control cycles for unit ventilators are available. The principal difference in the various cycles pertains to the amount of outdoor air delivered to the room. Usually, a room thermostat simultaneously controls a coil valve, damper, or step controller to regulate the heat supply and a damper to regulate the outdoor air supply. An airstream thermostat in the unit prevents the discharge of air below a desired minimum temperature. Unit ventilator control cycles provide the proper sequence for the following stages:

1. *Warm-up*—All standard unit ventilator cycles automatically close the outdoor air damper whenever maximum heating capacity is required. Then, only room air is circulated, thereby providing full unit capacity for rapid warm-up until the room temperature approaches the desired level.

2. *Heating and ventilating*—As the room temperature rises into the operating range of the thermostat, ventilation is accomplished through the partial or complete opening of the outdoor air damper, depending on the cycle used. Auxiliary heating equipment (fin tube, if used) is shut off. As the room temperature continues to rise, the heat supply to the coil is throttled.

3. *Cooling and ventilating*—When the room temperature rises above the set point level, cool air is discharged into the room. The room thermostat accomplishes this by throttling the heat supply, finally shutting it off, and then opening the outdoor air damper, as required, to prevent the room from overheating. The airstream thermostat frequently takes control during this stage to keep the discharge air temperature from falling below a set level. The unit cooling capacity is determined by the air delivery and difference between the outdoor air and room air temperatures. Cooling obtained in this manner is called "free cooling," "ventilation cooling," or "natural cooling."

ASHRAE control cycles: There are three basic cycles used, with variation, for unit ventilator control. They are referred to as ASHRAE Cycles I, II, and III.

ASHRAE Cycle I—During warm-up, the outdoor air damper is closed, and the unit handles 100% recirculated air. As room temperature approaches the thermostat set point, the outdoor air damper opens fully, and the unit handles 100% outdoor air. Unit capacity is then controlled by modulating the heating coil capacity.

ASHRAE Cycle II—This is the most widely used heating and ventilating cycle. During warm-up, the outdoor air damper is closed, and the unit handles 100% recirculated air. As the room temperature approaches the thermostat set point, the outdoor air damper opens to admit a predetermined minimum amount of outdoor air (usually between 25% and 50%). This minimum is determined by local code requirements and good engineering practice to provide adequate ventilation. Unit capacity is controlled by varying the heating coil output. If the room starts to overheat, the heating coil is first turned off, and then a gradually increasing amount of outdoor air is admitted until only outdoor air is being delivered.

This cycle usually incorporates a minimum discharge air temperature sensor that overrides the other controls to maintain an acceptable discharge temperature. When the outdoor air is very cold, the minimum air temperature control modulates the amount of outdoor air being delivered. This keeps the mixture temperature delivered to the room at 55°F to 60°F.

Figure 16-34 illustrates schematically the control system components for ASHRAE Cycle II.

ASHRAE Cycle III—During warm-up, the outdoor air damper is closed, and the unit handles 100% recirculated air. As room temperature approaches thermostat set point, a thermostat on the entering side of the heating coil modulates the outdoor air damper to maintain a constant mixed-air temperature entering the heating coil. The unit heating coil heats as needed to maintain the room thermostat set point.

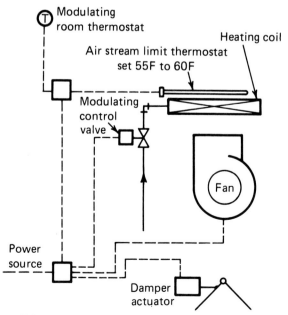

FIGURE 16-34 Control system components for ASHRAE Cycle II.

Cooling: Air conditioning unit ventilators can utilize any of the ASHRAE control cycles; however, when cooling, they normally operate with a fixed minimum of outdoor air as determined by code requirements or good engineering practice. This fixed minimum can either be the same or different from that used when heating. When cooling, the amount should equal the minimum required to give good ventilation, as excessive amounts of outdoor air greatly increase the air conditioning load and operating costs. Hydronic (chilled water) air conditioning unit ventilators can have capacity control with either a modulating two- or three-way valve or face-and-bypass dampers.

Unit ventilators using direct expansion (DX) cooling coils are controlled by cycling the condensing unit in response to a room air thermostat. In addition, a frost-protection device prevents frost buildup on the cooling coil during greatly decreased unit air delivery. This is a requirement because operation with a sizable portion of cool, wet outdoor air might have a tendency to form frost even though the unit ventilator and condensing unit are properly selected for the application design temperature.

Unit ventilator cooling control cycles should include provisions for cooling down a hot room on start-up by closing the outdoor air damper and operating on 100% recirculated air.

Changeover: Changeover from heating–ventilating to cooling can be controlled a number of ways. Changeover to mechanical cooling should occur when ventilation cooling on 100% outdoor air does not overcome the design heat loads in the room. Usually, this is in the neighborhood of 58°F outdoor air temperature.

Changeover can be done in individual units, as is common in four-pipe hydronic, hydronic cooling/electric heat, or direct expansion/electric heat applications. It can also be done by zones or the entire system on two-pipe hydronic systems.

Night cycle: For maximum economy, it is general practice to maintain the building at reduced temperature (heating) at night, over weekends, and vacations. At such times, the outdoor air dampers close, and the unit handles only recirculated air. Changeover is usually done with a manual switch centrally located in the building. Some systems are arranged so that individual units may be manually returned to the occupied cycle by providing an override switch for such units. The remaining units would stay on the night cycle.

The unit capacity on the night cycle can be controlled by cycling the unit fan or by modulation of coil valve or face-and-bypass dampers. Occasionally, a system is designed to utilize the natural convective capacity of the unit ventilator along with auxiliary direct radiation, if available. The natural convective capacity of unit ventilators is approximately 10% of their normal capacity.

Reset water schedule: Unit ventilators with face-and-bypass control should be selected for minimum heat pickup in the bypass position. However, all unit ventilators using face-and-bypass control require water temperature reset systems if full ventilation cooling capacity is to be assured.

The reset schedule should provide water at approximately 100°F when the outside air temperature is 60°F. This will provide adequate heating capacity for mild temperature conditions yet prevent any significant loss of ventilation cooling capacity.

If a reset water schedule is not provided, automatic valves should be installed to shut off water to the coil when maximum ventilation cooling capacity is required.

Auxiliary radiation: Auxiliary radiation (usually fin pipe) is sometimes recommended to prevent window downdrafts when (1) single-glazed windows are used, (2) windows extend over 40% or more of the window sill length, and (3) outdoor air temperatures are anticipated to be below 30°F for extended periods of room occupancy.

When all three conditions exist, heat loss through the windows at design conditions should be calculated and sufficient radiation installed below window sills to offset it. In heat loss calculations, infiltration can be ignored as unit ventilator systems normally cause a room to be at a slightly positive pressure.

Careful control of auxiliary radiation is important if room overheating is to be prevented. During certain periods, sun load, combined with heat given off by lights and occupants, will more than offset the total heat loss even though the outdoor air temperature is low enough to require downdraft protection. Unless properly sized and controlled, auxiliary radiation heat, combined with heat from other sources, may be so great that it exceeds the ventilation cooling capacity of the unit ventilator, thus making it impossible to maintain comfortable room temperatures.

If full ventilation cooling capacity is required, auxiliary radiation should be separately controlled, either by temperature scheduling or water flow modulation. This can be done either on the basis of individual rooms or zones, although individual room control is preferred.

Electric heat: Electric heat, in either heating only or cooling–heating units, is normally controlled by a pneumatic or an electric step controller. This device individually energizes the electric heat elements as re-

quired to maintain comfort conditions. Thus, the steps of modulation equal the number of electric elements provided. Standard heating and ventilating cycles are available with all electric heating units.

16-6 CONTROL OF COMMERCIAL CENTRAL-FAN HEATING SYSTEMS

In Section 16-2, we discussed some of the basic features of the central-fan system. Central-fan systems for heating and air conditioning consist of many separate pieces of apparatus, designed to provide any or all of the following functions: (1) heating, (2) humidification, (3) ventilation, (4) distribution, (5) air cleaning, and (6) cooling by use of cool outdoor air whenever necessary.

Many variations in the physical arrangement of equipment are used to accomplish these functions. Selection of a method that will prove most satisfactory for a specific installation depends upon local practice, local climatic and economic factors, and good engineering judgment.

When more than one function is to be accomplished by a single system, particular care must be exercised in arranging and interlocking the equipment so as to provide a completely coordinated sequence of operation. The use of automatic temperature control provides a satisfactory and economical means of accomplishing this coordination.

Central-fan winter heating and ventilating units may be divided into two general types, namely, (1) *tempering systems* for ventilation only, in which the mixture of outdoor air and return air is delivered at space temperature for ventilation only, the actual heat losses of the space being provided for by means of direct radiation; and (2) *heating and ventilating systems* that utilize sufficient heat-transfer surface in the unit to provide heat not only to temper the mixture of outdoor and return air, but also to provide for the heat losses from the conditioned space.

Automatic control: Because of limited space here, we shall consider only one of many possible control arrangements for a central-fan heating system. Pneumatic or electric–electronic systems may be used for central-fan systems.

Figure 16-35 illustrates a system for winter control of outdoor air quantity, with notable features of economy. Provisions have also been made for (1) closing the outdoor air damper when the fan stops, (2) a manually adjustable minimum quantity of outdoor air when the fan is in operation, (3) use of additional outdoor air for ventilation cooling when required, and (4) humidification. For purposes of illustration, a heating coil with

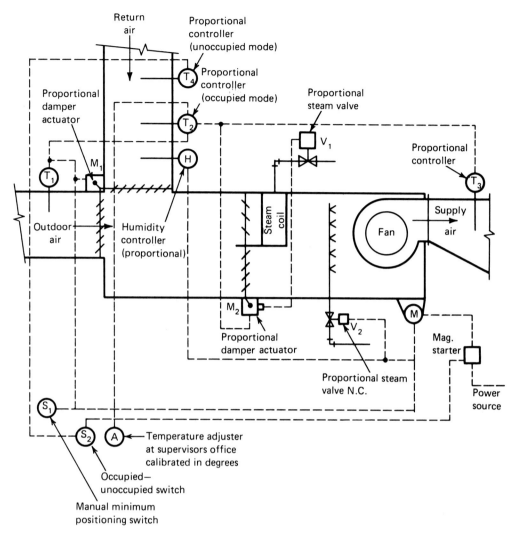

FIGURE 16-35 Control system for central-fan heating and ventilating system.

face-and-bypass dampers has been shown. This method of control may be used with any other method of providing heat. The sequence of operation is as follows:

1. When the fan is not operating, the outdoor air damper is automatically kept in the closed position.

2. During the pickup period, with the fan running, the outdoor air damper opens to a minimum position, which may be manually adjusted by a minimum-position switch S_1. This switch may be located at any remote point desired for easy adjustment.

3. The steam valve V_1 and the face-and-bypass dampers are modulated to provide the necessary amounts of heat called for by the proportional return air controller T_2, which is adjustable by temperature adjuster A.

4. As the return air temperature rises, the return air controller T_2 gradually throttles the steam valve V_1

and opens the bypass damper to provide less heat to the space.

5. Should the return air thermostat T_2 become satisfied and thus call for a closed steam valve V_1, the discharge controller T_3 modulates the valve and the face-and-bypass dampers to provide just enough heat to maintain a minimum discharge temperature of, say, 65°F. This temperature should be sufficiently low to provide cooling when required. If, however, the temperature of the mixed outdoor and return air should be 65°F or higher, so that no heat is required to maintain a discharge temperature of 65°F or more, the proportional controller T_3 allows the steam valve V_1 to close tight.

6. If the return air temperature rises no further, the system then operates with the outdoor air damper at its minimum position and with all steam to valve V_1 shut off.

7. If the internal heat in the space increases sufficiently to cause overheating, the outdoor air damper actuator M_1 is operated at the command of the proportional return air controller T_2. The set point of T_1 is adjusted for gradual opening of the outdoor air damper on a rise in outdoor air temperature and to return the damper to minimum position as the outdoor air temperature falls to a point where the minimum percentage of outdoor air would be sufficient to maintain a discharge temperature of 65°F.

For example, in a system designed for a minimum of 20% outdoor air, and with the return air at 70°F, the mixture of outdoor and return air would be at 65°F when the outdoor air temperature reached 45°F. However, as the temperature rises above 45°F, it is necessary to use more than the minimum amount of outdoor air if a 65°F discharge is to be maintained. For this condition, the outdoor air controller T_1 would be set to have the outdoor air damper wide open at 65°F and to close it gradually to the minimum position as the outdoor air temperature falls from 65°F to 45°F.

8. As long as the return air temperature does not indicate an overheated condition, the outdoor air damper is held at the minimum opening as set by the manual minimum-position switch S_1.

9. A proportional humidity controller H operates a proportional steam valve V_2 to control the steam flow for humidification. The valve is interconnected with the fan motor circuit so that it is tightly closed when the fan is stopped to prevent condensation of moisture in the ducts.

10. The manually operated occupied–unoccupied switch S_2 places the system under control of proportional controller T_4 for the unoccupied mode. Controller T_4 cycles the unit to maintain a reduced temperature in the space.

16-7 CONTROL OF YEAR-ROUND (SUMMER–WINTER) SYSTEMS

Where air conditioning equipment is provided for year-round air conditioning, it is desirable to exercise particular care in the selection of automatic control equipment and to pay some additional attention to the automatic control aspect of the equipment design problem.

Changeover: The combination of winter and summer conditioning equipment into one installation introduces an additional control problem. If all elements of the air conditioning system perform single functions, for example, if separate heating and cooling coils are used, the changeover from heating to cooling

or from cooling to heating might conceivably be accomplished by allowing the separate heating and cooling controls to function independently. On a rise in temperature, say, the heating thermostat would shut off the flow of steam to the heating coil, and, on a further rise, the cooling thermostat would start the operation of the cooling equipment.

But it is not quite as simple as that in most cases. For one thing, it is often necessary or desirable to utilize one piece of equipment in both the heating and the cooling cycle. A coil is sometimes used both for heating, with hot water as the medium, and for cooling, with chilled water. Means must then be provided for changing over from one source to the other. Again, the outdoor and return air dampers are normally used in both cycles, but the damper actuator must be positioned according to different schedules during the heating and cooling cycles. Provision must then be made for transferring command of the damper actuator from one controller schedule to the other.

In addition, even the single-purpose equipment elements may require special consideration of changeover means. Even if it is practical to have both heating and cooling capacity available at all times, suitable schedules of control sequence for heating and for cooling may not permit leaving the choice of heating or cooling operation to the independent operation of separate heating and cooling controllers. It is usually necessary, therefore, to provide for definite changeover, by means of suitable valves or switches, from one mode of operation to the other.

Manual versus automatic changeover: If it were always possible to determine that today is the last day of the heating season and tomorrow the cooling season begins, manual means of changeover would be entirely satisfactory. However, during late spring and early fall periods, there are times when the demands made upon the system may change, from day to day or even during the day, from a heating to a cooling load and vice versa. During the mild seasons, it is often necessary that heat be supplied during the morning and evening hours and at night if the building is occupied then; whereas temperature and humidity conditions during the day require cooling operation or at least dehumidification. In some applications, in fact, where very large internal heat gains may occur intermittently, this condition may occur during mild weather periods in the winter.

Even though a manual changeover can be accomplished easily, if it must be left to the judgment and alertness of the building engineer, some inefficiency and discomfort will often result. For these reasons, it is generally desirable to provide for automatic changeover. Electric or pneumatic switches and valves may be operated automatically by thermostats or humidity controllers located within the conditioned space or outdoors according to the requirements of the specific application. These changeover controls may reverse the action of actuators and valves, transfer command of actuators or valves from one set of controllers to another, or divert the flow of a medium such as water in order to convert the mode of operation from heating to cooling or vice versa.

If both the air conditioning system and the control system are designed with automatic changeover in view, the inclusion of this feature involves only slight additional investment. In view of the gains in operating economy and in comfort of occupants, this investment may be considered self-liquidating.

Year-Round Air Conditioning System (Electronic). Figure 16-36 illustrates only one of the many possible year-round air conditioning systems that includes heating, ventilating and cooling and the associated controls that could be used for a large space within a building (say, a school auditorium). The heating is provided by a hot water coil controlled by a three-way valve. Cooling is provided by a direct expansion (DX) coil with two stages of cooling. The control system provides for certain special features for the specific application, such as the following:

1. Manually selected occupied–unoccupied operation.

2. Fan operates continuously during occupied control sequence to provide good air circulation.

3. Fan cycles on–off during unoccupied control sequence to maintain lower space temperature.

4. Provides for cooling by using outdoor air (economy), which means a third cooling stage is available without the required use of refrigeration.

5. Starting, stopping, occupied–unoccupied, and temperature control by single supervisor.

The control sequence for the system shown in Figure 16-36 would be as follows:

1. *Occupied*—Occupied switch starts fan, which runs continuously. Mixed-air dampers move to maintain 65°F mixed-air temperature. Return air sensor T_1 positions three-way valve to maintain room temperature at setting of remote adjuster A. Three-way valve has a flat cam to provide dead-band between heating and cooling. When return air relative humidity rises above 50%, return air humidistat will energize one stage of DX cooling. When space requires cooling, first

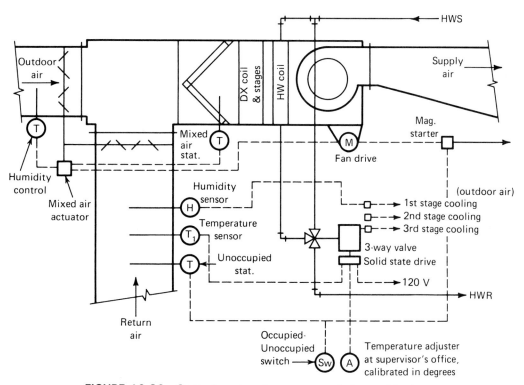

FIGURE 16-36 Control system for year-round air conditioning system (electronic).

stage will open outdoor air damper subject to total heat of entering air (enthalpy control). A further call for cooling will energize the first and second DX cooling stages as required.

2. *Unoccupied*—Unoccupied switch energizes the unoccupied return air thermostat, which cycles the unit to maintain space temperature at 60°F.

16-8   SELECTION OF CONTROL VALVES[4]

In the preceding sections, we have seen the use of control valves and dampers in air conditioning systems. The selection of valves and dampers must be done with care and knowledge of their operating characteristics if we are to have a satisfactorily, controllable air conditioning system. We shall first discuss the fundamentals of valve selection for heating–cooling applications. The benefits derived from proper selection of type of valve and its size are (1) greater *comfort* by providing even temperatures, (2) greater *economy* by using energy more efficiently, and (3) greater *life of equipment* by proper cycling.

A control valve regulates the flow of a liquid or gas in a system. This regulation is accomplished by varying the resistance that the valve introduced into the system as it is stroked. The increasing resistance causes the total system pressure drop to shift to the valve and in this way vary the flow in the system.

As the important function of the control valves in a system is becoming more fully realized, valves are receiving much more attention regarding their proper sizing and application. In the past, many valves were installed line size, and no consideration was given to the controllability of the valve. This usually produced oversized valves that were only performing correctly when they were fully open or fully closed. This, in addition to valves installed backwards, is one of the reasons for the shock waves in older heating systems that produce the annoying hammering and knocking that is all too familiar.

Theory of Coils.   A control valve is used to vary the flow quantity of heating fluid to a heating coil. As the flow through the coil is varied, the heat output of the coil does not vary in a linear manner. The reason for this is different in steam coils than it is in water coils.

Water coils: The heat output of a water coil varies with flow in a manner as shown in Figure 16-37. This is true because at full rated flow the water temperature drop (Δt) as it passes through the coil has a certain set value. But as less heat is required, the control valve is partially closed, and the flow through the coil is decreased. This decreases the water velocity through the coil tubes; the water then remains in the coil for a longer period of time and consequently has a larger temperature drop (Δt). For this reason, each gallon of water flowing gives off more heat, and the total output of the coil is not linearly related to the flow of water through it. This nonlinearity can be compensated for by selecting a control valve characteristic that will offset the characteristic of the coil.

An *equal-percentage valve* has this type of characteristic. When the coil and equal-percentage valve are used together, the combined characteristic approaches linearity (see Figure 16-38). This combination characteristic curve is arrived at by combining the individual characteristics of the heating coil and the equal-percentage plug. This characteristic approaches linear but is not quite so. When the effect of the increasing pressure drop across the valve is also added to this plot, the total characteristic goes slightly beyond linear. However, this characteristic is close enough to be considered linear from the controller's point of view.

What has been said about hot-water coils can also be said for chilled-water coils, except that the heat transfer is in the other direction.

Steam coils: Steam coils can act the same way as water coils but for a different reason. A steam coil producing its maximum heat output is full of steam that condenses and gives off heat. As less heat is required, the control valve begins to throttle back and decrease

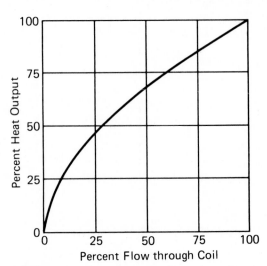

FIGURE 16-37   Nonlinear characteristic of a heating coil.

[4]The majority of the following discussion of control valves and their application is from *Engineering Data,* "Control Valves," Section Vb, Johnson Controls, Inc., Milwaukee, WI, 1980.

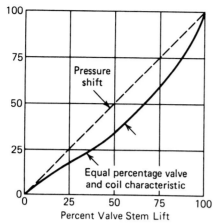

FIGURE 16-38 Total heat-transfer characteristic for control valve and coil combination.

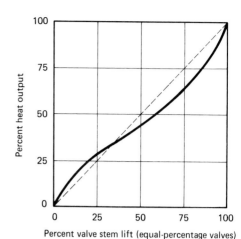

FIGURE 16-40 Heat output of steam coils when controlled by equal-percentage valves.

the flow of steam. The steam that is in the coil condenses and the pressure in the coil drops; this causes a greater pressure drop across the valve and the steam velocity increases. Because of this, the quantity of steam supplied to the coil is almost as great as it was before, and the heat output has only decreased a small percentage for a large movement of the control valve. This action will continue until the critical pressure drop is attained across the valve. From this point on, any further increase in pressure drop across the valve will not cause an increase in steam velocity; the maximum velocity was attained at the critical pressure drop. Because of the constant velocity under these conditions, the valve's performance will follow its inherent characteristic, and the heat output of the coil will resemble this characteristic. A *linear valve* on a steam coil will produce a total characteristic curve as shown in Figure 16-39, and an equal-percentage valve will produce the curve of Figure

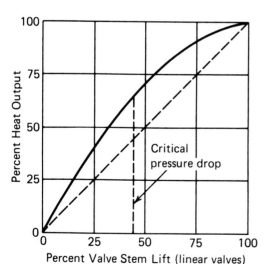

FIGURE 16-39 Heat output of steam coils when controlled by linear valves.

16-40. The equal-percentage plug more closely resembles a linear characteristic. Because the desired total characteristic of the control valve and coil is linear, the equal-percentage plug would be more desirable.

The preferred way to match a steam control valve and coil would be to size the valve so that it has critical pressure drop across it under full-load conditions. In this way, the flow of steam to the coil will depend entirely upon the free area between the valve plug and its seat. This area is determined by the characteristic of the valve and the position of the plug. Under these linear conditions, the use of a linear valve characteristic would produce the desired system characteristic. Unfortunately, this can sometimes require a larger pressure drop than is allowable for the control valve.

The majority of heating systems must operate most of the time in a stage of low heating capacity. For this part of the valve's stroke, the equal-percentage valve is quite nearly linear and possesses a relatively small valve gain. These facts make an equal-percentage valve very controllable and just as desirable as a linear valve for this type of application.

Valve Sizing. Control valves must be sized correctly to perform the job for which they are intended. Undersized valves cannot deliver sufficient quantities for maximum-load conditions, and oversized valves attempt to perform correctly but must do so at the very end of their strokes where hunting or cycling is difficult to avoid. Oversizing is definitely the most prevalent because in every step of putting the system together safety factors are used and the final result is a system that is well oversized—control valve included. Correct sizing, however, is not very complicated. Once the correct characteristic is chosen, a few simple equations will give the desired flow coefficient for sizing the valve.

Valve flow coefficient: The first step in finding the size of a valve is to determine the flow coefficient (C_v) that is required for the system. Again, repeating the definition as the number of gallons per minute of 60°F water that will flow through a *fully open valve* with a 1.0 psi pressure drop across it, the flow coefficient is determined by the construction of the valve and will not change. Identical valve sizes may have different flow coefficients if the body style or valve trim is different. The flow coefficient is probably the most useful piece of information necessary to size a valve.

Sizing water valves: The required flow coefficient (C_v) for a water control valve may be found from the following relationship:

$$C_v = \frac{\text{gpm}}{\sqrt{\Delta p}} \qquad (16\text{-}1)$$

where

    gpm = water flow required (gallons per minute)
    C_v = valve flow coefficient
    Δp = pressure drop between inlet and outlet of valve (psi)

The required water flow (gpm) for the coil should be available in the specifications or calculated from the heating or cooling requirements of the system.

Pressure drop across the valve (Δp): A valve and coil should be sized to produce a combined characteristic that is as close as possible to linear so that the controller can do an efficient job of controlling, as previously noted. A valve is matched with a coil based on the characteristic and flow coefficient of the valve. These valve parameters are determined at constant pressure drop, and, consequently, the valve should be operated at as close to a constant pressure drop as is possible. However, as the valve closes, the total system pressure drop shifts to the valve. This hardly approaches a constant pressure drop. The best that can be done is to keep the *relative change* in pressure as low as possible.

For example, in a 20-psi system (Δp_2), if the open valve has a 7-psi drop (Δp_1) initially, then the change in Δp as the valve closes would be ($\Delta p_2 - \Delta p_1$)(100)/Δp_1 or $(20 - 7)(100)/7 = 186\%$. A valve with an initial drop of 3 psi would have a relative change of $(20 - 3)(100)/3 = 567\%$, which would produce a much larger shift in characteristic. Therefore, the valve pressure drop at maximum flow should be as large as is *practical* for the system. This means that other system components require certain values of pressure drop to operate efficiently. Often, a maximum valve pressure drop is stated, and this value cannot be exceeded. When it is permissible to choose the pressure drop for the control valve, a value equal to 50% or greater of the pressure between the supply and return mains should be selected.

For liquids other than water, a correction for the difference in specific gravity of the liquid is necessary, and Eq. (16-1) becomes

$$C_v = \frac{\text{gpm}\sqrt{\text{SG}}}{\sqrt{\Delta p}} \qquad (16\text{-}2)$$

where SG is the specific gravity of the liquid.

ILLUSTRATIVE PROBLEM 16-1

What is the flow capacity of a 1-in. water valve used in a water system where the design pressure drop across the valve is 15 psi and the valve manufacturer quotes a flow coefficient of 8.6 for the valve?

Solution: By Eq. (16-1),

$$C_v = \frac{\text{gpm}}{\sqrt{\Delta p}}$$

$$\text{gpm} = C_v\sqrt{\Delta p} = 8.6\sqrt{15} = 33.3$$

ILLUSTRATIVE PROBLEM 16-2

Water is to flow through a control valve at the rate of 98 gpm with a pressure drop of 4.0 psi. What flow coefficient would be required for the valve?

Solution: By Eq. (16-1),

$$C_v = \frac{\text{gpm}}{\sqrt{\Delta p}} = \frac{98}{\sqrt{4.0}} = 49$$

Select a valve size having a $C_v = 49$ from a valve manufacturer's catalog.

Sizing steam valves: The procedure for sizing steam valves is very similar to that for water valves. The formulas are almost the same except for a few slight variations. It is recommended that steam control valves be sized for full pressure drop when the inlet pressure is 10.0 psig or below. When the inlet pressure is above 10.0 psig, the critical pressure drop should be used in sizing the valve. The critical pressure drop is equal to 45% of the absolute inlet pressure. Therefore, the pressure drop would be

$$\Delta p = 0.45 p_1$$

where

    Δp = pressure drop used to determine flow coefficient
    p_1 = absolute inlet pressure (psia)

Although the specific guide here recommends specific pressure drops to use in determining the C_v for a
steam valve, it would be better to look at a more general
set of equations for other pressure drops, and, if the
steam is superheated (rarely in heating systems), we
would have the following two relationships:

1. *When p_2 is 0.45 of p_1 or less,*

$$C_v = \frac{\dot{m}_s[1 + 0.0007(\text{deg. F superheat})]}{1.8p_1} \quad (16\text{-}3)$$

2. *When p_2 is more than 0.45 of p_1,*

$$C_v = \frac{\dot{m}_s[1 + 0.0007(\text{deg. F superheat})]}{2.1(p_1 - p_2)(p_1 + p_2)} \quad (16\text{-}4)$$

where

$$\dot{m}_s = \text{steam flow (lb/hr)}$$
$$\text{deg. F superheat} = \text{degrees of superheat}$$
$$p_1 = \text{absolute inlet pressure (psia)}$$
$$p_2 = \text{absolute outlet pressure (psia)}$$

If the steam contains moisture (wet steam), it has
a quality less than 100%, and the C_v value should be
corrected for this condition. If the percent moisture is
known, say, 2%, then the quality is $1.0 - 0.2 = 0.98$
or 98%. The C_v value obtained from either Eq. (16-3)
or Eq. (16-4) should be corrected by multiplying by the
square root of the quality.

Other saturated vapors to be controlled in HVAC
systems include refrigerants. Equations (16-3) and
(16-4) may be used if the appropriate constant appear-
ing in the denominator is selected for the particular vapor. Such constants, K, appear in Table 16-1 and may
be used in the following two relationships:

1. *When p_2 is 0.45 of p_1 or less,*

$$C_v = \frac{\dot{m}_v}{0.86K(p_1)} \quad (16\text{-}5)$$

2. *When p_2 is more than 0.45 of p_1,*

$$C_v = \frac{\dot{m}_v}{K\sqrt{(p_1 - p_2)(p_1 + p_2)}} \quad (16\text{-}6)$$

where

$$\dot{m}_v = \text{mass of saturated vapor (lb/hr)}$$
$$K = \text{constant from Table 16-1}$$

TABLE 16-1

Constants, K, for saturated vapors

Vapors	K
Freon 12	7.1
Freon 11	7.4
Freon 14	8.4
Freon 114	8.3
Dowtherm A	5.6
Ammonia	2.7

ILLUSTRATIVE PROBLEM 16-3

A preheat steam coil requires 800 lb/hr of saturated steam
at a system pressure of 20.0 psig. What should be the flow
coefficient for the control valve?

Solution: Assume that p_2 will be 0.45 of p_1 (critical pressure
drop). With no superheat, we have, by Eq. (16-3),

$$C_v = \frac{\dot{m}_s}{1.8p_1} = \frac{800}{1.8(20 + 14.7)} = 12.8$$

ILLUSTRATIVE PROBLEM 16-4

A steam control valve has a flow coefficient (C_v) of 20. It will
be installed in a steam system where the initial pressure is 20
psig, and downstream from the valve the pressure is 15 psig.
What maximum flow rate of saturated steam can be expected?

Solution: Since p_2 is more than 0.45 of p_1, we use Eq.
(16-4). By Eq. (16-4),

$$C_v = \frac{\dot{m}_s}{2.1\sqrt{(p_1 - p_2)(p_1 + p_2)}}$$

or

$$\begin{aligned}\dot{m}_s &= C_v(2.1)\sqrt{p_1^2 - p_2^2}\\ &= 20(2.1)\sqrt{(20 + 14.7)^2 - (15 + 14.7)^2}\\ &= 754 \text{ lb/hr}\end{aligned}$$

ILLUSTRATIVE PROBLEM 16-5

Determine the required flow coefficient (C_v) for a steam control valve having a maximum capacity of 1000 lb/hr of saturated steam with initial and final steam pressures of 20 psig
and 15 psig, respectively.

Solution: Since p_2 is more than 0.45 of p_1, we use Eq.
(16-4). By Eq. (16-4),

$$C_v = \frac{\dot{m}_s}{2.1\sqrt{p_1^2 - p_2^2}}$$

$$= \frac{1000}{2.1\sqrt{(20 + 14.7)^2 - (15 + 14.7)^2}}$$

$$= 26.5$$

ILLUSTRATIVE PROBLEM 16-6

Determine the flow coefficient (C_v) for a steam control valve to pass 1221 lb/hr of saturated steam when the inlet pressure is 20 psig.

Solution: Since the pressure drop is not specified, we will select the critical pressure drop, or $p_2 = 0.45p_1$, and use Eq. (16-3). By Eq. (16-3),

$$C_v = \frac{\dot{m}_s}{1.8p_1} = \frac{1221}{1.8(20 + 14.7)} = 19.5$$

Three-Way Valves. There are two basic types of three-way valves, as discussed previously—the mixing valve with two inlets and one outlet and the bypass valve with one inlet and two outlets.

There are also two types of applications to which these valves may be applied—a mixing application or a bypass application. The mixing valve can perform either application, but the bypass valve can be used *only* in a bypass application. These various combinations of three-way valve and piping arrangements for a coil are shown in Figures 16-41 through 16-43.

The type of application that is chosen for a certain job will depend on the required fail-safe conditions and the way in which the rest of the system is piped. It is

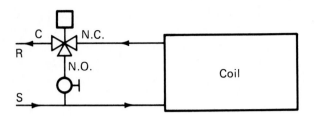

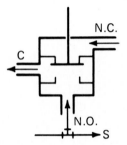

FIGURE 16-42 Mixing valve in bypass application piped NC (normally closed) to coil.

desirable to maintain as close as possible a constant system loss and pressure head in the system. This is possible if constant-flow and constant-friction losses are maintained. Three-way valves used in a bypass application are well suited for this because the flow through the valve is not stopped but just diverted through an alternate route back into the overall water system.

If this alternate route has an adjustable resistance (such as a balancing cock), it can be made to have the same resistance as the coil. In this way, the resistance to system flow will be the same for full bypass flow as it is for full flow through the coil. However, for all conditions between these two extremes, there exists two parallel paths for the system flow, and this greatly reduces the resistance. The maximum system flow will occur when the valve is in its midway position and one-half

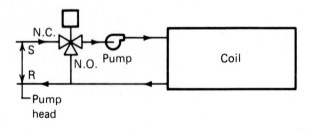

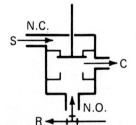

FIGURE 16-41 Mixing valve in mixing application piped NC (normally closed) to coil.

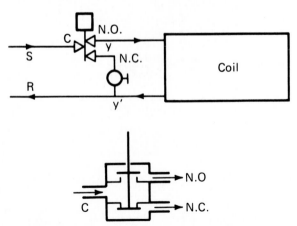

FIGURE 16-43 Parallel circuit: *y* through coil to *y'* bypass valve NO (normally open) to coil.

the flow is in each path. As the valve strokes in either direction from this middle position, the system resistance will increase, and the flow will decrease and approach the minimum flow conditions that are present at either extreme of the valve stroke (see Figure 16-44).

Again, the desired combined characteristic of the valve, coil, and bypass is linear as it was for the two-part valve and coil. But three-way valve characteristics will also shift from their inherent characteristics when they are subjected to this varying pressure drop across them. As before, the best way to combat this problem is to select the valve with as large a pressure drop as possible. This will keep the relative change in pressure across the valve small, resulting in less shift in the valve characteristic.

Mixing valves used in a mixing application (Figure 16-41), incorporate a pump that supplies water to the coil. This keeps a constant flow in the coil and thus eliminates the coil characteristic problem. Because of the constant flow through the coil, the heat output (heating coil) is directly related to the temperature of the water being supplied to the coil. The water temperature is regulated by mixing supply water with return water that has already been through the coil. This condition provides for very effective temperature control by the controller. But from the viewpoint of attempting to maintain constant overall system flow, this application is no better than two-way valve applications. At no-load conditions, the control valve completely closes off to overall system flow, and the pump is merely continuously circulating the same water through the coil. This causes great variations in system flow between full-load and no-load conditions.

Because the system pressure drop will shift to the valve as it is closed, the valve characteristic will also shift as it did for other valves. This cannot be eliminated, but the effects can be minimized. The same approach should be taken for mixing valves in mixing applications that we applied to all other valves; namely, the valve pressure drop must be as large as is practical so that the relative change in pressure across the valve will be kept as small as possible. This will make the shift of the valve characteristic relatively small.

Valve Selection. This discussion has presented some of the basic theories behind control valves and their applications in systems. The topics covered should provide a better understanding of valves and also aid in their selection. Unfortunately, it is very seldom that the perfect valve for a certain application is available, and the next best valve must be selected.

As a general rule of thumb, because most system components are oversized, the next smaller flow coefficient (C_v) than that calculated will provide satisfactory results. Good judgment must be used, however, when selecting a smaller valve so that the maximum capacity of the coil is not greatly reduced. On the other hand, choosing too large a valve can have very poor effects on controllability of the system, as was discussed earlier.

16-9 SELECTION OF CONTROL DAMPERS[5]

A damper is a device that controls the flow of air in an air conditioning system. The damper accomplishes airflow control by varying resistance to flow, just as a control valve does in a liquid circuit. This section deals with the multiple-blade damper in its basic design configurations—the parallel blade and the opposed blade. The same considerations that apply to valve selection—that is, correct size, flow characteristics, rangeability, and required pressure drop—also apply to damper selection.

Much has been published on the subject of valve selection and sizing. Instead of installing a valve the same size as the pipe, as was common practice not too many years ago, valves are now selected with regard to the function they are to perform. This was discussed in the preceding section. Manufacturers have given careful consideration to inner-valve construction. Most of them now provide a choice of flow characteristics ranging

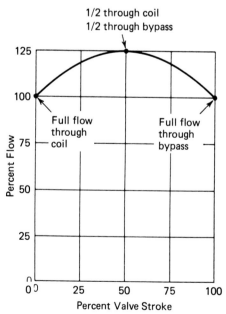

FIGURE 16-44 Three-way valve bypass application.

[5]Figures 16-45 through 16-63 and the associated descriptive matter are from *Damper Manual* and from *Engineering Data,* "Dampers," Section Db, Johnson Controls, Inc., Milwaukee, WI, 1980.

from quick-opening to linear to equal-percentage, or modifications of these, to meet various application requirements. They have published data regarding these characteristics, making it possible for the consulting engineer or control engineer in the field to make the best valve selection.

This has not been true with dampers. Damper design has changed very little. Published data regarding their performance characteristics have been extremely meager. Dampers are almost always the same size as the duct in which they are installed. As a result, performance of airflow control systems often leaves much to be desired.

Present Damper Design. Dampers are manufactured in two basic styles—those with blades that rotate parallel to one another (see Figure 16-45) and those with blades that operate in an opposed manner (see Figure 16-46). Each has different flow characteristics. The application determines which should be used in each case for best control. Their design provides a very limited choice in characteristics.

The characteristic built into a damper is not necessarily the characteristic of the damper when installed in an actual system. The characteristic designed into the damper, here called the *inherent flow characteristic,* is represented by the curve "percent flow versus operator travel," which is produced with a constant pressure drop maintained across the damper (see Figures 16-45 and 16-46). These are the basic characteristics designed into the dampers by virtue of the design of driving linkage, damper blade of leaf configuration, and other factors.

The *effective flow characteristic* is the actual flow

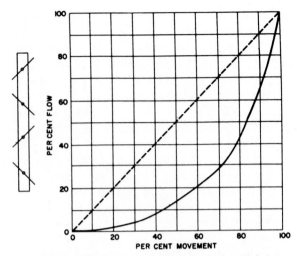

FIGURE 16-46 Opposed-blade damper inherent characteristic curve (dashed line is hypothetical linear characteristic curve)—constant pressure drop.

characteristic of the damper obtained when the damper is applied to a given air-handling system; that is, it is the inherent flow characteristic as altered by the varying pressure conditions in the system and other considerations. To select the inherent characteristic best suited to a particular flow control application, it is important to first decide what effective characteristic is most desirable for good control. Then, knowing how the system pressure variations will alter the inherent characteristic to produce an effective characteristic, the proper selection can be made.

The Ideal Effective Characteristic. Maximum control stability in a system occurs when there is a nearly linear relationship between the change in the controlled variable (error) as measured by the controller and the corrective action produced by the controlled device (damper). If, for example, a thermostat controls the temperature of a space by varying the volume of warm air admitted, each equal increment of space temperature change should result in an equal increment of change in air volume admitted to the space. Figure 16-47 shows how a linear flow characteristic produces airflow control through the entire travel of the damper (from closed to 100% open).

Since the relationship between the controller and damper operator, whether the control system is pneumatic, electric, or electronic, is essentially linear, it remains only to ensure that the effective damper characteristic be relatively linear to have the entire control loop linear.

The way in which a damper is used in a system to a great extent determines how much the system will alter the inherent damper characteristic. It is this altered in-

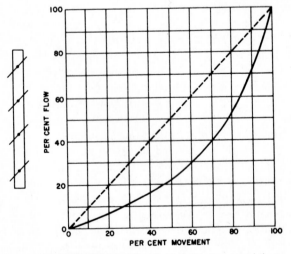

FIGURE 16-45 Parallel-blade damper inherent characteristic curve (dashed line is hypothetical linear characteristic curve)—constant pressure drop.

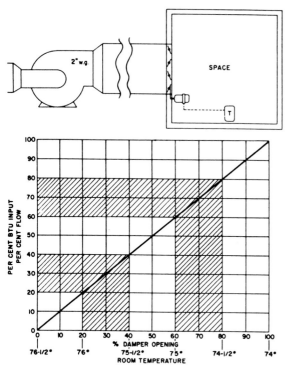

FIGURE 16-47 Variable-volume temperature control system.

herent characteristic, or effective characteristic, that is of importance to the end result. The following examples will illustrate why this is so.

Throttling control: For purposes of explanation only, assume a system as shown in Figure 16-47 wherein the space temperature is controlled by varying the volume of constant-temperature air admitted. Each equal increment in demand should result in equal increments of air volume change. The relationship between room temperature change and damper operator movement is essentially linear, as was mentioned previously. It remains to be seen how the relationship between operator movement and flow of air through the damper can be made as nearly linear as possible.

At maximum flow, assume the fan pressure is 2.0 in. WG (inches water gauge) and pressure drop across the wide-open damper is 0.01 in. WG. Resistance of the duct system is then 1.99 in. WG. At the other extreme, when the damper is completely closed, there is no flow and no duct loss, and the pressure drop across the damper is the full fan pressure of 2.0 in. WG, or 200 times what it was when wide open. (If the fan performance curve is taken into account, this ratio would be greater than 200 to 1.) Thus, as the damper gradually moves from wide open to closed, pressure drop across the damper gradually increases. Closing the damper to reduce flow is partially offset by this increase in pressure drop. With the increasing pressure drop partially

offsetting each increment of damper closing, the inherent characteristic curve will shift upward. Figure 16-48 shows how a near linear inherent characteristic would shift upward in this application. The damper characteristic curves in Figures 16-45 and 16-46, however, are for *constant pressure drop*. If a damper were constructed with an inherent linear characteristic, as shown by the dashed lines of Figures 16-45 and 16-46, variable pressure drop would bend the curves upward, and the effective characteristic would be far from linear in the system described (Figure 16-48). Knowing this, it is best to select a damper with an inherent characteristic that falls well below the linear curve so that the increasing pressure drop, as the damper closes, will bend the curve upward toward linear rather than away from it.

A system with several volume-control dampers, instead of the single damper used in the example, would behave exactly the same as shown in Figure 16-47. Whenever the volume of air handled by a duct system or even a portion of a duct system varies, a variable pressure drop across the dampers results unless the systems is specifically designed and controlled to prevent it. Variation of pressure drop will usually be quite large so that, normally, a damper with an inherent characteristic curve that is the farthest below the linear curve should be selected. This is the opposed-blade damper of Figure 16-46.

Face-and-bypass control: A face-and-bypass application must be treated much like the preceding example when dampers are applied. Although a constant pressure drop is maintained between points *A* and *B* of Figure 16-49, causing it to resemble a mixing or constant-pressure-drop application, other factors must be

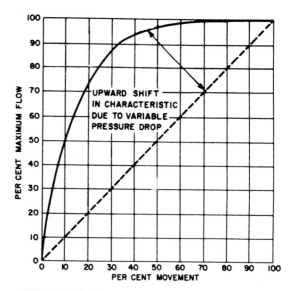

FIGURE 16-48 Shift in damper characteristic due to increasing pressure drop.

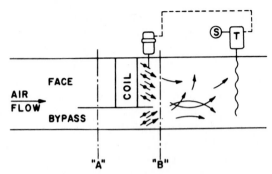

FIGURE 16-49   Face-and-bypass system.

analyzed. Looking at the face section alone, assume a pressure drop through the coil of 0.25 in. WG. Also assume a 0.10 in. WG pressure drop through the face damper in the wide-open position. When the face damper is wide open, the majority of the pressure drop in the face section is across the coil. But, as the face damper closes, the pressure drop shifts from the coil to the damper. At the near-closed position of the damper, the majority of the pressure drop in the face section is across the face damper. In the preceding volume-control example, the pressure drop shifted from the ductwork to the damper; in this example, it shifts from the coil to the face damper. This shift in pressure drop makes the problem of selection of the face damper much like the variable-volume damper.

When analyzing the bypass section by itself, a study of Figure 16-57 should be made. Figure 16-57 shows the resistance that is added by installing dampers smaller than the duct size. When the face-and-bypass dampers are each in a 50% open position, air is flowing through both the face-and-bypass section (airflow through 100% of duct area). When either damper is closed (100% face or 100% bypass), the pressure drop through the section is determined by the amount of total duct area being used for airflow. For example, with the face section closed, all the air must flow through the bypass section. The pressure drop added is that which is added by the blank-off section (closed face damper). Figure 16-57 shows that blanking off 70% of the total duct area at an approach velocity of 500 fpm will add approximately 0.34 in. WG. On the other hand, with the bypass closed, 30% of the total duct area is blanked off and 0.02 in. WG pressure drop is added. Although the pressure drop across the entire face-and-bypass section is relatively constant, the preceding example shows how the pressure drop varies within the section. This makes it impossible to consider a face-and-bypass damper in the same manner as a constant-pressure-drop or mixing application.

Since the pressure drop across the dampers varies as the dampers are throttled, the inherent characteristic

of the damper will be shifted. Truly linear dampers are therefore not the best choice. Figure 16-50 shows the total flow variations that can be expected if the proper characteristic is selected.

Mixing dampers: Application of mixing dampers often presents a considerably different problem from that described in the previous section. An outdoor air, return air, and exhaust damper combination could be this type of application. Figure 16-51 shows a combination of outdoor, return, and exhaust dampers in very close proximity to one another. No weather louvers, preheat coils, and so forth, are considered. Though not typical of many systems, it demonstrates one of the few constant-pressure-drop applications. The minimum outdoor air quantity provides the required ventilation and keeps the building under a slightly positive pressure.

With the maximum outdoor air and exhaust dampers closed and the return air damper open, assume that the pressure at A (mixing chamber) is -0.10 in. WG and that the system capacity is at the desired value of 10,000 cfm, of which 2500 cfm is minimum outdoor air and 7500 cfm is recirculated air. If the pressure at B is $+0.10$ in. WG, the pressure drop from B to A is 0.20 in. WG. When the maximum outdoor air and exhaust dampers are open and the return air damper is closed, it is required that the system capacity remain at 10,000 cfm. It is then necessary to select the maximum outdoor air damper to handle 7500 cfm, with 0.10 in. WG pressure drop, thus achieving the desired capacity. Knowing the pressure losses in ducts and dampers for various air velocities, it is possible to design a system that will essentially meet these constant-volume requirements.

Proper functioning of this type of air conditioning system depends upon maintaining a constant rate of flow for all positions of the maximum outdoor air, return air, and exhaust dampers, not just the extreme positions mentioned. If the pressure drops through the dampers can be held at a constant value by correct design of the system and proper selection of damper sizes, as described above, and if the dampers have linear inherent characteristics, nearly constant air volumes should be expected for all damper positions. Hence, a mixing application appears to require dampers of a relatively linear *inherent* characteristic if there are no appre-

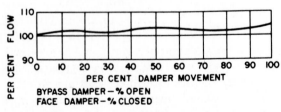

FIGURE 16-50   Percent flow versus percent damper position (face-and-bypass system).

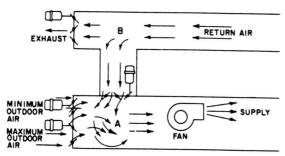

FIGURE 16-51 Typical outdoor, return, and exhaust air mixing application.

ciable changes in pressure drop to alter this inherent damper characteristic.

Unfortunately, very few systems are designed in the above manner. Abrupt expansion, abrupt conversion losses, minor duct losses, entrance protective device losses, and so forth, all provide losses that will tend to change the pressure drop across the damper as it modulates. Therefore, each of the dampers (outdoor air, return air, and exhaust) is studied separately and within its own system (as was done in the face-and-bypass example). It will be seen that truly linear characteristics for mixing applications are not necessary.

Importance of Proper Damper Sizing. For many years, it has been common practice to select the size of an automatic control valve to handle the required maximum flow at a pressure drop that will produce good controllability. An automatic damper is also a flow-control device, and, like a valve, it accomplishes its function by changing the flow area. It follows that it also should be sized for a pressure drop that will produce good controllability.

The damper, however, has not received this consideration. Much of this can be blamed on the fact that few manufacturers of dampers have had sufficient information about their damper characteristics and pressure drops to permit consulting engineers or control engineers to select dampers on this basis. For this reason, it has been extremely unusual for dampers to be selected and sized with system controllability in mind. Instead, the damper is selected to fit the duct at a point where it is most convenient to install it. This has created serious problems because these dampers are usually oversized and overall system controllability suffers.

Why size dampers? It is important that the damper be sized to use a reasonable portion of the total system pressure drop in the open position so as to enable it to perform its control functions. If this is done, it is then possible to determine the change in pressure drop that will occur as the damper closes and thereby

predict the shift in the inherent characteristics of the damper. Dampers with the proper characteristics can then be chosen so that the predictable change in pressure drop across the damper will shift the inherent characteristic curve to an effective characteristic curve, which will prevent cycling and result in stable control.

To correctly size a damper, it is important to know the following information about the system in which it will be installed:

1. *Total pressure drop* in the portion of the system in which the damper is to be located or total pressure drop that can be expected across the damper when in the closed position.
2. *Total volume* of air that the damper is expected to pass in the wide-open position.
3. *Inherent characteristics* of the damper to be selected for this system.

Figures 16-52 and 16-53 show the effect of variable pressure drops on the damper characteristics. These are the variations in pressure drop that occur, for example,

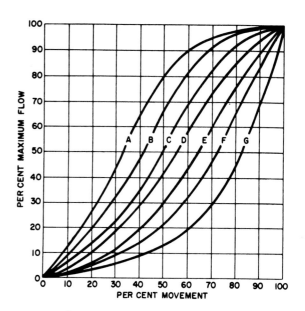

A—1
B—2
C—5
D—8
E—22
F—50
} PER CENT OF TOTAL SYSTEM DROP THROUGH THE DAMPER IN THE WIDE OPEN POSITION

G— INHERENT CHARACTERISTIC OF THE DAMPER AT A CONSTANT PRESSURE DROP

FIGURE 16-52 Characteristics of an opposed-blade damper (showing effect of varying percent of system pressure drop across damper).

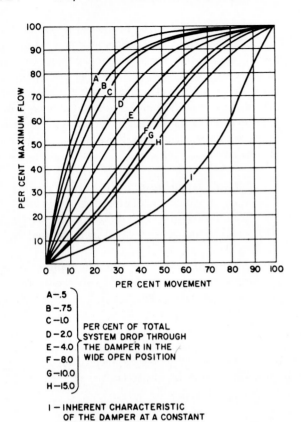

A —.5
B —.75
C —1.0
D —2.0
E —4.0
F —8.0
G —10.0
H —15.0

PER CENT OF TOTAL
SYSTEM DROP THROUGH
THE DAMPER IN THE
WIDE OPEN POSITION

I — INHERENT CHARACTERISTIC
OF THE DAMPER AT A CONSTANT
PRESSURE DROP

FIGURE 16-53 Characteristics of a parallel-blade damper (showing effect of varying percent of system pressure drop across damper).

when a damper is modulating to its closed position and the system resistance decreases, thus increasing the pressure drop across the damper. A study of the tables and the accompanying curves indicates the need for selecting the "wide-open" pressure drop by proper damper sizing. These curves are based on a particular range of damper sizes. As damper sizes change, these curves will vary slightly.

Curve G of Figure 16-52 shows the inherent characteristic of an opposed-blade damper. It is the curve that results when the pressure drop across the damper remains constant regardless of damper position. For a throttling control application, it was shown in Figure 16-47 that the pressure drop across the damper increases from its wide-open value, at full flow, to full fan pressure when the damper is closed. Thus, if the wide-open pressure drop is 0.01 in. WG and the full fan pressure at no flow is 2.0 in. WG, the open damper resistance in percent of system resistance is 0.5%. Figure 16-52 shows the effect on the damper characteristic for various values of open-damper resistance in percent of system resistance. Since a linear effective damper characteristic is desired for good control, curves C and D are best. Thus, a damper with an equal-percentage inherent characteris-

tic can be made essentially linear, in a throttling application, if it is sized so that its wide-open resistance falls in the range of 5% to 8% of the total system resistance.

For example, curve A of Figure 16-52 is decidedly nonlinear. When 50% open, it has 80% of full-open capacity. Most of its control is accomplished in less than one-half of its travel. It would likely produce unstable control when near its closed position.

Figure 16-53 shows similar curves for a parallel-blade damper. For a throttling application, its wide-open resistance would have to equal 10% to 15% of the total system resistance (curves G and H) to obtain a nearly linear effective characteristic. It is obvious that a damper with inherent equal-percentage characteristics (opposed-blade) requires less wide-open pressure drop to produce the desired effective characteristic under actual operation. As mentioned previously, Figure 16-52 indicates that its wide-open resistance should be 5% to 8% of the total system resistance. Although this is a small percentage of the total system resistance and should not be objectionable from the standpoint of added fan horsepower, it is a great deal more than can normally be expected if the dampers are sized the same as the duct.

To obtain the wide-open resistance required for proper control, the damper must usually be smaller than the duct. Correct damper sizing offers many advantages in addition to obtaining good control, such as the following:

1. *Initial cost*—Correct damper sizing will generally mean dampers smaller than the duct. Dampers are usually priced by the square foot. Thus, the correctly sized damper will be less expensive. The smaller damper also requires less operating power and can reduce the cost of the damper operators.
2. *Leakage*—For a given damper construction and pressure drop, leakage is proportional to the damper area. A smaller damper will result in less leakage.
3. *Rangeability*—This is a term commonly applied to valves but is just as applicable to dampers. It is expressed as a ratio of maximum flow to minimum controllable flow. Reduced leakage decreases the uncontrollable flow and therefore increases the rangeability. This improves the controllability of the system.

Example of poor damper selection: A look at Figure 16-53 shows why a parallel-blade damper is generally not suitable for throttling-type applications where the pressure drop across the damper increases appreciably as the damper (curve *I*) is more nearly linear than the inherent characteristics of an opposed-blade damper. To obtain the most nearly linear curve *H*, the damper would have to be sized to use up approximately 15% of the total system pressure drop in its wide-open position. If sized for 1% of the system pressure drop, a common occurrence, the damper would accomplish nearly 95% of its flow control in 50% of its travel (curve *C*). This would tend to produce unstable control and, since only about 50% of the total damper range is really effective, would multiply the effect of hysteresis and other losses.

Constant-pressure-drop applications: If the damper selected above is applied to a system where the pressure drop across the damper remains relatively constant, it can provide satisfactory control. Some mixing dampers are for this type of application. Since the pressure drop across the damper does not change appreciably, its inherent characteristics and effective actual characteristic are essentially the same. Sizing the wide-open damper to utilize a fair percentage of the total system pressure drop in this case is not as important as selecting the proper characteristic to produce constant total system flow.

Damper Sizing Methods. Figures 16-52 and 16-53 indicate the percent of total system pressure that should be utilized by the damper in the wide-open position in order to avoid shifting the inherent characteristics to anything but the desired (near linear) characteristics. Figures 16-54 and 16-55 show the approach velocities that produce the pressure drops illustrated by the curves in Figures 16-52 and 16-53. The approach velocity information is easier to use since it can be directly related to the duct area and, therefore, duct size if the cfm is known and no detailed pressure drop curves need be referred to.

Figure 16-56 shows the pressure drop that can be expected through a wide-open damper for various approach velocities and free area ratios. The free area ratio is the free area (total open area between blades and inside damper frame) divided by the total area of the damper. (Such data is usually available from the manufacturer.) The approach velocity is the velocity of the air in the duct approaching the complete damper area. Pressure drops in Figure 16-56 are based on a duct and damper of the same size. From the information above, it is obvious that selecting dampers to produce the desired characteristics, in most cases, adds very little pressure drop to the entire system. This drop should not be

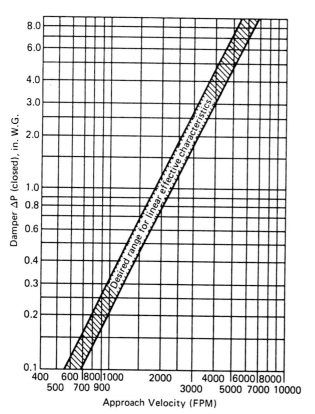

FIGURE 16-54 Desired pressure drops for opposed-blade damper.

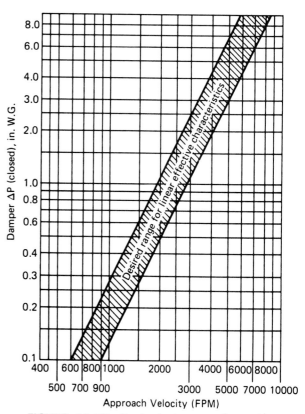

FIGURE 16-55 Desired pressure drops for parallel-blade damper.

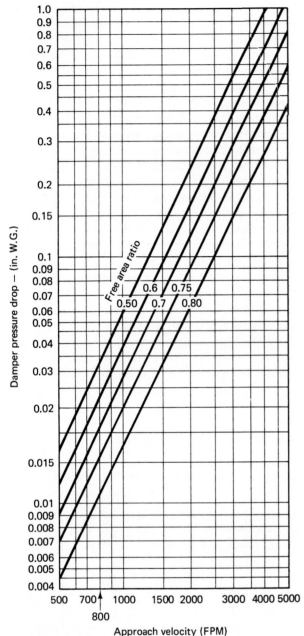

FIGURE 16-56 Pressure drops through wide-open damper.

objectionable and will be offset by the superior system operation that can be expected.

If dampers are properly sized, a common question will be, "where should they be located?" In most cases, there is a conversion section within the duct system in which the properly sized damper could be located. In practically all cases, it will mean moving the damper away from the coil (velocities of 400 to 600 fpm) to a location in the system where the velocities are higher. Moving the damper away from the coil will also produce better distribution of airflow across the coil. A

more controversial answer to the question might be that it is often possible to size the duct, or install conversion duct sections, to fit the damper. There are also inexpensive and convenient means for making the properly sized damper fit the duct while adding little pressure drop.

Application of Properly Sized Dampers. Selecting the damper with proper characteristics to do a good control job is relatively easy if characteristic curves are published by the damper manufacturers. Selecting a damper of proper size to match the selected characteristic to the system in which it is to be installed becomes feasible as damper manufacturers begin to take an engineering approach to the subject and publish the necessary test data. The only problem that remains is what to do if the properly sized damper is smaller than the duct in which it is to be installed. There are a number of approaches to this problem.

1. *Adapting the duct to fit the damper*—It is common practice for ducts to be designed to fit coils, filters, spray banks, and even "bug screens." In many cases, these components are of no more importance to the proper functioning of the system than the dampers. Realizing this, the designing engineer will find cases where the duct can be made to conform to the dimensions of a properly sized damper.

2. *Damper location*—The usual practice of locating a face damper near the coil has some disadvantages. Air distribution through the coil is poor. When the damper is throttling, there is a tendency to produce a number of relatively high-velocity air jets through portions of the coil. This is a major cause of freezing of preheat coils. In many cases, there is a conversion section to reduce the duct size downstream from the coil. In this conversion section, there is sometimes a location that provides the desired area for a correctly sized damper. If not, the damper may sometimes be located in the smaller distribution duct beyond the conversion section.

3. *Adapting the damper to fit the duct*—There will be a large number of cases where, regardless of the preceding suggestions, the properly sized damper will be smaller than the duct into which it is to be installed and some arrangements will be needed to accommodate the damper to the duct. In most installations, it will be possible to make one dimension of the duct the same as that of the damper.

A simple blank-off plate (see Figure 16-57) is an easy way of filling the duct. While this arrangement creates more turbulence and, therefore, a greater pressure loss than more elaborate arrangements, it has the advantage of being inexpensive. As Figure 16-57 indicates,

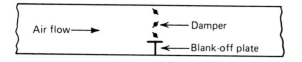

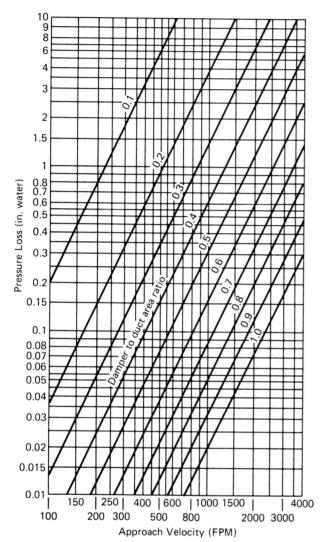

FIGURE 16-57 Effect of blank-off plate.

how the inherent characteristics are shifted upward when a damper is installed in the system in which the pressure drop varies as the damper throttles the airflow. In order to shift the inherent characteristic to the desired nearly linear effective characteristic, the damper must be sized so that its wide-open pressure drop is a certain percent of the pressure drop across the damper when it is closed. It can be seen that curves C and D of Figure 16-52 provide the most nearly linear effective characteristics. Figure 16-54 shows the damper velocities that produce the wide-open pressure drops necessary to create curves C and D in Figure 16-52. This damper velocity information is to be used for sizing dampers for throttling applications. It can be directly related to the duct area and duct size if the cfm is known, and no detailed pressure drop curves need be referred to.

The desired range band in Figure 16-54 shows the approach velocities necessary to produce an effective characteristic curve that approaches curve C or D of Figure 16-54. The vertical scale of Figure 16-54 shows Δp damper (closed) in in. WG. (Δp is the differential pressure or pressure drop.) This is the maximum static pressure expected on the entering side of the damper when closed or the anticipated pressure drop across it when closed. In many systems, this will be the maximum static pressure the fan can develop.

ILLUSTRATIVE PROBLEM 16-7

Assume a system that has a total closeoff pressure of 1.0 in. WG and the system flow is 15,000 cfm. What size damper should be used?

Solution: From Figure 16-54 for a closeoff pressure of 1.0 in. WG, the desired damper velocity equals 1800 to 2300 (fpm (use 2000 fpm for convenience).

Required damper area is

$$A = \frac{Q}{V} = \frac{15000}{2000} = 7.5 \text{ ft}^2$$

By referring to manufacturers' data, the rectangular dimensions of the damper may be determined. For example, a 30-by-36-in. damper has a nominal area of 7.5 ft². This damper could be selected for this application, and the duct should be sized to match the damper size. If the manufacturer of the above damper also quotes the free area as 5.73 ft², the free area ratio would be 5.73/7.50 = 0.76. Referring to Figure 16-56, the pressure drop through the wide-open damper would be approximately 0.085 in. WG.

Smaller-than-duct-size dampers: When a damper is properly sized, it will often be smaller than duct size, as previously stated. A blank-off plate can be used in the majority of applications to adapt the smaller damper to the duct size. Figure 16-58 shows that, when

the pressure drops created by this arrangement are not excessive if the blank-off plate is less than 30% of the duct area. If greater reductions in duct area are required, more elaborate conversion sections must be used, unless damper sizes are being reduced purposely to introduce excessive pressure drops.

While adapting the damper to the duct system presents some problems, there are usually only a few dampers in even a fairly elaborate central air conditioning system.

Throttling Application (Opposed-Blade Damper with Equal-Percentage Linkage).   Figure 16-52 shows

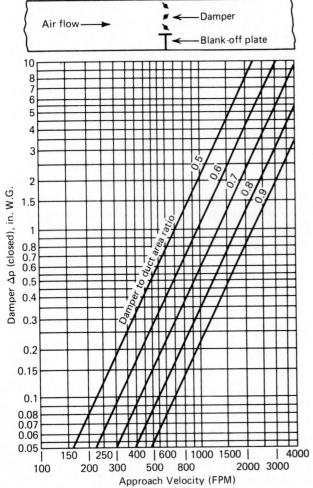

FIGURE 16-58 Desired pressure drops for opposed-blade damper with blank-off.

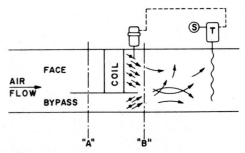

FIGURE 16-59 Face-and-bypass system (repeat of Figure 16-49).

plication because of the better mixing effect downstream. Figure 16-59 shows the proper method of installing the parallel-blade damper to obtain the best mixing.

The face-and-bypass damper is treated like a throttling application because of the internal pressure changes within the face-and-bypass section. As stated previously, the pressure drop through the coil is transferred to the face damper as the face damper closes. The pressure drop across the bypass also changes due to the smaller-than-duct-size effect. The face damper is sized to equal the area of the coil. By sizing the bypass damper from Figure 16-60, 100% flow will be maintained through the face-and-bypass section for all positions of the face-and-bypass damper (see Figure 16-50).

reducing the duct area by 30% or less, a blank-off plate can be used.

If the duct area is to be reduced by more than 30%, the pressure drop added by the blank-off is usually objectionable (see Figure 16-57). Since the blank-off plate affects the characteristics of the damper, a larger percent of the total closeoff pressure must be used for the damper and blank-off combination. This percent is larger than is normally used for a duct-size damper (see Figure 16-62). Figure 16-58 shows the proper wide-open velocity for the dampers installed in ducts with various percents of blank-off. Using this chart will produce an actual effective characteristic close to curves *C* and *D* of Figure 16-52 used for duct-size dampers.

Face-and-Bypass Application (Parallel-Blade Damper with Equal-Percentage Linkage). The parallel-blade damper lends itself to the face-and-bypass ap-

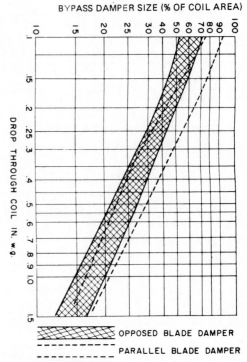

FIGURE 16-60 Bypass damper sizing curve.

ILLUSTRATIVE PROBLEM 16-8

Assume a system of 5000-cfm total flow, a specified coil face velocity of 600 fpm, and a pressure drop of 0.25 in. WG through the coil. Determine the size of the face-and-bypass damper sections.

Solution: The face damper will be coil size or as close to coil size as possible. If the actual duct size is not known, size it for the specified velocity through the coil.

Required face area is

$$A = \frac{Q}{V} = \frac{5000}{600} = 8.33 \text{ ft}^2$$

By checking manufacturers' data, we could select a standard damper, say, 30 by 42 in., giving a nominal area of 8.75 ft² for the face damper.

Using the parallel-blade damper, the bypass section and damper size should be about 40% of the face damper section (refer to Figure 16-60). Therefore, the bypass damper area is

$$A = 0.40 \times 8.75 = 3.50 \text{ ft}^2$$

The bypass damper size would be 12 by 42 in., and its blade axis dimension (42 in.) is the same as the face damper for ease of installation.

Mixing Applications. To properly size dampers for mixing applications such as outdoor air, return air, and exhaust dampers, each damper must be considered separately. Referring to Figure 16-61 and assuming that the dampers will be selected to produce a constant system flow and, therefore, a constant-mixing plenum pressure, each damper must be considered separately in its own area of operation. The outdoor air damper must be considered as a throttling damper in the system from point *A* (atmospheric pressure) to point *B* (constant-mixing plenum pressure). The pressure drops in this system are the duct and entrance losses, the drop through the weather louver, and the drop through the damper. These pressure drops will shift to the damper as the flow is reduced and, therefore, shift the inherent characteristics of the outdoor air damper.

The return air damper in the system must be considered from point *C* (a constant-pressure point in the return system) to point *B*. The losses in this section are the entrance and exit losses to that branch of the return system and the minor duct loss in that section, plus the loss of the return air damper. The exhaust damper is considered from point *C* to point *D* with similar pressure losses, which transfer to the damper as it closes.

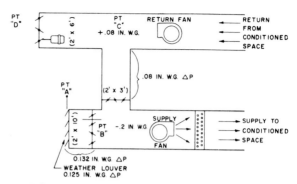

FIGURE 16-61 Typical constant-pressure-drop application.

ILLUSTRATIVE PROBLEM 16-9

Assume a system as illustrated in Figure 16-61. Total system volume delivered to the conditioned space is 10,000 cfm. The minimum outdoor air requirement for ventilation is 25%, or 2500 cfm. The return air fan has been selected to deliver 7500-cfm return air. The return fan is sized with sufficient discharge static pressure to overcome the duct friction between the fan discharge and the exhaust damper or the return damper, whichever is the greater. Determine size and characteristics of the outdoor air, return air, and exhaust dampers. (Typical pressure drops will be assumed as the problem progresses.)

Solution:

Outdoor air damper—Reduced-size dampers are usually difficult to use on outdoor air intakes. Tapered conversion sections cannot be used since it is usually difficult to install them between the weather louver and the outdoor air damper due to space limitations in the fan room. Reducing the outdoor air damper with blank-off plates is not permitted in a close-coupled arrangement since the velocity will increase in certain areas of the weather louver and snow and rain will be drawn into the outdoor air intake. The outdoor air damper must, by necessity, be sized to match the weather louver size.

For Figure 16-61, assume a normal velocity of 500 fpm through the weather louver at a pressure drop of 0.125 in. WG. To accommodate the 10,000 cfm at a velocity of 500 fpm, the area of the weather louver would be 10,000/500 = 20.0 ft². Assume the selected size of the weather louver to be 10.0 ft long by 2.0 ft high; the outdoor air damper will be selected to fit these dimensions. For this example, the following modules are selected. One module will be a nominal 30 in. parallel to the blade axis by 24 in. perpendicular to the blade axis, and two modules will be 48 in. and 42 in. parallel to the blade axis by 24 in. perpendicular to the blade axis. The 30-by-24-in. module will be used as the 25% minimum outdoor air damper, and the remaining modules will be used as a maximum outdoor air damper.

Characteristics—Consider the system *A*–*B* as a small throttling system. The total pressure drop in this system is the sum of the loss in the outdoor weather louver, the losses due to the abrupt conversion at the entrance, and some small duct loss. The drop across the weather louver is 0.125 in. WG. The pressure drop across the outdoor air damper must be determined by referring to Figure 16-56. Assume that a typical free area ratio of this damper to be 0.75, and, with the approach velocity of 500 fpm, the pressure drop across the wide-open damper is about 0.007 in. WG. The total pressure drop between points A and B will then be $(0.125 + 0.007) = 0.132$ in. WG, plus the drop caused by conversion losses mentioned above, for a total of approximately 0.20 in. WG. This 0.20 in. WG will be the drop across the damper when it is closed or the total system resistance of system A–B. Figure 16-54 shows that, with a system resistance of 0.20 in. WG, damper velocities between 775 and 1000 fpm will produce desired linear effective characteristics. The damper approach velocity in this problem is only 500 fpm.

A full-size damper with equal-percentage characteristics will therefore not produce the desired linear characteristics but will approach an effective characteristic curve between curves A and B in Figure 16-52. Since the effective characteristics of the outdoor air damper are not completely linearized, the return air damper must be selected and sized to produce a characteristic complementary to that produced by the outdoor air damper.

Return air damper—Constant-mixing plenum pressure will assure a relatively constant pressure at the suction of the fan and therefore do a great deal to produce constant system volume. Proper sizing of the return air damper can produce this required constant-mixing plenum pressure.

The pressure drop in the system in which the return air damper throttles is from B (-0.20 in. WG) to C ($+0.08$ in. WG), of 0.28 in. WG. In an installation, the exact operating pressure at points B and C are difficult to determine. The pressure drops of the components in the outdoor air section are available. In this example, they are the damper and weather louver, which produce a drop of 0.132 in. WG. The return air damper should be sized to produce a similar drop.

The velocity in the return duct is $7500/(2 \times 3) = 1250$ fpm. Figure 16-57 indicates that a duct velocity of 1250 fpm will produce a drop of 0.132 in. WG if the damper is approximately 70% of the duct area. The damper area should therefore be $6 \times 0.70 = 4.2$ ft^2. A 2-by-2-ft damper with a nominal free area of 4.0 ft^2 would be selected.

Since the return fan was selected and the system balanced to overcome the losses in the return duct, the $+0.08$ in. WG is lost in moving the return air to the return damper. The pressure at the entrance to the damper is atmospheric, or 0.0 in. WG. Therefore, the pressure in the mixing plenum will be equal to the pressure drop at the damper and the exit losses due to the abrupt expansion of the relative high-velocity air. The damper pressure drop (wide-open) as figured above is 0.132 in. WG. The abrupt expansion losses for the velocity ratio involved and the 1250-fpm return duct velocity (0.10 in. WG velocity pressure) is approximately 0.65 velocity heads, or 0.065 in. WG. The total drop is $0.132 + 0.065 = 0.197$ in. WG, which is very close to the 0.20 in. WG in the outdoor air duct.

The drop through the wide-open damper is 0.132 in. WG. The drop across the closed damper is 0.28 in. WG. The percent of the total system drop through the damper in the wide-open position is $0.132/0.28 = 0.47$, or 47%. The curves is Figure 16-62 show that a wide-open pressure drop of 47% will produce a curve similar to curve F. Adding the effective characteristics of the outdoor and return air dampers, as is done in Figure 16-63, produces a total system volume of near 100% at all positions of the dampers. The pressure at point B will also remain relatively constant at all damper positions, further assuring that total system volume will remain at a near constant value.

In the many systems tested, evaluated, and investigated, the return air damper sized in this manner produced from 45% to 75% of the total resistance of its return air section. This variation in resistance ratio has little effect on the actual effective characteristics of the return air damper (see Figure 16-62)

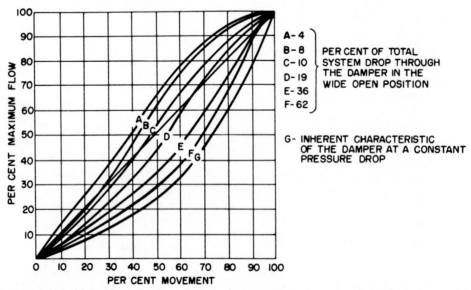

A-4
B-8
C-10
D-19
E-36
F-62

PER CENT OF TOTAL
SYSTEM DROP THROUGH
THE DAMPER IN THE
WIDE OPEN POSITION

G- INHERENT CHARACTERISTIC
OF THE DAMPER AT A CONSTANT
PRESSURE DROP

FIGURE 16-62 System characteristics for opposed-blade damper with blank-off.

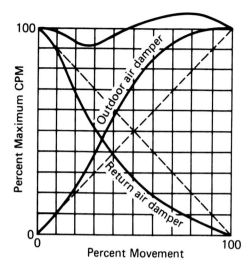

FIGURE 16-63 Desired effective characteristic curves for mixing damper application.

and hence the total volume of the system. In all cases where return air dampers were sized to produce the same wide-open pressure drop as is produced by the components (damper, weather louver, and at times, a preheat coil) in the outdoor air system at full flow, the total system flow varied less than 10% from the desired design flow.

If the return air damper is sized in the way described above, the velocity through it will be relatively high (about 1875 fpm). If the return duct at the entrance to the plenum covers the full width of the plenum, the high-velocity airstream from the return air will do an effective job of decreasing stratification when it interrupts the low-velocity stream from the outdoor air section. Two advantages are therefore gained by using this sizing method—constant total system volume and reduction in stratification problems.

Exhaust damper—When the return air damper is in the fully closed position, a 0.08 in. WG static pressure is available to overcome duct friction, the loss in the exhaust damper, and exit loss due to protection screens, expansion of the airstream, and so forth. (The duct losses and exit losses are assumed to be approximately 0.070 in. WG.) The exhaust damper must be sized to utilize the remaining available pressure, which is

less than 0.01 in. WG. The exhaust damper will handle 7500-cfm maximum outdoor air while positive building pressure will be maintained with 2500-cfm minimum outdoor air.

For this example, assume a duct-sized damper. Applying a duct-sized damper with nominal dimensions of 2.0 by 6.0 ft, and by referring to Figure 16-56, it is found that the pressure drop through the wide-open damper of this size with an approach velocity of 7500/(2 × 6) = 625 fpm (duct velocity when damper is wide open) would be 0.007 in. WG for a free area ratio of 0.80. The duct-sized damper would be acceptable for this application. Again considering the damper within its own little throttling system (point *C* to point *D* of Figure 16-61 and referring to Figure 16-54, desired approach velocities), it can be seen that this damper will produce near linear effective characteristics. The damper Δp (closed) will be 0.08 in. WG, and the desired approach velocity for this duct-sized damper should be between 475 and 625 fpm. This means that the pressure drop should change the equal-percentage inherent characteristic to a near linear effective characteristic. With the outdoor, return, and exhaust dampers sized and operated as described above, the damper and system flow characteristic would very closely approximate those of Figure 16-63. By applying dampers as described, the total system volume is maintained very close to a constant 100%.

Damper Leakage. Manufacturers of control dampers quote performance data in different ways, and it is necessary to carefully check such data to enable us to select properly. Data concerning damper leakage may be quoted as actual cfm leakage when the damper closes against a given static pressure, or it may be quoted as percent leakage. The former method is preferred. The percent leakage may be defined differently by different manufacturers, and percent leakage means nothing unless the total volume on which this leakage is based is known. Some manufacturers refer to percent leakage as a percentage of the volume that could theoretically pass through the damper at a wide-open pressure drop equal to the closeoff pressure. Others may refer to it as the percent of volume that will pass through the damper for a given application.

REVIEW PROBLEMS

16.1. Air conditioning involves the control of what four conditions of the air?

16.2. What is the basic difference between unitary or packaged systems and central-fan systems?

16.3. Define the following terms: (**a**) ventilation, (**b**) discharge air, (**c**) face-and-bypass damper system, (**d**) unitary system, (**e**) high-velocity double-duct, and (**f**) packaged multizone.

16.4. What is the function of the outdoor air sensor in an economizer system?

16.5. What are three types of controlled mediums normally used in heating systems?

16.6. Name three typical locations for primary sensors.

16.7. What kind of space controller could you use to operate two compressors in a central-fan system?

16.8. When more than two stages of cooling are used, what device is used to operate the stages?

16.9. What are four basic types of humidifiers commonly used in central-fan systems?

16.10. Determine the flow coefficient (C_v) of a water control valve for a flow rate of 9000 gpm and a pressure loss of 4 psi.

16.11. Determine the pressure drop across a water control valve that has a flow coefficient (C_v) of 4500 and where the flow rate is 9000 gpm.

16.12. A water control valve has a flow coefficient (C_v) of 4500. What will be the flow rate in gpm through the valve for a pressure drop of 9.0 psi?

16.13. Benzine (SG = 0.69) is to flow through a control valve at the rate of 9000 gpm with an allowed pressure drop of 4.0 psi. What is the required flow coefficient of the valve?

16.14. A steam control valve has a flow coefficient (C_v) of 20. The inlet steam pressure is 20.0 psig and the outlet pressure is 18 psig. What will be the flow rate through the valve?

16.15. A steam control valve is to be selected for an application requiring a flow rate of 675 lb/hr of saturated steam. The initial steam pressure is 15 psig. What should be the flow coefficient (C_v) for the valve?

BIBLIOGRAPHY

16.1. *Commercial Air Conditioning Controls,* "Basic Control Systems," Honeywell, Inc., Milwaukee, WI, 1972.

16.2. *Engineering Data,* "Control Valves," Section Vb, Johnson Controls, Inc., Milwaukee, WI, 1980.

16.3. *Damper Manual,* Johnson Controls, Inc., Milwaukee, WI, 1979.

Appendix A

Thermodynamic Properties of Steam and Moist Air

TABLE A-1 *

Saturated steam: temperature

		Specific Volume		Internal Energy			Enthalpy			Entropy		
Temp. Fahr.	.Press. Lbf. Sq. In.	Sat. Liquid	Sat. Vapor	Sat. Liquid	Evap.	Sat. Vapor	Sat. Liquid	Evap.	Sat. Vapor	Sat. Liquid	Evap.	Sat. Vapor
t	p	v_f	v_g	u_f	u_{fg}	u_g	h_f	h_{fg}	h_g	s_f	s_{fg}	s_g
32	.08859	.016022	3305.	.01	1021.2	1021.2	.01	1075.4	1075.4	.00003	2.1870	2.1870
32.018	.08866	.016022	3302.	.00	1021.2	1021.2	.01	1075.4	1075.4	.00000	2.1869	2.1869
35	.09992	.016021	2948.	2.99	1019.2	1022.2	3.00	1073.7	1076.7	.00607	2.1704	2.1764
40	.12166	.016020	2445.	8.02	1015.8	1023.9	8.02	1070.9	1078.9	.01617	2.1430	2.1592
45	.14748	.016021	203.7	13.04	1012.5	1025.5	13.04	1068.1	1081.1	.02618	2.1162	2.1423
50	.17803	.016024	1704.2	18.06	1009.1	1027.2	18.06	1065.2	1083.3	.03607	2.0899	2.1259
60	.2563	.016035	1206.9	28.08	1002.4	1030.4	28.08	1059.6	1087.7	.05555	2.0388	2.0943
70	.3632	.016051	867.7	38.09	995.6	1033.7	38.09	1054.0	1092.0	.07463	1.9896	2.0642
80	.5073	.016073	632.8	48.08	988.9	1037.0	48.09	1048.3	1096.4	.09332	1.9423	2.0356
90	.6988	.016099	467.7	58.07	982.2	1040.2	58.07	1042.7	1100.7	.11165	1.8966	2.0083
100	.9503	.016130	350.0	68.04	975.4	1043.5	68.05	1037.0	1105.0	.12963	1.8526	1.9822
110	1.2763	.016166	265.1	78.02	968.7	1046.7	78.02	1031.3	1109.3	.14730	1.8101	1.9574
120	1.6945	.016205	203.0	87.99	961.9	1049.9	88.00	1025.5	1113.5	.16465	1.7690	1.9336
130	2.225	.016247	157.17	97.97	955.1	1053.0	97.98	1019.8	1117.8	.18172	1.7292	1.9109
140	2.892	.016293	122.88	107.95	948.2	1056.2	107.96	1014.0	1121.9	.19851	1.6907	1.8892
150	3.722	.016343	96.99	117.95	941.3	1059.3	117.96	1008.1	1126.1	.21503	1.6533	1.8684
160	4.745	.016395	77.23	127.94	934.4	1062.3	127.96	1002.2	1130.1	.23130	1.6171	1.8484
170	5.996	.016450	62.02	137.95	927.4	1065.4	137.97	996.2	1134.2	.24732	1.5819	1.8293
180	7.515	.016509	50.20	147.97	920.4	1068.3	147.99	990.2	1138.2	.26311	1.5478	1.8109
190	9.343	.016570	40.95	158.0	913.3	1071.3	158.03	984.1	1142.1	.27866	1.5146	1.7932
200	11.529	.016634	33.63	168.04	906.2	1074.2	168.07	977.9	1145.9	.29400	1.4822	1.7762
210	14.125	.016702	27.82	178.10	898.9	1077.0	178.14	971.6	1149.7	.30913	1.4508	1.7599
220	17.188	.016772	23.15	188.17	891.7	1079.8	188.22	965.3	1153.5	.32406	1.4201	1.7441
230	20.78	.016845	19.386	198.26	884.3	1082.6	198.32	958.8	1157.1	.33880	1.3901	1.7289
240	24.97	.016922	16.327	208.36	876.9	1085.3	208.44	952.3	1160.7	.35335	1.3609	1.7143
250	29.82	.017001	13.826	218.49	869.4	1087.9	218.59	945.6	1164.2	.36772	1.3324	1.7001
260	35.42	.017084	11.768	228.64	861.8	1090.5	228.76	938.8	1167.6	.38193	1.3044	1.6864
270	41.85	.017170	10.066	238.82	854.1	1093.0	238.95	932.0	1170.9	.39597	1.2771	1.6731
280	49.18	.017259	8.650	249.02	846.3	1095.4	249.18	924.9	1174.1	.40986	1.2504	1.6602
290	57.53	.017352	7.467	259.25	838.5	1097.7	259.44	917.8	1177.2	.42360	1.2241	1.6477
300	66.98	.017448	6.472	269.52	830.5	1100.0	269.73	910.4	1180.2	.43720	1.1984	1.6356
310	77.64	.017548	5.632	279.81	822.3	1102.1	280.06	903.0	1183.0	.45067	1.1731	1.6238
320	89.60	.017652	4.919	290.14	814.1	1104.2	290.43	895.3	1185.8	.46400	1.1483	1.6123
330	103.00	.017760	4.312	300.51	805.7	1106.2	300.84	887.5	1188.4	.47722	1.1238	1.6010
340	117.93	.017872	3.792	310.91	797.1	1108.0	311.30	879.5	1190.8	.49031	1.0997	1.5901
350	134.53	.017988	3.346	321.35	788.4	1109.8	321.80	871.3	1193.1	.50329	1.0760	1.5793
360	152.92	.018108	2.961	331.84	779.6	1111.4	332.35	862.9	1195.2	.51617	1.0526	1.5688
370	173.23	.018233	2.628	342.37	770.6	1112.9	342.96	854.2	1197.2	.52894	1.0295	1.5585
380	195.60	.018363	2.339	352.95	761.4	1114.3	353.62	845.4	1199.0	.54163	1.0067	1.5483
390	220.2	.018498	2.087	363.58	752.0	1115.6	364.34	836.2	1200.6	.55422	.9841	1.5383
400	247.1	.018638	1.8661	374.27	742.4	1116.6	375.12	826.8	1202.0	.56672	.9617	1.5284
425	325.6	.019014	1.4249	401.24	717.4	1118.6	402.38	802.1	1204.5	.59767	.9066	1.5043
450	422.1	.019433	1.1011	428.6	690.9	1119.5	430.2	775.4	1205.6	.6282	.8523	1.4806
475	539.3	.019901	.8594	456.6	662.6	1119.2	458.5	746.4	1204.9	.6586	.7985	1.4571
500	680.0	.02043	.6761	485.1	632.3	1117.4	487.7	714.8	1202.5	.6888	.7448	1.4335
525	847.1	.02104	.5350	514.5	599.5	1113.9	517.8	680.0	1197.8	.7191	.6906	1.4007
550	1044.0	.02175	.4249	544.9	563.7	1108.6	549.1	641.6	1190.6	.7497	.6354	1.3851
575	1274.0	.02259	.3378	576.5	524.3	1100.8	581.9	598.6	1180.4	.7808	.5785	1.3593
600	1541.0	.02363	.2677	609.9	480.1	1090.0	616.7	549.7	1166.4	.8130	.5187	1.3317
625	1849.7	.02494	.2103	645.7	429.4	1075.1	654.2	492.9	1147.0	.8467	.4544	1.3010
650	2205.	.02673	.16206	685.0	368.7	1053.7	695.9	423.9	1119.8	.8831	.3820	1.2651
675	2616.	.02951	.11952	731.0	289.3	1020.3	745.3	332.9	1078.2	.9252	.2934	1.2186
700	3090.	.03666	.07438	801.7	145.9	947.7	822.7	167.5	990.2	.9902	.1444	1.1346
705.44	3204.	.05053	.05053	872.6	0	872.6	902.5	0	902.5	1.0580	0	1.0580

* Tables A-1 through A-3 are abridged from J. H. Keenan, P. G. Hill, and J. G. Moore, *Steam Tables—Thermodynamic Properties of Water Including Vapor and Liquid Phases,* John Wiley & Sons, Inc., New York, 1969, with permission. From Irving Granet, *Thermodynamics and Heat Power* (Reston, Va.: Reston Publishing Company, 1974).

TABLE A-2
Saturated steam: pressure

Press. Lbf. Sq.In. p	Temp. Fahr. t	Specific Volume Sat. Liquid v_f	Sat. Vapor v_g	Internal Energy Sat Liquid u_f	Evap. u_fg	Sat. Vapor u_g	Enthalpy Sat. Liquid h_f	Evap. h_fg	Sat. Vapor h_g	Entropy Sat. Liquid s_f	Evap. s_fg	Sat. Vapor s_g
.50	79.56	.016071	641.5	47.64	989.2	1036.9	47.65	1048.6	1096.2	.09250	1.9443	2.0368
1.0	101.70	.016136	333.6	69.74	974.3	1044.0	69.74	1036.0	1105.8	.13266	1.8453	1.9779
1.5	115.65	.016187	227.7	83.65	964.8	1048.5	83.65	1028.0	1111.7	.15714	1.7867	1.9438
2.0	126.04	.016230	173.75	94.02	957.8	1051.8	94.02	1022.1	1116.1	.17499	1.7448	1.9198
3.0	141.43	.016300	118.72	109.38	947.2	1056.6	109.39	1013.1	1122.5	.20089	1.6852	1.8861
4.0	152.93	.016358	90.64	120.88	939.3	1060.2	120.89	1006.4	1127.3	.21983	1.6426	1.8624
5.0	162.21	.016407	73.53	130.15	932.9	1063.0	130.17	1000.9	1131.0	.23486	1.6093	1.8441
7.5	179.91	.016508	50.30	147.88	920.4	1068.3	147.90	990.2	1138.1	.26297	1.5481	1.8110
10	193.19	.016590	38.42	161.20	911.0	1072.2	161.23	982.1	1143.3	.28358	1.5041	1.7877
14.696	211.99	.016715	26.80	180.10	897.5	1077.6	180.15	970.4	1150.5	.31212	1.4446	1.7567
15	213.03	.016723	26.29	181.14	896.8	1077.9	181.19	969.7	1150.9	.31367	1.4414	1.7551
20	227.96	.016830	20.09	196.19	885.8	1082.0	196.26	960.1	1156.4	.33580	1.3962	1.7320
25	240.08	.016922	16.306	208.44	867.9	1085.3	208.52	952.2	1160.7	.35345	1.3607	1.7142
30	250.34	.017004	13.748	218.84	869.2	1088.0	218.93	945.4	1164.3	.36821	1.3314	1.6996
35	259.30	.017078	11.900	227.93	862.4	1090.3	228.04	939.3	1167.4	.38093	1.3064	1.6873
40	267.26	.017146	10.501	236.03	856.2	1092.3	236.16	933.8	1170.0	.39214	1.2845	1.6767
45	274.46	.017209	9.403	243.37	850.7	1094.0	243.51	928.8	1172.3	.40218	1.2651	1.6673
50	281.03	.017269	8.518	250.08	845.5	1095.6	250.24	924.2	1174.4	.41129	1.2476	1.6589
55	287.10	.017325	7.789	256.28	840.8	1097.0	256.46	919.9	1176.3	.41963	1.2317	1.6513
60	292.73	.017378	7.177	262.06	836.3	1098.3	262.25	915.8	1178.0	.42733	1.2170	1.6444
65	298.00	.017429	6.657	267.46	832.1	1099.5	267.67	911.9	1179.6	.43450	1.2035	1.6380
70	302.96	.017478	6.209	272.56	828.1	1100.6	272.79	908.3	1181.0	.44120	1.1909	1.6321
75	307.63	.017524	5.818	277.37	824.3	1101.6	277.61	904.8	1182.4	.44749	1.1790	1.6265
80	312.07	.017570	5.474	281.95	820.6	1102.6	282.21	901.4	1183.6	.45344	1.1679	1.6214
85	316.29	.017613	5.170	286.30	817.1	1103.5	286.58	898.2	1184.8	.45907	1.1574	1.6165
90	320.31	.017655	4.898	290.46	813.8	1104.3	290.76	895.1	1185.9	.46442	1.1475	1.6119
95	324.16	.017696	4.654	294.45	810.6	1105.0	294.76	892.1	1186.9	.46952	1.1380	1.6076
100	327.86	.017736	4.434	298.28	807.5	1105.8	298.61	889.2	1187.8	.47439	1.1290	1.6034
105	331.41	.017775	4.234	301.97	804.5	1106.5	302.31	886.4	1188.7	.47906	1.1204	1.5995
110	334.82	.017813	4.051	305.52	801.6	1107.1	305.88	883.7	1189.6	.48355	1.1122	1.5957
115	338.12	.017850	3.884	308.95	798.8	1107.7	309.33	881.0	1190.4	.48786	1.1042	1.5921
120	341.30	.017886	3.730	312.27	796.0	1108.3	312.67	878.5	1191.1	.49201	1.0966	1.5886
125	344.39	.017922	3.588	315.49	793.3	1108.8	315.90	875.9	1191.8	.49602	1.0893	1.5853
130	347.37	.017957	3.457	318.61	790.7	1109.4	319.04	873.5	1192.5	.49989	1.0822	1.5821
135	350.27	.017991	3.335	321.64	788.2	1109.8	322.08	871.1	1193.2	.50364	1.0754	1.5790
140	353.08	.018024	3.221	324.58	785.7	1110.3	325.05	868.7	1193.8	.50727	1.0688	1.5761
145	355.82	.018057	3.115	327.45	783.3	1110.8	327.93	866.4	1194.4	.51079	1.0624	1.5732
150	358.48	.018089	3.016	330.24	781.0	1111.2	330.75	864.2	1194.9	.51422	1.0562	1.5704
160	363.60	.018152	2.836	335.63	776.4	1112.0	336.16	859.8	1196.0	.52078	1.0443	1.5651
170	368.47	.018214	2.676	340.76	772.0	1112.7	341.33	855.6	1196.9	.52700	1.0330	1.5600
180	373.13	.018273	2.533	345.68	767.7	1113.4	346.29	851.5	1197.8	.53292	1.0223	1.5553
190	377.59	.018331	2.405	350.39	763.6	1114.0	351.04	847.5	1198.6	.53857	1.0122	1.5507
200	381.86	.018387	2.289	354.9	759.6	1114.6	355.6	843.7	1199.3	.5440	1.0025	1.5464
225	391.87	.018523	2.043	365.6	750.2	1115.8	366.3	834.5	1200.8	.5566	.9799	1.5365
250	401.04	.018653	1.8448	375.4	741.4	1116.7	376.2	825.8	1202.1	.5680	.9594	1.5274
275	409.52	.018777	1.6813	384.5	733.0	1117.5	385.4	817.6	1203.1	.5786	.9406	1.5192
300	417.43	.018896	1.5442	393.0	725.1	1118.2	394.1	809.8	1203.9	.5883	.9232	1.5115
350	431.82	.019124	1.3267	408.7	710.3	1119.0	409.9	795.0	1204.9	.6060	.8917	1.4978
400	444.70	.019340	1.1620	422.8	696.7	1119.5	424.2	781.2	1205.5	.6218	.8638	1.4856
450	456.39	.019547	1.0326	435.7	683.9	1119.6	437.4	768.2	1205.6	.6360	.8385	1.4746
500	467.13	.019748	.9283	447.7	671.7	1119.4	449.5	755.8	1205.3	.6490	.8154	1.4645
550	477.07	.019943	.8423	458.9	660.2	1119.1	460.9	743.9	1204.8	.6611	.7941	1.4551
600	486.33	.02013	.7702	469.4	649.1	1118.6	471.7	732.4	1204.1	.6723	.7742	1.4464
700	503.23	.02051	.6558	488.9	628.2	1117.0	491.5	710.5	1202.0	.6927	.7378	1.4305
800	518.36	.02087	.5691	506.6	608.4	1115.0	509.7	689.6	1199.3	.7110	.7050	1.4160
900	532.12	.02123	.5009	523.0	589.6	1112.6	526.6	669.5	1196.0	.7277	.6750	1.4027
1000	544.75	.02159	.4459	538.4	571.5	1109.9	542.4	650.0	1192.4	.7432	.6471	1.3903
1250	572.56	.02250	.3454	573.4	528.3	1101.7	578.6	603.0	1181.6	.7778	.5841	1.3619
1500	596.39	.02346	.2769	605.0	486.9	1091.8	611.5	557.2	1168.7	.8082	.5276	1.3359
1750	617.31	.02450	.2266	634.4	445.9	1080.2	642.3	511.4	1153.7	.8361	.4748	1.3109
2000	636.00	.02565	.1813	662.4	404.2	1066.6	671.9	464.4	1136.3	.8623	.4238	1.2861
2250	652.90	.02698	.15692	689.9	360.7	1050.6	701.1	414.8	1115.9	.8876	.3728	1.2604
2500	668.31	.02860	.13059	717.7	313.4	1031.0	730.9	360.5	1091.4	.9131	.3196	1.2327
2750	682.46	.03077	.10717	747.3	258.6	1005.9	763.0	297.4	1060.4	.9401	.2604	1.2005
3000	695.52	.03431	.08404	783.4	185.4	968.8	802.5	213.0	1015.5	.9732	.1843	1.1575
3203.6	705.44	.05053	.05053	872.6	0	872.6	902.5	0	902.5	1.0580	0	1.0580

TABLE A-3

Properties of superheated steam

p(t Sat.)	Vapor 14.696 (211.99)				20 (227.96)				30 (250.34)			
t	v	u	h	s	v	u	h	s	v	u	h	s
Sat.	26.80	1077.6	1150.5	1.7567	20.09	1082.0	1156.4	1.7320	13.748	1088.0	1164.3	1.6996
150	24.10	1054.5	1120.0	1.7090	17.532	1052.0	1116.9	1.6710	11.460	1047.3	1111.0	1.6185
160	24.54	1058.2	1125.0	1.7171	17.870	1056.0	1122.1	1.6795	11.701	1051.6	1116.5	1.6275
170	24.98	1062.0	1129.9	1.7251	18.204	1059.9	1127.2	1.6877	11.938	1055.8	1122.0	1.6364
180	25.42	1065.7	1134.9	1.7328	18.535	1063.8	1132.4	1.6957	12.172	1059.9	1127.5	1.6449
190	25.85	1069.5	1139.8	1.7405	18.864	1067.6	1137.4	1.7036	12.403	1064.0	1132.9	1.6533
200	26.29	1073.2	1144.7	1.7479	19.191	1071.4	1142.5	1.7113	12.631	1068.1	1138.2	1.6615
210	26.72	1076.9	1149.5	1.7553	19.515	1075.2	1147.5	1.7188	12.857	1072.1	1143.5	1.6694
220	27.15	1080.6	1154.4	1.7624	19.837	1079.0	1152.4	1.7762	13.081	1076.1	1148.7	1.6771
230	27.57	1084.2	1159.2	1.7695	20.157	1082.8	1157.4	1.7335	13.303	1080.0	1153.9	1.6847
240	28.00	1087.9	1164.0	1.7764	20.475	1086.5	1162.3	1.7405	13.523	1084.0	1159.0	1.6921
250	28.42	1091.5	1168.8	1.7832	20.79	1090.3	1167.2	1.7475	13.741	1087.9	1164.1	1.6994
260	28.85	1095.2	1173.6	1.7899	21.11	1094.0	1172.1	1.7543	13.958	1091.7	1169.2	1.7064
270	29.27	1098.8	1178.4	1.7965	21.42	1097.7	1177.0	1.7610	14.173	1095.6	1174.2	1.7134
280	29.69	1102.4	1183.1	1.8030	21.73	1101.4	1181.8	1.7676	14.387	1099.4	1179.2	1.7202
290	30.11	1106.0	1187.9	1.8094	22.05	1105.0	1186.6	1.7741	14.600	1103.2	1184.2	1.7269
300	30.52	1109.6	1192.6	1.8157	22.36	1108.7	1191.5	1.7805	14.812	1106.9	1189.2	1.7334
310	30.94	1113.2	1197.4	1.8219	22.67	1112.4	1196.3	1.7868	15.023	1110.7	1194.1	1.7399
320	31.36	1116.8	1202.1	1.8280	22.98	1116.0	1201.0	1.7930	15.233	1114.4	1199.0	1.7462
330	31.77	1120.4	1206.8	1.8340	23.28	1119.7	1205.8	1.7991	15.442	1118.2	1203.9	1.7525
340	32.19	1124.0	1211.6	1.8400	23.59	1123.3	1210.6	1.8051	15.651	1121.9	1208.8	1.7586
350	32.60	1127.6	1216.3	1.8458	23.90	1126.9	1215.4	1.8110	15.859	1125.6	1213.6	1.7646
360	33.02	1131.2	1221.0	1.8516	24.21	1130.6	1220.1	1.8168	16.067	1129.3	1218.5	1.7706
370	33.43	1134.8	1225.7	1.8574	24.51	1134.2	1224.9	1.8226	16.273	1133.0	1223.3	1.7765
380	33.84	1138.4	1230.5	1.8630	24.82	1137.8	1229.7	1.8283	16.480	1136.7	1228.1	1.7822
390	34.26	1142.0	1235.2	1.8686	25.12	1141.4	1234.4	1.8340	16.686	1140.3	1233.0	1.7880
400	34.67	1145.6	1239.9	1.8741	25.43	1145.1	1239.2	1.8395	16.891	1144.0	1237.8	1.7936
420	35.49	1152.8	1249.3	1.8850	26.03	1152.3	1248.7	1.8504	17.301	1151.4	1247.4	1.8047
440	36.31	1160.1	1258.8	1.8956	26.64	1159.6	1258.2	1.8611	17.709	1158.7	1257.0	1.8155
460	37.13	1167.3	1268.3	1.9060	27.25	1166.9	1267.7	1.8716	18.116	1166.1	1266.6	1.8260
480	37.95	1174.6	1277.8	1.9162	27.85	1174.2	1277.2	1.8819	18.523	1173.4	1276.2	1.8364
500	38.77	1181.8	1287.3	1.9263	28.46	1181.5	1286.8	1.8919	18.928	1180.8	1285.9	1.8465
520	39.59	1189.1	1296.8	1.9361	29.06	1188.8	1296.3	1.9018	19.333	1188.2	1295.5	1.8564
540	40.41	1196.5	1306.4	1.9457	29.66	1196.1	1305.9	1.9114	19.737	1195.5	1305.1	1.8661
560	41.22	1203.8	1315.9	1.9552	30.26	1203.5	1315.5	1.9210	20.140	1203.0	1314.8	1.8757
580	42.04	1211.2	1325.5	1.9645	30.87	1210.9	1325.2	1.9303	20.543	1210.4	1324.4	1.8851
600	42.86	1218.6	1335.2	1.9737	31.47	1218.4	1334.8	1.9395	20.95	1217.8	1334.1	1.8943
620	43.67	1226.1	1344.8	1.9827	32.07	1225.8	1344.5	1.9485	21.35	1225.3	1343.8	1.9034
640	44.49	1233.5	1354.5	1.9916	32.67	1233.3	1354.2	1.9575	21.75	1232.8	1353.6	1.9123
660	45.30	1241.0	1364.2	2.0004	33.27	1240.8	1363.9	1.9662	22.15	1240.4	1363.3	1.9211
680	46.12	1248.6	1374.0	2.0090	33.87	1248.3	1373.7	1.9749	22.55	1247.9	1373.1	1.9298
700	46.93	1256.1	1383.8	2.0175	34.47	1255.9	1383.5	1.9834	22.95	1255.5	1383.0	1.9384
720	47.75	1263.7	1393.6	2.0259	35.07	1263.5	1393.3	1.9918	23.35	1263.2	1392.8	1.9468
740	48.56	1271.4	1403.4	2.0342	35.66	1271.2	1403.2	2.0001	23.75	1270.8	1402.7	1.9551
760	49.37	1279.0	1413.3	2.0424	36.26	1278.8	1413.0	2.0082	24.15	1278.5	1412.6	1.9633
780	50.19	1286.7	1423.2	2.0504	36.86	1286.5	1423.0	2.0163	24.55	1286.2	1422.5	1.9714
800	51.00	1294.4	1433.1	2.0584	37.46	1294.3	1432.9	2.0243	24.95	1294.0	1432.5	1.9793
850	53.03	1313.9	1458.1	2.0778	38.96	1313.8	1457.9	2.0438	25.95	1313.5	1457.6	1.9988
900	55.07	1333.6	1483.4	2.0967	40.45	1333.5	1483.2	2.0627	26.95	1333.2	1482.8	1.0178
950	57.10	1353.5	1508.8	2.1151	41.94	1353.4	1508.6	2.0810	27.95	1353.2	1508.3	2.0362
1000	59.13	1373.7	1534.5	2.1330	43.44	1373.5	1534.3	2.0989	28.95	1373.3	1534.0	2.0541
1100	63.19	1414.6	1586.4	2.1674	46.42	1414.5	1586.3	2.1334	30.94	1414.3	1586.1	2.0886
1200	67.25	1456.5	1639.3	2.2003	49.41	1456.4	1639.2	2.1663	32.93	1456.2	1639.1	2.1215
1300	71.30	1499.3	1693.2	2.2318	52.39	1499.2	1693.1	2.1978	34.92	1499.1	1692.9	2.1530
1400	75.36	1543.0	1747.9	2.2621	55.37	1542.9	1747.9	2.2281	36.91	1542.8	1747.7	2.1833
1500	79.42	1587.6	1803.6	2.2912	58.35	1587.6	1803.5	2.2572	38.90	1587.5	1803.4	2.2125
1600	83.47	1633.2	1860.2	2.3194	61.33	1633.2	1860.1	2.2854	40.88	1633.1	1860.0	2.2407
1800	91.58	1727.0	1976.1	2.3731	67.29	1727.0	1976.0	2.3391	44.86	1726.9	1976.0	2.2944
2000	99.69	1824.4	2095.5	2.4237	73.25	1824.3	2095.4	2.3897	48.83	1824.2	2095.3	2.3450
2200	107.80	1924.8	2218.0	2.4716	79.21	1924.8	2218.0	2.4376	52.81	1924.7	2217.9	2.3929
2400	115.91	2028.1	2343.4	2.5170	85.17	2028.1	2343.3	2.4830	56.78	2028.0	2343.3	2.4383

TABLE A-3 (continued)

p (t Sat.)	40 (267.26)				50 (281.03)				60 (292.73)			
t	v	u	h	s	v	u	h	s	v	u	h	s
Sat.	10.501	1092.3	1170.0	1.6767	8.518	1095.6	1174.4	1.6589	7.177	1098.3	1178.0	1.6444
200	9.346	1064.6	1133.8	1.6243	7.370	1060.9	1129.1	1.5940	6.047	1057.1	1124.2	1.5680
210	9.523	1068.8	1139.3	1.6327	7.519	1065.4	1135.0	1.6029	6.178	1061.9	1130.5	1.5774
220	9.699	1073.0	1144.8	1.6409	7.665	1069.9	1140.8	1.6115	6.307	1066.6	1136.6	1.5864
230	9.872	1077.2	1150.3	1.6488	7.810	1074.2	1146.5	1.6198	6.432	1071.2	1142.6	1.5952
240	10.043	1081.3	1155.6	1.6565	7.952	1078.6	1152.1	1.6279	6.556	1075.7	1148.5	1.6036
250	10.212	1085.4	1161.0	1.6641	8.092	1082.8	1157.7	1.6358	6.677	1080.1	1154.3	1.6118
260	10.380	1089.4	1166.2	1.6714	8.231	1087.0	1163.1	1.6434	6.797	1084.5	1160.0	1.6198
270	10.546	1093.4	1171.4	1.6786	8.368	1091.1	1168.5	1.6509	6.915	1088.8	1165.6	1.6275
280	10.711	1097.3	1176.6	1.6857	8.504	1095.2	1173.9	1.6582	7.031	1093.0	1171.1	1.6351
290	10.875	1101.2	1181.7	1.6926	8.639	1099.2	1179.2	1.6653	7.146	1097.2	1176.5	1.6424
300	11.038	1105.1	1186.8	1.6993	8.772	1103.2	1184.4	1.6722	7.260	1101.3	1181.9	1.6496
310	11.200	1109.0	1191.9	1.7059	8.904	1107.2	1189.6	1.6790	7.373	1105.4	1187.3	1.6565
320	11.360	1112.8	1196.9	1.7124	9.036	1111.2	1194.8	1.6857	7.485	1109.5	1192.6	1.6634
330	11.520	1116.6	1201.9	1.7188	9.166	1115.1	1199.9	1.6922	7.596	1113.5	1197.8	1.6700
340	11.680	1120.4	-1206.9	1.7251	9.296	1119.0	1205.0	1.6986	7.706	1117.4	1203.0	1.6766
350	11.838	1124.2	1211.8	1.7312	9.425	1122.8	1210.0	1.7049	7.815	1121.4	1208.2	1.6830
360	11.996	1128.0	1216.8	1.7373	9.553	1126.7	1215.1	1.7110	7.924	1125.3	1213.3	1.6893
370	12.153	1131.7	1221.7	1.7432	9.681	1130.5	1220.1	1.7171	8.032	1129.2	1218.4	1.6955
380	12.310	1135.5	1226.6	1.7491	9.808	1134.3	1225.0	1.7231	8.139	1133.1	1223.5	1.7015
390	12.467	1139.2	1231.5	1.7549	9.935	1138.1	1230.0	1.7290	8.246	1136.9	1228.5	1.7075
400	12.623	1143.0	1236.4	1.7606	10.061	1141.9	1235.0	1.7348	8.353	1140.8	1233.5	1.7134
420	12.933	1150.4	1246.1	1.7718	10.312	1149.4	1244.8	1.7461	8.565	1148.4	1243.5	1.7249
440	13.243	1157.8	1255.8	1.7828	10.562	1156.9	1254.6	1.7572	8.775	1156.0	1253.4	1.7360
460	13.551	1165.2	1265.5	1.7934	10.811	1164.4	1264.4	1.7679	8.984	1163.6	1263.3	1.7469
480	13.858	1172.7	1275.2	1.8038	11.059	1171.9	1274.2	1.7784	9.192	1171.1	1273.2	1.7575
500	14.164	1180.1	1284.9	1.8140	11.305	1179.4	1284.0	1.7887	9.399	1178.6	1283.0	1.7678
520	14.469	1187.5	1294.6	1.8240	11.551	1186.8	1293.7	1.7988	9.606	1186.2	1292.8	1.7780
540	14.774	1194.9	1304.3	1.8338	11.796	1194.3	1303.5	1.8086	9.811	1193.7	1302.6	1.7879
560	15.078	1202.4	1314.0	1.8434	12.041	1201.8	1313.2	1.8183	10.016	1201.2	1312.4	1.7976
580	15.382	1209.9	1323.7	1.8529	12.285	1209.3	1323.0	1.8277	10.221	1208.8	1322.2	1.8071
600	15.685	1217.3	1333.4	1.8621	12.529	1216.8	1332.8	1.8371	10.425	1216.3	1332.1	1.8165
620	15.988	1224.8	1343.2	1.8713	12.772	1224.4	1342.5	1.8462	10.628	1223.9	1341.9	1.8257
640	16.291	1232.4	1353.0	1.8802	13.015	1231.9	1352.4	1.8552	10.832	1231.5	1351.7	1.8347
660	16.593	1239.9	1362.8	1.8891	13.258	1239.5	1362.2	1.8641	11.035	1239.1	1361.6	1.8436
680	16.895	1247.5	1372.6	1.8977	13.500	1247.1	1372.0	1.8728	11.237	1246.7	1371.5	1.8523
700	17.196	1255.1	1382.4	1.9063	13.742	1254.8	1381.9	1.8814	11.440	1254.4	1381.4	1.8609
720	17.498	1262.8	1392.3	1.9147	13.984	1262.4	1391.8	1.8898	11.642	1262.0	1391.3	1.8694
740	17.799	1270.5	1402.2	1.9231	14.226	1270.1	1401.7	1.8982	11.844	1269.7	1401.2	1.8778
760	18.100	1278.2	1412.1	1.9313	14.467	1277.8	1411.7	1.9064	12.045	1277.5	1411.2	1.8860
780	18.401	1285.9	1422.1	1.9394	14.708	1285.6	1421.7	1.9145	12.247	1285.2	1421.2	1.8942
800	18.701	1293.7	1432.1	1.9474	14.949	1293.3	1431.7	1.9225	12.448	1293.0	1431.2	1.9022
850	19.452	1313.2	1457.2	1.9669	15.551	1312.9	1456.8	1.9421	12.951	1312.7	1456.4	1.9218
900	20.202	1333.0	1482.5	1.9859	16.152	1332.7	1482.2	1.9611	13.452	1332.5	1481.8	1.9408
950	20.951	1352.9	1508.0	2.0043	16.753	1352.7	1507.7	1.9796	13.954	1352.5	1507.4	1.9593
1000	21.700	1373.1	1533.8	2.0223	17.352	1372.9	1533.5	1.9975	14.454	1372.7	1533.2	1.9773
1100	23.20	1414.2	1585.9	2.0568	18.551	1414.0	1585.6	2.0321	15.454	1413.8	1585.4	2.0119
1200	24.69	1456.1	1638.9	2.0897	19.747	1456.0	1638.7	2.0650	16.452	1455.8	1638.5	2.0448
1300	26.18	1498.9	1692.8	2.1212	20.943	1498.8	1692.6	2.0966	17.449	1498.7	1692.4	2.0764
1400	27.68	1542.7	1747.6	2.1515	22.138	1542.6	1747.4	2.1269	18.445	1542.5	1747.3	2.1067
1500	29.17	1587.4	1803.3	2.1807	23.332	1587.3	1803.2	2.1561	19.441	1587.2	1803.0	2.1359
1600	30.66	1633.0	1859.9	2.2089	24.53	1632.9	1859.8	2.1843	20.44	1632.8	1859.7	2.1641
1800	33.64	1726.8	1975.9	2.2626	26.91	1726.8	1975.8	2.2380	22.43	1726.7	1975.7	2.2179
2000	36.62	1824.2	2095.3	2.3132	29.30	1824.1	2095.2	2.2886	24.41	1824.0	2095.1	2.2685
2200	39.61	1924.7	2217.8	2.3611	31.68	1924.6	2217.8	2.3365	26.40	1924.5	2217.7	2.3164
2400	42.59	2028.0	2343.2	2.4066	34.07	2027.9	2343.1	2.3820	28.39	2027.8	2343.1	2.3618

TABLE A-3 (continued)

p(t Sat.) t	v	200 (381.86) u	h	s	v	300 (417.43) u	h	s	v	350 (431.82) u	h	s	v	400 (444.70) u	h	s
Sat.	2.289	1114.6	1199.3	1.5464	1.5442	1118.2	1203.9	1.5115	1.3267	1119.0	1204.9	1.4978	1.1620	1119.5	1205.5	1.4856
400	2.361	1123.5	1210.8	1.5600	1.4915	1108.2	1191.0	1.4967								
410	2.399	1128.2	1217.0	1.5672	1.5221	1114.0	1198.5	1.5054								
420	2.437	1132.9	1223.1	1.5741	1.5517	1119.6	1205.7	1.5136	1.2950	1112.0	1195.9	1.4875				
430	2.475	1137.5	1229.1	1.5809	1.5805	1125.0	1212.7	1.5216	1.3219	1117.9	1203.6	1.4962	1.1257	1110.3	1193.6	1.4723
440	2.511	1142.0	1234.9	1.5874	1.6086	1130.3	1219.6	1.5292	1.3480	1123.7	1211.0	1.5045	1.1506	1116.6	1201.7	1.4814
450	2.548	1146.4	1240.7	1.5938	1.6361	1135.4	1226.2	1.5365	1.3733	1129.2	1218.2	1.5125	1.1745	1122.6	1209.6	1.4901
460	2.584	1150.8	1246.5	1.6001	1.6630	1140.4	1232.7	1.5436	1.3979	1134.6	1225.2	1.5201	1.1977	1128.5	1217.1	1.4984
470	2.619	1155.2	1252.1	1.6062	1.6894	1145.3	1239.1	1.5505	1.4220	1139.9	1232.0	1.5275	1.2202	1134.1	1224.4	1.5063
480	2.654	1159.5	1257.7	1.6122	1.7154	1150.1	1245.3	1.5572	1.4455	1145.0	1238.6	1.5346	1.2421	1139.6	1231.5	1.5139
490	2.689	1163.7	1263.3	1.6181	1.7410	1154.8	1251.5	1.5637	1.4686	1150.0	1245.1	1.5415	1.2634	1144.9	1238.4	1.5212
500	2.724	1168.0	1268.8	1.6239	1.7662	1159.5	1257.5	1.5701	1.4913	1154.9	1251.5	1.5482	1.2843	1150.1	1245.2	1.5282
510	2.758	1172.2	1274.2	1.6295	1.7910	1164.1	1263.5	1.5763	1.5136	1159.7	1257.8	1.5546	1.3048	1155.2	1251.8	1.5351
520	2.792	1176.3	1279.7	1.6351	1.8156	1168.6	1269.4	1.5823	1.5356	1164.5	1263.9	1.5610	1.3249	1160.2	1258.2	1.5417
530	2.826	1180.5	1285.0	1.6405	1.8399	1173.1	1275.2	1.5882	1.5572	1169.1	1270.0	1.5671	1.3447	1165.0	1264.6	1.5481
540	2.860	1184.6	1290.4	1.6459	1.8640	1177.5	1281.0	1.5940	1.5786	1173.7	1276.0	1.5731	1.3642	1169.8	1270.8	1.5544
550	2.893	1188.7	1295.7	1.6512	1.8878	1181.9	1286.7	1.5997	1.5998	1178.3	1281.9	1.5790	1.3833	1174.6	1277.0	1.5605
560	2.926	1192.7	1301.0	1.6565	1.9114	1186.2	1292.3	1.6052	1.6207	1182.8	1287.7	1.5848	1.4023	1179.2	1283.0	1.5665
570	2.960	1196.8	1306.3	1.6616	1.9348	1190.5	1297.9	1.6107	1.6414	1187.2	1293.5	1.5904	1.4210	1183.8	1289.0	1.5723
580	2.993	1200.8	1311.6	1.6667	1.9580	1194.8	1303.5	1.6161	1.6619	1191.6	1299.3	1.5960	1.4395	1188.4	1294.9	1.5781
590	3.025	1204.9	1316.8	1.6717	1.9811	1199.0	1309.0	1.6214	1.6823	1196.0	1304.9	1.6014	1.4579	1192.9	1300.8	1.5837
600	3.058	1208.9	1322.1	1.6767	2.004	1203.2	1314.5	1.6266	1.7025	1200.3	1310.6	1.6068	1.4760	1197.3	1306.6	1.5892
620	3.123	1216.9	1332.5	1.6864	2.049	1211.6	1325.4	1.6368	1.7424	1208.9	1321.8	1.6172	1.5118	1206.1	1318.0	1.5999
640	3.188	1224.9	1342.9	1.6959	2.094	1220.0	1336.2	1.6467	1.7818	1217.4	1332.8	1.6274	1.5471	1214.8	1329.3	1.6103
660	3.252	1232.8	1353.2	1.7053	2.139	1228.2	1347.0	1.6564	1.8207	1225.8	1343.8	1.6372	1.5819	1223.4	1340.5	1.6203
680	3.316	1240.8	1363.5	1.7144	2.183	1236.4	1357.6	1.6658	1.8593	1234.2	1354.6	1.6469	1.6163	1231.9	1351.6	1.6301
700	3.379	1248.8	1373.8	1.7234	2.227	1244.6	1368.3	1.6751	1.8975	1242.5	1365.4	1.6562	1.6503	1240.4	1362.5	1.6397
720	3.442	1256.7	1384.1	1.7322	2.270	1252.8	1378.9	1.6841	1.9354	1250.8	1376.2	1.6654	1.6840	1248.8	1373.4	1.6490
740	3.505	1264.7	1394.4	1.7408	2.314	1261.0	1389.4	1.6930	1.9731	1259.1	1386.9	1.6744	1.7175	1257.2	1384.3	1.6581
760	3.568	1272.7	1404.7	1.7493	2.357	1269.1	1400.0	1.7017	2.0104	1267.3	1397.5	1.6832	1.7506	1265.5	1395.1	1.6670
780	3.631	1280.6	1415.0	1.7577	2.400	1277.3	1410.5	1.7103	2.0476	1275.6	1408.2	1.6919	1.7836	1273.8	1405.9	1.6758
800	3.693	1288.6	1425.3	1.7660	2.442	1285.4	1421.0	1.7187	2.085	1283.8	1418.8	1.7004	1.8163	1282.1	1416.6	1.6844
820	3.755	1296.6	1435.6	1.7741	2.485	1293.6	1431.5	1.7270	2.121	1292.0	1429.4	1.7088	1.8489	1290.5	1427.3	1.6928
840	3.818	1304.7	1446.0	1.7821	2.527	1301.7	1442.0	1.7351	2.158	1300.3	1440.0	1.7170	1.8813	1298.8	1438.0	1.7011
860	3.879	1312.7	1456.3	1.7900	2.569	1309.9	1452.5	1.7432	2.193	1308.5	1450.6	1.7251	1.9135	1307.1	1448.7	1.7093
880	3.941	1320.8	1466.7	1.7978	2.611	1318.1	1463.1	1.7511	2.231	1316.7	1461.2	1.7331	1.9456	1315.4	1459.4	1.7173
900	4.003	1328.9	1477.1	1.8055	2.653	1326.3	1473.6	1.7589	2.267	1325.0	1471.8	1.7409	1.9776	1323.7	1470.1	1.7252
920	4.064	1337.0	1487.5	1.8131	2.695	1334.5	1484.1	1.7666	2.303	1333.3	1482.5	1.7487	2.0094	1332.0	1480.8	1.7330
940	4.126	1345.2	1497.9	1.8206	2.736	1342.8	1494.7	1.7742	2.339	1341.6	1493.1	1.7563	2.0411	1340.4	1491.5	1.7407
960	4.187	1353.3	1508.3	1.8280	2.778	1351.1	1505.3	1.7817	2.375	1349.9	1503.7	1.7639	2.0727	1348.7	1502.2	1.7483
980	4.249	1361.6	1518.8	1.8353	2.819	1359.3	1515.8	1.7891	2.411	1358.2	1514.4	1.7713	2.1043	1357.1	1512.9	1.7558
1000	4.310	1369.8	1529.3	1.8425	2.860	1367.7	1526.5	1.7964	2.446	1366.6	1525.0	1.7787	2.136	1365.5	1523.6	1.7632
1020	4.371	1378.0	1539.8	1.8497	2.902	1376.0	1537.1	1.8036	2.482	1375.0	1535.7	1.7859	2.167	1373.9	1534.3	1.7705
1040	4.432	1386.3	1550.3	1.8568	2.943	1384.4	1547.7	1.8108	2.517	1383.4	1546.4	1.7931	2.198	1382.4	1545.1	1.7777
1060	4.493	1394.6	1560.9	1.8638	2.984	1392.7	1558.4	1.8178	2.553	1391.8	1557.1	1.8002	2.229	1390.8	1555.9	1.7849
1080	4.554	1403.0	1571.5	1.8707	3.025	1401.2	1569.1	1.8248	2.588	1400.2	1567.9	1.8072	2.261	1399.3	1566.6	1.7919
1100	4.615	1411.4	1582.2	1.8776	3.066	1409.6	1579.8	1.8317	2.624	1408.7	1578.6	1.8142	2.292	1407.8	1577.4	1.7989
1200	4.918	1453.7	1635.7	1.9109	3.270	1452.2	1633.8	1.8653	2.799	1451.5	1632.8	1.8478	2.446	1450.7	1631.8	1.8327
1300	5.220	1496.9	1690.1	1.9427	3.473	1495.6	1688.4	1.8973	2.974	1495.0	1687.6	1.8799	2.599	1494.3	1686.8	1.8648
1400	5.521	1540.9	1745.3	1.9732	3.675	1539.8	1743.8	1.9279	3.148	1539.3	1743.1	1.9106	2.752	1538.7	1742.4	1.8956
1500	5.822	1585.8	1801.3	2.0025	3.877	1584.8	1800.0	1.9573	3.321	1584.3	1799.4	1.9401	2.904	1583.8	1798.8	1.9251
1600	6.123	1631.6	1858.2	2.0308	4.078	1630.7	1857.0	1.9857	3.494	1630.2	1856.5	1.9685	3.055	1629.8	1855.9	1.9535
1800	6.722	1725.6	1974.4	2.0847	4.479	1724.9	1973.5	2.0396	3.838	1724.5	1973.1	2.0225	3.357	1724.1	1972.6	2.0076
2000	7.321	1823.0	2094.0	2.1354	4.879	1822.3	2093.2	2.0904	4.182	1822.0	2092.8	2.0733	3.658	1821.6	2092.4	2.0584
2200	7.920	1923.6	2216.7	2.1833	5.280	1922.9	2216.0	2.1384	4.525	1922.5	2215.6	2.1212	3.959	1922.2	2215.2	2.1064
2400	8.518	2026.8	2342.1	2.2288	5.679	2026.1	2341.4	2.1838	4.868	2025.8	2341.1	2.1667	4.260	2025.4	2340.8	2.1519

Thermodynamic properties of moist air standard barometric pressure, 14.696 psia (29.92 in. Hg)

Temp. (°F)	Humidity Ratio, lb$_w$/lb$_a$ (W_s)	Volume (ft³/lb dry air)			Enthalpy (Btu/lb dry air)			Entropy [Btu/lb(dry air)-°F]			Enthalpy, h_w (Btu/lb)	Entropy, s_w (Btu/lb-°F)	Vapor Pressure, p_s (in. Hg)
		v_a	v_{as}	v_s	h_a	h_{as}	h_s	s_a	s_{as}	s_s			
10	0.0013158	11.832	0.025	11.857	2.402	1.402	3.804	0.00517	0.00315	0.00832	−154.13	−0.3141	0.062901
12	0.0014544	11.883	0.028	11.910	2.882	1.550	4.433	0.00619	0.00347	0.00966	−153.17	−0.3120	0.069511
14	0.0016062	11.933	0.031	11.964	3.363	1.714	5.077	0.00721	0.00381	0.01102	−152.20	−0.3100	0.076751
16	0.0017724	11.984	0.034	12.018	3.843	1.892	5.736	0.00822	0.00419	0.01241	−151.22	−0.3079	0.084673
18	0.0019543	12.035	0.038	12.072	4.324	2.088	6.412	0.00923	0.00460	0.01383	−150.25	−0.3059	0.093334
20	0.0021531	12.085	0.042	12.127	4.804	2.303	7.107	0.01023	0.00505	0.01528	−149.27	−0.3038	0.102798
22	0.0023703	12.136	0.046	12.182	5.285	2.537	7.822	0.01123	0.00554	0.01677	−148.28	−0.3018	0.113130
24	0.0026073	12.186	0.051	12.237	5.765	2.793	8.558	0.01223	0.00607	0.01830	−147.30	−0.2997	0.124396
26	0.0028660	12.237	0.056	12.293	6.246	3.073	9.318	0.01322	0.00665	0.01987	−146.30	−0.2977	0.136684
28	0.0031480	12.287	0.062	12.349	6.726	3.378	10.104	0.01420	0.00728	0.02148	−145.31	−0.2956	0.150066
30	0.0034552	12.338	0.068	12.406	7.206	3.711	10.917	0.01519	0.00796	0.02315	−144.31	−0.2936	0.164631
31	0.0036190	12.363	0.072	12.435	7.447	3.888	11.335	0.01568	0.00832	0.02400	−143.80	−0.2926	0.172390
32	0.0037895	12.389	0.075	12.464	7.687	4.073	11.760	0.01617	0.00870	0.02487	−143.30	−0.2915	0.180479
33	0.003947	12.414	0.079	12.492	7.927	4.243	12.170	0.01665	0.00905	0.02570	1.03	0.0020	0.18791
34	0.004109	12.439	0.082	12.521	8.167	4.420	12.587	0.01714	0.00940	0.02655	2.04	0.0041	0.19559
35	0.004277	12.464	0.085	12.550	8.408	4.603	13.010	0.01763	0.00977	0.02740	3.05	0.0061	0.20356
36	0.004452	12.490	0.089	12.579	8.648	4.793	13.441	0.01811	0.01016	0.02827	4.05	0.0081	0.21181
37	0.004633	12.515	0.093	12.608	8.888	4.990	13.878	0.01860	0.01055	0.02915	5.06	0.0102	0.22035
38	0.004820	12.540	0.097	12.637	9.128	5.194	14.322	0.01908	0.01906	0.03004	6.06	0.0122	0.22920
39	0.005014	12.566	0.101	12.667	9.369	5.405	14.773	0.01956	0.01139	0.03095	7.07	0.0142	0.23835
40	0.005216	12.591	0.105	12.696	9.609	5.624	15.233	0.02004	0.01183	0.03187	8.07	0.0162	0.24784
41	0.005424	12.616	0.110	12.726	9.849	5.851	15.700	0.02052	0.01228	0.03281	9.08	0.0182	0.25765
42	0.005640	12.641	0.114	12.756	10.089	6.086	16.175	0.02100	0.01275	0.03375	10.08	0.0202	0.26781
43	0.005863	12.667	0.119	12.786	10.330	6.330	16.660	0.02148	0.01324	0.03472	11.09	0.0222	0.27831
44	0.006094	12.692	0.124	12.816	10.570	6.582	17.152	0.02196	0.01374	0.03570	12.09	0.0242	0.28919
45	0.006334	12.717	0.129	12.846	10.810	6.843	17.653	0.02244	0.01426	0.03669	13.09	0.0262	0.30042
46	0.006581	12.743	0.134	12.877	11.050	7.114	18.164	0.02291	0.01479	0.03770	14.10	0.0282	0.31206
47	0.006838	12.768	0.140	12.908	11.291	7.394	18.685	0.02339	0.01534	0.03873	15.10	0.0302	0.32408
48	0.007103	12.793	0.146	12.939	11.531	7.684	19.215	0.02386	0.01592	0.03978	16.10	0.0321	0.33651
49	0.007378	12.818	0.152	12.970	11.771	7.984	19.756	0.02433	0.01651	0.04084	17.10	0.0341	0.34937
50	0.007661	12.844	0.158	13.001	12.012	8.295	20.306	0.02480	0.01712	0.04192	18.11	0.0361	0.36264
51	0.007955	12.869	0.164	13.033	12.252	8.616	20.868	0.02528	0.01775	0.04302	19.11	0.0381	0.37636
52	0.008259	12.894	0.171	13.065	12.492	8.949	21.441	0.02575	0.01840	0.04415	20.11	0.0400	0.39054
53	0.008573	12.920	0.178	13.097	12.732	9.293	22.025	0.02622	0.01907	0.04529	21.11	0.0420	0.40518
54	0.008897	12.945	0.185	13.129	12.973	9.648	22.621	0.02668	0.01976	0.04645	22.11	0.0439	0.42030
55	0.009233	12.970	0.192	13.162	13.213	10.016	23.229	0.02715	0.02048	0.04763	23.11	0.0459	0.43592
56	0.009580	12.995	0.200	13.195	13.453	10.397	23.850	0.02762	0.02122	0.04884	24.11	0.0478	0.45205
57	0.009938	13.021	0.207	13.228	13.694	10.790	24.484	0.02808	0.02198	0.05006	25.11	0.0497	0.46870
58	0.010309	13.046	0.216	13.262	13.934	11.197	25.131	0.02855	0.02277	0.05132	26.11	0.0517	0.48589
59	0.010692	13.071	0.224	13.295	14.174	11.618	25.792	0.02901	0.02358	0.05259	27.11	0.0536	0.50363
60	0.011087	13.096	0.233	13.329	14.415	12.052	26.467	0.02947	0.02442	0.05389	28.11	0.0555	0.52193
61	0.011496	13.122	0.242	13.364	14.655	12.502	27.157	0.02994	0.02528	0.05522	29.12	0.0575	0.54082
62	0.011919	13.147	0.251	13.398	14.895	12.966	27.862	0.03040	0.02617	0.05657	30.11	0.0594	0.56032
63	0.012355	13.172	0.261	13.433	15.135	13.446	28.582	0.03086	0.02709	0.05795	31.11	0.0613	0.58041
64	0.012805	13.198	0.271	13.468	15.376	13.942	29.318	0.03132	0.02804	0.05936	32.11	0.0632	0.60113
65	0.013270	13.223	0.281	13.504	15.616	14.454	30.071	0.03178	0.02902	0.06080	33.11	0.0651	0.62252
66	0.013750	13.248	0.292	13.540	15.856	14.983	30.840	0.03223	0.03003	0.06226	34.11	0.0670	0.64454
67	0.014246	13.273	0.303	13.577	16.097	15.530	31.626	0.03269	0.03107	0.06376	35.11	0.0689	0.66725
68	0.014758	13.299	0.315	13.613	16.337	16.094	32.431	0.03315	0.03214	0.06529	36.11	0.0708	0.69065
69	0.015286	13.324	0.326	13.650	16.577	16.677	33.254	0.03360	0.03325	0.06685	37.11	0.0727	0.71479
70	0.015832	13.349	0.339	13.688	16.818	17.279	34.097	0.03406	0.03438	0.06844	38.11	0.0746	0.73966
71	0.016395	13.375	0.351	13.726	17.058	17.901	34.959	0.03451	0.03556	0.07007	39.11	0.0765	0.76528
72	0.016976	13.400	0.365	13.764	17.299	18.543	35.841	0.03496	0.03677	0.07173	40.11	0.0783	0.79167
73	0.017575	13.425	0.378	13.803	17.539	19.204	36.743	0.03541	0.03801	0.07343	41.11	0.0802	0.81882
74	0.018194	13.450	0.392	13.843	17.779	19.889	37.668	0.03586	0.03930	0.07516	42.11	0.0821	0.84684
75	0.018833	13.476	0.407	13.882	18.020	20.595	38.615	0.03631	0.04062	0.07694	43.11	0.0840	0.87567
76	0.019491	13.501	0.422	13.923	18.260	21.323	39.583	0.03676	0.04199	0.07875	44.10	0.0858	0.90533

Source: Abridged from *ASHRAE Handbook 1989 Fundamentals,* American Society of Heating, Refrigerating, and Air Conditioning Engineers, Atlanta, GA.

Temp. (°F)	Humidity Ratio, lb$_w$/lb$_a$ (W_s)	Volume (ft³/lb dry air)			Enthalpy (Btu/lb dry air)			Entropy [Btu/lb(dry air)-°F]			Enthalpy, h_w (Btu/lb)	Entropy, s_w (Btu/ lb-°F)	Vapor Pressure, p_s (in. Hg)
		v_a	v_{as}	v_s	h_a	h_{as}	h_s	s_a	s_{as}	s_s			
77	0.020170	13.526	0.437	13.963	18.500	22.075	40.576	0.03721	0.04339	0.08060	45.10	0.0877	0.93589
78	0.020871	13.551	0.453	14.005	18.741	22.851	41.592	0.03766	0.04484	0.08250	46.10	0.0896	0.96733
79	0.021594	13.577	0.470	14.046	18.981	23.652	42.633	0.03811	0.04633	0.08444	47.10	0.0914	0.99970
80	0.022340	13.602	0.487	14.089	19.222	24.479	43.701	0.03855	0.04787	0.08642	48.10	0.0933	1.03302
81	0.023109	13.627	0.505	14.132	19.462	25.332	44.794	0.03900	0.04945	0.08844	49.10	0.0951	1.06728
82	0.023902	13.653	0.523	14.175	19.702	26.211	45.913	0.03944	0.05108	0.09052	50.10	0.0970	1.10252
83	0.024720	13.678	0.542	14.220	19.943	27.120	47.062	0.03988	0.05276	0.09264	51.09	0.0988	1.13882
84	0.025563	13.703	0.561	14.264	20.183	28.055	48.238	0.04033	0.05448	0.09481	52.09	0.1006	1.17608
85	0.026433	13.728	0.581	14.310	20.424	29.021	49.445	0.04077	0.05626	0.09703	53.09	0.1025	1.21445
86	0.027329	13.754	0.602	14.356	20.664	30.017	50.681	0.04121	0.05809	0.09930	54.09	0.1043	1.25388
87	0.028254	13.779	0.624	14.403	20.905	31.045	51.949	0.04165	0.05998	0.10163	55.09	0.1061	1.29443
88	0.029208	13.804	0.646	14.450	21.145	32.105	53.250	0.04209	0.06192	0.10401	56.09	0.1080	1.33613
89	0.030189	13.829	0.669	14.498	21.385	33.197	54.582	0.04253	0.06392	0.10645	57.09	0.1098	1.37893
90	0.031203	13.855	0.692	14.547	21.626	34.325	55.951	0.04297	0.06598	0.10895	58.08	0.1116	1.42298
91	0.032247	13.880	0.717	14.597	21.866	35.489	57.355	0.04340	0.06810	0.11150	59.08	0.1134	1.46824
92	0.033323	13.905	0.742	14.647	22.107	36.687	58.794	0.04384	0.07028	0.11412	60.08	0.1152	1.51471
93	0.034433	13.930	0.768	14.699	22.347	37.924	60.271	0.04427	0.07253	0.11680	61.08	0.1170	1.56248
94	0.035577	13.956	0.795	14.751	22.588	39.199	61.787	0.04471	0.07484	0.11955	62.08	0.1188	1.61154
95	0.036757	13.981	0.823	14.804	22.828	40.515	63.343	0.04514	0.07722	0.12237	63.08	0.1206	1.66196
96	0.037972	14.006	0.852	14.858	23.069	41.871	64.940	0.04558	0.07968	0.12525	64.07	0.1224	1.71372
97	0.039225	14.032	0.881	14.913	23.309	43.269	66.578	0.04601	0.08220	0.12821	65.07	0.1242	1.76685
98	0.040516	14.057	0.912	14.969	23.550	44.711	68.260	0.04644	0.08480	0.13124	66.07	0.1260	1.82141
99	0.041848	14.082	0.944	15.026	23.790	45.198	69.988	0.04687	0.08747	0.13434	67.07	0.1278	1.87745
100	0.043219	14.107	0.976	15.084	24.031	47.730	71.761	0.04730	0.09022	0.13752	68.07	0.1296	1.93492
101	0.044634	14.133	1.010	15.143	24.271	49.312	73.583	0.04773	0.09306	0.14079	69.07	0.1314	1.99396
102	0.046090	14.158	1.045	15.203	24.512	50.940	75.452	0.04816	0.09597	0.14413	70.06	0.1332	2.05447
103	0.047592	14.183	1.081	15.264	24.752	52.621	77.373	0.04859	0.09897	0.14756	71.06	0.1349	2.11661
104	0.049140	14.208	1.118	15.326	24.993	54.354	79.346	0.04901	0.10206	0.15108	72.06	0.1367	2.18037
105	0.050737	14.234	1.156	15.390	25.233	56.142	81.375	0.04944	0.10525	0.15469	73.06	0.1385	2.24581
106	0.052383	14.259	1.196	15.455	25.474	57.986	83.460	0.04987	0.10852	0.15839	74.06	0.1402	2.31297
107	0.054077	14.284	1.236	15.521	25.714	59.884	85.599	0.05029	0.11189	0.16218	75.06	0.1420	2.38173
108	0.055826	14.309	1.279	15.588	25.955	61.844	87.799	0.05071	0.11537	0.16608	76.05	0.1438	2.45232
109	0.057628	14.335	1.322	15.657	26.195	63.866	90.061	0.05114	0.11894	0.17008	77.05	0.1455	2.52473
110	0.059486	14.360	1.367	15.727	26.436	65.950	92.386	0.05156	0.12262	0.17418	78.05	0.1473	2.59891
111	0.061401	14.385	1.414	15.799	26.677	68.099	94.776	0.05198	0.12641	0.17839	79.05	0.1490	2.67500
112	0.063378	14.411	1.462	15.872	26.917	70.319	97.237	0.05240	0.13032	0.18272	80.05	0.1508	2.75310
113	0.065411	14.436	1.511	15.947	27.158	72.603	99.760	0.05282	0.13433	0.18716	81.05	0.1525	2.83291
114	0.067512	14.461	1.562	16.023	27.398	74.964	102.362	0.05324	0.13847	0.19172	82.04	0.1543	2.91491
115	0.069676	14.486	1.615	16.101	27.639	77.396	105.035	0.05366	0.14274	0.19640	83.04	0.1560	2.99883
116	0.071908	14.512	1.670	16.181	27.879	79.906	107.786	0.05408	0.14713	0.20121	84.04	0.1577	3.08488
117	0.074211	14.537	1.726	16.263	28.120	82.497	110.617	0.05450	0.15165	0.20615	85.04	0.1595	3.17305
118	0.076586	14.562	1.784	16.346	28.361	85.169	113.530	0.05492	0.15631	0.21122	86.04	0.1612	3.26335
119	0.079036	14.587	1.844	16.432	28.601	87.927	116.528	0.05533	0.16111	0.21644	87.04	0.1629	3.35586
120	0.081560	14.613	1.906	16.519	28.842	90.770	119.612	0.05575	0.16605	0.22180	88.04	0.1647	3.45052
121	0.084169	14.638	1.971	16.609	29.083	93.709	122.792	0.05616	0.17115	0.22731	89.04	0.1664	3.54764
122	0.086860	14.663	2.037	16.700	29.323	96.742	126.065	0.05658	0.17640	0.23298	90.03	0.1681	3.64704
123	0.089633	14.688	2.106	16.794	29.564	99.868	129.432	0.05699	0.18181	0.23880	91.03	0.1698	3.74871
124	0.092500	14.714	2.176	16.890	29.805	103.102	132.907	0.05740	0.18739	0.24480	92.03	0.1715	3.85298
125	0.095456	14.739	2.250	16.989	30.045	106.437	136.482	0.05781	0.19314	0.25096	93.03	0.1732	3.95961
126	0.098504	14.764	2.325	17.090	30.286	109.877	140.163	0.05823	0.19907	0.25729	94.03	0.1749	4.06863
127	0.101657	14.789	2.404	17.193	30.527	113.438	143.965	0.05864	0.20519	0.26382	95.03	0.1766	4.18046
128	0.104910	14.815	2.485	17.299	30.767	117.111	147.878	0.05905	0.21149	0.27054	96.03	0.1783	4.29477
129	0.108270	14.840	2.569	17.409	31.008	120.908	151.916	0.05946	0.21800	0.27745	97.03	0.1800	4.41181
130	0.111738	14.865	2.655	17.520	31.249	124.828	156.076	0.05986	0.22470	0.28457	98.03	0.1817	4.53148

TABLE A-4 *(continued)*

Temp. (°F)	Humidity Ratio, lb$_w$/lb$_a$ (W_s)	Volume (ft³/lb dry air)			Enthalpy (Btu/lb dry air)			Entropy [Btu/lb(dry air)-°F]			En- thalpy, h_w (Btu/lb)	En- tropy, s_w (Btu/ lb-°F)	Vapor Pressure, p_s (in. Hg)
		v_a	v_{as}	v_s	h_a	h_{as}	h_s	s_a	s_{as}	s_s			
131	0.115322	14.891	2.745	17.635	31.489	128.880	160.370	0.06027	0.23162	0.29190	99.02	0.1834	4.65397
132	0.119023	14.916	2.837	17.753	31.730	133.066	164.796	0.06068	0.23876	0.29944	100.02	0.1851	4.77919
133	0.122855	14.941	2.934	17.875	31.971	137.403	169.374	0.06109	0.24615	0.30723	101.02	0.1868	4.90755
134	0.126804	14.966	3.033	17.999	32.212	141.873	174.084	0.06149	0.25375	0.31524	102.02	0.1885	5.03844
135	0.130895	14.992	3.136	18.127	32.452	146.504	178.957	0.06190	0.26161	0.32351	103.02	0.1902	5.17258
136	0.135124	15.017	3.242	18.259	32.693	151.294	183.987	0.06230	0.26973	0.33203	104.02	0.1919	5.30973
137	0.139494	15.042	3.352	18.394	32.934	156.245	189.179	0.06271	0.27811	0.34082	105.02	0.1935	5.44985
138	0.144019	15.067	3.467	18.534	33.175	161.374	194.548	0.06311	0.28678	0.34989	106.02	0.1952	5.59324
139	0.148696	15.093	3.585	18.678	33.415	166.677	200.092	0.06351	0.29573	0.35924	107.02	0.1969	5.73970
140	0.153538	15.118	3.708	18.825	33.656	172.168	205.824	0.06391	0.30498	0.36890	108.02	0.1985	5.88945
141	0.158552	15.143	3.835	18.978	33.897	177.857	211.754	0.06431	0.31456	0.37887	109.02	0.2002	6.04256
142	0.163748	15.168	3.967	19.135	34.138	183.754	217.892	0.06471	0.32446	0.38918	110.02	0.2019	6.19918
143	0.169122	15.194	4.103	19.297	34.379	189.855	244.233	0.06511	0.33470	0.39981	111.02	0.2035	6.35898
144	0.174694	15.219	4.245	19.464	34.620	196.183	230.802	0.06551	0.34530	0.41081	112.02	0.2052	6.52241
145	0.180467	15.244	4.392	19.637	34.860	202.740	237.600	0.06591	0.35626	0.42218	113.02	0.2068	6.68932
146	0.186460	15.269	4.545	19.815	35.101	209.550	244.651	0.06631	0.36764	0.43395	114.02	0.2085	6.86009
147	0.192668	15.295	4.704	19.999	35.342	216.607	251.949	0.06671	0.37941	0.44611	115.02	0.2101	7.03435
148	0.199110	15.320	4.869	20.189	35.583	223.932	259.514	0.06710	0.39160	0.45871	116.02	0.2118	7.21239
149	0.205792	15.345	5.040	20.385	35.824	231.533	267.356	0.06750	0.40424	0.47174	117.02	0.2134	7.39413
150	0.212730	15.370	5.218	20.589	36.064	239.426	275.490	0.06790	0.41735	0.48524	118.02	0.2151	7.57977
151	0.219945	15.396	5.404	20.799	36.305	247.638	283.943	0.06829	0.43096	0.49925	119.02	0.2167	7.76958
152	0.227429	15.421	5.596	21.017	36.546	256.158	292.705	0.06868	0.44507	0.51375	120.02	0.2184	7.96306
153	0.235218	15.446	5.797	21.243	36.787	265.028	301.816	0.06908	0.45973	0.52881	121.02	0.2200	8.16087
154	0.243309	15.471	6.005	21.477	37.028	274.245	311.273	0.06947	0.47494	0.54441	122.02	0.2216	8.36256
155	0.251738	15.497	6.223	21.720	37.269	283.849	321.118	0.06986	0.49077	0.56064	123.02	0.2233	8.56871
156	0.260512	15.522	6.450	21.972	37.510	293.849	331.359	0.07025	0.50723	0.57749	124.02	0.2249	8.77915
157	0.269644	15.547	6.686	22.233	37.751	304.261	342.012	0.07065	0.52434	0.59499	125.02	0.2265	8.99378
158	0.279166	15.572	6.933	22.505	37.992	315.120	353.112	0.07104	0.54217	0.61320	126.02	0.2281	9.21297
159	0.289101	15.598	7.190	22.788	38.233	326.452	364.685	0.07143	0.56074	0.63216	127.02	0.2297	9.43677
160	0.29945	15.623	7.459	23.082	38.474	338.263	376.737	0.07181	0.58007	0.65188	128.02	0.2314	9.6648
161	0.31027	15.648	7.740	23.388	38.715	350.610	389.325	0.07220	0.60025	0.67245	129.02	0.2330	9.8978
162	0.32156	15.673	8.034	23.707	38.956	363.501	402.457	0.07259	0.62128	0.69388	130.03	0.2346	10.1353
163	0.33336	15.699	8.341	24.040	39.197	376.979	416.175	0.07298	0.64325	0.71623	131.03	0.2362	10.3776
164	0.34572	15.724	8.664	24.388	39.438	391.095	430.533	0.07337	0.66622	0.73959	132.03	0.2378	10.6250
165	0.35865	15.749	9.001	24.750	39.679	405.865	445.544	0.07375	0.69022	0.76397	133.03	0.2394	10.8771
166	0.37220	15.774	9.355	25.129	39.920	421.352	461.271	0.07414	0.71535	0.78949	134.03	0.2410	11.1343
167	0.38639	15.800	9.726	25.526	40.161	437.578	477.739	0.07452	0.74165	0.81617	135.03	0.2426	11.3965
168	0.40131	15.825	10.117	25.942	40.402	454.630	495.032	0.07491	0.76925	0.84415	136.03	0.2442	11.6641
169	0.41698	15.850	10.527	26.377	40.643	472.554	513.197	0.07529	0.79821	0.87350	137.04	0.2458	11.9370
170	0.43343	15.875	10.959	26.834	40.884	491.372	532.256	0.07567	0.82858	0.90425	138.04	0.2474	12.2149
171	0.45079	15.901	11.414	27.315	41.125	511.231	552.356	0.07606	0.86058	0.93664	139.04	0.2490	12.4988
172	0.46905	15.926	11.894	27.820	41.366	532.138	573.504	0.07644	0.89423	0.97067	140.04	0.2506	12.7880
173	0.48829	15.951	12.400	28.352	41.607	554.160	595.767	0.07682	0.92962	1.00644	141.04	0.2521	13.0823
174	0.50867	15.976	12.937	28.913	41.848	577.489	619.337	0.07720	0.96707	1.04427	142.04	0.2537	13.3831
175	0.53019	16.002	13.504	29.505	42.089	602.139	644.229	0.07758	1.00657	1.08416	143.05	0.2553	13.6894
176	0.55294	16.027	14.103	30.130	42.331	628.197	670.528	0.07796	1.04828	1.12624	144.05	0.2569	14.0010
177	0.57710	16.052	14.741	30.793	42.572	655.876	698.448	0.07834	1.09253	1.17087	145.05	0.2585	14.3191
178	0.60274	16.078	15.418	31.496	42.813	685.260	728.073	0.07872	1.13943	1.21815	146.05	0.2600	14.6430
179	0.63002	16.103	16.139	32.242	43.054	716.524	759.579	0.07910	1.18927	1.26837	147.06	0.2616	14.9731
180	0.65911	16.128	16.909	33.037	43.295	749.871	793.166	0.07947	1.24236	1.32183	148.06	0.2632	15.3097
181	0.69012	16.153	17.730	33.883	43.536	785.426	828.962	0.07985	1.29888	1.37873	149.06	0.2647	15.6522
182	0.72331	16.178	18.609	34.787	43.778	823.487	867.265	0.08023	1.35932	1.43954	150.06	0.2663	16.0014

Source: Abridged from *ASHRAE Handbook—1985 Fundamentals,* American Society of Heating, Refrigerating and Air-Conditioning Engineers, Atlanta, Ga.

Appendix B

Weather Data

TABLE B-1*

Outdoor design conditions (winter and summer)[a,b]

STATE AND STATION	WINTER		SUMMER			STATE AND STATION	WINTER		SUMMER		
	Lati-tude	DB 97½%	DB 2½%	WB 2½%	Outdoor Daily Range		Lati-tude	DB 97½%	DB 2½%	WB 2½%	Outdoor Daily Range
ALABAMA						Monterey	36	37	79	63	20
Alexander City	33	20	94	78	21	Napa	38	34	92	68	30
Anniston AP	33	19	94	78	21	Needles AP	34	37	110	75	27
Auburn	32	25	96	79	21	Oakland AP	37	37	81	63	19
Birmingham AP	33	22	94	78	21	Oceanside	33	40	81	68	13
Decatur	34	19	95	78	22	Ontario	34	34	97	71	36
Dothan AP	31	27	95	80	20	Oxnard AFB	34	37	80	69	19
Florence AP	34	17	95	78	22	Palmdale AP	34	27	101	68	35
Gadsden	34	20	94	77	22	Palm Springs	33	36	108	78	35
Huntsville AP	34	17	95	77	23	Pasadena	34	39	93	70	29
Mobile AP	30	29	93	79	18	Petaluma	38	32	90	68	31
Mobile CO	30	32	94	79	16	Pomona CO	34	34	96	72	36
Montgomery	32	26	95	79	21	Redding AP	40	35	101	69	32
Selma-Graig AFB	32	27	96	80	21	Redlands	34	37	96	71	33
Talladega	33	19	95	78	21	Richmond	38	38	81	64	17
Tuscaloosa AP	33	23	96	80	22	Riverside-March AFB	33	34	96	71	37
						Sacramento AP	38	32	97	70	36
ALASKA						Salinas AP	36	35	85	65	24
Anchorage AP	61	−20	70	61	15	San Bernadino, Norton AFB	34	33	98	73	38
Barrow	71	−42	54	51	12	San Diego AP	32	44	83	70	12
Fairbanks AP	64	−50	78	63	24	San Fernando	34	37	97	72	38
Juneau AP	58	− 4	71	64	15	San Francisco AP	37	37	79	63	20
Kodiak	57	12	66	60	10	San Francisco CO	37	44	77	62	14
Nome AP	64	−28	62	56	10	San Jose AP	37	36	88	67	26
						San Luis Obispo	35	37	85	64	26
ARIZONA						Santa Ana AP	33	36	89	71	28
Douglas AP	31	22	98	69	31	Santa Barbara CO	34	36	84	66	24
Flagstaff AP	35	5	82	60	31	Santa Cruz	37	34	84	65	28
Fort Huachuca AP	31	28	93	68	27	Santa Maria AP	34	34	82	64	23
Kingman AP	35	29	100	69	30	Santa Monica CO	34	45	77	68	16
Nogales	31	24	98	71	31	Santa Paula	34	36	89	71	36
Phoenix AP	33	34	106	76	27	Santa Rosa	38	32	93	68	34
Prescott AP	34	19	94	66	30	Stockton AP	37	34	98	70	37
Tuscon AP	33	32	102	73	26	Ukiah	39	30	96	69	40
Winslow AP	35	13	95	65	32	Visalia	36	36	100	72	38
Yuma AP	32	40	109	78	27	Yreka	41	17	94	66	38
						Yuba City	39	34	100	70	36
ARKANSAS						**COLORADO**					
Blytheville AFB	36	17	96	79	21	Alamosa AP	37	−13	82	61	35
Camden	33	23	97	80	21	Boulder	40	8	90	63	27
El Dorado AP	33	23	96	80	21	Colorado Springs AP	38	4	88	62	30
Fayetteville AP	36	13	95	76	23	Denver AP	39	3	90	64	28
Fort Smith AP	35	19	99	78	24	Durango	37	4	86	63	30
Hot Springs Nat. Pk.	34	22	97	78	22	Fort Collins	40	− 5	89	62	28
Jonesboro	35	18	96	79	21	Grand Junction AP	39	11	94	63	29
Little Rock AP	34	23	96	79	22	Greeley	40	− 5	92	64	29
Pine Bluff AP	34	24	96	80	22	La Junta AP	38	− 2	95	71	31
Texarkana AP	33	26	97	79	21	Leadville	39	− 4	73	55	30
						Pueblo AP	38	− 1	94	67	31
CALIFORNIA						Sterling	40	− 2	93	66	30
Bakersfield AP	35	33	101	71	32	Trinidad AP	37	5	91	65	32
Barstow AP	34	28	102	72	37						
Blythe AP	33	35	109	77	28	**CONNECTICUT**					
Burbank AP	34	38	94	70	25	Bridgeport AP	41	8	88	76	18
Chico	39	33	100	70	36	Hartford, Brainard Field	41	5	88	76	22
Concord	38	36	92	67	32	New Haven AP	41	9	86	76	17
Covina	34	41	97	72	31	New London	41	8	86	75	16
Crescent City AP	41	36	69	60	18	Norwalk	41	4	89	76	19
Downey	34	38	90	71	22	Norwich	41	2	86	76	18
El Cajon	32	34	95	73	30	Waterbury	41	4	88	76	21
El Centro AP	32	35	109	80	34	Windsor Locks, Bradley Field	42	2	88	75	22
Escondido	33	36	92	72	30						
Eureka/Arcata AP	41	35	65	59	11	**DELAWARE**					
Fairfield-Travis AFB	38	34	94	69	34	Dover AFB	39	15	90	78	18
Fresno AP	36	31	99	72	34	Wilmington AP	39	15	90	77	20
Hamilton AFB	38	35	85	68	28						
Laguna Beach	33	39	80	68	18	**DISTRICT OF COLUMBIA**					
Livermore	37	30	97	69	24	Andrews AFB	38	16	91	77	18
Lompoc, Vandenburg AFB	34	38	79	63	20	Washington National AP	38	19	92	77	18
Long Beach AP	33	38	84	70	22						
Los Angeles AP	34	43	83	68	15	**FLORIDA**					
Los Angeles CO	34	44	90	70	20	Belle Glade	26	39	91	79	16
Merced-Castle AFB	37	32	99	72	36	Cape Kennedy AP	28	40	89	80	15
Modesto	37	36	98	71	36	Daytona Beach AP	29	36	92	80	15

Source: Reproduced with permission from *Handbook of Air Conditioning, Heating, and Ventilating,* 3rd ed., Industrial Press, New York, 1979.

[a]See notes at end of table.

[b]Temperature in degrees F.

TABLE B-1 *(continued)*

STATE AND STATION	WINTER		SUMMER			STATE AND STATION	WINTER		SUMMER		
	Lati-tude	DB 97½%	DB 2½%	WB 2½%	Outdoor Daily Range		Lati-tude	DB 97½%	DB 2½%	WB 2½%	Outdoor Daily Range
Fort Lauderdale	26	45	90	80	15	Joliet AP	41	− 1	92	77	20
Fort Myers AP	26	42	92	80	18	Kankakee	41	1	92	77	21
Fort Pierce	27	41	91	80	15	LaSalle/Peru	41	1	93	77	22
Gainesville AP	29	32	94	79	18	Macomb	40	1	93	78	22
Jacksonville AP	30	32	94	79	19	Moline AP	41	− 3	91	77	23
Key West AP	24	58	89	79	9	Mt. Vernon	38	10	95	78	21
Lakeland CO	28	39	93	79	17	Peoria AP	40	2	92	77	22
Miami AP	25	47	90	79	15	Quincy AP	40	2	95	79	22
Miami Beach CO	25	48	89	79	10	Rantoul, Chanute AFB	40	3	92	77	21
Ocala	29	33	94	79	18	Rockford	42	− 3	90	76	24
Orlando AP	28	37	94	79	17	Springfield AP	39	4	92	78	21
Panama City, Tyndall AFB	30	35	91	80	14	Waukegan	42	− 1	90	76	21
Pensacola CO	30	32	90	81	14	**INDIANA**					
St. Augustine	29	35	92	80	16	Anderson	40	5	91	77	22
St. Petersburg	28	42	91	80	16	Bedford	38	7	93	78	22
Sanford	28	37	93	79	17	Bloomington	39	7	92	78	22
Sarasota	27	39	91	80	17	Columbus, Bakalar AFB	39	7	92	78	22
Tallahassee AP	30	29	94	79	19	Crawfordsville	40	2	93	77	22
Tampa AP	28	39	91	80	17	Evansville AP	38	10	94	78	22
West Palm Beach AP	26	44	91	80	16	Fort Wayne AP	41	5	91	76	24
GEORGIA						Goshen AP	41	0	90	76	23
Albany, Turner AFB	31	30	96	79	20	Hobart	41	0	91	76	21
Americus	32	25	96	79	20	Huntington	40	2	92	76	23
Athens	34	21	94	77	21	Indianapolis AP	39	4	91	77	22
Atlanta AP	33	23	92	77	19	Jeffersonville	38	13	94	78	23
Augusta AP	33	23	95	79	19	Kokomo	40	4	92	76	22
Brunswick	31	31	95	80	18	Layfayette	40	3	92	77	22
Columbus, Lawson AFB	32	26	96	79	21	La Porte	41	0	91	76	22
Dalton	34	19	95	77	22	Marion	40	2	91	76	23
Dublin	32	25	96	79	20	Muncie	40	2	91	77	22
Gainesville	34	20	92	77	21	Peru, Bunker Hill AFB	40	1	89	76	22
Griffin	33	22	93	78	21	Richmond AP	39	3	91	77	22
La Grange	33	20	94	78	21	Shelbyville	39	6	92	77	22
Macon AP	32	27	96	79	22	South Bend AP	41	3	89	76	22
Marietta, Dobbins AFB	34	21	93	77	21	Terre Haute AP	39	7	93	78	22
Moultrie	31	30	95	79	20	Valparaiso	41	− 2	90	76	22
Rome AP	34	20	95	77	23	Vincennes	38	9	94	78	22
Savannah-Travis AP	32	27	94	80	20	**IOWA**					
Valdosta-Moody AFB	31	31	94	79	20	Ames	42	− 7	92	78	23
Waycross	31	28	95	79	20	Burlington AP	40	0	92	78	22
HAWAII						Cedar Rapids AP	41	− 4	90	76	23
Hilo AP	19	61	83	73	15	Clinton	41	− 3	90	77	23
Honolulu AP	21	62	85	74	12	Council Bluffs	41	− 3	94	78	22
Kaneohe	21	61	83	73	12	Des Moines AP	41	− 3	92	77	23
Wahiawa	21	61	84	74	14	Dubuque	42	− 7	90	76	22
IDAHO						Fort Dodge	42	− 8	92	77	23
Boise AP	43	10	93	66	31	Iowa City	41	− 4	91	77	22
Burley	42	8	93	66	35	Keokuk	40	1	93	78	22
Coeur d'Alene AP	47	7	91	65	31	Marshalltown	42	− 6	91	77	23
Idaho Falls	43	− 6	88	64	38	Mason City AP	43	− 9	88	75	24
Lewiston AP	46	12	96	66	32	Newton	41	− 5	93	77	23
Moscow	46	1	89	63	32	Ottumwa AP	41	− 2	93	78	22
Mountain Home AFB	43	9	96	66	36	Sioux City AP	42	− 6	93	77	24
Pocatello AP	43	− 2	91	63	35	Waterloo	42	− 8	89	76	23
Twin Falls AP	42	8	94	64	34	**KANSAS**					
ILLINOIS						Atchison	39	2	95	78	23
Aurora	41	− 3	91	77	20	Chanute AP	37	7	97	78	23
Belleville, Scott AFB	38	10	95	78	21	Dodge City AP	37	7	97	73	25
Bloomington	40	3	92	78	21	El Dorado	37	8	99	77	24
Carbondale	37	11	96	79	21	Emporia	38	7	97	77	25
Champaign/Urbana	40	4	94	78	21	Garden City AP	38	3	98	73	28
Chicago, Midway AP	41	1	92	76	20	Goodland AP	39	4	96	70	31
Chicago, O'Hare AP	42	0	90	75	20	Great Bend	38	6	99	76	28
Chicago, CO	41	1	91	76	15	Hutchinson AP	38	6	99	76	28
Danville	40	4	94	78	21	Liberal	37	8	100	73	28
Decatur	39	4	93	78	21	Manhatten, Fort Riley	39	4	98	78	24
Dixon	41	− 3	91	77	23	Parsons	37	9	97	78	23
Elgin	42	− 4	90	76	21	Russell AP	38	4	100	76	29
Freeport	42	− 6	90	77	24	Salina	38	7	99	76	26
Galesburg	41	0	92	78	22	Topeka	39	6	96	78	24
Greenville	39	7	94	78	21	Wichita	37	9	99	76	23

TABLE B-1 (continued)

STATE AND STATION	WINTER Lati-tude	WINTER DB 97½%	SUMMER DB 2½%	SUMMER WB 2½%	Outdoor Daily Range
KENTUCKY					
Ashland	38	10	92	76	22
Bowling Green AP	37	11	95	78	21
Corbin AP	37	9	91	77	23
Covington AP	39	8	90	76	22
Hopkinsville, Campbell AFB	36	14	95	78	21
Lexington AP	38	10	92	77	22
Louisville AP	38	12	93	78	23
Madisonville	37	11	94	78	22
Owensboro	37	10	94	78	23
Paducah AP	37	14	95	79	20
LOUISIANA					
Alexandria AP	31	29	95	80	20
Baton Rouge AP	30	30	94	80	19
Bogalusa	30	28	94	79	19
Houma	29	33	92	80	15
Layfayette AP	30	32	93	81	18
Lake Charles AP	30	33	93	79	17
Minden	32	26	96	80	20
Monroe AP	32	27	96	81	20
Natchitoches	31	26	97	80	20
New Orleans AP	30	35	91	80	16
Shreveport AP	32	26	96	80	20
MAINE					
Augusta AP	44	− 3	86	73	22
Bangor, Dow AFB	44	− 4	85	73	22
Caribou AP	46	−14	81	70	21
Lewiston	44	− 4	86	73	22
Millinocket AP	45	−12	85	72	22
Portland AP	43	0	85	73	22
Waterville	44	− 5	86	73	22
MARYLAND					
Baltimore AP	39	15	91	78	21
Baltimore CO	39	20	92	78	17
Cumberland	39	9	92	75	22
Frederick AP	39	11	92	77	22
Hagerstown	39	10	92	76	22
Salisbury	38	18	90	78	18
MASSACHUSETTS					
Boston AP	42	10	88	74	16
Clinton	42	2	85	74	17
Fall River	41	9	86	74	18
Framingham	42	3	89	74	17
Gloucester	42	6	84	73	15
Greenfield	42	− 2	87	74	23
Lawrence	42	1	88	74	22
Lowell	42	3	89	74	21
New Bedford	41	13	84	73	19
Pittsfield AP	42	− 1	84	72	23
Springfield, Westover AFB	42	2	88	74	19
Taunton	41	0	86	75	18
Worcester AP	42	1	87	73	18
MICHIGAN					
Adrian	41	4	91	75	23
Alpena AP	45	− 1	85	73	27
Battle Creek AP	42	5	89	74	23
Benton Harbor AP	42	3	88	74	20
Detroit Met. CAP	42	8	88	75	20
Escanaba	45	− 3	80	71	17
Flint AP	43	3	87	75	25
Grand Rapids AP	42	6	89	74	24
Holland	42	6	88	74	22
Jackson AP	42	4	89	75	23
Kalamazoo	42	5	89	75	23
Lansing AP	42	6	87	75	24
Marquette CO	46	− 4	86	71	18
Mt. Pleasant	43	1	87	74	24
Muskegon AP	43	8	85	74	21
Pontiac	42	4	88	75	21

STATE AND STATION	WINTER Lati-tude	WINTER DB 97½%	SUMMER DB 2½%	SUMMER WB 2½%	Outdoor Daily Range
Port Huron	43	3	88	74	21
Saginaw AP	43	3	86	75	23
Sault Ste. Marie AP	46	− 8	81	71	23
Traverse City AP	44	4	86	73	22
Ypsilanti	42	5	89	74	22
MINNESOTA					
Albert Lea	43	−10	89	76	24
Alexandria AP	45	−15	88	74	24
Bemidji AP	47	−28	84	72	24
Brainerd	46	−20	85	73	24
Duluth AP	46	−15	82	71	22
Fairbault	44	−12	88	75	24
Fergus Falls	46	−17	89	74	24
International Falls AP	48	−24	82	69	26
Mankato	44	−12	89	75	24
Minneapolis/St. Paul AP	44	−10	89	75	22
Rochester AP	44	−13	88	75	24
St. Cloud AP	45	−16	88	75	24
Virginia	47	−21	83	71	23
Willmar	45	−14	88	75	24
Winona	44	− 8	89	76	24
MISSISSIPPI					
Biloxi, Keesler AFB	30	32	92	81	16
Clarksdale	34	24	96	80	21
Columbus AFB	33	22	95	79	22
Greenville AFB	33	24	96	80	21
Greenwood	33	23	96	80	21
Hattiesburg	31	26	95	79	21
Jackson AP	32	24	96	78	21
Laurel	31	26	95	79	21
McComb AP	31	26	94	79	18
Meridian AP	32	24	95	79	22
Natchez	31	26	94	80	21
Tupelo	34	22	96	79	22
Vicksburg CO	32	26	95	80	21
MISSOURI					
Cape Girardeau	37	12	96	79	21
Columbia AP	39	6	95	78	22
Farmington AP	37	8	95	78	22
Hannibal	39	4	94	78	22
Jefferson City	38	6	95	78	23
Joplin AP	37	11	95	78	24
Kansas City AP	39	8	97	77	20
Kirksville AP	40	− 3	94	78	24
Mexico	39	3	94	78	22
Moberly	39	2	94	78	23
Poplar Bluff	36	13	96	79	22
Rolla	38	7	95	78	22
St. Joseph AP	39	3	95	78	23
St. Louis AP	38	8	95	78	21
St. Louis CO	38	11	94	78	18
Sedalia, Whiteman AFB	38	9	94	77	22
Sikeston	36	14	96	79	21
Springfield AP	37	10	94	77	23
MONTANA					
Billings AP	45	− 6	91	66	31
Bozeman	45	−11	85	60	32
Butte AP	46	−16	83	59	35
Cut Bank AP	48	−17	86	63	35
Glasgow AP	48	−20	93	67	29
Glendive	47	−16	93	69	29
Great Falls AP	47	−16	88	63	28
Havre	48	−15	87	64	33
Helena AP	46	−13	87	63	32
Kalispell AP	48	− 3	84	63	34
Lewiston AP	47	−14	86	63	30
Livingston AP	45	−13	88	62	32
Miles City AP	46	−15	94	69	30
Missoula AP	46	− 3	89	63	36

TABLE B-1 (continued)

STATE AND STATION	WINTER		SUMMER		
	Lati-tude	DB 97½%	DB 2½%	WB 2½%	Outdoor Daily Range
NEBRASKA					
Beatrice	40	1	97	77	24
Chadron AP	42	− 9	95	70	30
Columbus	41	− 3	96	76	25
Fremont	41	− 3	97	77	22
Grand Island AP	41	− 2	95	75	28
Hastings	40	1	96	75	27
Kearney	40	− 2	95	75	28
Lincoln CO	40	0	96	77	24
McCook	40	0	97	72	28
Norfolk	42	− 7	95	76	30
North Platte AP	41	− 2	94	73	28
Omaha AP	41	− 1	94	78	22
Scottsbluff AP	41	− 4	94	69	31
Sidney AP	41	− 2	92	69	31
NEVADA					
Carson City	39	7	91	61	42
Elko AP	40	− 7	92	62	42
Ely AP	39	− 2	88	59	39
Las Vegas AP	36	26	106	71	30
Lovelock AP	40	11	96	64	42
Reno AP	39	7	92	62	45
Reno CO	39	17	92	62	45
Tonopah AP	38	13	92	63	40
Winnemucca AP	40	5	95	62	42
NEW HAMPSHIRE					
Berlin	44	−15	85	71	22
Claremont	43	− 9	87	73	24
Concord AP	43	− 7	88	73	26
Keene	43	− 8	88	73	24
Laconia	43	−12	87	73	25
Manchester, Grenier AFB	43	1	89	74	24
Portsmouth, Pease AFB	43	3	86	73	22
NEW JERSEY					
Atlantic City CO	39	18	88	77	18
Long Branch	40	13	91	76	18
Newark AP	40	15	91	76	20
New Brunswick	40	12	89	76	19
Paterson	40	12	91	76	21
Phillipsburg	40	10	91	76	21
Trenton CO	40	16	90	77	19
Vineland	39	16	90	77	19
NEW MEXICO					
Alamagordo, Holloman AFB	32	22	98	69	30
Albuquerque AP	35	17	94	65	27
Artesia	32	19	99	70	30
Carlsbad AP	32	21	99	71	28
Clovis AP	34	17	97	69	28
Farmington AP	36	9	93	65	30
Gallup	35	− 1	90	63	32
Grants	35	− 3	89	63	32
Hobbs AP	32	19	99	71	29
Las Cruces	32	23	100	69	30
Los Alamos	35	9	86	63	32
Raton AP	36	2	90	65	34
Roswell, Walker AFB	33	19	99	70	33
Santa Fe CO	35	11	88	63	28
Silver City AP	32	18	93	67	30
Socorro AP	34	17	97	66	30
Tucumcari AP	35	13	97	70	28
NEW YORK					
Albany AP	42	0	88	74	23
Albany CO	42	5	89	74	20
Auburn	43	2	87	73	22
Batavia	43	3	87	74	22
Binghamton CO	42	2	89	72	20
Buffalo AP	43	6	86	73	21
Cortland	42	− 1	88	73	23
Dunkirk	42	8	86	74	18

STATE AND STATION	WINTER		SUMMER		
	Lati-tude	DB 97½%	DB 2½%	WB 2½%	Outdoor Daily Range
Elmira AP	42	5	90	73	24
Geneva	42	2	89	73	22
Glens Falls	43	− 7	86	72	23
Gloversville	43	− 2	87	73	23
Hornell	42	− 5	85	72	24
Ithaca	42	0	88	73	24
Jamestown	42	5	86	73	20
Kingston	42	2	90	74	22
Lockport	43	6	85	74	21
Massena AP	45	−12	84	74	20
Newburgh-Stewart AFB	41	6	89	76	21
NYC - Central Park	40	15	91	76	17
NYC - Kennedy AP	40	21	87	76	16
NYC - LaGuardia AP	40	16	90	76	16
Niagara Falls AP	43	7	86	74	20
Olean	42	− 3	85	72	23
Oneonta	42	− 3	87	72	24
Oswego CO	43	6	84	74	20
Plattsburg AFB	44	− 6	84	73	22
Poughkeepsie	41	3	90	75	21
Rochester AP	43	5	88	74	22
Rome-Griffiss AFB	43	− 3	87	74	22
Schenectady	42	− 1	88	73	22
Suffolk County AFB	40	13	84	75	16
Syracuse AP	43	2	87	74	20
Utica	43	− 2	87	73	22
Watertown	44	−10	84	74	20
NORTH CAROLINA					
Asheville AP	35	17	88	74	21
Charlotte AP	35	22	94	77	20
Durham	36	19	92	77	20
Elizabeth City AP	36	22	91	79	18
Fayetteville, Pope AFB	35	20	94	79	20
Goldsboro, Seymour-Johnson AFB	35	21	92	79	18
Greensboro AP	36	17	91	76	21
Greenville	35	22	93	80	19
Henderson	36	16	92	78	20
Hickory	35	18	91	76	21
Jacksonville	34	25	92	80	18
Lumberton	34	22	93	80	20
New Bern AP	35	22	92	80	18
Raleigh/Durham AP	35	20	92	78	20
Rocky Mount	36	20	93	79	19
Wilmington AP	34	27	91	81	18
Winston-Salem AP	36	17	91	76	20
NORTH DAKOTA					
Bismarck AP	46	−19	91	72	27
Devil's Lake	48	−19	89	71	25
Dickinson AP	46	−19	93	70	25
Fargo AP	46	−17	88	74	25
Grand Forks AP	48	−23	87	72	25
Jamestown AP	47	−18	91	73	26
Minot AP	48	−20	88	70	25
Williston	48	−17	90	69	25
OHIO					
Akron/Canton AP	41	6	87	73	21
Ashtabula	42	7	87	75	18
Athens	39	7	91	76	22
Bowling Green	41	3	91	75	23
Cambridge	40	4	89	76	23
Chillicothe	39	9	91	76	22
Cincinnati CO	39	12	92	77	21
Cleveland AP	41	7	89	75	22
Columbus AP	40	7	88	76	24
Dayton AP	39	6	90	75	20
Defiance	41	1	91	76	24
Findlay AP	41	4	90	76	24
Fremont	41	3	90	75	24
Hamilton	39	8	92	77	22
Lancaster	39	5	91	76	23
Lima	40	4	91	76	24

TABLE B-1 *(continued)*

STATE AND STATION	WINTER Lati-tude	WINTER DB 97½%	SUMMER DB 2½%	SUMMER WB 2½%	Outdoor Daily Range
Mansfield AP	40	3	89	75	22
Marion	40	6	91	76	23
Middletown	39	7	91	76	22
Newark	40	3	90	76	23
Norwalk	41	3	90	75	22
Portsmouth	38	9	92	76	22
Sandusky CO	41	8	90	75	21
Springfield	40	7	90	76	21
Steubenville	40	9	89	75	22
Toledo AP	41	5	90	75	25
Warren	41	4	88	74	23
Wooster	40	3	88	75	22
Youngstown AP	41	6	86	74	23
Zanesville AP	40	3	89	76	23
OKLAHOMA					
Ada	34	16	100	78	23
Altus AFB	34	18	101	76	25
Ardmore	34	19	101	78	23
Bartlesville	36	9	99	78	23
Chickasha	35	16	101	76	24
Enid-Vance AFB	36	14	100	77	24
Lawton AP	34	16	101	77	24
McAlester	34	17	100	78	23
Muskogee AP	35	16	99	78	23
Norman	35	15	99	77	24
Oklahoma City AP	35	15	97	77	23
Ponca City	36	12	100	77	24
Seminole	35	16	100	77	23
Stillwater	36	13	99	77	24
Tulsa AP	36	16	99	78	22
Woodward	36	8	101	74	26
OREGON					
Albany	44	27	88	67	31
Astoria AP	46	30	76	60	16
Baker AP	44	1	92	65	30
Bend	44	4	87	62	33
Corvallis	44	27	88	67	31
Eugene AP	44	26	88	67	31
Grants Pass	42	26	92	66	33
Klamath Falls AP	42	5	87	62	36
Medford AP	42	23	94	68	35
Pendleton AP	45	10	94	65	29
Portland AP	45	24	85	67	23
Portland CO	45	29	88	68	21
Roseburg AP	43	29	91	67	30
Salem AP	45	25	88	67	31
The Dalles	45	17	91	68	28
PENNSYLVANIA					
Allentown AP	40	5	90	75	22
Altoona CO	40	5	87	73	23
Butler	40	2	89	74	22
Chambersburg	40	9	92	75	23
Erie AP	42	11	85	74	18
Harrisburg AP	40	13	89	75	21
Johnstown	40	5	87	73	23
Lancaster	40	6	90	76	22
Meadville	41	4	86	73	21
New Castle	41	4	89	74	23
Philadelphia AP	39	15	90	77	21
Pittsburgh AP	40	9	87	74	22
Pittsburgh CO	40	11	88	74	19
Reading CO	40	9	90	76	19
Scranton/Wilkes-Barre	41	6	87	74	19
State College	40	6	87	73	23
Sunbury	40	7	89	75	22
Uniontown	39	8	88	74	22
Warren	41	1	87	73	24
West Chester	40	13	90	76	20
Williamsport AP	41	5	89	75	23
York	40	8	91	76	22

STATE AND STATION	WINTER Lati-tude	WINTER DB 97½%	SUMMER DB 2½%	SUMMER WB 2½%	Outdoor Daily Range
RHODE ISLAND					
Newport	41	11	84	74	16
Providence AP	41	10	86	75	19
SOUTH CAROLINA					
Anderson	34	22	94	76	21
Charleston AFB	32	27	92	80	18
Charleston CO	32	30	93	80	13
Columbia AP	34	23	96	79	22
Florence AP	34	25	94	79	21
Georgetown	33	26	91	80	18
Greenville AP	34	23	93	76	21
Greenwood	34	23	95	77	21
Orangeburg	33	25	95	79	20
Rock Hill	35	21	95	77	20
Spartanburg AP	35	22	93	76	20
Sumter-Shaw AFB	34	26	94	79	21
SOUTH DAKOTA					
Aberdeen AP	45	−18	92	75	27
Brookings	44	−15	90	75	25
Huron AP	44	−12	93	75	28
Mitchell	43	−11	94	76	28
Pierre AP	44	− 9	96	74	29
Rapid City AP	44	− 6	94	71	28
Sioux Falls AP	43	−10	92	75	24
Watertown AP	45	−16	90	74	26
Yankton	43	− 7	94	76	25
TENNESSEE					
Athens	33	18	94	76	22
Bristol-Tri City AP	36	16	90	75	22
Chattanooga AP	35	19	94	78	22
Clarksville	36	16	96	78	21
Columbia	35	17	95	78	21
Dyersburg	36	17	96	79	21
Greenville	35	14	91	75	22
Jackson AP	35	17	95	79	21
Knoxville AP	35	17	92	76	21
Memphis AP	35	21	96	79	21
Murfreesboro	35	17	94	78	22
Nashville AP	36	16	95	78	21
Tullahoma	35	17	94	78	22
TEXAS					
Abilene AP	32	21	99	75	22
Alice AP	27	34	99	80	20
Amarillo AP	35	12	96	71	26
Austin AP	30	29	98	78	22
Bay City	29	33	93	80	16
Beaumont	30	33	94	80	19
Beeville	28	32	97	80	18
Big Spring AP	32	22	98	73	26
Brownsville AP	25	40	92	80	18
Brownwood	31	25	100	75	22
Bryan AP	30	31	98	78	20
Corpus Christi AP	27	36	93	80	19
Corsicana	32	25	100	78	21
Dallas AP	32	24	99	78	20
Del Rio, Laughlin AFB	29	31	99	77	24
Denton	33	22	100	78	22
Eagle Pass	28	31	104	79	24
El Paso AP	31	25	98	69	27
Fort Worth AP	32	24	100	78	22
Galveston AP	29	36	89	81	10
Greenville	33	24	99	78	21
Harlingen	26	38	95	80	19
Houston AP	29	32	94	80	18
Houston CO	29	33	94	80	18
Huntsville	30	31	97	79	20
Killeen–Gray AFB	31	26	99	77	22
Lamesa	32	18	98	73	26
Laredo AFB	27	36	101	78	23
Longview	32	25	98	80	20

TABLE B-1 (continued)

STATE AND STATION	WINTER Lati-tude	WINTER DB 97½%	SUMMER DB 2½%	SUMMER WB 2½%	SUMMER Outdoor Daily Range
Lubbock AP	33	15	97	72	26
Lufkin AP	31	28	96	80	20
McAllen	26	38	100	79	21
Midland AP	32	23	98	73	26
Mineral Wells AP	32	22	100	77	22
Palestine CO	31	25	97	79	20
Pampa	35	11	98	72	26
Pecos	31	19	100	71	27
Plainview	34	14	98	72	26
Port Arthur AP	30	33	92	80	19
San Angelo, Goodfellow AFB	31	25	99	75	24
San Antonio AP	29	30	97	77	19
Sherman-Perrin AFB	33	23	99	78	22
Snyder	32	19	100	74	26
Temple	31	27	99	78	22
Tyler AP	32	24	97	79	21
Vernon	34	18	101	76	24
Victoria AP	28	32	96	79	18
Waco AP	31	26	99	78	22
Wichita Falls AP	34	19	100	76	24
UTAH					
Cedar City AP	37	6	91	64	32
Logan	41	7	91	65	33
Moab	38	16	98	65	30
Ogden CO	41	11	92	65	33
Price	39	7	91	64	33
Provo	40	6	93	66	32
Richfield	38	3	92	65	34
St. George CO	37	26	102	70	33
Salt Lake City AP	40	9	94	66	32
Vernal AP	40	− 6	88	63	32
VERMONT					
Barre	44	−13	84	72	23
Burlington AP	44	− 7	85	73	23
Rutland	43	− 8	85	73	23
VIRGINIA					
Charlottesville	38	15	90	77	23
Danville AP	36	17	92	77	21
Fredericksburg	38	14	92	78	21
Harrisonburg	38	9	90	77	23
Lynchburg AP	37	19	92	76	21
Norfolk AP	36	23	91	78	18
Petersburg	37	18	94	79	20
Richmond AP	37	18	93	78	21
Roanoke AP	37	18	91	75	23
Staunton	38	12	90	77	23
Winchester	39	10	92	76	21
WASHINGTON					
Aberdeen	47	27	80	61	16
Bellingham AP	48	18	74	65	19
Bremerton	47	29	81	66	20
Ellensburg AP	47	6	89	65	34

STATE AND STATION	WINTER Lati-tude	WINTER DB 97½%	SUMMER DB 2½%	SUMMER WB 2½%	SUMMER Outdoor Daily Range
Everett-Paine AFB	47	24	78	65	20
Kennewick	46	15	96	68	30
Longview	46	24	86	66	30
Moses Lake, Larson AFB	47	− 1	93	66	32
Olympia AP	47	25	83	65	32
Port Angeles	48	29	73	58	18
Seattle-Boeing Fld.	47	27	80	65	24
Seattle CO	47	32	79	65	19
Seattle-Tacoma AP	47	24	81	64	22
Spokane AP	47	4	90	64	28
Tacoma-McChord AFB	47	24	81	66	22
Walla Walla AP	46	16	96	68	27
Wenatchee	47	9	92	66	32
Yakima AP	46	10	92	67	36
WEST VIRGINIA					
Beckley	37	6	88	73	22
Bluefield AP	37	10	86	73	22
Charleston AP	38	14	90	75	20
Clarksburg	39	7	90	75	21
Elkins AP	38	5	84	73	22
Huntington CO	38	14	93	76	22
Martinsburg AP	39	10	94	77	21
Morgantown AP	39	7	88	74	22
Parkersburg CO	39	12	91	76	21
Wheeling	40	9	89	75	21
WISCONSIN					
Appleton	44	− 6	87	74	23
Ashland	46	−17	83	71	23
Beloit	42	− 3	90	76	24
Eau Claire AP	44	−11	88	74	23
Fond du Lac	43	− 7	87	74	23
Green Bay AP	44	− 7	85	73	23
La Crosse AP	43	− 8	88	76	22
Madison AP	43	− 5	88	75	22
Manitowoc	44	− 1	86	74	21
Marinette	45	− 4	86	72	20
Milwaukee AP	43	− 2	87	75	21
Racine	42	0	88	75	21
Sheboygan	43	0	87	74	20
Stevens Point	44	−12	87	73	23
Waukesha	43	− 2	89	75	22
Wausau AP	44	−14	86	72	23
WYOMING					
Casper AP	42	− 5	90	62	31
Cheyenne AP	41	− 2	86	62	30
Cody AP	44	− 9	87	60	32
Evanston	41	− 8	82	57	32
Lander AP	42	−12	90	62	32
Laramie AP	41	− 2	80	59	28
Newcastle	43	− 5	89	67	30
Rawlins	41	−11	84	61	40
Rock Springs AP	41	− 1	84	57	32
Sheridan AP	44	− 7	92	65	32
Torrington	42	− 7	92	67	30

EXPLANATION OF DESIGN CONDITIONS:

WINTER — 97½% indicates that the temperature will be at or above the design temperature shown 97½% of the time.

SUMMER — 2½% indicates that the temperature will exceed the design temperature shown only 2½% of the time.

OUTDOOR DAILY RANGE — The outdoor daily range of DB temperatures is the difference between the average maximum and average minimum temperatures during the warmest month at each location.

TABLE B-2*

Normal degree-days by months for U.S. cities[a]

City	Jul	Aug	Sep	Oct	Nov	Dec	Jan	Feb	Mar	Apr	May	Jun	Total
ALABAMA													
Anniston–A	0	0	17	118	438	614	614	485	381	128	25	0	2820
Birmingham–A	0	0	13	123	396	598	623	491	378	128	30	0	2780
Mobile–A	0	0	0	28	219	376	416	304	222	47	0	0	1612
Mobile–C	0	0	0	23	198	357	412	290	209	40	0	0	1529
Montgomery–A	0	0	0	69	304	491	517	388	288	80	0	0	2137
Montgomery–C	0	0	0	55	267	458	483	360	265	66	0	0	1954
ARIZONA													
Flagstaff–A	49	78	243	586	876	1135	1231	1014	949	687	465	212	7525
Phoenix–A	0	0	0	22	223	400	474	309	196	74	0	0	1698
Phoenix–C	0	0	0	13	182	360	425	275	175	62	0	0	1492
Prescott–A	0	0	34	261	582	843	921	717	626	368	164	17	4533
Tuscon–A	0	0	0	24	222	403	474	330	239	84	0	0	1776
Winslow–A	0	0	20	274	663	946	1001	706	605	335	144	8	4702
Yuma–A	0	0	0	0	105	259	318	167	88	14	0	0	951
ARKANSAS													
Fort Smith–A	0	0	9	131	435	698	775	571	418	127	24	0	3188
Little Rock–A	0	0	10	110	405	654	719	543	401	122	18	0	2982
Texarkana–A	0	0	0	69	317	527	600	441	324	84	0	0	2362
CALIFORNIA													
Bakersfield–A	0	0	0	41	273	505	561	350	259	105	21	0	2115
Beaumont–C	0	0	18	103	298	487	574	473	437	286	146	18	2840
Bishop–A	0	0	55	253	564	803	840	664	546	319	140	38	4222
Blue Canyon–A	36	41	105	369	633	822	893	809	815	597	397	202	5719
Burbank–A	0	0	11	59	186	324	396	308	265	152	85	22	1808
Eureka–C	267	248	264	335	411	508	552	465	493	432	375	282	4632
Fresno–A	0	0	0	86	345	580	629	400	304	145	43	0	2532
Los Angeles–A	31	22	56	87	200	301	378	305	273	185	121	56	2015
Los Angeles–C	0	0	17	41	140	253	328	244	212	129	68	19	1451
Mount Shasta–C	37	46	165	434	705	939	998	787	722	549	357	174	5913
Oakland–A	84	77	76	157	336	508	552	400	360	282	212	119	3163
Red Bluff–A	0	0	0	59	319	564	617	423	336	177	51	0	2546
Sacramento–A	0	0	22	98	357	595	642	428	348	222	103	7	2822
Sacramento–C	0	0	17	75	321	567	614	403	317	196	85	5	2600
Sandberg–C	0	0	26	211	465	701	781	678	629	435	261	56	4243
San Diego–A	11	7	24	52	147	255	317	247	223	151	97	43	1574
San Francisco–A	144	136	101	174	318	487	530	398	378	327	264	164	3421
San Francisco–C	189	177	110	128	237	406	462	336	317	279	248	180	3069
San Jose–C	7	11	26	97	270	450	487	342	308	229	137	46	2410
Santa Catalina–A	21	11	24	77	168	311	375	328	344	264	205	121	2249
Santa Maria–A	98	94	111	157	262	391	453	370	341	276	229	152	2934

**Source:* Reproduced with permission from *Handbook of Air Conditioning, Heating, and Ventilating*, 3rd ed., Industrial Press, New York, 1979.

[a]The accompanying monthly normal number of degree-days for 335 weather stations are based on records for the 30-year period, 1921 to 1950, inclusive, and are those calculated by the U.S. Weather Bureau and released for publication late in 1953. In the table, A indicates airport weather station; C, city office station. The degree-day normals are derived from the values for the monthly normal maximum and minimum temperature, the monthly mean temperature being the sum of the maximum and minimum divided by 2. They are computed on the standard base of 65 degrees. The seasonal total is the total of the monthly figures. The 30-year period covered is consistent with the term of years accepted by the World Meteorological Organization for climate normal.

TABLE B-2 (continued)

City	Jul	Aug	Sep	Oct	Nov	Dec	Jan	Feb	Mar	Apr	May	Jun	Total
COLORADO													
Alamosa-A	64	121	309	648	1065	1414	1491	1176	1029	699	440	203	8059
Colorado Springs-A	8	21	124	422	777	1039	1122	930	874	555	307	75	6254
Denver-A	5	11	120	425	771	1032	1125	924	843	525	286	65	6132
Denver-C	0	5	103	385	711	958	1042	854	797	492	260	60	5673
Grand Junction-A	0	0	36	333	792	1132	1271	924	738	402	145	23	5796
Pueblo-A	0	0	74	383	771	1051	1104	805	775	456	203	27	5709
CONNECTICUT													
Bridgeport-A	0	0	66	334	645	1014	1110	1008	871	561	249	38	5896
Hartford-A	0	14	101	384	699	1082	1178	1050	871	528	201	31	6139
New Haven-A	0	18	93	363	663	1026	1113	1005	865	567	261	52	6026
DELAWARE													
Wilmington-A	0	0	47	282	585	927	983	876	698	396	110	6	4910
DISTRICT OF COLUMBIA													
Washington-A	0	0	37	237	519	837	893	781	619	323	87	0	4333
Washington-C	0	0	32	231	510	831	884	770	606	314	80	0	4258
Silver Hill Obs.	0	0	53	270	549	865	918	798	632	347	107	0	4539
FLORIDA													
Apalachicola-C	0	0	0	17	154	304	352	263	184	33	0	0	1307
Daytona Beach-A	0	0	0	0	83	205	245	187	137	11	0	0	868
Fort Myers-A	0	0	0	0	25	101	124	95	60	0	0	0	405
Jacksonville-A	0	0	0	16	148	309	331	247	169	23	0	0	1243
Jacksonville-C	0	0	0	11	129	276	303	226	154	14	0	0	1113
Key West-A	0	0	0	0	0	22	34	25	8	0	0	0	89
Key West-C	0	0	0	0	0	18	28	24	7	0	0	0	77
Lakeland-C	0	0	0	0	60	167	185	142	95	0	0	0	649
Melbourne-A	0	0	0	0	44	127	169	121	76	0	0	0	537
Miami-A	0	0	0	0	8	52	58	48	12	0	0	0	178
Miami-C	0	0	0	0	5	48	57	48	15	0	0	0	173
Miami Beach	0	0	0	0	0	37	43	34	9	0	0	0	123
Orlando-A	0	0	0	0	61	161	188	148	92	0	0	0	650
Pensacola-C	0	0	0	18	177	334	383	275	203	45	0	0	1435
Tallahassee-A	0	0	0	31	209	366	385	287	203	38	0	0	1519
Tampa-A	0	0	0	0	60	163	201	148	102	0	0	0	674
W. Palm Beach-A	0	0	0	0	7	62	85	61	33	0	0	0	248
GEORGIA													
Albany-A	0	0	0	39	242	427	446	333	236	40	0	0	1763
Athens-A	0	0	5	100	390	614	629	515	404	128	15	0	2800
Atlanta-A	0	0	8	110	393	614	632	512	404	133	20	0	2826
Atlanta-C	0	0	8	107	387	611	632	515	392	135	24	0	2811
Augusta-A	0	0	0	59	282	494	521	412	308	62	0	0	2138
Columbus-A	0	0	0	78	326	547	563	437	348	97	0	0	2396
Macon-A	0	0	0	63	280	481	497	391	275	62	0	0	2049
Rome-A	0	0	8	140	435	673	700	560	436	159	27	0	3138
Savannah-A	0	0	0	38	225	412	424	330	238	43	0	0	1710
Valdosta-A	0	0	0	32	203	366	386	290	210	38	0	0	1525
IDAHO													
Boise-A	0	0	135	389	762	1054	1169	868	719	453	249	92	5890
Lewiston-A	0	0	133	406	747	961	1060	815	663	408	222	68	5483
Pocatello-A	0	0	183	487	873	1184	1333	1022	880	561	317	136	6976
Salmon	22	55	292	592	996	1380	1513	1103	905	561	334	109	7922

TABLE B-2 (continued)

City	Jul	Aug	Sep	Oct	Nov	Dec	Jan	Feb	Mar	Apr	May	Jun	Total
ILLINOIS													
Cairo-C	0	0	28	161	492	784	856	683	523	182	47	0	3756
Chicago-A	0	0	90	350	765	1147	1243	1053	868	507	229	58	6310
Joliet-A	0	16	114	390	798	1190	1277	1084	893	519	233	64	6578
Moline-A	0	8	96	363	786	1181	1296	1075	862	453	199	45	6364
Peoria-A	0	11	86	330	759	1128	1240	1028	828	435	192	41	6087
Springfield-A	0	6	83	315	723	1066	1166	958	769	404	171	32	5693
Springfield-C	0	0	56	259	666	1017	1116	907	713	350	127	14	5225
INDIANA													
Evansville-A	0	0	59	215	570	871	939	770	589	251	90	6	4360
Fort Wayne-A	0	17	107	377	759	1122	1200	1036	874	516	226	53	6287
Indianapolis-A	0	0	79	306	705	1051	1122	938	772	432	176	30	5611
Indianapolis-C	0	0	59	247	642	986	1051	893	725	375	140	16	5134
South Bend-A	5	13	101	381	789	1153	1252	1081	908	531	248	62	6524
Terre Haute-A	0	5	77	295	681	1023	1107	913	725	371	145	24	5366
IOWA													
Burlington-A	0	0	83	336	765	1150	1271	1036	822	425	179	34	6101
Charles City-C	17	30	151	444	912	1352	1494	1240	1001	537	256	70	7504
Davenport-C	0	7	79	320	756	1147	1262	1044	834	432	175	35	6091
Des Moines-A	5	12	99	355	798	1203	1330	1092	868	438	201	45	6446
Des Moines-C	0	6	89	346	777	1178	1308	1072	849	425	183	41	6274
Dubuque-A	8	28	149	444	882	1290	1414	1187	983	543	267	76	7271
Sioux City-A	8	17	128	405	885	1290	1423	1170	930	474	228	54	7012
KANSAS													
Concordia-C	0	0	55	277	687	1029	1144	899	725	341	146	20	5323
Dodge City-A	0	0	40	262	669	980	1076	840	694	347	135	15	5058
Goodland-A	0	0	95	413	825	1128	1215	974	884	534	241	58	6367
Topeka-A	0	8	59	271	672	1017	1125	885	694	326	137	15	5209
Topeka-C	0	0	42	242	630	977	1088	851	669	295	112	13	4919
Wichita-A	0	0	32	219	597	915	1023	778	619	280	101	7	4571
KENTUCKY													
Bowling Green-A	0	0	47	215	558	840	890	739	601	286	98	5	4279
Lexington-A	0	0	56	259	636	933	1008	854	710	368	140	15	4979
Louisville-A	0	0	51	232	579	871	933	778	611	285	94	5	4439
Louisville-C	0	0	41	206	549	849	911	762	605	270	86	0	4279
LOUISIANA													
Baton Rouge-A	0	0	0	27	215	373	424	293	215	48	0	0	1595
Burrwood	0	0	0	0	81	225	303	226	169	29	0	0	1033
Lake Charles-A	0	0	0	22	218	353	416	284	210	40	0	0	1543
New Orleans-A	0	0	0	7	169	308	364	248	190	31	0	0	1317
New Orleans-C	0	0	0	5	141	283	341	223	163	19	0	0	1175
Shreveport-A	0	0	0	53	305	490	550	386	272	61	0	0	2117
MAINE													
Caribou-A	85	133	354	710	1074	1562	1745	1546	1342	909	512	201	10173
Eastport-C	141	136	261	521	798	1206	1333	1201	1063	774	524	288	8246
Portland-A	15	56	199	515	825	1237	1373	1218	1039	693	394	117	7681

TABLE B-2 *(continued)*

City	Jul	Aug	Sep	Oct	Nov	Dec	Jan	Feb	Mar	Apr	May	Jun	Total
MARYLAND													
Baltimore–A	0	0	43	256	558	884	942	820	651	360	97	0	4611
Baltimore–C	0	0	29	207	489	812	880	776	611	326	73	0	4203
Frederick–A	0	0	47	270	588	930	1001	865	673	368	106	0	4854
MASSACHUSETTS													
Boston–A	0	7	77	315	618	998	1113	1002	849	534	236	42	5791
Nantucket–A	22	34	111	372	615	924	1020	949	880	642	394	139	6102
Pittsfield–A	25	63	213	543	843	1246	1358	1212	1060	690	336	105	7694
MICHIGAN													
Alpena–C	50	85	215	530	864	1218	1358	1263	1156	762	437	135	8073
Detroit-Willow Run–A	0	10	96	393	759	1125	1231	1089	915	552	244	55	6469
Detroit–A	0	8	96	381	747	1101	1203	1072	927	558	251	60	6404
Escanaba–C	62	95	247	555	933	1321	1473	1327	1203	804	471	166	8657
Grand Rapids–A	14	29	144	462	822	1169	1287	1154	1008	606	301	79	7075
Grand Rapids–C	0	20	105	394	756	1107	1215	1086	939	546	248	58	6474
Lansing–A	13	33	140	455	813	1175	1277	1142	986	591	287	70	6982
Marquette–C	69	87	236	543	933	1299	1435	1291	1181	789	477	189	8529
Muskegon–A	26	48	152	462	795	1110	1243	1134	1011	642	350	116	7089
Sault Ste. Marie–A	109	126	298	639	1005	1398	1587	1442	1302	846	499	224	9475
MINNESOTA													
Duluth–A	56	91	298	651	1140	1606	1758	1512	1327	846	474	178	9937
Duluth–C	66	91	277	614	1092	1550	1696	1448	1252	801	487	200	9574
Internat'l Falls–A	70	118	356	716	1230	1733	1922	1618	1395	834	437	171	10600
Minneapolis–A	8	17	157	459	960	1414	1562	1310	1057	570	259	80	7853
Rochester–A	24	38	182	499	975	1426	1572	1316	1073	600	298	92	8095
Saint Cloud–A	32	53	225	570	1068	1535	1690	1439	1181	663	331	106	8893
Saint Paul–A	12	21	154	459	951	1401	1553	1305	1051	564	256	77	7804
MISSISSIPPI													
Jackson–A	0	0	0	69	310	503	535	405	299	81	0	0	2202
Meridian–A	0	0	0	90	338	528	561	413	309	85	9	0	2333
Vicksburg–C	0	0	0	51	268	456	507	374	273	71	0	0	2000
MISSOURI													
Columbia–A	0	6	62	262	654	989	1091	876	698	326	135	14	5113
Kansas City–A	0	0	44	240	621	970	1085	851	666	292	111	8	4888
Saint Joseph–A	0	5	49	265	681	1048	1175	930	716	326	127	14	5336
Saint Louis–A	0	0	45	233	600	927	1017	820	648	297	101	11	4699
Saint Louis–C	0	0	38	202	570	893	983	792	620	270	94	7	4469
Springfield–A	0	8	61	249	615	908	1001	790	632	295	118	16	4693
MONTANA													
Billings–A	8	20	194	497	876	1172	1305	1089	958	564	304	119	7106
Butte–A	115	174	450	744	1104	1442	1575	1294	1172	804	561	325	9760
Glasgow–C	14	30	244	574	1086	1510	1683	1408	1119	597	312	113	8690
Great Falls–A	24	50	273	524	894	1194	1311	1131	1008	621	359	166	7555
Havre–A	20	38	270	564	1023	1383	1513	1291	1076	597	313	125	8213
Helena–A	36	66	320	617	999	1311	1469	1165	1017	654	399	197	8250
Helena–C	51	78	359	598	969	1215	1438	1114	992	660	427	225	8126
Kalispell–A	47	83	326	639	990	1249	1386	1120	970	639	391	215	8055
Miles City–A	6	11	187	525	966	1373	1516	1229	1048	570	285	106	7822
Missoula–A	22	57	292	623	993	1285	1414	1100	939	609	365	176	7873

TABLE B-2 (continued)

City	Jul	Aug	Sep	Oct	Nov	Dec	Jan	Feb	Mar	Apr	May	Jun	Total
NEBRASKA													
Grand Island-A	0	6	84	369	822	1178	1302	1044	849	423	195	39	6311
Lincoln-A	0	12	82	340	774	1144	1271	1030	822	401	190	38	6104
Lincoln-C	0	7	79	310	741	1113	1240	1000	704	377	172	32	5865
Norfolk-A	0	17	122	422	903	1280	1417	1159	933	501	251	60	7065
North Platte-A	7	11	120	425	846	1172	1271	1016	887	489	243	59	6546
Omaha-A	0	5	88	331	783	1166	1302	1058	831	389	175	32	6160
Scottsbluff-A	0	0	137	456	867	1178	1287	1030	933	567	305	81	6841
Valentine-C	11	10	145	461	891	1212	1361	1100	970	543	288	83	7075
NEVADA													
Elko-A	6	28	229	546	915	1181	1336	1025	896	612	378	183	7335
Ely-A	22	44	228	561	894	1181	1302	1033	921	639	418	200	7443
Las Vegas-C	0	0	0	61	344	564	653	423	288	92	0	0	2425
Reno-A	27	61	165	443	744	986	1048	804	756	519	318	165	6036
Tonopah	0	5	96	422	723	995	1082	860	763	504	272	91	5813
Winnemucca-A	0	17	180	508	822	1085	1153	854	794	546	299	111	6369
NEW HAMPSHIRE													
Concord-A	11	57	192	527	849	1271	1392	1226	1029	660	316	82	7612
NEW JERSEY													
Atlantic City-C	0	0	29	230	507	831	905	829	729	468	189	24	4741
Newark-A	0	0	47	301	603	961	1039	932	760	450	148	11	5252
Trenton-C	0	0	55	285	582	930	1004	904	735	429	133	11	5068
NEW MEXICO													
Albuquerque-A	0	0	10	218	630	899	970	714	589	289	70	0	4389
Clayton-A	0	0	68	318	678	927	995	795	729	420	184	24	5138
Raton-A	17	36	148	431	798	1104	1203	924	834	543	292	87	6417
Roswell-A	0	0	8	156	501	750	787	566	443	185	28	0	3424
NEW YORK													
Albany-A	0	24	139	443	780	1197	1318	1179	989	597	246	50	6962
Albany-C	0	6	98	388	708	1113	1234	1103	905	531	202	31	6319
Bear Mountain-C	0	25	119	409	753	1110	1212	1098	921	561	244	59	6511
Binghamton-A	16	63	192	518	834	1228	1342	1215	1051	672	318	88	7537
Binghamton-C	0	36	141	428	735	1113	1218	1100	927	570	240	48	6556
Buffalo-A	16	30	122	433	753	1116	1225	1128	992	636	315	72	6838
New York-La Guard.-A	0	0	28	250	546	908	992	907	760	447	141	10	4989
New York-C	0	0	39	263	561	908	995	904	753	456	153	18	5050
New York-Central Pk.	0	0	31	250	552	902	1001	910	747	435	130	7	4965
Oswego-C	20	39	139	430	738	1132	1249	1134	995	654	355	90	6975
Rochester-A	9	34	133	440	759	1141	1249	1148	992	615	289	54	6863
Schenectady-C	0	19	137	456	792	1212	1349	1207	1008	597	233	40	7050
Syracuse-A	0	29	117	396	714	1113	1225	1117	955	570	247	37	6520
NORTH CAROLINA													
Asheville-C	0	0	50	262	552	769	794	678	572	285	105	5	4072
Charlotte-A	0	0	7	147	438	682	704	577	449	172	29	0	3205
Greensboro-A	0	0	29	202	510	772	806	672	528	241	50	0	3810
Hatteras-C	0	0	0	63	244	481	527	487	394	171	25	0	2392
Raleigh-A	0	0	16	149	438	701	732	613	477	202	41	0	3369
Raleigh-C	0	0	10	118	387	651	691	577	440	172	29	0	3075
Wilmington-A	0	0	0	73	288	508	533	463	347	104	7	0	2323
Winston-Salem-A	0	0	28	182	492	756	797	666	519	232	49	0	3721

TABLE B-2 (continued)

City	Jul	Aug	Sep	Oct	Nov	Dec	Jan	Feb	Mar	Apr	May	Jun	Total
NORTH DAKOTA													
Bismarck–A	29	37	227	598	1098	1535	1730	1464	1187	657	355	116	9033
Devils Lake–C	47	61	276	654	1197	1668	1866	1576	1314	750	394	137	9940
Fargo–A	25	41	215	586	1122	1615	1795	1518	1231	687	338	101	9274
Williston–C	29	42	261	605	1101	1528	1705	1442	1194	663	360	138	9068
OHIO													
Akron-Canton–A	0	17	83	378	738	1082	1166	1033	884	537	235	50	6203
Cincinnati–A	0	6	77	295	648	973	1029	871	732	392	149	23	5195
Cincinnati–C	0	0	42	222	567	880	942	812	645	314	108	0	4532
Cincinnati-Abbe Obs.	0	0	56	263	612	930	989	846	682	347	132	13	4870
Cleveland–A	0	10	75	340	699	1057	1132	1019	874	531	223	46	6006
Cleveland–C	0	9	60	311	636	995	1101	977	846	510	223	49	5717
Columbus–A	0	8	69	337	693	1032	1094	946	781	444	180	31	5615
Columbus–C	0	0	59	299	654	983	1051	907	741	408	153	22	5277
Dayton–A	0	5	74	324	693	1032	1094	941	781	435	179	39	5597
Sandusky–C	0	0	66	327	684	1039	1122	997	853	513	217	41	5859
Toledo–A	0	12	102	387	756	1119	1197	1056	905	555	245	60	6394
Youngstown–A	0	19	83	355	732	1085	1163	1030	877	534	241	53	6172
OKLAHOMA													
Oklahoma City–A	0	0	14	154	480	769	865	650	490	182	40	0	3644
Oklahoma City–C	0	0	12	149	459	747	843	630	472	169	38	0	3519
Tulsa–A	0	0	18	152	462	750	856	644	485	173	44	0	3584
OREGON													
Astoria–A	138	111	146	338	537	691	772	613	611	459	357	222	4995
Baker–C	25	47	255	518	852	1138	1268	972	837	591	384	200	7087
Burns–C	10	37	219	552	855	1156	1274	946	809	552	349	159	6918
Eugene–A	33	34	144	381	591	756	831	624	567	423	270	125	4779
Meacham–A	88	102	294	605	903	1113	1243	1008	961	717	527	327	7888
Medford–A	0	0	77	326	624	822	862	627	552	381	207	69	4547
Pendleton–A	0	0	104	353	717	921	1066	795	614	386	197	51	5204
Portland–A	25	22	116	319	585	750	856	658	570	396	242	93	4632
Portland–C	13	14	85	280	534	701	791	594	515	347	199	70	4143
Roseburg–C	14	10	98	288	531	694	744	563	508	366	223	83	4122
Salem–A	21	23	113	326	588	744	825	622	564	408	249	91	4574
Sexton Summit	88	69	169	456	714	877	905	801	797	621	450	270	6217
Troutdale–A	33	31	131	335	591	766	874	664	574	405	256	115	4775
PENNSYLVANIA													
Allentown–A	0	9	89	366	690	1051	1132	1019	840	495	164	25	5880
Erie–C	0	17	76	352	672	1020	1128	1039	911	573	273	55	6116
Harrisburg–A	0	0	69	308	630	964	1051	921	750	423	128	14	5258
Park Place–C	14	57	173	484	807	1200	1277	1142	998	648	290	85	7175
Philadelphia–A	0	0	47	269	573	902	986	879	704	402	104	0	4866
Philadelphia–C	0	0	33	219	516	856	933	837	667	369	93	0	4523
Pittsburgh-Allegheny–A	0	6	78	336	678	1004	1073	955	784	447	167	27	5555
Pittsburgh-Gr. Pitt.–A	0	20	94	377	720	1057	1116	986	818	486	195	36	5905
Pittsburgh–C	0	0	56	298	612	924	992	879	735	402	137	13	5048
Reading–C	0	5	57	285	588	930	1017	902	725	411	123	11	5060
Scranton–C	0	18	115	389	693	1057	1141	1028	859	516	196	35	6047
Williamsport–A	0	16	101	377	699	1057	1132	1005	828	477	181	25	5898

TABLE B-2 *(continued)*

City	Jul	Aug	Sep	Oct	Nov	Dec	Jan	Feb	Mar	Apr	May	Jun	Total
RHODE ISLAND													
Block Island–A	6	21	88	330	591	927	1026	955	865	603	335	96	5843
Providence–A	0	26	107	381	672	1035	1125	1019	874	570	258	58	6125
Providence–C	0	7	68	330	624	986	1076	972	809	507	197	31	5607
SOUTH CAROLINA													
Charleston–A	0	0	0	52	270	456	472	379	281	63	0	0	1973
Charleston–C	0	0	0	34	214	410	445	363	260	43	0	0	1769
Columbia–A	0	0	0	82	338	558	566	468	340	83	0	0	2435
Columbia–C	0	0	0	76	308	524	538	443	318	77	0	0	2284
Florence–A	0	0	0	94	347	574	588	487	334	83	0	0	2507
Greenville–A	0	0	10	131	411	648	673	552	442	161	32	0	3060
Spartanburg–A	0	0	7	136	414	654	670	549	436	152	26	0	3044
SOUTH DAKOTA													
Huron–A	10	16	149	472	975	1407	1597	1327	1032	558	279	80	7902
Rapid City–A	32	24	103	500	891	1218	1361	1151	1045	615	357	148	7535
Sioux Falls–A	16	21	155	472	984	1414	1575	1274	1023	558	276	80	7848
TENNESSEE													
Bristol–A	0	0	58	239	576	815	818	697	576	274	95	0	4148
Chattanooga–A	0	0	24	169	477	710	725	588	467	179	45	0	3384
Knoxville–A	0	0	33	179	498	744	760	630	500	196	50	0	3590
Memphis–A	0	0	17	126	432	673	725	574	427	139	24	0	3137
Memphis–C	0	0	13	98	392	639	716	574	423	131	20	0	3006
Nashville–A	0	0	22	154	471	725	778	636	498	186	43	0	3513
TEXAS													
Abilene–A	0	0	5	98	350	595	673	479	344	113	0	0	2657
Amarillo–A	0	0	37	240	594	859	921	711	586	298	99	0	4345
Austin–A	0	0	0	30	214	402	484	322	211	50	0	0	1713
Big Springs–A	0	0	0	75	316	577	639	454	314	105	0	0	2480
Brownsville–A	0	0	0	0	59	159	219	106	74	0	0	0	617
Corpus Christi–A	0	0	0	0	113	252	330	192	118	6	0	0	1011
Dallas–A	0	0	0	53	299	518	607	432	288	75	0	0	2272
Del Rio–A	0	0	0	26	188	371	419	235	147	21	0	0	1407
El Paso–A	0	0	0	70	390	626	670	445	330	110	0	0	2641
Fort Worth–A	0	0	0	58	299	533	622	446	308	90	5	0	2361
Ft. Worth–A. Carter Fld.	0	0	0	57	299	524	619	432	326	81	0	0	2338
Galveston–A	0	0	0	0	132	286	362	249	176	28	0	0	1233
Galveston–C	0	0	0	0	131	271	356	247	176	30	0	0	1211
Houston–A	0	0	0	7	181	321	394	265	184	36	0	0	1388
Houston–C	0	0	0	0	162	303	378	240	166	27	0	0	1276
Laredo–A	0	0	0	0	91	215	270	134	71	0	0	0	781
Lubbock–A	0	0	23	173	492	756	812	613	481	201	36	0	3587
Palestine–C	0	0	0	45	260	440	531	368	265	71	0	0	1980
Port Arthur–A	0	0	0	20	218	349	406	274	211	39	0	0	1517
Port Arthur–C	0	0	0	8	170	315	381	258	181	27	0	0	1340
San Angelo–A	0	0	0	72	280	502	556	378	257	62	0	0	2107
San Antonio–A	0	0	0	25	201	374	462	293	190	34	0	0	1579
Victoria–A	0	0	0	0	131	277	352	209	143	14	0	0	1126
Waco–A	0	0	0	44	251	459	557	385	263	66	0	0	2025
Wichita Falls–A	0	0	5	115	404	657	756	538	394	140	16	0	3025

TABLE B-2 *(continued)*

City	Jul	Aug	Sep	Oct	Nov	Dec	Jan	Feb	Mar	Apr	May	Jun	Total
UTAH													
Blanding	0	0	100	409	792	1079	1190	913	800	510	272	73	6138
Milford-A	0	0	114	462	828	1147	1277	955	800	516	269	77	6445
Salt Lake City-A	0	0	88	381	771	1039	1194	885	741	453	233	81	5866
Salt Lake City-C	0	0	61	330	714	995	1119	857	701	414	208	64	5463
VERMONT													
Burlington-A	19	47	172	521	858	1308	1460	1313	1107	681	307	72	7865
VIRGINIA													
Cape Henry-C	0	0	0	120	366	648	698	636	512	267	60	0	3307
Lynchburg-A	0	0	49	236	531	808	846	722	584	289	82	5	4153
Norfolk-A	0	0	9	152	408	688	729	644	500	265	59	0	3454
Norfolk-C	0	0	5	118	354	636	679	602	464	220	41	0	3119
Richmond-A	0	0	33	210	498	791	828	708	550	271	66	0	3955
Richmond-C	0	0	31	181	456	750	787	675	529	254	57	0	3720
Roanoke-A	0	0	50	233	543	806	840	722	588	289	81	0	4152
WASHINGTON													
Ellensburg-A	13	17	176	496	849	1116	1268	949	753	504	296	105	6542
Kelso-A	85	84	186	409	636	784	856	652	605	453	316	173	5239
Northhead	239	205	234	341	486	636	704	585	598	492	406	285	5211
Olympia-A	91	83	207	434	645	794	868	700	660	498	338	183	5501
Omak	0	46	194	533	921	1212	1352	1061	781	453	222	59	6834
Port Angeles-A	233	226	303	459	603	719	772	652	645	519	422	297	5850
Seattle-C	49	45	134	329	540	679	753	602	558	396	246	107	4438
Seattle-Tacoma-A	75	70	192	412	633	781	862	675	636	477	307	155	5275
Spokane-A	17	28	205	508	879	1113	1243	988	834	561	330	140	6852
Stampede Pass	251	260	414	701	1002	1203	1280	1064	1063	837	636	438	9149
Tacoma-C	66	62	177	375	579	719	797	636	595	435	282	143	4866
Tattosh Island-C	295	288	315	406	528	648	713	610	629	525	437	330	5724
Walla Walla-C	0	0	93	308	675	890	1023	748	564	338	171	38	4848
Yakima-A	0	7	150	446	807	1066	1181	862	660	408	205	53	5845
WEST VIRGINIA													
Charleston-A	0	0	60	250	576	834	887	750	632	310	110	8	4417
Elkins-A	9	31	122	412	726	995	1017	910	797	477	224	53	5773
Huntington-C	0	0	35	210	549	803	837	728	570	251	85	5	4073
Parkersburg-C	0	0	56	272	600	896	949	826	672	347	119	13	4750
Petersburg-C	0	5	72	308	654	942	967	820	667	384	133	14	4966
WISCONSIN													
Green Bay-A	32	58	183	515	945	1392	1516	1336	1132	696	347	107	8259
LaCrosse-A	11	20	152	447	921	1380	1528	1280	1035	552	250	74	7650
Madison-A	13	31	150	459	891	1302	1423	1207	1008	579	272	82	7417
Madison-C	10	30	137	419	864	1287	1417	1207	1011	573	266	79	7300
Milwaukee-A	20	32	134	428	831	1218	1336	1142	983	621	351	109	7205
Milwaukee-C	11	24	112	397	795	1184	1302	1117	961	606	335	100	6944
WYOMING													
Casper-A	13	24	231	577	951	1225	1324	1095	1011	660	381	146	7638
Cheyenne-A	33	39	241	577	897	1125	1225	1044	1029	717	462	173	7562
Lander-A	7	23	244	632	1050	1383	1494	1179	1045	687	396	163	8303
Rock Springs-A	20	32	266	648	1038	1349	1457	1182	1110	735	443	193	8473
Sheridan-A	27	41	239	578	957	1271	1392	1170	1035	645	387	161	7903

Source: Reproduced with permission from C. Strock and R. L. Koral, *Handbook of Air Conditioning, Heating, and Ventilating,* 2nd ed. (New York: Industrial Press, 1965).

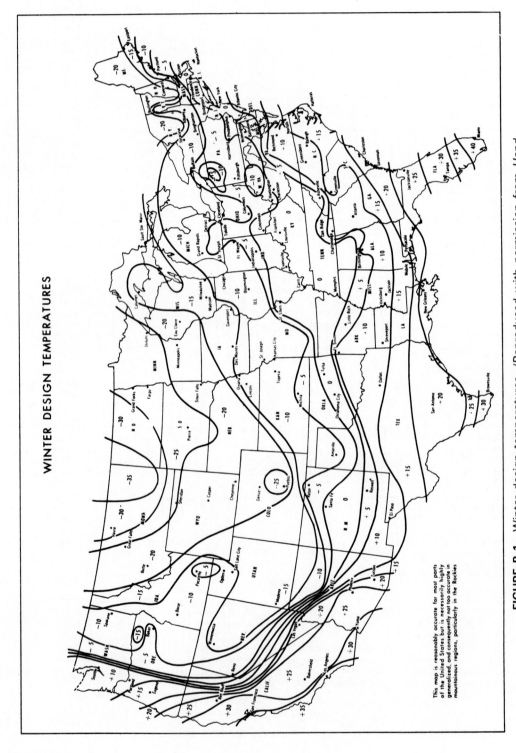

WINTER DESIGN TEMPERATURES

This map is reasonably accurate for most parts of the United States but is necessarily highly generalized, and consequently not too accurate in mountainous regions, particularly in the Rockies.

FIGURE B-1 Winter design temperatures. (Reproduced with permission from *Handbook of Air Conditioning, Heating, and Ventilating*, 3rd ed., Industrial Press, New York, 1979)

DATE NORMAL HEATING SEASON BEGINS

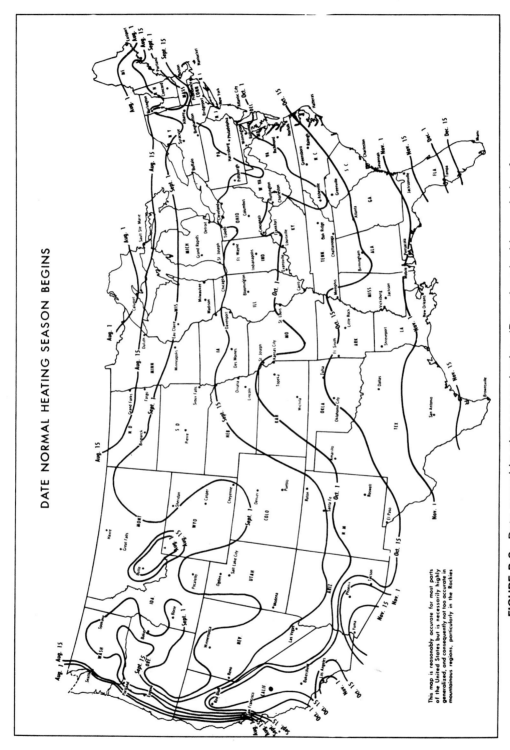

FIGURE B-2 Date normal heating season begins. (Reproduced with permission from *Handbook of Air Conditioning, Heating, and Ventilating*, 3rd ed., Industrial Press, New York, 1979)

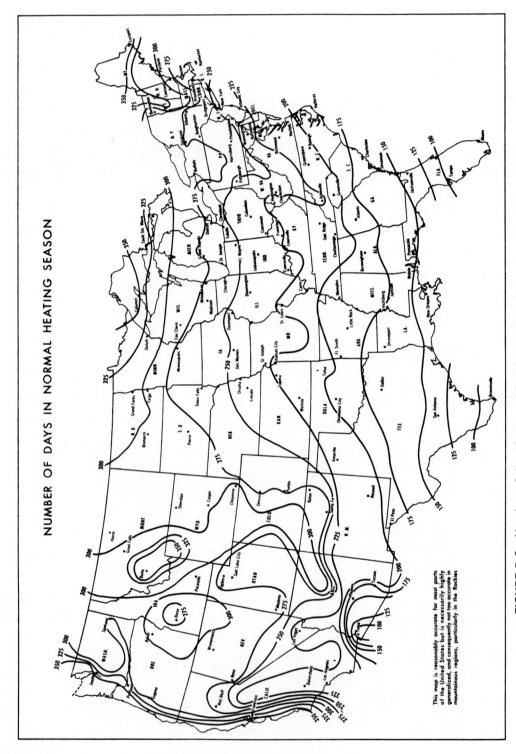

NUMBER OF DAYS IN NORMAL HEATING SEASON

This map is reasonably accurate for most parts of the United States but is necessarily highly generalized, and consequently not too accurate in mountainous regions, particularly in the Rockies

FIGURE B-3 Number of days in normal heating season. (Reproduced with permission from *Handbook of Air Conditioning, Heating, and Ventilating*, 3rd ed., Industrial Press, New York, 1979)

NORMAL NUMBER OF DEGREE-DAYS PER YEAR

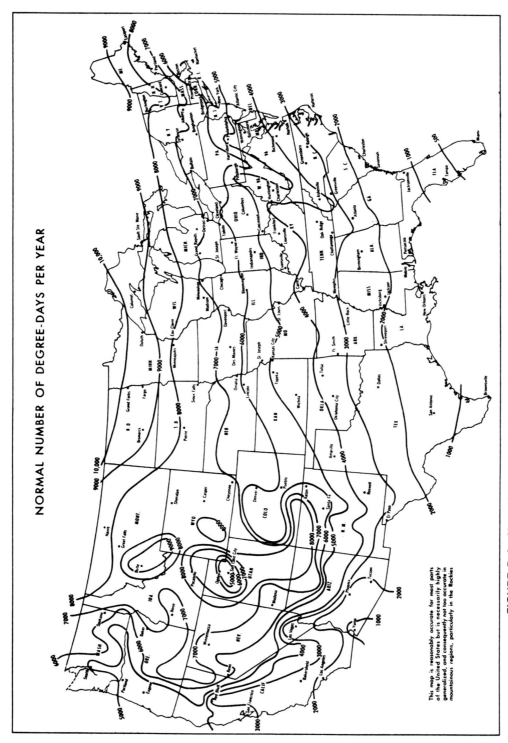

This map is reasonably accurate for most parts of the United States but is necessarily highly generalized, and consequently not too accurate in mountainous regions, particularly in the Rockies

FIGURE B-4 Normal number of degree-days per year. (Reproduced with permission from *Handbook of Air Conditioning, Heating, and Ventilating*, 3rd ed., Industrial Press, New York, 1979)

Appendix C

Thermal Properties of Building Components

TABLE C-1*

Thermal properties of typical building and insulating materials—design values[a]

Description	Density (lb/ft³)	Conductivity,[b] k (Btu-in./ hr-ft²-°F)	Conductance, C (Btu/hr- ft²-°F)	Resistance R Per Inch Thickness, $1/k$ (hr-ft²- °F/Btu)	Resistance R For Thickness Listed, $1/C$ (hr-ft²- °F/Btu)	Specific Heat (Btu/lb-°F)
Building Board (boards, panels, subflooring, sheathing, woodboard panel products)						
Asbestos-cement board	120	4.0	—	0.25	—	0.24
0.125 in.	120	—	33.00	—	0.03	
0.25 in.	120	—	16.50	—	0.06	
Gypsum or plaster board						
0.375 in.	50	—	3.10	—	0.32	0.26
0.5 in.	50	—	2.22	—	0.45	
0.625 in.	50	—	1.78	—	0.56	
Plywood (Douglas fir)[d]	34	0.80	—	1.25	—	0.29
0.25 in.	34	—	3.20	—	0.31	
0.375 in.	34	—	2.13	—	0.47	
0.5 in.	34	—	1.60	—	0.62	
0.625 in.	34	—	1.29	—	0.77	
Plywood or wood panels 0.75 in.	34	—	1.07	—	0.93	0.29
Vegetable fiber board sheathing, regular density[e]						

**Source:* Extracted from *ASHRAE Handbook 1989 Fundamentals* with permission from American Society of Heating, Refrigerating, and Air Conditioning Engineers, Atlanta, GA.

TABLE C-1 (continued)

Description	Density (lb/ft³)	Conductivity,[b] k (Btu-in./ hr-ft²-°F)	Conductance, C (Btu/hr-ft²-°F)	Resistance[c]R Per Inch Thickness, 1/k (hr-ft²-°F/Btu)	Resistance[c]R For Thickness Listed, 1/C (hr-ft²-°F/Btu)	Specific Heat (Btu/lb-°F)
0.5 in.	18	—	0.76	—	1.32	0.31
0.78125 in.	18	—	0.49	—	2.06	
Sheathing, intermediate density (0.5 in.)[e]	22	—	0.82	—	1.22	0.31
Nail base sheathing (0.5 in.)[e]	25	—	0.88	—	1.14	0.31
Shingle backer						
0.375 in.	18	—	1.06	—	0.94	0.31
0.3125 in.	18	—	1.28	—	0.78	
Sound-deadening board (0.5 in.)	15	—	0.74	—	1.35	0.30
Tile and lay-in panels, plain or acoustic	18	0.40	—	2.50	—	0.14
0.5 in.	18	—	0.80	—	1.25	
0.75 in.	18	—	0.53	—	1.89	
Laminated paperboard	30	0.50	—	2.00	—	0.33
Homogeneous board from repulped paper	30	0.50	—	2.00	—	0.28
Hardboard[e]						
Medium density	50	0.73	—	1.37	—	0.31
High density, service temp. service underlay	55	0.82	—	1.22	—	0.32
Particleboard[e]						
Low density	37	0.54	—	1.85	—	0.31
Medium density	50	0.94	—	1.06	—	0.31
High density	62.5	1.18	—	0.85	—	0.31
Underlayment (0.625 in.)	40	—	1.22	—	0.82	0.29
Wood subfloor (0.75 in.)		—	1.06	—	0.94	0.33

Building Membrane

Description	Density (lb/ft³)	Conductivity,[b] k (Btu-in./ hr-ft²-°F)	Conductance, C (Btu/hr-ft²-°F)	Resistance[c]R Per Inch Thickness, 1/k (hr-ft²-°F/Btu)	Resistance[c]R For Thickness Listed, 1/C (hr-ft²-°F/Btu)	Specific Heat (Btu/lb-°F)
Vapor-permeable felt	—	—	16.70	—	0.06	
Vapor-seal, two layers of mopped 15-lb felt	—	—	8.35	—	0.12	
Vapor-seal, plastic film	—	—	—	—	Negl.	

Finish Flooring Materials

Description	Density (lb/ft³)	Conductivity,[b] k (Btu-in./ hr-ft²-°F)	Conductance, C (Btu/hr-ft²-°F)	Resistance[c]R Per Inch Thickness, 1/k (hr-ft²-°F/Btu)	Resistance[c]R For Thickness Listed, 1/C (hr-ft²-°F/Btu)	Specific Heat (Btu/lb-°F)
Carpet and fibrous pad	—	—	0.48	—	2.08	0.34
Carpet and rubber pad	—	—	0.81	—	1.23	0.33
Cork tile (0.125 in.)	—	—	3.60	—	0.28	0.48
Terrazzo (1 in.)	—	—	12.50	—	0.08	0.19
Tile: asphalt, linoleum, vinyl, rubber vinyl asbestos	—	—	20.00	—	0.05	0.30 / 0.24 / 0.19
Wood, hardwood finish (0.75 in.)	—	—	1.47	—	0.68	

Insulating Materials

Description	Density (lb/ft³)	Conductivity,[b] k (Btu-in./ hr-ft²-°F)	Conductance, C (Btu/hr-ft²-°F)	Resistance[c]R Per Inch Thickness, 1/k (hr-ft²-°F/Btu)	Resistance[c]R For Thickness Listed, 1/C (hr-ft²-°F/Btu)	Specific Heat (Btu/lb-°F)
Blanket and batt:[f,g] mineral fiber, fibrous form process from rock, slag, or glass						
Approx. 3–4 in.	0.3–2.0	—	0.091	—	11	
Approx. 3.5 in.	0.3–2.0	—	0.077	—	13	
Approx. 5.5–6.5 in.	0.3–2.0	—	0.053	—	19	

TABLE C-1 (continued)

Description	Density (lb/ft³)	Conductivity,[b] k (Btu-in./ hr-ft²-°F)	Conductance, C (Btu/hr- ft²-°F)	Resistance[c] R		Specific Heat (Btu/lb-°F)
				Per Inch Thickness, 1/k (hr-ft²-°F/Btu)	For Thickness Listed, 1/C (hr-ft²-°F/Btu)	
Approx. 6–7.5 in.	0.3–2.0	—	0.045	—	22	
Approx. 9–10 in.	0.3–2.0	—	0.033	—	30	
Approx. 12–13 in.	0.3–2.0	—	0.026	—	38	
Board and slabs						
Cellular glass	8.5	0.35	—	2.86	—	0.18
Glass fiber, organic bonded	4–9	0.25	—	4.00	—	0.23
Expanded perlite, organic bonded	1.0	0.36	—	2.78	—	0.30
Expanded rubber (rigid)	4.5	0.22	—	4.55	—	0.40
Expanded polystyrene extruded						
Cut cell surface	1.8	0.25	—	4.00	—	0.29
Smooth skin surface	1.8–3.5	0.20	—	5.00	—	0.29
Expanded polystyrene, molded beads	1.0	0.26	—	3.85	—	—
Cellular polyurethane[h] (R-11 exp.) (unfaced)	1.5	0.16	—	6.25	—	0.38
Cellular polyisocyanurate[h,i] (R-11 exp.) (foil-faced, glass fiber-reinforced core)	2.0	0.14	—	7.20	—	0.22
Mineral fiberboard, wet felted						
Core or roof insulation	16–17	0.34	—	2.94	—	
Acoustical tile	18.0	0.35	—	2.86	—	0.19
Acoustical tile	21.0	0.37	—	2.70	—	
Mineral fiberboard, wet molded						
Acoustical tile	23.0	0.42	—	2.38	—	0.14

Loose Fill

Description	Density (lb/ft³)	Conductivity,[b] k	Conductance, C	Per Inch Thickness, 1/k	For Thickness Listed, 1/C	Specific Heat
Cellulosic insulation (milled paper or wood pulp)	2.3–3.2	0.27–0.32	—	3.70–3.13	—	0.33
Sawdust or shavings	8.0–15.0	0.45	—	2.22	—	0.33
Wood fiber, softwoods	2.0–3.5	0.30	—	3.33	—	0.33
Perlite, expanded	2.0–4.1	0.27–0.31	—	3.70–3.13	—	0.26
Mineral fiber (rock, slag, or glass)[g]						
Approx. 3.75–5 in.	0.6–2.0	—	—		11.0	0.17
Approx. 6.5–8.75 in.	0.6–2.0	—	—		19.0	
Approx. 7.5–10 in.	0.6–2.0	—	—		22.0	
Approx. 10.25–13.75 in.	0.6–2.0	—	—		30.0	
Mineral fiber (rock, slag, or glass) approx. 3.5 in. (closed sidewall application)[b]	2.0–3.5	—	—	—	12.0–14.0	
Vermiculite, exfoliated	7.0–8.2	0.47	—	2.13		0.32

Plastering Materials

Description	Density (lb/ft³)	Conductivity,[b] k	Conductance, C	Per Inch Thickness, 1/k	For Thickness Listed, 1/C	Specific Heat
Cement plaster, sand aggregate	116	5.0	—	0.20	—	0.20
0.375 in.	—	—	13.3	—	0.08	0.20
0.75 in.	—	—	6.66	—	0.15	0.20
Gypsum plaster						

TABLE C-1 (continued)

Description	Density (lb/ft³)	Conductivity,[b] k (Btu-in./ hr-ft²-°F)	Conductance, C (Btu/hr-ft²-°F)	Resistance[c] R Per Inch Thickness, 1/k (hr-ft²-°F/Btu)	For Thickness Listed, 1/C (hr-ft²-°F/Btu)	Specific Heat (Btu/lb-°F)
Lightweight aggregate						
0.5 in.	45	—	3.12	—	0.32	
0.625 in.	45	—	2.67	—	0.39	
Lightweight aggregate on metal lath (0.75 in.)	—	—	2.13	—	0.47	
Perlite aggregate	45	1.5	—	0.67	—	0.32
Sand aggregate	105	5.6	—	0.18	—	0.20
0.5 in.	105	—	11.10	—	0.09	
0.625 in.	105	—	9.10	—	0.11	
Sand aggregate on metal lath (0.75 in.)	—	—	7.70	—	0.13	
Vermiculite aggregate	45	1.7	—	0.59	—	

Masonry Materials

Description	Density (lb/ft³)	k	C	1/k	1/C	Specific Heat
Cement mortar	116	5.0	—	2.0	—	
Gypsum-fiber concrete 87.5% gypsum, 12.5% wood chips	51	1.66	—	0.60	—	0.21
Lightweight aggregates	120	5.2	—	0.19	—	
including expanded shale,	100	3.6	—	0.28	—	
clay or slate; expanded	80	2.5	—	0.40	—	
slags; cinders; pumice;	60	1.7	—	0.59	—	
vermiculite; also cellular	40	1.15	—	0.86	—	
concretes	30	0.90	—	1.11	—	
	20	0.70	—	1.43	—	
Perlite, expanded	40	0.93	—	1.08	—	
	30	0.71	—	1.41	—	
	20	0.50	—	2.00	—	0.32
Sand and gravel or stone aggregate						
Oven dried	140	9.0	—	0.11	—	
Not dried	140	12.0	—	0.08	—	0.22
Stucco	116	5.0	—	0.20	—	

Masonry Units

Description	Density (lb/ft³)	k	C	1/k	1/C	Specific Heat
Brick, common	120	5.0	—	0.20	—	0.19
Brick, face	130	9.0	—	0.11	—	
Clay tile, hollow						
1 cell deep (3 in.)	—	—	1.25	—	0.80	0.21
1 cell deep (4 in.)	—	—	0.90	—	1.11	
2 cells deep (6 in.)	—	—	0.66	—	1.52	
2 cells deep (8 in.)	—	—	0.54	—	1.85	
2 cells deep (10 in.)	—	—	0.45	—	2.22	
3 cells deep (12 in.)	—	—	0.40	—	2.50	
Concrete block, three-oval core:[k]						
Sand-and-gravel aggregate						
4 in.	—	—	1.40	—	0.71	0.22
8 in.	—	—	0.90	—	1.11	
12 in.	—	—	0.78	—	1.28	
Cinder aggregate						
3 in.	—	—	1.16	—	0.86	0.21
4 in.	—	—	0.90	—	1.11	
8 in.	—	—	0.58	—	1.72	
12 in.	—	—	0.53	—	1.89	

TABLE C-1 (continued)

Description	Density (lb/ft³)	Conductivity,[b] k (Btu-in./hr-ft²-°F)	Conductance, C (Btu/hr-ft²-°F)	Resistance[c] R		Specific Heat (Btu/lb-°F)
				Per Inch Thickness, $1/k$ (hr-ft²-°F/Btu)	For Thickness Listed, $1/C$ (hr-ft²-°F/Btu)	
Lightweight aggregate (expanded shale, clay, slate, or slag; pumice)						
3 in.	—	—	0.79	—	1.27	0.21
4 in.	—	—	0.67	—	1.50	
8 in.	—	—	0.50	—	2.00	
12 in.	—	—	0.44	—	2.27	
Roofing						
Asbestos-cement shingles	120	—	4.76	—	0.21	0.24
Asphalt roll roofing	70	—	6.50	—	0.15	0.36
Asphalt shingles	70	—	2.27	—	0.44	0.30
Built-up roofing (0.375 in.)	70	—	3.00	—	0.33	0.35
Slate (0.5 in.)	—	—	20.00	—	0.05	0.30
Wood shingles, plain and plastic film faced	—	—	1.06	—	0.94	0.31
Siding Materials (on flat surface)						
Shingles						
Asbestos-cement	120	—	4.75	—	0.21	
Wood, 16 in., 7.5 in. exposure	—	—	1.15	—	0.87	0.31
Wood, double, 16 in., 12 in. exp.	—	—	0.84	—	1.19	0.28
Wood, plus insulated backer board (0.3125 in.)	—	—	0.71	—	1.40	0.31
Siding						
Asbestos-cement, 0.25 in. lapped	—	—	4.76	—	0.21	0.24
Asphalt roll siding	—	—	6.50	—	0.15	0.35
Asphalt insulating siding (0.5-in. bed)	—	—	0.69	—	1.46	0.35
Hardboard siding (0.4375 in.)	40	1.49	—	0.67		0.28
Wood, drop (1–8 in.)	—	—	1.27	—	0.79	0.28
Wood, bevel, 0.5–8 in., lapped	—	—	1.23	—	0.81	0.28
Wood, bevel, 0.75–10 in., lapped	—	—	0.95	—	1.05	0.28
Wood, plywood, 0.375 in., lapped	—	—	1.59	—	0.59	0.29
Aluminium or steel, over sheathing[l]						
Hollow-backed	—	—	1.61	—	0.61	0.29
Insulating-board backed nominal 0.375 in.	—	—	0.55	—	1.82	0.32
Insulating-board backed nominal 0.375 in., foil backed			0.34		2.96	
Architectural glass	—	—	10.00	—	0.10	0.20
Woods (12% moisture content)[e,m]						
Hardwoods						0.39[n]
Oak	41.2–46.8	1.12–1.25	—	0.89–0.80	—	
Birch	42.6–45.4	1.16–1.22	—	0.87–0.82	—	
Maple	39.8–44.0	1.09–1.19	—	0.94–0.88	—	
Ash	38.4–41.9	1.06–1.14	—	0.94–0.88	—	

TABLE C-1 (continued)

Description	Density (lb/ft³)	Conductivity,[b] k (Btu-in./ hr-ft²-°F)	Conductance, C (Btu/hr- ft²-°F)	Resistance[c] R Per Inch Thickness, 1/k (hr-ft²- °F/Btu)	Resistance[c] R For Thickness Listed, 1/C (hr-ft²- °F/Btu)	Specific Heat (Btu/lb-°F)
Softwoods						0.39
Southern pine	35.6–41.2	1.00–1.12	—	1.00–0.89	—	
Douglas fir–larch	33.5–36.3	0.95–1.01	—	1.06–0.99	—	
Southern cypress	31.4–32.1	0.90–0.92	—	1.11–1.09	—	
Hem–fir, spruce–pine–fir	24.5–31.4	0.74–0.90	—	1.35–1.11	—	
West coast woods, cedars	21.7–31.4	0.68–0.90	—	1.48–1.11	—	
California redwood	24.5–28.0	0.74–0.82	—	1.35–1.22	—	

[a]Values are for a mean temperature of 75°F. Representative values for dry materials are intended as design (not specification) values for materials in normal use. Thermal values of insulating materials may differ from design values depending on their in-site properties (e.g., density, moisture content, orientation, etc.) and variability experienced during manufacture. For properties of a particular product, use the value supplied by the manufacturer or by unbiased tests.

[b]To obtain thermal conductivities in Btu/hr-ft-°F, divide the k factor by 12 in./ft.

[c]Resistance values are the reciprocals of C before rounding off C to two decimal places.

[d]Lewis (1967).

[e]U.S. Department of Agriculture (1974).

[f]Does not include paper backing and facing, if any. Where insulation forms a boundary (reflective or otherwise) of an air space, see Appendix Tables C-2b and C-3 for the insulating value of an air space with the appropriate effective emittance and temperature conditions of the space.

[g]Conductivity varies with fiber diameter. (Refer to *ASHRAE Handbook 1989 Fundamentals,* Chapter 20, "Factors That Affect Thermal Performance.") Batt, blanket, and loose-fill mineral fiber insulations are manufactured to achieve specified R-values, the most common of which is listed in the table. Due to differences in manufacturing processes and materials, the product thicknesses, densities, and thermal conductivities vary over considerable ranges for a specified R-value.

[h]For additional information, see Society of Plastics Engineers (SPI) Bulletin U108. Values are for aged, unfaced board stock. For changes in conductivity with age of expanded polyurethane/polyisocyanurate, refer to note g for cited reference.

[i]Values are for aged products with gas-impermeable facers on the two major surfaces. An aluminum foil facer of 0.001 in. thickness or greater is generally considered impermeable to gases. For change in conductivity with age of expanded polyisocyanurate, refer to note g for cited reference and to SPI Bulletin U108.

[j]Insulating values of acoustical tile vary depending on density of the board and on type, size, and depth of perforation.

[k]Values for fully grouted block may be approximated using values for concrete with similar unit weight.

[l]Values for metal siding applied over flat surfaces vary widely depending on amount of ventilation of air space beneath the siding; whether air space is reflective or nonreflective; and on thickness, type, and application of insulating backing-board used. Values given are averages for use as design guides and were obtained from several guarded hotbox tests (ASTM C236) or calibrated hotbox (ASTM C976) on hollow-backed types and types made using backing-boards of wood fiber, foamed plastic, and glass fiber. Departures of ±50% or more from the values may occur.

[m]See Adams (1971), MacLean (1941), and Wilkes (1979). The conductivity values listed are for heat transfer across the grain. The thermal conductivity of wood varies linearly with the density, and the density ranges listed are those normally found for the wood species given. If the density of the wood species is not known, use the mean conductivity value. For extrapolation to other moisture contents, the following empirical equation developed by Wilkes (1979) may be used:

$$k = 0.1791 + \frac{(1.874 \times 10^{-2} + 5.753 \times 10^{-4}\, M)\rho}{1 + 0.01M}$$

where ρ is the density based on oven-dry mass in lb/ft³ and M is the moisture content in percent.

[n]From Adams (1971), an empirical equation for the specific heat of moist wood at 75°F is as follows:

$$C_p = \frac{0.299 + 0.01M}{1 + 0.01M} + \Delta C_p$$

where ΔC_p accounts for the heat of absorption and is denoted by

$$\Delta C_p = M(1.921 \times 10^{-3} - 3.168 \times 10^{-5} \times M)$$

where M is the moisture content in percent by mass.

TABLE C-2a*

Surface conductances and resistances for air[a]

Position of Surface	Direction of Heat Flow	Non-reflective ε = 0.90		Reflective ε = 0.20		Reflective ε = 0.05	
Still air		f_i	R_a	f_i	R_a	f_i	R_a
Horizontal	Upward	1.63	0.61	0.91	1.10	0.76	1.32
Sloping (45)	Upward	1.60	0.62	0.88	1.14	0.73	1.37
Vertical	Horizontal	1.46	0.68	0.74	1.35	0.59	1.70
Sloping (45)	Downward	1.32	0.76	0.60	1.67	0.45	2.22
Horizontal	Downward	1.08	0.92	0.37	2.70	0.22	4.55
Moving Air (Any position)							
15-mph wind (for winter)	Any	6.00	0.17				
7.5-mph wind (for summer)	Any	4.00	0.25				

*Source: Extracted from *ASHRAE Handbook 1989 Fundamentals* with permission from American Society of Heating, Refrigerating, and Air Conditioning Engineers, Atlanta, GA.

[a]All conductance values expressed in Btu per (hour) (square foot) (degree F temperature difference). A surface cannot take credit for both an air-space resistance value and a surface resistance value. No credit for an air-space value can be taken for any surface facing an air space of less than 0.5 in.

TABLE C-2b*

Reflectivity and emittance values of various surfaces and effective emittances of air spaces

Surface	Reflectivity in Percent	Average Emittance ε	Effective Emittance E of Air Space	
			One Surface Emittance ε the Other 0.90	Both Surfaces Emittances ε
Aluminum foil, bright	92 to 97	0.05	0.05	0.03
Aluminum sheet	80 to 95	0.12	0.12	0.06
Aluminum coated paper, polished	75 to 84	0.20	0.20	0.11
Steel, galvanized, bright	70 to 80	0.25	0.24	0.15
Aluminum paint	30 to 70	0.50	0.47	0.35
Building materials: wood, paper, masonry, nonmetallic points	5 to 15	0.90	0.82	0.82
Regular glass	5 to 15	0.84	0.77	0.72

*Source: Extracted from *ASHRAE Handbook 1989 Fundamentals* with permission from American Society of Heating, Refrigerating, and Air Conditioning Engineers, Atlanta, GA.

TABLE C-3a*

Thermal resistance (R_a) of ¾-in. plane air spaces[a]

Position of Air Space	Direction of Heat Flow	Mean Temp. (°F)	Temp. Diff. (°F)	¾-in. Air Space			
				Value of Effective Emittance *E*			
				0.03	0.05	0.20	0.82
Horizontal	Up	90	10	2.34	2.22	1.61	0.75
		50	30	1.71	1.66	1.35	0.77
		50	10	2.30	2.21	1.70	0.87
		0	20	1.83	1.79	1.52	0.93
		0	10	2.23	2.16	1.78	1.02
45° Slope	Up	90	10	2.96	2.78	1.88	0.81
		50	30	1.99	1.92	1.52	0.82
		50	10	2.90	2.75	2.00	0.94
		0	20	2.13	2.07	1.72	1.00
		0	10	2.72	2.62	2.08	1.12
Vertical	Horizontal	90	10	3.50	3.24	2.08	0.84
		50	30	2.91	2.77	2.01	0.94
		50	10	3.70	3.46	2.35	1.01
		0	20	3.14	3.02	2.32	1.18
		0	10	3.77	3.59	2.64	1.26
45° Slope	Down	90	10	3.53	3.27	2.10	0.84
		50	30	3.43	3.23	2.24	0.99
		50	10	3.81	3.57	2.40	1.02
		0	20	3.75	3.57	2.63	1.26
		0	10	4.12	3.91	2.81	1.30
Horizontal	Down	90	10	3.55	3.28	2.10	0.85
		50	30	3.77	3.52	2.38	1.02
		50	10	3.84	3.59	2.41	1.02
		0	20	4.18	3.96	2.83	1.30
		0	10	4.25	4.02	2.87	1.31

*Source: Extracted from *ASHRAE Handbook 1989 Fundamentals* with permission from American Society of Heating, Refrigerating, and Air Conditioning Engineers, Atlanta, GA.

[a]All resistance values expressed in (hour) (square foot) (degree F temperature difference) per Btu.

TABLE C-3b *

Thermal resistance (R_a) of 1½-in. plane air spaces[a]

Position of Air Space	Direction of Heat Flow	Mean Temp. (°F)	Temp. Diff. (°F)	1½-in. Air Space			
				Value of Effective Emittance E			
				0.03	0.05	0.20	0.82
Horizontal	Up	90	10	2.55	2.41	1.71	0.77
		50	30	1.87	1.81	1.45	0.80
		50	10	2.50	2.40	1.81	0.89
		0	20	2.01	1.95	1.63	0.97
		0	10	2.43	2.35	1.90	1.06
45° Slope	Up	90	10	2.92	2.73	1.86	0.80
		50	30	2.14	2.06	1.61	0.84
		50	10	2.88	2.74	1.99	0.94
		0	20	2.30	2.23	1.82	1.04
		0	10	2.79	2.69	2.12	1.13
Vertical	Horizontal	90	10	3.99	3.66	2.25	0.87
		50	30	2.58	2.46	1.84	0.90
		50	10	3.79	3.55	2.39	1.02
		0	20	2.76	2.66	2.10	1.12
		0	10	3.51	3.35	2.51	1.23
45° Slope	Down	90	10	5.07	4.55	2.56	0.91
		50	30	3.58	3.36	2.31	1.00
		50	10	5.10	4.66	2.85	1.09
		0	20	3.85	3.66	2.68	1.27
		0	10	4.92	4.62	3.16	1.37
Horizontal	Down	90	10	6.09	5.35	2.79	0.94
		50	30	6.27	5.63	3.18	1.14
		50	10	6.61	5.90	3.27	1.15
		0	20	7.03	6.43	3.91	1.49
		0	10	7.31	6.66	4.00	1.51

Source: Extracted from *ASHRAE Handbook 1989 Fundamentals* with permission from American Society of Heating, Refrigerating, and Air Conditioning Engineers, Atlanta, GA.

[a]All resistance values expressed in (hour) (square foot) (degree F temperature difference) per Btu.

TABLE C-3c*

Thermal resistance (R_a) of 3½-in. plane air spaces[a, b]

Position of Air Space	Direction of Heat Flow	Mean Temp. (°F)	Temp. Diff. (°F)	3½-in. Air Space Value of Effective Emittance E			
				0.03	0.05	0.20	0.82
Horizontal	Up	90	10	2.84	2.66	1.83	0.80
		50	30	2.09	2.01	1.58	0.84
		50	10	2.80	2.66	1.95	0.93
		0	20	2.25	2.18	1.79	1.03
		0	10	2.71	2.62	2.07	1.12
45° Slope	Up	90	10	3.18	2.96	1.97	0.82
		50	30	2.26	2.17	1.67	0.86
		50	10	3.12	2.95	2.10	0.96
		0	20	2.42	2.35	1.90	1.06
		0	10	2.98	2.87	2.23	1.16
Vertical	Horizontal	90	10	3.69	3.40	2.15	0.85
		50	30	2.67	2.55	1.89	0.91
		50	10	3.63	3.40	2.32	1.01
		0	20	2.88	2.78	2.17	1.14
		0	10	3.49	3.33	2.50	1.23
45° Slope	Down	90	10	4.81	4.33	2.49	0.90
		50	30	3.51	3.30	2.28	1.00
		50	10	4.74	4.36	2.73	1.08
		0	20	3.81	3.63	2.66	1.27
		0	10	4.59	4.32	3.02	1.34
Horizontal	Down	90	10	10.07	8.19	3.41	1.00
		50	30	9.60	8.17	3.86	1.22
		50	10	11.15	9.27	4.09	1.24
		0	20	10.90	9.52	4.87	1.62
		0	10	11.97	10.32	5.08	1.64

Source: Extracted from *ASHRAE Handbook 1989 Fundamentals* with permission from American Society of Heating, Refrigerating, and Air Conditioning Engineers, Atlanta, GA.

[a]All resistance values expressed in (hour) (square foot) (degree F temperature difference) per Btu.

[b]Value in Tables C-3 apply only to air spaces of uniform thickness bounded by plane, smooth, parallel surfaces with no leakage of air to or from the space. Thermal resistance values for multiple air spaces must be based on careful estimates of mean temperature difference for each air space.

TABLE C-4*

Overall coefficients of heat transmission of various fenestration products

	Glass Only		Aluminum Frame no thermal break ($U_f = 1.9$)		Aluminum Frame thermal break ($U_f = 1.0$)		Wood or Vinyl Frame ($U_f = 0.4$)	
			Product[d] Type		Product[d] Type		Product[d] Type	
Glazing Type[b]	Center of Glass	Edge[c] of Glass	R	C	R	C	R	C
Single glazing								
glass	(1.11)	(n/a)	1.31	1.23	1.09	1.10	0.90	0.98
1/8 in. acrylic	(1.03)	(n/a)	1.26	1.16	1.02	1.03	0.84	0.92
Double glass								
1/4 in. air space	(0.57)	(0.66)	0.92	0.78	0.70	0.65	0.54	0.55
3/8 in. air space	(0.52)	(0.62)	0.88	0.74	0.66	0.60	0.50	0.51
1/2 in. and greater air space	(0.49)	(0.59)	0.87	0.72	0.64	0.59	0.49	0.49
Double glass, $\epsilon = 0.40$ on surface 2 or 3								
1/4 in. air space	(0.50)	(0.60)	0.87	0.73	0.65	0.59	0.49	0.50
3/8 in. air space	(0.43)	(0.55)	0.83	0.67	0.60	0.54	0.45	0.45
1/2 in. and greater air space	(0.41)	(0.54)	0.81	0.65	0.58	0.52	0.43	0.42
Double glass, $\epsilon = 0.15$ on surface 2 or 3								
1/4 in. air space	(0.45)	(0.56)	0.84	0.68	0.61	0.55	0.46	0.46
3/8 in. air space	(0.36)	(0.51)	0.78	0.62	0.56	0.48	0.41	0.39
1/2 in. air space	(0.34)	(0.50)	0.76	0.60	0.54	0.46	0.39	0.37
Double glass								
1/4 in. argon space	(0.52)	(0.62)	0.88	0.74	0.66	0.61	0.50	0.51
3/8 in. argon space	(0.48)	(0.59)	0.86	0.71	0.63	0.57	0.48	0.48
1/2 in. and greater argon space	(0.46)	(0.57)	0.82	0.69	0.62	0.56	0.47	0.47
Double glass, $\epsilon = 0.40$ on surface 2 or 3								
1/4 in. argon space	(0.43)	(0.55)	0.83	0.67	0.60	0.54	0.45	0.45
3/8 in. argon space	(0.38)	(0.52)	0.79	0.63	0.57	0.49	0.42	0.40
1/2 in. and greater argon space	(0.36)	(0.51)	0.78	0.62	0.56	0.48	0.41	0.39
Double glass, $\epsilon = 0.15$ on surface 2 or 3								
1/4 in. argon space	(0.36)	(0.51)	0.78	0.62	0.56	0.48	0.41	0.39
3/8 in. argon space	(0.30)	(0.48)	0.74	0.57	0.51	0.43	0.37	0.34
1/2 in. and greater argon space	(0.28)	(0.47)	0.73	0.55	0.50	0.42	0.36	0.33
Double glazing, 1/8 in. acrylic or polycarbonate								
1/4 in. air space	(0.52)	(0.62)	0.89	0.74	0.67	0.61	0.51	0.51
3/8 in. air space	(0.48)	(0.59)	0.86	0.71	0.64	0.57	0.48	0.48
1/2 in. and greater air space	(0.46)	(0.57)	0.85	0.69	0.62	0.56	0.47	0.47
Double glazing, 1/4 in. acrylic or polycarbonate								
1/4 in. air space	(0.48)	(0.59)	0.86	0.71	0.64	0.57	0.48	0.48
3/8 in. air space	(0.44)	(0.56)	0.84	0.68	0.61	0.54	0.46	0.45
1/2 in. and greater air space	(0.42)	(0.54)	0.82	0.66	0.60	0.53	0.45	0.43
Triple glass								
1/4 in. air space	(0.38)	(0.52)	0.79	0.64	0.57	0.50	0.42	0.41
3/8 in. air space	(0.34)	(0.50)	0.76	0.60	0.54	0.46	0.39	0.38
1/2 in. and greater air space	(0.32)	(0.49)	0.75	0.58	0.53	0.45	0.38	0.36
Triple glass, $\epsilon = 0.40$ on surface 2, 3, 4 or 5								
1/4 in. air spaces	(0.35)	(0.50)	0.77	0.61	0.55	0.48	0.40	0.39
3/8 in. air spaces	(0.30)	(0.48)	0.74	0.57	0.52	0.44	0.37	0.35
1/2 in. and greater air spaces	(0.28)	(0.47)	0.72	0.55	0.50	0.41	0.36	0.33
Triple glass or double glass with polyester film suspended in between, $\epsilon = 0.15$ on surface 2, 3, 4, or 5								
1/4 in. air spaces	(0.33)	(0.49)	0.76	0.59	0.53	0.45	0.39	0.37
3/8 in. air spaces	(0.27)	(0.46)	0.72	0.54	0.50	0.41	0.35	0.32
1/2 in. and greater air spaces	(0.24)	(0.45)	0.70	0.52	0.48	0.39	0.34	0.30

(continued)

TABLE C-4 (continued)

	Glass Only		Aluminum Frame no thermal break ($U_f = 1.9$) Product[d] Type		Aluminum Frame thermal break ($U_f = 1.0$) Product[d] Type		Wood or Vinyl Frame ($U_f = 0.4$) Product[d] Type	
Glazing Type[b]	Center of Glass	Edge[c] of Glass	R	C	R	C	R	C

Part A: U-Values for Vertical Installation[a], Btu/h·ft²·°F

Glazing Type[b]	Center of Glass	Edge[c] of Glass	R	C	R	C	R	C
Triple glass or double glass with polyester film suspended in between, $\epsilon = 0.15$ on surfaces 2 or 3 and 4 or 5								
1/4 in. air spaces	(0.28)	(0.47)	0.73	0.55	0.53	0.42	0.36	0.33
3/8 in. air spaces	(0.22)	(0.45)	0.69	0.51	0.47	0.37	0.32	0.29
1/2 in. and greater air spaces	(0.19)	(0.44)	0.67	0.48	0.45	0.35	0.31	0.26
Triple glass								
1/4 in. argon spaces	(0.34)	(0.50)	0.77	0.60	0.54	0.46	0.39	0.38
3/8 in. argon spaces	(0.31)	(0.48)	0.74	0.57	0.52	0.44	0.37	0.35
1/2 in. and greater argon spaces	(0.29)	(0.47)	0.73	0.56	0.51	0.42	0.36	0.34
Triple glass, $\epsilon = 0.40$ on surface 2, 3, 4, or 5								
1/4 in. argon spaces	(0.30)	(0.48)	0.74	0.57	0.52	0.44	0.37	0.35
3/8 in. argon spaces	(0.26)	(0.46)	0.72	0.54	0.49	0.41	0.35	0.32
1/2 in. and greater argon spaces	(0.25)	(0.46)	0.71	0.53	0.48	0.39	0.34	0.31
Triple glass or double glass with polyester suspended in between, $\epsilon = 0.15$ on surface 2, 3, 4, or 5								
1/4 in. argon spaces	(0.27)	(0.46)	0.72	0.54	0.50	0.41	0.35	0.32
3/8 in. argon spaces	(0.22)	(0.45)	0.69	0.51	0.47	0.37	0.33	0.29
1/2 in. and greater argon spaces	(0.20)	(0.44)	0.68	0.50	0.46	0.36	0.31	0.28
Triple glass or double glass with polyester film suspended in between, $\epsilon = 0.15$ on surfaces 2 or 3 and 4 or 5								
1/4 in. argon spaces	(0.22)	(0.45)	0.69	0.51	0.47	0.37	0.32	0.29
3/8 in. argon spaces	(0.17)	(0.43)	0.66	0.47	0.44	0.34	0.30	0.25
1/2 in. and greater argon spaces	(0.15)	(0.43)	0.65	0.46	0.43	0.32	0.29	0.24

Part B: U-Value Conversion Table for Sloped and Horizontal Glazing for Upward Heat Flow U-Value, Btu/h·ft²·°F

Slope													
90° (vertical)	0.10	0.20	0.30	0.40	0.50	0.60	0.70	0.80	0.90	1.00	1.10	1.20	1.30
45°	0.14	0.25	0.36	0.47	0.57	0.68	0.79	0.90	1.00	1.11	1.22	1.33	1.44
0 (horiz.)	0.19	0.29	0.40	0.51	0.61	0.72	0.82	0.93	1.04	1.14	1.25	1.35	1.46

*Source: Extracted from *ASHRAE Handbook 1989 Fundamentals* with permission from American Society of Heating, Refrigerating, and Air Conditioning Engineers, Atlanta, GA.

[a]All *U*-values are based on standard ASHRAE winter conditions of 70°F indoor and 0°F outdoor air temperature with 15-mph outdoor air velocity and zero solar flux. The outside surface coefficient at these conditions is approximately 5.1 Btu/hr-ft²-°F, depending on the glass surface temperature. With the exception of single glazing, small changes in the interior and exterior temperatures do not significantly affect overall *U*-values.

[b]Glazing layer surfaces are numbered from the outside to the inside. Double and triple refer to the number of glazing lites. All data are based on 1/8-in. glass unless otherwise noted. Thermal conductivities are 0.53 Btu/hr-ft-°F for glass and 0.11 Btu/hr-ft-°F for acrylic and polycarbonate.

[c]Based on aluminum spacers data. Edge-of-glass effect assumed to extend over the 2.5-in band around perimeter of each glazing unit as seen in Figure C-1.

[d]Product types described in Figure C-1.

Supplementary Notes for Tables C-4 and C-5:

In the absence of sunlight, air infiltration, and moisture condensation, the rate of heat transfer through a window system is proportional to the inside-to-outside air temperature difference. Heat is transferred by the combined effects of conduction, convention, and radiation. Usually, heat flows in parallel paths through the center of glass, edge of glass, and framing members. The general equation used to estimate the total rate of heat transfer, q_w, through a fenestration system from inside to outside is

$$\dot{q}_w = U_w A_w (t_i - t_o) \tag{a}$$

where

U_w = overall heat-transmission coefficient, *U*-value (Btu/hr-ft²-°F)
A_w = combined glazing plus frame area projected to a plane parallel to glass as viewed from outside (ft²)
t_i = temperature of warm space air (°F)
t_o = temperature of cold space air (°F)

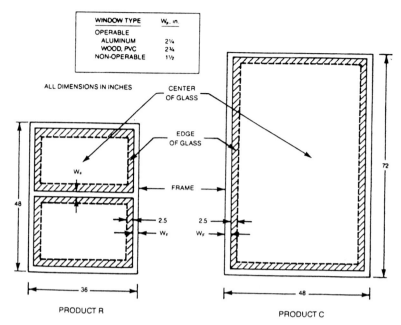

FIGURE C-1 *R* and *C* categories of frame dimensions, in inches.

The total rate of heat transfer through a fenestration system can be calculated knowing the separate heat-transfer contribution of the glazing and the frame. The glazing heat transfer includes a one-dimensional center-of-glass contribution and a two-dimensional edge-of-glass contribution. The frame contribution is primarily two-dimensional. The overall *U*-value is estimated using area-weighted *U*-values for each contribution by

$$U_w = \frac{U_{cg} A_{cg} + U_{eg} A_{eg} + U_f A_f}{A_{cg} + A_{eg} + A_f} \tag{b}$$

where the subscripts *cg, eg,* and *f* refer to the center of glass, edge of glass, and frame, respectively.

Computer programs have been set up that provide rapid estimates of glazing-unit heat transfer for a wide range of glazing construction.

Insulating glass units usually have continuous members around the glass perimeter to separate the glazing lites and provide an edge seal. These spacers are often constructed of aluminum, thereby greatly increasing conductive heat transfer between the contacted inner and outer lites and degrading the thermal performance of the glazing unit locally. Laboratory measurements have shown this conductive region to be limited to a 2.5-in.-wide band around the perimeter of the glazing unit.

Estimation of the rate of heat transfer through the framing elements of a fenestration system is complicated by (1) the variety of frame configurations for operable windows, (2) the different combinations of materials used for sash and frames, and (3) the different sizes available in residential and commercial applications. Available windows are usually framed in either wood, aluminum, or vinyl; and many manufacturers combine materials for structural or aesthetic purposes. Some aluminum-framed units have *thermal breaks* that reduce the conductive heat transfer through the framing element. Computer programs have been developed that analyze frame construction.

Appendix Table C-4 can be used to select a particular fenestration product to obtain a desired *U*-value. Part A of this table presents the computed overall *U*-values for a wide range of glazing units and framing materials installed vertically and based on component *U*-values shown in parentheses. Part B presents approximate *U*-values for nonvertical installation. The overall *U*-values in part A are appropriate for calculating compliance with the requirements of ASHRAE Standard 90. However, the values apply only to the specific design conditions described in the footnotes and are typically used to determine peak load conditions for sizing heating equipment. While these *U*-values have been determined for winter conditions,they can be used to estimate heat gain during peak cooling conditions since conductive gain, which is one of several factors, is usually a small portion of the total heat gain for windows in direct sunlight. Glazing designs and framing materials may be compared to choose a fenestration system that needs a specific winter design *U*-value.

The overall *U*-values in Table C-4 are based on the rough opening area for representative generic glazing products. Test data, when available, may provide more accurate results for specific products.

The multiple-glazing categories are appropriate for sealed glass units, acrylic domes, and the addition of storm sash to other glazing units. No distinction is made between flat and domed units such as skylights. For acrylic domes, use average air-space width to determine *U*-value. Unless otherwise noted, all multiple-glazed units are filled with dry air.

Table C-4, part A lists data for three values of hemispherical emittance. Uncoated glass has an emittance of 0.84. Emittances of various low-emittance glasses vary considerably between manufacturers and processes. When the emittance is known to be between the listed values, interpolation is recommended. When manufacturers' data are not available for low-emittance glass, assume glass with a pyrolytic (hard) coating has an emittance of 0.40 and glass with a sputtered (soft) coating has an emittance of 0.15. Some coated glazings have lower values of emittance. The use of tinted glass does not change the winter *U*-value. Also, some reflective glass may have an emittance less than 0.84.

Frame type refers to the primary unit. Thus, when storm sash is added over to other glazing, use the values for nonstorm frame. For glazing with a steel frame, use aluminum-frame values. The wood-frame category represents wood frame with and without cladding as well as vinyl frames.

Product types *R* (residential) and *C* (commercial) refer to the proportion of frame to glass area as shown in Figure C-1. The *U*-values in Table C-4 and the definition of these two products are based on frame sizes for operative windows. The type *R* category is based on products in which the glazing units are about 6 ft² in area and the overall size corresponds to a 3-by-4-ft window. For aluminum windows, this corresponds to 25% frame area, 27% edge of glass, and 48% center of glass; for wood or vinyl, the frame, edge, and center of glass areas are 30%, 26%, and

44%, respectively. The type *C* category is based on products in which the panes of glazing are about 24 ft^2 in area and overall size corresponds to a 4-by-6-ft window. For aluminum windows, type *C* category corresponds to 15% frame, 15% edge, and 70% center of glass; for wood and vinyl, the breakdown is 18%, 15%, and 67%, respectively. In this case, the key determinant for selecting the appropriate product is glazing-unit size, *not* overall window size. For products with glazing-unit sizes 16 ft^2 or less, use type *R* category. For products with glazing-unit sizes greater than 16 ft^2, such as sliding glass doors, use type *C* category. For large sealed glass units with mullions or grilles that form a thermal bridge in the air space between the glazing, use type *R* category.

Part B of Table C-4 lists approximate *U*-values for simple single- and double-glazed units installed in nonvertical installations where heat flows upward. After determining a *U*-value for vertical installation from part A, enter part B on the 90° slope line (vertical) and find that value. Then, read down and find the *U*-value corresponding to the desired slope. For simplicity, use the 90° value for products installed 60° to 90° above horizontal, the 45° value for installation 30° to 60° above horizontal, and the 0° value (horizontal) for installation within 30° of horizontal.

To estimate the overall *U*-value of a fenestration product that differs significantly from the assumption of Table C-4, Figure C-1, or both, first determine the percentage area that is frame/sash, center of glass, and edge of glass (based on a 2.5-in. band around the perimeter of each glazing unit). Next, determine the appropriate component *U*-values. These can be either from the standard values listed in parentheses in the table or from some other source such as test data or computed values. Finally, multiply the percent area and the corresponding *U*-value and sum these products to obtain the overall *U*-value, U_w.

Fenestration system performance is likely to be different from the data of Table C-4 when products are installed in buildings due to the combined effects of solar gain, wind, ambient temperature, and cloud cover. Since all of these factors influence the rate of heat transfer through fenestration, seasonal or annual energy flows cannot be accurately determined solely on the basis of the design-day *U*-values. Where it is desired to compare *U*-values at different wind speeds, Appendix Table C-5 provides approximate data to convert the overall *U*-value data at one specific wind condition to a *U*-value at another wind condition. Performance data in Table C-4, part A is based on the assumption that the effective sky temperature for radiative heat transfer is equal to the outdoor air temperature. That assumption gives reasonable results with 7.5- and 15-mph wind speeds. However, with a clear sky and nearly still wind conditions, significant heat transfer occurs by radiation and the data in Table C-4 may underestimate the actual *U*-values.

TABLE C-5*

Glazing *U*-value conversion from 15-mph wind
to 7.5-mph wind and still air

	Wind speed, mph	
15	7.5	0
	U-Value, Btu/h·ft^2·°F	
0.10	0.10	0.10
0.20	0.20	0.19
0.30	0.29	0.28
0.40	0.38	0.37
0.50	0.47	0.45
0.60	0.56	0.53
0.70	0.65	0.61
0.80	0.74	0.69
0.90	0.83	0.78
1.00	0.92	0.86
1.10	1.01	0.94
1.20	1.10	1.02
1.30	1.19	1.10

**Source:* Reprinted from *ASHRAE Handbook 1989 Fundamentals* with permission from American Society of Heating, Refrigerating, and Air Conditioning Engineers, Atlanta, GA.

TABLE C-6*

Transmission coefficients (U) for wood and steel doors, Btu/hr·ft².°F

Nominal Door Thickness, in.	Description	No Storm Door	Wood Storm Door[c]	Metal Storm Door[d]
Wood Doors[a, b]				
1-3/8	Panel door with 7/16-in. panels[e]	0.57	0.33	0.37
1-3/8	Hollow core flush door	0.47	0.30	0.32
1-3/8	Solid core flush door	0.39	0.26	0.28
1-3/4	Panel door with 7/16-in. panel[e]	0.57	0.33	0.36
1-3/4	Hollow core flush door	0.46	0.29	0.32
1-3/4	Panel door with 1-1/8-in. panels[e]	0.39	0.26	0.28
1-3/4	Solid core flush door	0.33	0.28	0.25
2-1/4	Solid core flush door	0.27	0.20	0.21
Steel Doors[b]				
1-3/4	Fiberglass or mineral wood core with steel stiffeners, no thermal break[f]	0.60	—	—
1-3/4	Paper honeycomb core without thermal break[f]	0.56	—	—
1-3/4	Solid urethane foam core without thermal break[a]	0.40	—	—
1-3/4	Solid fire rated mineral fiberboard core without thermal break[f]	0.38	—	—
1-3/4	Polystyrene core without thermal break (18 gage commercial steel)[f]	0.35	—	—
1-3/4	Polyurethane core without thermal break (18 gage commercial steel)[f]	0.29	—	—
1-3/4	Polyurethane core without thermal break (24 gage commercial steel)[f]	0.29	—	—
1-3/4	Polyurethane core with thermal break and wood perimeter (24 gage residential steel)[f]	0.20	—	—
1-3/4	Solid urethane foam core with thermal break[a]	0.19	0.16	0.17

Source: Reprinted from *ASHRAE Handbook 1989 Fundamentals* with permission from American Society of Heating, Refrigerating, and Air Conditioning Engineers, Atlanta, GA.

Note: All U-factors for exterior doors in this table are for doors with no glazing, except for the storm doors which are in addition to the main exterior door. Any glazing area in exterior doors should be included with the appropriate glass type and analyzed (see Appendix Table C-4). Interpolation and moderate extrapolation are permitted for door thicknesses other than those specified.

[a]Values are based on a nominal 32 by 80 in. door size with no glazing.
[b]Outside air conditions: 15 mph wind speed, 0°F air temperature; inside air conditions: natural convection, 70°F air temperature.
[c]Values for wood storm door are for approximately 50% glass area.
[d]Values for metal storm door are for any percent glass area.
[e]55% panel area.
[f]ASTM C 236 hotbox data on a nominal 3-by-7-ft door size with no glazing.

TABLE C-7*

Coefficients of transmission (U) for typical building sections*

CONSTRUCTION DETAIL	HEAT TRANSMISSION COEFFICIENT "U" Btuh per sq ft per °F.				
	INSULATION ADDED TO BASIC CONSTRUCTION				
	None	Blanket or Batt Type Thickness			
		1-1/2"	2"	3"	
EXPOSED FRAME AND VENEER WALLS					
1 Wood Siding or Wood Shingle Exterior					
a. Building paper, 25/32 in. wood sheathing, with wood or 3/8 in. gypsum lath and 1/2 in. plaster interior	.24	.10	.09	.07	
b. Same as (1a), but with 1/2 in. insulating board or 1/2 in. insulating lath and 1/2 in. plaster interior	.19	.10	.08	.06	
c. 25/32 in. insulating sheathing with wood or 3/8 in. gypsum lath and 1/2 in. plaster interior	.19	.09	.08	.06	
d. 3/8 in. plywood sheathing with wood or 3/8 in. gypsum lath and 1/2 in. plaster interior	.28	.11	.09	.07	
2 Asbestos-Cement Siding or Shingles (1/4 in. thick, tapped) or Stucco (1" over bldg. paper) on Frame					
a. Building paper, 25/32 in. wood sheathing, with wood or 3/8 in. gypsum lath and 1/2 in. plaster interior	.29	.11	.09	.07	
b. Same as (2a), but with 1/2 in. insulating board or 1/2 in. insulating lath and 1/2 in. plaster interior	.22	.10	.08	.06	
c. 25/32 in. insulating sheathing with wood or 3/8 in. gypsum lath and 1/2 in. plaster interior	.22	.10	.08	.06	
d. 3/8 in. plywood sheathing with wood or 3/8 in. gypsum lath and 1/2 in. plaster interior	.34	.12	.10	.07	
3 Panel Walls, Steel or Aluminum Skin					
a. Blown polyurethane core		*.16	.08	.05	
b. Polystyrene or fiber glass core		*.21	.12	.08	
c. Cellular glass core		*.30	.17	.12	
d. Corkboard, or, mineral wool with resin binder core		*.24	.14	.09	
4 Insulating Siding (1/2 in.), or Wood Shingles over Insulating Backer Board (5/16 in.)					
a. Building paper, 25/32 in. wood sheathing, with wood or 3/8 in. gypsum lath and 1/2 in. plaster interior	.21	.10	.08	.06	
b. Same as (4a), but with 1/2 in. insulating board or 1/2 in. insulating lath and 1/2 in. plaster interior	.18	.09	.08	.06	
c. 25/32 in. insulating sheathing with wood or 3/8 in. gypsum lath and 1/2 in. plaster interior	.18	.09	.08	.06	
d. 3/8 in. plywood sheathing with wood or 3/8 in. gypsum lath and 1/2 in. plaster interior	.24	.10	.09	.07	

*1" Insulation Core

TABLE C-7 *(continued)*

CONSTRUCTION DETAIL	HEAT TRANSMISSION COEFFICIENT "U" Btuh per sq ft per ° F.				
	INSULATION ADDED TO BASIC CONSTRUCTION				
	None	Blanket or Batt Type Thickness			
		1-1/2"	2"	3"	
5 Veneer (4 in. Face Brick or 4 in. Stone)					
a. Building paper, 25/32 in. wood sheathing, with wood or 3/8 in. gypsum lath and 1/2 in. plaster interior	.27	.11	.09	.07	
b. Same as (5a), but with 1/2 in. insulating board or 1/2 in. insulating lath and 1/2 in. plaster interior	.21	.10	.08	.06	
c. 25/32 in. insulating sheathing with wood or 3/8 in. gypsum lath and 1/2 in. plaster interior	.21	.10	.08	.06	
d. 3/8 in. plywood sheathing with wood or 3/8 in. gypsum lath and 1/2 in. plaster interior	.31	.11	.09	.07	
6 Sheet Metal Siding					
a. Aluminum sheet on studs, no sheathing or interior finish	1.18	.15	.12	.08	
b. Same as (6a), but with 1/2 in. insulating board interior expanded Polyesterene	.39	.12	.10	.07	
c. Galvanized sheet (plain or corrugated) on studs, no sheathing or interior finish	1.18	.15	.12	.08	
d. Same as (6c), but with 3/8 in. dry-wall interior	.47	.13	.11	.08	
e. Aluminum or galvanized sheet on 25/32 in. insulating sheathing, no interior finish	.46	.13	.11	.08	
7 Unsheathed Frame (old construction)					
a. Clapboard or wood siding, studs, lath and plaster interior	.33	.12	.10	.07	
8 Frame Interior Partitions on Studs					
a. Wood or 3/8 in. gypsum lath and 1/2 in. plaster finish, one side	.57	.14	.11	.08	
b. Metal lath and 3/4 in. plaster finish, one side	.68	.14	.11	.08	
c. Same as (8b), but with finish, both sides	.40	.12	.10	.07	
STANDARD MASONRY AND TILE WALLS (ABOVE GRADE ONLY)					
9 Brick, Face and Common					
a. 8 in. brick, no interior finish	.48				
b. Same as (9a), but with 5/8 in. plaster on interior surface of brick	.45				
c. Same as (9a), but with 3/8 in. gypsum lath and 1/2 in. plaster on furring	.29	.12	.11	.09	
d. 10 in. brick, no interior finish	.40				
e. Same as (9d), but with 5/8 in. plaster on interior surface of brick	.39				
f. Same as (9d), but with 3/8 in. gypsum lath and 1/2 in. plaster on furring	.26	.11	.10	.09	

TABLE C-7 (continued)

CONSTRUCTION DETAIL	HEAT TRANSMISSION COEFFICIENT "U" Btuh per sq ft per ° F.				
	INSULATION ADDED TO BASIC CONSTRUCTION				
	None	Blanket or Batt Type Thickness			
		1-1/2"	2"	3"	
9 Brick, Face and Common (Cont'd)					
g. 12 in. brick, no interior finish	.35				
h. Same as (9g), but with 5/8 in plaster on interior surface of brick	.33				
i. Same as (9g), but with metal lath and 3/4 in. plaster on furring	.25	.12	.10	.09	
j. Same as (9g), but with 3/8 in. gypsum lath and 1/2 in. plaster on furring	.24	.11	.10	.09	
10 Poured Concrete — Gravel Aggregate (140 lb. per cu. ft. density)					
a. 6 in. poured concrete — gravel aggregate — no interior finish	.75				
b. Same as (10a), but with 5/8 in. plaster on interior surface of concrete	.70				
c. Same as (10a), but with 3/8 in. gypsum lath and 1/2 in. plaster on furring	.37	.13	.11	.10	
d. 8 in. poured concrete — gravel aggregate — no interior finish	.67				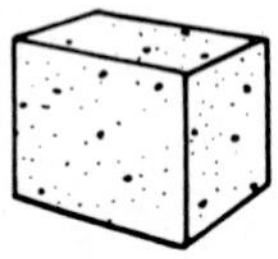
e. Same as (10d), but with 5/8 in. plaster on interior surface of concrete	.63				
f. Same as (10d), but with 3/8 in. gypsum lath and 1/2 in. plaster on furring	.35	.13	.11	.10	
g. 10 in. poured concrete — gravel aggregate — no interior finish	.61				
h. Same as (10g), but with 5/8 in. plaster on interior surface of concrete	.57				
i. Same as (10g), but with 3/8 in. gypsum lath and 1/2 in. plaster on furring	.33	.13	.11	.10	
11 Poured Concrete — Insulating (30 lb. per cu. ft. density)					
a. 6 in. poured concrete — insulating — no interior finish	.13				
b. Same as (11a), but with 5/8 in. plaster on interior surface of concrete	.13				
c. Same as (11a), but with 3/8 in. gypsum lath and 1/2 in. plaster on furring	.11	.09	.09	.08	
d. 8 in. poured concrete — insulating — no interior finish	.10				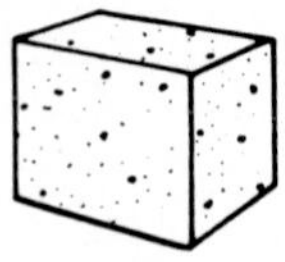
e. Same as (11d), but with 5/8 in. plaster on interior surface of concrete	.10				
f. Same as (11d), but with 3/8 in. gypsum lath and 1/2 in. plaster on furring	.09	.09	.09	.08	
g. 10 in. poured concrete — insulating — no interior finish	.08				
h. Same as (11g), but with 5/8 in. plaster on interior surface of concrete	.08				
i. Same as (11g), but with 3/8 in. gypsum lath and 1/2 in. plaster on furring	.08	.08	.08	.08	

Brick

Concrete

Concrete

TABLE C-7 *(continued)*

CONSTRUCTION DETAIL	HEAT TRANSMISSION COEFFICIENT "U" Btuh per sq ft per ° F.				
		INSULATION ADDED TO BASIC CONSTRUCTION			
	None	Blanket or Batt Type Thickness			
		1-1/2"	2"	3"	
12 Concrete Block — Gravel Aggregate					
a. 8 in. concrete block, no interior finish	.51				
b. Same as (12a), but with 5/8 in. plaster on interior surface of block	.48				
c. Same as (12a), but with 3/8 in. gypsum lath and 1/2 in. plaster on furring	.30	.13	.11	.09	
d. 12 in. concrete block, no interior finish	.47				
e. Same as (12d), but with 5/8 in. plaster on interior surface of blocks	.45				Blocks
f. Same as (12d), but with 3/8 in. gypsum lath and 1/2 in. plaster on furring	.29	.12	.11	.09	
13 Concrete Block — Cinder Aggregate					
a. 8 in. concrete block, no interior finish	.39				
b. Same as (13a), but with 5/8 in. plaster on interior surface of block	.38				
c. Same as (13a), but with 3/8 in. gypsum lath and 1/2 in. plaster on furring	.25	.12	.10	.09	
d. 12 in. concrete block, no interior finish	.36				
e. Same as (13d), but with 5/8 in. plaster on interior surface of block	.35				Blocks
f. Same as (13d), but with 3/8 in. gypsum lath and 1/2 in. plaster on furring	.24	.12	.10	.09	
14 Concrete Block — Light-Weight Aggregate					
a. 8 in. concrete block, no interior finish	.35				
b. Same as (14a), but with 5/8 in. plaster on interior surface of block	.34				
c. Same as (14a), but with 3/8 in. gypsum lath and 1/2 in. plaster on furring	.24	.12	.10	.09	
d. 12 in. concrete block, no interior finish	.32				
e. Same as (14d), but with 5/8 in. plaster on interior finish surface of block	.31				Blocks
f. Same as (14d), but with 3/8 in. gypsum lath and 1/2 in. plaster on furring	.22	.11	.10	.09	
15 4 in. Face Brick, Stone or Precast Concrete Backed by Gravel Aggregate Concrete Block					
a. 4 in. concrete block backing, no interior finish	.48				
b. Same as (15a), but with 5/8 in. plaster on interior surface of block	.46				
c. Same as (15a), but with 3/8 in. gypsum lath and 1/2 in. plaster on furring	.29	.12	.11	.09	
d. 8 in. concrete block backing, no interior finish	.40				
e. Same as (15d), but with 5/8 in. plaster on interior surface of block	.39				Brick & Block or Tile
f. Same as (15d), but with 3/8 in. gypsum lath and 1/2 in. plaster on furring	.26	.12	.10	.09	

TABLE C-7 *(continued)*

CONSTRUCTION DETAIL	HEAT TRANSMISSION COEFFICIENT "U" Btuh per sq ft per ° F.				
	INSULATION ADDED TO BASIC CONSTRUCTION				
	None	Blanket Batt Type Thickness			
		1-1/2"	2"	3"	
16 4 in. Face Brick, Stone, or Precast Concrete Backed by Cinder Aggregate Concrete Block					
a. 4 in. concrete block backing, no interior finish	.40				
b. Same as (16a), but with 5/8 in. plaster on interior surface of block	.39				
c. Same as (16a), but with 3/8 in. gypsum lath and 1/2 in. plaster on furring	.26	.12	.10	.09	Brick & Block or Tile
d. 8 in. concrete block backing, no interior finish	.33				
e. Same as (16d), but with 5/8 in. plaster on interior surface of block	.32				
f. Same as (16d), but with 3/8 in. gypsum lath and 1/2 in. plaster on furring	.23	.11	.10	.09	
17 4 in. Face Brick, Stone, or Precast Concrete Backed by Light-Weight Aggregate Concrete Block					
a. 4 in. concrete block backing, no interior finish	.35				
b. Same as (17a), but with 5/8 in. plaster on interior surface of block	.34				
c. Same as (17a), but with 3/8 in. gypsum lath and 1/2 in. plaster on furring	.24	.12	.10	.09	Brick & Block or Tile
d. 8 in. concrete block backing, no interior finish	.30				
e. Same as (17d), but with 5/8 in. plaster on interior surface of block	.29				
f. Same as (17d), but with 3/8 in. gypsum lath and 1/2 in. plaster on furring	.21	.11	.10	.09	
18 4 in. Face Brick, Stone, or Precast Concrete Backed by Gravel-Aggregate Poured Concrete					
a. 6 in. poured concrete backing, no interior finish	.54				
b. Same as (18a), but with 5/8 in. plaster on interior surface of concrete	.51				
c. Same as (18a), but with 3/8 in. gypsum lath and 1/2 in. plaster on furring	.31	.13	.11	.09	Brick, Stone or Precast Concrete Backed by Poured Concrete
d. 8 in. poured concrete backing, no interior finish	.49				
e. Same as (18d), but with 5/8 in. plaster on interior surface of concrete	.47				
f. Same as (18d), but with 3/8 in. gypsum lath and 1/2 in. plaster on furring	.30	.13	.11	.09	
MASONRY CAVITY WALLS					
19 4 in. Face Brick Exterior Construction Masonry Cavity Wall					
a. 4 in. gravel aggregate concrete block or 4 in. common brick inner section, no interior finish	.33				
b. Same as (19a), but with 5/8 in. plaster on interior surface of block or brick	.32				

TABLE C-7 (continued)

CONSTRUCTION DETAIL	HEAT TRANSMISSION COEFFICIENT "U" Btuh per sq ft per °F.				
	INSULATION ADDED TO BASIC CONSTRUCTION				
	None	Blanket or Batt Type Thickness			
		1-1/2"	2"	3"	
19 4 in. Face Brick Exterior Construction Masonry Cavity Wall (Cont'd)					
c. Same as (19a), but with 3/8 in. gypsum lath and 1/2 in. plaster on furring	.23	.11	.10	.09	
d. 4 in. cinder aggregate concrete block or 4 in. clay tile inner section, no interior finish	.30				
e. Same as (19d), but with 5/8 in. plaster on interior surface of block or tile	.29				
MASONRY INTERIOR PARTITIONS					
20 Gravel-Aggregate Hollow Concrete Block Partitions					
a. 8 in. block, no finish either side	.40				
b. Same as (20a), but with 5/8 in. plaster finish one side	.39				
c. Same as (20a), but with gypsum lath and 1/2 in. plaster on furring one side	.26	.12	.10	.09	
22 Cinder-Aggregate Hollow Concrete Block Partitions					
a. 8 in. block, no finish either side	.33				
b. Same as (21a), but with 5/8 in. plaster finish one side	.32				
c. Same as (21a), but with gypsum lath and 1/2 in. plaster on furring one side	.23	.11	.10	.09	

CONSTRUCTION DETAIL	HEAT TRANSMISSION COEFFICIENT "U" Btuh per sq. ft. per °F.						
	INSULATION ADDED TO BASIC CONSTRUCTION						
	Blanket, Batt or Fill Type Thickness						
	None		2		3	4	6
	Clg.	Htg.	Clg.	Htg.	Clg. & Htg.	Clg. & Htg.	Clg. & Htg.
WOOD AND CONCRETE FLOORS							
22 WOOD FRAME FLOORS							
a. With subfloor, hardwood flooring. No ceiling. Over untreated, open, occupied, space, or kitchen, laundry, etc.	.28	.34	.09	.09	.07	.05	.04
b. Same as (20a) but with 1" accoustical tile on 3/8" gypsum board or ½" insulating board only.	.17	.19	.07	.08	.06	.05	.04
c. Same as (20a) but with metal lath and 3/8" plaster	.21	.26	.08	.09	.07	.05	.04
23 CONCRETE FLOORS							
a. 4" to 6" concrete floor, tile or linoleum, plaster below.	.42	.57	.10	.10	.08	.06	.04
b. Same as (23a) but with suspended ½" tile on 3/8" gypsum board ceiling.	.21	.25	.08	.09	.06	.05	.04
c. Concrete floor on ground with vertical edge insulation. (Btu/per linear ft. of perimeter/degree temperature difference.)		.83		.59			

TABLE C-7 (continued)

CONSTRUCTION DETAIL	HEAT TRANSMISSION COEFFICIENT "U" Btuh per sq ft per °F.						
	INSULATION ADDED TO BASIC CONSTRUCTION						
	Blanket, Batt or Fill Type Thickness						
	None		2		3	4	6
	Clg.	Htg.	Clg.	Htg.	Clg. & Htg.	Clg. & Htg.	Clg. & Htg.
CEILINGS WITH ATTIC SPACE ABOVE (VENTED)							
24 Wood Joists, no floor							
a. 3/8 in. dry-wall interior	.46	.65	.10	.11	.08	.06	.04
b. 3/8 in. gypsum lath and 1/2 in. plaster interior	.44	.61	.10	.11	.08	.06	.04
c. 1/2 in. acoustical tile on 3/8 in. gypsum board interior	.30	.37	.09	.10	.07	.06	.04
d. 3/4 in. acoustical tile on 3/8 in. gypsum board interior	.25	.30	.09	.09	.07	.06	.04
25 Wood Joists, with floor Above (Vented)							
a. 25/32 in. wood or 3/4 in. plywood floor in attic, 3/8 in. dry-wall interior	.24	.30	.09	.09	.07	.06	.04
b. Same as (25a), but with 3/8 in. gypsum lath and 1/2 in. plaster interior	.24	.29	.09	.09	.07	.06	.04
c. Same as (25a), but with 1/2 in. acoustical tile on 3/8 in. gypsum board interior	.19	.22	.08	.08	.07	.05	.04
d. Same as (25a), but with 1/2 in. acoustical tile on furring interior	.20	.24	.08	.09	.07	.05	.04

CONSTRUCTION DETAIL	HEAT TRANSMISSION COEFFICIENT "U" Btuh per sq ft per °F.						
	PREFORMED INSULATION BETWEEN DECK & ROOF Nominal Thickness, in.						
	None		1/2"		1"	2"	3"
	Clg.	Htg.	Clg.	Htg.	Clg. & Htg.	Clg. & Htg.	Clg. & Htg.
FLAT ROOFS – BUILT-UP ON DECK WITH & WITHOUT SUSPENDED CEILING							
26 4 in. Gravel-Aggregate Concrete Slab Deck Under Built-up Roof							
a. No ceiling	.55	.70	.31	.35	.22	.14	.10
b. 3/8 in. dry-wall ceiling	.32	.38	.20	.22	.16	.10	.07
c. 3/8 in. gypsum lath and 1/2 in. plaster ceiling	.31	.37	.20	.22	.15	.10	.07
d. 1/2 in. acoustical tile ceiling on 3/8 in. gypsum board	.23	.26	.16	.18	.13	.09	.07
e. 1/2 in. acoustical tile ceiling on furring or channels	.25	.29	.17	.19	.14	.09	.07
27 8 in. Gravel-Aggregate Concrete Slab Deck Under Built-Up Roof							
a. No ceiling	.47	.57	.28	.32	.20	.14	.10
b. 3/8 in. dry-wall ceiling	.29	.34	.19	.21	.15	.10	.07
c. 3/8 in. gypsum lath and 1/2 in. plaster ceiling	.28	.33	.18	.20	.15	.10	.07
d. 1/2 in. acoustical tile ceiling on 3/8 in. gypsum board	.22	.24	.16	.17	.13	.09	.07
e. 1/2 in. acoustical tile ceiling on furring or channels	.23	.26	.16	.18	.13	.09	.07

TABLE C-7 (continued)

CONSTRUCTION DETAIL	HEAT TRANSMISSION COEFFICIENT "U" Btuh per sq ft per °F. PREFORMED INSULATION BETWEEN DECK & ROOF Nominal Thickness, in.							
	None		1/2"		1"	2"	3"	
	Clg.	Htg.	Clg.	Htg.	Clg. & Htg.	Clg. & Htg.	Clg. & Htg.	
28 2 in. Light-Weight Aggregate Concrete Slab Deck Under Built-Up Roof								
a. Deck poured on corrugated metal, or 1/4 in. asbestos-cement board, no ceiling	.27	.30	.20	.21	.15	.11	.08	
b. Same as (28a), but with 3/8 in. gypsum board ceiling	.20	.22	.15	.16	.12	.08	.06	
c. Same as (28a), but with 3/8 in. gypsum lath and 1/2 in. plaster ceiling	.20	.22	.15	.16	.12	.08	.06	
d. Same as (28a), but with 1/2 in. acoustical tile ceiling on 3/8 in. gypsum board	.16	.18	.12	.13	.11	.08	.06	
e. Same as (28a), but with 1/2 in. acoustical tile ceiling on furring or channels	.17	.19	.13	.14	.11	.08	.06	
f. Deck poured on 1 in. insulation board, no ceiling	.15	.16	.13	.13	.11	.09	.07	
g. Same as (28f), but with 3/8 in. dry-wall ceiling	.13	.14	.10	.11	.09	.07	.05	
h. Deck poured on 1 in. glass fiber board or 1-1/2 in. insulation board, no ceiling	.13	.14	.10	.11	.09	.07	.05	
i. Same as (28h), but with 3/8 in. dry-wall ceiling	.11	.12	.09	.10	.08	.06	.05	
29 4 in. Light-Weight Aggregate Concrete Slab Deck Under Built-Up Roof								
a. Deck poured on corrugated metal, or 1/4 in. asbestos-cement board, no ceiling	.17	.18	.14	.14	.11	.09	.07	
b. Same as (29a), but with 3/8 in. gypsum board ceiling	.14	.15	.11	.12	.10	.07	.05	
c. Same as (29a), but with 3/8 in. gypsum lath and 1/2 in. plaster ceiling	.14	.15	.11	.12	.10	.07	.05	
d. Same as (29a), but with 1/2 in. acoustical tile ceiling on 3/8 in. gpysum board	.12	.13	.10	.10	.09	.07	.05	
e. Same as (29a), but with 1/2 in. acoustical tile on ceiling furring or channels	.12	.13	.10	.10	.09	.07	.05	
f. Deck poured on 1 in. insulation board, no ceiling	.11	.12	.10	.10	.09	.07	.06	
g. Same as (29f), but with 3/8 in. dry-wall ceiling	.10	.11	.09	.09	.08	.06	.05	
h. Same as (29f), but with 3/8 in. gypsum lath and 1/2 in. plaster ceiling	.10	.10	.09	.09	.08	.06	.05	
i. Deck poured on 1 in. glass fiber board or 1-1/2 in. insulation board, no ceiling	.10	.10	.09	.09	.08	.07	.05	
j. Same as (29i), but with 3/8 in. dry-wall ceiling	.09	.09	.08	.08	.07	.05	.04	
k. Same as (29i), but with 3/8 in. gypsum lath and 1/2 in. plaster ceiling	.09	.09	.08	.08	.07	.05	.04	
l. Same as (29i), but with 1/2 in. acoustical tile ceiling on 3/8 in. gpysum board	.08	.08	.07	.07	.06	.05	.04	
m. Same as (29i), but with 1/2 in. acoustical tile ceiling on furring or channels	.08	.09	.07	.08	.07	.05	.04	

TABLE C-7 (continued)

CONSTRUCTION DETAIL	HEAT TRANSMISSION COEFFICIENT "U" Btuh per sq ft per °F.						
	PREFORMED INSULATION BETWEEN DECK & ROOF Nominal Thickness, in.						
	None		1/2"		1"	2"	3"
	Clg.	Htg.	Clg.	Htg.	Clg. & Htg.	Clg. & Htg.	Clg. & Htg.
30 2 in. Gypsum Deck Slab Under Built-up Roof							
a. Deck slab on 1/4 in. asbestos-cement board, no ceiling	.34	.40	.23	.26	.18	.12	.09
b. Same as (30a), but with 3/8 in. dry-wall ceiling	.25	.27	.17	.18	.13	.09	.07
c. Same as (30a), but with 3/8 in. gypsum lath and 1/2 in. plaster ceiling	.24	.26	.17	.18	.13	.09	.07
d. Same as (30a), but with 1/2 in. acoustical tile ceiling on 3/8 in. gypsum board	.19	.20	.14	.15	.12	.09	.06
e. Same as (30a), but with 1/2 in. acoustical tile ceiling on furring or channels	.20	.22	.15	.16	.12	.09	.06
f. Deck slab on 1 in. insulation board, no ceiling	.18	.20	.15	.15	.12	.09	.07
g. Same as (30f), but with 3/8 in. dry-wall ceiling	.15	.16	.11	.12	.10	.07	.06
h. Same as (30f), but with 3/8 in. gypsum lath and 1/2 in. plaster Ceiling	.15	.16	.12	.12	.10	.07	.06
i. Same as (30f), but with 1/2 in. acoustical tile ceiling on 3/8 in. gypsum board	.13	.13	.10	.10	.09	.07	.05
j. Same as (30f), but with 1/2 in. acoustical tile ceiling on furring or channels	.13	.14	.10	.11	.09	.07	.05
k. Deck slab on 1 in. glass fiber board or 1-1/2 in. insulation board, no ceiling	.15	.16	.12	.13	.11	.08	.07
l. Same as (30k), but with 3/8 in. dry-wall ceiling	.13	.13	.10	.10	.09	.07	.05
m. Same as (30k), but with 3/8 in. gpysum lath and 1/2 in. plaster ceiling	.12	.13	.10	.10	.09	.07	.05
n. Same as (30k), but with 1/2 in. acoustical tile ceiling on 3/8 in. gypsum board	.11	.12	.09	.10	.08	.06	.05
o. Same as (30K), but with 1/2 in. acoustical tile ceiling on furring or channels	.11	.12	.09	.10	.08	.06	.05
31 Wood Deck Under Built-Up Roof							
a. 25/32 in. wood deck, no ceiling	.40	.48	.26	.29	.19	.13	.09
b. Same as (31a), but with 3/8 in. dry-wall ceiling	.26	.31	.18	.19	.14	.09	.07
c. Same as (31a), but with 3/8 in. gypsum lath and 1/2 in. plaster ceiling	.26	.30	.18	.19	.14	.09	.07
d. Same as (31a), but with 1/2 in. acoustical tile ceiling on 3/8 in. gpysum board	.20	.22	.15	.16	.12	.08	.06

Roofing   Insulation
Roof Deck
Ceiling

Roofing   Insulation
Ceiling

TABLE C-7 (continued)

CONSTRUCTION DETAIL	HEAT TRANSMISSION COEFFICIENT "U" Btuh per sq ft per ° F.							
	PREFORMED INSULATION BETWEEN DECK & ROOF Nominal Thickness, in.							
	None		1/2"		1"	2"	3"	
	Clg.	Htg.	Clg.	Htg.	Clg. & Htg.	Clg. & Htg.	Clg. & Htg.	
31 Wood Deck Under Built-Up Roof (Cont'd)								
e. Same as (31a), but with 1/2 in. acoustical tile ceiling on furring or channels	.22	.24	.16	.17	.13	.09	.07	
f. 1-5/8 in. wood deck, no ceiling	.28	.32	.20	.22	.16	.11	.08	
g. Same as (31f), but with 3/8 in. dry-wall ceiling	.21	.23	.15	.16	.12	.08	.06	
h. Same as (31f), but with 3/8 in. gypsum lath and 1/2 in. plaster ceiling	.20	.23	.15	.16	.12	.08	.06	
i. Same as (31f), but with 1/2 in. acoustical tile ceiling on 3/8 in. gypsum board	.17	.18	.13	.13	.11	.08	.06	
j. Same as (31f), but with 1/2 in. acoustical tile ceiling on furring or channels	.18	.19	.13	.14	.11	.08	.06	
32 Flat Metal Deck Built-Up Roof								
a. No ceiling	.67	.90	.35	.40	.23	.15	.10	
b. 3/8 in. gypsum-board ceiling	.36	.44	.22	.24	.17	.10	.08	
c. 3/8 in. gypsum lath and 1/2 in. plaster ceiling	.34	.42	.20	.23	.16	.10	.07	
d. 1/2 in. acoustical tile ceiling on 3/8 in. gypsum board	.25	.29	.17	.19	.14	.09	.07	
e. 1/2 in. acoustical tile ceiling on furring or channels	.27	.32	.18	.20	.14	.09	.07	

CONSTRUCTION DETAIL	HEAT TRANSMISSION COEFFICIENT "U" Btuh per sq ft per ° F.				
	INSULATION ADDED TO BASIC CONSTRUCTION				
		RIGID INSULATION APPLIED TO INTERIOR OF FOUNDATION, THICKNESS			
		Insulating 1/2"	Sheathing 25/32"	Insulation Board	
	None			1"	2"
BASEMENT AND CRAWL-SPACE FOUNDATION WALLS					
33 Poured Concrete (No Interior Finish)					
a. 6 in. poured concrete — gravel-aggregate — above grade 140 lbs./cu. ft.	.75	.38	.30	.24	.16
b. Same as (33a), but with light-weight aggregate concrete 80 lbs./cu. ft.	.31	.22	.20	.16	.13
c. Same as (33a), but with insulating concrete 30 lbs./cu. ft.	.13	.13	.12	.11	.10
d. 8 in. poured concrete — gravel-aggregate — above grade 140 lbs./cu. ft.	.67	.36	.29	.23	.16
e. Same as (33d), but with light-weight aggregate concrete 80 lbs./cu. ft.	.25	.19	.17	.15	.12
f. 6 in., 8 in., or 10 in. poured concrete below grade	.06				

TABLE C-7 (continued)

CONSTRUCTION DETAIL	HEAT TRANSMISSION COEFFICIENT "U" Btuh per sq ft per °F.				
	INSULATION ADDED TO BASIC CONSTRUCTION				
	RIGID INSULATION APPLIED TO INTERIOR OF FOUNDATION, THICKNESS				
		Insulating	Sheathing	Insulation Board	
	None	1/2"	25/32"	1"	2"
34 Concrete Block (No Interior Finish)					
a. 8 in. concrete block — gravel-aggregate — above grade	.52	.31	.26	.21	.15
b. Same as (34a), but with cinder-aggregate block	.39	.26	.22	.18	.14
c. Same as (34a), but with light-weight aggregate block	.35	.25	.21	.18	.13
d. 10 in. concrete block — gravel-aggregate — above grade	.50				
e. Same as (34d), but with light-weight aggregate block	.33	.23	.20	.17	.13
f. 12 in. concrete block — gravel-aggregate — above grade	.47	.30	.25	.20	.14
g. Same as (34f), but with cinder-aggregate block	.36	.25	.22	.18	.13
h. 8 in., 10 in., or 12 in. block below grade	.06				

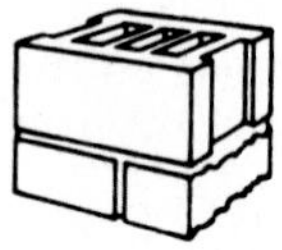

Blocks

Source: Load Calculation Digest, Commercial/Industrial Air Conditioning, General Electric Company, 1971; constructed from procedures and data contained in *ASHRAE Handbook and Product Directory 1977 Fundamentals.*

TABLE C-8

Heat loss below grade in basement walls

Depth ft	Path Length Through Soil, ft	Heat Loss, Btu/h·ft·°F							
		Uninsulated		R = 4.17		R = 8.34		R = 12.5	
0–1	0.68	0.410		0.152		0.093		0.067	
1–2	2.27	0.222	0.632	0.116	0.268	0.079	0.172	0.059	0.126
2–3	3.88	0.155	0.787	0.094	0.362	0.068	0.240	0.053	0.179
3–4	5.52	0.119	0.906	0.079	0.441	0.060	0.300	0.048	0.227
4–5	7.05	0.096	1.002	0.069	0.510	0.053	0.353	0.044	0.271
5–6	8.65	0.079	1.081	0.060	0.570	0.048	0.401	0.040	0.311
6–7	10.28	0.069	1.150	0.054	0.624	0.044	0.445	0.037	0.348

TABLE C-9

Heat loss through basement floors, Btu/hr·ft^2·°F

Depth of Foundation Wall Below Grade	Shortest Width of House, ft			
	20	24	28	32
5 ft	0.032	0.029	0.026	0.023
6 ft	0.030	0.027	0.025	0.022
7 ft	0.029	0.026	0.023	0.021

Note: $\Delta F = (t_a - A)$

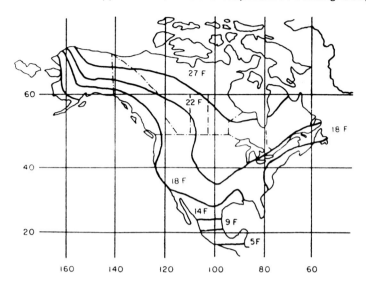

FIGURE C-2 Lines of constant amplitude, A.

TABLE C-10

Heat loss coefficient F_2 of slab floor construction, Btu/hr·°F per ft of perimeter

Construction	Insulation[a]	Degree Days (65°F Base)		
		2950	5350	7433
8-in. block wall,	Uninsulated	0.62	0.68	0.72
brick facing	R = 5.4 from edge to footer	0.48	0.50	0.56
4-in. block wall,	Uninsulated	0.80	0.84	0.93
brick facing	R = 5.4 from edge to footer	0.47	0.49	0.54
Metal stud wall,	Uninsulated	1.15	1.20	1.34
stucco	R = 5.4 from edge to footer	0.51	0.53	0.58
Poured concrete wall	Uninsulated	1.84	2.12	2.73
with duct near perimeter[b]	R = 5.4 from edge to footer, 3 ft under floor	0.64	0.72	0.90

[a]R-value units in °F·ft²·hr/Btu·in.

[b]Weighted average temperature of the heating duct was assumed at 110°F during the heating season (outdoor air temperature less than 65°F).

TABLE C-11*

Estimated U-values for insulated and uninsulated crawl spaces

Component	Uninsulated (Btu/hr-°F per ft of perimeter)	Insulated[a] (Btu/hr-°F per ft of perimeter)
16-in. exposed concrete blocks	0.7	0.18
7.5-in. sill box	0.188	0.071
First 12-in. block wall below grade	0.355	0.127
Second 12-in. block wall below grade	0.22	0.14
Third 12-in. block wall below grade	0.133	0.1
Total for perimeter wall	1.6	0.62
	$\dfrac{Btu}{hr\text{-}ft^2\text{-}°F}$	$\dfrac{Btu}{hr\text{-}ft^2\text{-}°F}$
Ground	0.077	0.077
Floor above crawl space	0.25	0.076[a]

**Source: ASHRAE Handbook 1989 Fundamentals* with permission from American Society of Heating, Refrigerating, and Air Conditioning Engineers, Atlanta, GA.

[a]Perimeter walls are insulated with R = 5.4; floor is insulated with R = 11 blankets or batts.

Appendix D

Heat Losses Through Pipe

TABLE D-1 *

Heat losses from bare steel pipe

HORIZONTAL PIPES

Diameter of Pipe, Inches	Temperature of Pipe, Deg. F.										
	100	120	150	180	210	240	270	300	330	360	390
	Temperature Difference, Pipe to Air, Deg. F.										
	30	50	80	110	140	170	200	230	260	290	320
	Heat Loss per Lineal Foot of Pipe, Btu per Hour										
½	13	22	40	60	82	106	133	162	193	227	265
¾	15	27	50	74	100	131	163	199	238	280	325
1	19	34	61	90	123	160	199	243	292	343	399
1¼	23	42	75	111	152	198	248	302	362	427	496
1½	27	48	85	126	173	224	280	343	410	483	563
2	33	59	104	154	212	275	344	420	503	594	692
2½	39	70	123	184	252	327	410	502	600	709	827
3	46	84	148	221	303	393	493	601	721	852	994
3½	52	95	168	250	342	444	556	680	816	964	1125
4	59	106	187	278	381	496	621	759	911	1076	1257
5	71	129	227	339	464	603	755	924	1109	1311	1532
6	84	151	267	398	546	709	890	1088	1306	1544	1806
8	107	194	341	509	697	906	1137	1391	1671	1977	2312
10	132	238	420	626	857	1114	1399	1714	2060	2437	2852
12	154	279	491	732	1003	1305	1640	2009	2415	2860	3346
14	181	326	575	856	1173	1527	1918	2350	2826	3347	3918
16	203	366	644	960	1314	1711	2149	2634	3168	3753	4395
18	214	385	678	1011	1385	1802	2266	2777	3339	3958	4635
20	236	426	748	1115	1529	1990	2501	3066	3690	4373	5123

VERTICAL PIPES

Diameter of Pipe, Inches	Temperature of Pipe, Deg. F.										
	100	120	150	180	210	240	270	300	330	360	390
	Temperature Difference, Pipe to Air, Deg. F.										
	30	50	80	110	140	170	200	230	260	290	320
	Heat Loss per Lineal Foot of Pipe, Btu per Hour										
½	11	20	35	52	71	93	116	142	170	201	235
¾	14	25	44	65	89	116	145	177	213	252	294
1	17	31	55	81	111	145	181	222	266	315	368
1¼	22	39	69	103	141	183	230	281	337	398	465
1½	25	45	79	118	161	210	263	321	386	456	532
2	31	56	99	147	201	262	328	401	481	569	665
2½	37	68	120	178	244	317	397	486	583	687	805
3	46	83	146	217	297	386	484	592	710	839	980
3½	52	94	166	248	339	440	552	676	810	958	1119
4	59	106	187	279	382	496	622	760	912	1078	1259
5	72	131	231	344	472	612	768	939	1126	1331	1555
6	86	156	275	410	562	729	915	1119	1342	1587	1853
8	112	203	358	534	731	950	1191	1456	1747	2065	2412
10	140	254	447	667	913	1186	1487	1818	2181	2578	3012
12	166	301	530	790	1081	1404	1761	2154	2584	3054	3567
14	195	354	624	930	1273	1653	2073	2536	3042	3596	4200
16	221	400	705	1051	1438	1868	2343	2865	3437	4063	4745
18	234	425	748	1115	1526	1982	2486	3040	3648	4311	5036
20	260	472	831	1239	1696	2203	2763	3378	4053	4791	5596

Source: Reprinted with permission from *Handbook of Air Conditioning, Heating, and Ventilating,* 3rd ed., Industrial Press, New York, 1979.

*TABLE D-2**

Heat losses from bare tarnished copper tube

HORIZONTAL TUBES

Nominal Diameter of Tube, Inches	Temperature of Tube, Deg. F.										
	100	120	150	180	210	240	270	300	330	360	390
	Temperature Difference, Tube to Air, Deg. F.										
	30	50	80	110	140	170	200	230	260	290	320
	Heat Loss per Lineal Foot of Tube, Btu per Hr.										
1/4	4	8	14	21	29	37	46	56	66	77	88
3/8	6	10	18	28	37	48	60	72	85	99	114
1/2	7	13	22	33	45	59	72	88	104	121	139
5/8	8	15	26	39	53	68	85	102	121	141	163
3/4	9	17	30	45	61	79	97	117	139	162	187
1	11	21	37	55	75	97	120	146	173	201	232
1 1/4	14	25	45	66	90	117	145	175	207	242	279
1 1/2	16	29	52	77	105	135	167	203	241	281	324
2	20	37	66	97	132	171	212	257	305	356	411
2 1/2	24	44	78	117	160	206	255	310	367	429	496
3	28	51	92	136	186	240	297	360	428	501	578
3 1/2	32	59	104	156	212	274	340	412	490	573	662
4	36	66	118	174	238	307	381	462	550	644	744
5	43	80	142	212	288	373	464	561	669	783	905
6	51	93	166	246	336	432	541	656	776	915	1059
8	66	120	215	317	435	562	699	848	1010	1184	1372
10	80	146	260	387	527	681	848	1031	1227	1442	1670
12	94	172	304	447	621	802	999	1214	1446	1699	1969

VERTICAL TUBES

Nominal Diameter of Tube, Inches	Temperature of Tube, Deg. F.										
	100	120	150	180	210	240	270	300	330	360	390
	Temperature Difference, Tube to Air, Deg. F.										
	30	50	80	110	140	170	200	230	260	290	320
	Heat Loss per Lineal Foot of Tube, Btu per Hr.										
1/4	3	6	10	15	21	27	34	41	49	57	66
3/8	4	8	14	21	28	36	45	55	65	77	88
1/2	5	10	17	26	35	46	57	69	82	96	111
5/8	6	12	21	31	42	54	68	82	98	114	132
3/4	7	14	24	36	49	64	79	96	114	134	155
1	10	18	31	46	63	82	102	123	147	172	198
1 1/4	12	21	38	57	77	100	125	151	180	210	243
1 1/2	14	25	45	67	91	118	147	178	212	248	287
2	18	33	59	88	120	155	192	233	277	325	375
2 1/2	22	41	73	109	148	191	238	288	343	402	464
3	27	49	87	129	176	227	283	343	408	478	552
3 1/2	31	57	101	150	204	264	328	398	474	554	641
4	35	64	114	171	232	300	374	453	539	631	729
5	43	80	142	212	288	373	464	561	669	783	905
6	52	96	170	253	344	445	554	670	798	934	1080
8	69	127	226	337	458	592	737	892	1063	1244	1438
10	86	158	281	419	570	737	917	1110	1322	1548	1789
12	103	189	336	501	682	881	1097	1328	1582	1851	2140

*Source: Reprinted with permission from *Handbook of Air Conditioning, Heating, and Ventilating,* 3rd ed., Industrial Press, New York, 1979.

TABLE D-3*

Heat loss through pipe insulation, various sizes of pipe

Insulation Conductivity k	Temperature Difference, Pipe to Air, Deg. F										
	30	50	80	110	140	170	200	230	260	290	320
	Heat Loss per Lineal Foot of Bare Pipe, Btu per Hour										
	15	27	50	74	100	131	163	199	238	280	325
	Heat Loss per Lineal Foot of Insulated Pipe, Btu per Hour										
1 Inch Thick Insulation — ¾ Inch Pipe											
0.20	2	6	7	10	13	16	18	21	24	27	29
0.25	3	6	9	12	16	19	23	26	29	33	36
0.30	4	7	11	15	19	23	27	31	35	39	43
0.35	5	8	12	17	21	26	30	35	40	44	49
0.40	5	9	14	19	24	29	34	39	44	50	55
0.45	6	10	15	21	27	32	38	44	49	55	61
0.50	6	10	17	23	29	35	42	48	54	60	67
0.55	7	11	18	25	32	38	45	52	59	65	72
0.60	7	12	19	27	34	41	48	56	63	70	77
1½ Inch Thick Insulation — ¾ Inch Pipe											
0.20	2	4	6	8	10	12	14	16	19	21	23
0.25	3	5	7	10	13	16	18	21	24	27	29
0.30	3	5	9	12	15	19	22	25	28	32	35
0.35	4	6	10	14	18	21	25	29	33	37	40
0.40	4	7	11	16	20	24	29	33	37	41	46
0.45	5	8	13	17	22	27	32	37	41	46	51
0.50	5	9	14	19	25	30	35	40	46	51	56
0.55	6	10	15	21	27	32	38	44	49	55	61
0.60	6	10	16	23	29	35	41	47	53	59	66
2 Inch Thick Insulation — ¾ Inch Pipe											
0.20	2	3	5	7	9	11	13	15	17	19	21
0.25	2	4	6	9	11	14	16	18	21	23	26
0.30	3	5	8	11	13	16	19	22	25	28	31
0.35	3	6	9	12	16	19	22	26	29	32	36
0.40	4	6	10	14	18	21	25	29	33	37	40
0.45	4	7	11	16	20	24	28	32	37	41	45
0.50	5	8	12	17	22	26	31	36	40	45	50
0.55	5	8	14	19	24	29	34	39	44	49	54
0.60	5	9	15	20	26	31	37	42	48	53	59
2½ Inch Thick Insulation — ¾ Inch Pipe											
0.20	2	3	5	6	8	10	12	13	15	17	19
0.25	2	3	5	7	9	11	13	15	17	19	21
0.30	3	4	7	10	12	15	17	20	23	25	28
0.35	3	5	8	11	14	17	20	23	26	29	32
0.40	3	6	9	13	16	20	23	26	30	33	37
0.45	3	6	10	14	18	22	26	29	33	37	41
0.50	4	7	11	16	20	24	28	33	37	41	45
0.55	5	8	12	17	22	26	31	36	40	45	50
0.60	5	8	13	18	24	29	34	39	44	49	54

Source: Reprinted with permission from *Handbook of Air Conditioning, Heating, and Ventilating,* 3rd ed., Industrial Press, New York, 1979.

TABLE D-3 (continued)

Insulation Conductivity k	Temperature Difference, Pipe to Air, Deg. F										
	30	50	80	110	140	170	200	230	260	290	320
	Heat Loss per Lineal Foot of Bare Pipe, Btu per Hour										
	19	34	61	90	123	160	199	243	292	343	399
	Heat Loss per Lineal Foot of Insulated Pipe, Btu per Hour										
1 Inch Thick Insulation — 1 Inch Pipe											
0.20	3	5	8	12	15	18	21	24	27	30	34
0.25	4	6	10	14	18	23	26	30	34	37	41
0.30	5	8	12	17	21	26	30	35	40	44	49
0.35	5	9	14	19	24	30	35	40	45	50	56
0.40	6	10	16	22	27	33	39	45	51	57	63
0.45	7	11	17	24	30	37	43	50	56	63	69
0.50	7	12	19	26	33	40	47	55	62	69	76
0.55	8	13	20	28	36	44	51	59	67	74	82
0.60	8	14	22	30	39	47	55	63	72	80	88
1¼ Inch Thick Insulation — 1 Inch Pipe											
0.20	2	4	7	9	12	14	17	19	22	24	27
0.25	3	5	8	11	15	18	21	24	27	30	33
0.30	4	6	10	14	17	21	25	29	32	36	40
0.35	4	7	11	16	20	24	29	33	37	41	46
0.40	5	8	13	18	23	27	32	37	42	47	52
0.45	5	9	14	20	25	31	36	41	47	52	58
0.50	6	10	16	22	28	33	39	45	51	57	63
0.55	6	11	17	24	30	37	43	49	56	62	69
0.60	7	12	19	26	32	39	46	53	60	67	74
2 Inch Thick Insulation — 1 Inch Pipe											
0.20	2	4	6	8	10	13	15	17	19	21	24
0.25	3	5	7	10	13	15	18	21	23	26	29
0.30	3	5	9	12	15	18	21	25	28	31	34
0.35	4	6	10	14	17	21	25	29	32	36	40
0.40	4	7	11	15	20	24	28	32	37	41	45
0.45	5	8	13	17	22	27	31	36	41	46	50
0.50	5	9	14	19	24	29	35	40	45	50	55
0.55	6	9	15	21	26	32	38	43	49	55	60
0.60	6	10	16	23	29	35	41	47	53	59	66
2½ Inch Thick Insulation — 1 Inch Pipe											
0.20	2	3	5	7	9	11	13	15	17	19	21
0.25	3	4	7	10	12	15	17	20	23	25	28
0.30	3	5	8	11	14	16	19	22	25	28	31
0.35	3	6	9	12	16	19	22	26	29	32	36
0.40	4	6	10	14	18	22	26	29	33	37	41
0.45	4	7	11	16	20	24	29	33	37	41	46
0.50	5	8	13	17	22	27	32	36	41	46	51
0.55	5	9	14	19	24	29	34	40	45	50	55
0.60	6	9	15	21	26	32	37	43	49	54	60

594   *Appendix D / Heat Losses Through Pipe*

TABLE D-3 (continued)

Insulation Conductivity k	Temperature Difference, Pipe to Air, Deg. F										
	30	50	80	110	140	170	200	230	260	290	320
	Heat Loss per Lineal Foot of Bare Pipe, Btu per Hour										
	27	48	85	126	173	224	280	343	410	483	563
	Heat Loss per Lineal Foot of Insulated Pipe, Btu per Hour										
1 Inch Thick Insulation — 1½ Inch Pipe											
0.20	4	7	11	15	19	23	27	31	35	39	43
0.25	5	8	13	18	23	28	32	37	42	47	52
0.30	6	10	16	21	27	33	39	45	50	56	62
0.35	7	11	18	24	31	38	44	51	58	64	71
0.40	7	12	20	27	35	42	50	57	65	72	80
0.45	8	14	22	30	39	47	55	63	72	80	88
0.50	9	15	24	33	42	51	60	69	78	87	96
0.55	10	16	26	36	46	55	65	75	85	94	104
0.60	10	17	28	38	49	59	70	80	90	100	111
1½ Inch Thick Insulation — 1½ Inch Pipe											
0.20	3	5	8	12	15	18	21	24	27	30	34
0.25	4	7	10	14	18	22	26	30	34	38	42
0.30	5	8	12	17	22	26	31	35	40	45	49
0.35	5	9	14	19	25	30	35	41	46	51	57
0.40	6	10	16	22	28	34	40	46	52	58	64
0.45	7	11	18	25	31	38	45	51	58	65	71
0.50	7	12	20	27	34	41	49	56	63	71	78
0.55	8	13	21	29	37	45	53	61	69	77	85
0.60	9	14	23	32	40	49	57	66	75	83	92
2 Inch Thick Insulation — 1½ Inch Pipe											
0.20	3	4	7	10	12	15	18	20	23	25	28
0.25	3	6	9	12	15	19	22	25	29	32	35
0.30	4	7	10	14	18	22	26	30	34	38	42
0.35	5	8	12	17	21	26	30	35	44	49	53
0.40	5	9	14	19	24	29	34	40	45	50	55
0.45	6	10	15	21	27	33	38	44	50	56	61
0.50	6	11	17	23	30	36	43	49	55	62	68
0.55	7	12	18	25	32	39	46	53	60	67	74
0.60	8	13	20	28	35	43	50	58	65	73	80
2½ Inch Thick Insulation — 1½ Inch Pipe											
0.20	2	4	6	9	11	13	16	18	21	23	25
0.25	3	5	8	11	14	17	20	23	25	28	31
0.30	4	6	9	13	16	20	23	27	30	34	37
0.35	4	7	11	15	19	23	27	31	35	39	44
0.40	5	8	12	17	22	26	31	35	40	45	49
0.45	5	9	14	19	24	29	34	40	45	50	55
0.50	6	10	15	21	27	32	38	44	49	55	61
0.55	6	10	17	23	29	35	42	48	54	60	67
0.60	7	11	18	25	32	38	45	52	59	65	72

TABLE D-3 *(continued)*

Insulation Conductivity *k*	Temperature Difference, Pipe to Air, Deg. F										
	30	50	80	110	140	170	200	230	260	290	320
	Heat Loss per Lineal Foot of Bare Pipe, Btu per Hour										
	33	59	104	154	212	275	344	420	503	594	692
	Heat Loss per Lineal Foot of Insulated Pipe, Btu per Hour										
1 Inch Thick Insulation — 2 Inch Pipe											
0.20	5	8	13	17	22	27	32	36	41	46	51
0.25	6	10	15	21	27	32	39	44	50	56	62
0.30	7	11	18	25	32	39	45	52	59	66	73
0.35	8	13	21	29	36	44	52	60	68	75	83
0.40	9	15	23	32	41	49	58	67	76	84	93
0.45	10	16	26	35	45	55	64	74	83	93	103
0.50	10	18	28	39	49	60	70	81	91	102	112
0.55	11	19	30	42	53	64	76	87	99	110	121
0.60	12	20	32	45	57	69	81	93	106	118	130
1½ Inch Thick Insulation — 2 Inch Pipe											
0.20	4	6	10	13	17	21	24	28	32	35	39
0.25	5	8	12	17	21	26	30	35	39	44	48
0.30	5	9	14	20	25	30	36	41	46	52	57
0.35	6	10	16	23	29	35	41	47	53	59	66
0.40	7	12	18	25	32	39	46	53	60	67	74
0.45	8	13	21	28	36	44	51	59	67	75	82
0.50	9	15	23	32	41	49	58	67	75	84	93
0.55	9	15	25	34	43	52	61	71	80	89	98
0.60	10	17	26	36	46	55	66	76	83	96	106
2 Inch Thick Insulation — 2 Inch Pipe											
0.20	3	5	8	11	14	17	20	23	27	30	33
0.25	4	6	10	14	18	22	25	29	33	37	41
0.30	4	8	12	17	21	26	30	35	39	44	48
0.35	5	9	14	19	24	30	35	40	45	50	56
0.40	6	10	16	22	28	33	39	45	51	57	63
0.45	7	11	18	24	31	37	44	51	57	64	70
0.50	7	12	19	27	34	41	48	56	63	70	77
0.55	8	13	21	29	37	45	53	61	69	77	84
0.60	9	14	23	31	40	49	57	66	74	83	92
2½ Inch Thick Insulation — 2 Inch Pipe											
0.20	3	5	7	10	13	15	18	21	23	26	29
0.25	3	6	9	12	16	19	22	26	29	32	36
0.30	4	7	11	15	19	23	27	31	35	39	43
0.35	5	8	12	17	22	26	31	35	40	45	49
0.40	5	9	14	19	25	30	35	40	46	51	56
0.45	6	10	16	22	27	33	39	45	51	57	63
0.50	6	11	17	24	30	37	43	50	56	63	69
0.55	7	12	19	26	33	40	48	54	61	68	76
0.60	8	13	21	29	37	45	53	61	69	77	85

***TABLE D-3** (continued)*

Insulation Conductivity k	Temperature Difference, Pipe to Air, Deg. F										
	30	50	80	110	140	170	200	230	260	290	320
	Heat Loss per Lineal Foot of Bare Pipe, Btu per Hour										
	39	70	123	184	252	327	410	502	600	709	827
	Heat Loss per Lineal Foot of Insulated Pipe, Btu per Hour										
1 Inch Thick Insulation — 2½ Inch Pipe											
0.20	5	9	14	20	25	31	36	42	47	52	58
0.25	7	11	18	24	31	38	44	51	57	64	71
0.30	8	13	21	29	36	44	52	60	68	75	83
0.35	9	15	24	33	42	50	59	68	77	86	95
0.40	10	17	27	37	47	57	67	77	87	97	107
0.45	11	18	29	40	52	63	74	85	96	107	118
0.50	12	21	33	45	57	70	82	94	107	119	131
0.55	13	22	35	48	61	74	87	100	113	126	139
0.60	14	23	37	51	65	79	93	106	120	134	148
1½ Inch Thick Insulation — 2½ Inch Pipe											
0.20	4	7	11	15	19	24	28	32	36	40	44
0.25	5	9	14	19	24	29	34	39	44	50	55
0.30	6	10	16	22	28	35	41	47	53	59	65
0.35	7	12	19	26	33	40	47	54	61	68	75
0.40	8	13	21	29	37	45	53	61	69	77	84
0.45	9	15	23	32	41	50	59	67	76	85	94
0.50	10	16	26	35	45	55	64	74	83	93	103
0.55	10	17	28	38	49	59	70	80	91	101	112
0.60	11	19	30	41	53	64	75	86	98	109	120
2 Inch Thick Insulation — 2½ Inch Pipe											
0.20	3	6	9	13	16	20	23	26	30	33	37
0.25	4	7	11	16	20	24	29	33	37	41	46
0.30	5	9	14	19	24	29	34	39	44	49	54
0.35	6	10	16	22	27	33	39	45	51	57	63
0.40	7	11	18	24	31	38	44	51	58	64	71
0.45	7	12	20	27	35	42	50	57	64	72	79
0.50	8	14	22	30	38	46	55	63	71	79	87
0.55	9	15	24	33	42	50	59	68	77	86	95
0.60	10	16	26	35	45	55	64	74	84	93	103
2½ Inch Thick Insulation — 2½ Inch Pipe											
0.20	3	5	8	11	14	17	20	23	26	29	32
0.25	4	6	10	14	18	21	25	29	33	36	40
0.30	4	7	12	16	21	25	30	34	39	43	48
0.35	5	9	14	19	24	29	35	40	45	50	55
0.40	6	10	16	22	27	33	39	45	51	57	63
0.45	7	11	18	24	31	37	44	50	57	64	70
0.50	7	12	19	27	34	41	48	55	63	70	77
0.55	8	13	21	29	37	45	53	61	69	77	84
0.60	9	14	23	31	40	49	57	66	74	83	92

TABLE D-3 (continued)

Insulation Conductivity k	Temperature Difference, Pipe to Air, Deg. F										
	30	50	80	110	140	170	200	230	260	290	320
	Heat Loss per Lineal Foot of Bare Pipe, Btu per Hour										
	46	84	148	221	303	393	493	601	721	852	994
	Heat Loss per Lineal Foot of Insulated Pipe, Btu per Hour										
1 Inch Thick Insulation — 3 Inch Pipe											
0.20	6	11	17	23	30	36	42	49	55	61	68
0.25	8	13	21	28	36	44	52	60	67	75	83
0.30	9	15	24	33	43	52	61	70	79	88	97
0.35	10	17	28	38	49	59	69	80	90	101	111
0.40	12	19	31	43	54	66	78	89	101	113	124
0.45	13	21	34	47	60	73	86	99	111	124	137
0.50	14	23	37	51	65	79	93	107	121	135	149
0.55	15	25	40	55	71	86	101	116	131	146	161
0.60	16	27	43	59	76	92	108	124	140	156	173
1½ Inch Thick Insulation — 3 Inch Pipe											
0.20	5	8	13	18	22	27	32	37	41	46	51
0.25	6	10	16	22	27	33	39	45	51	57	63
0.30	7	12	19	26	33	39	46	53	60	67	74
0.35	8	13	22	30	39	46	54	62	70	78	86
0.40	9	15	24	33	42	51	60	69	78	87	96
0.45	10	17	27	37	47	57	67	77	87	97	107
0.50	11	18	29	40	51	62	73	84	95	106	117
0.55	12	20	32	44	56	68	80	92	104	116	128
0.60	13	21	34	47	60	73	86	99	112	124	137
2 Inch Thick Insulation — 3 Inch Pipe											
0.20	4	7	11	15	18	22	26	30	34	38	42
0.25	5	8	13	18	23	28	33	38	43	47	52
0.30	6	10	15	21	27	33	39	45	51	56	62
0.35	7	11	18	25	31	38	45	52	58	65	72
0.40	8	13	20	28	36	43	51	58	66	74	81
0.45	8	14	23	31	40	48	57	65	74	82	91
0.50	9	16	25	34	44	53	62	72	81	90	100
0.55	10	17	27	37	48	58	68	78	88	98	109
0.60	11	18	29	40	51	62	73	84	95	106	117
2½ Inch Thick Insulation — 3 Inch Pipe											
0.20	3	6	9	12	16	19	23	26	29	33	36
0.25	4	7	11	16	20	24	28	33	37	41	45
0.30	5	8	13	19	24	29	34	39	44	49	54
0.35	6	10	16	22	27	33	39	45	51	57	63
0.40	7	11	18	24	31	38	44	51	58	64	71
0.45	7	12	20	27	35	42	50	57	64	72	79
0.50	8	14	22	30	38	46	55	63	71	79	87
0.55	9	15	24	33	42	51	60	69	78	87	96
0.60	10	16	26	36	45	55	65	74	84	94	103

TABLE D-3 (continued)

Insulation Conductivity *k*	Temperature Difference, Pipe to Air, Deg. F										
	30	50	80	110	140	170	200	230	270	300	320
	Heat Loss per Lineal Foot of Bare Pipe, Btu per Hour										
	59	106	187	278	381	496	621	759	911	1076	1257
	Heat Loss per Lineal Foot of Insulated Pipe, Btu per Hour										
1 Inch Thick Insulation — 4 Inch Pipe											
0.20	8	13	20	28	33	43	51	59	66	74	82
0.25	10	16	25	35	44	54	63	73	82	92	101
0.30	11	19	30	41	52	63	74	86	97	108	119
0.35	13	21	34	47	59	72	85	98	110	123	136
0.40	14	24	38	52	66	81	95	109	123	137	152
0.45	16	26	42	58	73	89	105	121	136	152	168
0.50	17	28	45	62	80	97	114	131	148	165	182
0.55	18	31	49	67	86	104	123	141	159	178	196
0.60	20	33	53	72	92	112	132	151	171	191	211
1½ Inch Thick Insulation — 4 Inch Pipe											
0.20	6	10	15	21	27	33	38	44	50	56	61
0.25	7	12	19	26	33	40	47	55	62	69	76
0.30	8	14	22	28	39	48	56	64	73	81	90
0.35	10	16	26	36	45	55	65	74	84	94	103
0.40	11	18	29	40	51	62	73	84	95	106	116
0.45	12	20	32	44	56	69	81	93	105	117	129
0.50	13	22	35	49	62	75	88	101	115	128	141
0.55	14	24	38	53	67	81	96	110	124	139	153
0.60	15	26	41	57	72	88	103	118	134	149	165
2 Inch Thick Insulation — 4 Inch Pipe											
0.20	5	8	13	17	22	27	31	36	41	46	50
0.25	6	10	16	21	27	33	39	45	51	57	62
0.30	7	12	19	26	32	39	46	53	60	67	74
0.35	8	13	21	29	37	45	53	61	69	77	85
0.40	9	15	24	33	42	51	60	69	79	88	97
0.45	10	17	27	37	47	57	67	78	88	98	108
0.50	11	19	30	41	52	63	74	85	96	107	118
0.55	12	20	32	44	56	69	81	93	105	117	129
0.60	13	22	35	48	61	74	87	101	114	127	140
2½ Inch Thick Insulation — 4 Inch Pipe											
0.20	4	7	11	15	19	23	27	31	35	39	44
0.25	5	8	13	18	24	29	34	39	44	49	54
0.30	6	10	16	22	28	34	40	46	52	58	64
0.35	7	12	19	26	32	39	46	53	60	67	74
0.40	8	13	21	29	37	45	52	60	68	76	84
0.45	9	15	24	32	41	50	59	68	76	85	94
0.50	10	16	26	36	45	55	65	75	84	94	104
0.55	11	18	28	39	49	60	71	81	92	102	113
0.60	11	19	31	42	53	65	76	88	99	111	122

Index